Fundamentals of Electroheat

Sergio Lupi

Fundamentals of Electroheat

Electrical Technologies for Process Heating

Sergio Lupi
University of Padua
Padua
Italy

ISBN 978-3-319-83420-7 ISBN 978-3-319-46015-4 (eBook)
DOI 10.1007/978-3-319-46015-4

Softcover reprint of the hardcover 1st edition 2016

Printed on acid-free paper

This Springer imprint is published by Springer Nature
The registered company is Springer International Publishing AG
The registered company address is: Gewerbestrasse 11, 6330 Cham, Switzerland

To my wife Manu

Preface

Electroheat is the technology of converting electrical energy into heat, mostly in industrial applications.

Electroheat technologies have greatly contributed to the industrial development during the twentieth century. In fact, the first applications date back to the early decades of that century, regarding arc furnaces for manufacture of calcium carbide and electric steel melting and, later on, induction and resistance furnaces. But the development became particularly rapid, in terms of both applications and research, especially in the last 50 years.

In particular, the continuous increase of energy and labor costs has further increased the importance of these technologies making them play a crucial role in the industrial activity of all industrial countries.

Due to the variety of applications, in the past each technology was considered a single independent engineering discipline which was often known only within its own field of application and therefore in a small circle of users.

In reality, the characteristics of all electroheating applications in production processes arise from their common origin, i.e., the transformation of electrical energy into heat, and soon it appeared more appropriate to treat them as a homogeneous topic. Such opportunity emerged progressively in the last 50 years with the increase of applications in the industry of traditional electroheat techniques and the introduction of new technological processes [1–7].

This book is intended as a basic teaching text for students and teachers dealing with industrial applications of electroheat. It is based on a series of lectures given by the author and his colleagues for many years at the engineering faculty of Padua University in courses for electrical engineering, material science engineering, and chemical engineering. It is hoped that it will also be useful to industrialists, engineers, scientists, and technicians who are more generally interested in industrial applications, efficiency of heating processes, and energy saving.

This book is subdivided into two introductory chapters dealing with heat transfer and electromagnetic phenomena in electrotechnologies and five chapters where the traditional most widespread industrial electroheat technologies are presented.

With this subdivision, this textbook may be useful to different teachers either adopting it as reference text for a basic comprehensive teaching program, or for organizing shorter courses at undergraduate or higher technicians level by selecting the topics of their interest.

Since the field of electroheat is extremely wide, taking into account the necessity of limiting the extent of a general basic course, some highly specialized technologies, such as laser, plasma, electron beam, heat pumps, infrared heating, and others have been omitted and the reader must address to specialized books and papers.

Special thanks are due to Prof. A. Aliferov and his colleagues of the State Technical University NGTU of Novosibirsk for their contribution on the chapter on Submerged Arc Furnaces, to my colleague Prof. M. Forzan who read and made numerous useful comments on parts of the manuscript, and to Mr. L. Fauliri for his assistance in preparation of drawings and figures.

Finally, my greatest thank goes to my wife who patiently supported me during the many hours I devoted to the preparation of this volume, often infringing upon or restricting planned family activities.

In our familiar slang: "*Grazie Pippuni!*"

Basic Electroheat Books

1. UIE—International Union for Electro-heat: History of Electroheat, Vulkan-Verlag GmbH, Essen, Germany, 202 p. (1968), ISBN 978-3-8027-2946-1
2. Lauster, F. (ed.): Manuel d'Électrothermie industrielle, 315 p. Dunod, Paris (1968) (in French)
3. U.I.E. (International Union for Electroheat): Elektrowärme - Theorie und Praxis. In: Girardet, E. (ed.) Verlag W, 902 p., ISBN 3-7736-0355-X (1974) (in German)
4. Orfeuil, M. (ed.): Electrothermie industrielle, 803 p., ISBN 2-04-012179-X. Dunod, Paris (1981) (in French)
5. Barber, H. (ed.): Electroheat. Granada, London, 308 p. (1983)
6. Mühlbauer, A. (Herausgeber) (ed.): Industrielle Elektrowärme-technik, 400 pp., ISBN 3-8027-2903-X. Vulkan-Verlag Essen (1992) (in German)
7. Metaxas, A.C. (ed.): Foundations of Electroheat—A Unified Approach, 500 p., ISBN 0-471-95644-9. Wiley (1996)

Padua, Italy
July 2016

Sergio Lupi

Contents

Symbols

div	*Divergence*—Scalar operator (–)
$grad$	*Gradient*—Vector operator (–)
∇	*Nabla* operator (–)
∇^2	*Laplacian* operator (–)
x, y, z	Cartesian coordinates system (–)
r,φ,z	Azimuthal coordinates system (–)
j	Imaginary unit (–)
$\dot{X}$	Complex quantity (–)
X^*	Complex conjugate value (–)
$\overline{X}$	Vector quantity (–)
$\dot{\overline{X}}$	Complex vector quantity (–)
J_0, J_1	Bessel functions of first kind and zero and first order (–)
Y_0, Y_1	Bessel functions of second kind of zero and first order (–)
T	Absolute temperature (K)
ϑ	Celsius temperature (°C)
Q	Heat (J)
q	Heat flux (W)
p	Heat flux density (W/m^2)
t	Time (s)
α_s, h	Surface heat transfer coefficient (W/m^2 °C)
τ	Time constant (s)
c	Specific heat capacity (J/(°C kg))
λ	Thermal conductivity (W/(m K))
k	Thermal diffusivity (m^2/s)
ρ	Electrical resistivity (Ω m)
$\sigma = 1/\rho$	Conductivity (S/m)
μ	(Absolute) magnetic permeability (H/m)
$\mu_0 = 4\pi \times 10^{-7}$	Magnetic permeability of vacuum (H/m)
$\mu_r = \mu/\mu_0$	Relative (magnetic) permeability (–)
ε	(Absolute) Dielectric constant (F/m)

$\varepsilon_0 = 8.86 \times 10^{-12}$	Dielectric constant of vacuum (F/m)
$\varepsilon_r = \varepsilon/\varepsilon_0$	Relative dielectric constant (–)
$\in$	Emissivity
H	Magnetic field intensity (A/m)
E	Electric field intensity (V/m)
D	Electric flux density (As/m^2)
B	Magnetic flux density (T)
Φ	Magnetic flux (Wb)
A	Magnetic vector potential (V s m^{-1})
J	Current density (A/m^2)
I	Current (A)
V	Electric potential (V)
U	Electric voltage (V)
P	Active power (W)
p_u	Active power per unit axial length (W/m)
p	Active power per unit surface (W/m^2)
w	Volumetric power density (W/m^3)
Q	Reactive power (VAr)
q	Reactive power per unit surface (VAr/m^2)
w_r	Reactive power per unit volume (VAr/m^3)
P_a	Apparent power (VA)
$\overline{S}$	Poynting vector (W/m^2)
W_E, E_w	Energy (J)
w_E	Energy volume density (J/m^3)
$\cos\phi$	Power factor (–)
$k = 1/\delta$	Attenuation coefficient of electromagnetic wave (1/m)
$\delta = \sqrt{\frac{2\rho}{\omega\mu}}$	Penetration depth (m)
f	Frequency (Hz)
ω	Angular velocity, angular frequency (rad/s)
R	Resistance (Ohm)
r	Resistance referred to unit surface (Ohm/m^2)
X	Reactance (Ohm)
x	Reactance referred to unit surface (Ohm/m^2)
Z	Impedance referred to unit surface (Ohm)
z	Impedance, referred to unit surface (Ohm/m^2)
R_{ac}	AC resistance (Ohm)
R_{dc}	DC resistance (Ohm)
$k_r = R_{ac}/R_{dc}$	Skin effect coefficient of resistance (–)
$k_x = X/R_{dc}$	Skin effect coefficient of reactance (–)
v	Volume (m^3)
A	Area (m^2)
Π	Perimeter (m)
l	Length (m)
w	Width of body of rectangular cross section (m)

$h = 2g$	Thickness of body of rectangular cross section (m)
r_i	Internal radius of a cylindrical body (m)
r_e	External radius of a cylindrical body (m)
$d = 2r_e$	Diameter of a cylindrical body (m)
η_e	Electrical efficiency (–)
η	Efficiency (–)
η_T	Thermal efficiency (–)
$\xi = \frac{r}{r_e}$	Dimensionless radial coordinate (–)
$m = \frac{\sqrt{2}\, r_e}{\delta}$	
N	Number of inductor turns (–)
n	Speed of rotation (r/s)
F	Force (N)
f_v	Specific force (N/m^3)
p_m	Magnetic pressure (N/m^2)

Chapter 1
Heat Transfer

Abstract The study of Electro-heat requires a multidisciplinary scientific approach and knowledge in different fields such as heat transfer, electromagnetic fields, electrical engineering, materials science, etc. In this book, before the treatment of the heating processes most applied in industry, there are two introductory chapters with hints concerning heat transfer problems in stationary or transient regime and particular electromagnetic phenomena that occur in applications. After a short theoretical outline of different modes of heat transfer (conduction, radiation and convection), the solution of a number of practical problems is discussed. In particular, the first part of the chapter deals with steady conduction heat transfer problems in bodies of different geometries, with or without internal heat generation. Then, in the second part, the transient heating and cooling temperature distributions in bodies of different geometries are analyzed with various initial and boundary conditions. In this chapter numerical examples are included to allow the student to become familiar with thermal units and the orders of magnitude typical in these applications.

1.1 Introduction

The *Temperature Field* describes the temperature values at all points of a given space at a given instant. A temperature field is in *unsteady-state conditions* when the temperatures values depend on spatial coordinates and time; it is in *steady-state conditions*, if temperatures are independent of time. In many practical cases 2D and 1D temperature fields are considered.

A temperature field is represented graphically by *isothermal surfaces*, each of which connects all points of the field having the same temperature. 1D and 2D fields are represented by means of isothermal curves. The distance between isotherms is inversely proportional to the *temperature gradient*; in this case, the scalar temperature field is associated with the vector field of temperature gradients.

Heat transfer indicates the spontaneous and irreversible processes of the transfer of heat in a region of space where the temperature field is non-uniform.

S. Lupi, *Fundamentals of Electroheat*, DOI 10.1007/978-3-319-46015-4_1

Heat is the quantity characterizing this process; it measures the amount of thermal energy exchanged in the process between physical systems (the body or the system of bodies considered).

Joule is the unit to quantify energy, work, or the amount of heat Q.

Heat transfer always occurs from a region at high temperature to another region at lower temperature. The basic requirement for heat transfer is the presence of a temperature difference. The temperature difference is the driving force of heat transfer, just as the voltage difference for electrical current.

The *heat flux* (or *heat flow rate)* is the rate of heat energy transfer through a given surface, per unit time, (W). *Heat flux surface density* is the heat rate per unit area. In SI units, heat flux density is measured in (W/m^2). Heat rate is a scalar quantity, while heat flux is a vector quantity. The average heat flux density through a surface is expressed as

$$p_t = \frac{q_t}{A}, \ (\mathrm{W/m^2}) \tag{1.1}$$

The total amount of heat transfer Q_t through the surface A during a time interval Δt can be determined from:

$$Q_t = \int_0^{\Delta t} p_t \cdot dt, \ (\mathrm{J}) \tag{1.2}$$

Heat transfer changes the internal energy of the systems involved according to the First Law of Thermodynamics.

1.2 Modes of Heat Transfer

The fundamental modes of heat transfer are *conduction*, *convection*, *radiation* (Fig. 1.1) [1].

In the great majority of electro-technology applications, all three modes of heat transfer (conduction, radiation and convection) occur simultaneously. In some cases, the increase of a certain heat transfer mode and reduction of others can improve process effectiveness. However, in other applications such an attempt could lead just to the opposite effect making a negative impact on the process.

1.2.1 Heat Conduction

Conduction is the most significant means of heat transfer within a solid or between solid objects in thermal contact.

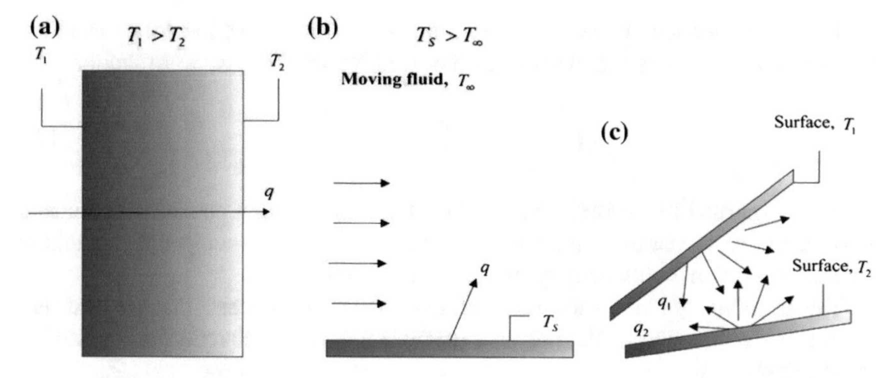

Fig. 1.1 Modes of heat transfer: **a** conduction through solid or stationary fluid; **b** convection from surface to moving fluid; **c** net radiation heat exchange between two surfaces [2]

On a microscopic scale *heat conduction* may be associated with the diffusion of energy due to the interaction of rapidly moving or vibrating atoms and molecules with neighboring atoms and molecules. The energy of particles depends on their random translational motion as well as the internal rotational and vibrational motions of molecules (Fig. 1.2).

In this way the energy (*heat*) is transferred from the more energetic particles to the less energetic ones as result of the interactions between them.

Steady state conduction (based on *Fourier's law*) is a condition that occurs when the temperature difference producing the conduction is constant, so that after an equilibration time, the spatial distribution of temperatures in the conducting object does not change further. In steady state conduction, the amount of heat entering a section is equal to amount of heat coming out.

Transient conduction occurs when the temperature distribution within a body changes as a function of time. The analysis of transient conduction temperature distributions is a complex problem which often requires theoretical approximations or the use of numerical solutions by computer.

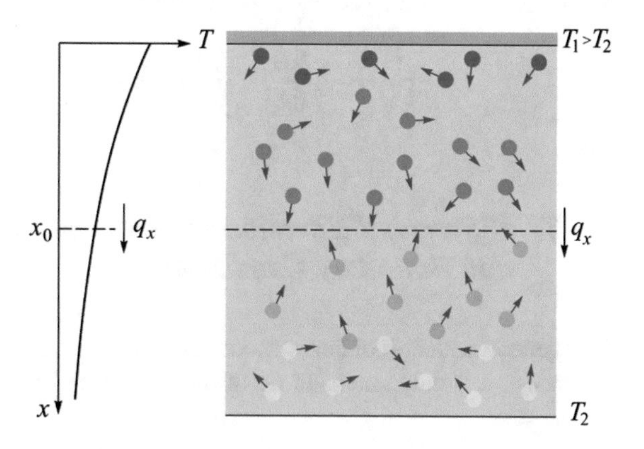

Fig. 1.2 Conductive heat flux q_x between heated walls at temperatures T_1 and T_2 due to motion of gas molecules [1, 2]

The *heat conduction law* (in differential form) states that the local heat flux surface density p_{cond} is proportional to the negative local temperature gradient

$$p_{cond} = -\lambda \frac{d\vartheta}{d\bar{n}} = -\lambda \,\mathrm{grad}\,\vartheta \tag{1.3}$$

where: $\bar{n}$—normal to the isothermal surface through the point considered, oriented in direction of increasing temperatures; $\mathrm{grad}\,\vartheta = \nabla\vartheta$—local temperature gradient (°C/m); λ—thermal conductivity of the material (W/m°C).

The negative sign in the second member of Eq. (1.3) indicates that the heat flux is positive in direction of the negative gradient, i.e. in the direction of decreasing temperatures.

The heat conduction law (1.3) which is also known as *Fourier's law* was empirically discovered by *Joseph Fourier* in 1822.

Thermal conductivity λ is a measure of the material's ability to conduct heat. The thermal conductivity is defined as the rate of heat transfer through a unit thickness of material per unit area and unit temperature difference.

In many calculations regarding electro-technologies, thermal conductivity of metals, melts, alloys and ceramics can assumed to be constant, $\lambda \approx$ *const.,* without taking into account its dependence on temperature. However, in other cases, especially when dealing with some types of oxides and alloys or when the temperature gradient is particularly strong, the dependence of thermal conductivity on temperature $\lambda = \lambda(\vartheta)$ cannot be neglected. In these cases the heat diffusion equation becomes non-linear.

Thermal conductivity values of different types of materials and their dependence on temperature can be found in Ref. [3].

Since the heat flux per unit area is a vector quantity, it can be represented as the sum of its components.

In a Cartesian coordinate system (*x*, *y*, *z*), it can be written as:

$$\bar{p}_t = \bar{u}_x p_x + \bar{u}_y p_y + \bar{u}_z p_z \tag{1.4}$$

with: $\bar{u}_x, \bar{u}_y, \bar{u}_z$—unit vectors in the directions *x*, *y*, *z*; and

$$p_x = -\lambda \frac{\partial\vartheta}{\partial x}; \quad p_y = -\lambda \frac{\partial\vartheta}{\partial y}; \quad p_z = -\lambda \frac{\partial\vartheta}{\partial z}. \tag{1.5}$$

1.2.2 *Differential Equation of Heat Conduction and Boundary Conditions*

The general equation of heat conduction represents in mathematical form the energy balance in a volume element *dv* of the body (see Fig. 1.3).

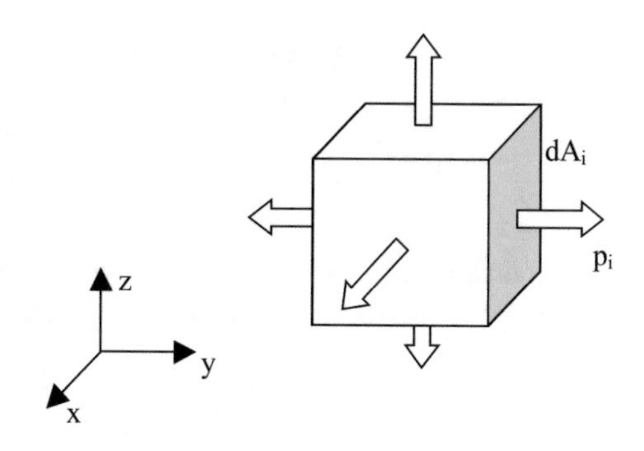

Fig. 1.3 Volume element dv with side surfaces dA_i, with the associated heat flows p_i taken with reference to the outgoing normal to surfaces

It establishes that the sum of the conduction heat flows through the external surfaces of the element:

$$\sum_i dq_i = \sum_i p_i \, dA_i \tag{1.6a}$$

and the power converted into heat inside the element:

$$dP_w = w(P, t) \cdot dv \tag{1.6b}$$

(with: w (W/m^3)—power per unit volume at point P and time t) produces an increase of internal energy of the element in the unit time, which is expressed by the relationship:

$$c\gamma \frac{\partial \vartheta}{\partial t} dv \tag{1.7}$$

where c, γ—specific heat (J/(°C kg)) and mass density (kg/m^3) of the material in the volume element *dv*.

The above sum can be therefore written as:

$$c\gamma \frac{\partial \vartheta}{\partial t} = \mathrm{div}(\lambda \, \mathrm{grad} \, \vartheta) + w \tag{1.8}$$

This is a differential equation of second order known as *Fourier differential equation*.

For practical cases, Eq. (1.8) can be rewritten as follows:

(a) in Cartesian coordinates (x, y, z)

$$c\gamma \frac{\partial \vartheta}{\partial t} = \frac{\partial}{\partial x}(\lambda \frac{\partial \vartheta}{\partial x}) + \frac{\partial}{\partial y}(\lambda \frac{\partial \vartheta}{\partial y}) + \frac{\partial}{\partial z}(\lambda \frac{\partial \vartheta}{\partial z}) + w(x, y, z, t) \tag{1.9a}$$

(b) in cylindrical coordinates (r, φ, z)

$$c\gamma(\frac{\partial \vartheta}{\partial t}) = \frac{1}{r}\frac{\partial}{\partial r}(r\lambda\frac{\partial \vartheta}{\partial r}) + \frac{1}{r}\frac{\partial}{\partial \varphi}(\frac{1}{r}\lambda\frac{\partial \vartheta}{\partial \varphi}) + \frac{\partial}{\partial z}(\lambda\frac{\partial \vartheta}{\partial z}) + w(r, \varphi, z, t) \quad (1.9b)$$

In particular, when dealing with homogeneous and isotropic bodies, where λ, c, γ are independent of coordinates and temperature, Eqs. (1.9a) and (1.9b) take the form:

$$\frac{\partial \vartheta}{\partial t} = k\left[\frac{\partial^2 \vartheta}{\partial x^2} + \frac{\partial^2 \vartheta}{\partial y^2} + \frac{\partial^2 \vartheta}{\partial z^2}\right] + \frac{w}{c\gamma} \quad (1.10a)$$

$$\frac{\partial \vartheta}{\partial t} = k\left[\frac{\partial^2 \vartheta}{\partial r^2} + \frac{1}{r}\frac{\partial \vartheta}{\partial r} + \frac{1}{r^2}\frac{\partial^2 \vartheta}{\partial \varphi^2} + \frac{\partial^2 \vartheta}{\partial z^2}\right] + \frac{w}{c\gamma} \quad (1.10b)$$

with: $k = \frac{\lambda}{c\gamma}$—thermal diffusivity ($m^2/s$).

Equations (1.8)–(1.10a) and (1.10b), with the condition $\partial\vartheta/\partial t = 0$, give as solution steady state conduction fields.

The solutions of Fourier equation are derived in analytical form, starting from Eqs. (1.10a) and (1.10b), when the material thermal properties are assumed to be constant, or from Eqs. (1.9a) and (1.9b) when they depend only on the coordinates of the point.

When variations with temperature of material properties must be taken into account, the solutions become more complex and, given the non-linearity of the problem, can be obtained only by numerical methods.

The solution of Fourier equation for a specific problem requires specification of the initial temperature distribution within the body and the boundary conditions that define the mode of heat transfer between the surface of the body and the surrounding environment.

Of particular interest are the following boundary conditions:

(a) ***Boundary conditions of first type***: *surface temperature constant (or zero)*

$$\vartheta_s = \text{const.} \quad (1.11a)$$

(b) ***Boundary conditions of second type***: *surface heat flux constant (or zero)*

$$\left[\frac{\partial \vartheta}{\partial n}\right]_s = \frac{p_S}{\lambda} \quad (1.11b)$$

with: p_S—surface specific power (W/m^2); *n*—normal to the surface oriented towards outside.

(c) ***Boundary condition of third type***: *surface heat losses proportional to the surface temperature*

$$\lambda \frac{\partial \vartheta}{\partial n} + \alpha_s(\vartheta - \vartheta_a) = 0 \rightarrow \frac{\partial \vartheta}{\partial n} + h(\vartheta - \vartheta_a) = 0 \tag{1.11c}$$

with: α_S (W/m^2 °C)—coefficient of heat transmission from the surface of the body to the surrounding medium; ϑ_a—temperature of the surrounding medium; $h = \alpha_s/\lambda$ (m^{-1}).

1.2.3 Steady State Conduction Heat Transfer in Bodies Without Internal Heat Generation (w = 0)

1.2.3.1 Heat Conduction Through a Large Plane Wall

From Eqs. (1.10a) and (1.10b), the Fourier equation, for steady conduction through a constant area plane wall (Fig. 1.4a), can be written:

$$\frac{\partial^2 \vartheta}{\partial x^2} = 0. \tag{1.12}$$

The solution of this DE with the boundary conditions $\vartheta = \vartheta_1$ at $x = 0$ and $\vartheta = \vartheta_2$ at $x = \Delta x$ gives:

$$\vartheta = \vartheta_1 - \frac{\vartheta_1 - \vartheta_2}{\Delta x} x \tag{1.13}$$

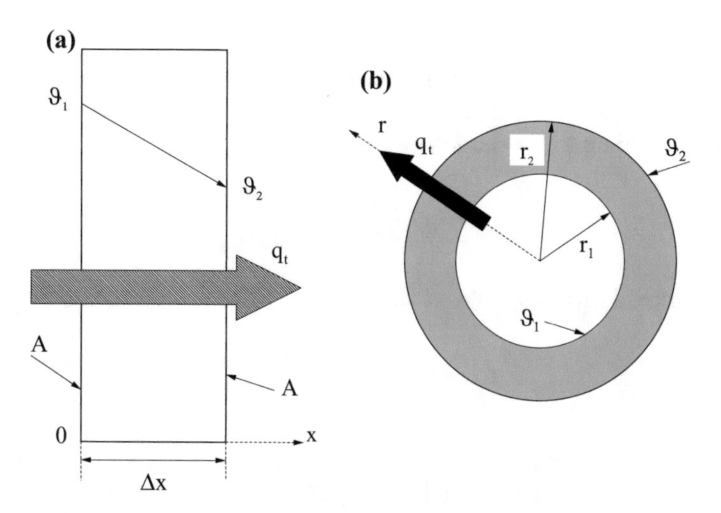

Fig. 1.4 **a** Steady state heat conduction in a large plane wall; **b** Steady state heat conduction in a cylindrical layer [3]

Since the temperature depends only on the coordinate *x*, the isothermal surfaces are planes *x* = *const.* parallel to the wall surfaces and the vector of thermal flux is directed along *x*. Introducing Eq. (1.13) in Eq. (1.5), the module of q_t can be written as follows:

$$q_{tx} = -\lambda A \frac{d\vartheta}{dx} = \lambda A \frac{\vartheta_1 - \vartheta_2}{\Delta x} = \frac{\vartheta_1 - \vartheta_2}{R_{th}}, \text{ (W)} \tag{1.14}$$

where: $R_{th} = \Delta x/(\lambda A)$ (°C/W)—is the *thermal resistance* of the wall against heat conduction or simply the *conduction resistance* of the wall.

1.2.3.2 Steady State Heat Transfer Through the Wall of a Pipe

Consider now the steady state heat transfer through the wall of a pipe in normal direction to the wall surface, neglecting heat transfer in other directions (Fig. 1.4b). Also in this case the heat transfer is one-dimensional, and the temperature depends only on the radial position, $\vartheta = \vartheta(r)$.

If we assume that there is no heat generation in the wall and that thermal conductivity is constant, the Fourier DE equation becomes:

$$\frac{d^2\vartheta}{dr^2} + \frac{1}{r}\frac{d\vartheta}{dr} = 0 \tag{1.15}$$

and the solution is:

$$\vartheta = \vartheta_1 - (\vartheta_1 - \vartheta_2)\frac{\ln(r/r_1)}{\ln(r_2/r_1)}. \tag{1.16}$$

Equation (1.16) shows that the isothermal surfaces are the concentric coaxial cylinders r = const., the vector of flux density is directed along the radius and its module—according to Fourier law—is given by:

$$p_{tr} = -\lambda \frac{d\vartheta}{dr} = \lambda \frac{\vartheta_1 - \vartheta_2}{r \ln(r_2/r_1)} \tag{1.17}$$

The thermal flux through the surface $A_\ell = 2\pi r\ell$, of axial length ℓ, therefore is:

$$q_{tr} = p_{tr} \cdot A_\ell = \pi\ell \frac{\vartheta_1 - \vartheta_2}{\frac{\ln(r_2/r_1)}{2\lambda}} \tag{1.18}$$

Equation (1.18) highlights that the thermal flux through every cylindrical surface is constant, i.e. it does not depend on the radius. Its value for unit axial length is therefore given by the formula:

$$q_{tu} = \frac{q_{tr}}{\ell} = \pi \frac{\vartheta_1 - \vartheta_2}{\frac{\ln(r_2/r_1)}{2\lambda}} = \frac{\pi(\vartheta_1 - \vartheta_2)}{R_{th}} \quad (1.18a)$$

with $R_{th} = (\ln r_2/r_1)/2\lambda$—thermal resistance to conduction of the cylindrical wall with temperature difference $(\vartheta_1 - \vartheta_2)$ between its surfaces.

1.2.3.3 Steady State Heat Transfer Through Multi-layered Walls

With reference to the multi-layered plane wall of Fig. 1.5a, we assume that no heat is lost through the edges and then in steady state the same heat flux q_t must flow through each layer.

With the symbols of the figure, according to Eq. (1.14), we can write:

$$\left.\begin{aligned} \vartheta_1 - \vartheta_2 &= q_t \cdot R_{th1} \\ \vartheta_2 - \vartheta_3 &= q_t \cdot R_{th2} \\ \vartheta_3 - \vartheta_4 &= q_t \cdot R_{th3} \end{aligned}\right\} \quad (1.19)$$

By adding Eq. (1.19) and expanding the result to a wall containing *n* layers we can write:

$$q_{tx} = \frac{\vartheta_1 - \vartheta_{n+1}}{R_{th1} + R_{th2} + R_{th3} + \cdots + R_{thn}} \quad (1.20a)$$

with $R_{thi} = \Delta x_i/(\lambda_{im} A)$ (°C/W)—thermal resistance of the *i*-th layer having average thermal conductivity λ_i (W/m °C).

Equation (1.20a) allows to evaluate the heat transmitted through a composite plane wall when the temperatures ϑ_1 and ϑ_{n+1} of external surfaces are known. The temperatures $\vartheta_2, \vartheta_3, \ldots, \vartheta_n$ between layers, may be calculated from Eq. (1.19) when Eqs. (1.20a) and (1.20b) has been solved for q_{tx}.

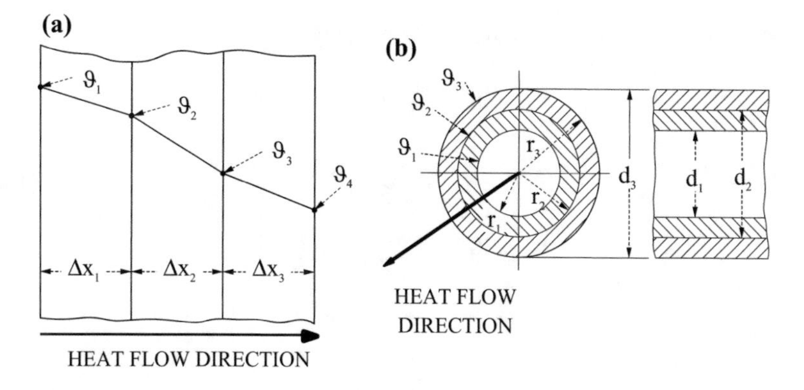

Fig. 1.5 Schematic of multi-layered plane (**a**) and cylindrical (**b**) wall (unidirectional flow)

Note: In many problems, instead of the temperatures of the internal and external surfaces, the temperatures ϑ_i, ϑ_e of the media lapping on such surfaces and the corresponding heat transfer coefficients α_{si}, α_{se} are known. In these cases to the total thermal resistance must be added the terms of resistances between each external surface and the surrounding medium and Eq. (1.20a), written in a general form becomes:

$$q_{tx} = \frac{\vartheta_i - \vartheta_e}{\frac{1}{\alpha_{si} A_i} + \frac{\Delta X_1}{\lambda_{1m} A_{1m}} + \frac{\Delta X_2}{\lambda_{2m} A_{2m}} + \cdots + \frac{\Delta X_n}{\lambda_{nm} A_{nm}} + \frac{1}{\alpha_{se} A_e}} \tag{1.20b}$$

with: $A_{km} = \sqrt{A_k \cdot A_{k+1}}$, $(k = 1, 2, \ldots, n)$; A_k, A_{km}—internal and mean values of surfaces of layers.

From a practical point of view, the most important cases are those of simple walls (n = 1), insulated walls (n = 2) and composite insulated walls (n = 3).

Using Eq. (1.18a), the same procedure can be used for a multilayered cylindrical wall like the one of Fig. 1.5b.

The corresponding equations in the case of n layers are:

$$\left.\begin{aligned} q_{tu} &= \frac{\pi(\vartheta_1 - \vartheta_2)}{R_{th\,1}} \\ &\cdots \\ q_{tu} &= \frac{\pi(\vartheta_i - \vartheta_{i+1})}{R_{thi}} \\ &\cdots \\ q_{tu} &= \frac{\pi(\vartheta_n - \vartheta_{n+1})}{R_{thn}} \end{aligned}\right\} \tag{1.21}$$

$$q_{tu} = \frac{\pi(\vartheta_1 - \vartheta_{i+1})}{\sum\limits_{i=1}^{n} \frac{\ln(r_{i+1}/r_i)}{2\lambda_i}} = \frac{\pi(\vartheta_1 - \vartheta_{i+1})}{R_{th,tot}} \tag{1.22}$$

with:

$R_{th,tot} = \sum\limits_{i=1}^{n} \frac{\ln(r_{i+1}/r_i)}{2\lambda_i}$—total thermal resistance of the wall.

The temperatures $\vartheta_2, \vartheta_3, \ldots, \vartheta_n$ between layers are:

$$\left.\begin{aligned} \vartheta_2 &= \vartheta_1 - \frac{q_{tu}}{2\pi\lambda_1} \ln\frac{r_2}{r_1} \\ &\cdots \\ \vartheta_i &= \vartheta_1 - \frac{q_{tu}}{\pi} \sum_{k=1}^{i-1} \frac{1}{2\lambda_k} \ln\frac{r_{k+1}}{r_k} \\ &\cdots \\ \vartheta_n &= \vartheta_1 - \frac{q_{tu}}{\pi} \sum_{k=1}^{n-1} \frac{1}{2\lambda_k} \ln\frac{r_{k+1}}{r_k} \end{aligned}\right\} \tag{1.23}$$

Example 1.1 Calculate specific surface losses through a simple plane wall of Schamotte with the following data:

$$\lambda_1 = 0.7\ \mathrm{W/m\ ^\circ C};\quad \Delta x = 0.275\ \mathrm{m};\quad A = 1\ \mathrm{m}^2;$$

surface temperatures: $\vartheta_1 = 1000\ ^\circ\mathrm{C}$, $\vartheta_2 = 50\ ^\circ\mathrm{C}$ (Fig. 1.6).

$$R_{th} = \frac{\Delta x}{\lambda A} = \frac{0.275}{0.7 \cdot 1} = 0.393\ ^\circ\mathrm{C/W}$$

$$q_t = \frac{\vartheta_1 - \vartheta_2}{R_{th}} = \frac{950}{0.393} = 2.417\ \mathrm{W}$$

Example 1.2 Specific losses through an insulated plane wall with two layers: $A = 1\ \mathrm{m}^2$; layer (1): Schamotte, $\lambda_1 = 0.7\ \mathrm{W/m\ ^\circ C}$, $\Delta x_1 = 0.125\ \mathrm{m}$, $\vartheta_1 = 1000\ ^\circ\mathrm{C}$; layer (2): Kieselgur, $\lambda_2 = 0.15\ \mathrm{W/m\ ^\circ C}$, $\Delta x_2 = 0.15\ \mathrm{m}$, $\vartheta_2 = 50\ ^\circ\mathrm{C}$.

$$R_{th1} = \frac{\Delta x_1}{\lambda_1 \cdot A} = \frac{0.125}{0.7 \cdot 1} = 0.179\ ^\circ\mathrm{C/W}$$

$$R_{th2} = \frac{\Delta x_2}{\lambda_2 \cdot A} = \frac{0.15}{0.15 \cdot 1} = 1.00\ ^\circ\mathrm{C/W}$$

$$q_t = \frac{\vartheta_1 - \vartheta_2}{R_{th1} + R_{th2}} = \frac{950}{0.179 + 1.00} = 806\ \mathrm{W}$$

$$\vartheta_{12} = \vartheta_1 - q_t \cdot R_{th1} = 1.000 - 806 \cdot 0.179 = 856\ ^\circ\mathrm{C}$$

Note—The results of the above examples show that the use of multi-layered insulated walls having the same overall thickness allows to drastically reduce heat conduction losses (in the examples from 2417 to 806 W/m²).

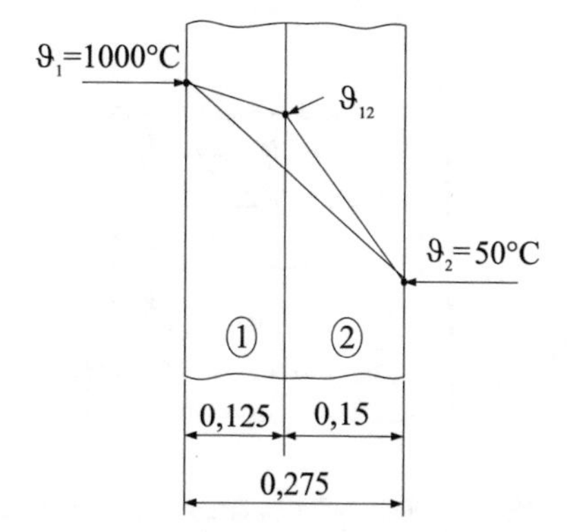

Fig. 1.6 Schematic of the wall of Examples 1.1 and 1.2 and temperature distributions

1.2.3.4 Thermal Insulation of a Cylindrical Tube—Critical Radius

Figure 1.7a represents a tube with internal and external radii r_1 and r_2 covered by an insulating layer of thickness δ and external radius r_3.

The values ϑ_1, ϑ_2 of temperatures of the internal and external surfaces, α_1, α_2 of the heat transfer coefficients at internal and external surfaces and λ_1, λ_2 of thermal conductivities of materials of the tube walls are assumed to be constant.

For this cylindrical two-layer system, the total thermal resistance for unit axial length can be written as follows:

$$R_{th,tot} = \frac{1}{2r_1\alpha_1} + \frac{1}{2\lambda_1}\ln\frac{r_2}{r_1} + \frac{1}{2\lambda_2}\ln\frac{r_3}{r_2} + \frac{1}{2r_3\alpha_2} \tag{1.24}$$

with: $r_3 = r_2 + \delta$.

If now we increase the thickness of the insulating layer, increasing r_3 while keeping constant $\lambda_1, \lambda_2, r_1, r_2$, it occurs an increase of the thermal resistance $(1/2\lambda_2)\ln(r_3/r_2)$ but at the same time a decrease of the term $1/(2r_3\alpha_2)$.

As a consequence, in Eq. (1.24) the total thermal resistance $R_{th,tot} = f(r_3)$ will show a minimum at the so-called "*critical radius of thermal insulation*", which is given by [5]:

$$r_3 = r_{cr} = \frac{\lambda_2}{\alpha_2}. \tag{1.25}$$

If $r_{cr} \leq r_2$, the increase of the thickness of the insulation layer will produce a continuous decrease of the losses $q_{tu} = (\pi\,\Delta\vartheta)/R_{th,tot}$ to the surrounding medium, as shown by the curve 1 of Fig. 1.7b. On the contrary, if $r_{cr} > r_2$, the heat transfer to the ambient will first increase increasing δ, it reaches a maximum for $r_{cr} = r_2$ and, for a further increase of δ, starts to decrease after the maximum.

Therefore, in order to reduce thermal losses, the choice of the insulating material should satisfy the condition:

$$\lambda_{ins} < \alpha_2 r_2$$

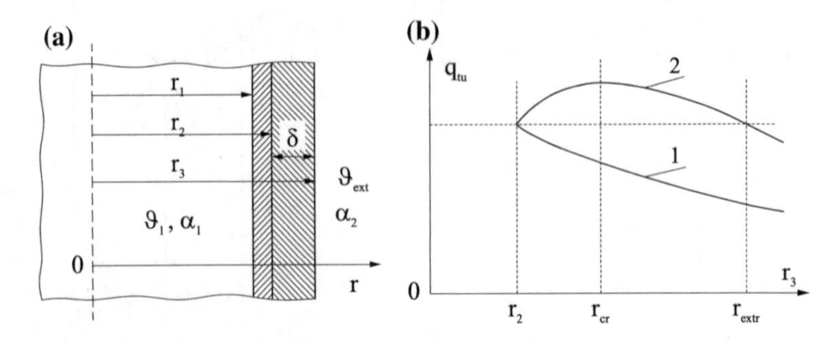

Fig. 1.7 Tube with external insulation layer: determination of critical radius [3, 4]

1.2.4 *Steady State Conduction Heat Transfer in Bodies with Internal Heat Generation (w ≠ 0)*

1.2.4.1 Plane Wall with Uniform Heat Sources and Given Temperature on External Surfaces

An example of this type of process occurs in conductors where the flow of a DC electrical current produces Joule heat sources uniformly distributed in the wall thickness ($w = const.$).

With reference to the plane wall and the coordinate system of Fig. 1.8a, we suppose constant the material thermal conductivity λ and given the temperature ϑ_S on the wall external surfaces *(boundary condition of first type)*.

Since the wall is infinite along *y* and *z*, in steady-state conditions ($\partial\vartheta/\partial t = 0$) with $w = const.$, the temperature field varies only along *x* (i.e. $\partial\vartheta/\partial y = \partial\vartheta/\partial z = 0$) and Eq. (1.10a) becomes:

$$\frac{\partial \vartheta}{\partial x^2} + \frac{w}{\lambda} = 0 \tag{1.26}$$

The boundary conditions with temperature imposed at the external surfaces are:

$$\vartheta = \vartheta_S \quad \text{for } x = \pm\delta \tag{1.27a}$$

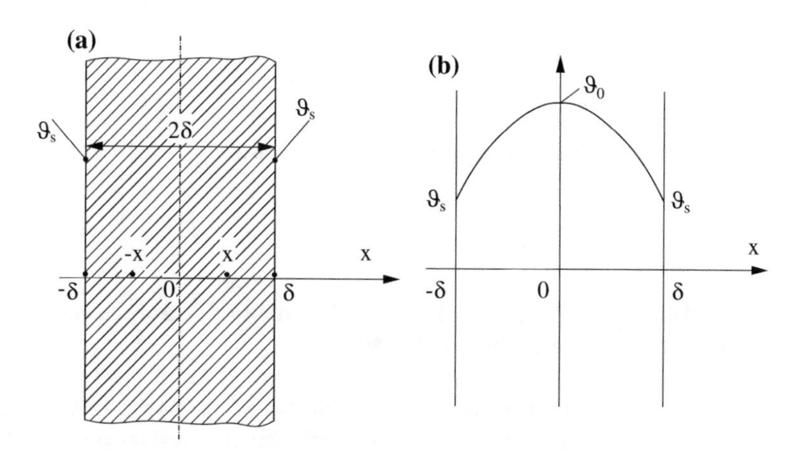

Fig. 1.8 a Plane wall with uniform distribution of heat sources w = const. and boundary conditions of first type; **b** Temperature distribution in plane wall with w = const. and boundary conditions of first type

For the equality of the external surface temperatures and the uniform distribution of w, the temperature field will be symmetric to the plane $x = 0$ and the heat flux will have the same values but opposite directions on the right and left side, i.e. it is:

$$\vartheta(x) = \vartheta(-x) \quad \text{and} \quad q(x) = q(-x) \quad \text{for} \; -\delta \leq x \leq \delta.$$

As a consequence, at the wall center plane the heat flux density and the temperature gradient are equal to zero, i.e. it is:

$$\frac{\partial \vartheta}{\partial x} = 0 \quad \text{for } x = 0. \tag{1.27b}$$

The solution of Eq. (1.26) with the boundary conditions (1.27a) at $x = \delta$ and (1.27b) at $x = 0$, gives the following temperature distribution in the semi-plane $x > 0$:

$$\vartheta = \vartheta_S + \frac{w}{2\lambda}\left(\delta^2 - x^2\right) \tag{1.28}$$

The maximum of this temperature distribution (shown in Fig. 1.8b) occurs at $x = 0$, and has the value:

$$\vartheta_0 = \vartheta(x = 0) = \vartheta_S + \frac{w\delta^2}{2\lambda}$$

The heat flux density on a generic plane $x = const.$ therefore is:

$$p_t(x) = -\lambda \frac{\partial \vartheta}{\partial x} = w \cdot x, \tag{1.29}$$

it has the maximum value at the surface of the plane wall:

$$p_t(\delta) = w \cdot \delta. \tag{1.29a}$$

1.2.4.2 Cylindrical Body with Uniform Distribution of Heat Sources and Given Temperature at the Surface

The body considered here is a "long" cylindrical bar of radius r_e, with constant thermal conductivity λ and uniform distribution of internal heat sources $w = \text{const.}$ (Fig. 1.9a).

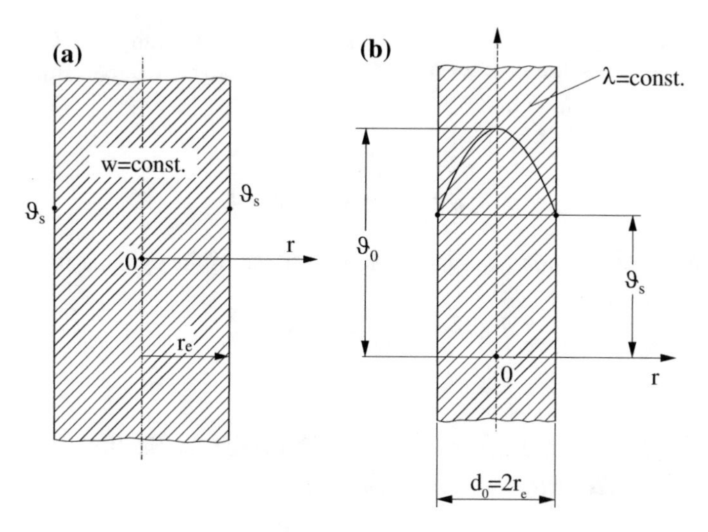

Fig. 1.9 **a** Cylindrical bar with uniform distribution of heat sources w = const. and boundary conditions of first type; **b** Temperature distribution in cylindrical bar with w = const. and boundary conditions of first type

It is required to evaluate the distribution of temperature ϑ and heat flux density p in steady-state condition, when the bar surface is maintained at a given constant temperature ϑ_S (boundary condition of first type).

For the previous assumptions the temperature field is a function only of radius, $\vartheta = f(r)$, and is symmetric to the axis of the bar.

With the condition $\partial\vartheta/\partial t = 0$, Eq. (1.10b) becomes:

$$\frac{\partial^2\vartheta}{\partial r^2} + \frac{1}{r}\frac{\partial\vartheta}{\partial r} + \frac{w}{\lambda} = 0 \tag{1.30}$$

The boundary conditions therefore are:

- at the surface: $\vartheta = \vartheta_S \quad \text{for } r = r_e$ (1.31a)
- on the axis: $\partial\vartheta/\partial r = 0 \quad \text{for } r = 0$ (1.31b)

By integration of Eq. (1.30) with the boundary conditions (1.31a) and (1.31b), one obtains:

$$\left.\begin{aligned} \vartheta &= \vartheta_S + \tfrac{w}{4\lambda}\left(r_e^2 - r^2\right) \\ \vartheta_0 &= \vartheta(r=0) = \vartheta_S + \tfrac{w}{4\lambda} r_e^2 \end{aligned}\right\} \tag{1.32}$$

Figure 1.9b shows the temperature distribution in the bar.

From Eqs. (1.3), (1.32) it is easy to obtain the corresponding heat flux densities through a generic cylindrical surface of radius r and the bar external surface, where the heat flux is maximum:

$$\left.\begin{aligned} p &= -\lambda \frac{\partial \vartheta}{\partial r} = \frac{w}{2} r \\ p_S &= p(r = r_e) = \frac{w}{2} r_e \end{aligned}\right\} \tag{1.33}$$

The heat flux per unit axial length at the surface of the cylinder therefore is:

$$q_u = p_S \cdot 2\pi r_e = w \cdot \pi r_e^2 \tag{1.34}$$

1.2.5 Transient Conduction Heat Transfer in Bodies Without Internal Heat Generation (w = 0)

Heating and cooling processes where the temperature in the body varies with position and time are called transient heat transfer processes. Typical examples are the heating or cooling of workpieces in a furnace.

The solution of transient heat transfer processes is obtained as solution of the differential Eqs. (1.10a) and (1.10b) in the considered coordinates system with the relevant boundary conditions (1.11a) and (1.11b).

1.2.5.1 Convective Cooling (or Heating) of an Infinite Slab

We make reference to the infinite slab of Fig. 1.10, which is initially at a constant temperature ϑ_{i0} and is cooled by convection in an external medium at constant temperature ϑ_e. The convective heat transfer coefficient α at the surface of the slab and the thermal conductivity of the slab material are supposed to be constant and independent on time.

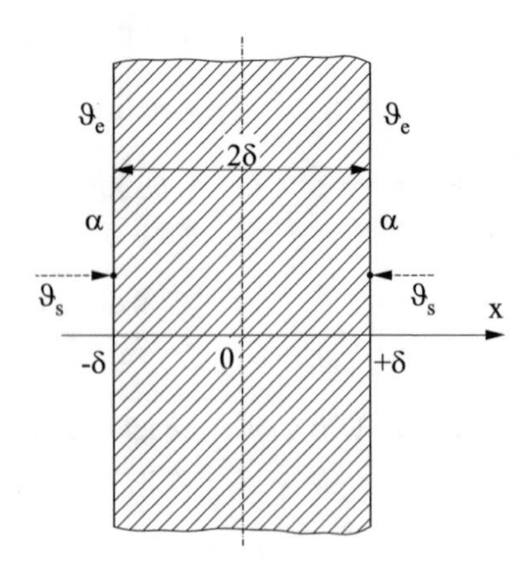

Fig. 1.10 Geometry considered in the cooling (or heating) of an infinite slab

In this system the temperature is a function only of the position x and the time t; therefore it is $\partial\vartheta/\partial y = \partial\vartheta/\partial z = 0$.

As a consequence, in the case here considered with $w = 0$, Eq. (1.10a) reduces to:

$$\frac{\partial\vartheta}{\partial t} = \alpha\frac{\partial^2\vartheta}{\partial x^2}. \tag{1.35}$$

Moreover, the temperature field must be symmetric with respect to the symmetry plan, i.e. it is:

$$\vartheta(x) = \vartheta(-x), \quad \text{for } -\delta \le x \le +\delta,\ t > 0 \tag{1.36}$$

from which it follows also that is:

$$\partial\vartheta/\partial x = 0, \quad \text{for } x = 0,\ t > 0. \tag{1.37a}$$

Taking into account Eq. (1.36), it's sufficient to evaluate the temperature distribution only in half slab, e.g. on its right part, with boundary and initial conditions given by Eq. (1.37a) and

$$\vartheta = \vartheta_{io}, \quad \text{for } 0 \le x \le \delta,\ t = 0. \tag{1.37b}$$

$$-\lambda\frac{\partial\vartheta}{\partial x} = \alpha(\vartheta_s - \vartheta_e), \quad \text{for } x = \delta,\ t > 0 \tag{1.37c}$$

with ϑ_s—temperature of the slab surface (°C).

To obtain the solution in a general form, it is convenient to introduce dimensionless quantities for coordinates, time and temperature respectively:

$$\xi = \frac{x}{\delta}; \quad \tau = \frac{\alpha \cdot t}{\delta^2}; \quad \Theta = \frac{\vartheta - \vartheta_{i0}}{\vartheta_e - \vartheta_{io}} \tag{1.38a}$$

Note: in the literature τ is known as *Fourier number*, and is often indicated with the symbol F_o or N_{Fo} [3, 4].

In the new variables, the derivatives can be written as follows:

$$\left.\begin{aligned} \frac{\partial\vartheta}{\partial x} &= \frac{\partial\Theta}{\partial\xi}\cdot\frac{(\vartheta_e - \vartheta_{i0})}{\delta} \\ \frac{\partial^2\vartheta}{\partial x^2} &= \frac{\partial^2\Theta}{\partial\xi^2}\cdot\frac{(\vartheta_e - \vartheta_{i0})}{\delta^2} \\ \frac{\partial\vartheta}{\partial t} &= \frac{\partial\Theta}{\partial\tau}\cdot\frac{\alpha(\vartheta_e - \vartheta_{i0})}{\delta^2} \end{aligned}\right\}, \tag{1.38b}$$

and Eq. (1.35) becomes:

$$\frac{\partial \Theta}{\partial \tau} = \frac{\partial^2 \Theta}{\partial \xi^2}. \tag{1.39}$$

The initial and boundary conditions (1.37a) and (1.37b), are rewritten in the form:

$$\left.\begin{aligned} &\Theta = 1 && \text{for } \tau = 0 \text{ and } 0 \le \xi \le 1 \\ &\frac{\partial \Theta}{\partial \xi} = 0 && \text{for } \tau > 0 \text{ and } \xi = 0 \\ &\frac{\partial \Theta}{\partial \xi} = \frac{\alpha\delta}{\lambda} \cdot \Theta_s && \text{for } \tau > 0 \text{ and } \xi = 1 \end{aligned}\right\} \tag{1.39b}$$

with: $\Theta_s = (\vartheta_s - \vartheta_{i0})/(\vartheta_e - \vartheta_{i0})$—dimensionless temperature of the slab surface.

Note: the dimensionless quantity $(\alpha\delta/\lambda)$ is know in the literature as *Biot number* and is often indicated with the symbol *Bi* or N_{Bi} [3, 4].

The solution of Eq. (1.39) is conveniently obtained by the method of separation of variables [3, 5], expressing the function $\Theta(\xi, \tau)$ as the product of two functions, one dependent on the variable ξ only, the second one only on the variable τ, i.e. $\Theta(\xi, \tau) = X(\xi) \cdot T(\tau)$. In this way from Eq. (1.39) are obtained two separate ordinary differential equations in the variables *X* and *T*, whose integration constants can be evaluated applying the initial and boundary conditions.

The solution for $\Theta(\xi, \tau)$ is finally written in the form [3]:

$$\Theta(\xi, \tau) = 1 - \sum_{n=1}^{\infty} A_n \cdot \cos(\nu_n \xi) \cdot e^{-\nu_n^2 \tau} \tag{1.40}$$

where:

$A_n = \frac{2 \sin \nu_n}{\nu_n + \sin \nu_n \cdot \cos \nu_n}$, $n = 1, 2, \ldots$

ν_n—roots of the equation: $\cot(\nu) = (\nu\lambda)/(\alpha\delta)$.

The values of ν and ν^2 are given as a function of $(\alpha\delta/\lambda)$ in Table A.1 of appendix.

Equation (1.40) calculated for $\xi = 1$ and $\xi = 0$ gives the values of temperature at the surface and the center of the slab:

$$\Theta_s = \Theta(1, \tau) = 1 - \sum_{n=1}^{\infty} A_n \cdot \cos \nu_n \cdot e^{-\nu_n^2 \tau} \tag{1.41a}$$

$$\Theta_a = \Theta(0, \tau) = 1 - \sum_{n=1}^{\infty} A_n \cdot e^{-\nu_n^2 \tau} \tag{1.41b}$$

Equation (1.40) can be used also for calculating the heating of a slab placed in a medium at temperature $\vartheta_e > \vartheta_{io}$. In this case the solution gives at any instant and any point the difference $\vartheta - \vartheta_{io}$, i.e. the increase of temperature above ϑ_{io}.

It must be noticed that in practical calculations, for $\tau > 0.3$ (i.e. after the initial period of the thermal transient) all exponential terms, except the one for $n = 1$, tend to zero and, with an accuracy within few percent, can be used the approximation:

$$\Theta(\xi, \tau) \approx \frac{2 \sin \nu_1}{\nu_1 + \sin \nu_1 \cdot \cos \nu_1} \cos(\nu_1 \xi) \cdot e^{-\nu_1^2 \tau} \tag{1.42}$$

1.2.5.2 Thermal Transients in Cylindrical Loads—General Solution

We refer now to a cylindrical workpiece, infinitely long in axial direction, with known temperature distribution along radius at $t = 0$, and specified conditions of heat exchange between surface and environment during the thermal transient.

The consideration of a "long" system allows us to neglect edge effects, so that the temperature ϑ is a function only of position along radius and time: $\vartheta = \vartheta(r, t)$.

The study of thermal transient requires the solution of Fourier Eq. (1.10b), with the relevant initial and boundary conditions.

We will use the notations:

r_e, r radius of the cylinder and radius of a generic radial internal position (m)
ϑ temperature (°C) at radius r and time t (s)
ϑ_s, ϑ_a values of ϑ at surface and axis
$\vartheta_0(r)$ initial distribution of temperature along radius.

In cylindrical coordinates (r, φ, z), taking into account that the heat transfer takes place only along r and the boundary conditions are independent on φ and z, Eq. (1.10b) takes the form:

$$\frac{\partial \vartheta}{\partial t} = k \cdot \left[\frac{\partial^2 \vartheta}{\partial r^2} + \frac{1}{r}\frac{\partial \vartheta}{\partial r}\right], \quad 0 \le r \le r_e \tag{1.43}$$

Assuming now that the initial distribution of temperature along radius $\vartheta_0(r)$ is known and the surface temperature ϑ_s is maintained at zero during the thermal transient, Eq. (1.43) is solved with the initial and boundary conditions:

$$\left.\begin{array}{ll} \vartheta = \vartheta_0(r), & \text{for } t = 0 \\ \vartheta = \vartheta_s = 0, & \text{for } t > 0; r = r_e \end{array}\right\} \tag{1.44}$$

Applying as in the previous paragraph the method of separation of variables, suppose now that the solution of (1.34) has the form:

$$\vartheta = u(r) \cdot e^{-k\alpha^2 t} \tag{1.45}$$

where $u(r)$ is a function of the coordinate r only.

Then, from Eq. (1.45) we have:

$$\left.\begin{aligned}\frac{\partial\vartheta}{\partial t} &= -k\alpha^2\cdot u\cdot e^{-k\alpha^2 t}\\ \frac{\partial\vartheta}{\partial r} &= \frac{du}{dr}\cdot e^{-k\alpha^2 t}\\ \frac{\partial^2\vartheta}{\partial r^2} &= \frac{d^2u}{dr^2}\cdot e^{-k\alpha^2 t}\end{aligned}\right\}. \tag{1.46}$$

Equation (1.45) can be a solution of the DE (1.43), only if the function *u (r)* is in turn solution of:

$$\frac{\partial^2 u}{\partial r^2}+\frac{1}{r}\frac{\partial u}{\partial r}+\alpha^2 u=0 \tag{1.47}$$

which is obtained by substitution of Eqs. (1.46) into (1.43).

As known, this is the Bessel differential equation of order zero which has the general solution:

$$u = A\cdot J_o(\alpha r)+B\cdot Y_o(\alpha r) \tag{1.48a}$$

with: Jo (x), Yo (x)—respectively Bessel functions of order zero of the first and second kind; A, B—integration constants.

Since it is $Y_0(0)=-\infty$, and it is not physically possible that on the axis ($r = 0$) the temperature becomes infinite, it must be assumed $B = 0$ and the solution (1.48a) becomes:

$$\vartheta = A\cdot J_o(\alpha r)\cdot e^{-k\alpha^2 t} \tag{1.48b}$$

This solution can meet the boundary conditions (1.44) only if α is chosen so that it is:

$$J_o(\beta) = J_o(\alpha r_e) = 0 \tag{1.49}$$

This equation has an infinite number of positive real roots $\beta_1, \beta_2, \beta_3, \ldots$, whose values are given in Table A.2 of appendix for $C=\infty$.

Finally, assuming that the function $\vartheta_o(r)$ can be expanded in series of Bessel functions of the form:

$$\vartheta_o(r)=\sum_{n=1}^{\infty}A_n\cdot J_o(\alpha_n r) \tag{1.50}$$

the solution of the problem becomes:

$$\vartheta=\sum_{n=1}^{\infty}A_n\cdot J_o(\alpha_n r)\cdot e^{-k\alpha_n^2 t} \tag{1.51}$$

The constants A_n can be evaluated by multiplying both sides of Eq. (1.50) by $r\,J_o(\alpha_n r)$ and integrating between 0 and r_e:

$$\int_0^{r_e} r \cdot \vartheta_0(r) \cdot J_o(\alpha_n r)\,dr = A_n \int_0^{r_e} r \cdot \left[\sum_{k=1}^{\infty} J_o(\alpha_k r)\right] \cdot J_o(\alpha_n r)\,dr$$

Using the relations (A3.1) and (A3.3) given in Appendix A.3, we obtain:

$$\int_0^{r_e} r \cdot \vartheta_o(r) \cdot J_o(\alpha_n r)\,dr = A_n \frac{r_e^2}{2} J_1^2(\alpha_n r_e)$$

and:

$$A_n = \frac{2}{r_e^2 J_1^2(\alpha_n r_e)} \int_0^{r_e} r \cdot \vartheta_o(r) \cdot J_o(\alpha_n r)\,dr \tag{1.52}$$

with: J_1 (x)—Bessel function of the first kind and order 1.

Equation (1.51) thus can be written as follows:

$$\vartheta = \frac{2}{r_e^2} \sum_{n=1}^{\infty} e^{-k\alpha_n^2 t} \frac{J_o(\alpha_n r)}{J_1^2(\alpha_n r_e)} \int_0^{r_e} r \cdot \vartheta_o(r) \cdot J_o(\alpha_n r)\,dr \tag{1.53}$$

The development in series of Bessel functions (1.50) is "appropriate" for the solution of the thermal transient in a cylindrical workpiece with zero (or constant) surface temperature (boundary condition of the first type).

When the boundary condition for $r = r_e$ is different, it is assumed that the function $\vartheta_0(r)$ can be still described by a series of Bessel functions of the type (1.50), but with a different choice of the values α_n such that the solution satisfies the new boundary condition.

For example, in case of boundary condition of the second type (surface heat flux equal to zero), the condition

$$\left[\frac{\partial \vartheta}{\partial r}\right]_{r=r_e} = 0 \quad \text{for } t > 0 \tag{1.54a}$$

is satisfied by a solution of the type (1.51) only if the values α_n are chosen in such a way that it is:

$$J_1(\alpha_n r_e) = J_1(\beta_n) = 0 \tag{1.54b}$$

since it is $J_o'(z) = -J_1(z)$.

Taking into account Eq. (A3.4) of Appendix A.3, the determination of coefficients A_n and the temperature ϑ in this case must be carried out using the relationships:

$$A_n = \frac{2}{r_e^2 J_o^2(\alpha_n r_e)} \int_0^{r_e} r \cdot \vartheta_o(r) \cdot J_o(\alpha_n r)\, dr \tag{1.55a}$$

$$\vartheta = \frac{2}{r_e^2} \sum_{n=1}^{\infty} e^{-k\alpha_n^2 t} \frac{J_o(\alpha_n r)}{J_o^2(\alpha_n r_e)} \int_0^{r_e} r \cdot \vartheta_o(r) \cdot J_o(\alpha_n r)\, dr \tag{1.55b}$$

Similarly, when the heat losses from the surface of the cylinder are proportional to the temperature of the surface (boundary condition of the third type):

$$\left[\frac{\partial \vartheta}{\partial r} + h\,\vartheta\right]_{r=r_e} = 0 \quad \text{for } t > 0, \tag{1.56}$$

the coefficients α_n must be solution of equation:

$$\left.\begin{aligned} &-\alpha_n J_1(\alpha_n r_e) + h\, J_o(\alpha_n r_e) = 0 \\ &\beta_n J_1(\beta_n) - A J_o(\beta_n) = 0 \end{aligned}\right\} \tag{1.57}$$

with: $\beta_n = \alpha_n r_e$; $A = r_e h = r_e \alpha_s / \lambda$.
In this case it is:

$$A_n = \frac{2\alpha_n^2}{r_e^2 (h^2 + \alpha_n^2) J_o^2(\alpha_n r_e)} \int_0^{r_e} r \cdot \vartheta_o(r) \cdot J_o(\alpha_n r)\, dr \tag{1.58a}$$

$$\vartheta = \frac{2}{r_e^2} \sum_{n=1}^{\infty} e^{-k\alpha_n^2 t} \frac{\alpha_n^2 J_o(\alpha_n r)}{(h^2 + \alpha_n^2) J_0^2(\alpha_n r_e)} \int_0^{r_e} r \cdot \vartheta_o(r) \cdot J_o(\alpha_n r)\, dr \tag{1.58b}$$

The values of β_n for the above mentioned cases are given in Table A.2 of Appendix.

1.2.5.3 Convective Heating or Cooling of an Infinite Cylindrical Workpiece

- *Cylinder at initial temperature $\vartheta_0(r) = const.$ and surface temperature $\vartheta_s = 0$ during the cooling transient.*

The initial and boundary conditions are:

$$\left.\begin{aligned} &\vartheta = \vartheta_o = \text{const.}, && \text{for } t = 0 \\ &\vartheta_s = 0, && \text{for } t > 0;\ r = r_e \end{aligned}\right\} \tag{1.59}$$

From Eq. (A3.2), and Eqs. (1.52), (1.53) we have:

$$\frac{\vartheta}{\vartheta_o} = \frac{2}{r_e} \sum_{n=1}^{\infty} e^{-k\alpha_n^2 t} \frac{J_o(\alpha_n r)}{\alpha_n J_1(\alpha_n r_e)} \tag{1.60}$$

with: $J_o(\alpha_n r_e) = 0$.

Using the dimensionless quantities:

$$\xi = \frac{r}{r_e}; \quad \beta_n = \alpha_n r_e; \quad \Theta = \frac{\vartheta}{\vartheta_o}; \quad \tau = \frac{kt}{r_e^2},$$

Eq. (1.60) can be rewritten as follows:

$$\Theta = \frac{\vartheta}{\vartheta_o} = 2 \sum_{n=1}^{\infty} e^{-\beta_n^2 \tau} \frac{J_o(\beta_n \xi)}{\beta_n J_1(\beta_n)} \tag{1.61}$$

with: $J_o(\beta_n) = 0$.

- *Cylinder at initial temperature $\vartheta_0(r) = 0$ and surface temperature $\vartheta_s = const.$ during the heating transient.*

In this case the initial and boundary conditions are:

$$\left.\begin{array}{ll} \vartheta = \vartheta_o(r) = 0, & \text{for } t = 0 \\ \vartheta = \vartheta_s = \text{const.} & \text{for } t > 0;\ r = r_e \end{array}\right\} \tag{1.62}$$

Introducing the new variable:

$$x = \vartheta - \vartheta_s, \tag{1.63}$$

the conditions (1.62) can be written as follows:

$$\left.\begin{array}{ll} x = -\vartheta_s, & \text{for } t = 0 \\ x = 0, & \text{for } t > 0;\ r = r_e \end{array}\right\}. \tag{1.64}$$

With the exception of the sign of ϑ_s, they are identical to (1.59). According to Eqs. (1.61) and (1.53), the solution of the heating transient with the initial and boundary conditions (1.62) thus becomes:

$$\Theta = \frac{\vartheta}{\vartheta_s} = 1 - 2 \sum_{n=1}^{\infty} e^{-\beta_n^2 \tau} \frac{J_o(\beta_n \xi)}{\beta_n J_1(\beta_n)} \tag{1.65}$$

with $J_o(\beta_n) = 0$.

Using Eq. (1.65), the calculation of the temperature transient can be developed in tabular form for different values of ξ and τ, as shown as an example in Table 1.1, where the temperature ϑ_a of the axis is calculated with Eq. (1.65) for $\xi = 0$:

$$\Theta_a = \frac{\vartheta_a}{\vartheta_s} = 1 - 2\sum_{n=1}^{\infty} \frac{e^{-\beta_n^2 \tau}}{\beta_n J_1(\beta_n)} \tag{1.66}$$

The results of the table show that at the beginning of the transient a low number of terms of Eq. (1.66) is sufficient for obtaining a precision adequate to technical applications.

The curves of Fig. 1.11 give the same results in graphical form; they show that the use of dimensionless parameters allows to draw diagrams having a general character, which can be used for different geometrical dimensions and material characteristics.

The curves show that the temperatures of all internal points tend asymptotically to the surface temperature, but that the points near the surface approach this temperature faster than the internal ones.

As example of application of this diagram, consider the heating of a steel cylinder (diameter $2r_e$ = 100 mm; thermal diffusivity k = 0.0525 cm^2/s) with goal final temperatures $\vartheta_s = 1.000$ °C and $\vartheta_a = 881$ °C. Since from the diagram it is $\vartheta_s/\vartheta_a = 0.881$ at $\tau = 0.45$, the time required for the heating process is

$$t = \frac{\tau \cdot r_e^2}{k} = \frac{0.45 \cdot 25}{0.0525} \approx 214 \text{ s}$$

Table 1.1 Values of temperatures calculated with Eqs. (1.64) and (1.65)

$\beta_n \Rightarrow$	2.4048	5.5201	8.6537	11.7915	14.931	18.071	
$J_1(\beta_n) \Rightarrow$	0.51914	−0.34026	0.27145	−0.23245	0.20654	−0.1877	
$\tau \Downarrow$	$e^{-\beta_n^2\tau}/\beta_n J_1(\beta_n)$						ϑ_a/ϑ_s
0.025	0.6932	−0.2485	0.0655	−0.0113	0.0012	−0.0001	0.0000
0.050	0.5999	−0.1160	0.0101	−0.0003	–	–	0.0126
0.075	0.5191	−0.0542	0.0015	–	–	–	0.0672
0.10	0.4492	−0.0253	0.0002	–	–	–	0.1518
0.15	0.3364	−0.0055	–	–	–	–	0.3382
0.20	0.2520	−0.0012	–	–	–	–	0.4984
0.25	0.1887	−0.0003	–	–	–	–	0.6232
0.30	0.1413	−0.0001	–	–	–	–	0.7176
0.35	0.1058	–	–	–	–	–	0.7884
0.40	0.0793	–	–	–	–	–	0.8414
0.45	0.0594	–	–	–	–	–	0.8812
0.50	0.0444	–	–	–	–	–	0.9112
0.60	0.0249	–	–	–	–	–	0.9501
0.75	0.0105	–	–	–	–	–	0.9791

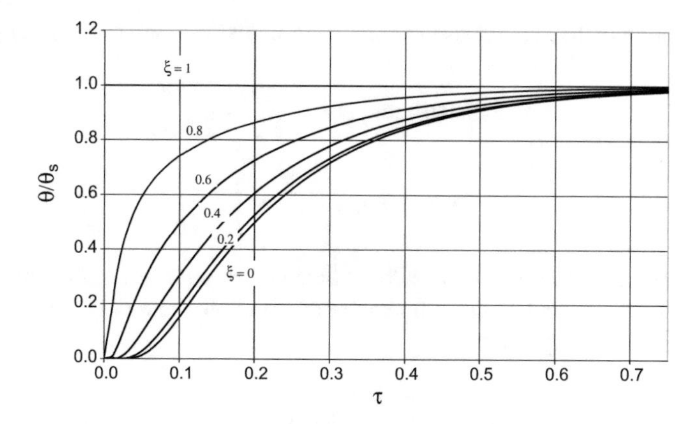

Fig. 1.11 Transient heating of a cylinder with zero initial temperature and surface maintained at a higher constant temperature ϑ_s

1.2.5.4 Heating with Constant Surface Power Density

In case of a cylinder being at zero initial temperature heated with constant surface power density p_o, (W/m^2), no internal heat generation and negligible surface losses, Eq. (1.43) is solved with the following conditions:

$$\left.\begin{array}{ll} \vartheta = \vartheta_o(r) = 0, & \text{for } t = 0 \\ \frac{\partial \vartheta}{\partial r} = \frac{p_o}{\lambda}, & \text{for } t > 0;\ r = r_e \end{array}\right\} \tag{1.67}$$

Introducing the dimensionless parameters of radius, temperature and time:

$$\xi = \frac{r}{r_e}; \quad \Theta = \frac{\lambda}{p_o r_e}\vartheta; \quad \tau = \frac{k}{r_e^2}t,$$

the Fourier equation (1.43) and the initial and boundary conditions (1.67) can be written in the form:

$$\frac{\partial \Theta}{\partial \tau} = \frac{\partial^2 \Theta}{\partial \xi^2} + \frac{1}{\xi}\frac{\partial \Theta}{\partial \xi} \tag{1.68a}$$

$$\left\{\begin{array}{ll} \Theta = \Theta_o(\xi) = 0, & \tau = 0 \\ \frac{\partial \Theta}{\partial \xi} = 1, & \tau > 0; \xi = 1 \end{array}\right. \tag{1.68b}$$

As known, in heating transients with constant surface power and negligible surface losses, after an initial period during which the temperature of each point varies with a different law, it is reached a kind of thermal equilibrium characterized by different radial temperature distributions which increase linearly in time at the same speed.

In the range of linear increase of temperatures the following conditions hold:

$$\bullet \text{ at a generic radius } r: \quad \frac{\partial \Theta}{\partial \xi} = \frac{1}{2}\xi \frac{\partial \Theta}{\partial \tau} \tag{1.69a}$$

$$\bullet \text{ at } r = r_e: \quad \frac{\partial \Theta}{\partial \tau} = 2 \tag{1.69b}$$

Combining Eqs. (1.69a) and (1.69b) is obtained an equation in the only variable ξ, which allows to calculate the temperature distribution along radius in the range of linear increase of temperatures:

$$\frac{\partial \Theta}{\partial \xi} - \xi = 0 \tag{1.70}$$

whose solution is:

$$\Theta = \frac{1}{2}\xi^2 + \mathrm{C}.$$

Evaluating the integration constant C with the condition:

$$\Theta = \Theta_{\mathrm{a}}, \quad \text{at } \xi = 0$$

the solution becomes:

$$\Theta - \Theta_{\mathrm{a}} = \frac{1}{2}\xi^2 \tag{1.71}$$

This equation highlights that, in the range of linear increase of temperatures, the distribution of Θ along radius is parabolic, with average value Θ_{av} which corresponds to the average temperature in the cylinder cross-section:

$$\Theta_{\mathrm{av}} = 2\int_0^1 \xi \cdot \Theta \cdot \mathrm{d}\xi = \Theta_{\mathrm{a}} + \frac{1}{4} \tag{1.72}$$

For Eq. (1.69b) we can also write:

$$\Theta_{\mathrm{av}} = 2\tau \tag{1.73}$$

Finally, from Eqs. (1.71) to (1.73), we obtain:

$$\Theta = 2\tau + \frac{1}{2}\xi^2 - \frac{1}{4} \tag{1.74}$$

Taking into account the initial transient, to this solution must be added a term which vanishes in the range of linear temperature increase and, on the other hand, must fulfil the conditions (1.68b).

For the first of these conditions, similarly to what done previously, we can write the solution of Eq. (1.68a) in the form:

$$\Theta = 2\tau + \frac{1}{2}\xi^2 - \frac{1}{4} - \sum_{n=1}^{\infty} A_n J_0(\beta_n \xi) \cdot e^{-\beta_n^2 \tau} \tag{1.75}$$

with: $\beta_n = \alpha_n r_e$—(= 3.83, 7.02, 10.17, 13.32 ….) positive roots of the equation $J_1(\beta) = 0$.

From the second of Eqs. (1.68b) and (1.75), for $\tau = 0$ it is:

$$0 = \frac{1}{2}\xi^2 - \frac{1}{4} - \sum_{n=1}^{\infty} A_n J_0(\beta_n \xi).$$

Then the calculation of coefficients A_n can be developed, as described in previous paragraphs, with reference to the function:

$$f(\xi) = \frac{1}{2}\xi^2 - \frac{1}{4} = \sum_{n=1}^{\infty} A_n J_0(\beta_n \xi). \tag{1.76}$$

Using relationships (A.3.1) and (A.3.4) of Appendix A.3, it is:

$$\int_0^1 \xi f(\xi) J_0(\beta_n \xi) d\xi = A_n \frac{1}{2} J_0^2(\beta_n)$$

and:

$$A_n = \frac{2}{J_0^2(\beta_n)} \int_0^1 \xi \cdot f(\xi) \cdot J_0(\beta_n \xi) d\xi = \frac{2}{\beta_n^2 J_0(\beta_n)} \tag{1.77}$$

Introducing this expression in Eq. (1.75), finally we obtain:

$$\Theta = 2\tau + \frac{1}{2}\xi^2 - \frac{1}{4} - 2\sum_{n=1}^{\infty} \frac{J_0(\beta_n \xi)}{\beta_n^2 J_0(\beta_n)} e^{-\beta_n^2 \tau} \tag{1.78}$$

The use of dimensionless parameters allows to represent the solution (1.78) in a single diagram of general use (Fig. 1.12a), valid for different materials, geometric dimensions of the cylinder and values of specific power. The ordinates multiplied by $(p_0 r_e/\lambda)$ represent temperatures, the abscissas multiplied by (r_e^2/k), times.

The diagrams of Fig. 1.12b shows the temperature distributions along the radius of the cylinder for values of τ equal to 0.0125, 0.025, 0.05, 0.1, 0.25, 0.50. For

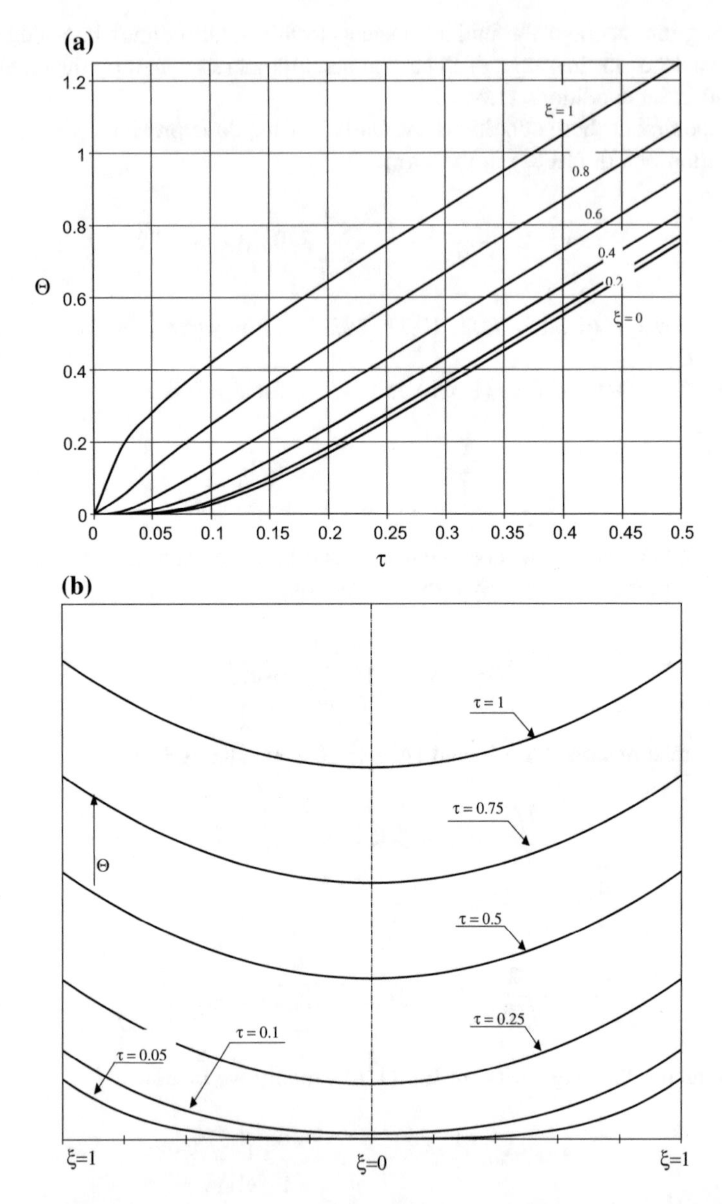

Fig. 1.12 **a** Heating transient of a cylinder heated with constant surface power density, no internal heat generation and negligible surface heat losses; **b** Radial temperature distributions at different time instants in the heating transient of Fig. 1.10a

$\tau \geq 0.25$ the summation in Eq. (1.78) practically vanishes and, as already said, until the end of heating the temperatures of all the points of the cylinder increase linearly in time with a parabolic radial distribution which remains unchanged in all instants.

As a consequence, at each instant after $\tau \geq 0.25$ the temperature difference $\Delta\Theta = \Theta_s - \Theta_a$ between surface and axis of the cylinder, evaluated through Eq. (1.74) for $\xi = 1$ and $\xi = 0$, is:

$$\Theta_s = 2\tau + \frac{1}{4}; \quad \Theta_a = 2\tau - \frac{1}{4} \tag{1.79a}$$

and:

$$\Delta\Theta = \frac{1}{2} \Rightarrow \Delta\vartheta = \vartheta_s - \vartheta_a = \frac{p_0 r_e}{2\lambda} \tag{1.79b}$$

Equation (1.79b) points out that relatively uniform heating can be obtained only in bodies of small geometrical dimensions, high thermal conductivity or with the use of low values of p_0.

From Eqs. (1.79a) and (1.79b), the temperatures of surface and axis for $\tau = 0.25$ are:

$$\vartheta_s' = \frac{p_0 r_e}{\lambda}\Theta_s' = 0.75\frac{p_0 r_e}{\lambda}; \quad \vartheta_a' = \frac{p_0 r_e}{\lambda}\Theta_a' = 0.25\frac{p_0 r_e}{\lambda}.$$

The heating time necessary to bring the surface of the cylinder at a given temperature (greater than ϑ_s'), can be obtained from Eqs. (1.79a) and (1.79b).

With the position:

$$\varepsilon = \frac{\vartheta_s - \vartheta_a}{\vartheta_s} = \frac{\Theta_s - \Theta_a}{\Theta_s} = \frac{1}{2\Theta_s},$$

it results:

$$\tau = \frac{\Theta_s}{2} - \frac{1}{8} = \frac{1}{4\varepsilon}\left(1 - \frac{\varepsilon}{2}\right)$$

or:

$$t = \frac{r_e^2}{4\varepsilon k}\left(1 - \frac{\varepsilon}{2}\right) \tag{1.80}$$

Example 1.3 Heating with constant surface power density a steel billet ($2r_e$ = 100 mm) to final surface temperature $\vartheta_s = 1200$ °C with temperature differential $\Delta\vartheta = 60$ °C. The data for steel are:

$$\lambda = 0.291\ \text{W/cm °C}; \quad c = 0.71\ \text{W s/g °C}; \quad k = 0.0525\ \text{cm}^2/\text{s}.$$

According to Eqs. (1.79b) and (1.80) it is:

$$p_0 = \frac{2\lambda\Delta\vartheta}{r_e} = \frac{2 \cdot 0.291 \cdot 60}{5} \approx 7\ \mathrm{W/cm^2}$$

$$t = \frac{25}{4 \cdot 0.05 \cdot 0.0525} 0.975 = 2321\ \mathrm{s} = 38.7\ \mathrm{min}$$

1.2.5.5 Heating a Cylindrical Charge at Zero Initial Temperature in a Chamber at Constant Temperature

Equation (1.43) must be solved with the following initial and boundary conditions:

$$\begin{cases} \vartheta = \vartheta_0(r) = 0, & \text{at } t = 0 \\ \frac{\partial \vartheta}{\partial r} - h(\vartheta_c - \vartheta) = 0, & \text{for } t > 0,\ r = r_e \end{cases} \tag{1.81}$$

Introducing the new variable $y = \vartheta - \vartheta_c$, the conditions (1.81) become:

$$\begin{cases} y_0(r) = -\vartheta_c, & \text{at } t = 0 \\ \frac{\partial y}{\partial r} + hy = 0, & \text{for } t > 0,\ r = r_e \end{cases} \tag{1.82}$$

The solution of Fourier equation in the variable y with these conditions is given by Eqs. (1.58a) and (1.58b).

According to Eq. (A3.5), it is:

$$\begin{aligned} A_n &= \frac{2\alpha_n^2}{r_e^2(h^2 + \alpha_n^2)J_o^2(\alpha_n r_e)} \int_0^{re} r \cdot \vartheta_o(r) \cdot J_o(\alpha_n r)\, dr = \\ &= -\frac{2\alpha_n^2 \vartheta_c}{r_e^2(h^2 + \alpha_n^2)J_o^2(\alpha_n r_e)} \frac{r_e}{\alpha_n} J_1(\alpha_n r_e) \end{aligned} \tag{1.83a}$$

Since α_n must satisfy the conditions (1.57), using the notations $\beta_n = \alpha_n r_e$ and $A = h\, r_e$, Eq. (1.83a) simplifies in the form:

$$A_n = -\frac{2A\vartheta_c}{(A^2 + \beta_n^2)J_o(\beta_n)} \tag{1.83b}$$

Thus the solution in the variable y is:

$$y = -2A\vartheta_c \sum_{n=1}^{\infty} \frac{J_0(\beta_n \xi)}{(A^2 + \beta_n^2)J_o(\beta_n)} e^{-\beta_n^2 \tau} \tag{1.84a}$$

while in the variable ϑ it becomes:

$$\frac{\vartheta}{\vartheta_c} = 1 - 2A\sum_{n=1}^{\infty}\frac{J_0(\beta_n\xi)}{(A^2+\beta_n^2)J_o(\beta_n)}e^{-\beta_n^2\tau} \quad (1.84b)$$

In particular, the temperatures ϑ_s of the surface and ϑ_a of the axis, and the temperature differential $\Delta\vartheta = \vartheta_s - \vartheta_a$ are:

$$\left.\begin{aligned} \frac{\vartheta_s}{\vartheta_c} &= 1 - 2A\sum_{n=1}^{\infty}\frac{e^{-\beta_n^2\tau}}{(A^2+\beta_n^2)} \\ \frac{\vartheta_a}{\vartheta_c} &= 1 - 2A\sum_{n=1}^{\infty}\frac{e^{-\beta_n^2\tau}}{(A^2+\beta_n^2)J_0(\beta_n)} \end{aligned}\right\} \quad (1.85a)$$

$$\frac{\Delta\vartheta}{\vartheta_c} = \frac{\vartheta_s - \vartheta_a}{\vartheta_c} = -2A\sum_{n=1}^{\infty}\frac{e^{-\beta_n^2\tau}}{(A^2+\beta_n^2)}\left[1 - \frac{1}{J_0(\beta_n)}\right] \quad (1.85b)$$

Finally, according to (1.84b), it is possible to calculate the variation during heating transient of the average temperature ϑ_{av} in the billet cross section. The result is:

$$\begin{aligned} \frac{\vartheta_{av}}{\vartheta_c} &= 2\int_0^1 \xi\left(\frac{\vartheta}{\vartheta_c}\right)d\xi = 1 - 4A\sum_{n=1}^{\infty}\frac{J_1(\beta_n)}{\beta_n(A^2+\beta_n^2)J_o(\beta_n)}e^{-\beta_n^2\tau} = \\ &= 1 - 4A^2\sum_{n=1}^{\infty}\frac{e^{-\beta_n^2\tau}}{\beta_n^2(A^2+\beta_n^2)} \end{aligned} \quad (1.85c)$$

1.2.5.6 Transient Temperature Equalization

In the transients previously considered, at the end of heating period there is a temperature differential $\Delta\vartheta' = \vartheta_s' - \vartheta_a'$ between surface and axis of the cylinder. In many practical applications, when this difference is too high for meeting the process requirements, it is necessary to provide a period of temperature equalization or, in continuous furnaces with progressive advancement of workpieces, an area where the equalization takes place.

In other cases the same technological process requires a maintenance period of the charge at high temperature.

This equalization process can be obtained either with an additional input of heat, for maintaining at constant temperature the surface of the workpiece, or at expenses of internal heat, theoretically without supplying any additional heat, in practice either keeping the body in a thermally insulated chamber or providing only the power necessary to compensate surface losses.

As an example of these processes we consider the transient temperature equalization in a heated body which has the following initial parabolic temperature distribution:

$$\vartheta_0(r) = \vartheta_s' - \Delta\vartheta'[1 - \left(\frac{r}{re}\right)^2]$$

We must solve the Fourier equation with the appropriate initial and boundary conditions.

(a) *equalization at constant surface temperature*

The relevant boundary conditions are:

$$\left.\begin{array}{ll} \vartheta_0(r) = \vartheta_s' - \Delta\vartheta'[1 - \left(\frac{r}{r_e}\right)^2], & \text{for } t = 0 \\ \vartheta = \vartheta_s', & \text{for } t > 0;\ r = r_e \end{array}\right\} \quad (1.86a)$$

Introducing the new variable $x = \vartheta - \vartheta_s'$, the above conditions become:

$$\left.\begin{array}{ll} x = f_0(r) = -\Delta\vartheta'[1 - \left(\frac{r}{r_e}\right)^2], & \text{for } t = 0 \\ x = 0, & \text{for } t > 0;\ r = r_e \end{array}\right\} \quad (1.86b)$$

As shown by Eqs. (1.51) and (1.52), the solution can be written in the form:

$$x = \sum_{n=1}^{\infty} A_n J_0(\alpha_n x) \cdot e^{-k\alpha_n^2 t} \quad (1.87)$$

with: $J_0(\alpha_n r_e) = 0$

$$\begin{aligned} A_n &= \frac{2}{r_e^2 J_1^2(\alpha_n r_e)} \int_0^{r_e} r J_0(\alpha_n r) f_0(r) dr = \\ &= \frac{-2\Delta\vartheta'}{r_e^2 J_1^2(\alpha_n r_e)} \left\{ \frac{r_e}{\alpha_n} J_1(\alpha_n r_e) - \frac{1}{r_e^2} [\frac{r_e^3}{\alpha_n} J_1(\alpha_n R) - \frac{2r_e^2}{\alpha_n^2} J_2(\alpha_n r_e)] \right\} = \\ &= -\frac{8\Delta\vartheta'}{\alpha_n^3 r_e^3 J_1(\alpha_n r_e)} \end{aligned}$$

Then we obtain:

$$\frac{\vartheta}{\vartheta_s'} = 1 + \frac{x}{\vartheta_s'} = 1 - 8\frac{\Delta\vartheta'}{\vartheta_s'} \sum_{n=1}^{\infty} \frac{J_0(\beta_n \xi)}{\beta_n^3 J_1(\beta_n)} \cdot e^{-\beta_n^2 \tau} \quad (1.88)$$

with: $J_0(\beta_n) = 0$

In particular, for $\xi = 0$, we have:

$$\frac{\vartheta_a}{\vartheta_s'} = 1 - 8\frac{\Delta\vartheta'}{\vartheta_s'}\sum_{n=1}^{\infty}\frac{1}{\beta_n^3 J_1(\beta_n)}e^{-\beta_n^2\tau} = 1 - \frac{\Delta\vartheta'}{\vartheta_s'}F_a' \tag{1.89}$$

(b) *equalization at expenses of stored heat*

In this case the initial and boundary conditions are:

$$\left.\begin{array}{ll} \vartheta_0(r) = \vartheta_s' - \Delta\vartheta'[1 - \left(\frac{r}{r_e}\right)^2], & \text{for } t = 0 \\ \frac{\partial\vartheta}{\partial r} = 0, & \text{for } t > 0;\ r = r_e \end{array}\right\} \tag{1.90}$$

The solution of this transient, given in Ref. [5], is the following:

$$\vartheta = \frac{2}{r_e^2}\left\{\int_0^{r_e} r\vartheta_0(r)dr + \sum_{n=1}^{\infty} e^{-k\alpha_n^2 t}\frac{J_0(\alpha_n r)}{J_0^2(\alpha_n r_e)}\int_0^{r_e} r\vartheta_0(r)J_0(\alpha_n r)\,dr\right\} \tag{1.91}$$

with: $J_1(\alpha_n r_e) = 0$.

The first term of this equation represents the average value of the initial temperature distribution, while the second one corresponds to Eq. (1.58b) with $h = 0$.

In fact, for $h \neq 0$ (i.e. when heat transfer to the environment occurs), the average temperature must tend to zero, whereas for $h = 0$ (i.e. when heat exchange with the environment is negligible), the temperature of all points of the workpiece cross-section must tend to the average temperature of the initial distribution.

Introducing in Eq. (1.91) the values the integrals:

$$\int_0^{r_e} r\vartheta_s' - \Delta\vartheta'[1 - (\frac{r}{r_e})^2]dr = (\vartheta_s' - \frac{\Delta\vartheta'}{2})\frac{r_e^2}{2}$$

$$\int_0^{r_e} r\vartheta_s' - \Delta\vartheta'[1 - (\frac{r}{r_e})^2]J_0(\alpha_n r)dr = \frac{2\Delta\vartheta'}{\alpha_n^2}J_0(\alpha_n r_e),$$

finally we obtain:

$$\vartheta = \frac{2}{r_e^2}\left\{(\vartheta_s' - \frac{\Delta\vartheta'}{2})\frac{r_e^2}{2} + 2\Delta\vartheta'\sum_{n=1}^{\infty}\frac{J_0(\alpha_n r)}{\alpha_n^2 J_0(\alpha_n r_e)}e^{-k\alpha_n^2 t}\right\} \tag{1.92a}$$

or, in dimensionless quantities:

$$\frac{\vartheta}{\vartheta_s'} = 1 - \frac{\Delta\vartheta'}{\vartheta_s'}\left\{\frac{1}{2} - 4\sum_{n=1}^{\infty}\frac{J_0(\beta_n\xi)}{\beta_n^2 J_0(\beta_n)}\cdot e^{-\beta_n^2\tau}\right\} \tag{1.92b}$$

with: $J_1(\beta_n) = 0$.

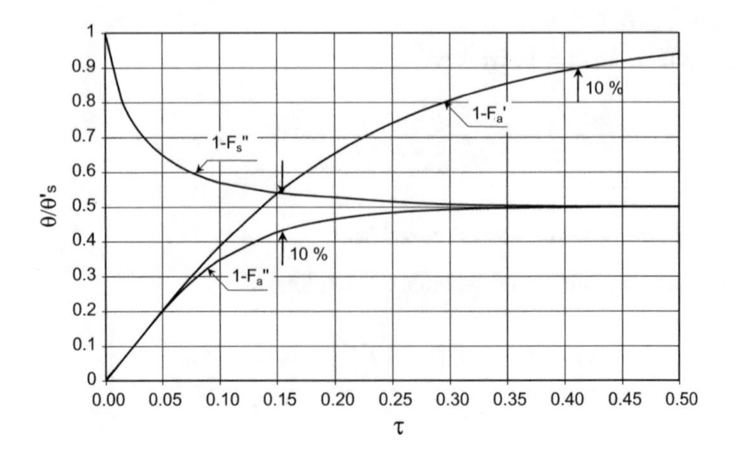

Fig. 1.13 Transient temperature equalization at expenses of stored heat or with constant surface temperature in the case $\Delta\vartheta'/\vartheta_s' = 1$ [6]

In particular, for $\xi = 1$ and $\xi = 0$, it is:

$$\frac{\vartheta_s}{\vartheta_s'} = 1 - \frac{\Delta\vartheta'}{\vartheta_s'}\left\{\frac{1}{2} - 4\sum_{n=1}^{\infty}\frac{1}{\beta_n^2}e^{-\beta_n^2\tau}\right\} = 1 - \frac{\Delta\vartheta'}{\vartheta_s'}F_s'' \tag{1.93a}$$

$$\frac{\vartheta_a}{\vartheta_s'} = 1 - \frac{\Delta\vartheta'}{\vartheta_s'}\left\{\frac{1}{2} - 4\sum_{n=1}^{\infty}\frac{1}{\beta_n^2 J_0(\beta_n)}e^{-\beta_n^2\tau}\right\} = 1 - \frac{\Delta\vartheta'}{\vartheta_s'}F_a'' \tag{1.93b}$$

Figure 1.13 shows a graphical representation of the functions F_a', F_a'', F_s'' used for the evaluation of the soaking time under the given boundary conditions and transient temperature distributions of Eqs. (1.89), (1.93a), (1.93b) as a function of the dimensionless time τ, assuming equal to 1 the initial surface-to-core dimensionless differential ($\Delta\vartheta'/\vartheta_s' = 1$).

The analysis of the curves shows that the equalization at the expense of stored heat takes place much more rapidly than the one with constant surface temperature. In particular, the reduction of the temperature differential between surface and axis to 30 and 10 % of the initial value is obtained in the first case with values of τ approximately equal to 0.075 and 0.15 respectively, while in the second case the same conditions occur for $\tau \approx 0.225$ and $\tau \approx 0.42$.

Example 1.4

(a) Heating with constant surface power density of a copper cylindrical billet, $r_e = 75$ mm, to the final average temperature $\vartheta_{av} = 850$ °C with temperature differential $\Delta\vartheta = \vartheta_s - \vartheta_a = 50$ °C.

Average values of material properties:

$\lambda = 360\ \mathrm{W/m\ ^\circ C}$; $c\gamma = 3.75 \cdot 10^6\ \mathrm{W\ s/m^3\ ^\circ C}$; $k = 96 \cdot 10^{-6}\ \mathrm{m^2/s}$.

- Required power density:

$$p_0 = \frac{2\lambda}{r_e}\Delta\vartheta = \frac{2 \cdot 360}{7.5 \cdot 10^{-2}} 50 \approx 48 \cdot 10^4\ \mathrm{W/m^2}$$

- Heating time:

$$t_0 = \frac{c\gamma\,\pi r_e^2}{2\pi r_e p_0}\vartheta_{av} = \frac{3.75 \cdot 10^6 \cdot \pi \cdot (7.5 \cdot 10^{-2})^2}{2\pi \cdot 7.5 \cdot 10^{-2} \cdot 48 \cdot 10^4} 850 \approx 250\,\mathrm{s}$$

- Dimensionless heating time: $\tau = \frac{kt_0}{r_e^2} = \frac{96 \cdot 10^{-6} \cdot 250}{(7.5 \cdot 10^{-2})^2} = 4.267$
- Dimensionless temperature of the axis:

$$\Theta_a = \Theta_{av} - \frac{1}{4} = 2\tau - \frac{1}{4} = 2 \cdot 4.267 - 0.25 = 8.283$$

- Temperature of the axis:

$$\vartheta_a = \frac{p_0 r_e}{\lambda}\Theta_a = \frac{48 \cdot 10^4 \cdot 7.5 \cdot 10^{-2}}{360} 8.283 \approx 828\ ^\circ\mathrm{C}$$

- Temperature of the surface: $\vartheta_s = \vartheta_a + \Delta\vartheta \approx 828 + 50 \approx 878\ ^\circ\mathrm{C}$.

(b) *reduction of the total processing time by using thermal equalization at expenses of stored heat*

- Heating power density: $p_0 = 150 \cdot 10^4\ \mathrm{W/m^2}$
- Heating time:

$$t_0 = \frac{c\gamma\,\pi r_e^2}{2\pi r_e p_0}\vartheta_{av} = \frac{3.75 \cdot 10^6 \cdot \pi \cdot (7.5 \cdot 10^{-2})^2}{2\pi \cdot 7.5 \cdot 10^{-2} \cdot 150 \cdot 10^4} 850 \approx 80\,\mathrm{s}$$

- Dimensionless heating time: $\tau_0 = \frac{kt_0}{r_e^2} = \frac{96 \cdot 10^{-6} \cdot 80}{(7.5 \cdot 10^{-2})^2} = 1.365$

- Dimensionless axial temperature:

$$\Theta_a = 2\tau - \frac{1}{4} = 2 \cdot 1.365 - 0.25 = 2.480$$

- Axial temperature:

$$\vartheta_a = \frac{p_0 r_e}{\lambda}\Theta_a = \frac{150 \cdot 10^4 \cdot 7.5 \cdot 10^{-2}}{360} 2.480 \approx 775\ ^\circ\mathrm{C}$$

- Temperature differential:

$$\Delta\vartheta = \vartheta_s - \vartheta_a = \frac{p_0 r_e}{2\lambda} = \frac{150 \cdot 10^4 \cdot 7.5 \cdot 10^{-2}}{2 \cdot 360} \approx 156\ ^\circ\mathrm{C}$$

- Surface temperature: $\vartheta_s = \vartheta_a + \Delta\vartheta = 775 + 156 = 931\ ^\circ\mathrm{C}$
- Additional equalization time necessary for reducing the temperature differential to about 33 % of the previous value:

$$\Delta t_0 = \frac{r_e^2}{k}\Delta\tau' = \frac{(7.5 \cdot 10^{-2})^2}{96 \cdot 10^{-6}} 0.08 \approx 4.7\ \mathrm{s},$$

with $\Delta\tau'$ given by the diagram of Fig. 1.13.

The diagrams of Fig. 1.14a, b show the heating transients of the above examples, calculated with the numerical program ELTA which allows to take into account the variations of material characteristics with temperature. [15] The comparison of analytical and numerical results, summarized in Table 1.2, shows that the analytical calculation allows to obtain results that are in very good agreement with the numerical calculations.

1.2.5.7 Cooling Transient with Heat Losses from the Cylinder Surface (Boundary Conditions of Third Type)

The cooling takes place with the following initial and boundary conditions:

$$\left.\begin{array}{ll} \vartheta_0(r) = \vartheta_s' - \Delta\vartheta'[1 - (\frac{r}{r_e})^2], & \text{for } t = 0 \\ \frac{\partial\vartheta}{\partial r} + h\vartheta = 0, & \text{for } t > 0;\ r = r_e \end{array}\right\} \qquad (1.94)$$

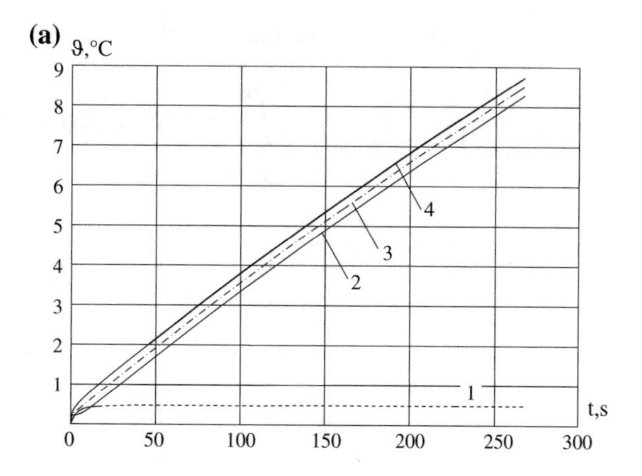

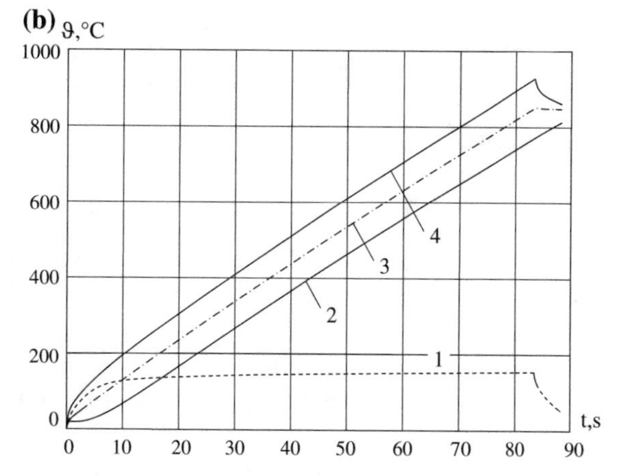

Fig. 1.14 **a** Heating with constant surface power density of a copper cylindrical billet, $r_e = 75$ mm, to the final average temperature $\vartheta_{av} =$ 850 °C with final temperature differential $\Delta\vartheta = 50$ °C. (*1* $\Delta\vartheta$; *2* ϑ_a; *3* ϑ_{av}; *4* ϑ_s); **b** Reduction of processing time by the using thermal equalization at expenses of stored heat for the same case of Fig. 1.14a. (*1* $\Delta\vartheta$; *2* ϑ_a; *3* ϑ_{av}; *4* ϑ_s)

Table 1.2 Comparison of analytical and numerical results

Case	p_o (W/cm^2)	t_0 (s)	Δt_0 (s)	ϑ_{av}(°C)	ϑ_a(°C)	ϑ_s (°C)	$\Delta\vartheta$ (°C)	Calculation
(a)	48	250	0	850	828	878	50	Analytical
	48	267	0	850	823	873	47	ELTA
(b)	150	80	0	850	775	931	156	Analytical
	150	83.5	0	850	771	925	154	ELTA
	150	80	5	850	–	–	47	Analytical
	150	83.5	5	850	810	865	55	ELTA

From Eqs. (1.57), (1.58a) and (1.58b) we have:

$$A_n = \frac{2\alpha_n^2}{r_e^2(h^2+\alpha_n^2)J_0^2(\alpha_n r_e)}\int_0^{r_e} r\cdot\vartheta_o(r)\cdot J_o(\alpha_n r)dr =$$

$$= \frac{2\alpha_n^2}{r_e^2(\alpha_n^2+h^2)J_0^2(\alpha_n r_e)}\left\{\begin{aligned} &(\vartheta_s' - \Delta\vartheta)\frac{r_e}{\alpha_n}J_1(\alpha_n r_e) + \\ &+\frac{\Delta\vartheta'}{r_e^2}\left[\frac{r_e^3}{\alpha_n}J_1(\alpha_n r_e) - \frac{2r_e^2}{\alpha_n^2}J_2(\alpha_n r_e)\right]\end{aligned}\right\} = \tag{1.95}$$

$$= \frac{2\alpha_n^2}{r_e^2(\alpha_n^2+h^2)J_0^2(\alpha_n r_e)}\left\{\begin{aligned} &\vartheta_s'\frac{r_e}{\alpha_n}J_1(\alpha_n r_e) - \\ &-\frac{2\Delta\vartheta'}{\alpha_n^2}\left[\frac{2}{\alpha_n r_e}J_1(\alpha_n r_e) - J_0(\alpha_n r_e)\right]\end{aligned}\right\} =$$

$$= \frac{2A\vartheta_s'}{(\beta_n^2+A^2)J_0(\beta_n)}\left\{1 - \frac{2\Delta\vartheta'}{\vartheta_s'}\left[\frac{2}{\beta_n^2} - \frac{1}{A}\right]\right\}$$

and the solution becomes:

$$\frac{\vartheta}{\vartheta_s'} = 2A\sum_{n=1}^{\infty}\frac{J_0(\beta_n\xi)}{(\beta_n^2+A^2)}\left\{1 - \frac{2\Delta\vartheta'}{\vartheta_s'}\left[\frac{2}{\beta_n^2} - \frac{1}{A}\right]\right\}e^{-\beta_n^2\tau} \tag{1.96}$$

with:

$$A = h\,r_e = \frac{\alpha_s}{\lambda}r_e; \quad \beta_n = \alpha_n r_e; \quad -\beta_n J_1(\beta_n) + AJ_0(\beta_n) = 0$$

In particular, the solution of the cooling transient of a cylinder from uniform initial temperature ϑ_s' is obtained for $\Delta\vartheta' = 0$.

In this case the solution is:

$$\frac{\vartheta}{\vartheta_s'} = 2A\cdot\sum_{n=1}^{\infty}\frac{J_0(\beta_n\xi)}{(\beta_n^2+A^2)}e^{-\beta_n^2\tau} \tag{1.97}$$

Example 1.5 Cooling of a steel billet ($2r_e = 39$ mm) from uniform temperature $\vartheta_s' = 1200\,^\circ C$, with radiation losses towards an environment at temperature $\vartheta_a = 0\,^\circ C$

Data: $\lambda = 0.290\ W/cm\,^\circ C$; $k = 0.0525\ cm^2/s$; $\varepsilon = 0.8$.

The average radiation losses p_{av} at the beginning of cooling in the temperature range 1200–1000 °C (calculated as shown in Appendix A.4) are:

$$\begin{aligned} p_{av} &= 5.67 \cdot 10^{-12} \cdot \frac{\varepsilon}{5} \cdot [\frac{T_2^5 - T_1^5}{T_2 - T_1}] = \\ &= 5.67 \cdot 10^{-12} \frac{0.8}{5} [\frac{1473^5 - 1273^5}{200}] = 16.3 \, (\frac{W}{cm^2}) \end{aligned} \tag{1.98}$$

From the equation:

$$p_{av} = \alpha_s(\vartheta_m - \vartheta_a) \tag{1.99}$$

we have:

$$\begin{aligned} \alpha_s &= \frac{16.3}{1100} = 0.01481; \\ h &= \frac{\alpha_s}{\lambda} = \frac{0.01481}{0.290} = 0.05107; \\ A &= h\, r_e = 0.05107 \cdot 1.95 \approx 0.1 \end{aligned}$$

From Table A.2, for A = 0.1 it is:

$$\beta_1 = 0.4417 \quad \beta_2 = 3.8577 \quad \beta_3 = 7.0298$$

The calculation of the surface temperature, made with Eq. (1.97) for $\xi = 1$, at $\tau = 0.1$ gives the following values of the terms of the summation and the overall result:

$$\frac{2A}{(\beta_1^2 + A^2)} e^{-\beta_1^2 \tau} = 0.8343 \qquad \frac{2A}{(\beta_2^2 + A^2)} e^{-\beta_2^2 \tau} = 0.0030$$

$$\frac{2A}{(\beta_3^2 + A^2)} e^{-\beta_3^2 \tau} = 0.0000$$

$$\begin{aligned} \frac{\vartheta_s}{\vartheta_s'} &= 0.8373; \\ \vartheta_s &= 1200 \cdot 0.8373 = 1005\,^\circ C; \\ t &= \frac{r_e^2}{k} \tau = \frac{1.95^2}{0.0525} 0.1 \approx 7.2 \text{ s} \end{aligned}$$

This means that in about 7 s the surface of the billet cools down of about 200 ° C, i.e. from 1200 to 1000 °C !!!

1.3 Heat Convection

1.3.1 Introduction

Convection heat transfer occurs whenever a fluid is in contact with a solid surface that is at a different temperature than the fluid. It also occurs between two fluids or gases that are at different temperatures and/or have different densities.

Due to the importance of this mode of heat transfer in the applications, in the following we will refer only to cases in which the fluid is in contact with a solid body.

The convective heat transfer may take the form of

- *Natural or free convection,*

  or

- *Forced or assisted convection.*

In the first case the fluid motion is produced by the buoyancy forces due to density differences caused by temperature variations in the fluid; the second one occurs when the fluid flow is induced by mechanical means, like a pump, a fan or a mixer.

The heat transfer per unit time through a surface by convection is described by the *Newton's Law of Cooling* which can be expressed by the relation:

$$q = h_c A(\vartheta_s - \vartheta_\infty) \tag{1.100}$$

where:

q	heat transfer per unit time (W)
A	area of surface in contact with the fluid (m^2)
$(\vartheta_s - \vartheta_\infty)$	temperature difference between solid surface and the bulk fluid (°C)
h_c	convective heat transfer coefficient W/(m^2 °C).

As regards the temperature ϑ_∞, it depends on the process specifics. In convection within fluids limited by a single flat wall, e.g. convection on a flat plate, ϑ_∞ is the temperature of the fluid outside the thermal boundary layer, where it is not affected anymore by the phenomenon of heat exchange. For the flow in a pipe, ϑ_∞ is the average temperature at a particular cross section.

The presence of a boundary layer near the solid surface is due to the tendency of a viscous fluid to adhere to the wall which delimits the fluid flow, thus reducing to zero the relative speed of the flow at the solid surface. In the boundary velocity layer near the surface there are therefore strong velocity gradients, while outside it the speed tends to the undisturbed value v_∞. The thickness δ of the boundary velocity layer is defined as the distance from the surface where the velocity v is $v = 0.99 \cdot v_\infty$.

To the velocity boundary layer corresponds a thermal boundary layer δ_t which also increases in the flow direction.

Figures 1.15 and 1.16 schematically show the shapes of the velocity and thermal boundary layers on a flat plane.

The thickness δ_t of the thermal boundary layer at any location along the surface is defined as the distance from the surface at which the temperature difference $(\vartheta - \vartheta_s)$ equals $0.99(\vartheta_\infty - \vartheta_s)$.

Since the heat transfer through the boundary layer in the direction normal to the surface is due to thermal conduction, which is directly related to the temperature gradient, the convection heat transfer depends from the shape of the temperature profile in the thermal boundary layer.

In conclusion, convective heat transfer is a complex phenomenon which is influenced by several different parameters, i.e.

- fluid properties
- fluid stream velocity
- fluid flow regime
- temperatures ϑ_s and ϑ_∞ of surface and free stream
- geometry of the solid surface
- surface roughness.

Due to these influences a precise determination of the convective heat transfer coefficient h_c is very difficult, and it could be obtained only through the solution of three-dimensional, time-dependent Partial Differential Equations of Mass, Momentum and Energy transfer governing fluid motion.

Therefore practical design values of h_c are mostly based on correlation of experimental data. These correlations, which are valid for particular geometries and flow conditions, are typically expressed in terms of universal dimensionless numbers, by using the following symbols:

d, ℓ characteristic length (e.g. height ℓ for vertical plate or diameter d for cylinder) (m)

c_f fluid specific heat (J/kg °C)

λ_f fluid thermal conductivity (W/m °C)

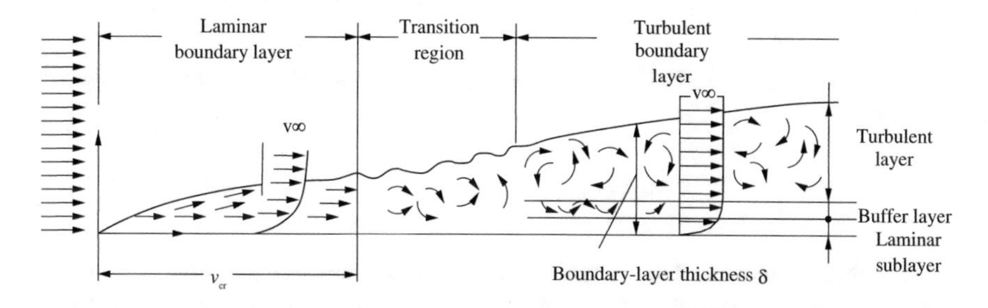

Fig. 1.15 Development of the velocity boundary layer on a flat plate at different flow regimes [7]

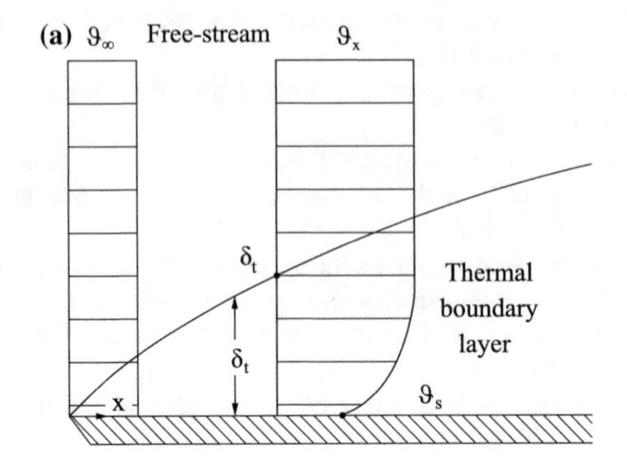

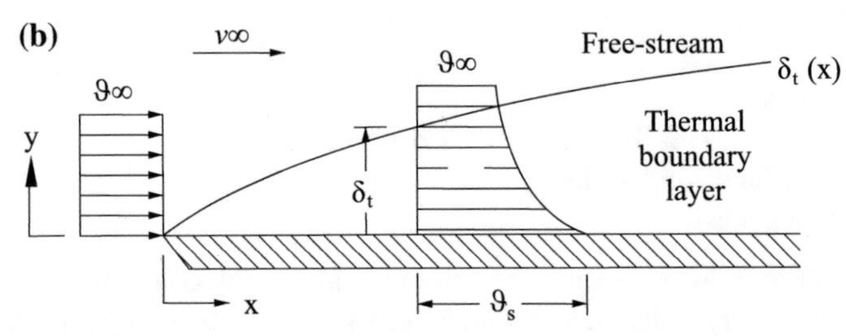

Fig. 1.16 Thermal boundary layer δ_t on an isothermal flat plate: **a** $\vartheta_s < \vartheta_\infty$; **b** $\vartheta_s > \vartheta_\infty$ (ϑ_s—temperature of the surface) [1, 7]

g	acceleration due to gravity (m/s^2)
v_m	mean velocity of the solid surface relative to the fluid (m/s)
$\Delta\vartheta = \vartheta_s - \vartheta_\infty$	temperature difference between solid surface and fluid (°C)
γ	density, mass per unit volume (kg/m^3)
β_f	coefficient of thermal expansion of fluid (K^{-1})
k	thermal diffusivity (m^2/s)
μ_f	fluid dynamic viscosity (kg/m s)
ν_f	fluid kinematic viscosity (m^2/s)

The fluid properties μ, v_f, β_f and k are typically evaluated at the film temperature $\vartheta_f = (\vartheta_s + \vartheta_\infty)/2$.

The definitions and the physical meaning of the dimensionless numbers are the following.

$$\bullet\ \textit{Nusselt number}: \mathrm{Nu} = \frac{h_c\, d}{\lambda_f} \tag{1.101a}$$

The Nusselt number Nu represents the ratio of convective to conductive heat transfer across the boundary layer. A value of Nu close to unity means that heat convection and thermal conduction are of similar magnitude (as it occurs in laminar flows), while large values are typical for turbulent flows (where convection is more efficient).

$$\bullet\ \textit{Reynolds number}\text{: } \mathrm{Re} = \frac{\nu_f\, d}{\mu_f} \tag{1.101b}$$

The Reynolds number Re represents a measure of the ratio of inertial to viscous forces acting in the fluid flow. High values of Re characterize the turbulent flow; low values the laminar one.

$$\bullet\ \textit{Prandtl number}\text{: } \mathrm{Pr} = \frac{c_f \mu_f}{\lambda_f} \tag{1.101c}$$

The Prandtl number P_r approximates the ratio of momentum diffusivity to thermal diffusivity. Small values of P_r (lower than 1) indicate that heat conduction is more effective than convection; the opposite occurs if its value is greater than one.

$$\bullet\ \textit{Grashof number}\text{: } \mathrm{Gr} = \frac{g\, \beta_f\, d^3 \Delta\, \vartheta}{\nu_f^2} \tag{1.101d}$$

The Grashof number Gr represents the ratio of buoyancy forces due to the spatial variation of fluid density caused by temperature differences, to the restraining forces due to fluid viscosity. When $\mathrm{Gr} \gg 1$, the viscous force is negligible compared to buoyancy and inertial forces. When buoyant forces overcome viscous forces, the flow undergoes a transition to turbulent flow, which occurs in the range $10^8 < \mathrm{Gr} < 10^9$ for natural convection on vertical flat plates. At higher Gr numbers, the boundary layer is turbulent; at lower Gr values, the boundary layer is laminar.

$$\bullet\ \textit{Rayleigh number}\text{: } \mathrm{Ra} = \frac{g\beta_f\, d^3 \Delta\, \vartheta}{\nu_f\, \alpha} = \mathrm{Gr} \cdot \mathrm{Pr} \tag{1.101e}$$

The Rayleigh number Ra gives an indication whether the natural convection boundary layer is laminar or turbulent. When Ra is below a critical value for that fluid, heat transfer is primarily due to conduction. When it exceeds the critical value, heat transfer occurs mainly by convection.

We give in the following some correlations available in literature for simple geometries with forced and natural convection and examples of evaluation of the convection heat transfer coefficient.

However it must be taken into account that different correlations can be found in the literature for the same heat transfer conditions and that, in any case, their use gives only a rough estimate of the convection coefficient h_c.

Typical ranges of the overall convective heat transfer coefficients are shown in Table 1.3.

Table 1.3 Range of heat transfer coefficients

Medium	Heat transfer coefficient (W/m^2 °C)	Medium	Heat transfer coefficient (W/m^2 °C)
Natural convection			
Air	5–25	Liquids	50–3000
Oil	60–1800	Water	300–6000
Boiling water	2.500–60.000		
Forced convection			
Air/superheated stem	20–300	Liquid metals	5.000–40.000
Steam (condensing)	6.000–120.000		

1.3.2 *Average Nusselt Number in Forced Convection Flow Parallel to a Flat Plane of Length ℓ*

(a) *in laminar regime* ($Re_\ell < 10^5$)

$$\overline{Nu}_\ell = \frac{h_c \cdot \ell}{\lambda_f} = 0.66 \cdot Re_f^{0.5} \cdot Pr_f^{0.33} \cdot \left(\frac{Pr_f}{Pr_s}\right)^{0.25} \tag{1.102a}$$

Equation (1.102a) simplifies for gases where it is $(Pr_f/Pr_s) \approx 1$; in particular, for air at ambient temperature, it becomes:

$$\overline{Nu}_\ell = \frac{h_c \cdot \ell}{\lambda_f} \approx 0.57 \cdot Re_f^{0.5} \tag{1.102b}$$

(b) *in turbulent regime* ($Re_\ell > 10^5$)

$$\overline{Nu}_\ell = \frac{h_c \cdot \ell}{\lambda_f} = 0.037 \cdot Re_f^{0.8} \cdot Pr_f^{0.43} \cdot \left(\frac{Pr_f}{Pr_s}\right)^{0.25} \tag{1.103a}$$

In particular for turbulent flow in air, Eq. (1.103a) becomes:

$$\overline{Nu}_\ell = \frac{h_c \cdot \ell}{\lambda_f} \approx 0.032 \cdot Re_f^{0.8} \tag{1.103b}$$

Example 1.6 Calculate the average convection heat transfer coefficient on a solid flat plane, $\ell = 2$ m length, at temperature $\vartheta_s = 120$ °C, on which flows air at temperature $\vartheta_f = 10$ °C with velocity $v = 5$ m/s.

Data for air at $\vartheta_f = 10\ °C$ (see Table A.5):

$$\nu_f = 14.16 \cdot 10^{-6}\ m^2/s; \quad \lambda_f = 2.51 \cdot 10^{-2}\ W/m\ K; \quad Pr = 0.705.$$

Solution:

- Reynolds number:

$$Re_\ell = \frac{v_f\,\ell}{\mu_f} = \frac{5 \cdot 2}{14.16 \cdot 10^{-6}} = 706\,214\ (>\ 10^5\ \text{turbulent regime})$$

- average Nusselt number, Eq. (1.103b):

$$\overline{Nu}_\ell = \frac{h_c \cdot \ell}{\lambda_f} \approx 0.032 \cdot (706\,214)^{0.8} = 1529$$

- average convection heat transfer coefficient:

$$h_c = \overline{Nu}_\ell \cdot \frac{\lambda_f}{\ell} = 1529 \cdot \frac{2.51 \cdot 10^{-2}}{2} = 19\ W/m^2\ K$$

1.3.3 Average Nusselt Number in Forced Convection Flow Normal to the Axis of a Cylinder

- *for* $5 < Re_f < 10^3$

$$\overline{Nu}_f = 0.5 \cdot Re_f^{0.5} \cdot Pr_f^{0.38} \cdot \left(\frac{Pr_f}{Pr_s}\right)^{0.25} \tag{1.104a}$$

- *for* $10^3 < Re_f < 2 \cdot 10^5$

$$\overline{Nu}_f = 0.26 \cdot Re_f^{0.6} \cdot Pr_f^{0.37} \cdot \left(\frac{Pr_f}{Pr_s}\right)^{0.25} \tag{1.104b}$$

- *for* $10^3 < Re_f < 2 \cdot 10^5$

$$\overline{Nu}_f = 0.023 \cdot Re_f^{0.8} \cdot Pr_f^{0.37} \cdot \left(\frac{Pr_f}{Pr_s}\right)^{0.25} \tag{1.104c}$$

Example 1.7 Calculate the forced convection average heat transfer coefficient due to the cross flow of dry air at 20 °C with velocity 3 m/s over a cylinder, 0.016 m diameter, at surface temperature of 100 °C.

Physical properties (from Table A.5 of appendix):

$$\nu_f = 15.06 \cdot 10^{-6}\ m^2/s; \quad \lambda_f = 2.59 \cdot 10^{-2}\ W/m\ K;$$

$$Pr_f = 0.703; \quad Pr_s = 0.688$$

Solution:

- Reynolds number:

$$Re_d = \frac{v_f d}{\mu_f} = \frac{3 \cdot 16 \cdot 10^{-3}}{15.06 \cdot 10^{-6}} = 3187$$

- average Nusselt number (Eq. 1.104b):

$$\overline{Nu}_d = 0.26 \cdot (3187)^{0.6} \cdot \left(\frac{0.703}{0.688}\right)^{0.25} = 33.06$$

- average convection heat transfer coefficient:

$$h_c = \overline{Nu}_d \cdot \frac{\lambda_f}{d} = 33.06 \cdot \frac{2.59 \cdot 10^{-2}}{16 \cdot 10^{-3}} = 53.52 \quad W/m^2\ K$$

1.3.4 Average Nusselt Number in Natural Convection from Isothermal Vertical Planes or Horizontal Cylinder

$$\overline{Nu}_f = C \cdot (Gr \cdot Pr)_f^n \cdot \left(\frac{Pr_f}{Pr_s}\right)^{0.25} \tag{1.105}$$

with the constants C and n given in Table 1.4 as a function of $(Gr \cdot Pr)_f$.

Table 1.4 .

Flow	$(Gr \cdot \mathrm{Pr})_f$	C	n
Along vertical plane	$10^3 \ldots 10^9$	0.75	0.25
Along vertical plane	$>6 \cdot 10^{10}$	0.15	0.33
On horizontal cylinder	$10^3 \ldots 10^9$	0.5	0.25

Example 1.8 Calculate natural convection heat transfer coefficient from a horizontal cylinder in air. Cylinder diameter d = 12 mm, surface temperature $\vartheta_s = 130$ °C, *air at* 20 °C.

Physical properties from Table A.5 in Appendix.
Solution:

- Film temperature $\vartheta_f = \frac{130+20}{2} = 75$ °C
- Grashof number at film temperature:

$$\mathrm{Gr_f} = \frac{g\beta_f d^3 \Delta\vartheta}{\nu_f^2} = \frac{9.81 \cdot 2.87 \cdot 10^{-3} \cdot 110 \cdot 0.012^3}{(20.56 \cdot 10^{-6})^2} = 12{,}660$$

- Prandtl numbers at film and surface temperature:

$$\mathrm{Pr}_f = 0.693; \quad \mathrm{Pr}_s = 0.685$$

- Rayleigh number:

$$\mathrm{Ra} = \mathrm{Gr_f} \cdot \mathrm{Pr_f} = 12{,}660 \cdot 0.693 = 8773$$

- Nusselt number (from Eq. 1.105):

$$\overline{\mathrm{Nu}}_f = 0.5 \cdot (\mathrm{Gr} \cdot \mathrm{Pr})_f^{0.25} \cdot \left(\frac{\mathrm{Pr_f}}{\mathrm{Pr_s}}\right)^{0.25} =$$

$$= 0.5 \cdot (8773)^{0.25} \cdot \left(\frac{0.693}{0.685}\right)^{0.25} = 4.85$$

- average convection heat transfer coefficient:

$$h_c = \overline{\mathrm{Nu}}_f \cdot \frac{\lambda_f}{d} = 4.85 \cdot \frac{3.01 \cdot 10^{-2}}{12 \cdot 10^{-3}} = 12.2\ \mathrm{W/m^2\,K}$$

1.4 Radiation Heat Transfer

Heat transfer by radiation is the phenomenon of energy transmission between two bodies due to their temperature difference. Unlike conduction and convection, there is no need of an intermediate medium between the bodies for heat transfer.

All bodies at elevated temperature radiate energy in the form of photons moving in random directions, with random phase and frequency.

The energy emitted per unit time from a surface A (m^2), which is at the absolute temperature T (K), is described by the *Stepan-Boltzman law*:

$$W = A \cdot \varepsilon \cdot \sigma \cdot T^4 \quad (1.106)$$

where: $\sigma = 5.67 \cdot 10^{-8}$ ($W/m^2\ K^4$) is the Stefan-Boltzman constant; ε—the *emissivity* or total emission factor of the surface, which depends on temperature and surface conditions (e.g. degree of roughness, oxidation).

When the photons radiated by one body reach another surface, they may be either absorbed, reflected or transmitted, as schematically shown in Fig. 1.17.

The behavior of a surface to the incident radiation can be described by the dimensionless quantities:

$\alpha = E_a/E_i$ *Absorptivity*—absorbed fraction of the incident radiation
$\rho = E_r/E_i$ *Reflectance*—reflected fraction of the incident radiation
$\tau = E_t/E_i$ *Transmittance*—fraction of the incident radiation transmitted.

From energy considerations it is

$$\alpha + \rho + \tau = 1.$$

In nature no bodies have reflectance ρ equal to zero or absorptivity α equal to unity.

A body which reflects only a very small fraction of the incident radiation is called "black".

By definition a *black-body*, has absorptivity $\alpha = 1$ and surface emissivity of 1. Such a body will absorb 100 % of the thermal radiation incident on it and will emit energy according Eq. (1.106) with $\varepsilon = 1$ (ideal radiator).

Most physical objects have surface emissivity less than 1 and hence do not have blackbody surface properties.

Values of emissivity for some commonly used metals are given in Table 1.5. As shown by the table, emissivity can have strong variations for the same material depending on surface conditions and temperature.

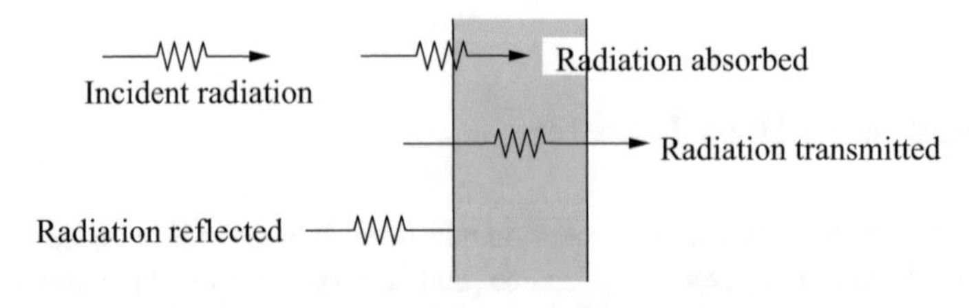

Fig. 1.17 Absorption, reflection and transmission of radiation incident on a body [8]

Table 1.5 Emissivity of selected materials [9]

Material	Temperature (°C)	Emissivity ε	Material	Temperature (°C)	Emissivity ε
Aluminium			**Graphite**		
Unoxidized	25–100	0.02–0.03	Natural	1000–2800	0.77–0.83
Unoxidized	500	0.06	**Iron**		
Oxidized	200–600	0.011–0.019	Polished	425–1020	0.144–0.377
Higly polished	225–575	0.039–0.057	**Silver**		
Brass			Pure, polished	20–800	0.019–0.046
Unoxidized	25–100	0.04	**Stainless steel**		
Polished	250–380	0.3–0.4	316, repeated heating	230–870	0.57–0.66
Oxidized	200–600	0.6	310, after furnace service	220–530	0.9–0.97
Carbon steel			**Titanium**		
Polished	740–1040	0.52–0.56		600–1200	0.217–0.286
Smooth oxidized	200–600	0.78–0.82		1600	0.323
Strongly oxidized	40–250	0.95	**Tungsten**		
Cast iron				100	0.039
Oxidized	200–600	0.64–0.78		700	0.070
Molten	1540	0.29		1300	0.195
Copper				1900	0.268
Electrolytically polished	20–800	0.025–0.061		3100	0.345
Highly polished	38	0.002			
Polished	38	0.003			
Oxidized	200–600	0.57–0.87			
Molten	1075–1275	0.16–0.13			

It is defined *gray-body* a body that emits only a fraction of the thermal energy emitted by an equivalent blackbody. By definition, a gray-body has surface emissivity less than 1, and surface reflectivity greater than zero.

It is defined *opaque body* a body which absorbs and reflects all the radiation falling upon it. For the opaque body it is $\tau = 0$ and $\alpha + \rho = 1$.

1.4.1 Radiation Between Surfaces of Solids Separated by a Non Absorbing Medium

The radiation heat transfer between two surfaces depends on the view the surfaces have of each other and the emitting and absorbing characteristics of the surfaces.

A general expression of the energy interchange between gray bodies with two surfaces A_1 and A_2 at different uniform temperatures $\vartheta_1 > \vartheta_2$ can be written in the form:

$$p_{12} = 5.67 \cdot 10^{-8} \cdot \varepsilon_{12} \cdot F_{12} \cdot \left[T_1^4 - T_2^4\right], \ \mathrm{W/m^2} \tag{1.107}$$

where: $\sigma = 5.67 \cdot 10^{-8}\ (\mathrm{W/m^2\,K^4})$—Stefan-Boltzman constant; ε_{12}—coefficient of mutual radiation between the surfaces; F_{12}—geometrical shape or view factor.

The *geometrical view factor* F_{12} is defined as the fraction of energy leaving the surface A_1 in all directions which reaches A_2. It depends only on shape and orientation of the two surfaces.

Analogously it is defined the view factor F_{21} as the fraction of energy leaving A_2 which is intercepted by A_1.

The view factors have symmetrical character, i.e. the terms A_1F_{12} and A_2F_{21} are identical.

The calculation of view factors is based on simple geometrical considerations. For the two infinitesimal surface areas dA_1 and dA_2 in arbitrary position and orientation shown in Fig. 1.18, the differential view factor is given by:

$$dF_{12} = \frac{\cos\beta_1 \cos\beta_2}{\pi r_{12}^2} \tag{1.108}$$

To obtain the total heat transferred from a finite area A_1, to a finite area A_2, it is necessary to integrate over both surfaces, and the view factor F_{12} becomes:

$$F_{12} = \frac{1}{A_1}\int_{A_1}\left(\int_{A_2}\frac{\cos\beta_1\cos\beta_2}{\pi r_{12}^2}dA_2\right)dA_1 \tag{1.109}$$

Then the computation of view factors is therefore a problem of mathematical integration, which can be not trivial, except for some simple geometries:

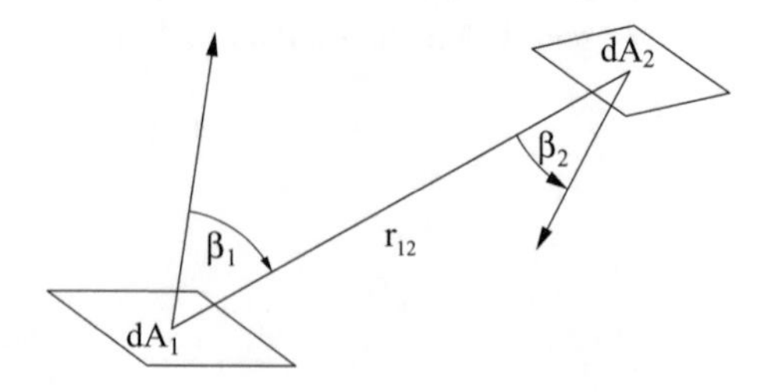

Fig. 1.18 Infinitesimal surface areas for definition of view factor [8]

1. *identical parallel strips of width W and separation H*

$$F_{12} = F_{21} = \sqrt{1+h^2} - h \tag{1.109a}$$

with: $h = H/W$ (see Fig. 1.19a)

2. *two parallel coaxial discs with separation H and diameters $d_1 = 2r_1$ and $d_2 = 2r_2$* (Fig. 1.19b):

$$\left.\begin{aligned} F_{12} &= \frac{1}{d_1^2}\left[\sqrt{H^2 + \frac{(d_1+d_2)^2}{2}} - \sqrt{H^2 + \frac{(d_2-d_1)^2}{2}}\right] \\ F_{21} &= F_{12}\cdot\frac{A_1}{A_2} = F_{12}\cdot\left(\frac{d_1}{d_2}\right)^2 \end{aligned}\right\} \tag{1.109b}$$

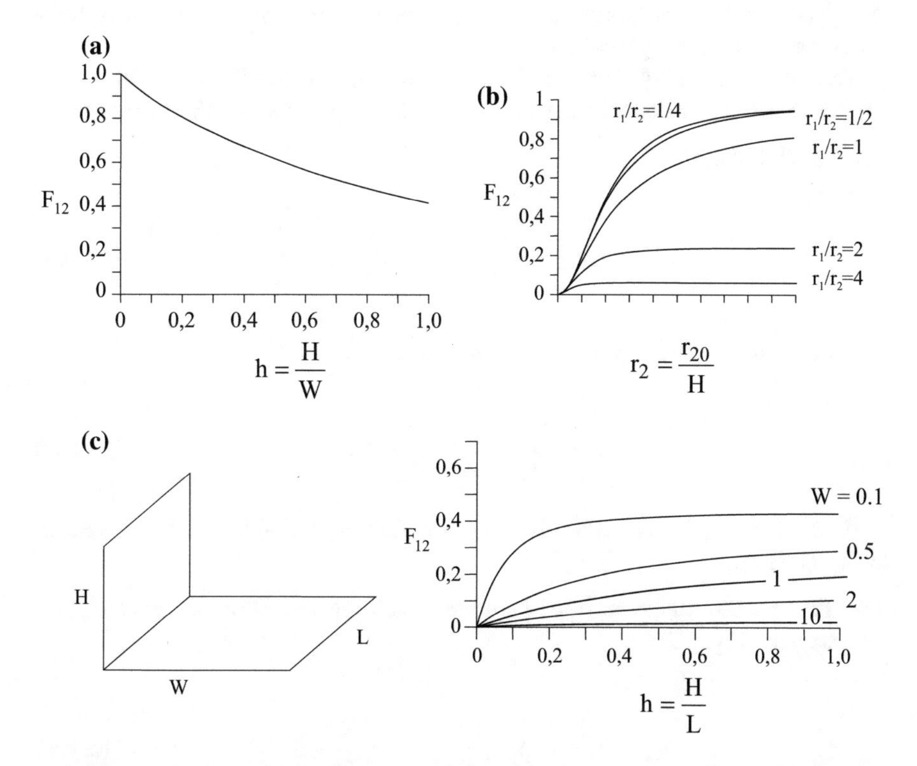

Fig. 1.19 **a** View factor F_{12} between two identical parallel strips of width W and separation H ($h = H/W$); **b** View factor F_{12} between a disc of radius r_{10} and a coaxial parallel disc of radius r_{20}, with separation H ($r_1 = r_{10}/H$ and $r_2 = r_{20}/H$); **c** View factor F_{12} from a horizontal rectangle to an unequal adjacent vertical rectangular plate

3. *perpendicular rectangles with a common edge* (Fig. 1.19c)

$$F_{12} = \frac{1}{\pi W}\left[h \cdot \tan^{-1}\left(\frac{1}{h}\right) + w \cdot \tan^{-1}\left(\frac{1}{w}\right) - \right.$$
$$\left. - \sqrt{h^2 + w^2} \cdot \tan^{-1}\left(\frac{1}{\sqrt{h^2 + w^2}}\right) + \frac{1}{4} \cdot \ln\left(a \cdot b^{w^2} \cdot c^{h^2}\right)\right] \tag{1.109c}$$

with:

$$h = H/L; \quad w = W/L;$$

$$a = \frac{(1 + h^2)(1 + w^2)}{1 + h^2 + w^2}; \quad b = \frac{w^2(1 + h^2 + w^2)}{(1 + w^2)(h^2 + w^2)}; \quad c = \frac{h^2(1 + h^2 + w^2)}{(1 + h^2)(h^2 + w^2)}$$

In the literature are available several catalogs which give the values of view factors for a large number of other geometries [10, 11]. Moreover, the so-called *view factor algebra* allows to extend such results to the case of several surfaces under the sight from a given one.

The coefficient ε_{12} of mutual radiation, which depends on the values ε_1 and ε_2, has particular expressions for the following relative positions of the surfaces:

(a) Parallel plates

$$\varepsilon_{12} = \left(\frac{1}{\varepsilon_1} + \frac{1}{\varepsilon_2} - 1\right)^{-1} \tag{1.110a}$$

(b) Body of surface A_1 placed inside the surface A_2

$$\varepsilon_{12} = \left[\frac{1}{\varepsilon_1} + \frac{A_1}{A_2}\left(\frac{1}{\varepsilon_2} - 1\right)\right]^{-1} \tag{1.110b}$$

For $A_2 >> A_1$, Eq. (1.110b) gives $\varepsilon_{12} = \varepsilon_1$. For $A_1 \approx A_2$, Eq. (1.110b) reduces to Eq. (1.110a).

(c) Two parallel plane surfaces between which are interposed *n* parallel plane screens (Fig. 1.20a)

$$\varepsilon_{12} = \left[\frac{1}{\varepsilon_1} + \sum_{i=1}^{n}\left(\frac{2}{\varepsilon_{is}} - 1\right) + \frac{1}{\varepsilon_2} - 1\right]^{-1} \tag{1.110c}$$

where ε_{is} is the emissivity of the *i*-th screen.

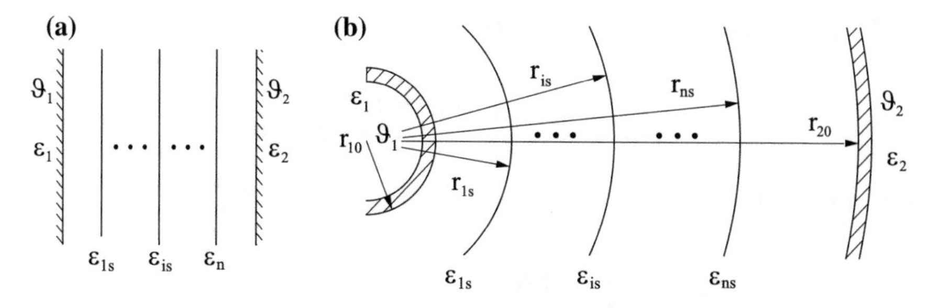

Fig. 1.20 **a** system of plane screens between two parallel plates; **b** system of coaxial cylindrical screens

Formula (1.110c) highlights that the heat transfer between the two surfaces does not depend on the distance between the screens.

(d) System of coaxial cylindrical screens of radii r_{is} between two coaxial surfaces at radii r_{10} and r_{20} (Fig. 1.20b)

$$\varepsilon_{12} = \left[\frac{1}{\varepsilon_1} + \sum_{i=1}^{n} \frac{r_{10}}{r_{is}}\left(\frac{2}{\varepsilon_{is}} - 1\right) + \frac{r_{10}}{r_{20}}\left(\frac{1}{\varepsilon_2} - 1\right)\right]^{-1} \tag{1.110d}$$

Example 1.9 A tube with external diameter d = 0.2 m and surface temperature $\vartheta_s = 400$ °C is placed horizontally in a large ambient at temperature $\vartheta_a = 30$ °C. Emissivity of the tube surface is $\varepsilon_s = 0.8$. Calculate the heat losses by radiation and convection per unit length of the tube.

Solution.

- Radiation losses—[see Eq. (1.98)]:

$$\begin{aligned} p_{ur} &= \pi d \cdot 5.67 \cdot 10^{-8} \cdot \varepsilon_s \cdot \left(T_s^4 - T_a^4\right) = \\ &= 3.14 \cdot 0.2 \cdot 5.67 \cdot 10^{-8} \cdot 0.8 \cdot \left(673^4 - 303^4\right) = 5604\,\mathrm{W/m} \end{aligned}$$

- Convection losses

Data for air at $\vartheta_a = 30$ °C (see Appendix, Table A.5) are:

$$\beta_a = 3.30 \cdot 10^{-3}\,\mathrm{K}^{-1}; \quad \nu_a = 16.00 \cdot 10^{-6}\,\mathrm{m^2/s};$$

$$\lambda_a = 2.67 \cdot 10^{-2}\,\mathrm{W/m\,K}; \quad \mathrm{Pr}_a = 0.701.$$

From Eq. (1.96) the Nusselt number is:

$$\overline{\mathrm{Nu}}_a = 0.5 \cdot (\mathrm{Gr} \cdot \mathrm{Pr})_a^{0.25} \cdot \left(\frac{\mathrm{Pr}_a}{\mathrm{Pr}_s}\right)^{0.25}$$

with: $\mathrm{Pr}_s = 0.678$

$$\mathrm{Gr}_a \cdot \mathrm{Pr}_a = \frac{g\beta_a d^3 \Delta\vartheta}{\nu_a} \cdot \mathrm{Pr}_a = \\ = \frac{9.81 \cdot 3.30 \cdot 10^{-3} \cdot 0.2^3 \cdot 370}{(16.00 \cdot 10^{-6})^2} 0.701 = 2.624 \cdot 10^8$$

Therefore we have:

$$\mathrm{Nu} = 0.5 \cdot \left(2.624 \cdot 10^8\right)^{0.25} \cdot \left(\frac{0.701}{0.678}\right)^{0.25} = 64.17$$

and

$$h_c = \mathrm{Nu}\frac{\lambda_a}{d} = 64.17 \cdot \frac{2.67 \cdot 10^{-2}}{0.2} = 8.57\ \mathrm{W/m^2\,K}$$

Finally, convection losses are:

$$p_{uc} = h_c \cdot \pi d \cdot (\vartheta_s - \vartheta_a) = 8.57 \cdot 3.14 \cdot 0.2 \cdot 370 = 1992\ \mathrm{W/m}$$

and total losses:

$$p_{u\_tot} = p_{ur} + p_{uc} = 5604 + 1992 = 7596\ \mathrm{W/m} \cdot$$

References

1. Incropera, F.P., de Witt, D.P., Bergman, T.L., Lavine, A.S.: Fundamentals of Heat and Mass Transfer, 6th edn. Wiley & Sons, New York (2007)
2. Nacke, B., Baake, E., Lupi, S., Forzan, M., et al.: Theoretical background and aspects of electrotechnologies. Physical principles and realization, p. 356. Pub. house ETU (2012). ISBN 978-5-7629-1237-2
3. Cherednichenko, V.S., Aliferov, A.I., et al.: Heat Transmission in Electrotechnologies, 2 ed., p. 571. NGTU (2011). ISBN 978-5-7782-1813-0 (in Russian)
4. McAdams, W.H.: Heat Transmission, 3rd edn, p. 532. McGraw-Hill Inc., New York (1954)
5. Carslaw, H.S., Jaeger J.C.: Conduction of Heat in Solids, p. 505. Clarendon University Press, Oxford (1959). ISBN 0 19 853303 9
6. Lupi, S.: Elettrotermia (Teaching Notes), p. 467. Libreria Progetto (Padova, Italy) (2005) (in Italian)

7. Talukdar, P.: Introduction to Convective Heat Transfer. IIT—Indian Institute of Technology, Dehli
8. Radiation Heat Transfer. http://web.mit.edu/19
9. Emissivity of common materials. http://www.omega.com/literature/transactions/volume1/emissivitya.html
10. Howell, J.R.: A catalog of radiation configuration factors. McGraw-Hill (1982)
11. Martinez, I.: Radiative View Factors. Webserver.dmt.upm.es/~isidoro/tc3/Radiation%20View%20factors.pdf
12. Bahrami, M.: Steady Conduction Heat Transfer. Simon Fraser University, ENSC 388 (F09)
13. Lupi, S., Rudnev, V.: Principles of Induction Heating—Heat Transfer Phenomena. ASM Handbook, Volume 4C, Induction Heating and Heat Treatment, pp. 6–14. ASM International (2014)
14. Cherednichenko, V.S., Aliferov, A.I., et al.: Heat Transmission. Part 1: p. 232; Part 2: p. 378, 2 edn., NGTU (2007). ISBN 978-5-7782-1386-9 (in Russian)
15. Vologdin, V., Vologdin, Vl., Jr., Bukanin, V., Ivanov, A: Practice of Computer-Assisted Design of induction installations. MEP—International Scientific Colloquium Modelling for Electromagnetic Processing Hannover, September 16–19, 2014

Chapter 2
Electromagnetic Fields in Electro-technologies

Abstract This chapter is the second introductory chapter of the book. It deals with the basic laws of electromagnetic fields and describes some specific phenomena occurring in electro-heat technologies. In particular, the electromagnetic wave diffusion in a conductive half-space is presented and the basic quantities that characterize the thermal processes based on internal heating sources are given. The penetration depth of the electromagnetic wave is a parameter that characterizes all the relevant phenomena in these applications. A qualitative description is provided of several effects influencing the distribution of the current density and, as a consequence, the heating sources in the heated workpieces: the proximity effect (that occurs between two conductors that carry electrical current), the ring effect (that occurs in bended conductors), the slot effect (that occurs in conductors placed in the slot of a magnetic yoke), and the end and edge effects (that occur due to the finite length of inductor and load and their relative position).

In electro-technological processes, electromagnetic phenomena are the basis for generation of heat and forces inside the body to be heated (in the following *workpiece*).

This chapter recalls the basic laws of electromagnetic fields and describes some relevant phenomena occurring in these technologies.

2.1 Basic Equations

Maxwell's equations is a set of equations which fully describe electromagnetic phenomena and, in particular, allow to calculate spatial distribution and direction of field vectors, including their time dependence, as a function of current density and characteristics of medium.

S. Lupi, *Fundamentals of Electroheat*, DOI 10.1007/978-3-319-46015-4_2

The field vectors are:

$\overline{E}$ electric field strength (V/m)
$\overline{D}$ electric flux density (As/m^2)
$\overline{H}$ magnetic field intensity (A/m)
$\overline{B}$ magnetic induction (magnetic flux density) (T)
$\overline{J}$ current density (A/m^2)

Maxwell's equations can be written in the form:

$$\mathrm{rot}\,\overline{H} = \overline{J} + \frac{\partial \overline{D}}{\partial t} \quad (\mathit{Ampere's\ law}) \tag{2.1}$$

$$\mathrm{rot}\,\overline{E} = -\frac{\partial \overline{B}}{\partial t} = -\mu \frac{\partial \overline{H}}{\partial t} \quad (\mathit{Induction\ law}) \tag{2.2}$$

$$\overline{D} = \varepsilon\,\overline{E} = \varepsilon_0\,\varepsilon_r\,\overline{E} \tag{2.3}$$

$$\overline{B} = \mu\,\overline{H} = \mu_0\,\mu_r\,\overline{H} \tag{2.4}$$

where the quantities ε and μ, characteristics of the material, are respectively *electrical permittivity* (or dielectric constant) and *magnetic permeability*.

The parameters $\varepsilon_r = \varepsilon/\varepsilon_0$ and $\mu_r = \mu/\mu_o$ denote relative permittivity and relative magnetic permeability.

Magnetic permeability and dielectric constant have known constant values in vacuum [$\mu_0 = 4 \cdot \pi \cdot 10^{-7}$ ($\mathrm{H\,m^{-1}}$); $\varepsilon_0 = c^2/\mu_0$ ($\mathrm{F\,m^{-1}}$); $c = 299792.458$ (km/s)—velocity of light], but in many practical cases they depend on process parameters, such as temperature or, for permeability of ferromagnetic materials, also on magnetic field intensity.

In Eq. (2.1), the first term of r.h.s. represents the conduction current density due to an electric field $\overline{E}$ in a medium:

$$\overline{J} = \sigma\,\overline{E} = \frac{\overline{E}}{\rho} \quad (\mathit{Ohm's\ law}), \tag{2.5}$$

while the second term $\partial\overline{D}/\partial t$ is the displacement current density which represents the dominant mechanism of electrical conduction in dielectric materials. The quantity $\sigma = 1/\rho$ ($\Omega^{-1}\ \mathrm{m}^{-1}$), denotes material's *electrical conductivity* and ρ its *electrical resistivity* (Ω m).

If an electrically conductive body is placed in an electric field, Joule heat will develop in the body. The electric power converted into heat within each volume element is the power density w:

$$w(t) = \sigma \cdot E^2(t) = \rho \cdot J^2(t) \ (W/m^3) \tag{2.6}$$

The total active power converted into heat is calculated by integration of w over the volume v of the body:

$$P(t) = \int_V \sigma E^2(t)\, dv = \rho \int_V J^2(t)\, dv \tag{2.7}$$

The *Poynting's vector* $\dot{\bar{S}}$ is defined as:

$$\dot{\bar{S}} = \dot{\bar{E}} \times \bar{H}^*. \tag{2.8}$$

It describes the change of energy density in space. It can also be interpreted as describing the flow of energy since it specifies at any point in space the surface power density (in VA m^{-2} or W m^{-2}) and the direction of flow.

Thus considering the rate of the energy flow out from a volume v enclosed by a closed surface A, and denoting with $\dot{P}_S$ the complex power, the rate of complex power flow is:

$$-\dot{P}_S = \oint_A \dot{\bar{S}} \cdot d\bar{A} = \int_V \mathrm{div}\, \dot{\bar{S}}\, dv \tag{2.9}$$

Expressing the term *div* $\dot{\bar{S}}$ by Eq. (2.8), on the basis of vector identities and Maxwell's equations it is possible to show that Eq. (2.9) can be re-written as follows:

$$\begin{aligned} \oint_A \dot{\bar{S}} \cdot d\bar{A} &= \int_V \mathrm{div}\, \dot{\bar{S}}\, dv = \\ &= -\frac{\partial}{dt} \int_V \left(\frac{1}{2}\mu H^2\right) dv - \frac{\partial}{\partial t} \int_V \left(\frac{1}{2}\varepsilon E^2\right) dv - \int_V \dot{\bar{J}} \cdot \bar{E}^*\, dv = \\ &= -\frac{\partial}{\partial t} W_m - \frac{\partial}{\partial t} W_e - \int_V \dot{\bar{J}} \cdot \bar{E}^*\, dv \end{aligned} \tag{2.10}$$

where is:

$W_m = \int_v \left(\frac{1}{2}\, \mu\, H^2\right) dv$ the energy in the magnetic field;

$W_e = \int_v \left(\frac{1}{2}\, \varepsilon E^2\right) dv$ the energy in the electric field;

$\int_v \dot{\bar{J}} \cdot \bar{E}^*\, dv$ a term which represents *either* the power dissipated as ohmic losses *or* the power generated by a source inside v

Therefore, the power delivered by a generator or absorbed by a consumer, can be evaluated by the integral of Poynting vector over a surface enclosing the generator or the consumer.

2.2 Equations for AC Fields

In most applications steady-state AC fields are used and all field quantities $\overline{H}$, $\overline{E}$, $\overline{B}$ vary sinusoidally in time:

$$\overline{H} = \overline{H}_m \cdot \sin \omega t, \quad \overline{B} = \overline{B}_m \cdot \sin \omega t, \quad \overline{E} = \overline{E}_m \cdot \sin(\omega t + \phi)$$

In this case, using the complex exponential form $(e^{j\omega t})$, the field quantities can be rewritten in the form:

$$\dot{\overline{H}} = \overline{H}_m \cdot e^{j\omega t}, \quad \dot{\overline{B}} = \overline{B}_m \cdot e^{j\omega t}, \quad \dot{\overline{E}} = \overline{E}_m \cdot e^{j(\omega t + \phi)}$$

and the general form of Maxwell's equations for harmonic processes becomes:

$$\text{rot}\,\dot{\overline{H}} = (\sigma + j\omega\varepsilon)\,\dot{\overline{E}} \tag{2.11a}$$

$$\text{rot}\,\dot{\overline{E}} = -j\,\omega\,\mu\,\dot{\overline{H}} \tag{2.11b}$$

with: $j = \sqrt{-1}$; ω—angular frequency (rad/s).

Equations (2.11a) and (2.11b) are written in a simpler form in the following cases of practical interest:

(a) *Electromagnetic field in electrical conductors*

In applications concerning conductive materials with high electrical conductivity or electromagnetic fields at relatively low frequency (e.g. less than 10 MHz), it is $\sigma \gg \omega\varepsilon$ and the conduction current density is much greater than displacement current density, so that Eqs. (2.11a) and (2.11b) have the form:

$$\text{rot}\,\dot{\overline{H}} = \sigma\,\dot{\overline{E}} \tag{2.12a}$$

$$\text{rot}\,\dot{\overline{E}} = -j\,\omega\,\mu\,\dot{\overline{H}} \tag{2.12b}$$

Applying the rotational operator to each member of Eqs. (2.12a) and (2.12b) and taking into account Eq. (2.5), we can write:

$$\mathrm{rot\,rot}\,\dot{\overline{\mathrm{E}}} + \mathrm{j}\,\omega\,\mu\,\sigma\,\dot{\overline{\mathrm{E}}} = 0 \tag{2.13a}$$

$$\mathrm{rot\;rot}\,\dot{\overline{\mathrm{H}}} + \mathrm{j}\,\omega\,\mu\,\sigma\,\dot{\overline{\mathrm{H}}} = 0 \tag{2.13b}$$

Using the rules of vector algebra and taking into account that the electrical field intensity $\dot{\overline{\mathrm{E}}}$ and the magnetic field intensity $\dot{\overline{\mathrm{H}}}$ satisfy zero divergence conditions, can be obtained the differential equations of field distributions in electrical conductors (passive conducting regions):

$$\nabla^2\,\dot{\overline{\mathrm{E}}} + \dot{\mathrm{k}}^2\,\dot{\overline{\mathrm{E}}} = 0 \tag{2.14a}$$

$$\nabla^2\,\dot{\overline{\mathrm{H}}} + \dot{\mathrm{k}}^2\,\dot{\overline{\mathrm{H}}} = 0 \tag{2.14b}$$

where: $\dot{\mathrm{k}}^2 = -\mathrm{j}\,\omega\,\mu\,\sigma = \frac{2}{\delta^2}(1-\mathrm{j})$—eddy current constant ($\mathrm{m}^{-2}$); δ—penetration depth of electromagnetic field (m); ∇^2—the Laplacian, which has different forms in different coordinates system.

(b) *Electromagnetic field in non-conducting materials*

In non-conducting materials, where $\sigma \approx 0$, it is $\sigma \ll \omega\varepsilon$ and the Maxewell's Eqs. (2.11a) and (2.11b) take the form:

$$\mathrm{rot}\,\dot{\overline{\mathrm{H}}} = \mathrm{j}\,\omega\,\varepsilon\,\dot{\overline{\mathrm{E}}} \tag{2.15a}$$

$$\mathrm{rot}\,\dot{\overline{\mathrm{E}}} = -\,\mathrm{j}\,\omega\,\mu\,\dot{\overline{\mathrm{H}}}. \tag{2.15b}$$

As in (a), we can obtain the wave equations:

$$\nabla^2\,\dot{\overline{\mathrm{E}}} + \dot{\mathrm{k}}_0^2\,\dot{\overline{\mathrm{E}}} = 0 \tag{2.16a}$$

$$\nabla^2\,\dot{\overline{\mathrm{H}}} + \dot{\mathrm{k}}_0^2\,\dot{\overline{\mathrm{H}}} = 0 \tag{2.16b}$$

with: $\mathrm{k}_0 = \omega\sqrt{\varepsilon\mu}$—real constant ($\mathrm{m}^{-1}$).

Equations (2.16a) and (2.16b) are used for calculating the field distribution in non-conducting materials, e.g. in dielectric or microwave heating applications.

2.3 Electromagnetic Wave Diffusion in Conductive Half-Space

2.3.1 Electromagnetic Field Distribution

To describe how the electromagnetic wave propagates inside an electrically conductive body, consider the homogeneous semi-infinite metal body of Fig. 2.1, infinitely extended along *x, y, z*, with interface in the plane *x-z* and depth along *y*.

We look for the solution of Eq. (2.14b) with the following assumptions:

- homogeneous material with constant electrical resistivity ρ and magnetic permeability μ;
- sinusoidal field quantities;
- exciting plane wave of magnetic field incident on the plane *xoz*.

Since in Cartesian coordinates the Laplace's operator has the form

$$\nabla^2 \dot{\overline{\mathrm{H}}} = \frac{\partial^2 \dot{\overline{\mathrm{H}}}}{\partial \mathrm{x}^2} + \frac{\partial^2 \dot{\overline{\mathrm{H}}}}{\partial \mathrm{y}^2} + \frac{\partial^2 \dot{\overline{\mathrm{H}}}}{\partial \mathrm{z}^2},$$

and the magnetic field intensity $\dot{\overline{\mathrm{H}}}$ is constant in any plane of coordinate *y*, then it is $\partial^2 \dot{\overline{\mathrm{H}}}/\partial \mathrm{x}^2 = \partial^2 \dot{\overline{\mathrm{H}}}/\partial \mathrm{z}^2 = 0$ and Eq. (2.14b) becomes:

$$\frac{\partial^2 \dot{\overline{\mathrm{H}}}}{\partial \mathrm{y}^2} = \mathrm{k}^2 \dot{\overline{\mathrm{H}}} \tag{2.17}$$

with

$$\mathrm{k}^2 = \mathrm{j}\,\omega\sigma\mu.$$

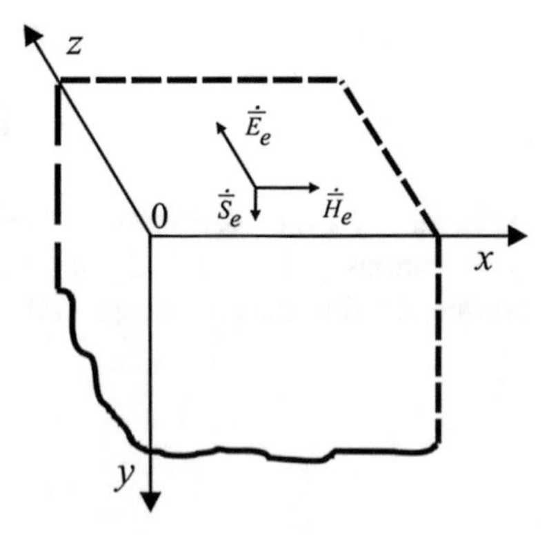

Fig. 2.1 Schematic of a semi-infinite homogeneous conductive body

The general solution of this linear differential equation of second order is:

$$\dot{\bar{H}} = \dot{\bar{C}}_1 \cdot e^{-ky} + \dot{\bar{C}}_2 \cdot e^{ky}, \tag{2.18}$$

and the spatial modulus of the magnetic field intensity vector is

$$\dot{H} = \dot{C}_1 \cdot e^{-ky} + \dot{C}_2 \cdot e^{ky}, \tag{2.18a}$$

with $\dot{\bar{C}}_1, \dot{\bar{C}}_2, \dot{C}_1, \dot{C}_2$—integration constants, $\dot{\bar{C}}_1, \dot{\bar{C}}_2$—complex vectors and $\dot{C}_1$, $\dot{C}_2$—moduli of the complex vectors.

With the assumptions made, the vector of magnetic field intensity is directed along *ox*, while its components along the axes *oz* and *oy* are equal to zero, i.e. $\dot{H}_z = \dot{H}_y = 0$.

Moreover it is:

$$\frac{\partial \dot{H}_z}{\partial y} = \frac{\partial \dot{H}_y}{\partial z} = \frac{\partial \dot{H}_x}{\partial z} = \frac{\partial \dot{H}_z}{\partial x} = \frac{\partial \dot{H}_y}{\partial x} = 0$$

and

$$\operatorname{rot} \dot{\bar{H}} = \bar{u}_z \cdot \operatorname{rot}_z \dot{\bar{H}} = -\bar{u}_z \cdot \frac{\partial \dot{H}_x}{\partial y}$$

with u_z—unit vector along the coordinate *x*.

We are able now to write the general solution for the electric field intensity since—omitting the subscript *x* of $\dot{H}_x$—from Eq. (2.12a) we have:

$$\dot{\bar{E}} = \rho \cdot \operatorname{rot} \dot{\bar{H}} = -u_z \cdot \rho \cdot \frac{\partial \dot{H}}{\partial y} \tag{2.19a}$$

Using Eq. (2.18a), finally we obtain:

$$\dot{\bar{E}} = \bar{u}_z \cdot k \cdot \rho \cdot \left(\dot{C}_1 \cdot e^{-ky} - \dot{C}_2 \cdot e^{ky}\right). \tag{2.19b}$$

The integration constant $\dot{C}_1$ and $\dot{C}_2$ of Eqs. (2.18a) and (2.19b) can be determined from the boundary conditions. In particular,

- the constant $\dot{C}_2$ must be equal to zero, since the field intensity cannot increase when *y* increases. In fact, the first term $\dot{C}_1 \cdot e^{-ky}$ of Eq. (2.19b) describes the incident wave, the second term $\dot{C}_2 \cdot e^{ky}$—the reflected one. But in the propagation of an electromagnetic wave inside a semi-infinite half space there is not any reflective surface and the reflected wave is absent.

- the constant $\dot{\bar{C}}_1$ is determined from the condition that on the surface $y = 0$ is known the amplitude of the exciting magnetic field vector $\dot{\bar{H}} = \dot{\bar{H}}_e$ tangential to the metal surface. In this hypothesis the magnetic field intensity does not vary at the crossing of the surface between the conductive medium and air.

Therefore, it results $\dot{\bar{C}}_1 = \dot{\bar{H}}_e$ and $\dot{C}_2 = 0$.
Finally we obtain:

$$\dot{\bar{H}} = \bar{u}_x \cdot \dot{\bar{H}}_e \cdot e^{-ky} \tag{2.20a}$$

$$\dot{\bar{E}} = \bar{u}_z \cdot k \cdot \rho \cdot \dot{H}_e \cdot e^{-ky} \tag{2.20b}$$

where k is the damping coefficient of the electromagnetic wave and δ the penetration depth:

$$k = \sqrt{\frac{j\omega\mu}{\rho}} = \sqrt{2j}\sqrt{\frac{\omega\,\mu}{2\,\rho}} = \frac{1+j}{\delta} = \frac{\sqrt{2}}{\delta}\,e^{j\frac{\pi}{4}}. \tag{2.21a}$$

$$\delta = \sqrt{\frac{2\,\rho}{\omega\,\mu}} = \sqrt{\frac{2\,\rho}{\pi\,f\,\mu_0\,\mu_r}} \tag{2.21b}$$

The penetration depth δ has dimension of a length; its value varies with the square root of the electrical resistivity and inversely with the square root of the relative magnetic permeability and the frequency. Therefore, it depends upon the electromagnetic material properties and the applied frequency and, for the same frequency, it has different values in different materials.

For example, at 50 Hz: $\delta_{Cu50} \approx 1$ cm; at 5 kHz: $\delta_{Cu5000} \approx 1$ mm.

In trigonometric form Eqs. (2.20a) and (2.20b) can be written as follows:

$$\dot{\bar{H}} = \bar{u}_x \cdot H_{em} \cdot e^{-\frac{y}{\delta}} \cdot \sin\left(\omega t - \frac{y}{\delta}\right) \tag{2.22a}$$

$$\dot{\bar{E}} = \bar{u}_z \cdot \frac{\sqrt{2}}{\delta} \cdot \rho \cdot H_{em} \cdot e^{-\frac{y}{\delta}} \cdot \sin\left(\omega t - \frac{y}{\delta} + \frac{\pi}{4}\right) \tag{2.22b}$$

and the amplitudes of the electric and magnetic field intensities are

$$\left.\begin{aligned} H_m &= H_{em} \cdot e^{-\frac{y}{\delta}} \\ E_m &= H_{em} \cdot \tfrac{\sqrt{2}}{\delta} \cdot \rho \cdot e^{-\frac{y}{\delta}} \end{aligned}\right\}. \tag{2.23}$$

As shown by Eqs. (2.20a), (2.20b), (2.22a), (2.22b) and (2.23), the amplitudes of the magnetic and electric field intensities decay exponentially as the electromagnetic wave penetrates in the conducting body, with a decay rate proportional to the damping factor $k = (1+j)/\delta$, which is an index of the phenomenon of skin effect in conductors.

The factor $e^{-y/\delta}$ describes the damping, while the second term $e^{-jy/\delta}$ says that the maximum field strength is achieved at different times in different places.

2.3.2 Current and Power Distributions

The instantaneous values of the current density vector $\dot{\bar{J}} = \dot{\bar{E}}/\rho$ in the thickness of the semi-infinite metal body, evaluated from Eq. (2.22b), are:

$$\dot{\bar{J}} = \bar{u}_z \cdot \frac{\sqrt{2}}{\delta} \cdot H_{em} \cdot e^{-\frac{y}{\delta}} \cdot e^{j\left(\omega t - \frac{y}{\delta} + \frac{\pi}{4}\right)}. \tag{2.24}$$

By introducing the notation

$$J_{em} = \frac{\sqrt{2}}{\delta} \cdot H_{em}, \tag{2.25a}$$

for the amplitude of the current density vector inside the semi-infinite metallic body we can write:

$$J = J_{em} \cdot e^{-\frac{y}{\delta}}. \tag{2.25b}$$

The r.m.s. value J of modulus of the current density vector therefore is:

$$J_m = \frac{J_{em}}{\sqrt{2}} \cdot e^{-\frac{y}{\delta}}. \tag{2.25c}$$

The maximum amplitude of the current density vector, given by Eq. (2.25a), occurs at the surface $y = 0$.

The distribution of the volume power density inside the semi-infinite body can be evaluated according Eq. (2.25c) by the relationship:

$$w(y) = \rho \cdot J^2 \cdot e^{-\frac{2y}{\delta}} = \frac{1}{2} \cdot J_{em}^2 \cdot e^{-\frac{2y}{\delta}} = w_e \cdot e^{-\frac{2y}{\delta}} \tag{2.26}$$

with w_e—value of w at $y = 0$.

At $y = \delta$ it is $w_\delta = w_e \cdot e^{-2} = 0.135 \cdot w_e$, i.e. the heat source density falls within one penetration depth to about 14 % of its surface value.

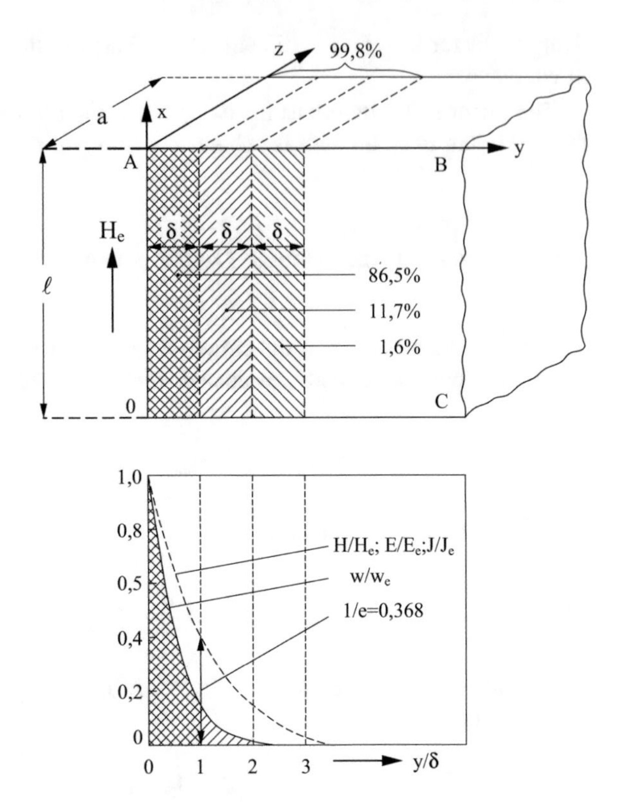

Fig. 2.2 Decay of magnetic field (H), electric field (E), current density (J) and power density (w) inside the conducting medium, as a function of distance from the surface

By integration of the heat source density along the y coordinate the following results are obtained:

- active power dissipated in a layer below the surface of length ℓ along x and depth a along z (see Fig. 2.2):

$$P_{wy} = \ell\, a \int_0^y w\, dy = \ell\, a\, w_e \int_0^y e^{-2y/\delta}\, dy = \ell\, a\, H_e^2 \frac{\rho}{\delta}\left(1 - e^{-2y/\delta}\right) \qquad (2.27a)$$

- total active power within an element of the semi-plane of length ℓ (along x) and depth a (along y):

$$P_{w\infty} = \ell\, a \int_0^\infty w\, dy = \ell\, a\, w_e \int_0^\infty e^{-2y/\delta} dy = \ell\, a\, H_e^2 \frac{\rho}{\delta} \qquad (2.27b)$$

- ratio $P_{wy}/P_{w\infty}$:

$$\frac{P_{wy}}{P_{w\infty}} = (1 - e^{-2y/\delta}) \tag{2.27c}$$

- power dissipated in the first layer of thickness δ below the surface

$$P_\delta = P_{w\infty}\left(1 - e^{-2}\right) = 0,865 \cdot P_{w\infty}. \tag{2.27d}$$

This means that more than 86 % of the total power is developed in the first layer of thickness δ. From the previous equations it's easy to demonstrate that in the subsequent two layers of thickness δ the power is 11.7 and 1.6 % of the total power, as shown in Fig. 2.2.

The diagram of the figure also shows that at a depth greater than 3 times δ, the values of all the quantities of interest (current density, volume specific power and magnetic field intensity) are practically negligible.

2.3.3 Wave Impedance

The electrical impedance of one square meter of the metallic semi-infinite body, which is called *wave impedance*, is equal to:

$$\dot{Z}_{e1} = R_{e1} + jX_{e1} = \frac{\dot{E}}{\dot{H}} = \frac{\dot{H}_e \cdot \rho \cdot \frac{1+j}{\delta} \cdot e^{-\frac{1+j}{\delta}y}}{\dot{H}_e \cdot e^{-\frac{1+j}{\delta}y}} = \frac{\rho}{\delta}(1+j) \tag{2.28}$$

The corresponding resistance and reactance therefore are:

$$R_{e1} = \frac{\rho}{\delta}; \quad X_{e1} = \frac{\rho}{\delta}.$$

As a consequence, the power factor of the semi-infinite body is:

$$\cos\varphi = \frac{R_{e1}}{\sqrt{R_{e1}^2 + X_{e1}^2}} = \frac{1}{\sqrt{2}} = 0.707 \tag{2.29}$$

It is noteworthy to note that the power factor is equal to 0.707 and does not depend on material characteristics and frequency.

2.4 Special Electromagnetic Phenomena in Electro-technologies

2.4.1 Skin-Effect

The skin-effect is the effect which produces a decrease of the current density from the surface towards the internal part of the body in conductors where AC current or magnetic flux in magnetic cores flow.

In a conductor supplied with DC current, the current density J is constant in the cross-section and is given by:

$$J = \frac{I}{A},$$

with: *I*—current in the conductor (A); *A*—conductor cross-section (m^2).

A different situation occurs in a conductor where AC current flows.

As example consider the infinitely long cylindrical conductor of Fig. 2.3a, where the AC current *I* flows.

The direction of the magnetic flux lines $\dot{\overline{B}}_I$ produced by the current I, is determined by the advancement of the right-hand screw rule.

This AC magnetic flux $\dot{\overline{B}}_I$ induces, in ideal rectangular paths inside the conductor, the eddy current *i* which in turn produces a reaction magnetic field, characterised by the vector $\dot{\overline{B}}_i$.

According to Lenz's law, the direction of the magnetic field $\dot{\overline{B}}_i$ produced by the current *i*, is opposite to that of the magnetic induction created by the current *I* and the direction of the induced current *i* in any path encircling the lines of magnetic induction also will obey the right-hand screw law.

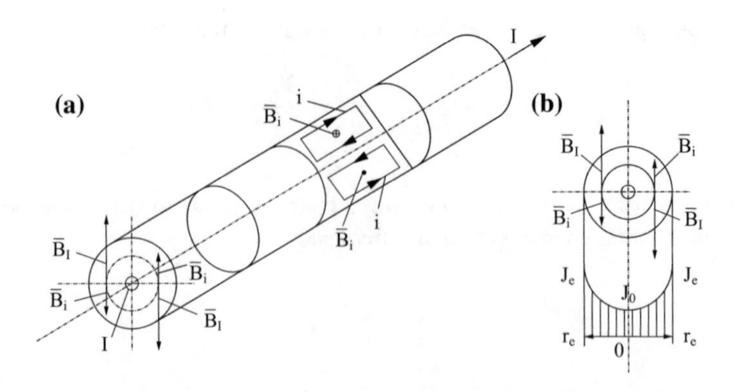

Fig. 2.3 Skin-effect in a cylindrical conductor [1, 2]

Therefore, the induced eddy currents *i* will increase the total value of the current near the surface and weaken it in the central part of the conductor, producing in this way an uneven distribution of the current density in the cross-section. This phenomenon is illustrated qualitatively in Fig. 2.3b.

2.4.2 Exponential Distributions and Equivalent Surface Active Layer

As shown by Eqs. (2.25a)–(2.25c), in the conducting semi-infinite body or, more generally, when the skin-effect is pronounced, the distribution of the current density from the surface towards the interior of the conductor follows the exponential law:

$$J = J_{em} e^{-\frac{y}{\delta}}, \tag{2.30a}$$

where J—current density at a distance y from the conductor's surface (A/m^2); J_{em}—corresponding value at the conductor's surface (A/m^2); δ—penetration depth of the electromagnetic wave (m).

Introducing in Eq. (2.21b) the values $\omega = 2\pi f$ and $\mu_0 = 4\pi \cdot 10^{-7}$ (H/m), the penetration depth can be expressed in the more convenient form:

$$\delta = 503 \sqrt{\frac{\rho}{\mu_r f}} \ \text{(m)}. \tag{2.30b}$$

From Eq. (2.30a), the penetration depth can be defined as the distance from the surface of the conductor at which the current density is reduced to a value $e = 2.718$ times lower than the current density at the surface.

It has been also shown that, in this case, the distribution of volumetric power density from the surface to the conductor interior is expressed by the relationship:

$$w = w_e e^{-\frac{2y}{\delta}} \tag{2.30c}$$

with: $w_e = \rho J_{em}^2$.

Considering that, as shown by Eq. (2.27d), nearly all active power is concentrated in the surface layer of thickness δ, can be sometimes useful in practical calculations to make reference to an *equivalent surface active layer* of thickness δ'. This equivalent surface layer is obtained by replacing the actual exponential current density distribution in the heated body by a fictitious distribution, such that in the equivalent active layer the current density is uniformly distributed with value J', while it is equal to zero in the remaining part of the cross-section (see Fig. 2.4).

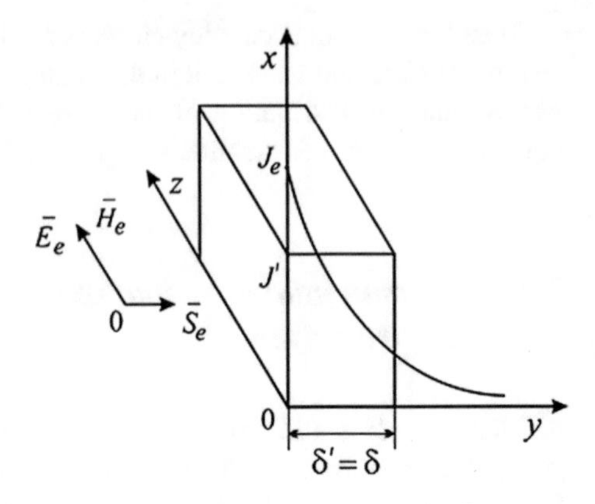

Fig. 2.4 Equivalent surface active layer

The thickness δ' of the equivalent layer and the value of the current density J' are determined by the condition that the total active power dissipated in the conductor by the real exponential current density distribution must be equal to the one dissipated in the equivalent active layer.

Choosing arbitrarily the value of δ' equal to the penetration depth δ, it's easy to demonstrate that the current density at the surface of the body with exponential distribution and the current density in the equivalent active layer are linked by the relationship $\mathrm{J_e} = \sqrt{2}\,\mathrm{J}'$ [1].

Note: The introduction of the active layer thickness δ' and the equivalent current density J' is an artificial operation without physical meaning.

2.4.3 Non-exponential Distributions of Current Density and Heat Sources

As stated before, the exponential distributions of the current density and heat sources described by Eqs. (2.30a)–(2.30c) occur in homogeneous and linear bodies in case of pronounced skin effect (e.g. at relative high frequency).

At lower frequencies, also in case of homogeneous and linear materials, the distributions differ from exponentials, depending on frequency, material characteristics and geometry of conductor.

A typical example occurs in the longitudinal flux induction heating of non-magnetic cylindrical work-pieces, where—with constant resistivity—the radial distributions of induced current density are those illustrated in Fig. 2.5. In this case the distribution becomes nearly exponential only for values of $\mathrm{m} = \sqrt{2}\mathrm{r_e}/\delta \geq 12$.

In many applications of rapid induction heating or direct resistance heating, during the thermal transient the temperature may have considerable different values in the work-piece cross-section and—as a consequence—the local resistivity, which is temperature dependent, is different from point to point.

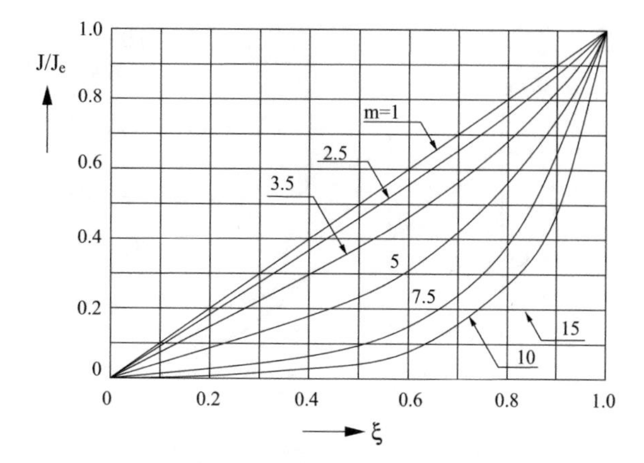

Fig. 2.5 Radial distributions of induced current density in cylindrical workpieces ($\xi = r/r_e$; $m = \sqrt{2}(r_e/\delta)$; r_e—external radius of cylinder) [3]

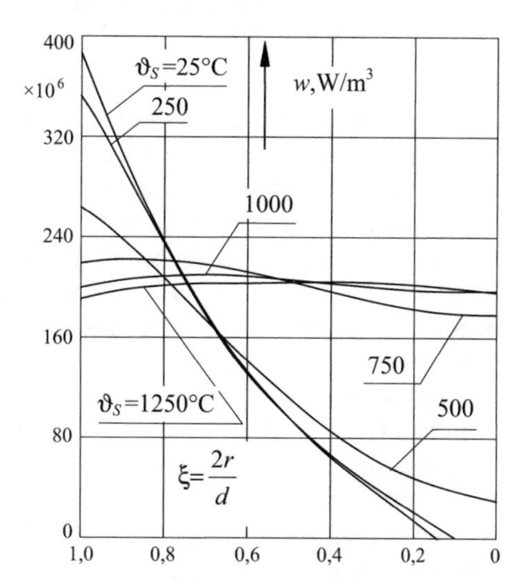

Fig. 2.6 Power density distributions along radius in DRH of a cylindrical rod, at different instants when surface temperature is ϑ_s ($r_e = d/2 = 30$ mm; $\xi = r/r_e$; $\ell = 6$ m; U = 100 V; steel C45) [3]

This reflects on the distribution in the cross-section of specific power per unit volume, i.e. the heat sources, which may differ significantly from the exponential one even in case of pronounced skin effect.

This phenomenon is more evident in ferromagnetic work-pieces, where the current density distribution is influenced not only by the temperature dependent resistivity, but also by the local values of magnetic permeability, which in turn depend not only on temperature but also on local magnetic field intensity.

Figure 2.6 shows an example of the specific power radial distribution in the direct resistance heating (DRH) of a magnetic steel rod. The curves correspond to different instants of the process, when surface temperatures were those specified in the figure. They show that at the beginning of the process the power distribution is nearly exponential. In subsequent instants of the heating transient—due to the variations of resistivity and permeability with temperature and local magnetic field

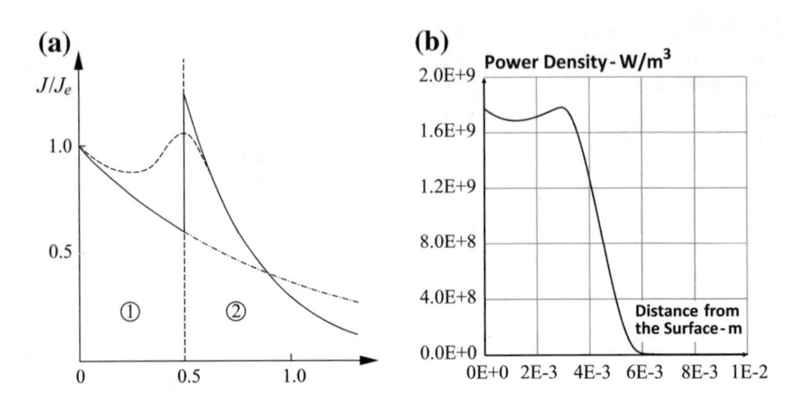

Fig. 2.7 **a** Schematic representation of the magnetic wave phenomenon; **b** radial distribution of power density in induction heating at 4 kHz of a carbon steel shaft, 100 mm diameter, with H_0 = 240 kA/m (r.m.s. value)

intensity—the power distributions remain highly non-uniform below Curie point, while in the final stages above Curie temperature they have a quite different shape, becoming more "flat" and showing the influence at the surface of radiation losses.

A particular phenomenon, known as "*magnetic wave*", occurs in induction surface hardening where the power density distribution inside the workpiece may show a peculiar shape, significantly different from the commonly assumed exponential distribution. In this case, as usual, the power density has its maximum value at the surface, and decreases toward the core. But, at a certain distance from the surface, it starts to increase again, reaching a maximum value before finally declining (Fig. 2.7) [3, 4].

This behavior can be explained considering that, during rapid induction heating, like in surface hardening of a carbon steel workpiece, it may occur a heating stage where the core has still magnetic properties (region 2 in Fig. 2.7a), while a surface layer (region 1) is non-magnetic having exceeded the Curie temperature.

In this situation, the current density distribution can be schematically represented by two exponentials corresponding to two different values of δ due to the material characteristics of the magnetic and non-magnetic regions, as it is illustrated in Fig. 2.7a.

In particular, in some cases, the values of the current and power density at the surface of separation of the two layers, can be even higher than those at the surface.

However, since in practice the values of ρ and μ change gradually from layer to layer (*natura non facit saltus*), the real distributions are more complex, as indicated qualitatively by the dashed curve in the same figure or as shown by the example of Fig. 2.7b, which gives the radial distribution of power density during the transition of Curie point in the induction heating of a carbon steel shaft.

2.4.4 Proximity Effect

The proximity effect consists in a re-distribution of the current density in the conductor cross-section as a result of the interaction of the electromagnetic field produced by the conductor itself and the fields of all other nearby current carrying conductors.

In a single conductor carrying AC current, the current is distributed unevenly over the cross-section, symmetrically about the plane of symmetry for conductors of rectangular cross section (Fig. 2.8a) or the axis for cylindrical conductors.

In case of two conductors in which alternating currents flow, the current density distribution in the cross-sections depends on the distance between them and the direction of currents.

For conductors with currents flowing in opposite directions (i.e. with phase difference 180°), the maximum of the current density occurs on the two sides placed in front to each other (Fig. 2.8b). For currents in the same direction (phase difference equal to 0°), the highest current density occurs on the far side of conductors (Fig. 2.8c).

This phenomenon is similar to the skin-effect where the current density distribution is higher at the surface of the conductor where the magnetic field intensity is maximum (see Eqs. 2.23 and 2.25a–2.25c) and decreases where the magnetic field is lower.

In this case the magnetic field H produced by one conductor is modified by the influence of the magnetic field H' produced by the current in the other conductor.

Consider for example the case of Fig. 2.8b where there are two conductors with opposite directed currents of the same intensity I. For the super-position principle, the total magnetic field intensity produced by the two conductors is the sum of H and H'.

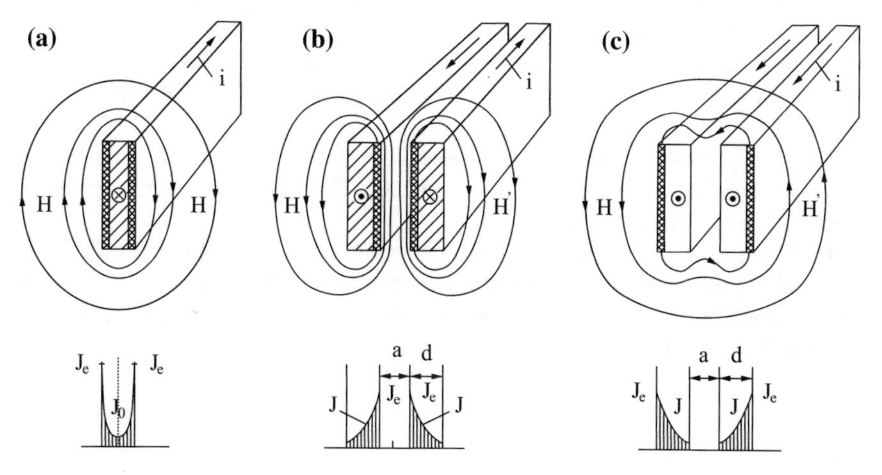

Fig. 2.8 Skin (**a**) and proximity effects (**b**, **c**) in rectangular straight conductors [5]

If the distance (a) between the conductors is much less than their height (h) and is comparable with their thickness (d), the magnetic field intensity H_i in the internal region between the two conductors is nearly the double of H_1, i.e. is

$$H_i = H + H' \approx 2 \cdot H = 2 \cdot (I/2h) = I/h,$$

while on the opposite sides of the conductors it is practically nil, i.e.

$$H_e = H - H' \approx 0.$$

Obviously the influence of the field H' on the total field depends on the value of the current producing it: it is maximum when the conductors are very close to each other and decreases with the increase of the distance (a).

In any case the current density inside the conductors concentrates towards the surfaces where the magnetic field is maximum, as schematically shown in the figure.

Similarly, we can explain the redistribution of the current density when the currents in the conductors are in the same direction (Fig. 2.8c).

Arguing similarly, in case of equal currents the magnetic field intensities in the internal and external regions respectively are:

$$H_i = H - H' \approx 0 \quad \text{and} \quad H_e = H + H' \approx 2 \cdot H$$

In this case the current density inside the conductors is maximum on the external sides of the conductors and decreases inside the cross sections as illustrated in the figure.

The proximity effect depends on the ratio a/d and is more pronounced as the distance between the conductors decreases (i.e., for smaller ratios a/d) and the greater is the thickness of the conductor in comparison with the penetration depth.

A special case of proximity effect occurs in induction heating where the conductor of the inductor, carrying the current I_1, is located near the surface of an electrically conductive body (the workpiece), in which the eddy current i_2 is induced by the exciting magnetic field produced by the inductor. (see the sketch of Fig. 2.9).

For the right-hand screw rule, the direction of the induced current i_2 is opposite to the direction of the current I_1 in the inductor. Thus, the phase difference of the currents I_1 and i_2 will be 180°.

In this situation, uneven distributions of current density will result both in the inductor and the heated body. This effect corresponds to the proximity effect in two parallel conductors, with the maximum of the current density located at the surfaces of the inductor and the heated body.

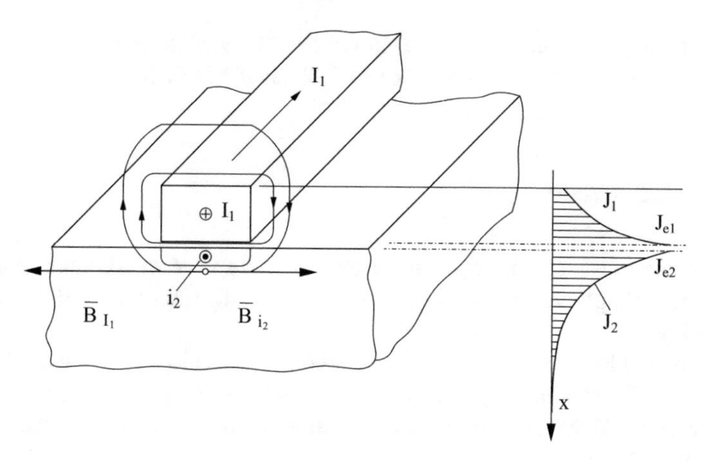

Fig. 2.9 Proximity effect in induction heating [1]

2.4.5 *Ring Effect*

In ring shaped or bent conductors in which DC or AC current flows, the current density tends to concentrate on the inner part of the ring or the bending radius, as shown in Fig. 2.10. This phenomenon is known as ring- or coil-effect.

Due to the asymmetrical pattern of the magnetic field, the current density will be higher in the section of the conductor having the smallest reactance (and resistance), i.e. near the inner surface of the ring, and lower on the opposite side of the cross-section.

The effect is stronger the greater is the ratio d/r_0—for a conductor of rectangular cross section (with d—radial thickness of the conductor and r_0—bending radius), or the ratio r_1/r_0—for a conductor of circular cross-section of radius r_1. Moreover, it is more pronounced the higher is the frequency of the current flowing in the

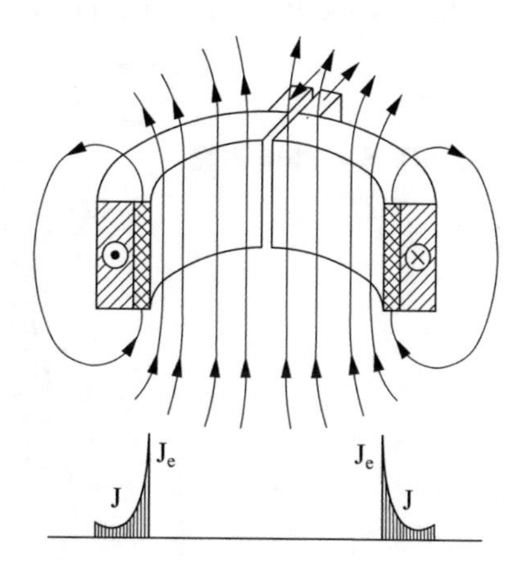

Fig. 2.10 Ring effect in a rectangular conductor and current density distribution in the cross-section [1, 5]

conductor. In particular, when the skin effect is very high, the current density is practically concentrated in the internal surface layer of thickness δ.

2.4.6 Slot Effect

The presence of a magnetic core or a magnetic yoke near an AC current carrying conductor, strongly modifies the current density distribution in the conductor cross-section.

As an example, we will illustrate this phenomenon, which is known as *slot effect,* in the case of a conductor placed in an open slot of a magnetic circuit made of magnetic laminations or other magnetic material suited for the frequency of the current (Fig. 2.11).

Since the magnetic material has permeability much higher than copper, the magnetic field lines tend to flow mainly in magnetic paths of lower magnetic reluctance. In particular, in the upper part of the slot the magnetic flux lines are confined in the magnetic material, while in the lower part they reclose partly through the conductor and partly in air in the lower air gap external to the conductor. This last part of flux lines is totally linked with the conductor cross-section.

For a qualitative description, the conductor cross-section can be ideally subdivided into several elementary layers. The layers closer to the magnetic material (at the bottom of the slot), are 'linked' with a magnetic flux density higher than those close to the open side of the slot.

Therefore, in the inner elements of the slot, stronger counter-electromotive forces will be induced in comparison with those near the open side. As a consequence the internal current paths are characterized by higher reactance values, and their current density will be of lower intensity in comparison with the elements in the regions of the conductor with lower reactance near the open side of the slot, where the maximum current density will occur.

The power density concentration near the opening of the slot will be greater the higher are the frequency and the slot depth.

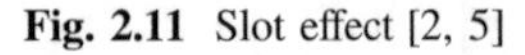

Fig. 2.11 Slot effect [2, 5]

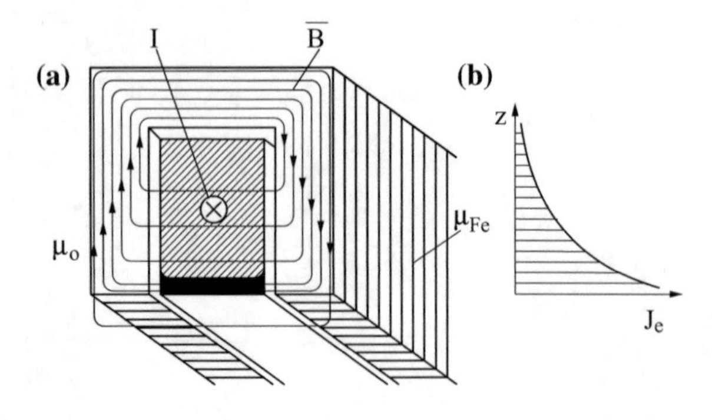

2.4.7 *Load and Coil End- and Edge-Effects*

As described in previous paragraphs, the current and power density distributions have a fundamental influence on the transient and final temperature pattern in the workpiece to be heated.

These distributions, in turn, are influenced not only by the above mentioned electromagnetic phenomena and the heat transfer conditions but also by the distortion of the magnetic field that occur at the end regions of the inductors and at the edges of workpieces in induction heating, or near the contact system in direct resistance heating, where the current lines have a complex two dimensional pattern also in axis-symmetric geometries.

These field distortions are known as *end-* and *edge-effects.*

End- and edge-effects act differently depending on the application, the type of load, magnetic or non-magnetic, cylindrical or with rectangular cross-section (slabs) and—in induction heating—relative dimensions and position of inductor and load.

A detailed discussion of this topic can be found in the bibliography [4, 6–10]. Here we will illustrate only some few examples in order to underline the importance of these phenomena.

2.4.7.1 Longitudinal End- and Edge-Effect

These effects occur typically in the induction heating of cylindrical workpieces with solenoidal coils or circular plates with pancake coils. They may be illustrated by the sketch of Fig. 2.12, which refers to the heating of a long homogeneous workpiece with a multi-turn inductor.

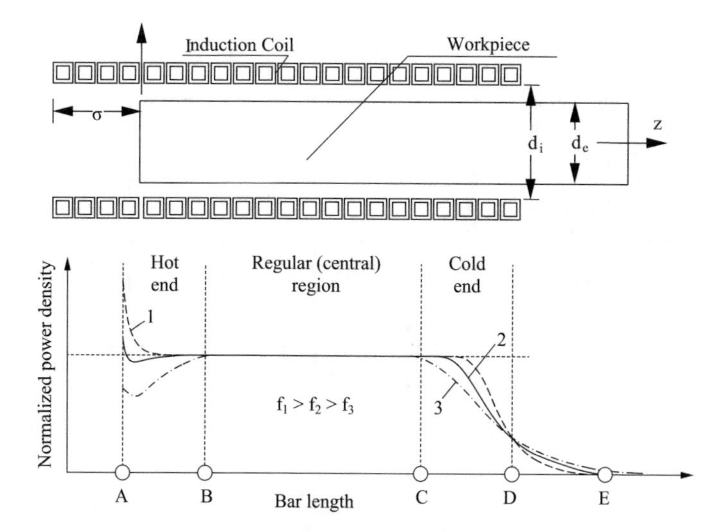

Fig. 2.12 Longitudinal end-effect (*zone CE*) and edge-effect (*zone AB*) in a workpiece

As regards the end-effect of the coil, it is known that the density of the induced current under the coil end (zone CE of Fig. 2.12) is approximately two times lower than in the central part. Accordingly, in this end zone of the workpiece the power density is equal to one quarter of that in the central part.

Depending on the work-piece and coil geometrical dimensions and the frequency, the inductor coil overhang σ and the material characteristics, the end effect (or load edge effect) can produce under heating or overheating of the work-piece end regions (zone *A-B* in Fig. 2.12).

Basically, this effect in non-magnetic workpieces is defined by three variables:

d_e/δ ratio charactering the skin effect;
σ/δ normalized coil overhang;
d_i/d_e coupling ratio between inductor and load

If the coil and workpiece lengths are two times larger than the length of the end effect zones, then the end effects at the opposite ends do not mix and in the "*central part*" there is a uniform magnetic field zone, which corresponds to the field in an infinitely long system.

Figure 2.13 shows, as an example, the normalized power density distributions referred to the value of the central part, along the length of a typical aluminium induction heating system. It shows that an appropriate choice of the above parameters allows to obtain a reasonable uniform power density distribution, like the one in the curve with overhang $\sigma/d_e = 0.7$.

At the workpiece butt-end there is a local surplus of power, while in the adjacent internal zone there is a power deficit. This power profile combined with material thermal conductivity and heat losses from the workpiece front end provides a thermal equalization which can produce a nearly uniform temperature distribution along the workpiece length.

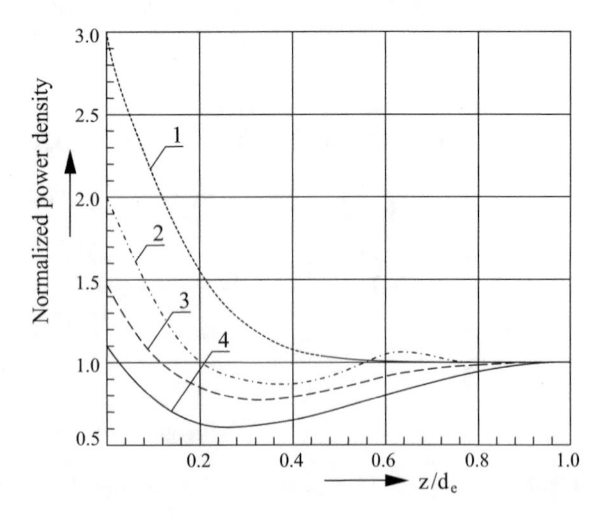

Fig. 2.13 Power distribution in the end zone along a typical aluminium heating system for different coil overhangs, normalized to the value of the central part ($d_e/\delta = 6$; *1* $\sigma/d_e = 1.5$; *2* $\sigma/d_e = 0.7$; *3* $\sigma/d_e = 0.3$; *4* $\sigma/d_e = 0$) [8]

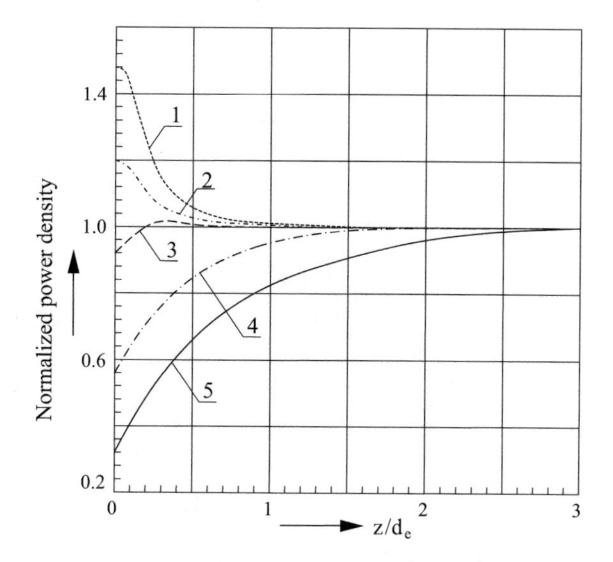

Fig. 2.14 Power distribution, normalized to the value of the central part, in the end zone of a ferromagnetic workpiece for different surface permeability in central part ($d_e/\delta = 6$; *1* $\mu = 10$; *2* $\mu = 15$; *3* $\mu = 30$; *4* $\mu = 90$; *5* $\mu = 180$) [8]

The end-effects have quite different features in magnetic workpieces, since they are affected by two opposing factors: on one side the demagnetizing effect of eddy currents which tends to force the magnetic field out of the workpiece, on the other side, the magnetizing effect of the material, which tends to gather the magnetic field lines within the work-piece.

The first factor tends to increase the power at the workpiece ends, like in non-magnetic materials; the second one, on the contrary, causes a power reduction. Therefore, unlike from non-magnetic material, the ends of ferromagnetic work-pieces, even in magnetic fields produced by coils with large overhangs, may be either overheated or underheated.

Figure 2.14 shows an example of the power distribution, normalized to the value of the central part, in the end zone of a ferromagnetic workpiece, for different surface permeability values in the central part. As reference is taken the surface permeability of the central part because μ varies along the workpiece length, and the skin-effect parameter d_e/δ accordingly varies in the axial direction also.

The diagram shows that the power deficit in the end zone is more pronounced with high values of μ, while a surplus of power can occur with the lower permeability values.

2.4.7.2 Longitudinal and Transverse End and Edge Effects in Rectangular Slabs

In the induction heating of rectangular slabs end- and edge-effects occur both in the transversal and longitudinal direction, due to the finite size of the workpiece.

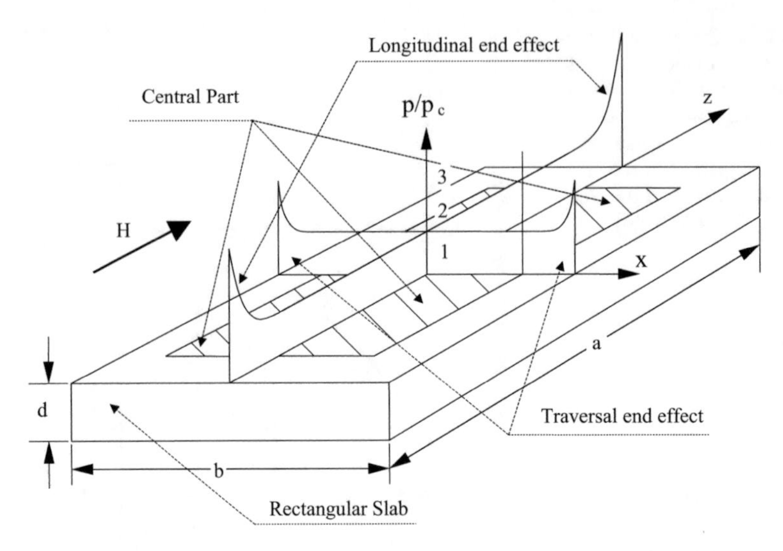

Fig. 2.15 Electromagnetic end and edge effects in induction heating of a slab heated in longitudinal uniform magnetic field

Suppose that the slab of Fig. 2.15, which has length (*a*) and width (*b*) much larger than its thickness (*d*), is placed in an initially uniform magnetic field, oriented along the z-axis.

As schematically shown in the figure, the electro-magnetic field distribution in the slab can be subdivided in three zones: the *central part*, the zone of *tranversal edge effect* and the zone of *longitudinal end effect*. In the central part the field distribution is the same of an infinite plate excited with the same surface magnetic field intensity.

Basically, end and edge-effects have a two dimensional character and are determined only by the parameter (d/δ). The only exception occurs at the three-edge corners, where the field distribution is three-dimensional and is the result of combined end and edge effects [4, 7].

The same figure shows the distributions along the axes x and z of the power density p induced on the upper surface of the slab, normalized to value p_c in the central point of the centre line of the longer side.

Figure 2.16 gives an example of such distributions along the slab width as a function of the relative distance (x'/d) from the edges of the slab, for different values of d/δ. They are characterised near the edges by a region with induced power values lower than the ones in the “central” zone (under-heated area) and an external area with higher power density values (possible overheating).

In conclusion, the figures underline on one hand the strong influence that edge- and end-effects may have on the power density distributions and, as a consequence, on the heating pattern in the workpiece; on the other hand, they suggest that—in

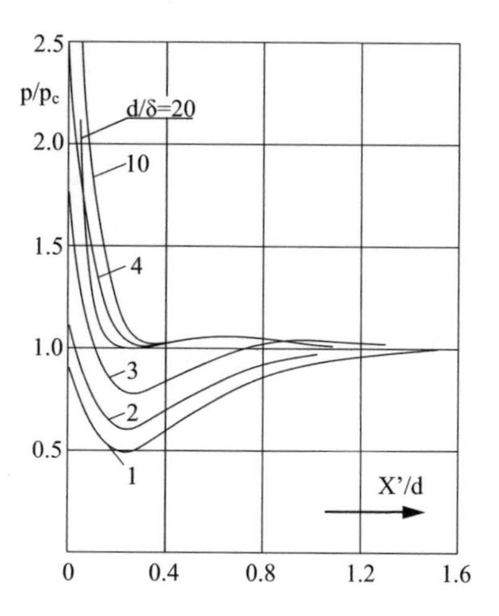

Fig. 2.16 Relative power density distribution along slab width for different values of d/δ (x′-distance from slab edge) [7]

some cases—a convenient choice of the frequency allows to obtain an induced power pattern where the overheated areas can compensate to a certain extent the under-heated ones, by taking advantage of the material's thermal conductivity.

References

1. Aliferov, A., Lupi, S.: Direct Resistance Heating of Metals, 223 pp. NGTU, Novosibirsk (2004). ISBN 5-7782-0475-2 (in Russian)
2. Aliferov, A., Lupi, S.: Induction and Direct Resistance Heating of Metals, 411 pp. NGTU, Novosibirsk (2010). ISBN 978-5-7782-1622-8 (in Russian)
3. Lupi, S.: Appunti di Elettrotermia (Teaching notes), 453 pp. Libreria Progetto, Padua (Italy), (2005) (in Italian)
4. Rudnev, V., Loveless, D., Cook, R., Black, M.: Handbook of Induction Heating, 777 pp. Marcel Dekker Inc., New York (2003). ISBN 0-8247-0848-2
5. Fomin, N.I., Satylobskij, L.M.: Electrical Furnaces and Induction Heating Installations, 247 pp. Metallurghia, Moscow (1979) (in Russian)
6. Slukhotskii, E., Nemkov, V., et al.: Induction Heating Installations, 328 pp. Energija, Leningrad (1981) (in Russian)
7. Nemkov, V., Demidovich, V.B.: Theory and Calculation of Induction Heating Installations, 280 pp. Energoatomiszdat, Leningrad (1988). ISBN 5-283-04409-2 (in Russian)
8. Rudnev, V., Loveless, D.: Longitudinal Flux Induction Heating of Slabs, Bars and Strips is no Longer "Black Magic. J. Melting/Forming/Joining (1995)
9. ASM Handbook: Volume 4C—Induction Heating and Heat Treatment, 820 pp. ASM International, Materials Park, OH 44073-0002 (2014). ISBN-10 1-62708-012-0
10. EDF: Induction, Conduction électrique dans l'industrie, 780 pp. CFE DOPEE85, Paris (France) (1996) (in French)

Chapter 3
Arc Furnaces

Abstract This chapter deals with Electric Arc Furnaces (EAFs) installations used for melting processes based on the heat produced by one or more arcs burning between the ends of one or more electrodes and the charge. The first part of the chapter deals with AC direct arc furnaces for production of steel, considering in particular the power supply, the furnace transformer, the secondary high-current circuit, the electrodes and the furnace vessel. Then the equivalent circuits and the operating characteristics of the installation are presented, making reference either to the approximation with sinusoidal quantities or the non-linear characteristic of the arc. The analysis of these characteristics allows the selection of the optimum operating point. Finally, the evolution of the steel production cycle and the main innovations introduced in modern furnaces in the last 40 years, like foaming slag practice, scrap preheating, CO post-combustion, intensive usage of oxygen and carbon, bottom stirring are dealt with. At the end of the first part the voltage fluctuations produced by EAFs on the supply network (the "Flicker" phenomenon) and the development of DC EAFs are shortly discussed. The second part of the chapter is devoted to Submerged Arc Furnaces used for production of calcium carbide, ferroalloys, other non-ferrous alloys and phosphorus by chemical processes of reduction of one or more oxides of the ore by a reducing agent loaded with the ore. The design and energetic characteristics and the special type of self-baking electrodes used in these furnaces are described.

3.1 Introduction

Arc heating is the heating technique in which heat is produced primarily by one or more arcs burning in the space between the ends of one or more electrodes and the charge. Heat is transferred to the material to be heated mainly by radiation and, partially, by convection and conduction.

Depending on whether the arc current flows directly or not in the material to be heated, we distinguish the cases of direct arc heating and indirect arc heating.

S. Lupi, *Fundamentals of Electroheat*, DOI 10.1007/978-3-319-46015-4_3

It is defined arc-resistance-heating (or submerged arc heating) the heating process in which the heat is developed partly by the arc and partly by the Joule effect of the current flowing in charge.

Corresponding to the above defined types of heating electric arc furnaces (EAFs) may be classified as follows:

- *Direct arc furnaces*
- *Indirect arc furnaces*
- *Arc-resistance furnaces (submerged arc furnaces).*

Taking into account different production processes, arc furnaces can be also subdivided into:

- *Furnaces for steel melting*
- *Reduction furnaces*
- *Vacuum re-melting furnaces*
- *Electroslag re-melting processes.*

For the importance of their application in the steel industry, the high value of installed unit power and the high energy consumption, arc furnaces can be considered the most important electro-heating installations from the technical and economical point of view.

Arc furnaces for steel melting have undergone many technological developments and refinements over the years, but their basic structure remained significantly unchanged.

Conventionally, these developments can be divided into four stages.

In the early stage (from last decades of XIX century till first decades of the XX century), many chemists and metallurgists had conducted small experiments with electric carbon arc furnaces, but only after the researches of Siemens in 1882, the electric arc furnace reached a level of technology that attracted commercial interests.

The first patent for the use of an indirect arc furnace was obtained by the Italian Stassano in 1898. [1] As a result of the work of Stassano and the successful production of ferro-alloys in its electric furnace, the process for production of high-quality steel from scrap was next commercialized in France by Héroult and in Sweden by Kjellin, under patents taken out in 1900 (Fig. 3.1) [1, 2].

The furnaces were commonly of few tons capacity (0.5–1.5 t) and the research was focused on the improvement of design solutions. In 1910 there were more than 100 units of different design operating worldwide. The subsequent combination of decreased costs of electricity and increased demand for steel during WWI transformed the arc furnace into the major mean for producing high-quality metal, with nearly 1000 unit at work in 1920.

The second period, from the twenties to the middle of the XX century, was characterized by the construction of a large number of furnaces, with capacity increased in the range 30–50 tons. The most important changes in the furnace design were driven by the development of the steelmaking technology.

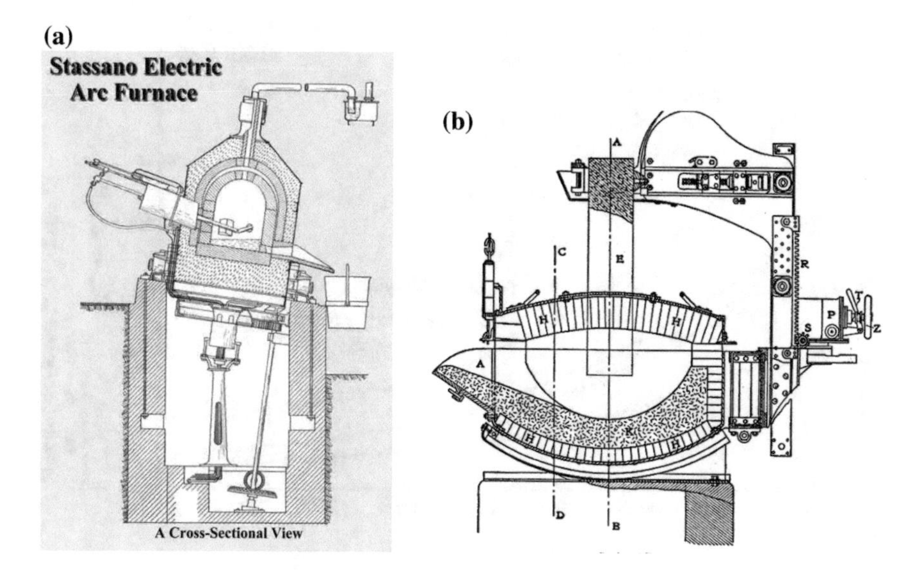

Fig. 3.1 **a** Stassano arc furnace (1898); **b** Héroult refining furnace (1908)

The third stage (from the 50s to the 80s of the last century) was characterized by a further increase in the furnace capacity (50–100 ton, 200–250 kVA/ton), the increased power of EAF transformers, the sharp shortening of tap-to-tap time and the use of oxygen burners. The main objectives of all furnace developments were maximal productivity and environmental, health and safety requirements.

For increasing productivity, in 1963 the first 135-ton UHP (Ultra-High-Power) furnaces were installed, equipped with 70–80 MVA transformers and specific power in the range 520–600 kVA/ton. Due to their successful operation, UHP furnaces became widespread rather quickly, reaching in the early 80s a specific power of about 1000 kVA/ton.

In the first period of UHP furnaces operation, the increase of productivity was limited by the rapid deterioration of sidewalls and roof refractory lining and subsequent increase in downtime for repairs. This technological drawback was eliminated by replacing up to 85 % of the total lining surface with water-cooled panels.

This phase ended with the use of the furnace only as a melting unit producing an intermediate semi-product. All the subsequent process operations, which require reduced power, were moved to the secondary ladle metallurgy unit for further processing.

Then in the fourth stage (last 30 years)—the stage of mass production of electrical steel—an impressive number of innovations, like the introduction of secondary metallurgy, the water-cooled panels, the water cooled oxygen lance, the foaming slag practice, the ladle furnace, the lance manipulation etc., have shortened the tap-to-tap time to 30–40 min in the best 100–130 tons furnaces operating with scraps.

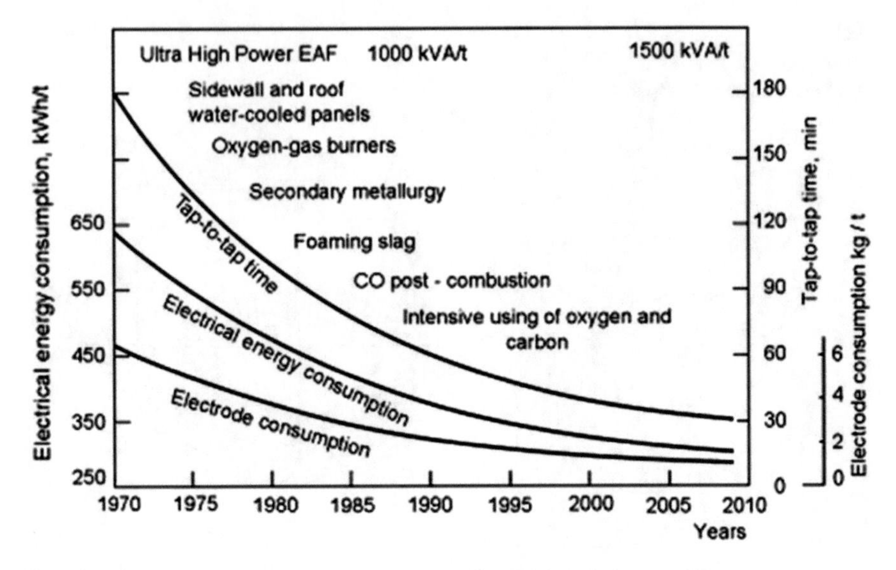

Fig. 3.2 Basic innovations and improvement in 120-t EAF performances [3]

The electrical energy consumption was reduced approximately to half, from 580–650 to 320–350 kWh/ton, and the electrical energy share in the overall energy consumption dropped to 50 % while the electrode consumption was reduced of 4–5 times (Fig. 3.2).

As regards some rough economic figures on steel production, the cost of raw materials, scrap and ferro-alloys amounts to about 75 % of the general costs of an EAF operating on scrap. The so-called operating costs constitute the remaining 25 %, while the costs of energy, electrodes and refractories account for about 60 % of the latter [3].

3.2 EAFs for Steel Production

3.2.1 Steel Production Process

Melting by recycling ferrous scrap in an Electric Arc Furnace (EAF) is the most common process for producing steel.

The EAF constitutes a chemical reactor that utilizes electricity to transform steel scrap to molten steel. However, even if the main source of energy is electricity, other types of energy inputs (combustion in furnace of natural gas, oil or coal, scrap preheating, melt injection, etc.) constitute a significant part of the total energy balance, as shown in Fig. 3.3.

The total theoretical energy required to melt and superheat the scrap to the typical tap temperatures is about 350–380 kWh/t-steel. This energy can be provided

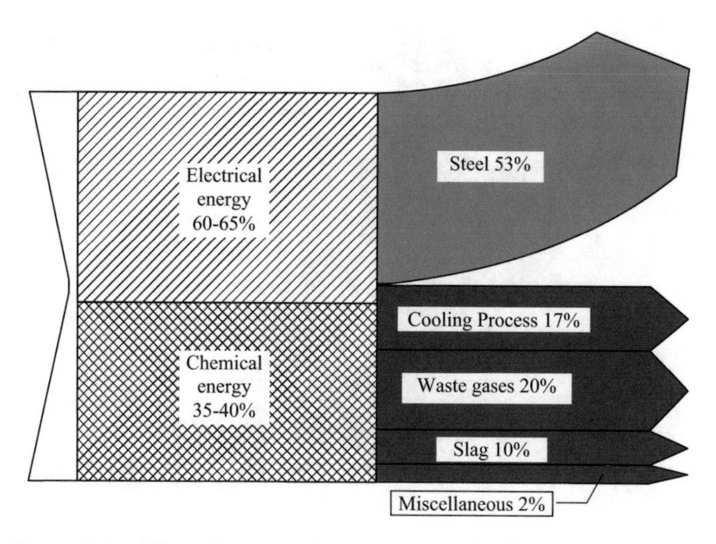

Fig. 3.3 Energy balance flow chart of an electric arc furnace

by the electric arc, from fossil fuel injection or oxidation of the scrap feedstock. In practice, the energy used is highly dependent on product mix, material and energy costs and is unique to the specific furnace operation.

Factors such as raw material composition, power input rates and operating practices—such as post-combustion, scrap preheating—can greatly influence the energy balance.

As shown in the schematic views of Fig. 3.4, the arc furnace consists of a shell with movable roof, through which three graphite electrodes can be raised and lowered.

At the start of the process, the scrap together with limestone for slag formation, is charged from a suspended basket into the furnace with upraised electrodes and roof swung clear. When charging is completed, the roof is closed, the electrodes are lowered into the furnace and an arc is created which generates the heat for melting the scrap.

The main task of the furnace is to convert as fast as possible the solid raw materials to liquid steel. Nevertheless, if time is available, metallurgical operations like oxygen and coal powder injection or alloys additions to reach the required chemical compositions may be performed during the flat bath operation period (after melting), thus constituting a pre-treatment before secondary steelmaking operations.

When the correct composition and temperature of the melt have been achieved, the furnace is tapped rapidly into a ladle. Final adjustments to precise steel grade specification can be made subsequently, in a secondary steel making unit.

Depending upon the grade of steel required, after the molten metal is tapped into a ladle some extra refining stages are made for chemical composition improvement,

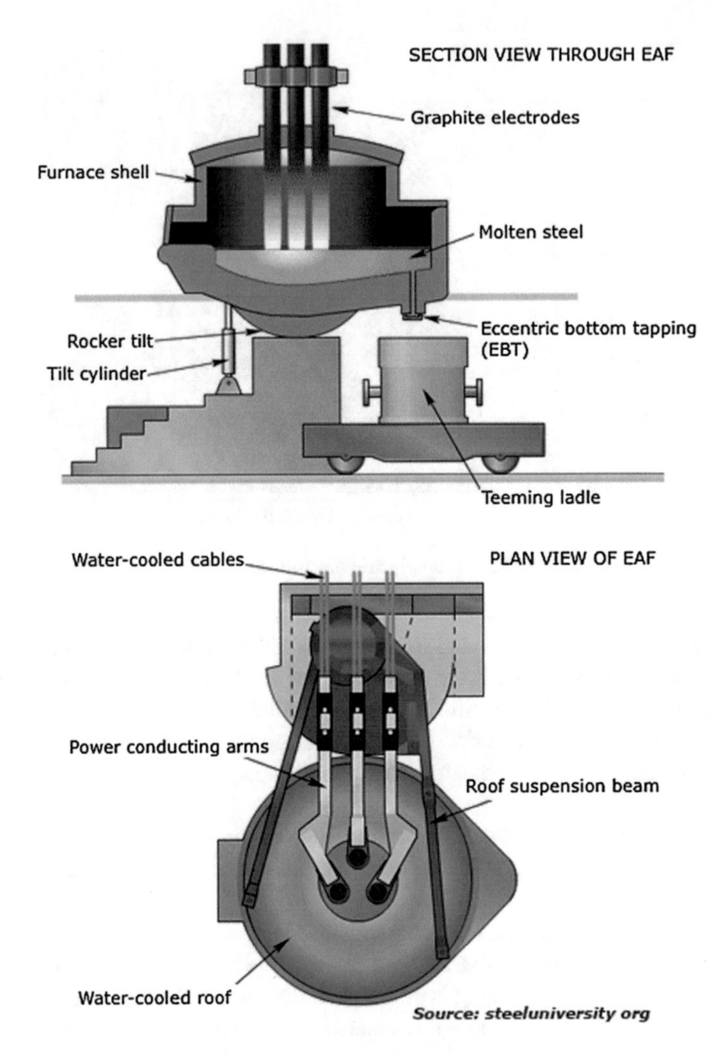

Fig. 3.4 Section and plan view of an electric arc furnace [4]

temperature homogenization, removing of unwanted gases or reduction of some elements to the required low level.

These refining stages are known as secondary steel making and can include ladle stirring with argon, powder or wire injection, vacuum degassing and ladle arc heating.

3.2.2 Installation Description

EAFs for steel production are direct arc furnaces in which the arcs burn between the tips of electrodes and the charge. They are characterised by big size and high installed power, and are currently built with nominal capacity between 10 and 150 tons and power ranging from 5 to 95 MVA. Modern units have capacity most commonly between 50 and 80 tons, while the maximum capacity is 400 tons (with installed power of approximately 160 MVA).

The new constructions are sometimes designed for working with sponge iron and/or pig iron (in form of "pellets or briquettes"), obtained from the ore by reducing oxides. They are not yet used on large scale, due to the still more favourable price of steel scrap, which constitutes the raw material most commonly used.

The typical design of a three-phase EAF comprises three graphite electrodes, a vessel with rounded non-conductive bottom (hearth) and cylindrical walls with openings, one for the tapping spout and one as working or slagging door for de-slagging and making additions, and a roof constituted by a self-supporting dome made of refractory bricks provided externally by a water-cooled iron ring, as shown in Fig. 3.5.

The electrodes are attached to the supporting arms by means of contact pads and clamps; the power conducting arms and clamps are water-cooled. The electrodes can move up and down for arc ignition and extinction and power control during operation. To this end, the arms are moved by hydraulic or electro-mechanical actuators.

Loading of the furnace is normally done by lifting and rotating roof, carrier arms and electrodes around a vertical axis.

Tapping is done first by removing the slag through the back door and then by pouring the molten metal through a front lower tapping spout. To this end a tilting device enables the furnace to be swung out of a certain angle around a horizontal rolling axis towards either the tapping or the slagging side. This is done by means of rockers and roller tracks.

3.2.3 Power Supply System of EAFs

Because of high levels of power and short-circuit power required, the electricity supply is normally ensured by connecting the furnace to a high-voltage power system. A step down transformer is therefore needed to provide the low voltage/high current supply to the electrodes. In fact, electrode voltages are typically in the range from 200 to 1500 V, while currents are of the order of some tens of kA.

An EAF installation and its power supply system include the following main elements (Fig. 3.6):

- high voltage equipment
- reactive power compensation devices

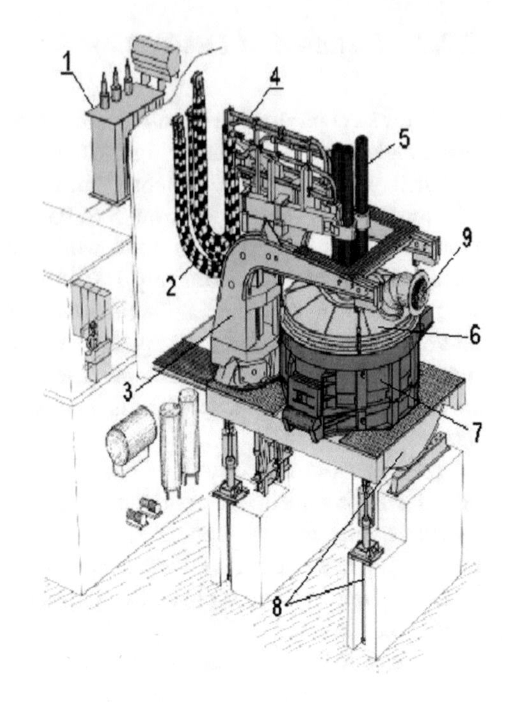

Fig. 3.5 Three-phase arc furnace: mechanical structure and main electrical parts (*1* trasformer and reactor; *2* flexible cables; *3* roof suspension arms with lift and swing mechanism; *4* electrode arms and high-current water-cooled bus tubes; *5* electrodes; *6* roof; *7* external mechanical structure of vessel; *8* tilting devices of the main support platform with toothed rockers and racks; *9* fume extraction duct) [5]

- tapped insulating step-down furnace transformer
- flexible cables
- bus-bars and electrode clamps
- electrodes
- furnace shell.

In small and medium-size furnace installations supply voltages of 6–30 kV may be used, while large furnaces of capacities up to 300 tons and supply power up to more than 100 MVA are connected either directly to the mains grid at 110–220 kV or even to higher voltages, via an intermediate transformer.

Given the high power involved and the low values of furnace transformer secondary voltages, to keep within reasonable values the transformation ratio of this transformer, the connection to the HV grid is done by an intermediate step-down transformer and its high-voltage equipment.

In this case the furnace transformer, with stepped voltage control, is fed from the intermediate transformer.

In small power installations, where the reactance of the circuit is not sufficient to ensure in normal operation arc stability and to limit short circuit currents, it may by installed an extra reactor (see point 5 of Fig. 3.6) between the furnace transformer and the supply network.

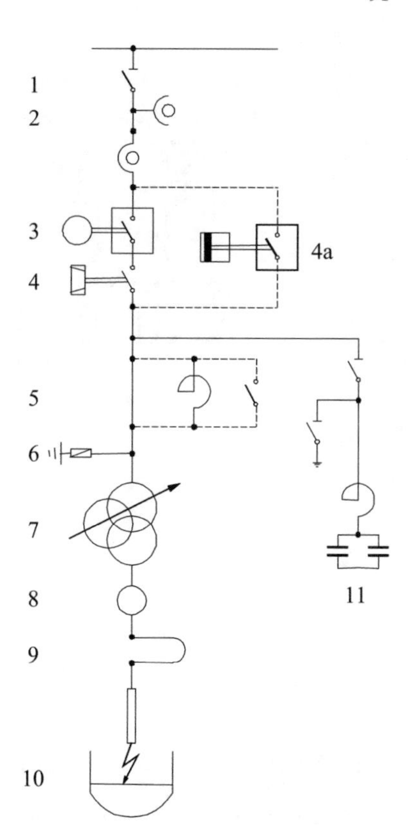

Fig. 3.6 Schematic circuit diagram of the EAF power supply. (*1* supply disconnector; *2* power supply current and voltage transformers for feeding instrumentation and protective devices; *3* oil circuit breaker (short circuit protective breaker); *4* vacuum contactor (furnace on/off switch); *4a* air-blast circuit breaker; *5* reactor with bypass switch; *6* surge diverter; *7* furnace tap change transformer; *8* Rogowsky coil for current measurement (electrode control); *9* furnace supply cables; *10* arc furnace; *11* reactive power and flicker compensation system) [6]

The installation's operating characteristics mainly depend on:

- size of furnace transformer (maximum rated power, voltage, current and voltage control range);
- sum of all impedances of the supply circuit, including extra reactor, intermediate circuit transformer and supply network;
- arc characteristics (i.e. its working conditions, length, voltage and current).

From the point of view of efficiency and safety, the design and construction of the electrical installation should be done by an unique manufacturer, since a strong mutual influence exists between the various circuit elements.

Let us now examine more in detail these elements.

3.2.4 High Voltage Installation

It constitutes the link between the high voltage grid and the furnace transformer. As already said, the furnace transformers are usually fed at voltages between 6 and 30 kV, and—more recently, for large units—directly with voltages up to 110 kV.

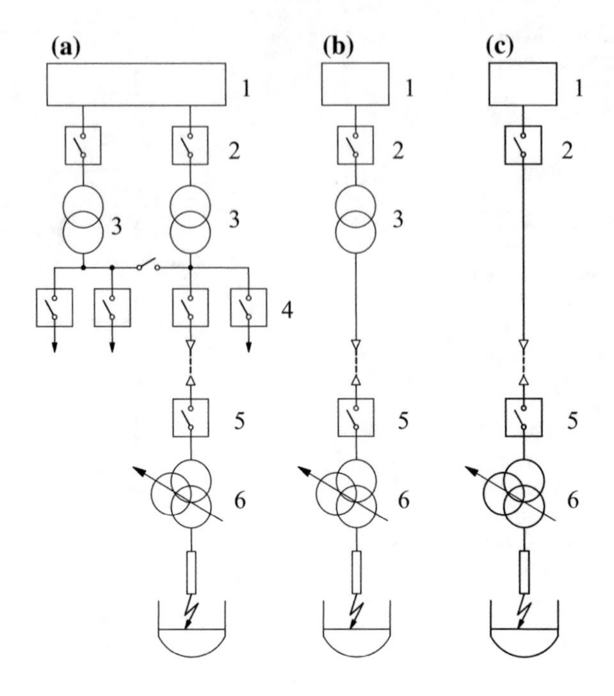

Fig. 3.7 Connections of arc furnace to the HV supply network. (*1* HV supply grid; *2* protection circuit breaker; *3* intermediate step-down transformer; *4* MV circuit breaker; *5* furnace on/off circuit breaker; *6* furnace transformer) [6, 7]

There are several schemes of connection to the high voltage grid; some of them are illustrated in Fig. 3.7.

The first solution (Fig. 3.7a) is used in high power installations: the bus bar system is subdivided into two sections. It provides the supply of the furnace which is characterized by highly variable energy absorption, the second one is connected to the loads of the plant characterised by more constant energy requirements.

The other solutions shown in the figure, provide a direct connection of the furnace to the supply grid (1) either via an intermediate step-down transformer (3) used only for feeding the furnace (Fig. 3.7b), or without intermediate transformer (Fig. 3.7c).

The step-down transformer reduces the grid voltage to a lower furnace bus voltage. This furnace bus can supply several furnaces and can be connected to some step-down transformers, as in the layout of a steel plant shown in Fig. 3.8.

An important component of the high-voltage installation is the furnace on/off switch which, in the case of the arc furnaces, works in particularly heavy duty conditions.

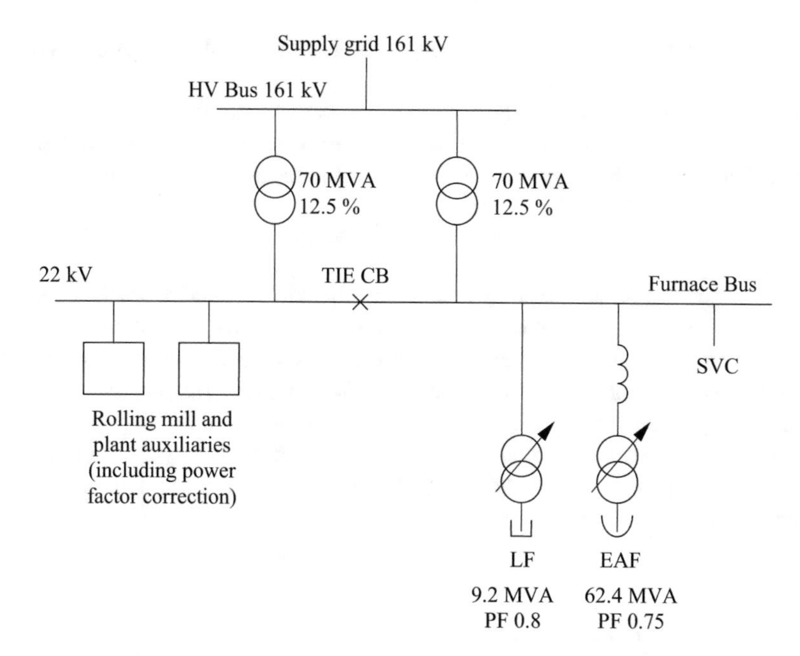

Fig. 3.8 Layout of a steel plant with an AC electric arc furnace, a ladle furnace and two rolling mills (*TIE CB* tie circuit breaker for emergency operation; *SVC* Static Var Compensator) [8]

Instead of the combination of a circuit breaker for short circuit protection and a vacuum contactor used as furnace on/off switch, as in Fig. 3.6, it is possible to employ a special air blast circuit breaker performing both functions.

This circuit breaker must ensure interruption of the current both in normal operation and in case of short-circuited electrodes (including three-phase short-circuits). Moreover, it should avoid creating excessive over-voltages and should have a reasonable service life under normal operating conditions, which include very frequent switching on/off operations, which can be even of some hundreds per day. Air blast circuit breakers are particularly suited to this type of service and meet the above requirements because of their ability to open the circuit under load or at no-load up to about 40,000 times before a major rebuilt is necessary.

The EAF usually works with power factor in the range 0.65–0.8. To meet the requirements of utility companies, the power factor must be kept at levels higher than 0.90–0.92 by compensating the inductive reactive power.

If the supply network is relatively strong, the easiest and cheapest way to improve power factor is the installation of fixed and/or incrementally switchable capacitor banks, which usually are connected in parallel with the HV side of the furnace transformer.

The calculation of the capacitors power rate is simple if the operational cos φ is known.

Denoting by: Qc—the reactive power of capacitors, P—the furnace active power, φ_1, φ_2—the phase angles of furnace current respectively with and without power factor correction, the reactive power of the capacitor bank is given by:

$$Q_c = P(tg\varphi_1 - tg\varphi_2).$$

The compensation devices may include an adjustable inductive reactance to form a filter for reducing injection of harmonics into the network, balancing asymmetric loads and compensating the reactive power fluctuations, which occur—especially in relatively weak supply grids—at frequencies from 1 to 20 Hz. These fluctuations, even if they remain within the power supply contractual limits, may produce disturbances of various kinds to other voltage sensitive users fed by the same network (like loss of synchronism of CRT TV, spurious pulses to digital equipment, light *flicker* on human eye, etc.).

3.2.5 Furnace Transformer

The furnace transformer is a special transformer designed specifically for arc furnace operation; it is used to transform the voltage level of the supply grid to the arc furnace voltage.

It is usually an oil three-phase transformer (with water cooling and oil/water heat exchanger), with primary star and secondary open-delta connections, subjected to heavy working conditions, much more severe than those of a normal transformer.

In fact, arc furnace transformers have to meet the following special requirements:

- high secondary current, up to 120 kA (short-circuit current requirements are higher)
- very high rated power (up to 160 MVA)
- low secondary voltage, in the range from 200 to 800–1000 V, and high primary voltage up to 110 kV
- adjustable furnace voltage to suit furnace operating demands at high switching rate (up to 800 switching operations per day)
- robust construction in order to withstand electro-dynamic stresses due to frequent short circuits that are normal during furnace operation
- equal reactance in the three phases (triangulated secondary system)
- relatively high leakage reactance, especially in low-power units
- separate adjustment of the secondary three phase voltages to limit the so called wild phase operating conditions.

In large power units, particularly in reduction furnaces, sometimes three independent single-phase transformers are used. They allow to obtain higher values of furnace power factor, due to a more rational layout of the high current circuit between transformer secondary windings terminals and furnace electrodes, and to adjust independently power and voltage of each phase.

They also offer the possibility to maintain the continuity of service in case of failure of one unit either by keeping, as a reserve, a fourth single-phase transformer or by using the remaining two at reduced power in the so-called V-connection. However, cost, size and weight of three single-phase transformers are significantly higher than those of one three-phase transformer with the same total power.

In three-phase units, the commonly used connection of primary and secondary windings is triangle, with the possibility to change it from triangle to star in order to modify the secondary voltage.

The delta connection of secondary windings has the advantage, compared to the star connection, to share over two phases the short-circuit current between two electrodes and gives the possibility to make, in the high current circuit, the so-called two-wire reactance compensation with go and return anti-phase conductors, and to obtain equal values of reactance in the three phases.

To widen the voltage adjustment range, the transformer is generally equipped with additional taps, situated on the primary side due to the small number of secondary turns (mostly 1–3) and the high furnace currents. The number of these taps usually ranges from four to ten depending on the transformer power.

The most common schemes for transformer voltage control are:

- *direct control,* which is carried out stepwise by changing the number of primary turns, and is used mostly for power lower than 10 MVA,
- *indirect control,* with an intermediate circuit transformer, generally used for furnace power above 10 MVA.

Figure 3.9 shows the schematic diagram of a furnace transformer with direct regulation.

By increasing the number of primary turns at constant supply voltage, the magnetic induction in the iron core is reduced and thereby also the voltage induced

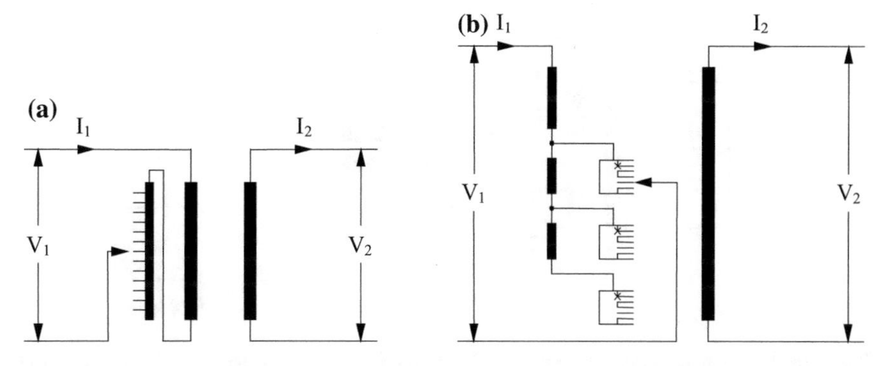

Fig. 3.9 Single-phase schematic circuit diagram of direct voltage control of furnace transformer (**a** with in phase control winding; **b** with coarse and fine voltage steps) [9]

in the secondary winding is reduced. The control range can be widened by delta-star or parallel-series connection of the high voltage winding. The tap change switching is generally done at no-load or, sometimes, with tap change-switch under load. High voltage values above 30 kV can create problems to the windings insulation and the tap change-switch. The main drawback of the direct control scheme is that the voltage steps are not equal.

The design power is a parameter indicative of the cost of transformer [10].

Denoting with:

V_{2M}, V_{2m} maximum and minimum values of secondary voltage at constant primary voltage V_1;

I_{1M}, I_{1m} maximum and minimum values of primary current with constant secondary current I_2,

$$x = \frac{V_{2M}}{V_{2m}} = \frac{I_{1M}}{I_{1m}} \quad ; I_{1av} = \frac{I_{1M} + I_{1m}}{2} = I_{1m}\left(\frac{1+x}{2}\right); P = V_{2M} I_2,$$

in the hypothesis of designing the coils of the tap change winding for the average current I_{av}, the design power of the transformer is:

$$P_d = P\left[1 + \frac{1}{4}\left(x - \frac{1}{x}\right)\right]. \tag{3.1}$$

For furnace transformers with primary voltage higher than 30 kV, it is generally used the indirect control circuit, schematically illustrated by the single-phase diagram of Fig. 3.10.

The voltage control with intermediate circuit transformer consists of a main transformer with a tertiary winding and an auxiliary regulating transformer (or "booster transformer"). The two transformers have secondary windings series

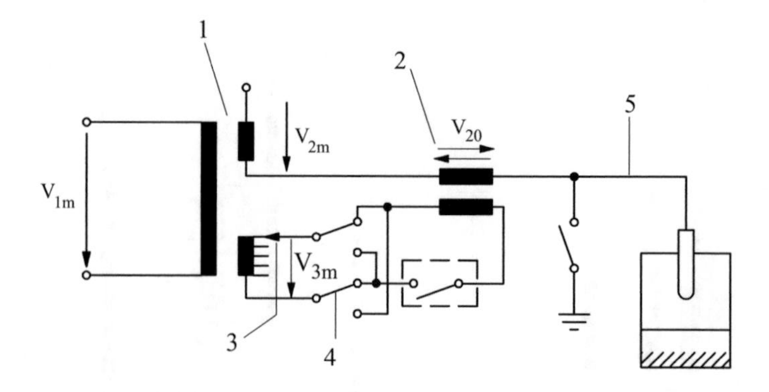

Fig. 3.10 Single-phase circuit diagram of voltage control with intermediate circuit transformer. (*1* main transformer; *2* auxiliary transformer; *3* load step switch; *4* change-over switch for adding or subtracting V_{3m} to V_{2m}; *5* high current circuit) [9]

connected and are integrated in one tank. According to the position of the change-over switch, the adjustable voltage of the auxiliary transformer V_{20} is added or subtracted to the fixed secondary voltage V_{2m}. In this scheme, the switching is done always under load and, given its very wide range of voltage control, it is not necessary, in general, the change of connection star/delta.

Advantages of this method are:

- wide control range of voltage with equal steps
- small variations of short-circuit voltage in the whole field of regulation, since the total reactance of the installation is independent of the position of the switch
- possibility of eliminating the intermediate transformer 220 (110)/30 kV
- possibility to select a convenient intermediate circuit voltage in order to reduce cost of capacitors and filters and facilitate measurement of the current in the intermediate circuit, which is proportional to the furnace current.

With the symbols previously indicated, the total design power of the main and auxiliary transformers is given by the expression:

$$P_d = P\left[1 + \frac{1}{2}\left(1 - \frac{1}{x}\right)\right]. \quad (3.2)$$

Assuming for example:

$$V_{2m} = 200\,(V), \quad V_{2M} = 600\,V; \quad I_2 = 30\,(kA),$$

it is x = 3 and the design power of transformers of the two schemes becomes:

- with direct control P_d = 30 MVA
- with indirect control P_d = 24 MVA.

However, it should be taken into account that the cost reduction, which theoretically is always obtained with the indirect control, is achieved in practice only when the design power is greater than 10 MVA, given the higher cost of construction per MVA of the scheme with intermediate circuit transformer.

Normal values for primary voltages are 6 and 30 kV or, for high power transformers, 110; the secondary voltage typically is in the range 100–250 V for furnaces up to 7 MVA, and can reach 800–1500 V in UHP (Ultra High Power) furnaces.

The transformer windings must be able to withstand frequent short-circuit currents of about 2.5–4 times the rated current, while the short-circuit voltage usually is in the range from 5 to 10 % of the rated voltage, with a value increasing with the power of transformer; it can be up to 10–14 % in UHP furnaces.

In EAF installations of power lower than 7 MVA, where the reactance of the furnace transformer is insufficient to limit the short-circuit current and to ensure arc stability, additional reactors may be installed with reactance of the order of 5–30 %. These reactors Z_a can be connected either in series with the supply circuit (as shown

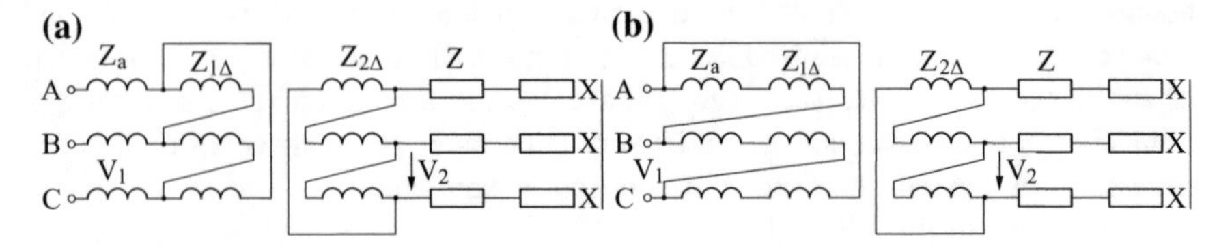

Fig. 3.11 Connection of additional reactors Z_a in series with the furnace transformer [11]

in Fig. 3.11a) or in series with the windings inside the transformer (as in Fig. 3.11b).

Taking into account their normal duty cycle, furnace transformers are also heavily overloaded from the thermal point of view; the relevant standards admit overloads of 100, 50 and 25 % for times of 5, 30 and 120 min respectively, after operation at nominal load.

3.2.6 *Secondary High-Current Circuit*

It is called high-current circuit (or "*secondary circuit*") the section of circuit between the secondary terminals of the furnace transformer and the arcs burning between electrodes and melt.

Taking into account that the intensity of currents flowing through this circuit is always very high (from 30 to more than 100 kA), the circuit layout and design must be carefully designed to reduce as much as possible voltage drops and losses. In fact, these values strongly affect electrical efficiency, installation power factor and utilisation coefficient of the furnace transformer.

Moreover, the circuit layout must give values of the mutual inductances between phases as much identical as possible, in order to reduce the phenomena of the "*dead*" and "*wild*" phases which lead to decrease the production and may give rise to faults.

Schematically the elements of the secondary high-current circuit are illustrated in Fig. 3.12. However, it must be taken into account that the design of the circuit has been drastically modified in modern furnaces.

In particular, the *sub-station bus bars* are the interconnections between transformer and flexible cables; they should be as short as possible.

The *connexions bars* on the sub-station wall are equipotential connections made of copper plates on which sub-station bus-bars and flexible cables are bolted.

The *flexible cables* are air- or water-cooled cables, made either by copper ropes or thin copper sheets, parallel connected; they establish electrical continuity between fixed blocks and electrode holding arms, allowing vertical movement of electrodes.

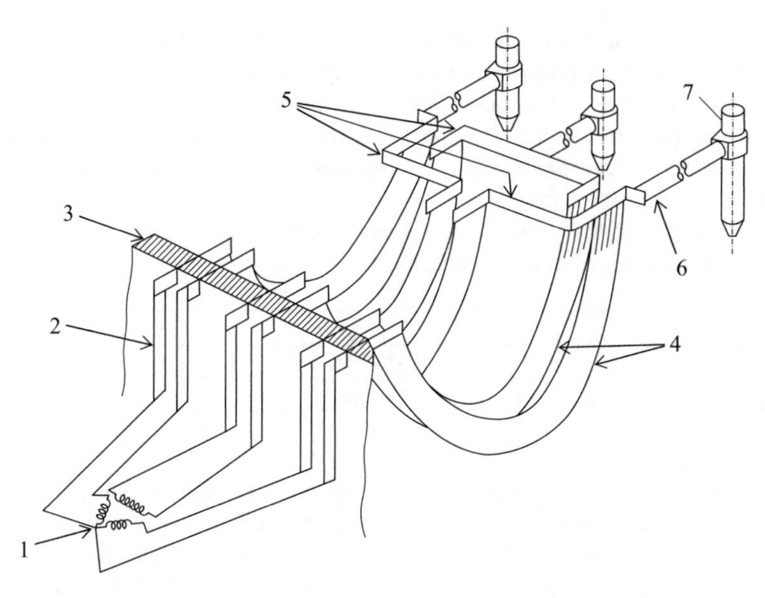

Fig. 3.12 Secondary high-current circuit (*1* three-phase transformer secondary; *2* sub-station bus-bars; *3* sub-station wall; *4* flexible cables; *5* delta connection bars; *6* high-current water-cooled bus tubes; *7* contact pad and electrode clamping ring) [12]

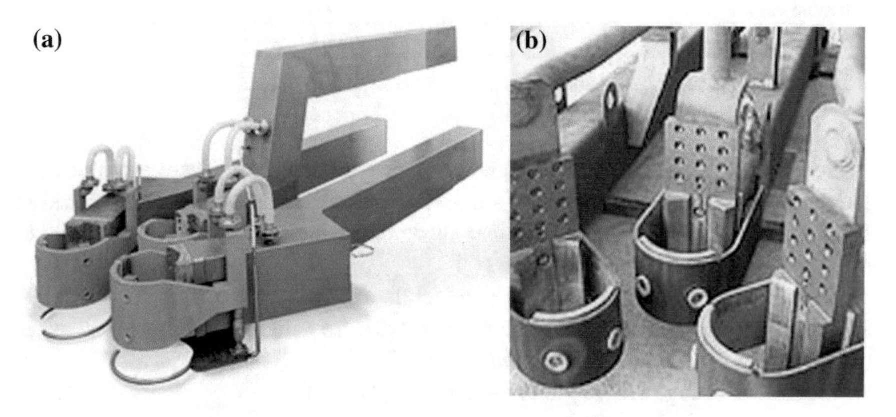

Fig. 3.13 **a** conducting electrode-carrier arms; **b** contact pads and clamping rings

The *delta connection bars* are movable equipotential connections, that can move jointly with the electrodes supporting arms. They are usually made of bronze, since are exposed to high thermal radiation from the furnace.

The *high-current water-cooled bus tubes* connect the delta connection bars with contact pads. They consist of water-cooled copper or aluminium tubes, placed on the electrode supporting arms. In modern design the supporting arms are made of bi-metallic plates of steel and copper, thus being able to carry the secondary high current without the need of bus tubes (see Fig. 3.13a).

The *contact pads* are water-cooled plates made of copper, which are pressed against the electrodes by the *clamping rings*, as shown in Fig. 3.13b. They carry the current from the bus tubes or the conducting arms to the electrodes assuring a low value of contact resistance.

Typical average values of the total reactance and resistance of the high-current circuit, variable with furnace power, are the following:

	r (mΩ)	x (mΩ)
Furnaces for steel	0.5–1.5	2.0–5.0
Submerged arc furnaces	0.1–0.3	0.5–1.0

It's easy to understand the difficulties and costs of construction for achieving such low values, especially for installations in the higher power range.

As an example, the different parts of the high current secondary circuit may contribute to the overall circuit reactance of a loaded furnace in the following proportions:

Transformer	12 %
Transformer bus bars	8 %
Flexible cables	29 %
High-current bus tubes	35 %
Electrodes	16 %

Several design layouts of the high-current circuit are used, where a progressive reduction of reactance is achieved—albeit at gradually increasing costs—making the delta connection of the transformer secondary windings as close as possible to the electrodes.

They allow, by arranging two conductors of the same phase with currents flowing in opposite directions close to each other, to achieve to a greater or lesser extent the so-called "*reactance compensation*".

As shown in Fig. 3.14, typical layouts are [12, 13]:

A. the *conventional connection* (non-compensated)—where the triangle is closed directly at the output terminals of transformer (Fig. 3.14a)
B. *partial interleave*—in which the triangle is closed immediately after the flexible cables and the reactance compensation is achieved only on a portion of the high current circuit (Fig. 3.14b)
C. *fully interleaved Van Roll system*—in which the triangle is closed on the electrodes and the reactance compensation is practically obtained on the whole secondary circuit (Fig. 3.14c).

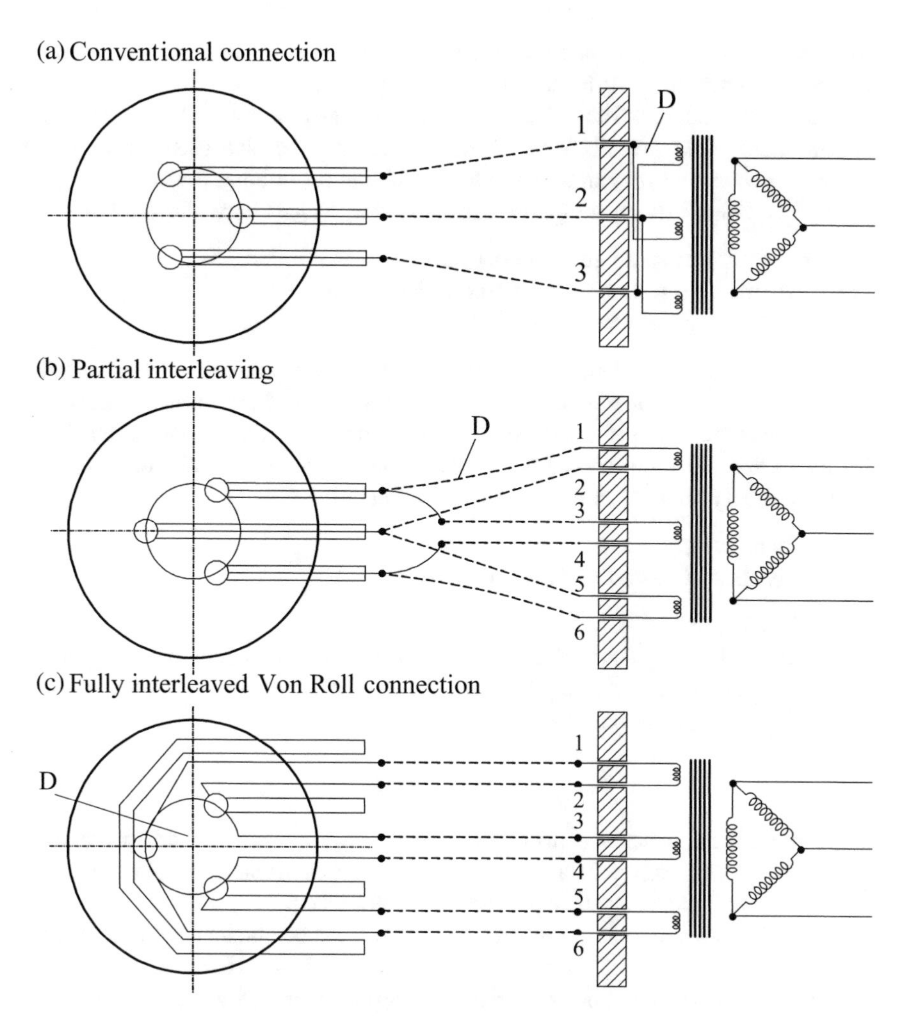

Fig. 3.14 High-current circuit layouts (*A* conventional; *B* with partial reactance compensation; *C* total compensation or Von Roll scheme; *D* closure of triangle)

In addition to the circuit layout, number, shape and mutual arrangement of bus-bars in each phase strongly affect the performance of the secondary circuit.

In fact, these factors will influence not only the DC resistance and the reactance of the circuit itself, but also the values of mutual inductances between phases. Mutual inductances, in turn, increase the equivalent resistance of the high current circuit because of the uneven distribution of current in conductors cross-sections due to skin and proximity effects between conductors and between conductors and parts of the furnace made of ferromagnetic steel.

The transformer bus-bar system is usually subdivided into a number of elementary bars in order to reduce skin effect losses and to facilitate heat dissipation.

The bars of one phase are parallel connected on one side at the transformer output and, on the other side, on the sub-station wall blocks.

The most used bus-bars have rectangular (in air) or tubular (water-cooled) cross-section; other special cross-section shapes may be also used for reducing reactance values, but they may introduce additional assembling costs.

In order to limit overheating, the current density values normally used are:

1,1–1,4 A/mm^2 in rectangular copper bars in air
0,6–0,8 A/mm^2 in rectangular aluminium bars in air
2–5 A/mm^2 in tubular water-cooled bars

The main design problem of the high current secondary circuit is to achieve the same reactance in all phases, in order to assure a balanced arc-furnace operation.

As regards the layout of the three-phase high current conductors (set of bus-bars, flexible cables and water-cooled bus tubes on carrier arms), two solutions were the most commonly adopted in older furnaces, namely the

- *co-planar arrangement*
- *arrangement at vertexes of equilateral triangle*

sketched in Fig. 3.15a, b.

However, the co-planar arrangement of Fig. 3.15a produces an asymmetric reactance distribution, with equal reactance in the outer phases and a smaller reactance in the middle one, i.e.

$$X_{1e} \approx X_{3e} > X_{2e}.$$

An unbalanced reactance distribution may cause asymmetric arc power and radiation factors, leading to a higher consumption of refractory lining near the so-called "wild" phase, if other provisions are not taken.

Theoretically a symmetric reactance distribution can be obtained by triangulating the conductors of the whole high-current secondary circuit. However, in comparison with the co-planar one, the triangular arrangement of conductors requires higher construction costs, more space in vertical direction and a greater distance between transformer and furnace in order obtain the triangle layout of flexible cables. Moreover, it gives rise to a greater unbalance of the three phases as a result of a relatively small displacement of one electrode arm in vertical direction (which is normal during normal operation of the furnace, where the electrodes vertical positions are independently regulated) [12].

Besides the two previously mentioned arrangements, other design layouts have been proposed for obtaining symmetrical reactance, e.g. the use of high current conductor tubes with different cross sections (Fig. 3.15c) or—in case of several conductors—separating the ones of the outer phases by calculated spaces and setting close together the centre-phase conductors (as in Fig. 3.15d, h).

A complex task is also the calculation of the circuit resistance and reactance, since it must take into account the variety of different layouts, the presence of flexible cables, which give a variable component to the total reactance for their

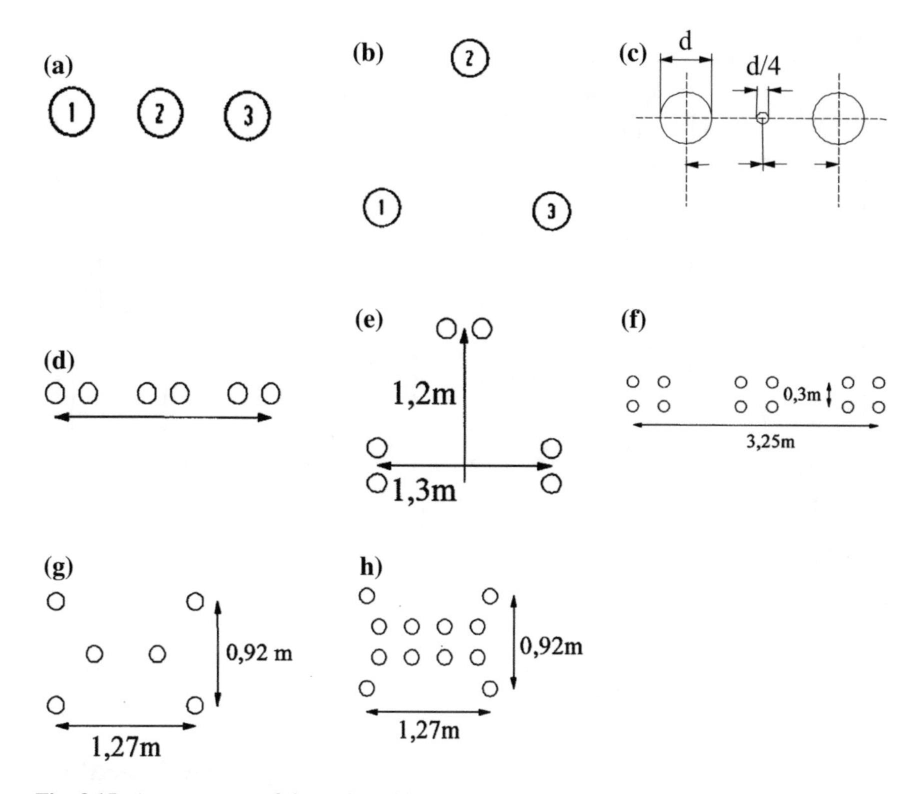

Fig. 3.15 Arrangements of three-phase high-current conductors [12]

oscillations due to electro-dynamic forces, the subdivision into elementary bus-bars, the presence of contact resistances at the joints of the different parts, etc.

The calculation of resistances is done taking into account the coefficients of increase of DC resistance due to skin and proximity effects and the so-called "*power transfer*" occurring in case of unbalanced mutual inductances between phases.

Values of such coefficients are usually given by graphs which can be found in Ref. [14, 15].

Approximate formulas are also given for the contact resistance of bolted connections of the secondary circuit [6, 11].

The experience shows that the increase of resistance due to the above mentioned phenomena, in some cases can be as high as 100 % of the DC value.

The total reactance of the secondary circuit is determined through the coefficients of self and mutual inductance of its different parts. In Ref. [2], a number of formulas allow to take into account the geometrical layout of conductors in the phases.

However, the calculation of inductances and resistances, taking into account layout of conductors and distribution of currents in the bar cross-sections due to proximity and skin effects can be done more conveniently today by numerical methods (e.g. FEM), for which many powerful commercial programs are available.

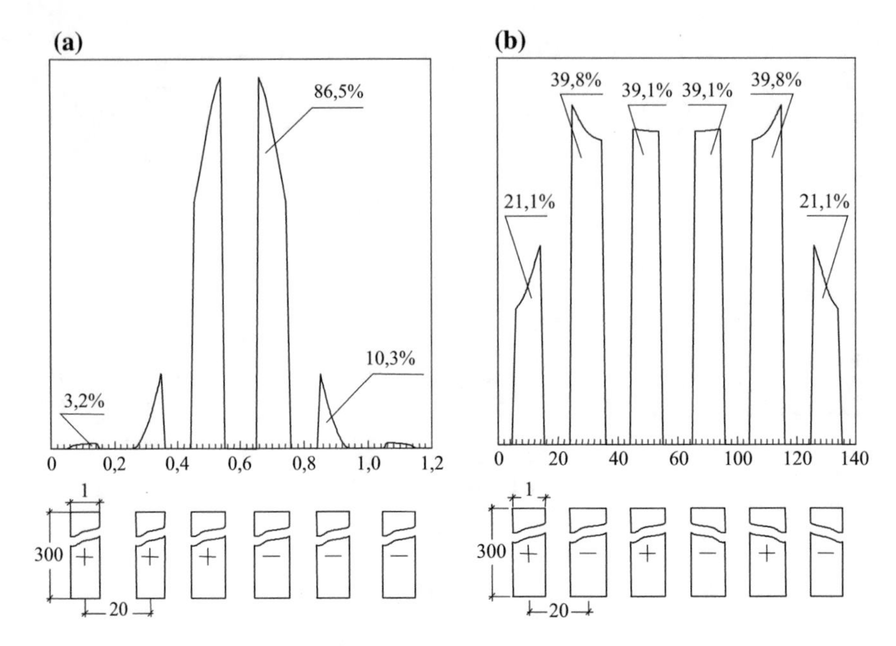

Fig. 3.16 Current distribution (in % of the total current) in a set of 6 bus-bars subdivided in two groups (**a** two separate groups; **b** interlaced) [16]

An example of results obtained with such programs, respectively for interlaced or separated groups of rectangular bus-bars, is shown in Fig. 3.16.

3.2.7 Electrodes

The electrodes are conductive cylindrical elements, made of amorphous carbon or graphite, having one end connected to the secondary high-current circuit through the clamping rings, while at the other end (the electrode tip) the arc burns towards the charge or another electrode.

Basically three types of electrodes are used: (1) prebaked electrodes made of amorphous carbon, (2) graphite electrodes used in steel making EAFs and (3) self-baking electrodes, know as "*Soderberg*" electrodes, used in ore-thermal reduction furnaces.

Average values of physical properties of graphite and amorphous carbon are shown in Table 3.1.

One of the most important properties for selection of electrode material is the electrical resistivity: its value determines the electrode diameter, given the arc current intensity.

Table 3.1 Typical data of graphite and amorphous carbon

	Graphite	Amorphous carbon
Resistivity (Ω cm)	$500\text{–}1300 \cdot 10^{-6}$	$3500\text{–}6500 \cdot 10^{-6}$
Tensile strength (kg/cm^2)	40	20
Compressive strength (kg/cm^2)	170	140–400
Bending strength (kg/cm^2)	100	65
Thermal conductivity (W/m K)	116–175	50
Specific heat (Wh/kg K)	0.186–0.465	0.21–0.26
Density (real) (gr/cm^3)	2.2–2.3	1.8–2.1
Density (bulk) (gr/cm^3)	1.5–1.7	1.45–1.6
Temperature of oxidation start (°C)	500	400
Current density (A/cm^2)	10–30	5–15

Taking into account that above 400–500 °C the electrode starts to burn in air, the balance of Joule losses in the electrode resistance and the heat transferred to the environment from the electrode surface must occur below this temperature.

Therefore, denoting with:

- α heat transfer coefficient from the electrode surface to the surroundings, in the portion of electrode exposed to air, $W/(m^2K)$
- $\Delta\vartheta$ temperature difference between electrode surface and surroundings, K
- $d = 2r_e$ electrode diameter, m
- ℓ length of the section considered, m
- ρ_o resistivity of material at the temperature of start of oxidation, Ω m
- I electrode current intensity, A,

it must be:

$$\alpha \cdot \Delta\vartheta \cdot \pi d\ell = \rho_o \frac{4\ell}{\pi d^2} \cdot I^2$$

from which the diameter d is: equation at the center of the line

$$d = \sqrt[3]{\frac{4\rho_0 I^2}{\pi^2 \alpha \Delta\vartheta}} \tag{3.3}$$

Commercial graphite electrodes have diameters in the range 200–800 mm, while those of amorphous carbon between 200 and 1300 mm.

The maximum size of graphite electrodes depends on skin effect which produces current density distributions in the cross-section of the type shown in Fig. 3.17a.

In selecting the electrode diameter, it must be also taken into account the influence of proximity effects which, as illustrated in Fig. 3.17b, introduce a further unevenness in the current density distribution, which may produce asymmetric electrode consumption or cracks formation due to thermal stresses.

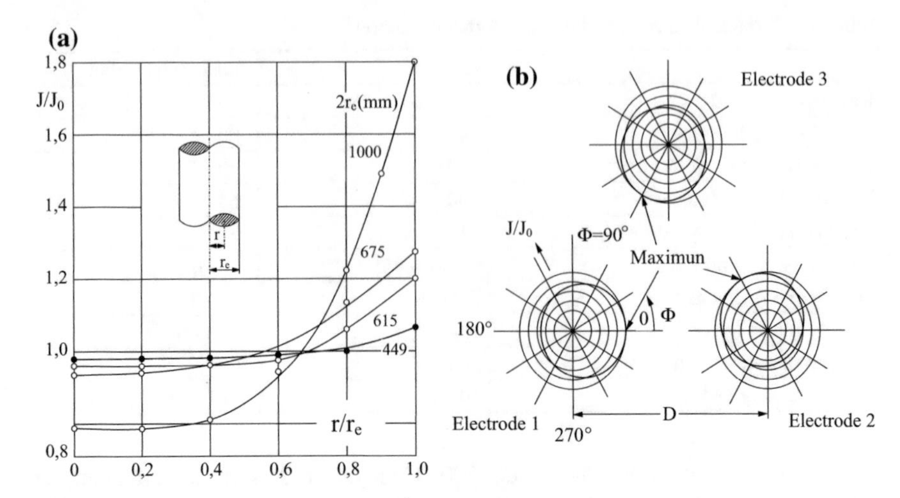

Fig. 3.17 **a** Radial current density distribution in graphite electrodes without proximity effect for different electrode diameters; **b** Instantaneous current density distributions in cylindrical electrodes of a three-phase electrode arrangement as a function of phase-angle φ (pitch circle diameter $D/2r_e = 2.7$, $2r_e = 615$ mm, phase sequence 1,2,3) [9]

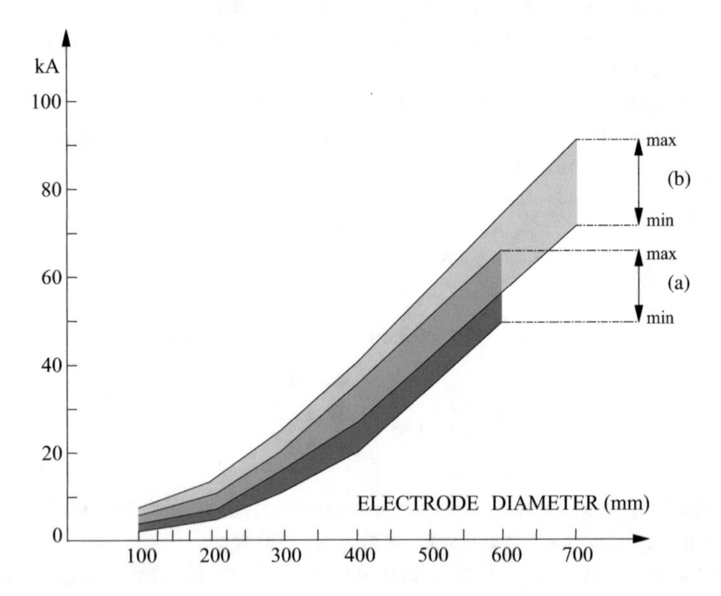

Fig. 3.18 Typical current carrying capacity of graphite electrodes as a function of diameter (**a** normal operating conditions; **b** heavy operating conditions)

Given the higher temperature at which starts oxidation of graphite, and the material's resistivity at this temperature (900–1000 μΩ cm for graphite, 4000–5000 μΩ cm for amorphous carbon), Eq. (3.3) says that—for the same cross section

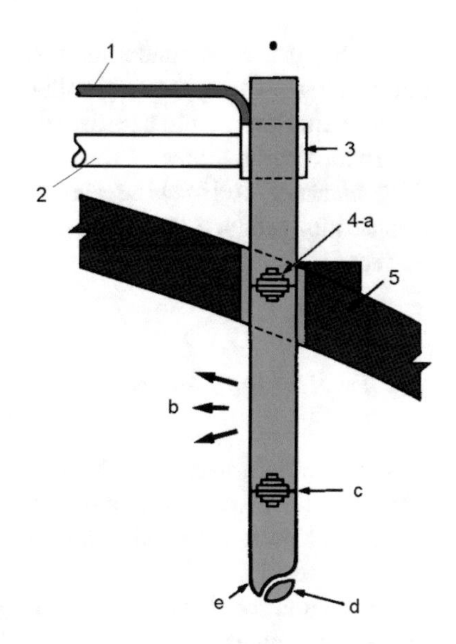

Fig. 3.19 Mechanism and relative importance of electrode consumption. (*1* bus-tube; *2* support arm; *3* clamp; *4* graphite nipples; *5* furnace roof; *a* top joint breakage: 5 %; *b* surface oxidation: 45 %; *c* bottom joint breakage: 5 %; *d* tip rupture due to thermal stress: 5 %; *e* tip volatilization and microspalling: 40 %) [17]

—the admissible current in graphite electrodes is about 2.5 times greater than in the amorphous carbon ones.

Typical values of the current carrying capacity of commercial graphite electrodes, evaluated through Eq. (3.3), is given by the diagrams of Fig. 3.18.

Another important factor is the electrode consumption that occurs continuously during operation of the furnace; this consumption is much higher for amorphous carbon electrodes than for the graphite ones. In furnaces with graphite electrodes for steel production this consumption is in the range of 1.0–2.0 kg/t. However, this value is influenced by many factors, such as the furnace working operating point, the method of performing the melting process and the use of electrodes with metallic or refractory coatings.

Figure 3.19 shows the main causes of electrode consumption: among them surface oxidation and volatilization of material in contact with the arc play a predominant role.

From the economic point of view, it must be remembered that today the cost of electrode material is comparable to that of the energy consumed during the melting process.

In order to ensure continuity of production cycle, the electrodes are produced in the form of sections that are added to the top of the electrode columns, and are connected to them by means of tapered threaded joints of graphite, called "*nipples*", as shown in Fig. 3.19. These joints must be carefully designed in order to ensure both the necessary mechanical strength and a good electrical contact, as most malfunctions of electrodes are due to them.

As an example, an arc furnace 65 MVA/180 t, has electrode columns constituted by 3 sections, 600 mm diameter, 2.4 m length each, with total weight of about 3 t.

Finally, it must be underlined that for their lower ash content, lower consumption and lower failure rates, graphite electrodes can better ensure the purity of melt, what justifies their wide use, despite a cost about three times higher than that of the amorphous carbon ones.

"*Soderberg*" self-baking electrodes, which are used in reduction continuous furnaces for production of ferro-alloys, Ca-carbide, phosphorous, will be described in Sect. 3.7.3.

3.2.8 Furnace Vessel

A direct arc furnace for steelmaking is generally composed of a cylindrical refractory-lined vessel with a rounded non-conductive bottom and a retractable roof constituted by a self-supporting dome, through which the electrodes enter in the furnace (see Fig. 3.5).

Vessel and roof have the functions of containing the melt and to limit heat losses to the surrounding.

The vessel is primarily split into three sections:

- the *furnace shell*, comprising the sidewalls and the lower steel "bowl"; it is provided with a door towards the tapping spout and a working door for de-slagging and making additions;
- the *hearth*, hemispherical in shape, constituted by the refractory lining of the lower bowl;
- a *domed roof*, which may be refractory-lined or water-cooled. It is provided externally with a water-cooled iron ring, with ports for the electrodes and a fourth hole for fume extraction. The roof is supported by suspension beams and is lifted up and swung aside for charging.

Separate from the furnace structure, there are the electrode supporting arms, the electrical system, and the tilting platform on which the furnace lays.

3.2.8.1 The Furnace Shell

The design of EAFs is constantly evolving, taking into account the changes in the melting technology, the production of more resistant refractory materials and the development of the technology of electromechanical construction.

Figure 3.20 shows a modern EAF in the assembly stand of the Russian company JSC "Sibelectroterm".

For the definition of the geometry of the vessel first must be chosen its shape and ratios of main dimensions, height of roof position above liquid metal, diameter of electrodes pitch circle, internal profile of lining and configuration of the housing metal structure.

Fig. 3.20 Arc furnace EAF-30N2 at the assembly stand of "Sibelectroterm": (*1* current-carrying bimetallic arms; *2* water-cooled panels; *3* working door; *4* gas extraction duct; *5* water-cooled roof; *6* bay window; *7* bottom pouring system; *8* tilting platform) [18]

The vessel consists of a cylindrical steel structure usually with rounded bottom. It is constructed with water cooled thick steel plates, welded together with appropriate stiffeners, or with welded steel rings.

The dimensions of the vessel cross-section depend on furnace capacity, the need to minimize heat losses to the environment during charging operations (through the choice of relatively small diameters), the diameters of electrodes and electrode pitch circle, the distance between it and the refractory lining and the thickness of lining.

Size and weight of the furnace, surface of heat transfer and specific power consumption depend from the ratio between vessel diameter and its height.

For large furnaces of normal power, the ratio of bath diameter to depth of molten metal varies from 4.2 to 5, and the bath depth is about 1100–1200 mm. For modern UHP furnaces the bath depth is greater and the above ratio is lower than 4 [19].

This change of bath dimensions has decreased both the area of the molten surface and of contact between slag and metal. This decrease of the bath surface was a logical consequence of the fact that in modern melting technology the processes of desulfurization and de-oxidation of the metal is done outside the furnace and that the metal de-phosphorization is facilitated by a significant increase of the metal-slag contact surface and the vigorous "boiling" of the bath due to the oxidation of carbon. The bath depth is limited only by the possibility of heat transfer from the arcs to the metal. If the basic arc heating is combined with the process of oxidation of carbon (pulverized coal injection and carbon charged into the furnace with the scrap), usual in modern melting processes, the stirring of the metal produced by the formation of bubbles of carbon monoxide (CO) contributes to the uniform heating of the bath.

The thickness of the slag layer floating on the melt is mostly in the range 100–300 mm and the height of the cylindrical portion of the vessel above the melt is the one necessary for containing one basket of scrap.

In fact, given the convenience of reducing the number of charges during a melting cycle, the latter size is chosen so that, with a scrap of normal density, the first charge corresponds to about 50–60 % of the furnace melting capacity, so that the melting cycle can be completed with a single intermediate charge.

This has led, in recent years, to increase the height of the vessel, with consequent increased protection of the hearth at the beginning of melting and the roof from excessive radiation at the end of it.

However, increasing the height of the vessel, increases also the length of the section of electrodes inside the furnace and thus the total surface of the electrodes exposed to oxidation and, as a consequence, the electrodes consumption due to oxidation of their lateral surface. In this condition, also the probability of breakage of electrodes increases.

From the above, it follows that cannot be given unique rules for the choice of dimensions of the furnace vessel.

Tables 3.2 give characteristics and corresponding capacity of two series of furnaces produced in the 70ties by EAF manufacturers.

In particular, comparing power and melting capacity of the two series, one can see that—for the same capacity and vessel diameter—the power is significantly different in normal type furnaces (Table 3.2a) and in so-called UHP (*Ultra High Power*) furnaces (Table 3.2b).

The corresponding specific power, as a function of furnace capacity, in modern UHP furnaces can reach values up to 1,000 kVA/t, as shown in Fig. 3.21.

Due to the high installed specific power, UHP furnaces are characterized by shorter melting times and, consequently, by a tap-to-tap time reduced to half in comparison with that of normal-type furnaces, higher production rates and lower specific energy consumption (see Fig. 3.2).

3.2.8.2 Vessel Lining

The lining of bottom and walls of the vessel is made with refractory materials which must have appropriate characteristics, i.e. withstand to high temperatures and temperature variations, low thermal and electrical conductivity, low coefficient of thermal expansion, maximum resistance to slag attack and to abrasion during charging, which is usually done with a drop-bottom charging basket.

Some characteristics of the most commonly used materials are shown in the diagrams of Fig. 3.22.

Depending on the material used, refractory linings can be classified into two groups: alkaline or acid.

The refractories of the first group, which are the most used for production of steel, use as raw materials Magnesite ($MgCO_3$, MgO), CaO, Chromite-Magnesite

Table 3.2 Characteristics of two series of EAFs produced in the 70ties (a-normal; b-UHP)

(a)

Nominal capacity (t)	Maximum capacity (t)	Vessel diameter (mm)	Transformer power (MVA)	Electrode diameter (mm)
50	60	5500	18–22	550
60	80	5700	22–28	550
80	90	6000	25–33	600
90	100	6200	28–38	600
100	110	6400	31–42	600
110	120	6600	34–46	600
120	130	6800	37–50	600
130	140	7000	40–52	600
140	150	7200	43–57	600
160	180	7400	50–68	600
180	200	7600	57–76	600
200	240	7800	68–92	600
240	280	8000	80–107	600
280	300	8200	86–115	600

(b)

Capacity (t)	Vessel diameter (mm)	Rated power/maximum power (MVA)	Capacity (t)	Vessel diameter (mm)	Rated power/maximum power (MVA)
10–12	3000	5.5/6.6	44–57	4900	30/36
12–15	3200	7.0/8.4	56–68	5200	35/42
15–18	3400	8.0/9.6	64–82	5500	40/48
18–22	3600	10/12	76–96	5800	45/54
22–26	3800	12.5/15	90–112	6100	52.5/63
26–31	4000	15/18	104–130	6400	62.5/75
32–39	4300	20/24	120–148	6700	72/86
40–49	4600	25/30	–	–	–

and Dolomite, either in the form of bricks or as granular compounds (3–15 mm granulation) to be rammed [6, 21].

The lining of the bottom hearth, in this case consists of three layers (see Fig. 3.23):

- a first external layer of thermal insulation, constituted by insulating sheets and insulating powder (about 10 mm thickness), and a layer of Schamotte, 40–60 mm thick;
- an intermediate layer of magnesite bricks, which is the so-called long duration lining;
- an internal layer, 200–400 mm thick, of rammed magnesite to be frequently renewed.

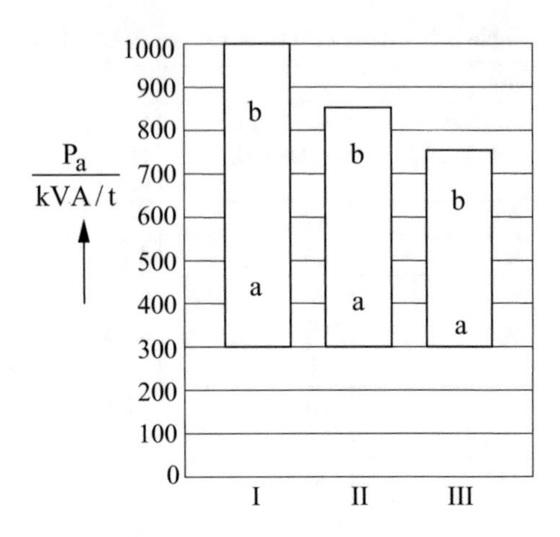

Fig. 3.21 Specific power for arc furnaces of different capacities (*I* capacity 35 t.; *II* capacity 75 t; *III* capacity 110 t; *a* normal type furnaces; *b* UHP furnaces; N.B.—values of IIb and IIIb today can be up to 900 and 850 respectively) [20]

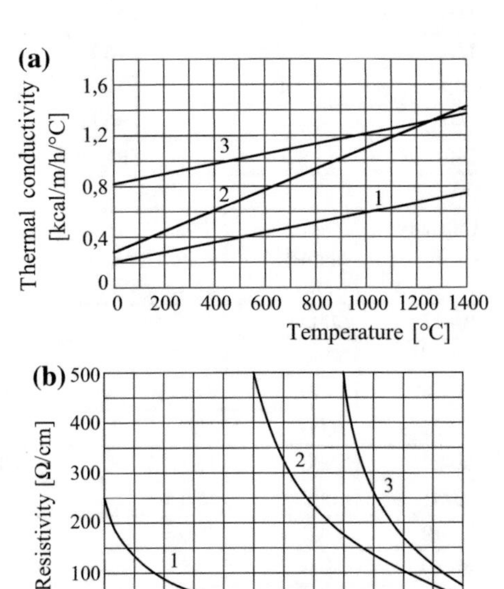

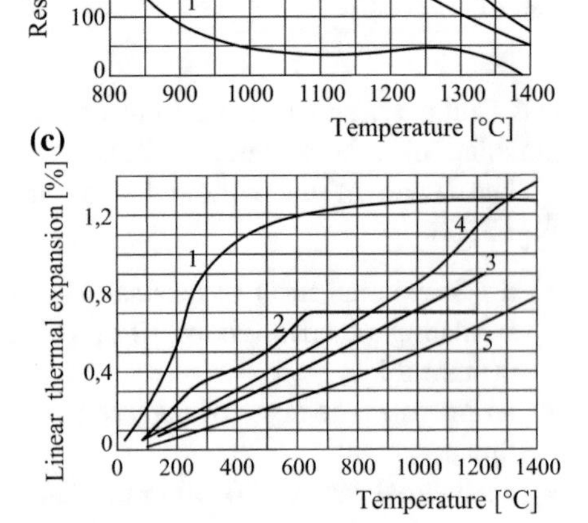

Fig. 3.22 Thermal conductivity (**a**), electrical resistivity (**b**) and thermal expansion coefficient (**c**) of refractory materials (*1* Bauxite; *2* Silika; *3* Magnesite; *4* acid Schamotte, *5* basic Schamotte; *6* Chromite; *7* Carborundum)

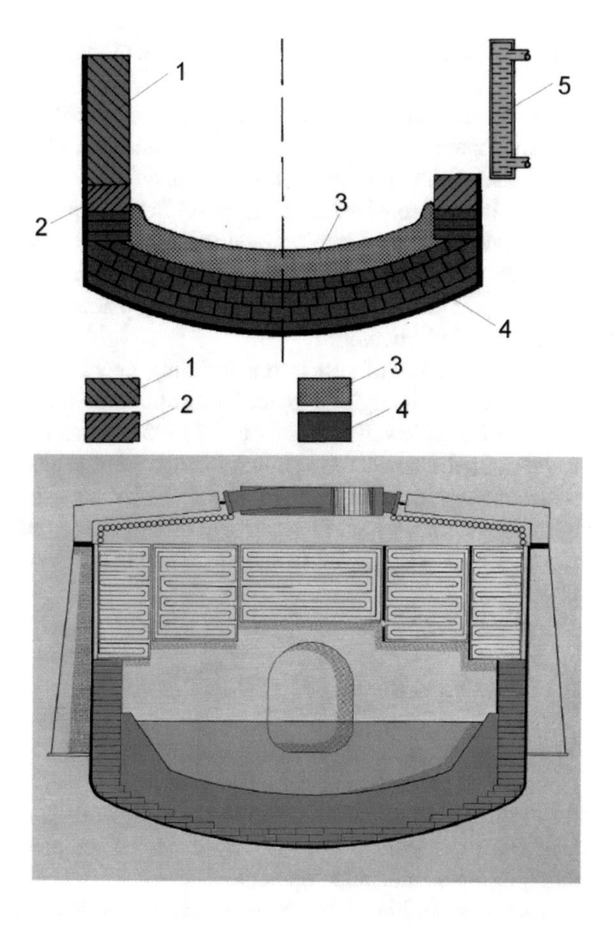

Fig. 3.23 Schematic arrangement of refractory lining for basic arc furnaces with normal sidewalls or water-cooled sidewalls (*1* Cromo-magnesite sidewall lining; *2* Magnesite sidewall lining; *3* rammed magnesite, short duration layer; *4* bricks of magnesite, long duration layer; *5* water cooled sidewall plate) [17, 22]

The sidewall refractory lining is made with a first insulating layer, similar to that of the hearth, a second layer of bricks of pre-fabricated blocks of magnesite or chrome-magnesite and a gap between the two layers, 40–50 mm thick, filled with fine granulated refractory with good thermal conductivity, to allow for thermal expansion [21, 23].

Instead of using bricks, most frequently the second layer is made with a hot-rammed mixture of dolomite, magnesium or chromium-magnesium granules, to which is added about 10 % of bituminous binder.

Refractory ramming or the use of prefabricated blocks allow important economies, because the sidewalls lining must be rebuilt on average every 100–300 castings (refractory consumption averages 0.8 kg/t of steel for bricks and 4.5 kg/t for gunned repair materials, i.e. an average consumption in the order of 2.7 kg/t in steel industry) [4, 23, 24].

Acids refractory lining is used for production of pig iron and steel from low-phosphorus and low-sulphur scrap. The main raw materials used are SiO_2, ZrO_2 and aluminium-silicate. They are used when the slag and the atmosphere are acid, and cannot be used under basic conditions; therefore selected scrap must be used.

Hearth and sidewalls are similar to those described for basic lining, but using as building materials quartz sand, SiO_2 (Quarzite), CaO (Silicate) and "Silica" (92–96 % SiO_2 with addition of lime as binder).

The hearth consists of a layer of Schamotte powder, one or two layers of Schamotte bricks and two or three layers of Silica bricks. Above them is rammed a layer constituted by a mixture of quartz with addition of boric acid.

Sidewalls are built with Silica bricks, placed over a thermal insulation layer made of light Schamotte.

For production of cast iron acid lining may allow up to 250–500 castings.

Before concluding this topic, it must be underlined that in all modern furnaces a large part of sidewall lining (about 75–85 %) has been replaced by pre-fabricated water-cooled elements (as shown in Fig. 3.23) which allow—albeit with an additional energy consumption of about 30 kWh/t—to reduce the consumption of refractory lining, which therefore can be used for a much higher number of castings (up to one thousand), and to shorten meltdown times due to a the higher meltdown power.

3.2.8.3 Vessel Roof

The vessel is enclosed on top by means of its roof.

The roof hangs from steel straps mounted on two suspension beams at the sides (see Fig. 3.4). The entire suspension frame work is designed to allow to raise the roof from the shell and swinging it out of an opening angle of about 60° for charging or maintenance operations.

As shown in Fig. 3.24, the roof is provided with three circular central holes for the electrodes and a fourth hole for fume-extraction.

The roof is almost always fitted with a water-cooled bezel ring which mates with the vessel upper flange. The external ring has diameter slightly larger than the vessel upper flange to reduce thermal stress on the metal and it is provided, in the internal part, with a circular 'knife' that goes to mate a groove on the vessel upper flange, filled with sand to act as a thermal 'seal'.

The roof lining was in the past arranged in the form of a self-supporting domed arch constituted of rings of domed side-arch silica bricks. The ratio between height of dome and roof diameter was usually 1:10–1:12, the thickness of the layer of bricks of about 200–300 mm.

Roofs of silica brick can withstand about 50–70 melting cycles in basic furnaces, 100–150 in the acid ones. These values can be much higher in entirely water-cooled roofs, which consist of a bricked-up section like in a normal roof, and a water-cooled ring-shaped section around the central part. The latter comprises nearly 85 % of the total roof surface.

The electrode holes have diameter 40–50 mm greater than the electrode diameter. The space between electrodes and roof is "sealed" with water-cooled

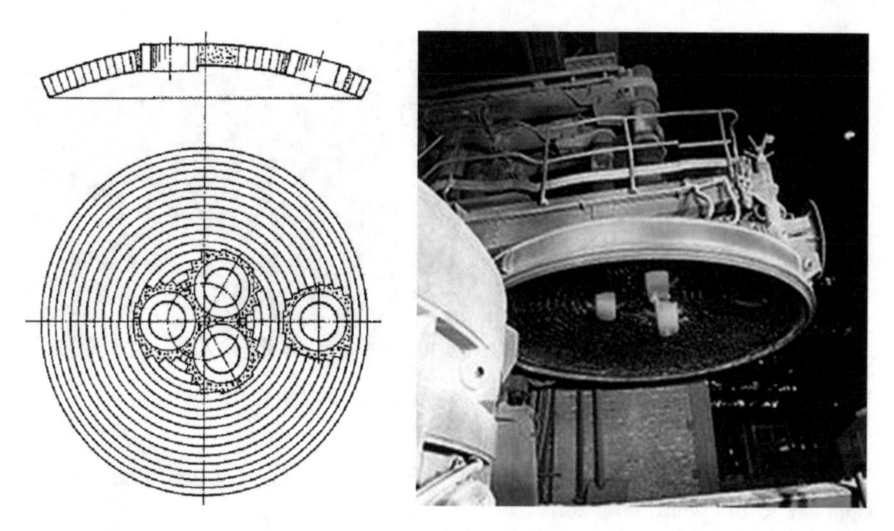

Fig. 3.24 Roof with ringed courses, electrodes and fume-extraction holes [9]

non-magnetic metal rings, subdivided into segments to prevent the flow of induced currents.

The diameter of the electrodes circle in UHP furnaces is significantly reduced in comparison with conventional furnaces. Its value must be adequate for accommodate the electrode holders without risk of contact during their movement. If the electrodes circle diameter is large, the strength of the construction of the central part of the roof increases and it is easier to accommodate the electrode holders. On the other hand, this solution worsens the heating of the central part of the bath surface and increases the unevenness of the temperature distribution along the circumference of the furnace inside wall. The reduction of the electrode circle diameter reduces the inductance of the secondary current circuit and produces the reduction and a more uniform distribution of thermal load of the walls.

In the past, the electrode circle diameter in furnaces of 100 tons capacity was typically 1800 mm, in modern furnaces about 1200–1400 mm.

3.2.8.4 Metal Tapping

Whilst older furnaces were equipped with discharge spouts, most of new EAFs are equipped with bottom tapping systems, which allow to decrease the volume of refractory material.

As shown in Fig. 3.25a, in old furnaces the discharge spout is made of steel and lined with rammed refractory Schamotte; it is fixed on the side wall of the vessel at the level of the tapping door, usually placed above the level of the slag. It should be as short as possible for reducing cooling and oxidation of molten metal.

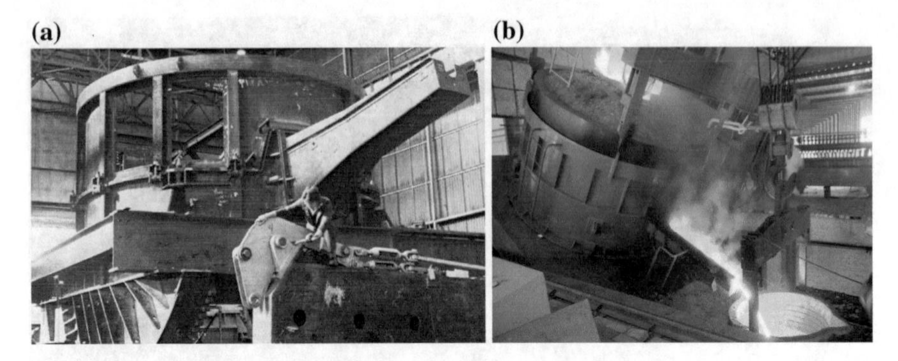

Fig. 3.25 Furnace metal shell, discharge spout and tilting rockers

The vessel is also provided with a working and de-slagging door diametrically opposite to the pouring spout.

The furnace can rotate around a horizontal axis tilting toward the pouring spout (with an angle of about 40° from the vertical) and, on the opposite side, towards the working and de-slagging door (with angle of 10°–15°).

The tilting movement is done by means of rockers, roller tracks and hydraulic cylinders. The rockers are rigidly fastened to the frame supporting the vessel; the roller tracks are imbedded in the foundations and incorporate either a row of holes or a toothed track, arranged in a way that—in case of failure—the container goes back to its upright position.

The traditional scheme of metal tapping (Fig. 3.26a) limits the area of the water-cooled walls and requires a higher consumption of expensive high quality refractory bricks because, for security reasons, above the pouring spout, the water-cooled panels must be placed substantially higher than in the rest of the furnace.

Although the change of pouring technology in UHP furnaces did not require and did not include metal de-slagging during tapping, with the development of the secondary metallurgy it became mandatory to cutoff the slag from the metal, and different schemes of metal tapping have been developed to this end (see Fig. 3.26).

The so-called siphon outlet (Fig. 3.26b) was used in classical heavy-duty furnaces with new pouring technology. Siphon pouring completely solves the problem of pouring the metal from the furnace without slag, since it allows to leave the slag in the furnace and to operate the furnace with residual melt. It does not provide significant benefits in the design of the furnace installation but it allows to increase the area of water-cooled panels.

To increase the surface of water-cooled panels, have been developed furnaces with bottom tapping (Fig. 3.26c). This design increases the area of water-cooled walls from 70 to 85 %. The tilting of a furnace with bottom tapping is performed only during de-slagging at an angle of 12°, thus converting the furnace to a near-static configuration. This allows to simplify significantly the design of the

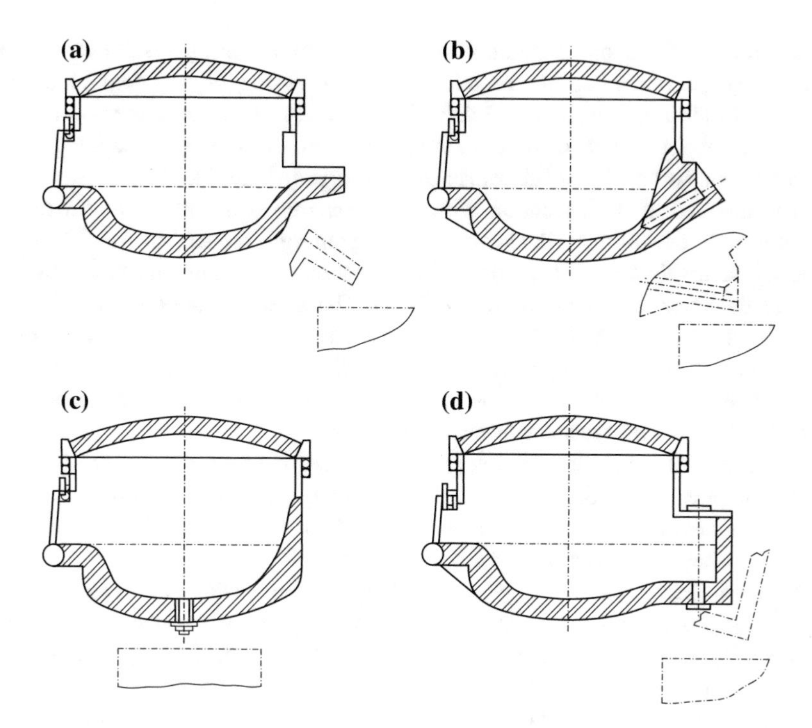

Fig. 3.26 Different schemes of metal tapping: *a* traditional, *b* with siphon; *c* bottom tapping; *d* eccentric bottom tapping [18]

furnace mechanical equipment and to reduce the length of the furnace secondary circuit.

The hole in the bottom of the furnace is rammed with a magnesite tube. After complete emptying the furnace, before loading a new charge, the pouring system is closed by a locking pneumatic valve located under the bottom of the furnace, and it is filled with a special refractory magnesite sealing compound. Before tapping the valve opens and the molten steel breaks through the sintered compound at the top of the nozzle. The magnesite tube can be used for about 100 heats.

In addition to reducing consumption of refractory walls, the advantages of this type of furnace are quick pouring of the melt, reduction of heat losses during this operation, reduction of gas inclusions in the metal during pouring, reduction of electrical losses in the secondary circuit and reduction of wear of the lining in the ladle.

Disadvantages are the impossibility of leaving the slag in the furnace, of adopting the operating mode in which part of the molten metal is left inside the furnace from a melt to the next, and the difficulty of maintenance of the bottom hole. Due to these shortcomings, furnaces with bottom pouring have not received wide diffusion.

In order to overcome these drawbacks, the bottom pouring system was moved from the center of the bottom to a side special protrusion ("*bay window*") located in the rear of the furnace (see Figs. 3.4, 3.20 and 3.26d). The level of the pouring hole is slightly above the level of the furnace bottom. Such type of tapping in the literature is also known as "*eccentric bottom tapping*" (EBT).

The practice of the operation with EBT has shown that it facilitates furnace maintenance. The tilting of the furnace during pouring is reduced to 10°–12°, which reduces the level of the lower edge of molten metal, thus allowing to increase the surface of water-cooled panels to 84–89 %. The steel production with EBT and separate de-slagging hole allow to produce steel virtually without slag, and if necessary to leave some metal in the furnace. Since the output flow of the metal is compact and short, metal oxidation is lower and its temperature drop is reduced to only 20–35 °C, with consequent energy savings.

Moreover, the reduction of tilting angle reduces the required length of flexible cable garlands, thus decreasing the total resistance and reactance of the current supply conductors. The reduction of length of the cable garlands also allows to bring the transformer nearer to the pouring spout of 1–1.5 m.

Eccentric tapping is performed as follows. A ladle with the additives on its bottom, is placed under the furnace before pouring. Then is opened the shut-off device (Fig. 3.27) and the melt is poured from the furnace. During pouring (about 2 min in large furnaces) the furnace is tilted slightly towards the ladle, to ensure a constant level of metal above the bottom hole. The tilting of the furnace is automatically locked when is reached the desired maximum tilting angle. When the ladle is filled with the required amount of metal, the furnace returns to its original position, while the outlet remains open. From the top of the working platform the

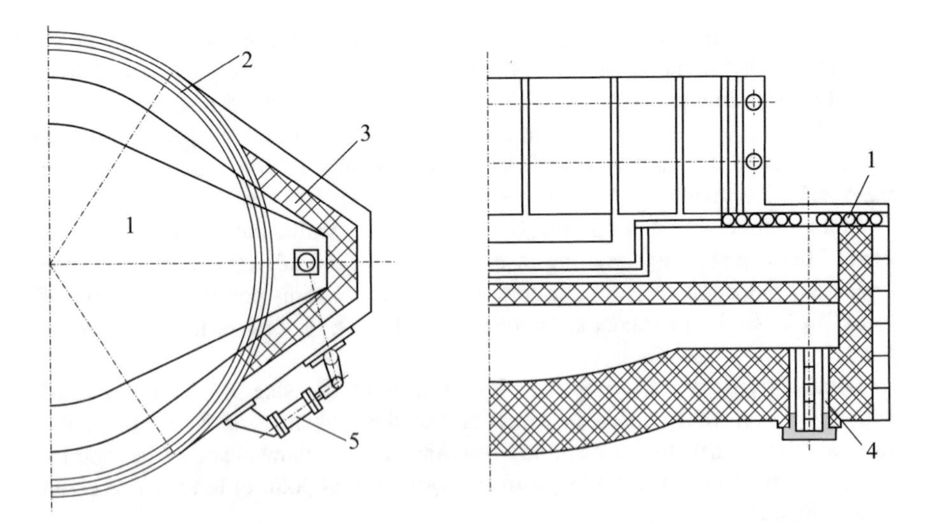

Fig. 3.27 Eccentric tapping arc furnace: *1* water cooled bay window, *2* wall panels, *3* lining of bay window, *4* bay window hole, *5* drive of the closing device [18, 25]

pouring hole is flushed with oxygen. The control of the locking device is done from the control panel located far from the furnace. From the same remote console are controlled tilting of the furnace, addition of additives into the ladle and all transport means necessary for the steel production. After maintenance operation, the valve is closed and the hole is filled from the top with a refractory compound. The maintenance operation of the pouring hole is no longer than 3 min.

3.2.8.5 Electrodes Arms and Column Assembly

The mechanical structure which supports the electrodes consists of three positioning columns which can be shifted up and down by hydraulic cylinders incorporated in each column. The horizontal electrode arms, mostly made with non-magnetic steel, are attached to the top ends of the positioning columns.

There are two types of design of electrode holder arms, i.e. the conventional one with conducting current tubes located on the carrier arms and the modern one with conductive arms made of bi-metallic plates of steel and copper, able to carry the high current to the contact pads without the need of bus tubes (see Figs. 3.5, 3.13a and 3.28).

The characteristic feature of the new design is the combination of electrical, mechanical and heat transfer functions in one unit—a water-cooled arm with conductive external surface.

Conductive arms have a number of advantages in comparison with conventional ones. The electrical resistance and reactance of the secondary circuit of an arc furnace with conductive arms are lower than in a furnace with conventional electrode holder. This allows, for a given power of the furnace transformer, to increase the arc power of each phase.

The thermal load of electrode arms of both types, which is determined by external factors, is concentrated at their ends and is nearly the same, but the heat losses due to the current flow are higher in conventional electrode arms which have a smaller current carrying section.

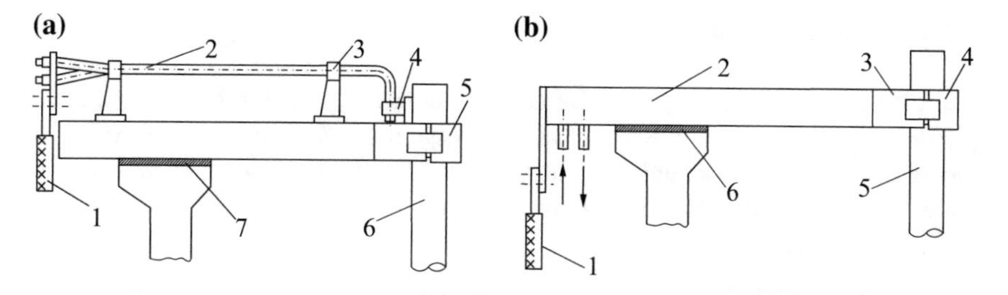

Fig. 3.28 Electrode holder arms—**a** conventional (*1* flexible cables, *2* tubular bus-bars, *3* supports, *4* contact pads, *5* clamping ring, *6* electrode, *7* electrical insulation); **b** conductive (*1* flexible cables, *2* conducting arm, *3*, *4* contact pads and clamping ring, *6* electrode, *7* electrical insulation) [25]

Moreover, in conventional electrode-holder arms there are additional losses as a result of currents induced in steel elements under the influence of the electromagnetic field, while in current-carrying arms such phenomenon is eliminated due to the screening action of the external conducting layer. This increases the life of electrode clamping mechanism and electrical insulation, thus decreasing costs of repairs and maintenance.

Mechanical loads of the two types of electrode arms are both static and dynamic. Static loads are determined by the gravity force, while the dynamic ones occur when tilting the furnace, during the electrodes movement, at the contact of electrodes with the solid charge, etc. In a first approximation for both types of electrode arms these loads can be considered identical, when copper current-carrying elements are used.

However, the electromagnetic forces occurring on the current carrying tubes of conventional arms, can cause vibrations and deterioration of the electrical insulation between them and the holder arm.

The most common design of conductive arms, used since the middle of the 80s by companies producing EAFs, is constituted by a tube of rectangular cross section, obtained by welding bimetallic (copper-steel) sheets. The thickness of the outer copper conductive layer is typically 8–10 mm. The bimetallic tube is water cooled inside, in order to assure normal temperature conditions of the current-carrying parts near the spring-hydraulic clamping mechanism of the electrode, as well as of the electrical insulation located between the bottom "face" of the arm and the supporting column.

The effectiveness of electrode conductive arms increases with the capacity of the furnace vessel, since the reduction of power losses and the increase arc power are proportional to the square of the current. This indicates the convenience of their use also in DC electric arc furnaces, where the intensity of the working current is two times higher then in AC three-phase furnaces of the same capacity.

Finally, the mechanical construction of electrodes arms and column assembly must be mechanically very robust to allow for fast adjustment vertical movements (speeds up to 150 mm/s) with minimum oscillations.

3.2.8.6 Loading System

The different types of direct arc furnaces differ, as regards mechanical structure, primarily in the loading system, which is always made from the upper opening of the vessel. There are several solutions, all aiming to reduce loading time and to allow the most beneficial working conditions of the furnace.

The main solutions are:

- A portal which holds and lifts-up the roof and the electrodes and allows to move them horizontally along a rail in order to open the vessel for loading. Disadvantages of this type of construction is that roof and electrodes can be

Fig. 3.29 Bucket charging

damaged during the movement due to vibrations and that the flexible cables of the secondary circuit must be relatively long;

- Lifting-up roof and electrode holder and moving the vessel horizontally;
- Lifting-up the roof and the electrode holder and swinging it around a vertical axis. This is the most widely used charging system today in furnaces with diameters up to about 5 m, as it allows the movement of roof and electrodes without appreciable vibrations and with limited length of the flexible cables.

The charge is normally charged with buckets (see Fig. 3.29). Before loading, the basket may pass to a scrap pre-heater, which uses hot furnace off-gases to heat the scrap and recover energy, increasing in this way the melting process efficiency. Scrap preheating can produce energy savings up to 50 kWh/ton of steel [4].

The scrap basket is then taken to the melt shop, the roof is swung off the furnace, and the furnace is charged with scrap from the basket.

The bottom of clamshell consists of two movable half-shell which, when are opened by the charging crane, frees virtually the entire bucket cross-section to permit the scrap to drop out.

After charging, the roof is swung back over the furnace and meltdown starts.

3.3 Equivalent Circuits of EAF

3.3.1 Single Phase Equivalent Circuit

As it was shown in Fig. 3.6, the electrical circuit of an arc furnace installation consists of several series connected elements, such as bus-bars and high-voltage equipment, reactors, transformers, secondary high-current circuit, the electrodes and —within the vessel itself—electrical arc(s) and molten metal.

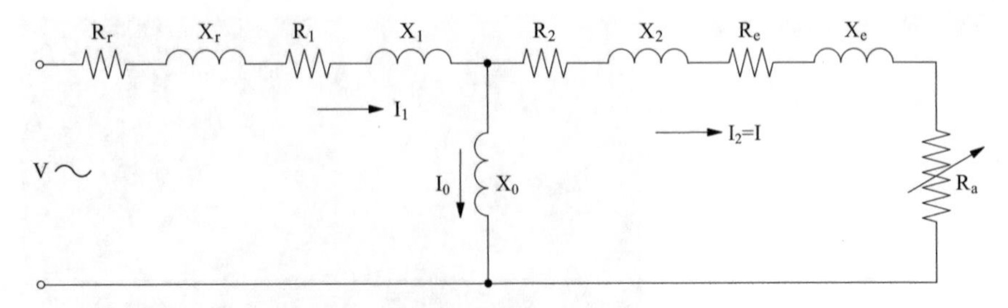

Fig. 3.30 Single-phase equivalent circuit of a direct arc furnace

Usually the design of the furnace, the analysis of its working conditions and the influence of the different elements is made with reference to equivalent circuits in which, with applied voltages equal to the secondary voltage of the furnace transformer, the absorbed current and power are the same as in the real furnace.

In the equivalent circuits the power that in the furnace is transformed into heat in the arc is dissipated in an equivalent arc resistance R_a, which is non-linear and variable from zero (when electrodes are in contact with the molten metal) to infinity (interruption of the arc).

Assuming ideally that the three phases are absolutely symmetrical (theoretically with triangulated secondary circuit and flexible cables) and the arc resistances of the three phases are equal, it is usual to refer to a single phase equivalent circuit.

As said before, in this equivalent circuit all circuit parameters must be referred to the low voltage side of the furnace transformer.

The secondary circuit and the electrode can be represented by an equivalent resistance R_e and a reactance X_e, the transformer (or transformers) with their equivalent circuit, the additional series reactor with a resistance and a series reactance, while the impedance of the high voltage conductors can be generally neglected.

This leads to draw for the single-phase equivalent circuit of a direct arc furnace the circuit diagram of Fig. 3.30,

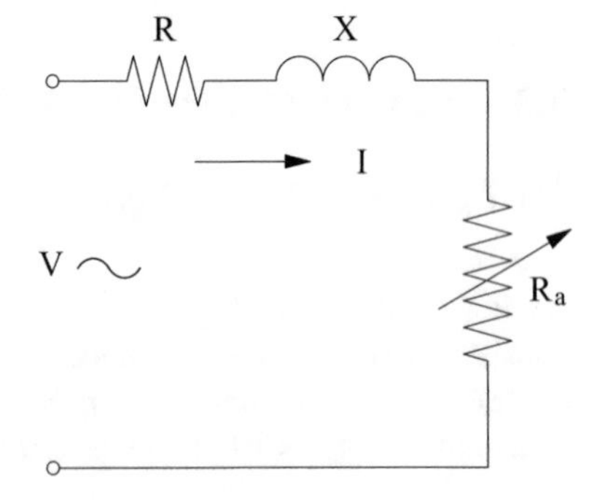

Fig. 3.31 Simplified equivalent circuit of a direct arc furnace

where is:

R_e, X_e	equivalent resistance and reactance of high-current secondary circuit and electrode,
R_1, X_1, R_2, X_2	resistance and leakage reactance of primary and secondary windings of furnace transformer,
X_o	reactance of the magnetizing branch of transformer,
R_r, X_r	resistance and reactance of the series reactor,
R_a	useful resistance which includes arc resistance and resistance of the portion of electrode which contributes to useful power,
V	secondary phase voltage of the furnace transformer

Assuming negligible the transformer no-load current $I_{0,}$ it is $I_1 = I_2 = I$ and the equivalent circuit can be simplified as shown in Fig. 3.31, with:

I arc current
X total supply circuit reactance
R resistance producing to the total losses in the supply circuit

Attention must be paid to the fact that for calculating resistance and reactance of the single-phase equivalent circuit, it must be considered the actual value of the transformation ratio of furnace transformer, which is different for the different control taps and, in case of a three-phase transformer with delta connected windings, first it must be replaced with an equivalent transformer Y/Y connected and then to refer all impedances to the secondary side taking into account the change with this substitution of resistance and reactance of the windings.

3.3.2 Circuit Analysis with Sinusoidal Quantities

For operating an arc furnace it is necessary to select the more convenient working conditions in order to minimize the power consumption and achieve the maximum production rate.

The different operating modes are characterized by the values of current, voltage, impedance, power and power factor.

The main operating characteristics provide the values of the above electrical parameters; other useful characteristics give copper losses, thermal losses, electrical efficiency, total efficiency, specific energy consumption, production rate, etc.

All these characteristics are plotted as a function of the current since the control of the furnace is based on this value, which is recorded continuously.

Other characteristics can be given as a function of different quantities (such as arc voltage, useful power or arc resistance), but are of minor practical interest because the measurement of these quantities is not usually carried out during the normal operation of the furnace.

However, as examples, also diagrams as a function of the arc voltage will be presented here.

Starting from the equivalent circuit of Fig. 3.31, it is possible to derive the theoretical operating characteristics of the furnace either analytically, by using the equations of the circuit, or graphically by drawing its circle diagram.

3.3.2.1 Analytical Characteristics

The analysis of the operating conditions of the furnace on the basis of the circuit of Fig. 3.31 is easy in the hypothesis of sinusoidal arc current and voltage.

Although this hypothesis does not correspond to reality, due to the non-linear characteristics of the arc which will be discussed later, the results of the analysis provides a simplified overview of the different operating conditions, and can be used for steel furnaces through an appropriate increase of the short-circuit reactance [26].

The well known equations of main electrical parameters of the circuit of Fig. 3.31 are:

Current:

$$\begin{aligned} I &= \frac{V}{\sqrt{(R+R_a)^2+X^2}} \\ &= \frac{V}{X}\frac{1}{\sqrt{1+[(R+R_a)/X]^2}} \end{aligned} \tag{3.4}$$

Short-circuit current:

$$I_{sc} = \frac{V}{X}\frac{1}{\sqrt{1+(R/X)^2}} \tag{3.4a}$$

Dimensionless current:

$$i = \frac{I}{I_{sc}} = \sqrt{\frac{1+(R/X)^2}{1+[(R+R_a)/X]^2}} \tag{3.4b}$$

Arc voltage:

$$V_a = R_a \cdot I \tag{3.5}$$

Dimensionless arc voltage:

$$v = \frac{V_a}{V} = \frac{R_a/X}{\sqrt{1+[(R+R_a)/X]^2}} \tag{3.5a}$$

Apparent power:

$$P_a = VI = \frac{V^2}{X}\sqrt{\frac{1}{1+[(R+R_a)/X]^2}} \tag{3.6}$$

Short-circuit apparent power:

$$P_{asc} = \frac{V^2}{X}\sqrt{\frac{1}{1+(R/X)^2}} \tag{3.6a}$$

Dimensionless apparent power and current:

$$\frac{P_a}{P_{asc}} = \frac{I}{I_c} = i \tag{3.6b}$$

Total active power:

$$P = (R+R_a)I^2 = \frac{V^2}{X}\frac{(R+R_a)/X}{1+[(R+R_a)/X]^2} \tag{3.7}$$

Arc power:

$$P_{arc} = R_a I^2 = \frac{V^2}{X}\frac{R_a/X}{1+[(R+R_a)/X]^2} \tag{3.7a}$$

Reactive power:

$$Q = XI^2 = \frac{V^2}{X}\frac{1}{1+[(R+R_a)/X]^2} \tag{3.8}$$

Short-circuit reactive power:

$$Q_{sc} = \frac{V^2}{X}\frac{1}{1+(R/X)^2} \tag{3.8a}$$

Power factor:

$$\cos\varphi = \frac{P}{P_a} = \frac{[(R+R_a)/X]}{\sqrt{1+[(R+R_a)/X]^2}} \tag{3.9}$$

Electrical efficiency:

$$\eta_e = \frac{P_{arc}}{P} = \frac{R_a}{R + R_a} \tag{3.10}$$

Power ratios:

$$\begin{cases} \dfrac{P}{P_{asc}} = \dfrac{[(R+R_a)/X]\sqrt{1+(R/X)^2}}{1+[(R+R_a)/X]^2} \\ \dfrac{P_{arc}}{P_{asc}} = \dfrac{(R_a/X)\sqrt{1+(R/X)^2}}{1+[(R+R_a)/X]^2} \\ \dfrac{Q}{Q_{sc}} = \dfrac{1+(R/X)^2}{1+[(R+R_a)/X]^2} = i^2 \end{cases} \tag{3.11}$$

With the positions:

$$Z^2 = (R+R_a)^2 + X^2, \cos\varphi = (R+R_a)/Z, \sin\varphi = X/Z,$$

the total active power can be written as follows:

$$P = \frac{V^2}{X} \sin\varphi \cos\varphi. \tag{3.12}$$

It varies, as a function of power factor, according to the curve 1 of Fig. 3.32.

For comparison, in the same figure are given the analogous curves for single-phase and three-phase circuits with arc, which will be discussed later.

As known, in the circuit of Fig. 3.31, the maximum active power is absorbed when it is

$$R + R_a = X$$

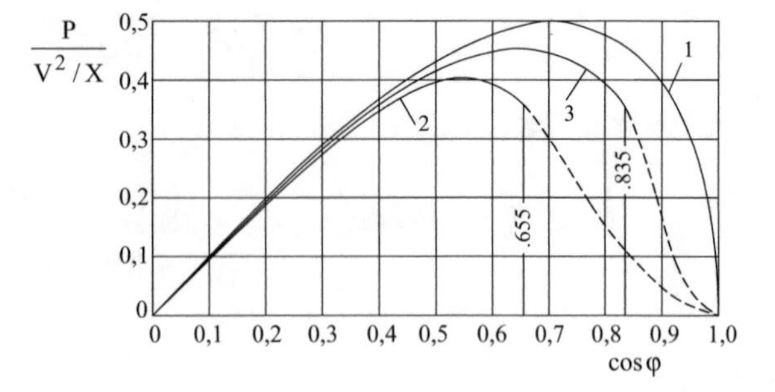

Fig. 3.32 Active power as a function of power factor (*1* single-phase circuit with sinusoidal quantities; *2* single-phase circuit with arc; *3* three-phase circuit with arc)

while the maximum arc power occurs for

$$R_a = \sqrt{R^2 + X^2}.$$

- *Single-phase circuit with R = 0*

If we assume $R = 0$, i.e. we neglect the losses of the furnace supply circuit, all above quantities become functions only of the ratio (R_a/X). From Eq. (3.5a) we can write:

$$v = \frac{V_a}{V} = \frac{R_a/X}{\sqrt{1 + (R_a/X)^2}} \tag{3.13}$$

and

$$\frac{R_a}{X} = \sqrt{\frac{v^2}{1 - v^2}}.$$

This relationship allows us to re-write Eqs. (3.4)–(3.10) in the compact form:

$$\left.\begin{aligned} & I_0 = \frac{V}{X}\sqrt{1 - v^2}; I_{sc0} = \frac{V}{X}; \frac{I_0}{I_{sc0}} = \sqrt{1 - v^2} \\ & P_{a0} = \frac{V^2}{X}\sqrt{1 - v^2}; P_{asc0} = \frac{V^2}{X}; \frac{P_{a0}}{P_{asc0}} = \frac{I_0}{I_{c0}} = \sqrt{1 - v^2} \\ & P_0 = P_{arc0} = \frac{V^2}{X} v\sqrt{1 - v^2}; \frac{P_0}{P_{asc0}} = \frac{P_{arc0}}{P_{asc0}} = v\sqrt{1 - v^2} \\ & \cos\varphi_0 = \frac{P_0}{P_{a0}} = v; \quad \frac{Q_0}{Q_{sc0}} = 1 - v^2 \end{aligned}\right\} \tag{3.14}$$

The ratios $\frac{I_0}{I_{sc0}}, \frac{P_{a0}}{P_{asc0}}, \frac{P_0}{P_{asc0}}$ and $\cos\varphi_0$ vary, as a function of *v*, as shown by the dashed and dotted curves in Fig. 3.33.

For v = 0.707 the active power has a maximum, corresponding to which it is:

$$P_0/P_{asc0} = 0.5; \quad I_0/I_{sc0} = 0.707; \quad \cos\varphi_0 = 0.707.$$

For R = 0 Eq. (3.4b) becomes:

$$i = \frac{I}{I_c} = \frac{1}{\sqrt{1 + \left(\frac{R_a}{X}\right)^2}}$$

from which we can write

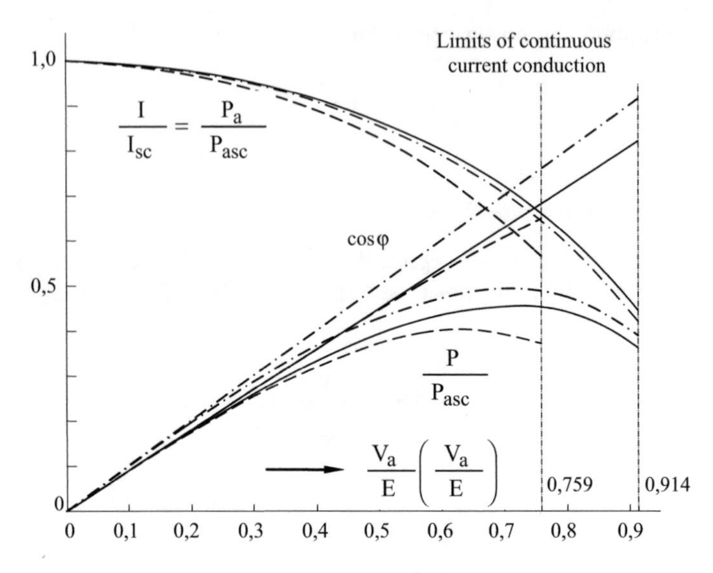

Fig. 3.33 Single-phase circuit characteristics as a function of $v = V_a/V$ with sinusoidal quantities (*dashed* and *dotted lines*); with arc (*dashed lines*); three-phase circuit with arc (*solid lines*)

$$\frac{R_a}{X} = \sqrt{\frac{1 - i^2}{i^2}}.$$

By substituting this relationship in Eqs. (3.5a), (3.6b) and (3.9) we obtain:

$$v = \frac{V_a}{V} = \sqrt{1 - i^2} \tag{3.15}$$

$$\left.\begin{aligned} \frac{P_a}{P_{asc}} = \frac{I}{I_{sc}} = i \quad &; \frac{P}{P_{asc}} = i\sqrt{1 - i^2} \\ \frac{Q}{Q_{sc}} = i^2 \quad &; \cos\varphi = \sqrt{1 - i^2} \end{aligned}\right\} \tag{3.16}$$

The variations of the above quantities as a function of *i* are represented by the dashed and dotted curves in Fig. 3.34.

They show the following:

- the active power increases with the current, reaches its maximum value for i = 0.707 (cos φ = 0.707) and then decreases to zero at i = 1 (short-circuit). At its maximum it is equal to 50 % of the short-circuit power;
- the reactive power increases with the increase of the current with a quadratic law, reaching the maximum value in short-circuit conditions (i = 1);
- the apparent power varies linearly from zero, for i = 0, to the short-circuit value for i = 1;

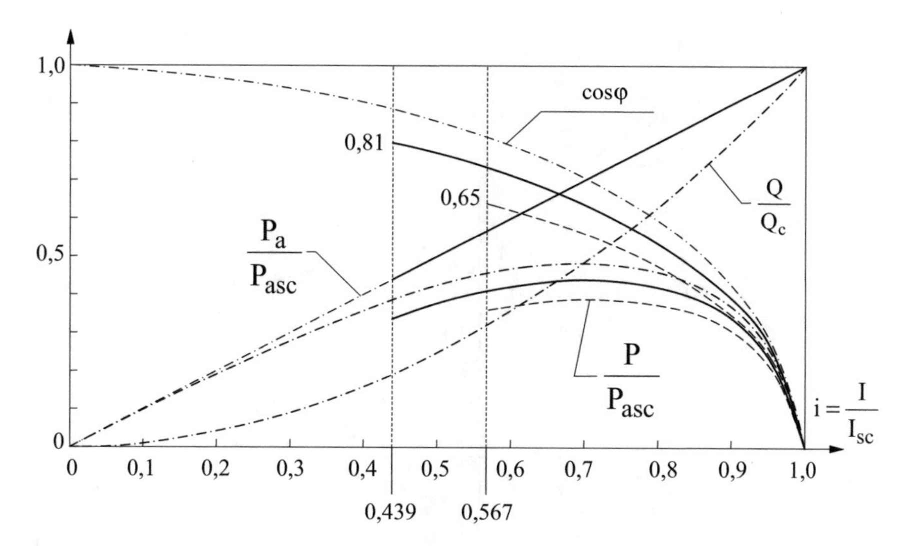

Fig. 3.34 Analytical characteristics as a function of current with sinusoidal quantities (*dashed* and *dotted lines*); with arc (*dashed curves*); three-phase circuit with arc (*solid lines*)

- the power factor decreases when the current increases and becomes zero in short-circuit.

3.3.2.2 The Circle Diagram

As known, the analysis of the simplified circuit of Fig. 3.31 with sinusoidal quantities can be conveniently developed graphically, through the construction of the circle diagram. On one hand it allows to consider the influence of the resistance of the supply circuit, on the other hand it has the limitations connected with a graphical construction.

The construction of the diagram is based on the equation of the circuit divided by jX:

$$-j\frac{\dot{V}}{X} = -j\left(\frac{R + R_a}{X}\right)\dot{I} + \dot{I} = \text{const.} \tag{3.17}$$

which highlights that—at constant furnace phase voltage—the geometrical sum of the vectors $j(R + R_a/X)\dot{I}$ and $\dot{I}$, phase displaced to each other of $\pi/2$, is equal to the vector $-j\dot{V}/X$ which has constant amplitude and lags the vector $\dot{V}$ of $\pi/2$. The tip of the vector $\dot{I}$ therefore moves on a circle of diameter $-j\dot{V}/X$.

The graphical construction can be done as illustrated in Fig. 3.35.

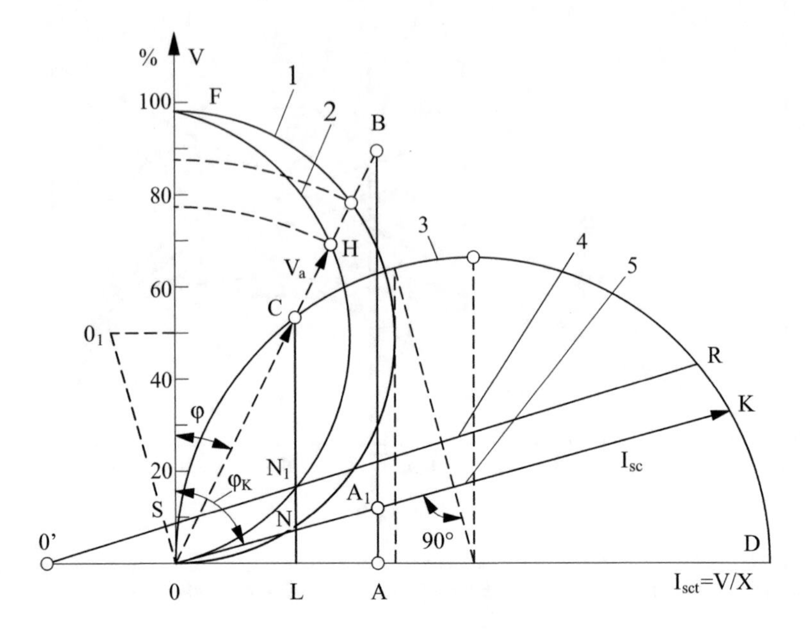

Fig. 3.35 Circle diagram of circuit of Fig. 3.31. (*1* supply voltage and cos φ; *2* arc voltage; *3* current; *4* thermal losses; *5* electrical losses) [11]

Starting from the origin O, it must be traced on the vertical axis—in the scale of voltages m_v—the segment OF = m_vV, and—on the horizontal one—in the scale of impedances m_z—the segment OA = m_zX.

If on the vertical line through the point A is set out the segment $AA_1 = m_zR$, then OA_1 represents the impedance of the circuit for $R_a \to 0$ (short-circuit impedance).

Now it must be drawn the semi-circle ODC passing through O, with diameter OD = (V/X)m_i, which—in the scale of the currents $m_i = m_v/m_z$—represents the theoretical short-circuit current I_{sct}, i.e. the short-circuit current for R = 0 and $R_a = 0$.

By varying the arc resistance R_{arc} from zero to infinity, the tip C of the current vector $\dot{I}$ moves along the circumference from the point K (actual point of short-circuit for $R_a = 0$) to the point O (point of no-load: $R_a \to \infty$)

The vector OK, obtained by extending the segment OA_1 to the intersection with the circle, represents the actual short-circuit current I_{sc}.

When the arc resistance varies from $R_a = 0$ to $R_a = \infty$, the phase angle φ between voltage and current varies from the value φ_k (for which tg φ_k = X/R) to zero.

To evaluate the components of the supply voltage $\dot{V}$ respectively in phase and in quadrature with the current $\dot{I}$, it must be drawn the semicircle OEF: the segment OE, determined by the intersection of this circle with the direction of the current, represents the active component, the segment EF the reactive one.

For easiness of reading the active component of voltage, on the segment OE the scale of voltages is indicated as percentage of the total voltage.

For the determination of the arc voltage $\dot{V}_a$ we can draw the circle OHF along which moves the tip of the vector representative of $\dot{V}_a$ when the resistance R_a varies; its center is the intersection of the perpendicular from O to OK with the perpendicular to OF from its midpoint. The vector OH is then the vector representative of the arc voltage $\dot{V}_a$ corresponding to the current $\dot{I}$, while the vector HE represents the voltage drop in the resistance of the supply circuit for the same current.

For the assumption V = const., the diagram of currents gives also the power values, since the segment OC represents both the current I (in the scale of currents) and the apparent power of the furnace P_a = VI (in the scale of power $m_p = m_i/V$).

From the foregoing the circle diagram shows on the ordinates the values of the phase active power; it gives also on the abscissae the corresponding reactive power, while the distance from zero indicates the apparent power.

Therefore it is:

$$\mathrm{CL} \equiv \mathrm{I}\cos\varphi \equiv \mathrm{VI}\cos\varphi = \mathrm{P}$$
$$\mathrm{OL} \equiv \mathrm{I}\sin\varphi \equiv \mathrm{VI}\sin\varphi = \mathrm{Q},$$

i.e. the total active power is proportional to the segment CL, while the segments LN and CN represent respectively the electrical power losses $P_e = R I^2$ and the arc power $P_{arc} = R_a I^2$.

The maximum absorbed power occurs on the perpendicular to the diameter through the center of the circle (Fig. 3.36), while the maximum arc power is determined by the perpendicular to the diameter passing through the point of tangent to the circle circumference parallel to the short circuit current.

The total power for the whole three-phase system is determined by multiplying by 3 the values of the circle diagram.

The electrical efficiency is given by the ratio:

$$\eta_e = \frac{P_{arc}}{P} = \frac{R_a}{R + R_a} = \frac{CN}{CL} \tag{3.18}$$

In EA furnaces always occur also thermal losses P_t which depend on furnace design, type of refractory material, operating temperature, duty cycle, etc.

Since a part of arc power is spent to compensate these losses, the actual useful power used in the melting process is the difference between arc power and total losses.

$$P_u = P_{arc} - (P_e + P_t) \tag{3.19}$$

If the power P_t is constant as a function of I, the useful power can be evaluated on the circle diagram with reference to a line SR parallel to the short-circuit current

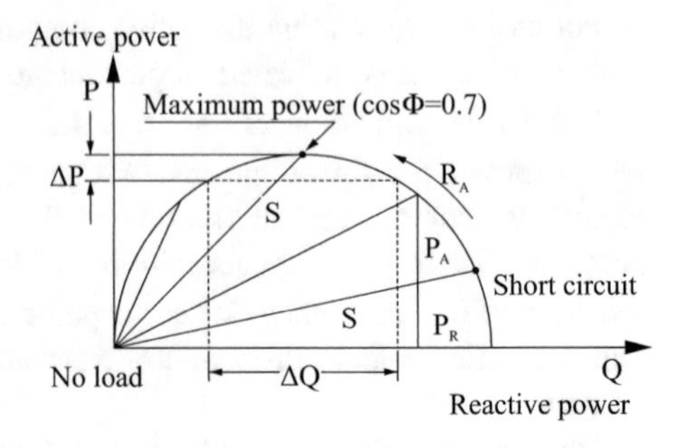

Fig. 3.36 Power on the circle diagram

I_c and shifted vertically from it of the amount NN_1 equal, in the power scale, to thermal losses.

In this case the useful power P_u is represented in Fig. 3.35 by the segment CN_1, the power P_t by the segment NN_1, the electrical power losses P_e by the segment LN.

On the contrary, if the power P_t increases with the arc current, then also in the circle diagram instead of the straight line SR must be drawn a curve whose points are moved from the short-circuit current line of quantities corresponding to the power P_t.

Taking into account thermal losses, the total efficiency of the furnace becomes:

$$\eta = \frac{P_u}{P} = \frac{P_{arc} - P_{loss}}{P} = \frac{P - (P_e + P_t)}{P} = \frac{CN_1}{CL} \tag{3.20}$$

Figure 3.37 shows typical operating characteristics obtained in this way, in case of resistance $R \neq 0$, negligible thermal losses and constant circuit parameters.

From the curves it can be observed that, with the increase of the current, the power factor decreases, the losses in the power supply circuit increase with the square of the current, while the arc and the total active power initially increase, reach a maximum and then decrease.

Note that the maximum arc power occurs at a current I' lower than the value I_M corresponding to the maximum absorbed active power and that radiation is higher with long arcs (with a maximum at $\cos \varphi = 0.85$). This is why, before the introduction of the technology of foaming slags, the melting was preferably carried out at low power factor using short arcs.

In addition to the above features, other technological parameters of performance of steel melting furnaces can be obtained from the circle diagram. In particular,

- *the production rate:*

$$M \approx \frac{P_u}{380} = \frac{P_{arc} - P_{loss}}{380}\,(t/h), \tag{3.21}$$

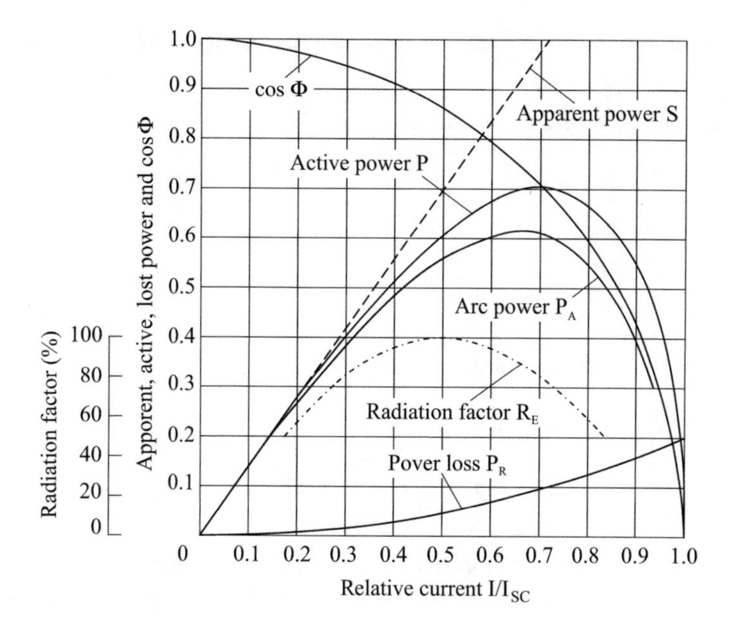

Fig. 3.37 EAF characteristics with $R \neq 0$ and constant equivalent circuit parameters

Fig. 3.38 Coefficient k_{op} for evaluation of operational reactance (*a* initial melting period; *b* main melting period; *c* melting with foaming slag) [27]

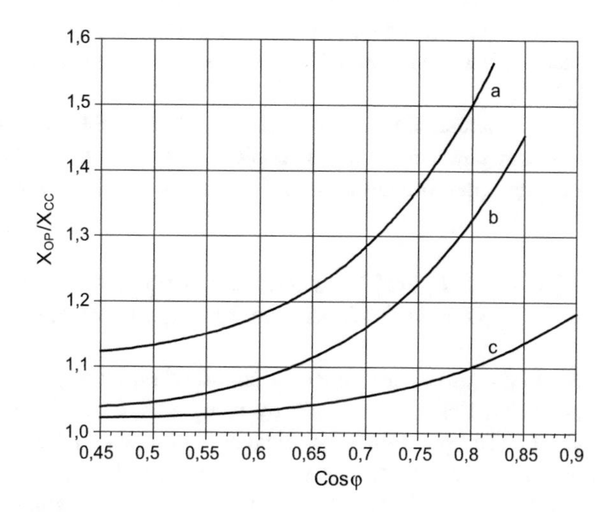

380 being the theoretical amount of energy, kWh, required for melting 1 ton of steel.[1]

- *the melting time of 1 ton of steel:*

$$T = \frac{1}{M}(h/t) \tag{3.22}$$

- *the specific electrical energy consumption*

$$\sigma = \frac{P}{M} = \frac{380}{\eta}(kWh/t). \tag{3.23}$$

The specific electrical consumption has a minimum at the maximum total efficiency, for a value of current I' lower than the one I_M corresponding to maximum arc power.

The productivity is maximum for a current $I'' > I'$, which occurs at the minimum melting time per tonne, i.e. when the arc power P_a is maximum.

The experience has shown that optimal furnace operation, for which the production cost of molten metal is minimum, is achieved with a current value intermediate between I' and I'', at a power factor in the range between 0.75 and 0.85 [11].

In conclusion, the characteristics based on the simplified equivalent circuit with constant parameters and sinusoidal quantities allow to estimate with good engineering approximation the real operating characteristics.

3.3.2.3 Analytical Characteristics with R ≠ 0, Considering Arc Current Deformation Through Correction Coefficients

Making again reference to the circuit of Fig. 3.31, the calculation of the operating characteristics can be made by using Eqs. (3.4)–(3.11), introducing—outside the short-circuit point—an empirical correction coefficient which takes into account the increase of reactance due to the time-dependent behavior of the arc as well as the non-linearity and fluctuations of the electrical parameters.

Since this behavior is different in the initial phase of melting and the subsequent working periods and depends on arc length, power factor, arc impedance and use of

[1]Some authors make reference to 340 kWh/ton, taking into account the heat accumulated in the lining and partially re-transferred to the charge.

foaming slags, also the correction factor has different values in the different operating conditions.

The calculation is performed introducing outside the short-circuit point, instead of the value X of the reactance, the so-called "*operational reactance*":

$$X_{op} = k_{op} \cdot X \tag{3.24}$$

Thus, the circuit impedance at all operating points—with the exception of the short-circuit one—is:

$$\dot{Z}_{op} = (R + R_a) + jX_{op} = (R + R_a) + jk_{op}X \tag{3.25}$$

The values of the coefficient k_{op} are given by the curves of Fig. 3.10 or by empirical equations; the curves highlight that the operational reactance can be up to 50 % higher than that of the short-circuit point.

Example 3.1 For further clarification, in the following is developed an example of calculation with reference to the diagram of the installation and the typical data of a three-phase furnace shown in Fig. 3.39, using the following symbols:

$\dot{Z}_{scs} = R_{scs} + jX_{scs}$	short-circuit impedance of the supply grid at point of connection of the furnace;
$\dot{Z}_{ti} = R_{ti} + jX_{ti}$	impedance of intermediate transformer;
$\dot{Z}_{tf} = R_{tf} + jX_{tf}$	impedance of furnace transformer;
$\dot{Z}_{scf} = R_{scf} + jX_{scf}$	short-circuit impedance of the furnace;
R_a	arc resistance

Moreover, we assume the following data:

$$X_{scs}/R_{scs} = 10; \quad X_{ti}/R_{ti} = 10; \quad X_{tf}/R_{tf} = 5$$
$$\dot{Z}_{scf} = 0.41 + j2.57\,(\text{m}\Omega).$$

The calculation procedure then develops as follows:

1. *Short-circuit impedance of the supply*

$$Z_{scs} = \frac{V^2}{P_{a_{scs}}} = \frac{635^2}{1800 \cdot 10^6} = 0.224\,\text{m}\Omega$$

$$R_{scs} = \frac{Z_{scs}}{\sqrt{1 + (X_{scs}/R_{scs})^2}} = \frac{0.224}{\sqrt{1 + 100}} = 0.0223\,\text{m}\Omega$$

$$X_{scs} = 10 \cdot R_{scs} = 0.223\,\text{m}\Omega$$

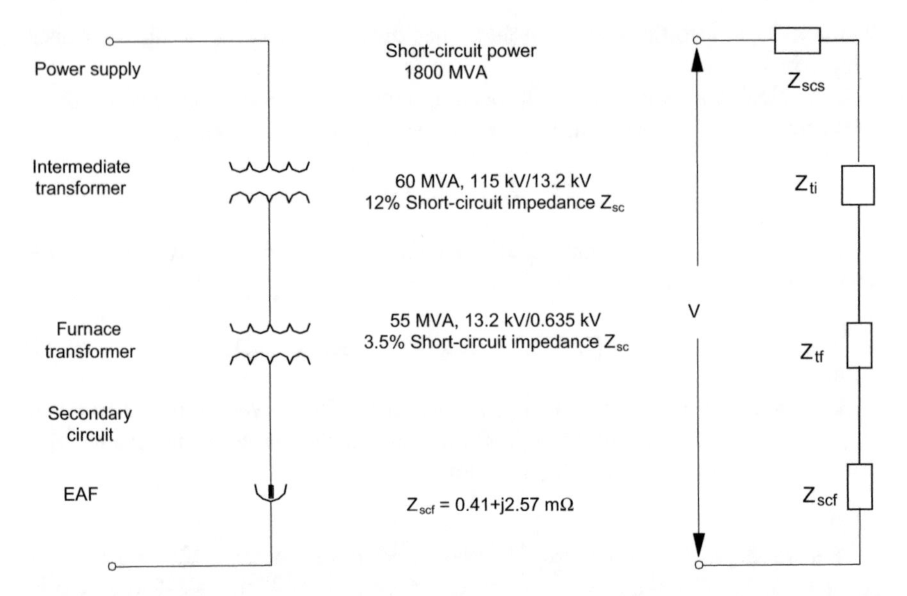

Fig. 3.39 Schematic circuit diagram and single-phase equivalent circuit of a three-phase furnace

2. *Impedance of intermediate transformer*

$$Z_{ti} = \frac{V^2 \cdot Z\%}{P_{ati}} = \frac{635^2 \cdot 0.12}{60 \cdot 10^6} = 0.806\,m\Omega$$

$$R_{ti} = \frac{Z_{ti}}{\sqrt{1 + (X_{ti}/R_{ti})^2}} = \frac{0.806}{\sqrt{1+100}} = 0.0802\,m\Omega$$

$$X_{ti} = 10 \cdot R_{ti} = 0.802\,m\Omega$$

3. *Impedance of furnace transformer*

$$Z_{tf} = \frac{V^2 \cdot Z\%}{P_{atf}} = \frac{635^2 \cdot 0.035}{55 \cdot 10^6} = 0.257 m\Omega$$

$$R_{tf} = \frac{Z_{tf}}{\sqrt{1 + (X_{tf}/R_{tf})^2}} = \frac{0.257}{\sqrt{1+25}} = 0.0503 m\Omega$$

$$X_{tf} = 5 \cdot R_{tf} = 0.252 m\Omega$$

4. *Short-circuit impedance*

$$\dot{Z}_{sc} = \dot{Z}_{scs} + \dot{Z}_{ti} + \dot{Z}_{tf} + \dot{Z}_{scf} = R + jX = 0.563 + j3.847\ (m\Omega)$$
$$Z_{sc} = \sqrt{R^2 + X^2} = 3.888\ m\Omega$$

5. *Short-circuit working conditions (R_a = 0)*

Apparent power (MVA)	$P_{asc} = \frac{V^2}{Z_{sc}} = \frac{635^2}{3.888 \cdot 10^{-3}} = 103.71$
Current (kA)	$I_{sc} = \frac{S_{sc}}{\sqrt{3} \cdot V} = \frac{103.71 \cdot 10^6}{\sqrt{3} \cdot 635} = 94.296$
Power factor	$\cos\varphi_{sc} = \frac{R_{sc}}{Z_{sc}} = \frac{0.563}{3.888} = 0.145$
Short-circuit power, (MW)	$P_{sc} = S_{sc} \cdot \cos\varphi_{sc} = 103.71 \cdot 0.145 = 15.038$
Reactive power, (MVAr)	$Q_{sc} = S_{sc} \cdot \sin\varphi_{sc} = 103.71 \cdot 0.989 = 102.61$

As indicated previously, for all operating conditions outside the short circuit point reference must be made to the operational impedance given by Eq. (3.25).

Taking into account that the values of coefficient k_{op} are usually given as a function of power factor, it is convenient to develop the calculation starting from this value.

Table 3.3 shows the equations used and the results obtained for different values of cos φ with the values of k_{op} given by curve b of Fig. 3.38.

In the table the usual notations are used: P_a—apparent power, P—active power from the supply, P_{arc}—arc power; P_{loss}—power losses in the supply circuit, Q—reactive power; η_e—electrical efficiency, V_a—arc voltage; ℓ_a—arc length, calculated with the relationship:

$$V_a^{(V)} \approx \ell_a^{(mm)} + 40$$

With the data of the Table 3.3 are plotted the characteristics of Fig. 3.40.

3.3.3 Arc Characteristics

For a more deep analysis of the furnace operating conditions it's necessary to take into account that in the circuit of Fig. 3.31 arc voltage and current are not sinusoidal.

For this purpose, in this paragraph is described the behavior of the arc resistance, which in turn depends on the electric arc characteristics.

Table 3.3 Parameters calculated for the EAF furnace of Fig. 3.39 with different power factors

$\cos\varphi$	0.5	0.6	0.65	0.7	0.8	0.85
$R = 0.5628\ (\text{m}\Omega)$	–	–	–	–	–	–
$X = 3.847\ (\text{m}\Omega)$	–	–	–	–	–	–
$X_{op} = k_{op} \cdot X\ (\text{m}\Omega)$	4.039	4.155	4.270	4.463	5.117	5.578
$R + R_a = X_{op}/\text{tg}\varphi\ (\text{m}\Omega)$	2.332	3.116	3.652	4.374	6.822	9.000
$Z_{op}\ (\text{m}\Omega)$	4.6643	5.1935	5.619	6.2487	8.5278	10.589
$P_a = V^2/Z_{op}\ (\text{MVA})$	86.450	77.640	71.765	64.529	47.284	38.080
$I = P_a/(\sqrt{3} \cdot V)\ (\text{kA})$	78.601	70.592	65.249	58.671	42.991	34.622
$P = P_a \cdot \cos\varphi\ (\text{MW})$	43.225	46.584	46.647	45.170	37.827	32.368
$P_u = 3 \cdot R_a \cdot I^2\ (\text{MW})$	32.793	38.171	39.454	39.359	34.705	30.342
$P_{loss} = P - P_u\ (\text{MW})$	10.431	8.414	7.193	5.812	3.121	2.024
$Q = P_a \cdot \sin\varphi\ (\text{MVAR})$	74.868	62.112	54.534	46.080	28.370	20.061
$\eta_e = 1 - (P_{loss}/P)$	0.759	0.819	0.846	0.871	0.918	0.937
$V_a = R_a \cdot I\ (\text{V})$	139	180	201	224	269	292
$\ell_a \approx (P_u/3 \cdot I) - 40\ (\text{mm})$	99.1	140	162	184	229	252

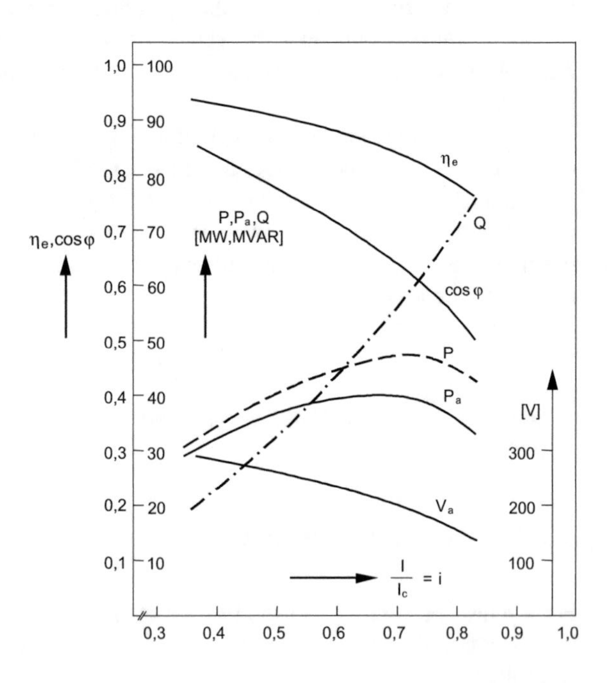

Fig. 3.40 Characteristics of the furnace of Fig. 3.39

As known, the arc is characterized by relatively high currents (usually not lower than several kA in industrial furnaces), low voltages (in the range of several tens or hundreds of volts) and high plasma temperatures in the region of the discharge (generally above 3000–4000 K).

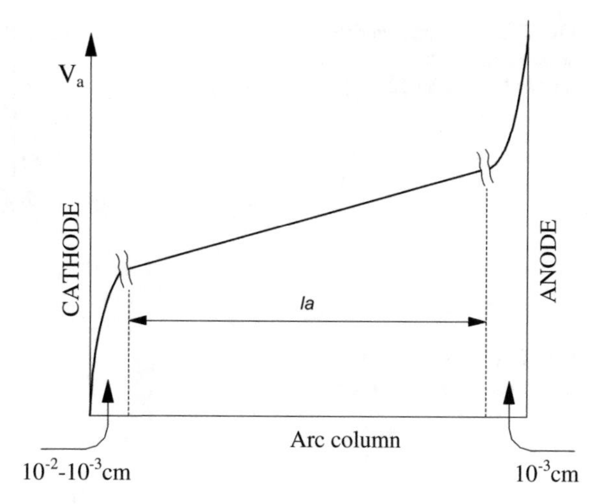

Fig. 3.41 Potential distribution in an electric arc in air (ℓ_a—length of arc column)

The voltage distribution in the discharge space of a DC arc burning between electrodes in air, is of the type shown in Fig. 3.41.

It is characterized by two voltage drops in the close proximity of anode and cathode (the latter, in general, relatively small) and a uniform longitudinal voltage gradient (of about 1 V/mm) along the arc column.

The total arc voltage V_a, sum of voltage drops in the three above mentioned regions, varies—as a function of the arc current and for constant gap between electrodes—according to the empirical equation of Ayrton [6]:

$$V_a = a + b \cdot \ell_a + \frac{c + d \cdot \ell_a}{I} \tag{3.26}$$

where:

a, b, c, d	constants depending on electrode material and medium in which the arc burns (e.g. for carbon electrodes a = 39, b = 2, c = 17, d = 10)
V_a, ℓ_a, I	arc voltage, V, length, mm, and current, A, respectively.

If the current is kept constant, the arc characteristics as a function of the arc length become:

$$V_a = \left(a + \frac{c}{I}\right) + \left(b + \frac{d}{I}\right)\ell_a \tag{3.26a}$$

and are represented by the diagrams of Fig. 3.42.

In particular, at the high currents values usual in industrial furnaces and constant arc length, the arc voltage V_a is constant and independent of the current; in this case the Eq. (3.26a) simplifies in the form:

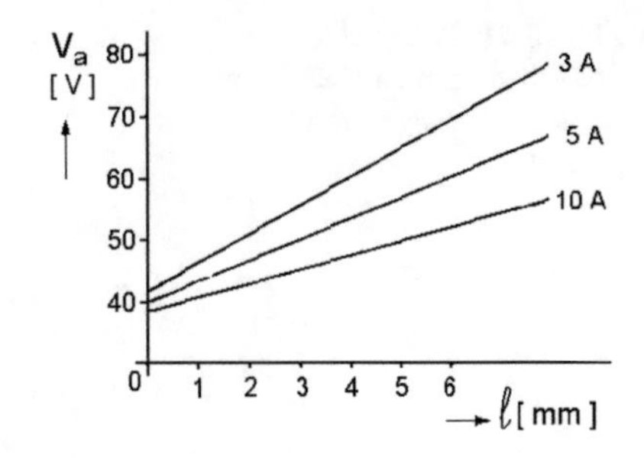

Fig. 3.42 Arc characteristics at constant arc current I as a function of arc length ℓ_a [6]

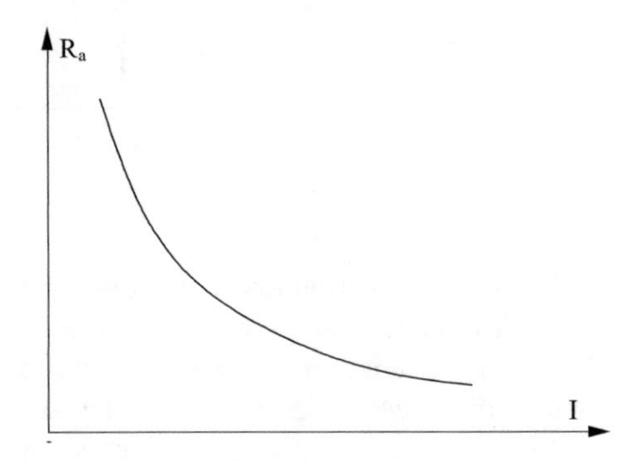

Fig. 3.43 Arc resistance R_a as a function of arc current [11]

$$V_a = a + b\ell_a \tag{3.27}$$

Equations (3.26), (3.26a) and Fig. 3.42 show that the arc voltage varies with the arc length according to straight lines, whose slope decreases when the current increases.

From Eq. (3.27) it follows also that, at constant arc length, the arc resistance R_a :

$$R_a = \frac{V_a}{I} = \frac{\text{const.}}{I} \tag{3.28}$$

is inversely proportional to the arc current, according to the diagram of in Fig. 3.43 (*non-linear resistance*).

This phenomenon of the decrease of R_a when the current I increases can be explained taking into account that, as I increases, increases the degree of ionization that determines the conductivity of the arc column and at the same time increases the cross-section of the ionized region through which the current flows.

If current and arc voltage variations are sufficiently low to allow the arc to adapt to the new conditions of temperature, ionization and emission from the electrodes, than the static characteristic of curve 1 in Fig. 3.44 holds; in this case arc voltages at ignition and extinction are practically the same.

But when voltage and current variations are fast, a "*hysteresis*" phenomenon occurs, for which when the current increases the arc voltage is higher than the static

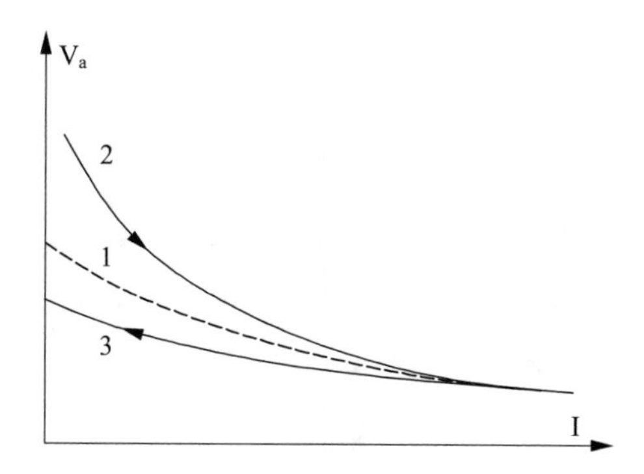

Fig. 3.44 Static and dynamic DC arc characteristics (*1* static; *2*, *3* dynamic) [6]

one, whereas when the current decreases it is lower, as illustrated by the dynamic characteristics of curves 2 and 3 in the same figure.

The higher arc voltage values occurring when the current increases are due to an ionization of the arc column lower than the normal one, so that a higher voltage gradient is needed to increase ionization to the value which allows the flow of the current; the opposite occurs when the current decreases.

Unlike what happens in DC, with AC current the arc values and temperature conditions vary at each instant, the arc must extinguish and restrike at each half cycle when the current passes through zero and the polarity of electrodes changes its sign at every half period.

Moreover, there is not a unique dependence of voltage on current, since the volt-ampere characteristic follows different curves at the increase and decrease of the current ("hysteresis" of AC arcs). The volt-ampere characteristic occurring when the current intensity varies so rapidly as to inhibit the establishment of equilibrium conditions, is called *arc dynamic characteristic*.

The arc dynamic characteristics are of the type shown in Figs. 3.45a, b, for relatively low and high currents values respectively.

These characteristic are markedly influenced by nature of electrodes, arc length and parameters of the external supply circuit.

In particular, Fig. 3.45a shows that: (1) when the current crosses zero, due to the gas deionization and cooling of electrodes, a voltage peek V_r is necessary for arc re-ignition; (2) owing to the above deionization and cooling, the voltage for burning the arc during the period of increasing current will be greater than that of the static characteristic (shown with dotted lines in the figure). It occurs a hysteresis phenomenon analogous to the one previously described for DC arcs; (3) after the maximum current is reached and the current starts to decrease, the ionization is in excess of that required for lower currents values, and the value of the required voltage is lower than the static one. When the current further decreases, the arc voltage tends to increase up to a peak value V_e (extinction peek voltage), in analogy to what happens in the static characteristics of DC arcs.

Since in arc furnaces for steel, Eq. (3.27) becomes:

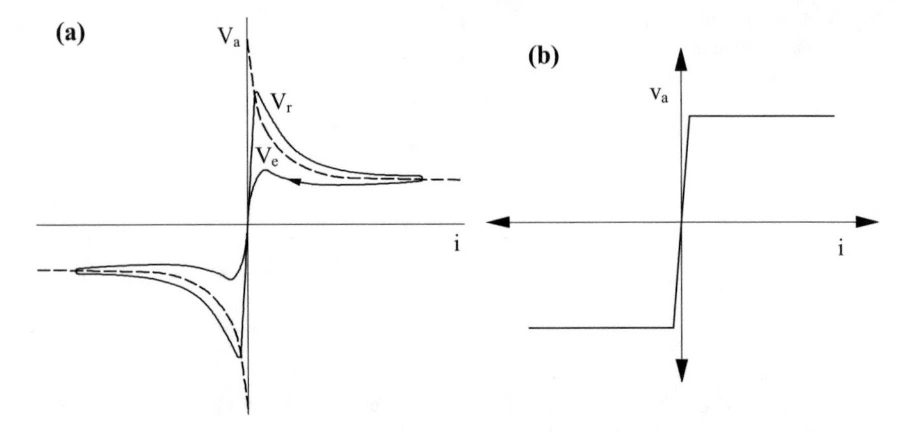

Fig. 3.45 AC arc volt-ampere dynamic characteristics (**a** for low arc currents; **b** for high currents) [6, 28]

$$V_a \approx 40 + 1 \cdot \ell_a,$$

with reference to the power $P_u = 3V_aI$ of a three-phase furnace, the arc length can be expressed as:

$$\ell_a \cong \frac{P_u}{3I} - 40.$$

To understand the mechanism of heat transfer from the arc to the vessel and the charge it's useful now to examine the balance of the energy transformed into heat in the arc column. The heat transfer from the arc occurs in the usual ways, i.e. conduction, convection and radiation. Taking into account the temperature of the arc, a considerable amount of heat, about 35–50 %, is transmitted by radiation; the remainder by convection, since conduction is negligible given the low thermal conductivity of plasma.

An example, in Fig. 3.46 is shown the energy balance in a low power arc. The figure highlights the fact that a considerable amount of heat is transferred by convection and radiation also towards the electrode and the walls of the vessel, amount that is greater the longer is the arc length.

Therefore the thermal efficiency of heat transfer from the arc to the bath, for arcs freely radiating on molten metal, decreases with the increase of arc length as shown, as an example, in the diagram of Fig. 3.47.

The experience has also shown that the arc column is not cylindrical, but tends to a conical shape, as illustrated in Fig. 3.48.

As a consequence, in the arc is present a radial component of the current density J_r, which interacts with the azimuthal component B_φ of the induction produced by the arc current; this produces a force $F_z = (J_r \wedge B_\varphi)$ acting on the conductive medium in the direction of the anode.

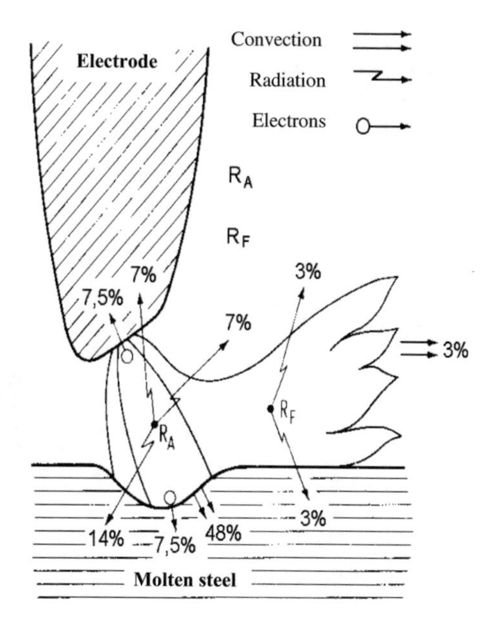

Fig. 3.46 Thermal energy balance of an arc 7 kA–143 V. (R_A radiation from the arc column; R_F radiation from the flame) [29]

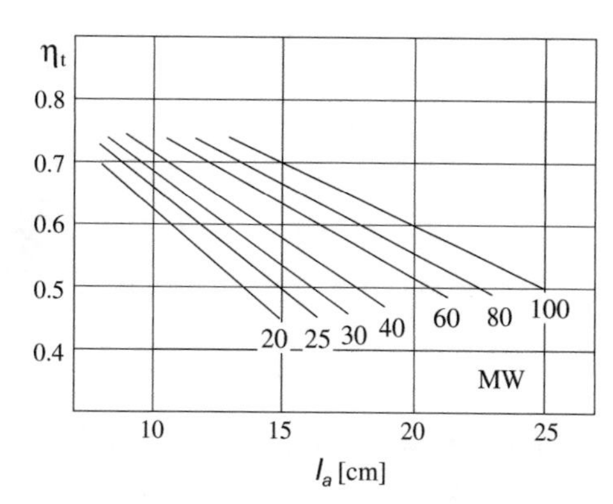

Fig. 3.47 Heat transfer efficiency from arc to molten bath as a function of the arc length ℓ_a [26]

Given that this force is proportional to J_r^2 and that the values of the current density are very high, it follows that an intense plasma jet is directed toward the surface of the molten metal.

In addition, on the arc current act also the components of the magnetic field created by the currents flowing in the other electrodes and in the melt (see Fig. 3.49).

The overall result is a deflection of the arc and a plasma jet in the direction of the sidewalls of the vessel accompanied by a rapid rotary motion at the electrode tip.

This phenomenon is less clearly defined in the half period in which the electrode is positive, but in any case, it gives rise to significant heat losses towards the vessel walls.

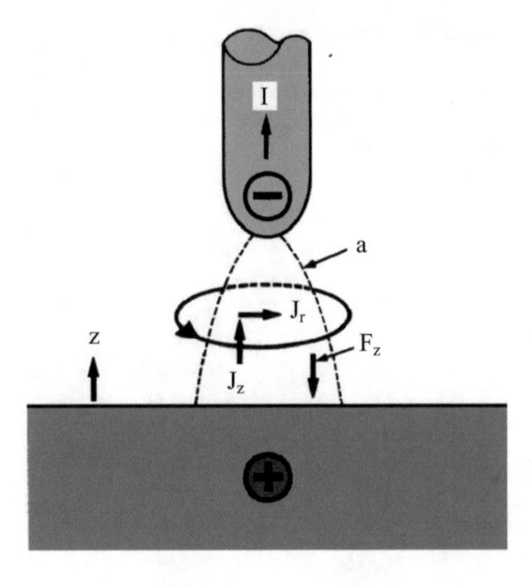

Fig. 3.48 Generation of arc jet in the arc column (*a* arc column) [17]

Fig. 3.49 Deflection of arcs towards sidewalls [17]

In recent years, this undesired phenomenon has been counter-balanced by the use of foaming slags which allow to work with "long" arcs and, at the same time, to improve heat transfer from the arc to the molten bath and reduce the damage of the vessel refractory lining.

3.3.4 Single-Phase and Three-Phase Symmetric Circuits with Arc

With AC supply, the wave forms of arc current and voltage are strongly influenced by the characteristics of the supply external circuit, as shown in the examples of Fig. 3.50 which refer to low-frequency and external circuits purely resistive or inductive.

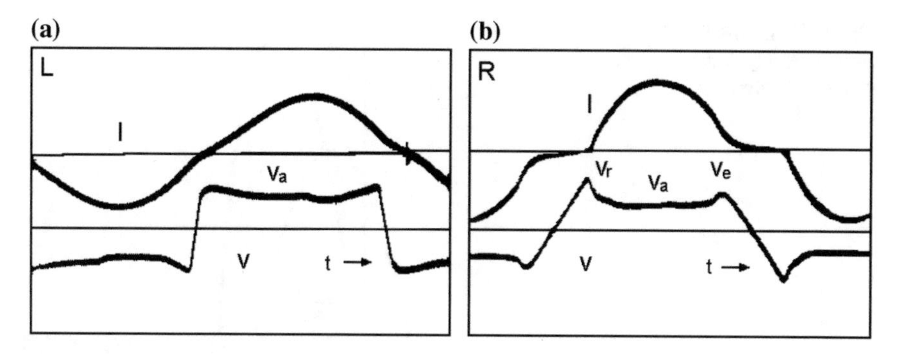

Fig. 3.50 Wave forms of AC arc voltage and current in case of current limitation by inductance (**a**) or by resistance (**b**) (v_r, v_e—see Fig. 3.45a) [28]

To analyze this influence and the waveforms of arc voltage and current, we make reference to the simplified equivalent circuit of Fig. 3.51, comprising a sinusoidal voltage supply $v = V_M \sin(\omega t + \varphi) = \sqrt{2} V \sin(\omega t + \varphi)$, a resistance R, an inductive reactance $X = \omega L$ and the arc gap.

Moreover, as shown in Fig. 3.52, as origin of times is taken the instant when the current is zero (i.e. $i = 0$ for $\omega t = 0$) and it is assumed that the arc voltage has an idealized rectangular wave shape which reverses its sign at each half period, neglecting the hysteresis phenomenon and the peek voltages at the arc re-ignition and extinction.

These conditions are typical for arcs with high thermal inertia as in arc furnaces for steel melting.

As known, with these hypotheses, in the circuit of Fig. 3.51 the arc voltage v_a must satisfy the differential equation:

$$L \frac{di}{dt} + R i = v - v_a \tag{3.29}$$

Fig. 3.51 Single-phase equivalent circuit with arc

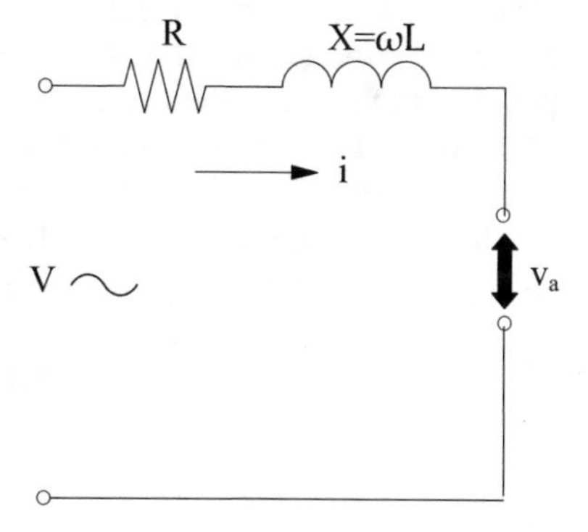

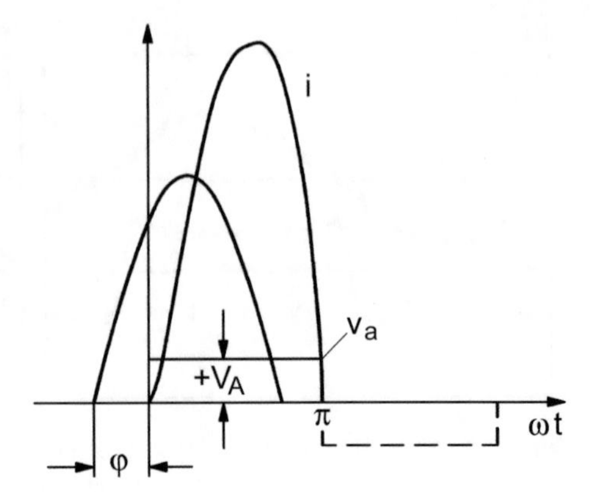

Fig. 3.52 Schematic of current, supply and arc voltage waveforms

3.3.4.1 Single-Phase Circuit with R = 0

With the further hypothesis that the resistance *R* is negligible in comparison with the reactance $X = \omega L$, for the current *i* the following relationship holds:

$$i = \frac{V_M}{L} \int_0^{\omega t} \sin(\omega t + \varphi) dt - \frac{v_a}{L} \int_0^t dt$$

from which, by integration, we obtain:

$$i = \frac{V_M}{X} [-\cos(\omega t + \varphi) + \cos\varphi] - \frac{v_a}{X} \omega t \tag{3.30}$$

Since in the spark gap acts a voltage equal to the sum of the supply voltage and the *e.m.f.* of self-induction, when the current passes through zero (i.e. at the arc extinction) the circuit becomes open, the self-induction e.m.f. goes to zero and the voltage across the spark abruptly jumps to the value of the supply voltage. According to the curves of Fig. 3.52, for a sufficiently large time interval from this instant to the instant of $v = 0$, i.e. a sufficiently large value of the angle φ, the supply voltage, which is increasing by a sinusoidal curve with reversed polarity, can be equal to or greater than $|v_a|$, and the arc immediately sparks in the reverse polarity. In this case, the current becomes continuous. Consequently, the presence of the inductance in the circuit tends to inhibit the interruption of the current and produces in the circuit the conditions of *continuous current flow*.

Then the arc ignition occurs only when $v \geq v_a$ (i.e. at time $t = 0$ in Fig. 3.52) at an angle φ greater than φ_o for which it is:

$$V_M \sin\varphi_0 = v_a \tag{3.31}$$

Moreover, assuming continuous current conditions, the subsequent current transition through zero occurs at $\omega t = \pi$.

By imposing the condition $i = 0$ for $\omega t = \pi$, Eq. (3.30) gives the phase angle between supply voltage and current:

$$\cos\varphi = \frac{\pi}{2}\frac{v_a}{V_M}; \quad \sin\varphi = \sqrt{1 - \frac{\pi^2}{4}\left(\frac{v_a}{V_M}\right)^2}; \tag{3.32}$$

Since it must be always $\varphi \geq \varphi_0$, from Eqs. (3.31) and (3.32) we can write:

$$\frac{v_a}{V_M} \leq \sqrt{1 - \frac{\pi^2}{4}\left(\frac{v_a}{V_M}\right)^2}$$

or:

$$\frac{v_a}{V_M} \leq 0.538 \quad \text{and} \quad \frac{v_a}{V} \leq 0.759 \tag{3.33}$$

These inequalities represent the conditions of current continuity, where the circuit is sufficiently inductive to allow the current flow also when the supply voltage is lower than the voltage v_a.

For $v_a/V_M > 0.538$, at the critical angle $\varphi_{cr} = 32°20'$, the conduction becomes discontinuous and the interval of conduction can be obtained from Eq. (3.30) by imposing the condition $i = 0$ for $\omega t = \delta$, i.e.

$$\cos(\varphi + \delta) = \cos\varphi - \frac{v_a}{V_M}\delta,$$

which highlights that the conduction angle δ decreases with the increase of the ratio v_a/V_M and the value of φ (which must always be greater than φ_o).

Typical waveforms, for continuous and discontinuous conduction, are shown in Fig. 3.53 for different values of v_a/V_M.

It's interesting to note that, due to the presence of the arc, the phase-angle between applied voltage and current is always lower than 90° and smaller the higher is the ratio of arc voltage to applied voltage.

In case of continuous conduction, introducing (3.32) in (3.30), the arc current can be written in the form:

$$i = -\frac{V_M}{X}\cos(\omega t + \varphi) + \frac{v_a}{X}\left(\frac{\pi}{2} - \omega t\right) = i' + i'', \tag{3.34}$$

which shows that it consists of two components (i′) and (i″): the first, sinusoidal, lagging in phase the supply voltage by $\pi/2$, with amplitude equal to the maximum value of the short circuit current; the second one, increases and decreases linearly with time at the same frequency of the applied voltage.

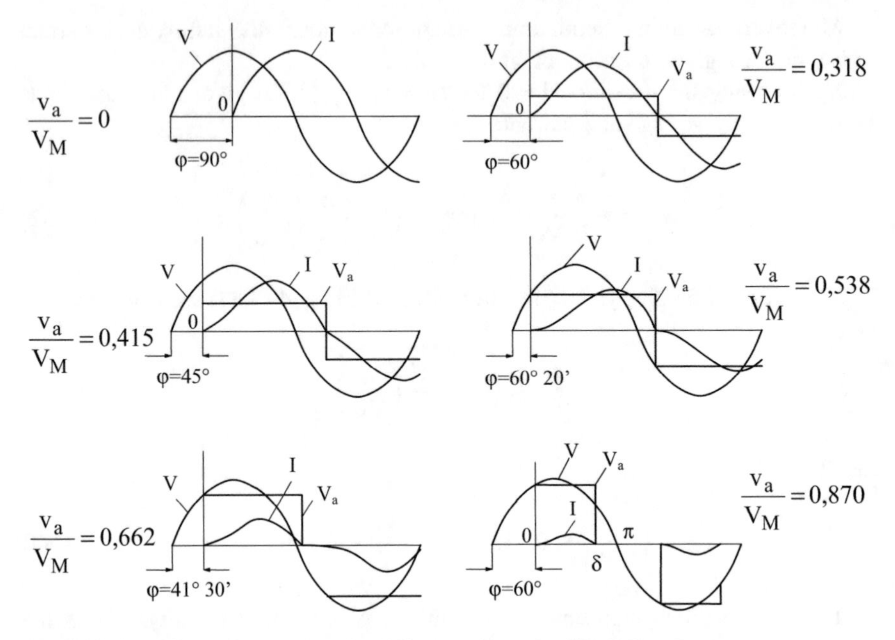

Fig. 3.53 Waveform of arc current for different values of v_a/V_M [30]

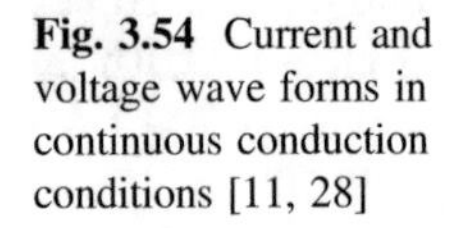

Fig. 3.54 Current and voltage wave forms in continuous conduction conditions [11, 28]

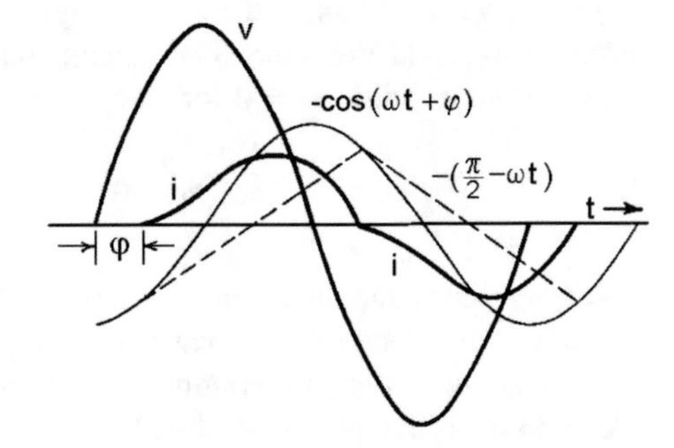

The arc current waveform with its two components and the applied voltage are plotted in Fig. 3.54.

From Eq. (3.34) it's easy to calculate the r.m.s. value of current and active power of the arc:

$$\begin{aligned} I &= \sqrt{\frac{1}{\pi}\int_0^{\pi} i^2\, d\omega t} = \frac{V}{X}\sqrt{1-\left(2-\frac{\pi^2}{12}\right)\left(\frac{V_a}{V}\right)^2} \\ &= \frac{V}{X}\sqrt{1-1.178\left(\frac{V_a}{V}\right)^2} \end{aligned} \tag{3.35}$$

$$P_{arc} = \frac{1}{\pi}\int_{0}^{\pi} v_a\, i d\omega t = 0.9\frac{V^2}{X}\left(\frac{v_a}{V}\right)\sqrt{1 - 1.234\left(\frac{v_a}{V}\right)^2} \tag{3.36}$$

Equations (3.35) and (3.36) show that in a given circuit, at constant supply voltage, it is possible to control both current and power through the variation the arc voltage, i.e. modifying the arc length. In particular, the current increases decreasing v_a.

Denoting by $I_{sc} = (V/X)$ the short-circuit current (i.e. electrode in contact with the charge: $v_a = 0$), the condition of continuous current flow (3.33) can be expressed as:

$$0.567 \le \frac{I}{I_c} \le 1.0 \tag{3.37}$$

According to Eq. (3.35), for the apparent and reactive power we can write:

$$\left.\begin{aligned} P_a &= VI = \frac{V^2}{X}\sqrt{1 - 1.178\left(\frac{v_a}{V}\right)^2} \\ Q &= XI^2 = \frac{V^2}{X}\left[1 - 1.178\left(\frac{v_a}{V}\right)^2\right] \end{aligned}\right\} \tag{3.38}$$

Denoting with $P_{asc} = Q_{sc} = V^2/X$ the values of P_a and Q with short-circuited electrode ($v_a = 0$), Eqs. (3.38) can be written in dimensionless form:

$$\left.\begin{aligned} \frac{P}{P_{asc}} &= 0.9\left(\frac{v_a}{V}\right)\sqrt{1 - 1.234\left(\frac{v_a}{V}\right)^2} \\ \frac{P_a}{P_{asc}} &= \frac{I}{I_c} = \sqrt{1 - 1.178\left(\frac{v_a}{V}\right)^2} \end{aligned}\right\} \tag{3.39}$$

The "power factor", calculated as the ratio between active and apparent power, therefore is:

$$\cos\varphi = \frac{P}{P_a} = 0.9\left(\frac{v_a}{V}\right)\sqrt{\frac{1 - 1.234\left(\frac{v_a}{V}\right)^2}{1 - 1.178\left(\frac{v_a}{V}\right)^2}} \tag{3.40}$$

The values of the ratios P/P_{asc}, P_a/P_{asc}, and P/P_a are given, as a function of v_a/V, by the dashed curves of Fig. 3.33.

They show that the active power has a maximum equal to 0.405 (in dimensionless units) for $v_a/V = 0.65$ and $\cos\varphi = 0.575$.

Further increasing $\cos\varphi$ from this value to 0.65 (limit beyond which starts the discontinuous conduction) the power decreases.

From Eq. (3.40), in continuous current flow condition, within an accuracy of few percent it is:

$$\cos\varphi \approx 0.9\left(\frac{v_a}{V}\right) \tag{3.41}$$

3.3.4.2 Single Phase Circuit with R ≠ 0

In a real supply circuit neither R nor X can be neglected. The analysis of such circuit with non-zero resistance and inductance leads to the following more complex equation for the current in case of continuous conduction [11]:

$$i = \frac{V_M}{Z}\sin(\omega t + \varphi - \Gamma) + \frac{v_a}{R}\left[\frac{2\,e^{-(R/X)\omega t}}{1 + e^{-(R/X)\pi}} - 1\right] \tag{3.42}$$

This equation shows that the arc current, plotted in Fig. 3.55, is non-sinusoidal and crosses zero later than in the circuit with R = 0. It consists of two components: the first one (*i′*), is a sine wave phase-shifted with respect to the arc voltage, with amplitude equal to the short circuit current, the second one (*i″*), is characterized by the coefficient (v_a/R) which changes sign at each crossing of zero of the current (*i*), and by exponentials with time constant (L/R).

Also in this case, as in the circuit with R = 0, if the ratio (v_a/V_M) increases above a critical value, the current becomes discontinuous. This critical value depends on the ratio X/R between the inductive reactance and resistance of the supply circuit, as shown by the curves 2 and 4 of Fig. 3.56, which refer respectively to the maximum value V_M and the r.m.s. value V of the supply voltage.

Curve 4 shows also that for values X/R > 3, the critical value $(v_a/V_M)_{cr}$ is very close to 0.537, i.e. the same value obtained in the previous paragraph for the circuit with R = 0 (or X/R → ∞).

Given that the ratio X/R is usually between 5 and 10, as first approximation, in most practical cases we can assume as continuity condition also for the circuit with R ≠ 0:

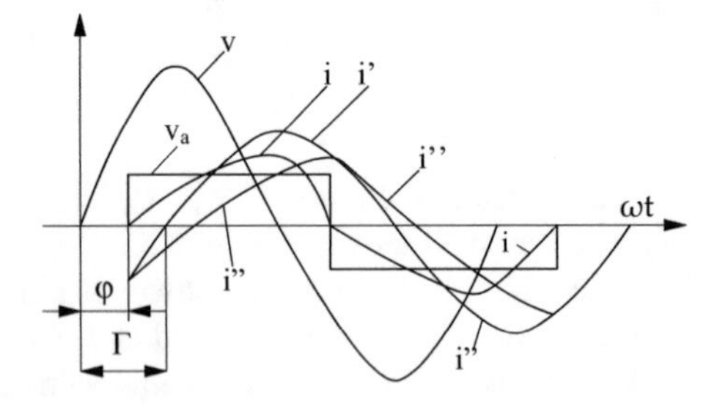

Fig. 3.55 Current and voltage wave forms in single-phase circuit with arc and R ≠ 0, in continuous current flow conditions [11]

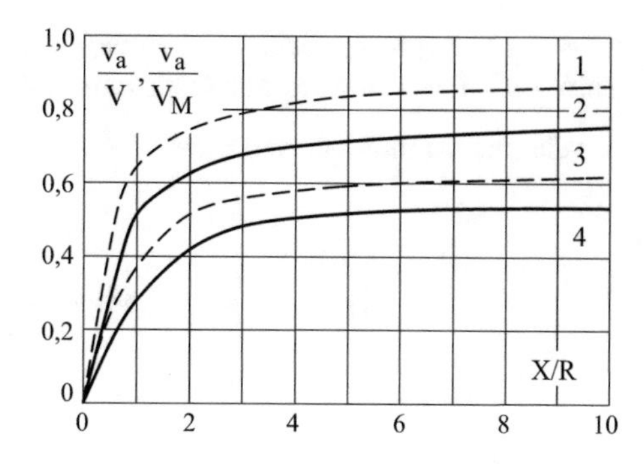

Fig. 3.56 Critical arc voltage of transition from continuous to discontinuous arc current, as a function of the ratio X/R (curves *1*, *3*: v_a/V and v_a/V_M for three-phase circuit; curves *2*, *4*: v_a/V and v_a/V_M for single-phase circuit) [11]

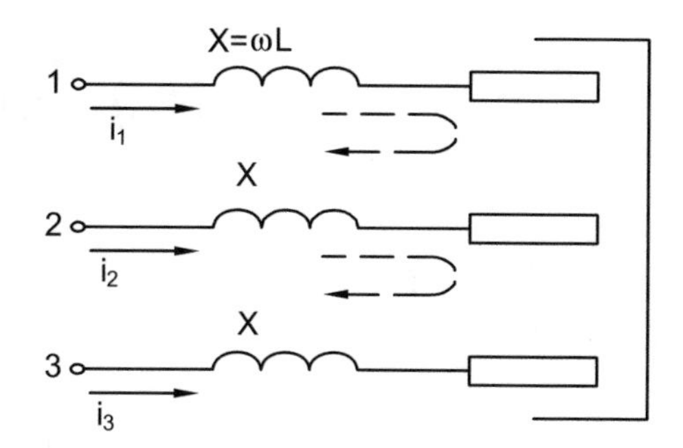

Fig. 3.57 Simplified equivalent circuit of three-phase furnace with $X \gg R$

$$\frac{V_a}{V_M} \leq 0.537$$

In some AC arc furnaces, where the arc burns with interruptions (like in small arc steel furnaces during the melting period), for reducing or eliminating the problem can be inserted in the supply circuit additional reactors (*chokes*), thus increasing the ratio X/R.

3.3.4.3 Three-Phase Circuit with R = 0

Similar results can be obtained for three-phase circuits. In particular, for the circuit of Fig. 3.57, in the hypotheses of three arcs burning between electrodes and molten pool, three phase symmetrical sinusoidal supply voltages, external supply circuit with identical reactance X in each phase, $R \ll X$, equal arc voltages

$v_a = v_{a1} = v_{a2} = v_{a3}$, and continuous arc currents in the three phases with phase shift 2π/3, the following equations can be obtained [16]:

- Current continuity condition:

$$\frac{v_a}{E_M} \leq 0.646; \quad \left(\frac{v_a}{E} \leq 0.914\right) \tag{3.43}$$

- Arc current r.m.s. value:

$$I = \frac{E}{X}\sqrt{1 - 0.966\left(\frac{v_a}{E}\right)^2} \tag{3.44}$$

- Total active power:

$$P = 3 \cdot P_1 = 3 \cdot 0.9\,\frac{E^2}{X}\left(\frac{v_a}{E}\right)\sqrt{1 - 0.975\left(\frac{v_a}{E}\right)^2} \tag{3.45}$$

with: $E = E_M/\sqrt{2}$—r.m.s. value of phase voltage.

The dashed curves 1 and 3 of Fig. 3.56, which refer to E_M and E respectively, are those valid for the circuit of Fig. 3.57. They show the change of the limit value for the continuity of the arc current, as a function of the ratio X/R, when the resistance R of each phase is not negligible with respect to the reactance X.

On the characteristics of Figs. 3.33 and 3.34 are shown the limits of operation with continuous current for single-phase and three-phase EAFs.

3.3.5 Non-symmetric Three Phase Circuits: Phasor Diagram, "Dead" and "Wild" Phases

During operation of three-phase arc furnaces, normally sudden and frequent load variations occur in the three phases, even for long periods of the working cycle, which are produced by the same melting process.

During these variations, different arc lengths, phase impedances and current intensities are present in each phase, thus making non-symmetric the supply power circuit and leading to a deterioration of the furnace operating characteristics.

The asymmetry may also be produced by inequality of the phase impedances of the secondary circuit, caused by different mutual inductances between the phases.

In fact, the *e.m.f.*'s of mutual induction and self induction and the active voltage drops due to the currents flowing in the phases, may produce a peculiar phenomenon by increasing or decreasing the impedance of each phase. This phenomenon modifies the values of currents and power in individual phases, reducing them in a certain phase and increasing in another.

In this case, even in ideal conditions of balanced load, there is an apparent "*transfer of power*" from one phase to another, which produces the phenomena of the so-called '*dead*' and '*wild*' phase. These phenomena will cause always a reduction of production rate and economy of operation and, sometimes, even furnace breakdowns.

The analysis of the influence of the *e.m.f.*'s of mutual inductance on the circuit impedances can be understood in a simple way through the vector diagram of Fig. 3.58a, which refers to the equivalent circuit of Fig. 3.58b and to the hypothesis of equal currents in the phases [11].

The vectors of the diagram represent the following quantities:

I_1, I_2, I_3	phase currents
R_1I_1, R_2I_2, R_3I_3	voltage drops in the resistances, in phase with the currents I_1, I_2, I_3
X_1I_1, X_2I_2, X_3I_3	voltage drops in the own reactances in phase quadrature with I_1,I_2,I_3
$\omega M_{12}I_2$, $\omega M_{13}I_3$,...	voltage drops due to mutual inductances, in phase quadrature with I_2,I_3
v_{a1}, v_{a2}, v_{a3}	arc voltages, in phase with I_1, I_2, I_3
V_{12}, V_{23}, V_{31}	phase to phase voltages

As it can be seen, in each phase the vector sum of the voltage drops in the mutual inductances can be decomposed into two components: one in phase with the phase current and one in quadrature to it.

The first component modifies the voltage drop in phase with the current, what is equivalent to the appearance of an additional resistance R_{add} (positive or negative).

The second one modifies the inductive voltage drop, which is equivalent to an additional reactance X_{add} in the phase (which has always a demagnetizing effect).

These additional impedances (which are also known as "*transfer impedances*") produce variations of phase currents and power, which is equivalent to a "transfer" of active power from one phase to another. This causes an uneven distribution of power between the three phases, which increases losses in the supply circuit and produces unbalance of the arc voltages.

With reference to the diagram of Fig. 3.58 and to the inequality of arc voltages that it highlights, it is called "*dead phase*" the phase with lower arc voltage and power, and "*wild phase*" the one with higher arc voltage and power.

The change of the phase sequence of the supply voltages, usually leads to the change in the direction of power transfer.

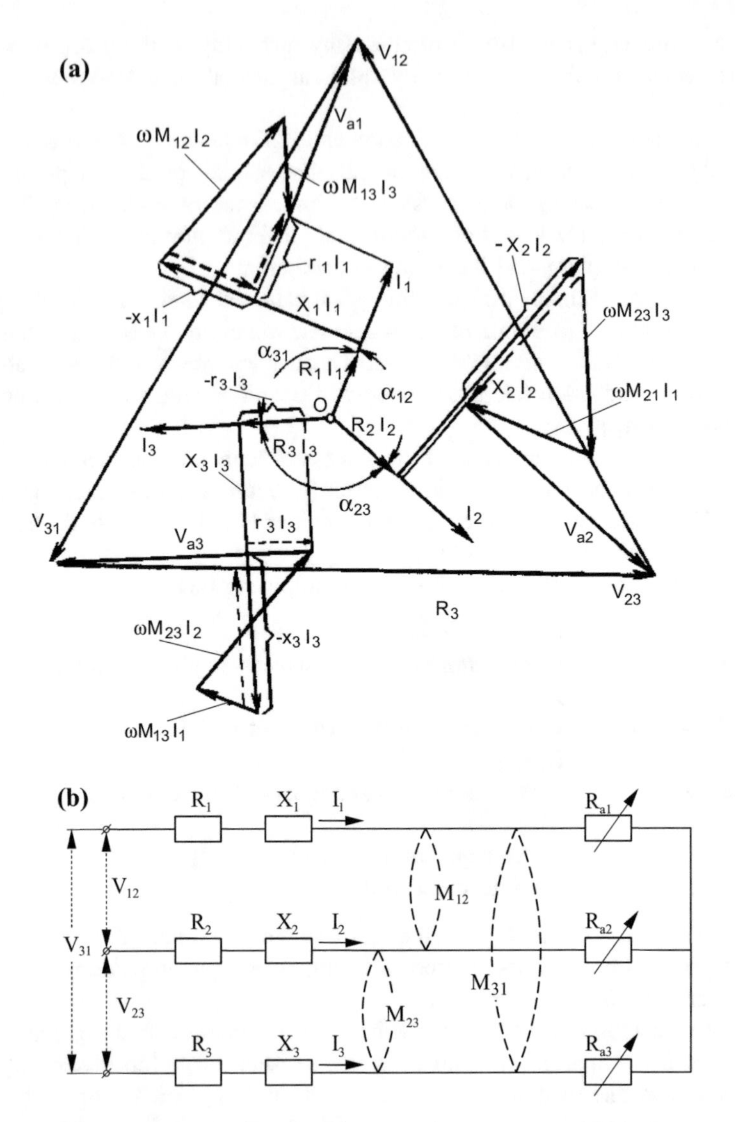

Fig. 3.58 a Vector diagram of voltages and currents in the equivalent circuit of figure; **b** Equivalent circuit of a three-phase electric arc furnace

Finally, even if the power transfer occurs only in some parts of the secondary circuit, it always causes abnormal operating conditions also in those parts where this transfer does not occur. Therefore, even though these phenomena theoretically do not reduce the total amount of energy released in the arcs of the three phases, they are always highly undesirable.

First, because the amount of metal molten in proximity of the "dead" phase is reduced much more than it is increased near the "wild" phase, so that the overall performance of the furnace is reduced.

Secondly, because the increase of power in the "wild" phase produces a more rapid wear of the adjacent refractory lining, leading to increased downtime with loss of production and repair costs for the renovation of the vessel refractory.

Elimination or reduction of these phenomena theoretically can be achieved with a symmetric circuit, with equal arc lengths, electrodes arranged at the vertices of an equilateral triangle, with the two-wire reactance compensation in the high-current secondary circuit and the closure of the triangle connection of three phases at the electrodes.

However, taking into account that the EAF power regulation is done by independent control of the arc lengths of the three phases, this ideal situation never occurs and the situation of normal operation is that of a dissymmetric circuit.

From the vector diagram of Fig. 3.58, the "*transfer*" resistance and reactance of phase 1 is:

$$\left.\begin{aligned} R_{1tr} &= \omega\left[M_{12}\,\frac{I_2}{I_1}\,\cos\left(\alpha_{12}-\frac{\pi}{2}\right) - M_{13}\,\frac{I_3}{I_1}\,\cos\left(\alpha_{13}-\frac{\pi}{2}\right)\right] \\ X_{1tr} &= -\omega\left[M_{12}\,\frac{I_2}{I_1}\,\sin\left(\alpha_{12}-\frac{\pi}{2}\right) + M_{13}\,\frac{I_3}{I_1}\,\sin\left(\alpha_{13}-\frac{\pi}{2}\right)\right] \end{aligned}\right\} \quad (3.46)$$

Similar equations can be written for phases 2 and 3 by rotation of indexes.

The corresponding "transfer" of active power and the variation of reactive power in each phase due to the demagnetizing action of the components of currents in adjacent phases are therefore given by the relationships:

$$\left.\begin{aligned} P_{itr} &= R_{itr}\,I_i^2 \\ Q_{itr} &= X_{itr}\,I_i^2 \end{aligned}\right\} \quad (i = 1, 2, 3). \quad (3.47)$$

The currents of different phase and their phase angles, can be obtained from the diagram of Fig. 3.58.

The following special cases of Eq. (3.46) can be of practical interest:

- *Symmetric circuit ($M_{12} = M_{23} = M_{31} = M$), phases at vertices of an equilateral triangle and unbalanced load:*

$$\left.\begin{aligned} R_{1tr} &= \omega M\left[\frac{I_2}{I_1}\,\cos\left(\alpha_{12}-\frac{\pi}{2}\right) - \frac{I_3}{I_1}\,\cos\left(\alpha_{13}-\frac{\pi}{2}\right)\right] \\ X_{1tr} &= -\omega M\left[\frac{I_2}{I_1}\,\sin\left(\alpha_{12}-\frac{\pi}{2}\right) + \frac{I_3}{I_1}\,\sin\left(\alpha_{13}-\frac{\pi}{2}\right)\right] \end{aligned}\right\} \quad (3.48)$$

Also in this case there is a "transfer" of power from one phase to another which gives rise to the phenomena of the "dead" and "wild" phase.

- *Non-symmetric circuit and balanced load ($I_1 = I_2 = I_3\, I$; $\alpha_{12} = \alpha_{23} = \alpha_{31} = 2\pi/3$):*

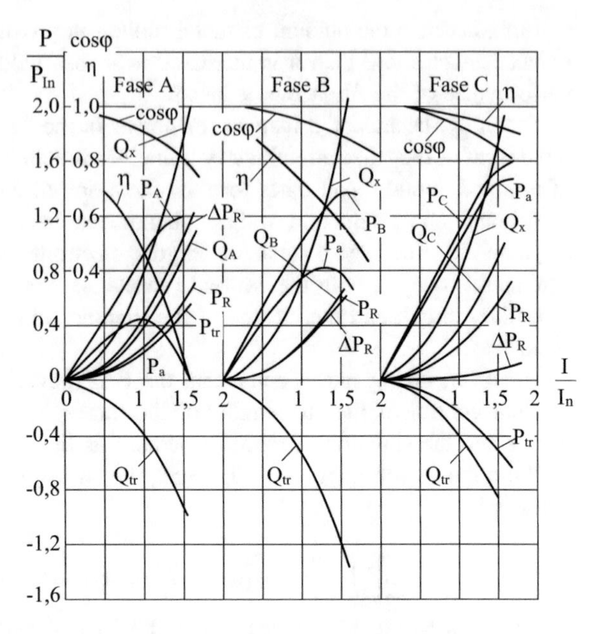

Fig. 3.59 Characteristics of a three-phase furnace with unequal mutual inductances and equal phase currents (P_a arc power; P_R power loss in the resistances R; P_{tr} 'transferred' power; $\Delta P_R = P_R + P_{tr}$; $P = \Delta P_R + P_a$; Q_X reactive power in the reactance X; Q_{tr} 'transferred' reactive power; $Q = \Delta Q_X + Q_{tr}$) [11]

$$\left.\begin{aligned} R_{1tr} &= \tfrac{\sqrt{3}}{2}\,\omega(M_{12} - M_{13}); \quad X_{1tr} = -\tfrac{1}{2}\,\omega(M_{12} + M_{13}) \\ R_{2tr} &= \tfrac{\sqrt{3}}{2}\,\omega(M_{23} - M_{12}); \quad X_{2tr} = -\tfrac{1}{2}\,\omega(M_{23} + M_{12}) \\ R_{3tr} &= \tfrac{\sqrt{3}}{2}\,\omega(M_{31} - M_{32}); \quad X_{3tr} = -\tfrac{1}{2}\,\omega(M_{31} + M_{32}) \end{aligned}\right\} \tag{3.49}$$

The characteristics of Fig. 3.59 refer to this case: they point out that, due to the power transfer, the values of power, efficiency and power factor, as a function of phase currents, can be very different in the three phases.

- *Non-symmetric circuit with balanced load and $M_{12} = M_{23} = M$ and $M_{13} = M'$ (co-planar arrangement of phases):*

$$\left.\begin{aligned} R_{1tr} &= \tfrac{\sqrt{3}}{2}\,\omega(M - M') \quad &;\quad X_{1tr} &= -\tfrac{1}{2}\,\omega(M + M') \\ R_{2tr} &= 0 \quad &;\quad X_{2tr} &= -\omega M \\ R_{3tr} &= -\tfrac{\sqrt{3}}{2}\,\omega(M - M') \quad &;\quad X_{3tr} &= -\tfrac{1}{2}\,\omega(M + M') \end{aligned}\right\} \tag{3.50}$$

In this case it is:

$$P_{1tr} = -P_{3tr}; \quad P_{2tr} = 0; \quad Q_{1tr} = Q_{3tr}.$$

Phase 1 is the 'wild' phase; phase 3 is the 'dead' one.

For the analytical solution of the unbalanced three-phase system with arcs of Fig. 3.58b the reader may refer to Ref. [25].

3.4 Power Control and Energy Balance of EAFs for Steel Production

3.4.1 Steel Production Cycle

In the last 40 years many improvements have been introduced in the construction of the EAFs and the production cycle in order to reduce tap-to-tap times and energy consumption. In order to better understand the influence of such innovations, the different phases of the production cycle will be described, even if nowadays, with the advent of the secondary metallurgy, some of them are no longer carried out in the furnace vessel or are performed in a different way.

Figure 3.60 shows the phases of a typical production cycle, commonly referred to as "tap-to-tap-cycle". We can distinguish the following main periods:

1. *Refractory lining maintenance*—Is the operation of re-lining and refractory maintenance of furnace bottoms and walls by gunning and fettling in order to assure that the refractory thickness is greater than the appropriate threshold value.
2. *Scrap charging*—as already mentioned, it is usually made by lifting and swinging away the suspended roof, the electrode arms and the electrodes together with their clamps around a vertical axis, to allow the loading of the scrap into the vessel from above.

In order to reduce thermal losses and increase production efficiency, the size of the vessel should be such as to contain the molten steel and the slag corresponding

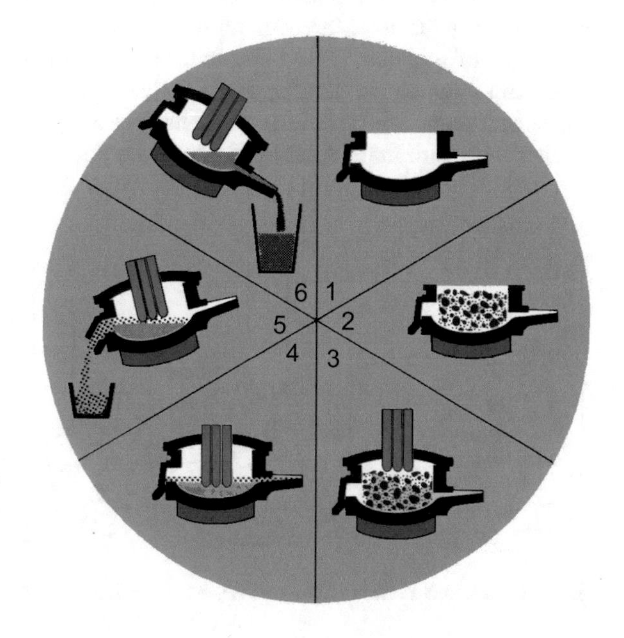

Fig. 3.60 Sequence of operations in a typical arc furnace melting cycle (*1* refractory maintenance; *2* scrap charging; *3* melting; *4* refining; *5* deslagging; *6* tapping) [17]

to the furnace melting capacity and to enable, with a scrap of normal density, to complete the melting cycle with a single intermediate charging operation; to this end the first charging "bucket" should correspond approximately to 50–60 % of the final melt.

In some modern furnaces of special design the whole melting cycle is completed with a single charge operation.

3. *Melting*—after the furnace is charged and the roof is in place, the operator lowers the electrodes, each controlled by its own regulator and mechanical drive. The melting process starts with low voltage (short arc) between electrodes and scrap.

During this period the arc is very unstable. In order to improve arc stability small pieces of scrap are placed in the upper layer of the charge.

This is the heaviest period of the cycle from the point of view of the electrical supply, since the arcs burn on a solid mass very heterogeneous and unstable; characteristic of this period are frequent and sudden variations of the power absorbed from the mains, which can be reduced but not eliminated by the power control system. Especially during the first minutes the situation is even more difficult for frequent short-circuits between electrodes.

After switching on, when the melting starts, the electrodes descend and penetrate into the scrap forming bores. The arc creates a crater on the surface and within the scrap volume (Figs. 3.61a, b); during this period the arc behavior is unstable and requires a rapid control of the energy input, through the vertical movement of the electrode arm holders which adjust arc lengths and therefore arc voltages

In this phase the scrap helps to protect the furnace lining from the high-intensity radiation from the arc. Moreover, during this period, it's very high the likelihood of collapse of scrap on the electrodes with formation of bridges that will short-circuit the electrodes; the rapid vertical movement necessary for arc length adjustment can result in electromechanical oscillations of the electrodes arm holders with undesirable perturbations on the power supply network.

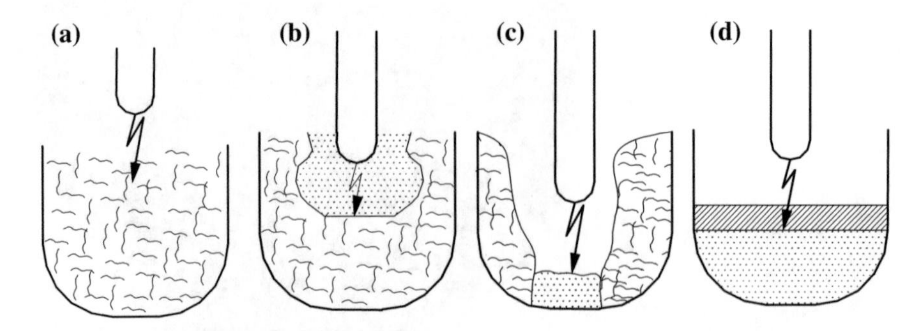

Fig. 3.61 **a** and **b** starting melting phase; **c** final melting phase; **d** refining and overheating

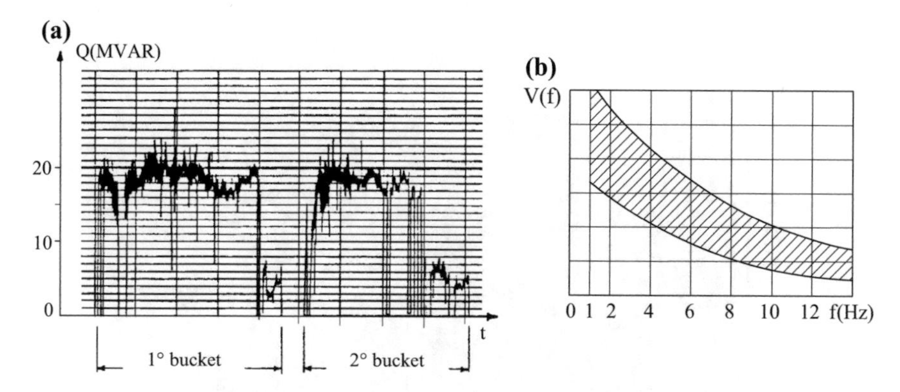

Fig. 3.62 **a** Reactive power absorption in melting two subsequent buckets; **b** Corresponding frequency spectrum

As it has been noted previously, when the electrodes are short-circuited, the active power is reduced to extremely low levels, while the reactive power increases dramatically.

The melting period is therefore characterized by:

- large and rapid variations of active and reactive power with very high peaks of the latter (Fig. 3.62a);
- large and rapid variations of absorbed current, with peaks consisting almost exclusively of reactive current (short-circuit between electrodes and scrap);
- large and rapid variations of power factor, which in certain moments tends to zero.

The frequency of variations of reactive power is different from instant to instant and from one type of furnace to another; it ranges generally between 1 and 20 Hz, with higher values in low power furnaces (Fig. 3.62b).

The current in this period can vary from zero to the short-circuit value.

In the final stage of melting a bucket, the arcs "burn" on a nearly complete metal pool, as shown in Fig. 3.61c, d, and the arc behavior is much more stable and calm.

In this phase, however, the walls of the vessel and the roof are not protected from the heat radiated by the arcs, as it was in the previous stage by the presence of the scrap. If foaming slags are not used in order to reduce the wear of refractory, it is necessary to shorten arc length and power to reduce radiation heat losses and avoid refractory damage and hot spots.

In the initial melting period the furnace works, on average, with active power equal to about 1.2 times that of normal operation and power factor between 0.70 and 0.85.

Figure 3.63 shows a typical diagram of active power absorbed in traditional old furnaces, where the refining process was carried out in the same melting vessel.

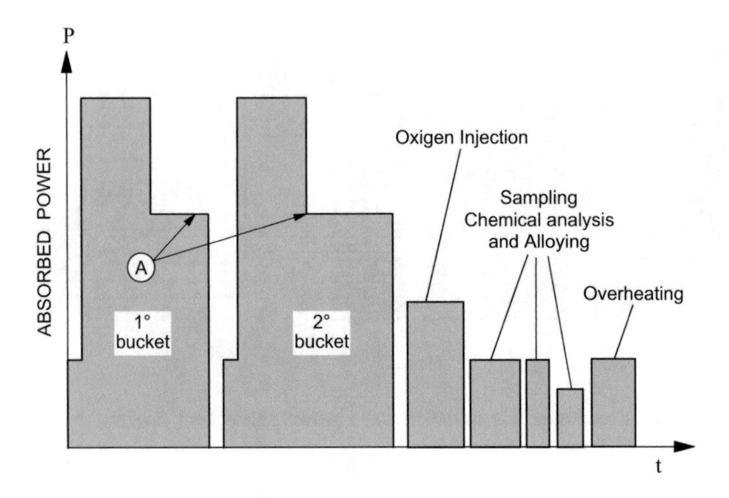

Fig. 3.63 Typical power input during a melting cycle for a traditional arc furnace (*A* reduction of arc voltage to avoid excessive heat radiation towards vessel walls) [17]

4. *refining*—during this period, are carried out the operations of reduction of carbon content by oxygen injection, sampling and analysis of the melt, addition of alloying elements to bring the steel to the desired final composition and overheating.

Oxygen is injected in the melt to oxidize the carbon in the steel or the charged carbon. The decarburization process is an important source of energy. It is now common that between 30 to 40 % of the total energy input to the EAF comes from oxy-fuel burners and oxygen lancing. In addition, carbon monoxide that evolves during this operation helps to flush nitrogen and hydrogen out of the metal. It also foams the slag, which helps to minimize heat loss and shield the arc, thereby reducing damage to refractories.

During the refining period the electrodes work on a flat molten pool, therefore there are not appreciable disturbances on the furnace operating conditions. The melt is sufficiently quiet and the control system can keep almost constant to the set value the instantaneous power developed by the arcs. The furnace works on average at 25–30 % of its rated active power with power factor between 0.8 and 0.9.

Taking into account the low utilisation factor during this period of the available installed power, in the last decades (see Fig. 3.2) the EAF has evolved into a fast low-cost melting unit of the scrap with the objective of improving product quality and reducing costs. To this end the most part of refining is carried out in a secondary Ladle Refining Furnace (LRF). This allows the EAF to concentrate on melting the scrap and removing impurities via oxidation reactions, while temperature and chemistry adjustments are carried out more optimally in a reduced power LRF.

During a melting cycle the two furnaces are operated simultaneously, thereby allowing an intensive use of the mains power and the reduction to about half of the tap to tap time.

The separation of melting and refining stages has led to a significant increase of installed capacity and hour production rate; in this way the production has been increased from about 1 to more than 2 t/h/MVA.

5. *De-slagging and Tapping*—For these operations, as already mentioned, in traditional furnaces the vessel is rotated around a horizontal tilting axis towards the working and de-slagging door (with an angle of about 15°) and, on the other side, towards the pouring spout (with an angle of about 40°). The movement is done by means of rockers and roller tracks, so designed that—in case of a failure—the vessel returns to its upright position automatically.

As mentioned in Sect. 3.2.8, different metal tapping schemes (bottom tapping and eccentric bottom tapping) have been developed which allow quick pouring of the melt and have strongly influenced the design of the furnace.

3.4.2 Innovations in Modern Furnaces

As shown in Fig. 3.2, a number of innovations have been introduced in the last 40 years in EAF operations in order to improve the performance and reduce the energy consumption of the steel production process.

Among them the most important are:

- *Foaming slag practice*

As illustrated in Fig. 3.61, at the start of meltdown the radiation from the arc to the side-walls is negligible because the electrodes are surrounded by the scrap. As the melting proceeds, the efficiency of heat transfer to the scrap and bath declines as more heat is radiated from the arc to the side walls.

By covering the arc in a layer of slag the arc is shielded and more energy is transferred to the bath; this has led to the introduction of the foaming slag practice. The slag is foamed by carbon monoxide (CO) which is produced by the injection of oxygen with granular coal or carbon. In some cases only carbon is injected, which reacts with the iron oxide in the slag to produce CO. In this way the foamy slag cover can be increased from 10 cm to about 30 cm thickness.

Slag foaming is usually carried out once a flat molten pool is achieved.

The use of the foaming slag allows to improve the thermal efficiency of energy transfer from arc to melt from 35–40 % without foamy slag to about 65–90 % (as shown in Fig. 3.64), depending on the thickness of the slag layer.

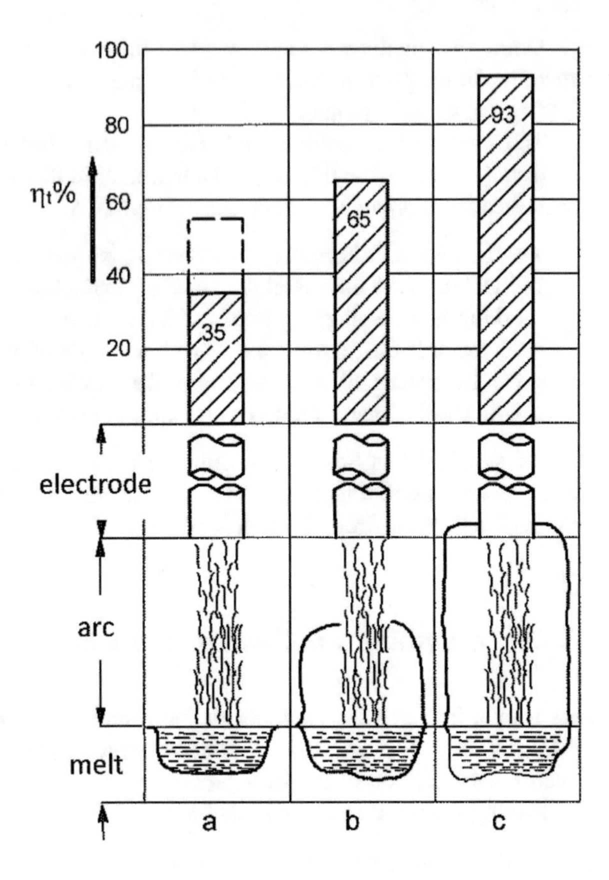

Fig. 3.64 Thermal efficiency of free burning arcs (*a*), arcs partially protected by foaming slag (*b*) and arcs fully protected (*c*) [27]

If a thick foaming slag layer is used, it is possible to increase the arc voltage considerably. This allows to operate with greater power input, lower electrode current and higher power factor.

- *Scrap preheating*

About 20 % of the total energy consumed in EAF steelmaking leaves the furnace in waste gases, with a loss of energy up to 130 kWh/ton.

A significant portion of this energy can be recuperated by using this gas to preheat the scrap.

Different technological solutions are available, which allow to reduce energy consumption by about 15 % over conventional EAF operations without scrap preheating.

- *CO post-combustion*

Currently, most of EAF furnaces used for steel production employ the post combustion technology to transfer heat to the molten steel bath. The technology involves injection of oxygen into the EAF to oxidize CO, which is already present in the furnace, to CO_2.

CO gas is produced in large quantities in EAFs both from oxygen lancing and slag foaming. If CO is not combusted in the furnace, typically it is burned in the gas collecting system end route to the bag-house. The purpose of post-combustion is to capture a significant amount of heat evolved from the oxidation of CO and transfer it into the melting process.

The heat of combustion of CO–CO_2 is three times greater than that of combustion of C–CO. Thus it represents a large potential energy source in EAFs. Electricity saving is estimated in the range of 50–100 kWh/ton of steel. Efficient transfer of the heat energy from post combustion gases to the molten steel bath can therefore reduce steel production costs and improve productivity [4].

There are many different systems of post combustion used in EAFs; all are based on the principle of burning CO (and H_2) above the bath when solid scrap is still high in the furnace. This because in this stage the heat transfer between hot combustion gases and scrap is more efficient, since the scrap is cold and the large surface area of the scrap promotes convective heat transfer, which is the most prominent mode of heat transfer from burners and post combustion.

- *Intensive usage of oxygen and carbon*

Much of EAF productivity gain achieved in the past decades is linked to the increased injection of oxygen and carbon into the furnace. It complements the electric arc energy input with additional exothermic energy created by the chemical reactions of fuel or gas, oxygen, and carbon injected into the furnace. About 30–40 % of the total energy input to the EAF is commonly given by oxy-fuel burners and oxygen lancing.

Multi-function tools for oxygen injection have become a standard for new EAFs. With these tools, oxygen can be injected in different ways, e.g. for decarburization with supersonic O_2 nozzles combined with carbon injection for foamy slag forming, in situ post-combustion with low velocity injectors, scrap melting boosting with oxy-fuel burners. These burners increase the effective capacity of the furnace by increasing the melting speed and reduce the consumption of electrode material. Additional temperature homogeneity benefits can be obtained by directing the heat to the “cold spots” in the furnace, which exist in the areas lying between electrodes on the periphery of the furnace bottom.

Moreover, the injection of oxygen helps to remove different elements from the steel bath, like phosphorus, silicon and carbon.

Typically all auxiliary equipment, that is burners, oxygen injectors and carbon powder injection for foamy slag are installed in the lower section of the side walls as shown in Fig. 3.65.

Oxygen lancing has also become an integral part of modern EAF melting operations; oxygen lances are used to cut scrap, decarburize (refine) the bath, and foam the slag. Energy savings due to oxygen lancing arise from both exothermic reactions (oxidation of carbon and iron) and the stirring of the bath, leading to its temperature and composition homogeneity.

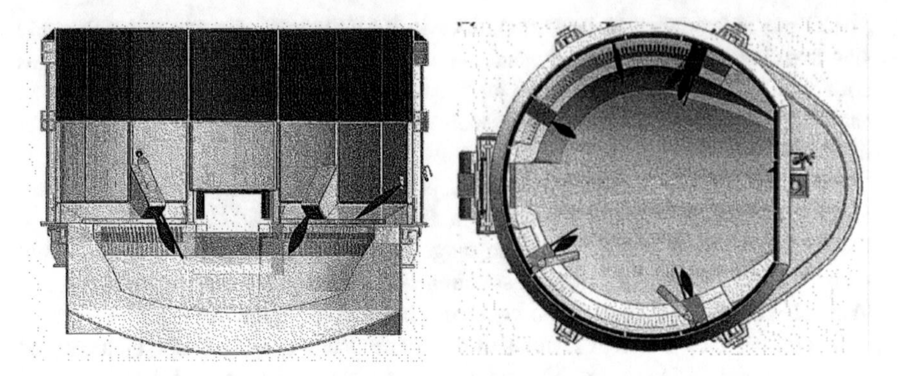

Fig. 3.65 Burners, oxygen injectors and carbon powder injection [31]

Fig. 3.66 Electrical energy versus oxygen consumption in EAFs [32]

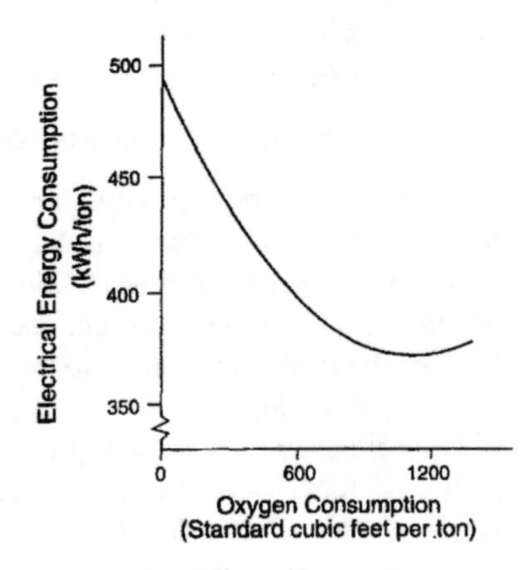

The relationship between electrical energy and oxygen consumption in EAFs is shown in the diagram of Fig. 3.66, while Fig. 3.67 gives a typical energy balance of a modern EAF.

Depending upon melt-shop operation, about 60–65 % of the total energy input is electric, the remainder being chemical energy arising from the oxidation of elements such as carbon, iron and silicon and the burning of natural gas with the oxy-fuel burners. About 53–57 % leaves the furnace in the liquid steel, while the remainder is lost into the slag, waste gases and cooling.

- *EAF bottom stirring*

In conventional arc furnaces there is little natural turbulence within the bath. Due to absence of stirring, large piece of scrap can take a long time to melt and may require the use of oxygen lancing. Inert-gas stirring, with argon or nitrogen, or bottom-stirring, agitates the bath, thus accelerating homogenization of chemical composition and temperature of the steel, as well as the chemical reactions between steel and slag.

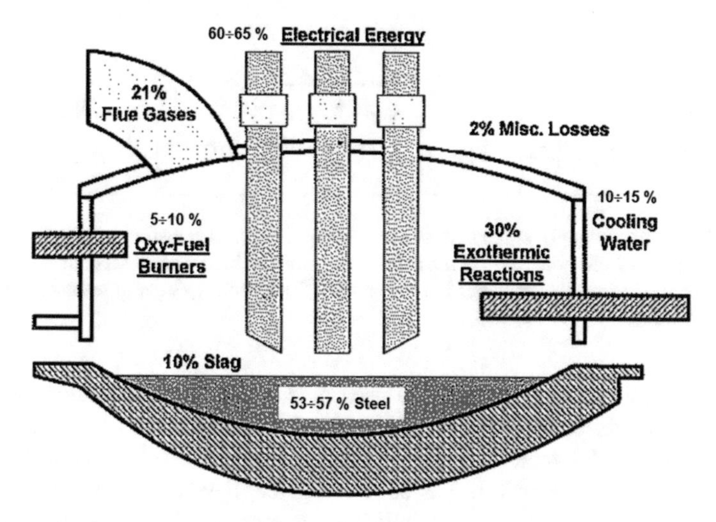

Fig. 3.67 Typical energy flow in steel a.c. EAFs [33]

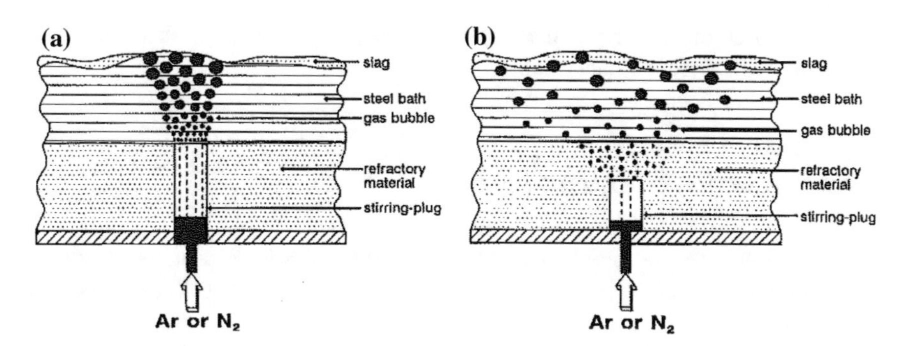

Fig. 3.68 Industrial bottom stirring systems in electric arc furnace: **a** direct-contact, **b** indirect-contact stirring-plug system [34]

Stirring with inert gas, argon or nitrogen, through the bottom of electric-arc furnaces allows:

- to eliminate temperature and concentration gradients
- shorten tap-to-tap times
- reduce refractory, electrode and power consumption
- improve the control of temperature and chemical composition of the steel.

Industrial systems for bottom stirring are either with direct contact plug or with indirect contact plug. Figure 3.68 illustrates these two systems.

In the direct-contact stirring-plug system, the stirring plugs extend through the refractory material of the hearth and are in direct contact with the steel bath into which the stirring gas is injected.

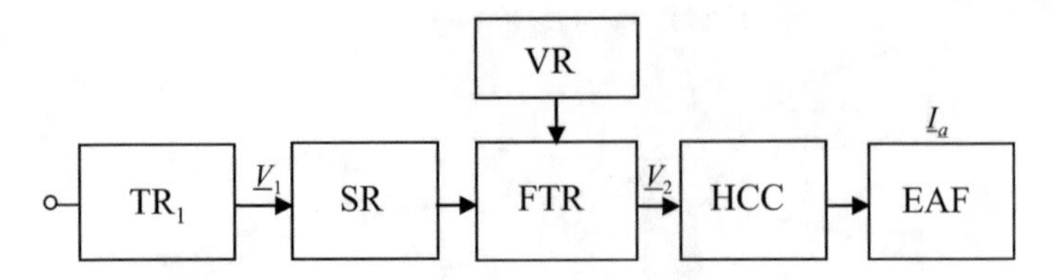

Fig. 3.69 Electrical supply system of an EAF. (*TR₁* step-down transformer; *SR* series reactor; *TRF* furnace transformer; *VR* voltage regulation system; *HCC* high current secondary circuit; *EAF* arc furnace; V_1 phase secondary voltage of step-down transformer; V_2 phase secondary voltage of furnace transformer; I_a arc current)

In the indirect-contact system, are used much shorter stirring plugs, which are covered by the refractory material of the hearth. This refractory material is a specially developed dry ramming mix, having a grain-size distribution such that the sintered hearth is permeable to gas but of sufficient density to prevent penetration by the liquid steel. As a result, the stirring plugs are not in contact with the steel bath, but the stirring gas can be introduced into the bath through the refractory hearth.

Note that in the indirect contact system a larger area of the bath is stirred as compared with direct contact one.

3.4.3 Selection of Operating Point

In order to analyze the problem of power control and selection of the operating point of the arc furnace it is necessary to consider the typical arc melting process of Fig. 3.60.

The main goals of the process optimization is to minimize the total melting time considered as the time period from loading the first bucket and tapping, and to increase process efficiency and productivity. The end of the melting and refining phases is based on temperature measurements and chemical analysis.

Let us consider again the electric supply system of a typical three-phase AC EAF and its simplified equivalent circuit (Figs. 3.69 and 3.31).

The working conditions of the EAF at any instant of the process are determined by the secondary voltage of the furnace transformer and the arc length.

The power control is obtained in two ways: the first—which allows only a step-wise regulation—by adjusting the position of the taps change switch or the star–delta connection, thus modifying the output voltage V_2 of the furnace transformer; the second one by changing the arc length (i.e. the arc voltage $V_a = R_a I_a$), through the vertical displacement of the electrodes in a convenient direction; this function is fulfilled by the automatic regulator.

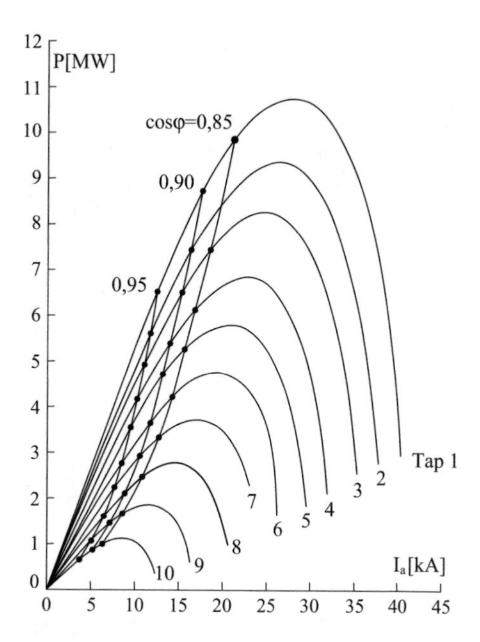

Fig. 3.70 Power characteristics of a 10 MVA furnace at different taps of furnace transformer [17]

However, as shown in Sects. 3.3.2 and 3.3.4, the characteristics of basic thermal and electric parameters of the furnace are usually given as a function of the arc current.

Therefore the current I_a has been chosen as basic parameter for continuous power regulation, selecting a different set value of it for each voltage step of the furnace transformer (Fig. 3.70).

Figure 3.71 shows the characteristics of a furnace for a given tap setting of the transformer. It highlights that to a disturbance consisting in a decrease of V_a corresponds an increase of the current I_a, which—for returning to the set power value—must be compensated by an increase of the arc length ℓ, i.e. by raising the electrodes.

The maximum of the arc current corresponds to the short-circuit condition when the electrodes are in direct contact with the charge.

$$I_{sc} = V_2/\sqrt{R^2 + X^2},$$

while the arc power reaches its maximum at the characteristic value I_{amax}.

Theoretically the operating point can be taken between I' and I'', and as indicated in previous paragraphs, and is usually selected close to the point of maximum arc power.

Given the shape of the characteristic, since the arc power undergoes relatively small variations in a relative large range of values near the maximum, it would be possible to choose an operating point either to the right or to the left of that point.

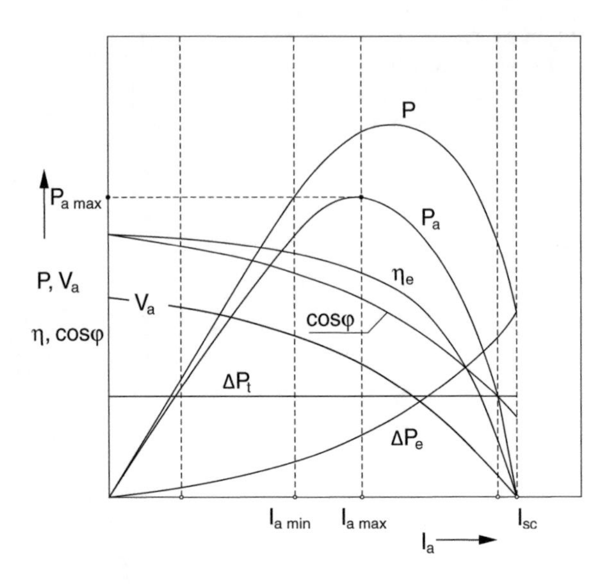

Fig. 3.71 Active power (P), arc power (P_a), power factor ($cos\ \varphi$), electrical efficiency (η_e), electrical and thermal losses (P_e, P_t) as a function of arc current (I_a) [35]

However it must be taken into account that to the right of the point of maximum:

- arc voltage and arc length are small and, as a consequence heat radiation towards the refractory is less intense;
- since the arc current is higher, it is higher also the consumption of electrodes;
- arc stability is greater since, due to the lower power factor, the arc re-striking voltage available after the current crosses zero is higher. Moreover, for the low value of power factor the arc current is higher for the same apparent power, and the heat capacity of the arc column is also higher, facilitating an easier re-striking of the arc because of the lower cooling of plasma.

On the other end, for points to the left of the maximum:

- the arc voltage is higher and arc length is longer. Then in the period during which the scrap is completely molten, if other measures—such as the use of foaming slag—are not taken, there is a higher heat radiation towards the refractory which will produce lower thermal efficiency;
- the power factor is higher, resulting in lower costs of the capacitor bank for power factor correction;
- the electrical efficiency is higher;
- the arc current is smaller, resulting in lower consumption of electrodes.

In conclusion, the main drawback of choosing an operating point to the left of point of maximum arc power, is the increased radiation towards the vessel walls and the resulting increased consumption of refractory.

This problem has been solved in modern furnaces, by the use of foaming slag and the lining of the side walls of the vessel with prefabricated, water-cooled panels, as shown in Fig. 3.23.

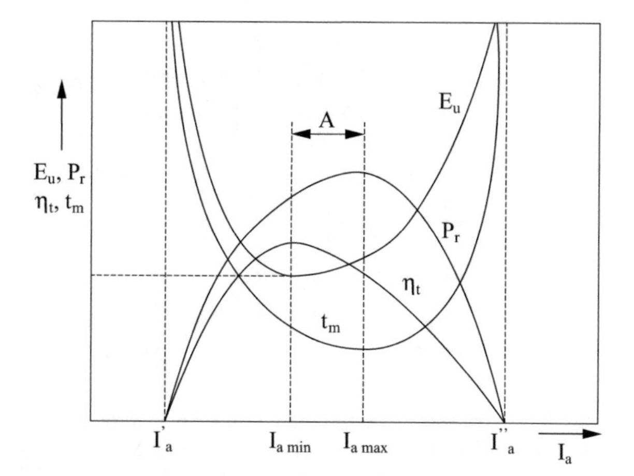

Fig. 3.72 Variations of specific energy consumption (E_u), productivity (P_r), total efficiency (η_t) and time for melting unit mass (t_m) as a function of arc current [35]

Therefore, for most arc furnaces the practical working area is selected on the left of maximum arc power, in the range between I_{amin}, where the process total efficiency η_t reaches its maximum, and I_{amax}, where at the maximum arc power also production rate is maximum. This range corresponds to the interval A in Fig. 3.72.

Only for some types of EAFs arc currents higher than I_{amax} are used.

However, the selection of the optimal operating point is even more complex, since it depends not only on the optimum value of arc current but also on other selected factors of the process, like for instance productivity *Pr,* melting time t_m per unit mass, electrical energy consumption, production costs, etc.

Generally, it is a good solution to take as optimum operating point the upper limit I_{amax} of the range A in Fig. 3.72. In fact, if the arc current increases from the value I_{amin} to the value I_{amax} the productivity increases and the melting time for unit mass decreases. Outside the interval $(I_{amin} - I_{amax})$ any variation of the arc current may cause unexpected simultaneous changes of all above mentioned factors.

3.4.4 Power Control Systems

Also the power control systems have undergone continued development over the years.

The first automatic regulators worked on the principle of maintaining constant the value of an electrical quantity, e.g. current, voltage or power.

A subsequent improvement of the arc furnace control systems was based on the measurement of the arc impedance to control the arc power, i.e. the electrodes are displaced varying the arc length in order to maintain constant, for each electrode, the ratio between arc voltage V_a (measured between electrode tip and melt, at the vessel bottom) and arc current (Fig. 3.73).

In this case, the controller works using the equation:

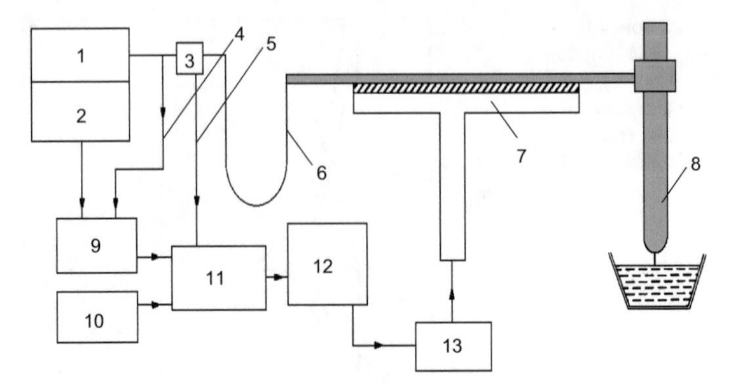

Fig. 3.73 Arc impedance controller (*1* main transformer; *2* tap changer; *3* TA; *4* voltage signal; *5* current signal; *6* flexible cables; *7* mast and arm; *8* electrode; *9* voltage tap compensator; *10* bias signal; *11* error signal detector and amplifier; *12* power amplifier; *13* prime mover) [17]

$$V_a - A \cdot I_a = 0 \tag{3.51}$$

with A—a constant, and must fulfil the following conditions:

- for $V_a - AI = 0$, the electrode remains still;
- for $V_a - AI > 0$, it moves downward and the current increases till fulfilment of the condition (3.50);
- for $V_a - AI < 0$, it moves upwards and the current decreases to a value fulfilling condition (3.50).

A further improvement was constituted by the so-called "*improved constant impedance control*" (or two constants control), based on the control equation:

$$V_a - A \cdot I_a - B = 0 \tag{3.52}$$

In this case, at the beginning of the melting period, in order to lower the electrodes without current ($I = 0$) it is sufficient to decrease the voltage V_a so that the difference $(V_a - B)$ goes close to zero and thus the lowering speed becomes small even if the value of V_a is still sufficiently high for the establishment of the arc.

Moreover, when the supply of the furnace is switched off ($I = 0$ and $V_a = 0$), it is $-B < 0$ and consequently the electrode does not remain stationary but moves upward to its upper position. This prevents the immersion of the electrode in the molten bath, thus avoiding cooling and pollution of the liquid steel.

Concerning the response of electrodes to a deviation from the set value, the control systems can be classified as proportional or non-proportional, depending on the fact that the speed of the movement response is proportional or non-proportional to the deviation from the set value.

Finally, a variety of electrode activators are used, which are based either on electric or electro-hydraulic systems. They must realize movement speeds of the electrode as high as 150 mm/s (9 m/min), and to be able to reverse the movement up and down in less than 150 ms.

However, the constant current or constant impedance controllers have the following disadvantages:

- when a variation of supply voltage occurs, the controller maintains the set point of the ratio between arc voltage and current, while the power of the furnace varies with the square of supply voltage variation;
- the controller reacts in the same way in both directions, for the same amplitude of deviation from the set value, while in practice values and number of deviations are not equal in both directions;
- the insensitivity of the controller may produce considerable differences in the controlled power, even with the same set value.

These phenomena may lead overall to variations of the energy supplied to the furnace during a melting cycle of about ±10–15 % and, consequently, to a great dispersion of melting times.

This drawback is eliminated by modern controllers which are operated under on-line computer control, which regulates the input of energy into the furnace in a way to satisfy the conditions:

$$\left.\begin{aligned} \Delta E_W &= \int_0^t (P - P_n)dt \rightarrow 0 \\ \Delta I \cdot t &= \int_0^t (I - I_n)dt \rightarrow 0 \end{aligned}\right\} \quad (3.53)$$

with: P, I, P_n, I_n—instantaneous values and set values of power and current respectively; ΔE_W—deviation of electrical energy from the set value.

With these controllers, which correspond to the block diagram of Fig. 3.74, the energy supplied to the furnace can be kept constant within ±2 % and the dispersion of tap-to-tap time decreases correspondingly.

3.4.5 Energy Balance of EAFs

In the energy balance of an electric arc furnace, some data are important reference numbers for EAF steelmakers. In particular:

- tap-to-tap time,
- electric energy consumption per ton of liquid steel,
- specific graphite electrode consumption
- specific consumption of refractory lining
- productivity, in tons per hour.

As already shown in Fig. 3.2, in the last decades impressive improvements of these parameters have been achieved, decreasing tap-to-tap time from an average level of more than 2 h in the middle of the sixties to about half an hour.

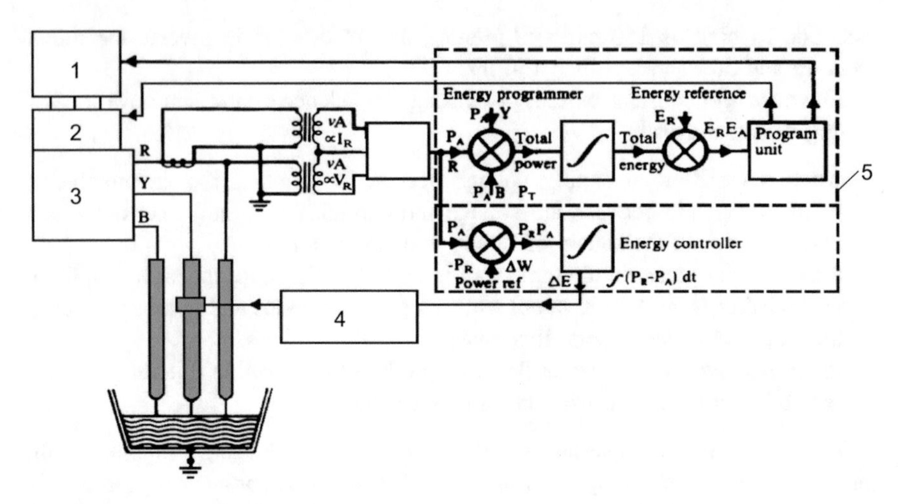

Fig. 3.74 Electrode position and furnace transformer controller with superimposed energy/power input control (*1* circuit breaker; *2* tap changer; *3* transformer; *4* electrode position controller; *5* energy/power input control unit) [17]

The electrical energy consumption was reduced more than two times from 630 to 300 kWh/t and electrode material consumption was also decreased from 6.5 to 1–2 kg/t of liquid steel (or even a lower value, i.e. about 0.9 kg/t, in DC arc furnaces) [3, 36].

All above factors justify the use of arc furnaces for steel production, which today cover approximately 40–45 % of the total annual steel production in the world.

Since high costs of scrap and energy are characteristic of the steelmaking process, in order to minimize total production costs, of particular importance in the production process are the costs of charge materials and energy.

The most important elements which influence the specific electric energy consumption of an electric arc furnace are the following:

- kind of charge (scrap only or mixture of scrap, pig iron and directly reduced iron)
- furnace characteristics (e.g. capacity, transformer apparent power, AC or DC current)
- heat recovery (from water or steam, scrap pre-heating)
- lining (consumption, surface of water cooled elements)
- additional chemical energy delivered to the electric arc furnace (fuel + oxygen, secondary burning of carbon oxide).

The total energy demand of an electric arc furnace can be described by different mathematical models. The most used model is the model of Kőhle, which is based upon linear regression of the average values from a big number of furnaces, taking into consideration a big number of process parameters [37, 38].

The reliability of results given by this model were checked for AC and DC EAFs with capacity in the range 64–147 t.

The evaluation of the melting process from the point of view of energy consumption and efficiency is done by the so-called utility factor η_U defined by the ratio

$$\eta_U = \frac{E_{wh}}{E_{wt}}$$

of the energy E_{wh} used for increasing the liquid steel enthalpy to the total energy E_{wt} delivered to the electric arc furnace from different sources in the same period of time.

In fact, in modern arc installations, besides basic electric energy, a number of additional energy components are used. In particular:

- energy from chemical reactions in the melt,
- from burning natural gas,
- from oxidation of graphite electrodes,
- heat of the pre-heated hot scrap,
- heat recovered from CO post-combustion.

In conventional AC electric arc furnaces the value of the utility factor is in the range η_U =0.5–0.6, while in modern EAFs it can reach slightly higher values.

A typical mass and energy balance for an AC electric arc furnace is shown in Fig. 3.75 [39].

Similar results for other types of electric arc furnaces can be found in Ref. [18, 40].

The analysis of different energy balances highlights once more that the steelmaking process in modern EAFs is not a pure arc heating process but a combined

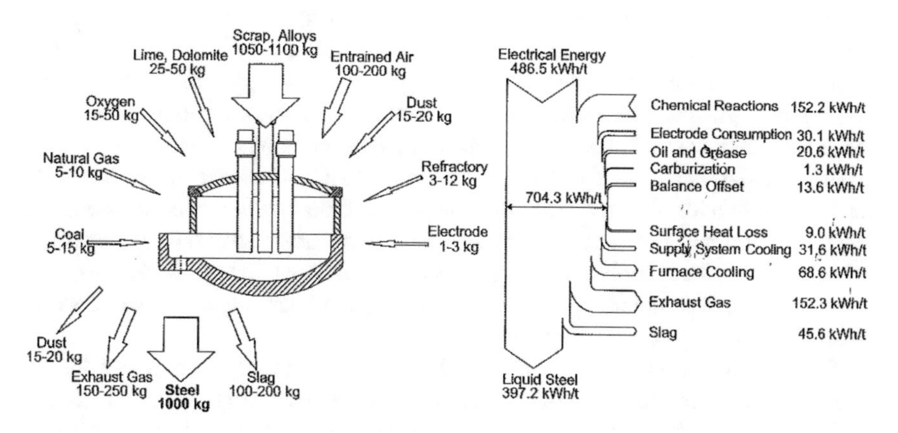

Fig. 3.75 Typical mass and energy balance of an EAF [39]

electric-fuel process, where electric energy is the main source with a percentage which can range from less that 50 to more than 65 % of the total energy consumption.

The per cent of the other sources on the total supplied energy in average are:

- 15–22 % from chemical reactions,
- 2–6 % from gas-oxygen burners,
- 22–24 % from coal burning
- about 10 % from burning organic compounds
- 3–4 % from electrode consumption.

However, the total energy demand varies from one furnace to the other and depends on type of scraps and methods of carrying out the process. Its value is usually in the range from 700 to 800 kWh/ton.

Only about half of the total supplied energy (sum of electrical and chemical energy) is used directly in the melting process for increasing the enthalpy of the charge.

The remaining part is generally lost in:

- gas emission (10–20 %),
- dust and slag (4–6 %),
- cooling systems (5–13 %),
- radiation losses (2–8 %)
- miscellaneous including electrical (10 %).

In some installations a small fraction of these losses can be recovered [41].

3.5 Interference of EAFs on the Supply Network ("Flicker")

AC and DC arc furnaces represent one of the most intensive disturbing loads in sub-transmission or transmission electric power systems; they are characterized by rapid changes of the absorbed power that occur especially during the melting-down initial phase when the electrodes bore into the scrap and the arc behavior may vary instantaneously from short circuit to open circuit conditions.

In particular, in DC arc furnaces the presence of AC/DC static converters and the random motion of the arc with nonlinear and time-varying characteristics are responsible for dangerous perturbations such as waveform distortions and repetitive fluctuations of the supply voltage (voltage fluctuations). These variations may cause many power quality problems on the high voltage supply network which affect the power system performance.

The diagrams of Figs. 3.36 and 3.76 show that even modest changes of the active power ΔP around the working point, produce very large variations ΔQ of reactive power, with consequent voltage drops in the supply network which, in turn, give rise to such fluctuations. The maximum instantaneous current absorption

Fig. 3.76 Active and reactive power and cos φ variations from operating point to short-circuit

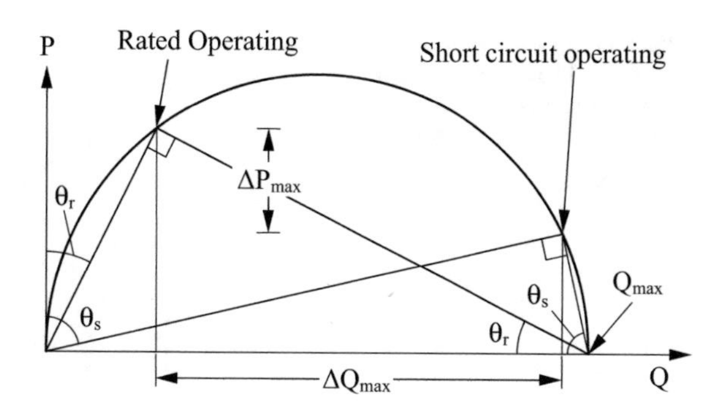

occurs in short-circuit conditions, which are frequent during the melt-down phase; it depends, in turn, on the electrical characteristics of the furnace supply circuit, in particular on its equivalent total reactance. Voltage drops and voltage fluctuations are therefore larger when the arc working conditions jump from the rated operating point to short circuit.

In Fig. 3.62a, b have been already given typical diagrams of reactive power absorption in a furnace during the melting period and the harmonic spectrum of the corresponding voltage fluctuations.

The above voltage fluctuations, even if they remain within contractual limits, lead to disturbances of various kinds because of their periodicity, which usually is in the range between 1 and 20 Hz.

Since the EAF is connected to a high-voltage supply network, these fluctuations have an impact on the whole distribution network, affecting other consumers fed from the same grid. In particular, they can cause sudden flashes of luminosity in fluorescent and filament lamps and disturbances in other electrical equipment.

It's normal practice to limit these fluctuations within limits that avoid inconveniences to other voltage-sensitive users (e.g. loss of synchronism of television, spurious pulses to digital equipment, light flicker on the human eye, etc.).

The appliances most sensitive to voltage fluctuations are filament lamps, since the resulting changes in their light intensity can cause an unpleasant physiological sensation (nuisance) to persons subjected to such light variations.

The filament lamps transform voltage fluctuations in variations of light intensity according to a law that, for small voltage fluctuations, can be expressed by the relationship:

$$\frac{\Delta I_L}{I_{nL}} = k\,\frac{\Delta V}{V_n} \tag{3.54}$$

with:

I_{nL} light intensity of the lamp at its nominal voltage V_n

ΔI_L, ΔV variations of light intensity and supply voltage

k coefficient ("*gain factor*"), depending on type of lamp and frequency of voltage fluctuations, shown as an example in Fig. 3.77

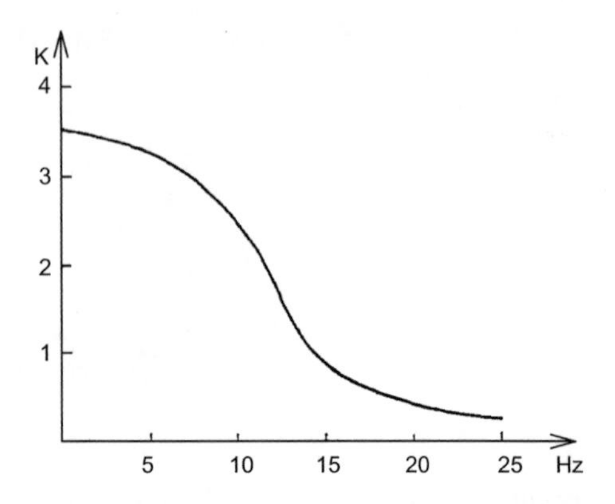

Fig. 3.77 Typical gain factor k of 60 W filament lamps as a function of frequency of fluctuations of supply voltage

In particular, in the lower frequency range, the values of k are typically of about 2.5–3.5; it follows that the filament lamp produces light intensity fluctuations considerably higher than voltage fluctuations, and that this kind of lamps is the appliance most sensitive to this type of variations of supply voltage.

The phenomenon associated with these rapid fluctuations of light flux, is named "*flicker*".

The problem of establishing limits of permissible level of flicker requires the evaluation of personal nuisance and the identification of the link between the parameters of the arc furnace installation and those of the high voltage supply network.

As regards the evaluation of nuisance, it should be noted that the determination of the permissible level of flicker cannot be absolute, being linked not only to the human element that can experience in different ways light fluctuations, but also to the type of lamps subjected to the voltage variations.

An indication of perceptibility of flicker is given by the curves of Fig. 3.78. They show that the human eye has a selective response dependent on the frequency of light periodic fluctuations, with maximum sensitivity around 8–10 Hz, frequencies at which also fluctuations of amplitude greater than 0.3 % can be annoying.

Results of researches and experiments performed varying independently frequency (f), amplitude (a) and duration (T) of the disturbance (flicker), have shown the following:

- if two types of flicker (a_1, f_1 and a_2, f_2) are adjusted to yield equal discomfort to one observer, than other observers would also experience the same discomfort;
- the intensity of discomfort perceived is proportional to the square of the amplitude (a) of fluctuations, i.e. the experiments have indicated equal discomfort when the integral

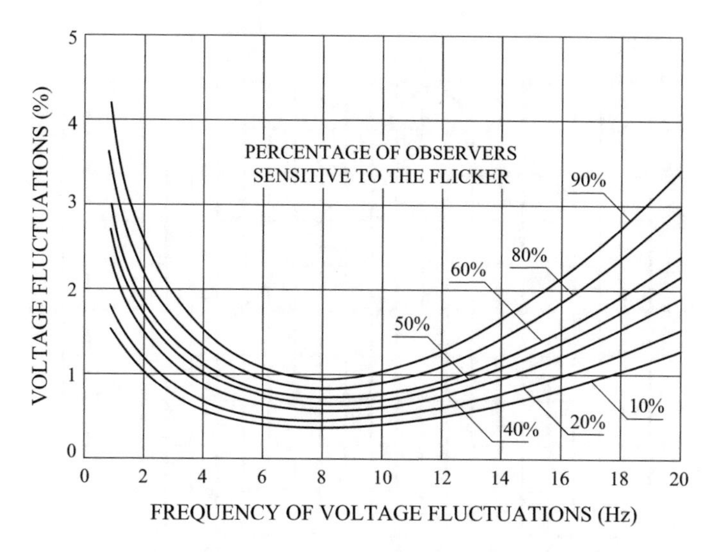

Fig. 3.78 Per cent periodic variations of voltage at which flicker becomes annoying as a function of frequency

$$\int (a)^2 dt$$

has the same value.

This has led to define a quantity D (*dose*) which expresses the cumulative amount of discomfort caused to an individual by the variation of luminous flux in a given time interval T, through the relationship:

$$D = \int_0^T (a_{10}\%)^2 dt \tag{3.55}$$

where:

$a_{10}\% = (a_f\%) \cdot k_f$ percentage voltage variation referred to the reference frequency of 10 Hz;

k_f coefficient defined as the ratio of the amplitude of a sinusoidal variation of luminous flux at frequency '*f*' to the amplitude of the variation at frequency 10 Hz producing equal discomfort (this coefficient is given in Fig. 3.79)

Since, as shown in Fig. 3.62, the voltage fluctuations caused by EAFs undergo strong variations during the melt-down stage, it's necessary to make a significant period of recording flicker for the evaluation of its statistical behavior.

Moreover, given that the fluctuations have very different characteristics during melt-down and refining periods, and may vary from one cycle to another, since they

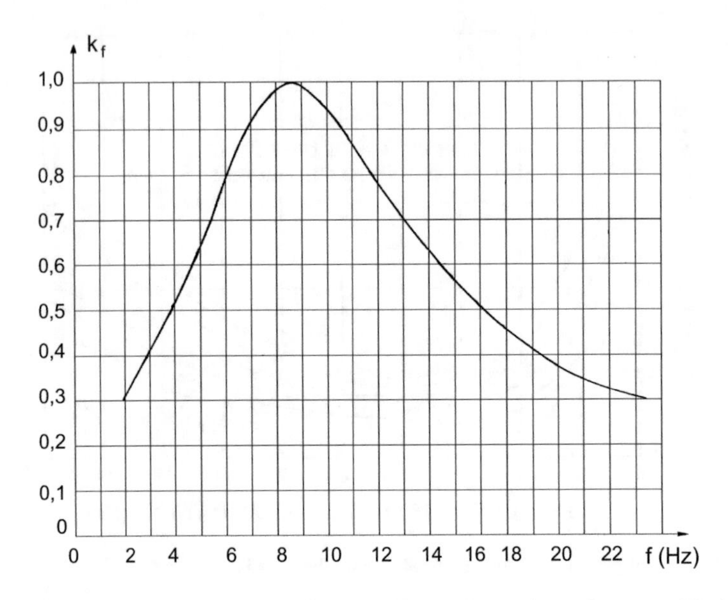

Fig. 3.79 Ratio of amplitudes of fluctuation at 10 Hz and fluctuation at frequency 'f' giving the same dose of discomfort D

Fig. 3.80 Simplified equivalent circuit of the supply of an AC/EAF for evaluation of voltage fluctuations

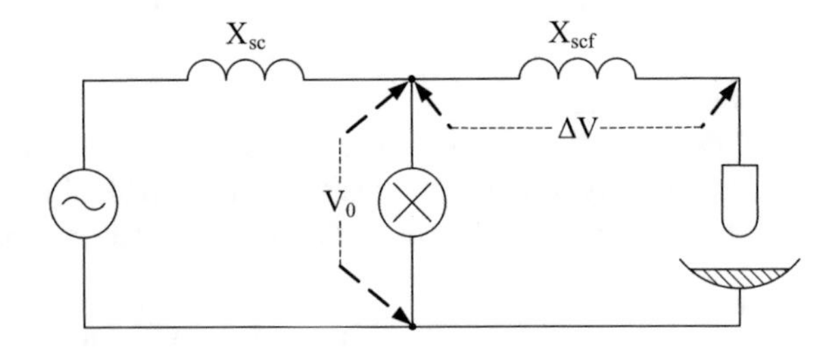

depend upon a number of parameters (e.g. quality and quantity of scrap used, operating point, quantity of injected oxygen, unpredictable consequences of collapse of the scrap during melting), the flicker evaluation should be done both during short periods (5–15 min) and long ones (at least some tens of process cycles).

The link between the parameters characterizing the furnace and those characterizing the power supply network is obtained in a simplified way by analyzing the maximum voltage fluctuation occurring at the transition from no-load to short-circuit conditions.

With reference to the circuit of Fig. 3.80, and having denoted with:

V_0	supply voltage
ΔV	variation of supply voltage from no-load operation to short-circuit
$\Delta V/V_0$	voltage fluctuation
P_{asc}	short-circuit power of supply network at the furnace connection point

P_{ascf} short-circuit power of the furnace (with short circuited electrodes),

it is:

$$\frac{\Delta V}{V_0} = \frac{P_{ascf}}{P_{asc}} \tag{3.56}$$

The experience has shown that, in case of a single furnace connected to a supply network, to keep the flicker below the limit of discomfort, the ratio P_{asc}/P_{ascf} at the point of connection should be greater than 40.

On this basis an empirical relation has been established that predicts the "*probable flicker value*" (p_{fl}) when connecting an arc furnace at a node A of a supply grid without creating flicker problems: [42]

$$p_{fl} = k_{fl} \frac{P_{ascf}}{P_{asc}} \tag{3.57}$$

with: $40 < k_{fl} < 80$.

More complex is the assessment of flicker when more furnaces are connected to the same node, since in these conditions the resulting flicker is due to statistical probability that the disturbances due to the individual furnaces will sum up [43, 44].

For flicker mitigation, the most used structures are reactive power compensators connected in parallel with the load, as schematically shown in Fig. 3.81.

In particular they are:

- rotating synchronous compensators—neglecting costs, their features are not entirely appropriate, due to their relatively long response times and inability to compensate unbalanced loads;
- saturable reactors;
- static var compensators (SVC);
- static synchronous compensators (STATCOM).

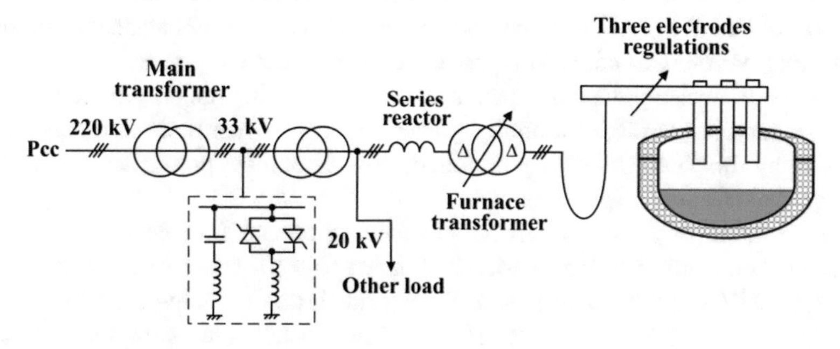

Fig. 3.81 Classic AC/EAF supply with reactive power compensator

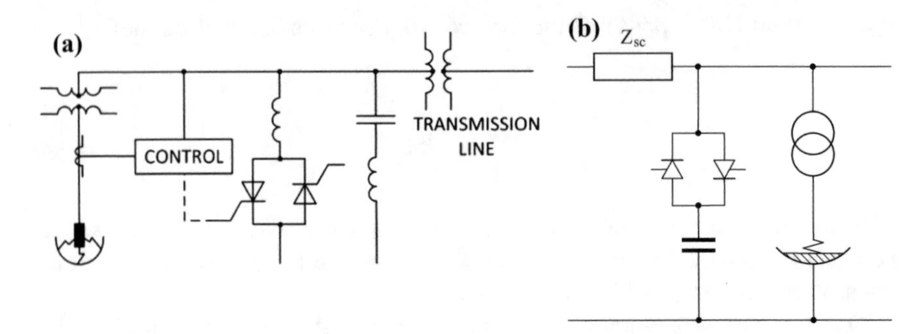

Fig. 3.82 Static Var Compensators (SVC) with: **a** thyristor controlled reactor; **b** thyristor switched capacitors

The last two solutions are nowadays the more used since, in addition to mitigate flicker, they also allows to control the power factor (averaged in short time intervals) and the unbalance of reactive load [45–47].

The shunt connected Static Var Compensator (SVC) is a well mature technology used extensively to control AC voltage. SVCs are arranged, as shown in Fig. 3.82, either as thyristor-controlled reactors (TCR) or thyristor-switched capacitors, or their combination, with fixed passive filters for eliminating dominant harmonics generated from electronic switching phenomena.

The susceptance of the TRC is controlled by varying the thyristor's firing angle, but their operational efficiency is inherently limited by the ability to respond rapidly to the fluctuating arc furnace load and the possibility of compensating only reactive power.

The main elements of a Static Syncronous Compensator (STATCOM) are schematically illustrated in Fig. 3.83.

Whereas SVCs operate by selectively connecting passive components (inductors or capacitors) to the power line, the STATCOM is essentially a controlled AC voltage source connected to the power line by a suitable tie reactance.

By controlling the STATCOM voltage source, any desired current can be forced to flow through the tie reactance.

Connected to the AC supply of an EAF, STATCOM can thus supply those components of the EAF load comprising non-sinusoidal, unbalanced, randomly fluctuating currents, in addition to the fundamental reactive power.

These are precisely the components associated with the flicker production when they are supplied through the utility power network. When these components are supplied by the STATCOM they no longer flow though the power network and the voltage flicker is drastically reduced.

An example of flicker reduction obtained in an arc furnace by STATCOM compensation is given in Fig. 3.84, which refers to a 100-ton furnace with nominal power 80 MVA, supplied from a 138 kV grid through a step-down transformer feeding the furnace at about 15 kV. The installation comprises a ±80 MVA STATCOM with a shunt coupling transformer connecting it to the 15 kV

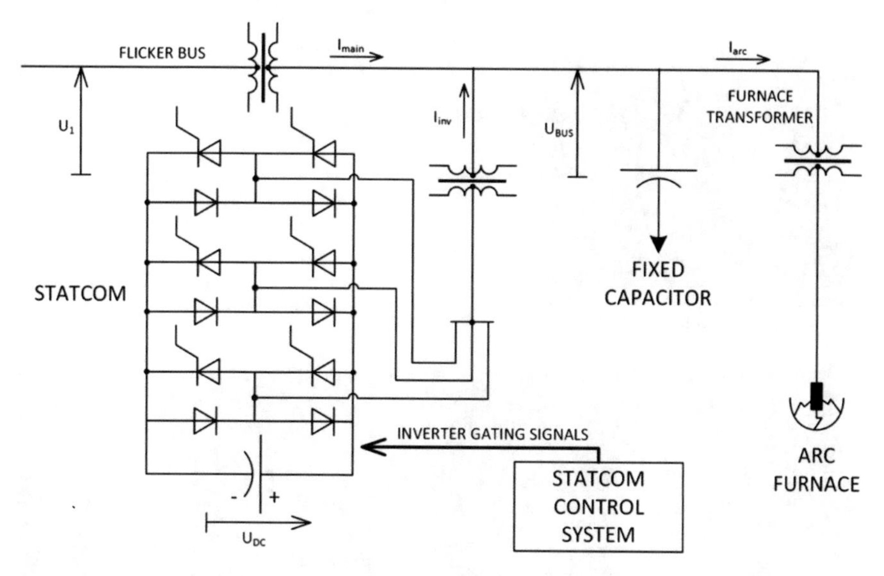

Fig. 3.83 Main elements of a STATCOM/EAF compensator [45]

furnace bus, and a fixed 60 MVA AC capacitor bank, also connected in shunt to the furnace bus [45].

The inverter is sized to supply all reactive power required by the furnace, the randomly fluctuating instantaneous real and reactive power high frequency components drawn by the furnace and those components of the current (including negative sequence fundamental) associated with unbalanced furnace operation.

In essence, the STATCOM control system acts to make the inverter operate as a controlled current source, delivering the desired current components to the EAF bus on command.

Finally, we must remember that the insertion of compensators modifies the frequency spectrum of fluctuations, reducing the frequency of components at lower frequencies and increasing those at the higher ones.

3.6 DC Arc Furnaces

The idea of reducing interferences of arc furnaces on the supply network through stabilization measures acting directly on the arc, has led to the development of DC arc furnaces.

After many attempts in the 60s and 70s, limited at that time by the insufficient development of power electronics, in 1982 was installed a first prototype of DC arc furnace for steel production, rated 12 t/6 MW.

(a)

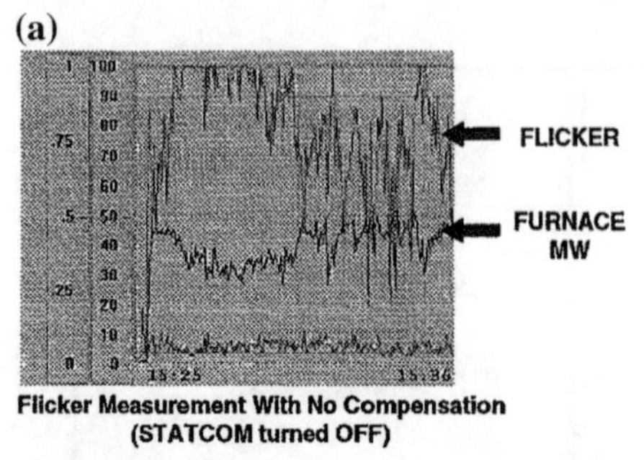

(b)

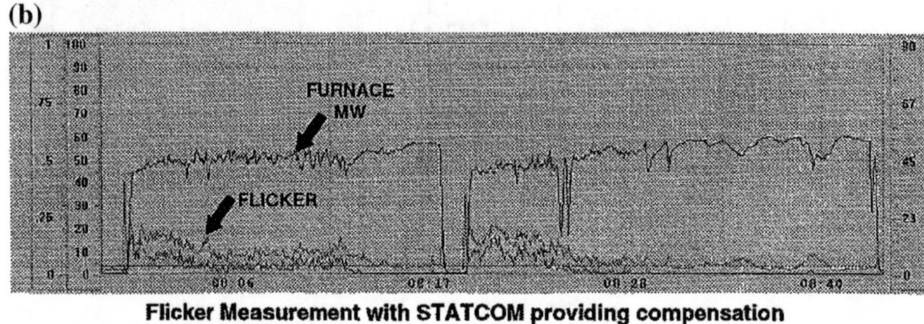

Fig. 3.84 Flicker reduction in an arc furnace by STATCOM compensation (**a** with no compensation; **b** with STATCOM compensation)

This furnace was characterized by a single graphite electrode in the center of the vessel, with function of cathode; between its tip and the charge was burning a DC arc, 300 V/20 kA. The anode was constituted by the molten charge, connected to the DC supply circuit through a bottom electrode, designed to be replaced after wear.

The positive experience of operation of a subsequent furnace 130 t/60 MW, installed in Japan since 1982, led to a boom of this new technique: in 1989 were operating worldwide 39 DC furnaces. Only 10 years after the construction of the first prototype, in 1992 in USA were put in operation two DC arc furnaces 150 t/80 MW. The installation of DC/EAFs reached in 1996 a quota of more than 40 %.

The circuit diagram of a DC furnace corresponds to the scheme of Fig. 3.85.

For the power supply of DC currents up to 80–100 kA are used two 6-pulse bridge rectifiers connected in parallel to form a scheme with 12 pulses and low ripple, according to the circuit of Fig. 3.86. For higher arc currents, four 6-pulse bridges connected in a 24-pulse scheme are used.

The usual twelve-pulse scheme creates current harmonics with mode 11–13, 23–25, etc., which must be compensated by the filter circuits shown in Fig. 3.85.

The cathode is represented by a single graphite electrode, which is connected to the transformer by a current-conducting electrode-arm and flexible cables. With the

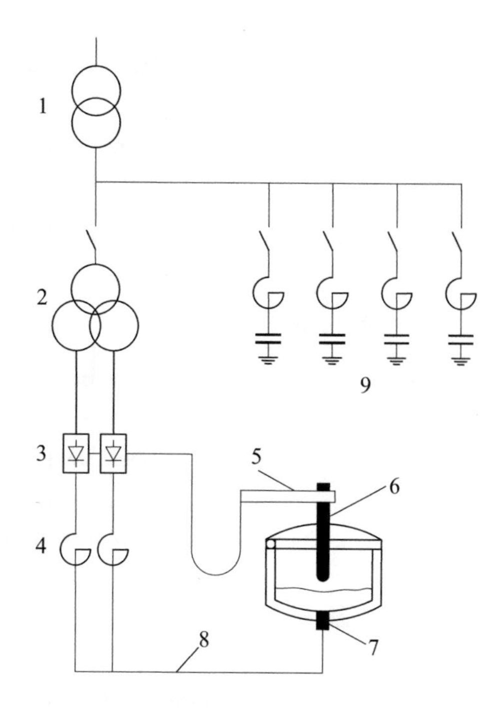

Fig. 3.85 Schematic circuit of a DC arc furnace [29] (*1* Step-down transformer; *2* Power converter transformer; *3* Power converter; *4* Smoothing choke; *5* Electrode arm; *6* Graphite cathode electrode; *7* Bottom anode electrode; *8* High-curret line; *9* Filter circuits)

introduction of graphite electrodes with diameter up to 1000 mm, nowadays it's possible to design DC furnaces up to 120–135 MW, 850–1000 V/160 kA.

However, for furnaces of increased size (greater than 150 tons), the electrode diameter and its consequent current carrying capacity and graphite consumption became a potential limiting factor; this has led to the development of the dual-electrode AC/EAF design. Instead of an upper graphite electrode and a lower conductive hearth, this type of EAF uses two upper graphite electrodes.

This design concept allows the utilization of lower diameter standard electrodes and lower density currents compared to the single electrode solution. It has been used in the construction of the 420 t largest DC/EAF built worldwide, shown in Fig. 3.87, which ensures a production of 360 t/h.

Data of the twin-electrode design are: sizes from 20 to 420 t, active power above 160 MW, tap-to-tap time 45 min, energy consumption <450 kWh/t, oxygen consumption <40 Nm^3/t. [48]

Unlike what occurs in AC/EAFs, the DC arc is not affected by any rotation, but the arc column moves around stochastically with a cone-shaped geometry. However, it can be influenced by the magnetic field produced by the large current loop of the high current circuit, which tends to deflect the arc away from the power supply. This deflection can be minimized by a suitable layout of the high current circuit.

Different phenomena occur in the twin electrode furnace, where the steady magnetic field causes the arcs to attract each other.

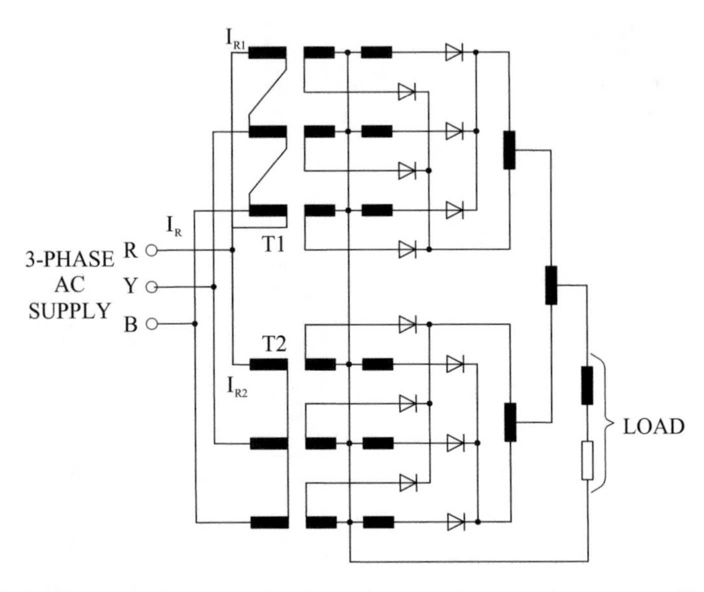

Fig. 3.86 Combination of two six-pulse circuits into a twelve-pulse single-way rectifier

Fig. 3.87 Fast DC/EAF 420-t (300-t tapping size) [48]

Finally, the operation with high arc voltages, which are needed at high MW ratings, requires arc lengths between 600 and 1000 mm, which in turn demand generation of a strong foaming slag. Therefore, these DC furnaces are unsuitable for melting processes which do not allow foaming slag operation.

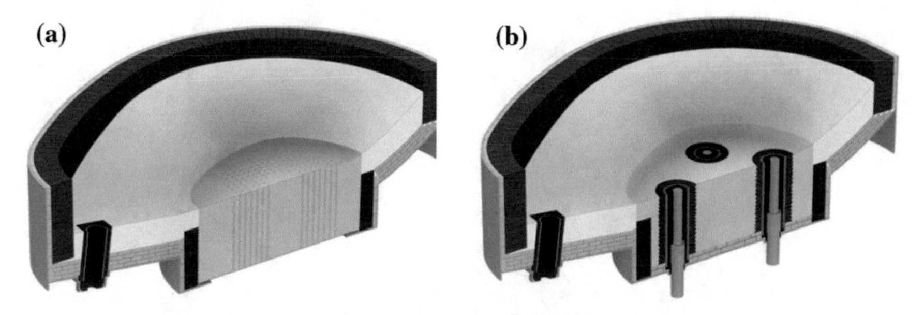

Fig. 3.88 Typical bottom electrodes: **a** pin-type, **b** billet type

The bottom electrode is an important technological component of the furnace for which several designs have been proposed and are still under development. However, bottom configurations are not longer, as in the past, a major deterrent to installing such furnaces, but could restrict bottom injection technology. Bottom electrodes should be replaced every 2000–3000 melts.

The main types of bottom electrodes are:

- *Pin-type*—(Fig. 3.88a) with several steel pins embedded in the central part of the refractory bottom, air cooled from underneath;
- *Fin-type*—is similar to the previous one, but is constituted by sheet steel rings embedded in the refractory and welded on the furnace bottom;
- *Billet-type*—(Fig. 3.88b) consisting of cylindrical water-cooled steel blocks incorporated in the refractory of the furnace bottom;
- *C-brick type*—where the furnace bottom is constituted by current conducting carbon-magnesite bricks, intensively air-cooled from underneath.

Unlike to what occurs in AC furnaces, in DC furnaces the control of arc current, which is stabilized by the series reactance of the DC circuit, is done acting on the switching-on angle α of thyristors, independently from the adjustment of arc voltage through the movement of the electrode.

Figure 3.89 shows the so-called furnace power control diagram, where the active power is plotted as a function of the reactive one, for different values of arc current, power factor and thyristor's ignition angle α (the diagram refers to a 95 MVA DC arc furnace).

It can be demonstrated, in a very simplified approximation, that the power diagram for constant values of arc current and voltage, describes a quarter of a circle.

Within the region indicated in the diagram, the operating point can be chosen freely: in case of a variation of the arc length, the current is first kept constant by acting on the switch-on angle α of thyristors and is stabilized by the DC reactor. At the same time acts the slower closed-loop control which adjusts the arc voltage independently by vertically moving the electrode and thus is again reached the original working set-point.

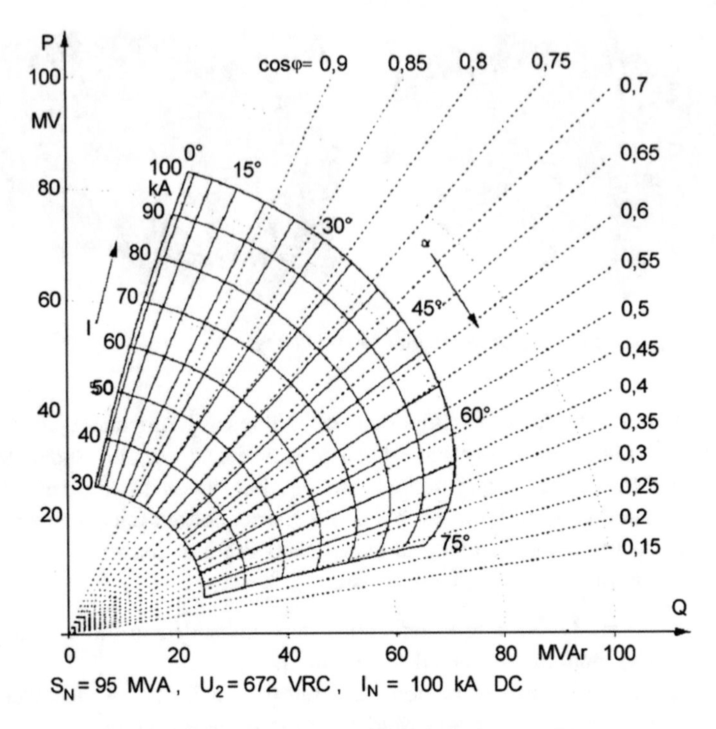

Fig. 3.89 Diagram for power control of a 95 MVA DC arc furnace [29]

Typical operating set-points of α lie between 30° and 45°.

During the melting period, within this range of ignition angles, there is a wide control range, so that variations of reactive power can be smaller than in AC furnaces with the same installed power. For this reason, in DC/EAFs the phenomenon of "flicker" can be reduced by 35–50 % in comparison with conventional AC "low impedance" furnaces. This technique is therefore an alternative solution in case of weak power supply grids, to the AC furnace with dynamic compensation of reactive current.

In practice, since this degree of flicker reduction cannot be met always, in big plants additional dynamic power factor correction systems can be installed in case of weak supply grids which, however, in turn can create higher frequency harmonics and consequent further disturbances [49].

In conclusion, main advantages of DC furnaces are: (a) reduction of flicker, mainly due to the operation with "long" arcs and limitation of the current variations obtained with thyristor control; (b) lower electrode consumption due to the fact that the high anode voltage drop occurs on the molten metal surface, thus producing lower overheating and lower oxidation of the electrode.

Electrode consumption is in the range of 1–1.5 kg/t.

3.7 Submerged Arc Furnaces[2]

3.7.1 Description

Among electrical arc furnaces, *submerged arc furnaces* play a special role. These installations are used for very different technological processes, which lead to specific requirements for the furnace design.

This explains the wide range of unit capacities ranging from few MVA to 100 MVA.

The conversion of electrical energy into heat occurs both in the electric arc and the processed materials, because the electrode tips are buried in the slag/charge, and the arc current flows through the slag, between matte and electrode.

Submerged arc furnaces are also known as *arc reduction furnaces* since they are used for production of calcium carbide, ferroalloys, other non-ferrous alloys and phosphorus by chemical processes of reduction of one or more oxides of the ore by a reducing agent loaded with the ore.

As reducing agents are used elements forming a more stable oxide than that of the reduced element, or oxides which are quickly removed from the area of reaction. The most common and cheapest reducing element is carbon; the process involving the reduction of metal oxides using carbon as reduction agent is called *carbo-thermic reaction.*

Another type of submerged arc furnaces are the so-called *ore melting furnaces*, where are performed technological processes of melting ores without development of reduction reactions or where these reactions occur in a very limited extent. These furnaces are used for melting refractories (e.g. mullite, bakor) or synthetic and welding fluxes, where the production process requires homogenization of the melt.

Other furnaces are those used for production of copper and copper-nickel matte, where the separation of components is obtained using the effect of their different weight in liquid state. The reduction of primary oxide impurities and their segregations is performed in electric furnaces for production of normal and alloyed electro-corundum, fused refractories (periclase).

Thus, submerged arc furnaces include electro-thermal installations where a great variety of processes are performed, but all have a unique main goal, i.e. the realization through the heating of reduction of the main elements of the ore minerals, or melting of the ore for homogenization or segregation of its constituents.

Furnaces for these services are physically different from steel-making furnaces and usually operate on a continuous, rather than batch, basis.

These furnaces are applied for production of two groups of alloys.

The first group for ferro-silicon, crystalline silicon, ferro-chrome, calcium-silicon, silicon-manganese, ferro-manganese, phosphorus, calcium carbide, copper and copper-nickel matte, corundum, silicon carbide, synthetic fluxes, and others.

---

[2]This paragraph is in large part taken from Ref. [54] with authorization of the authors.

The second is used for production of ferro-tungsten, ferro-vanadium, boron carbide, fused refractories and others.

The world consumption of alloys of the first group is in the range of tens to hundreds of thousands tons per year, while for the second group it amounts only to several thousand tons.

With the growth of production of steel, started to increase continuously also the production and consumption of ferroalloys. In fact, in 2001 the production of ferroalloys in the world was 19.2 million tons; in 2005—27.6 million tons, with an increase of more than 31 %. The average production in these years was about 2.5 % of the total world steel production [50].

The production rate required is the most important factor for the selection of power and type of the installation.

Table 3.4 gives typical data of some submerged arc installations.

3.7.2 Technological Processes and Influence on Design of Submerged Arc Furnaces

The most widespread group of submerged arc furnaces are the reduction furnaces for production of ferroalloys, in which ores-oxides of different elements are reduced and, as a rule, are melted with addition of iron in the charge.

The most common ferroalloy is ferrosilicon, which is used for de-oxidation and alloying of steel.

Several grades of ferrosilicon are produced, which mostly differ in silicon content (ranging from 18 to 90 %), and the so-called "metallic silicon", which is an alloy containing iron within the limits of 0.4–1.5 %.

As charge materials for ferrosilicon production are used quartzite (a siliceous sandstone consisting of 97–98 % of silica—SiO_2), coke as reducing agent, and iron —in the form of steel shavings. The reduction of silica with solid carbon is likely determined by a combination of several chemical reactions taking place in the furnace, involving intermediate products like gaseous silicon monoxide and solid silicon carbide.

The theoretical temperature of starting the reaction is 1514 °C [51].

A great influence has the presence of iron, which dissolves silicon and removes it from the reaction zone. Iron also destroys silicon carbides formed by reaction in the solid phase, which helps to speed up the total reaction of reduction of silica.

When silicon is dissolved into iron, heat is released, thus reducing electrical energy consumption. The presence of iron in the charge also reduces the starting temperature of reduction of silicon.

Melting of ferrosilicon is a continuous process with periodic loading of the charge on the open top of the furnace, or continuous loading in closed furnaces, with periodic pouring of alloy and slag, and continuous removal of furnace gases.

Table 3.4 Operating performance of some submerged arc electrical furnaces (For furnaces with round bath is given the collapse diameter of electrodes, for furnaces with rectangular baths, the distance between electrodes axes)

№	Type of furnace	Main product	Power of transformer (MVA)	Full operating capacity of the furnace (MVA)	Useful working capacity (MW)	Working current in the electrode (кА)	Working voltage (V)	Power factor (cos φ)	Electrical efficiency ($\eta_{эл}$)	Electrode diameter (mm)	Collapse diameter of electrodes* (mm)
3	RKO-20.0	Ferrosilicium, 75 %	20.0	16.8	13.5	53.0	183.0	0.90	0.87	1100	2700
4	RKO-29.0	Ferrosilicium, 65 %	29.0	24.5	19.0	72.3	195.8	0.89	0.87	1400	2900
5	RKZ-40.0	Ferrosilicium, 65 %	40.0	40.0	32.5	106	217.4	0.69	0.87	1500	3900
6	RKZ-21.0	Ferrosilicium, 45 %	21.0	21.7	14.9	69.0	182.0	0.79	0.87	1200	2900
7	RKZ-80.0	Ferrosilicium, 45 %	80.0	80.0	66.2	171	268.8	0.56	0.91	1900	5400
9	RKZ-27.5	Ferrosilicium, 25 %	27.5	27.4	23.5	79.8	198.7	0.78	0.90	1400	3700
11	RKO-16.5	Ferro-silicon-chrome	16.5	18.9	13.5	63.3	173.0	0.80	0.89	1200	300
13	RKO-16.5	Ferro-chrome	16.5	16.8	13.1	54.5	178.5	0.82	0.91	1200	2800
14	RKO-16.5	Silico-chrome	13.9	13.9	9.71	46.0	163.0	0.86	0.87	1100	2550
15	RPO-11.0	Ferro-manganese	11.2	10.2	6.83	42.0	141.0	0.79	0.84	1100	2100
17	RKZ-81	Silico-manganese	81.0	75.0	32.7	149	290.0	0.54	0.80	2000	5650
18	RPZ-80	Silico-manganese	80.0	80.0	66.8	101	263.5	0.76	0.91	2400 × 1200	3600
19	RKO-15.0	Silico-calcium	15.0	13.2	8.15	57.0	134.0	0.73	0.84	1050	2500
21	RPO-40к	Calcium carbide	40.0	36.8	30.0	86.0	247.0	0.88	0.92	2800 × 650	2400
22	RPO-60к	Calcium carbide	60.0	45.8	36.0	103	257.0	0.86	0.91	2800 × 650	2400
24	RKZ-48ф	Phosphorus	50.0	50.0	46.8	62.5	460.0	0.96	0.97	1400	4000
25	RKZ-80ф	Phosphorus	80.0	72.0	58.9	78.0	533.4	0.84	0.97	1700	4800
27	RKO-16.5	Normal corundum	16.5	14.8	12.4	39.6	217.0	0.90	0.92	1200	3000
28	RPZ-20	Copper nickel matte	20.0	16.1	14.0	28.4	327.0	0.92	0.95	1200	3400

The melting process occurs mainly near the electrodes. The main part of the current flows from the tip of the electrode to the melt, forming an electric arc. Due to this arc, the nearby space is heated to a temperature of about 1700–1750 °C, where the melting and reduction process occurs. In this space forms a certain amount of vapor of silicon and iron which, together with reaction gases, produces a gas chamber acting as a crucible.

Figure 3.90 schematically shows the workspace of an ore reduction furnace.

The lining of the vessel works at the high temperature of the melt and under the chemical and the mechanical effects of the movement of molten metal and slag. To ensure the correct pouring of metal and slag, the mutual positions of the crucibles must be relatively close. The dimensions of the vessel must be chosen such that, on one hand—a sufficiently high temperature of melt and slag is ensured for subsequent pouring and casting, on the other hand—to assure the lowest consumption of lining under the influence of the above factors.

According to the technology used, there is the tendency to ensure that the inner surface of the lining layer forms a skull—a protective frozen layer constituted by the charge, particles of the lining, slag and melt. The skull protects the lining from further deterioration thus ensuring to it a longer life. The greatest deterioration of the lining is due usually to the slag. Since in production of ferrosilicon is used a small amount of slag, the formation of skull is not a priority. Due these process characteristics, the refractory lining layer of the vessel of furnaces for ferrosilicon production is made with carbon blocks.

The pouring of metal and slag is made together through tap-holes, i.e. holes situated in the side wall lining at the level of the hearth. The pouring is made regularly 6–8 times per shift. Between subsequent metal tapping, the tap-hole openings are closed by special stoppers constituted by a mixture of coal mass and clay, which before pouring are cut by oxygen or electrical burnouts.

The reaction gases produced in the gas chamber, passing through the layers of the charge, release heat to them, and silicon and iron will condense on the colder layers. In open furnaces on the surface of the charge (at the furnace top) the gas emitted from the furnace is burned, creating severe temperature conditions for equipment and personnel, and the dust is carried away with the gas and fumes polluting the atmosphere and the surrounding area of the shop.

Therefore there is a tendency to cover the bath of the furnace with a cupola, and to collect the gas out from the furnace stand, to remove the dust and to send it to recycling.

As an example, a 21 MVA furnace produces during operation about 2200 Nm^3/h of reaction gases consisting of carbon monoxide (80–90 %), carbon dioxide, hydrogen, oxygen, nitrogen, hydrocarbons. In a closed furnace can be collected up to 95 % of these gases.

Since the reaction of reduction occurs mainly at the walls of the crucibles and each crucible has dimensions slightly larger than the diameter of the electrode, then in furnaces for ferrosilicon the loading of the charge is made in the electrode area. In open furnaces the charge is introduced and piled up near the electrodes, where it forms a cone with a base diameter equal to approximately twice the diameter of

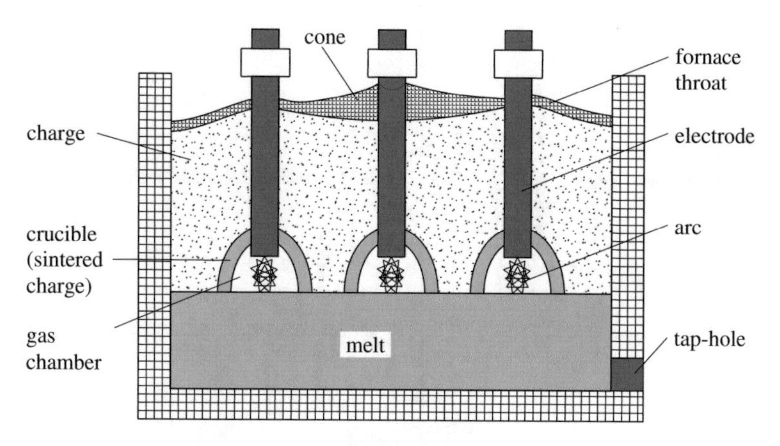

Fig. 3.90 Schematic drawing of an ore reduction furnace

electrodes (see Figs. 3.90 and 3.91). In closed furnaces the charge is loaded by hoppers, covering each electrode (Fig. 3.92).

A vault cover can be used in furnaces for smelting. The open top of furnaces for melting ferrosilicon with silicon content of more than 65 %, require a constant maintenance due to the sintering of the upper layers of the charge and by this reason cannot be covered with a cupola. Their operation is therefore done in open version, but in order to protect the shop structures and the personnel from thermal radiation, above the furnace is set a hood, which collects the throat gases for cleaning up the dust in the gas treatment unit.

In the last years, with the aim of recovering heat from the combustion of reaction gases, on the furnace top is built an opening, the so-called "low hood", which is a kind of heat-recovery unit. At the same time it facilitates the process of gas treatment, since it sharply reduces the volume of aspirated gases.

In order to obtain a more uniform thermal load of the lining, counteracting the formation of sintered regions at the top of the furnace, and a more uniform descent of the charge, the furnaces for production of ferrosilicon are equipped with a rotation mechanism of the vessel with revolution time of 60–80 h and angle of rotation of 80º–120º. It has been verified that the rotation is useful for processes with low quantity or without slag, while it is ineffective for processes with high quantity of slag.

Close to the ones previously described are the technology and the construction of furnaces for melting calcium carbide, silicon-calcium, carbon-ferromanganese, carbon-ferrochrome, silicon-manganese, silicon-aluminum, cast iron.

But each process imposes special requirements to the furnace design.

For example, in furnaces for production of carbon-ferrochrome the refractory lining is made of magnesite, which is not chemically reactive with the slag and the melt.

Furnaces for melting silicon manganese and carbon-ferromanganese are classified as "high slag" processes, since they are characterized by ratios of slag to melt of

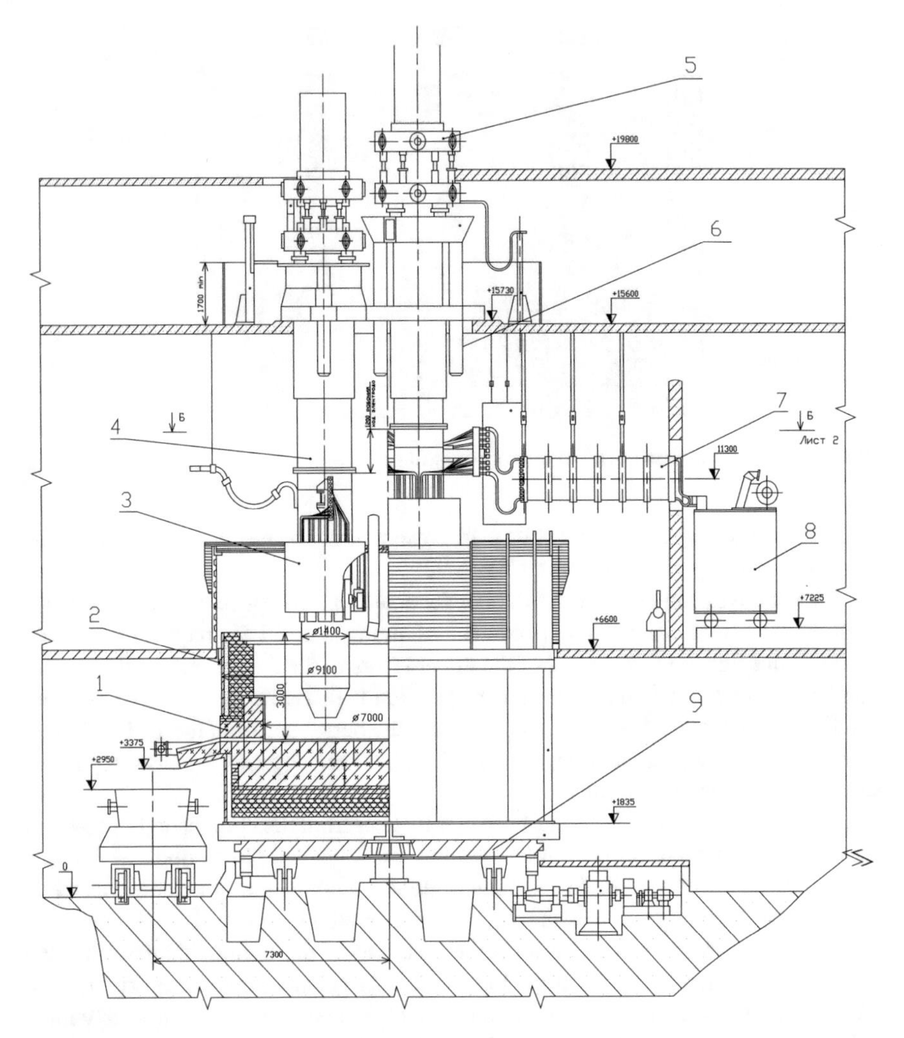

Fig. 3.91 Schematic of open ferroalloy reduction furnace RKO-25 (*1* lining, *2* bath cover, *3* clamping mechanism of electrode, *4* casing of self-baking electrode, *5* device for slipping the electrode; *6* hydraulically operated hoist; *7* secondary circuit, *8* transformer; *9* base plate of rotation mechanism)

0.7–0.8 and 1.1–1.2 respectively. This situation, in comparison with "low-slag" furnaces, leads to increase the reaction zone and the electrodes consumption.

This has led to the development of high-power 6-electrode furnaces for silicon manganese and carbon ferromanganese with rectangular vessels and arrangement of rectangular electrodes in line (as in Fig. 3.93) or in two concentric rows for circular electrodes.

Similar to round open furnaces for production of ferroalloy are the furnaces for production of normal fused corundum. Normal corundum is an abrasive material

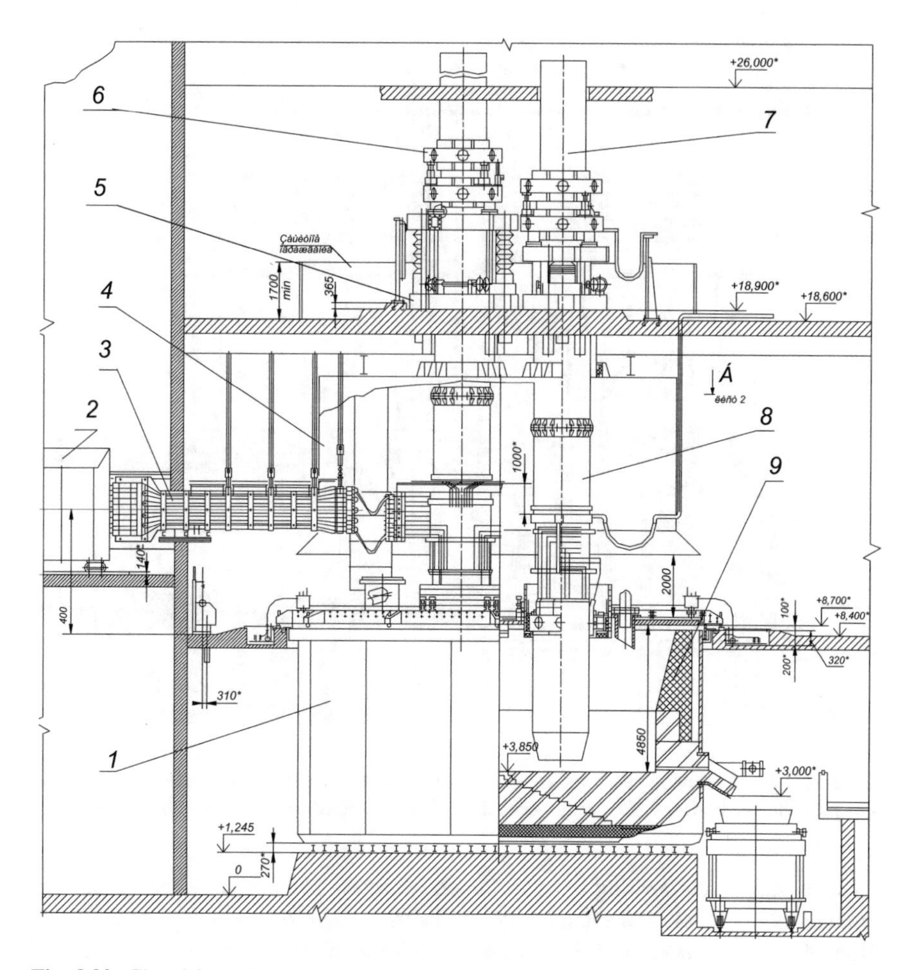

Fig. 3.92 Closed ferroalloy submerged arc furnace RKZ-33 (*1* vessel; *2* transformer; *3* secondary circuit; *4* fume hood; *5* hydraulically operated hoist; *6* device for slipping the electrode; *7* self-baking electrode; *8* supporting cylinder: Mantel; *9* lining)

containing 93–98 % of mineral corundum (which is constituted by aluminum oxide crystals) and some impurities contained in the mineral. It is produced by a semi-melting process of raw materials, constituted by bauxite mixed with a reducing agent. In this process the oxides of iron, silicon, part of titanium, are reduced to form ferroalloy.

Since the temperature of fused electro-corundum is about 2200–2300 °C, at which all known refractories would be destroyed, the vessel is lined with the same electro-corundum and effective lining is constituted by the skull which forms on the vessel walls.

In some cases, such as production of low-carbon ferrochrome and ferromanganese, the melting is done in tilting furnaces (Fig. 3.94), which are simplified

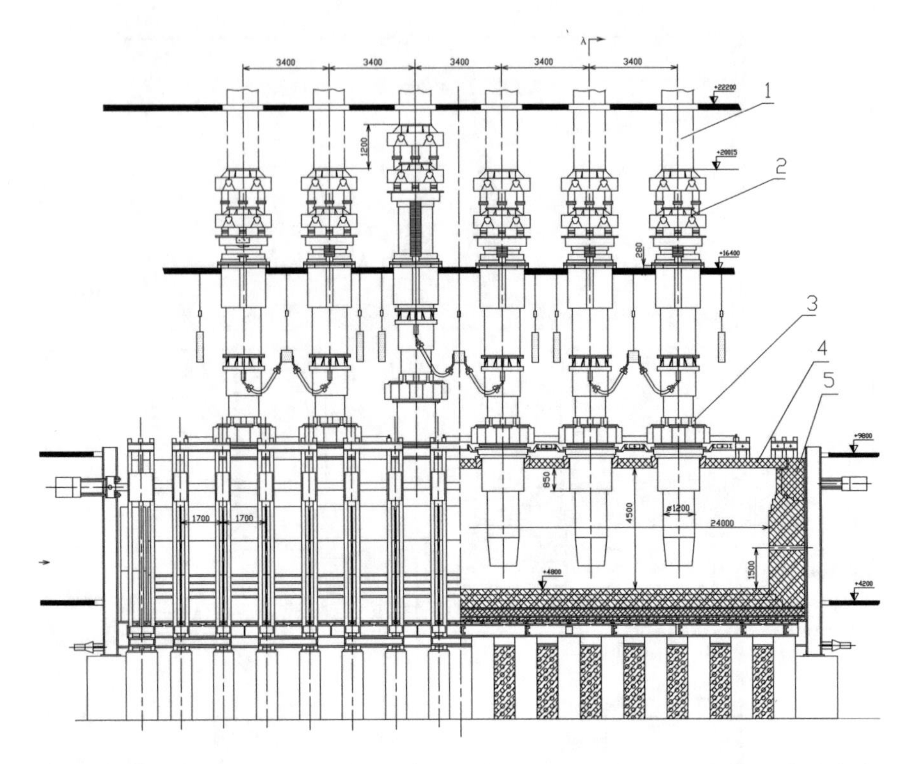

Fig. 3.93 Closed ferroalloy electric furnace RPG-48 with rectangular vessel (*1* electrode, *2* device to bypass the electrodes, *3* clamping mechanism of the electrode, *4* cover of bath, *5* lining)

versions of the steel-making furnaces. Such furnaces cannot be covered because of continuous operation and complexity of loading. Therefore, to simplify the operation of such furnaces, the must with the column of the electrode holder is not connected with the tilting mechanism and is placed on a separate foundation. The electrodes during pouring are moved upwards allowing to tilt the vessel at the required angle. The practice has shown that the most suitable material for the lining of these furnaces is constituted by magnesia bricks. In order to increase the life of the lining and to allow a more complete reaction, the vessel is rotated during the melting process.

For obtaining some molten products such as boron carbide, alloyed corundum or a number of refractories (periclase), the charge materials can be melted and kept in liquid state only in a very small area near the electrodes. In this case the pouring of the melt through a tap-hole opening is not possible, and the melting is done by the so-called “block-process,” which consists of the following: the charge, which is fed continuously or intermittently, is first melted; then, when a new quantity of charge is loaded, the melting zone shifts upwards, while the lower layers solidify. In this way one block of the desired product is melted layer-by-layer till to fill the crucible

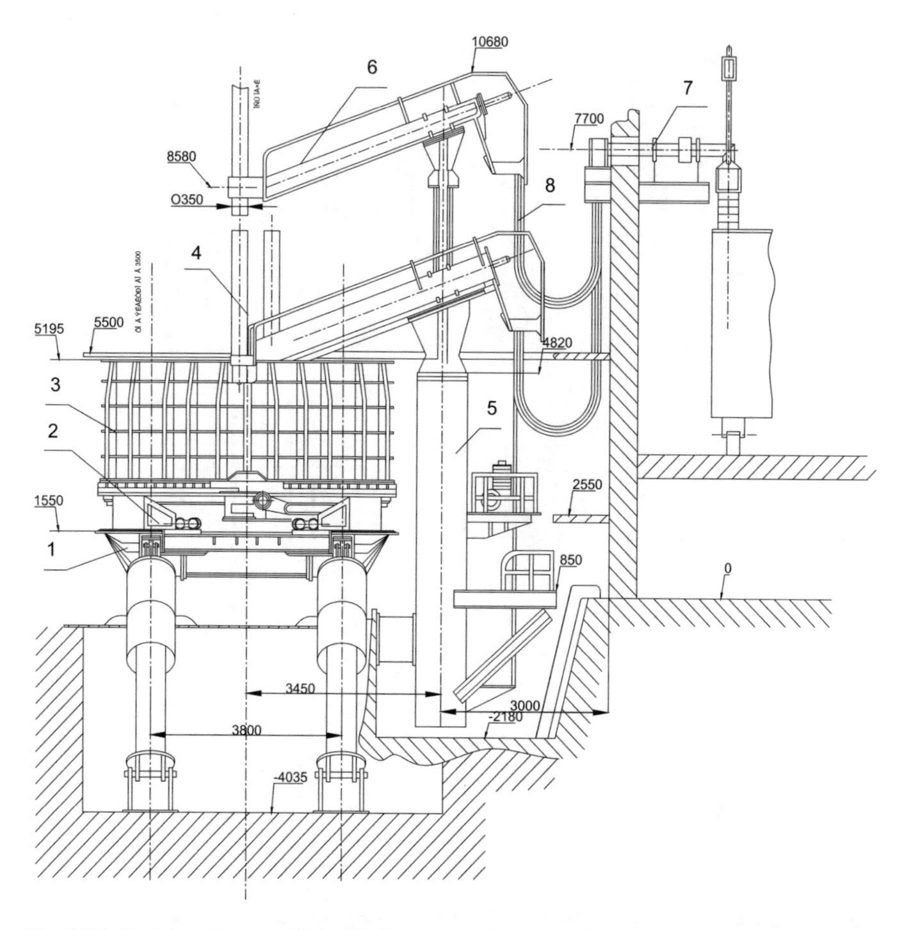

Fig. 3.94 Refining furnace RKO-3,5 (*1* furnace tilting mechanism; *2* mechanism of vessel rotation; *3* crucible of the furnace: shell and lining; *4* electrode; *5* mechanism of electrode movement; *6* electrode holder; *7* secondary circuit; *8* cables)

of the furnace. At this stage the furnace is switched-off, it is cooled and the block is extracted, i.e. is separated from the unreacted charge, located on the periphery.

In these furnaces, for thermal insulation of the hearth are usually used refractory bricks while, for the walls, the same charge.

In order to facilitate the cutting of blocks and to increase production are used interchangeable vessels, which can be moved on a trolley within the shop to the cooling and cutting unit.

The power of such furnaces does not exceed 1500–2000 kVA (Fig. 3.95).

There are some processes for production of silicon carbide or carborundum, where melting is not required since the chemical reactions occur when the charge is still solid. The electrical furnaces used for these processes are fundamentally

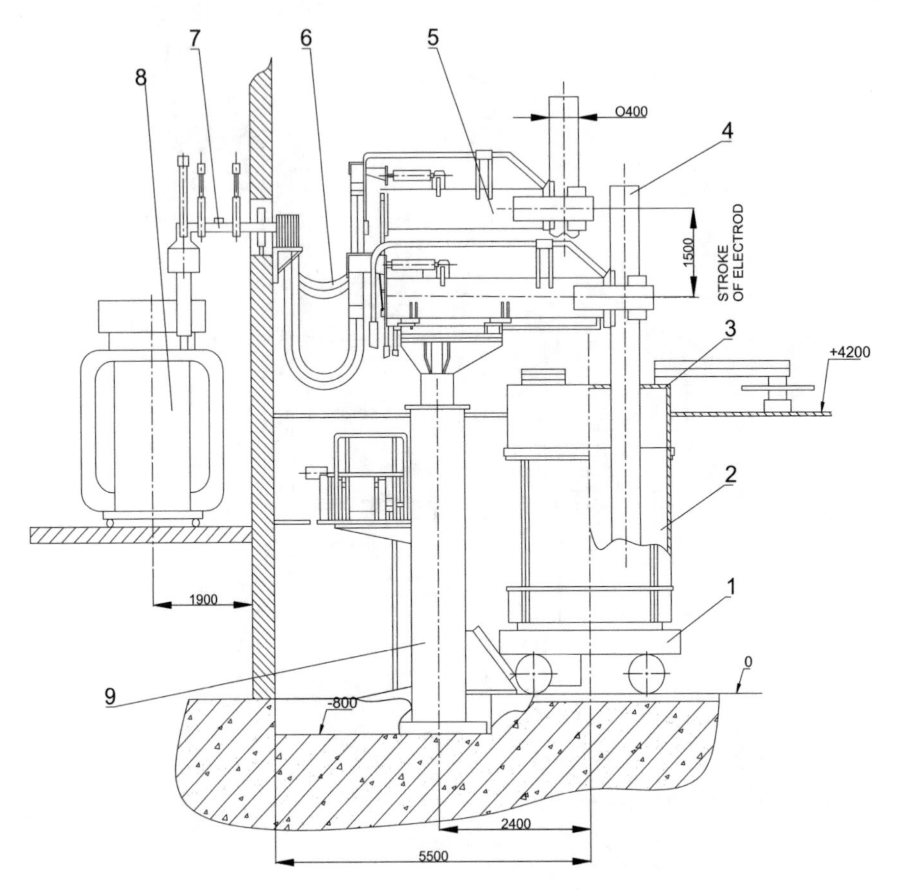

Fig. 3.95 Ore-thermal electric furnace RKZ-12 for melting refractory materials "on blocks" (*1* trolley; *2* bath cover; *3* hood; *4* graphite electrodes; *5* electrode holder; *6* flexible cables; *7* bus-bars; *8* transformer; *9* shaft moving mechanism of electrodes)

different from the previously considered ore-thermal furnaces, since they use direct resistance heating.

These furnaces are single-phase, with a reaction zone positioned horizontally and limited by lateral and hearth iron shields, inside which is laid a core of carbonaceous materials. The space between core and shields is filled with a charge consisting of pure silica, petroleum coke, old unreacted charge, fillings as gas barrier and salt.

The current supply is made though rectangular electrodes, located at the end walls.

In the production of silicon carbide, where the heat is produced by the electric current flowing partly through the core and partly through the charge, the core is heated to a temperature of about 2000–2250 K.

Under the influence of this temperature, chemical reactions occur with formation of crystalline SiC and products of incomplete reduction (as amorphous and aggregates of silicon). A significant part of the charge, located in the peripheral zones, is heated to a lower temperature, insufficient for producing these reactions, and acts as thermal insulation. Such products are located around the core in concentric circles.

After the end of melting the furnace is cooled. Then is done the so-called "hot maintenance" of the furnace, during which are taken off the side panels and the unreacted charge, while the sintered part of semi-reacted charge are removed from the top and the sides of the unit. The remaining part of the block is cooled further, and then the silicon carbide is separated from the rest. The whole process takes several days.

3.7.3 Self Baking Electrodes

Ore-reducing electrical furnaces usually work with self-baking electrodes which are constituted of a steel shell filled till a certain height by an electrode mass.

The use of shaped electrodes of amorphous carbon and less often of graphite is necessary only when it is impossible the introduction in the melt of the iron coming from self-baking electrodes, like for example in melting crystalline silicon, or when is needed a concentration of power bigger than the one allowed by self-baking electrodes. Moreover, shaped electrodes are much more expensive and the design of electrode holders for them is much more complex than the one for self-baking electrodes.

Data of amorphous carbon or graphite electrodes have been given at Sect. 3.2.7.

With the exception of the above cases, the use of shaped electrodes is undesirable in submerged arc furnaces because of their higher costs, more complex holding and movement mechanisms, complexity of increasing their length inside the furnace and ensuring the integrity of the joint.

The vast majority of ore-thermal furnaces are therefore equipped with high power continuous self-baking electrodes, which were invented in 1917 by the Norwegian engineer S. Söderbergh.

The self-baking electrode consists of a metal casing designed to wear during operation, filled with an electrode paste [6, 52].

The paste undergoes the self-baking process mainly within the furnace, as a result of the heat transmitted by the furnace ambient to the electrode and the one developed as Joule losses by the current flowing in it. In this way the baking, which for amorphous carbon electrodes is done in special furnaces with a very long and expensive process, in this case is completed directly in the reduction furnace during its normal operation.

As a consequence, the electrode is much cheaper and suitable for continuous production processes in high-power furnaces.

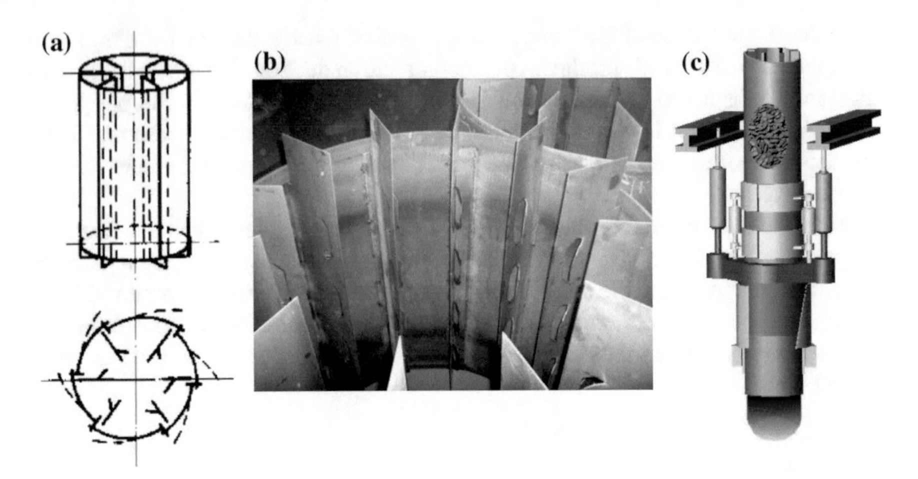

Fig. 3.96 Self-baking "Soderberg" electrodes (**a** external steel casing; **b** steel casing and internal fins; **c** electrode external view)

The casing of the electrode is constituted by a cylinder made of sheet steel, provided with a certain number of internal fins with bent triangular incisions, or—more recently—round holes. The fins have mechanical functions of reinforcement and electrical functions of improving the electrical contact between metal casing and electrode paste and the current distribution in the electrode cross-section (see Fig. 3.96).

The electrode consists of a number of sections; as the lower end of the electrode is combusted, on its top is connected a new section.

The thickness of the casing steel sheet is 3–6 mm, height and thickness of the ribs are calculated on the basis of the electrode current intensity and the conditions of retention of the electrode in the slipping device.

The admissible current density in the casing of self-baking electrodes should not exceed 2 A/mm^2.

The casing is filled with an electrode paste of composition similar to that used for graphite and amorphous carbon electrodes. Each company has its own recipe for producing the electrode paste, but the main components of the paste are about 50 % of granular anthracite with grain size up to 20 mm, 25–50 % of foundry coke powder with size up to 0.5 mm, the remaining part (25–27 % of the dry weight) of a binder consisting of a mixture of tar, pitch and coal tar.

Typical values of baked electrode paste are given in Table 3.5.

The paste becomes liquid at about 100 °C, condition which is achieved during baking at 2–2.5 m from the top contact clamps. The level of crushed paste must be higher of 200–500 mm.

Figure 3.97 shows an example of temperature distribution in the electrode [18].

Table 3.5 Physical properties of baked electrode paste

Electrical resistivity (Ω m 10^{-5})	6–8
Mechanical strength	
At compression (MPa)	14–22
At tension (MPa)	>1.5
Density	
Effective (Kg/m^3)	1900
Apparent (Kg/m^3)	1300–1400
Porosity (%)	>26–28
Volatile content (%)	15–16
Ash (%)	<4

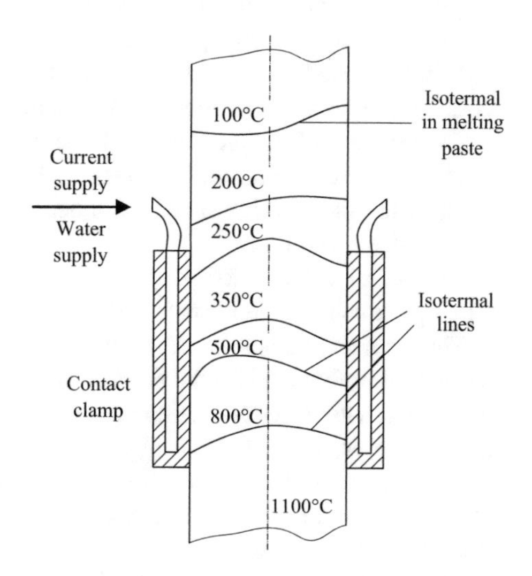

Fig. 3.97 Distribution of temperature in self-baking electrodes

Due to the effects of Joule heating and heat transfer by conduction, the electrode paste heats and becomes homogenous and creamy at 80–100 °C, then at 400 °C begins the process of sintering, which ends at 700–800 °C.

With a proper selection of cross-section of the electrode and operating conditions of the furnace, the baking rate of the paste is equal to consumption rate of the electrode, which burns in the lower part. To this end, it must be selected an appropriate electrode slipping rate, increasing the length of a quantity equal to its consumption.

Deviations from normal running an technological conditions, like reduced power, stop of the furnace, modification of the charge, fast slipping, deterioration of electrode paste, etc., all lead to modifications on the baking process.

Since the liquid paste has low electrical and thermal conductivity, the baking process must be completed within the lower third of the height of the contact clamps.

In this condition the contact clamps, which are pressed on the electrode casing, will deform it a little providing a reliable electrical and thermal contact.

If the electrode sintering is completed out of the contact clamps, the casing overheats since the whole current flows through it, it burns and the liquid electrode paste can leak, leading to a serious accident that requires a long furnace shut-down. A premature baking of the electrode paste is also undesirable, because the contact of the solid electrode with the slipping plates will degrade the electrical contact, leading to premature failure of the contact clamps.

During operation it is difficult to modify the vertical position of the baking zone, i.e. the mode of electrode baking. The change of temperature and flow rate of the air supplied to the electrode cooling system or the change of the flow rate of cooling water in the contact clamps give no significant results. If the electrodes are not completely baked the only solution is to reduce the electrode slipping rate and to reduce the power, i.e. to modify the running conditions of the process.

The position of the area where the electrode baking occurs is influenced by the width of contact plates, the diameter of the cooling channels of the plates and the depth of the bath, which modify the heat exchange conditions between electrode, contact plates and bath.

The set of devices for supplying current to the electrode, its casing and the slipping movement elements is indicated as "*electrode holder*" (Fig. 3.98).

The contact clamps work in very severe conditions, since electric currents flowing through them into the electrodes can have intensities up to 100 kA, they are in contact with electrode temperatures up to 800 °C, they work in a reducing atmosphere coming from the high temperature zone of the furnace top and containing a number of harmful active chemical substances. Thus the requirements for their safe operation in these heavy conditions can be satisfied only by clamps made of copper or its alloys.

Söderberg electrodes are available with diameters up to 2000 mm; their current carrying capacity varies with electrode diameter as shown in the diagram of Fig. 3.99.

3.7.4 Energetic Characteristics of Reduction Furnaces

Energy balance and energetic characteristics of ore-reduction furnaces are important parameters for metallurgists.

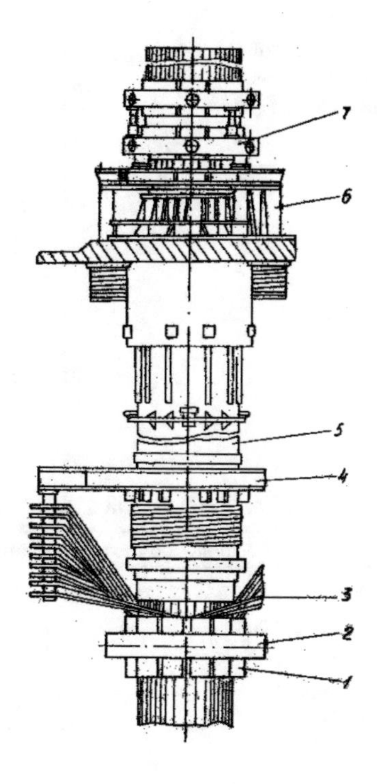

Fig. 3.98 Electrode holder of an ore-thermal furnace (*1* contact plate; *2* pressure ring; *3* current supply tube; *4* element of the current-supply; *5* hanging mantel; *6* hydraulically operated device; *7* unit for electrode slipping)

In order to obtain reliable data, the energy balance should include the actual mass flow of materials and consider a period of stable working conditions of the furnace with stable composition of the charge and minimal deviations from normal operation.

The realization of tests for obtaining these parameters requires a long period of data recording (usually at least 1 month) and the involvement of a team of researchers, during which the furnace should work continuously for several days at different levels of secondary voltage for the subsequent determination of the optimal voltage supply.

In the following some tables some tables give typical data of furnaces for melting ferrosilicon, production of phosphorus and refining furnaces [18].

Another technical-economic index of particular interest is the specific energy consumption, which depends on the final product and is determined from the energy balance of the furnace (Tables 3.6, 3.7, 3.8 and 3.9).

Table 3.10 gives typical values of specific energy consumption for production of various ferroalloys.

Fig. 3.99 Current carrying capacity and current density of Söderberg electrodes [53]

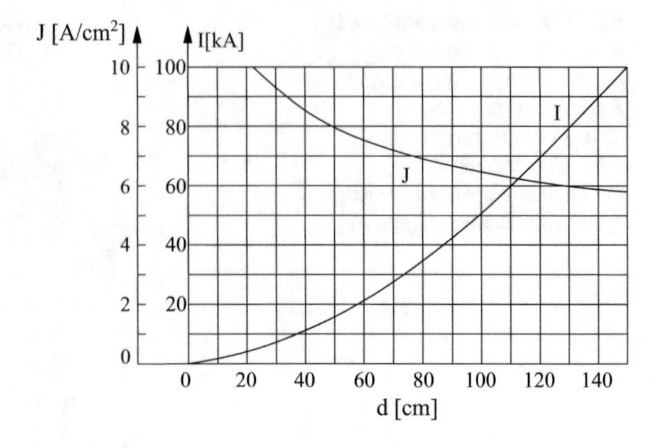

Table 3.6 Electric power distribution in a 21 MVA furnace

Item (%)	Ferro-silicon furnaces	Large phosphorus furnaces	Refining furnaces
Useful electrical power	90	96.5	87
Losses of furnace transformer	1.8	1	4.2
Losses in bus-bars	8.2	2.5	5.2
Losses in electrodes	–	–	3.6

Table 3.7 Thermal balance of a 16.5 MVA furnace for melting 45 % ferrosilicon (low-slag process)

Input	%	Output	%
Electrical energy	93.92	Electro-thermal reactions	75.40
Dissolution of silicon in iron	5.88	Melt	14.61
Formation of silicates	0.20	Slag	0.38
		Gas	5.29
		Thermal loss	4.23
Total	100	Total	100

Table 3.8 Energy balance of a 50 MVA phosphorus furnace (high-slag process)

Input	%	Output	%
Electrical energy	70.62	Endo-thermic reactions	72.35
Oxidation of carbon	18.29	Slag	19.32
Exothermic reactions	10.70	Ferroalloys	0.57
Sensible heat of the charge	0.39	Gases	1.94
		Thermal losses	1.91
		Electrical losses	3.96
		Residual	0.05
Total	100	Total	100

Table 3.9 Energy balance of a refining furnace, 3.5 MVA capacity (open arc)

Input	%	Output	%
Electrical energy	64.37	Alloys	11.01
Exo-thermal reactions for formation of slag	11.38	Slag	60.00
		Evaporation of humidity	0.91
		Decomposition of carbonate	2.95
Thermal reduction of silicon, chromium oxides, iron, etc.	21.32	Losses from the open furnace top	14.05
		Heat losses from construction elements	9.14
Oxidation of carbon	2.19	Residual	1.94
Sensible heat of the charge	0.74		
Total	100	Total	100

Table 3.10 Specific energy consumption for production of ferroalloys

Alloy	Electrical energy consumption (kWh/t)
Ferrosilicon, 18 %	2000
Ferrosilicon, 45 %	4500–4800
Ferrosilicon, 75 %	8300–8700
Ferrosilicon, 90 %	12,500
Crystalline silicon	12,000–12,300
Carbon ferrochrome	3500
Carbon ferromanganese	3090

References

1. Tiburzi, A.: La Pratica del Forno Elettrico. In: Hoepli, U. (ed.) Milano, 259 pp (1918) (in Italian)
2. Mainardis, M.: I forni elettrici. Ulrico Hoepli, Milano (1953) (in Italian)
3. Toulouevski, YuN, Zinurov, I.Y., Zinurov, IYu.: Innovation in Electric Arc Furnaces. Springer, Berlin (2010)
4. Institute for Industrial Productivity: Electric Arc Furnace—Industrial Efficiency Technology & Measures. Electric Arc Furnace Publications, 2010
5. Krupp 110-t-Hochleistungs – Lichtbogenofen. Krupp Industrie und Stahlbau publication (1980) (in German)
6. Union Internationale d'Electrothermie (UIE): Elektrowärme - Theorie und Praxis, 902 p. Verlag W. Girardet, Essen (1974) (in German)
7. Becker-Barbrok, U., Felix, W., Papachristos, G.: Installations de four à arc à grande puissance de fusion. Brown Boveri Rev. **2**, 103–113 (1978)
8. Silva, A., Hultqvist L., Wilk-Wilczynski, A.: Steel plant performance, power supply system design and power quality aspects. In: 54th Electric Furnace Conference, pp. 1–13, December 1996
9. Plockinger, E., Etterich, O.: Electric Furnace Steel Production, 622 pp. Wiley, New York (1985)

10. Coppadoro, F.: La regolazione della tensione dei trasformatori per l'alimentazione dei forni elettrici. L'Elettrotecnica, vol.LI, n.1, gennaio, 24–31 (1964) (in Italian)
11. Markov, N.A.: Electrical Circuits and Operating Regimes of Arc Furnaces Installations, 204 pp. Energhia, Moscow (1975) (in Russian)
12. McGee, L., Sparrow, J.O.: Optimum design of arc furnace secondary conductors. In: VII Congress International d'Electrothermie (UIE), Warszawa, n. 402 (1972)
13. Sundberg, Y.: The power circuit of arc furnaces. Elektro-wärme Int. **30**, B2, B93–B99 (1972)
14. Egidi, C.: Effetto pelle. L'Elettrotecnica, vol. XXXV, n. 4bis, 188–212 (1948) (in Italian)
15. Dwight, H.B.: Proximity effect in wires and thin tubes. J. A.I.E.E 961–970 (1923)
16. Lupi, S.: Appunti di Elettrotermia (Teaching notes). Libreria Progetto, Padova (Italy), 2005, 457 p., (in Italian)
17. Swinden, D.J.: The arc furnace, Teaching Monograph 1, The Electricity Council, 1986, 52 pp
18. Aliferov, A., Lupi, S., Forzan, M. et al.: Theory and practice of application of arc furnaces. Tempus European Project, Intensive Course Specific II, St. Petersburg, Publishing House of ETU, 2013, 234 pp. ISBN 978-57629-1418-5
19. Afanasev, V.V.: Dimensions and shape of circular vessels of arc furnaces. Elektrometallurgia, 2009, n.1, 1–12 (in Russian)
20. Becker-Barbrok, U., Felix, W., Papachristos, G.: Installations de four à arc à grande puissance de fusion. Brown Boveri Rev. **2**, 103–113 (1978). (in French)
21. Rhu, E.: Refractories—magnesia-carbon refractories, history, development, types and applications. Int. Ceram. Monogr. **1**(2), 772–793 (1994)
22. BBC: Furnace Systems "Arcmelt". BBC Publication No. CH-IH 512 484 E
23. Bases, G.: The Basics of Brick and Refractories for Ferrous Foundries. Foundry, Nov. 28, 2003
24. Raja, B.V.: Status & Outlook of Indian Refractory Industry, pp. 9–13. Steelworld (2006)
25. Gorieva L.L.: Electric Arc Installations, 111 p. NGTU, Novosibirsk (2008). ISBN 978-5-7782-1096-7 (in Russian)
26. Schwabe W.E.: Electrical and thermal factors in UHP arc furnace design operation. In: IX UIE International Congress, Cannes, 20–24 Oct., 1980, No. IICa4
27. Timm K.: Lichtbogenöfen. In: Industrielle Elektrowärme-technik, pp. 309–339. Vulkan-Verlag, Essen (1992). ISBN3-8027-2903-X
28. Cobine, J.D.: Gaseous Conductors—Theory and Engineering Applications. Dover Publications Inc., New York (1958)
29. Mühlbauer A.: Industrielle Elektrowärme-technik, 400 pp. Vulkan-Verlag Essen (1992). ISBN 3-8027-2903-X (in German)
30. Paschkis, V., Persson, J.: Industrial Electric Furnaces and Appliances. Interscience Publishers Inc., New York (1960)
31. Knapp, H., Hein, M.: Advanced EAF Technologies, pp. 1–5. Millenium Steel (2004)
32. Jones, J.: Understanding Electric Arc Furnaces Operations, pp. 1–6. EPRI Center for Materials Production (1997)
33. Wilson, E., Kan, M., Mirle, A.: Intelligent technologies for electric arc furnace optimization. ISS Technical Paper, Edward Wilson, EFC 98, pp. 1–6
34. Wijngaarden, M.J., Pieterse, A.T.: Bottom-stirring in an electric arc furnace. J. S. Afr. Inst. Min. Metall. 27–33 (1994)
35. Hering, M.: Basics of Electroheat, part I. WNT Warszawa (1998)
36. Szekely, J., Trapaga, G.: Zukunftsperspektiven für neue Technologien in der Stahlindustrie. Stahl und Eisen, vol. 114, Nr. 9, pp. 43–55. Düsseldorf (1994)
37. Köhle, S.: Recent improvements in modelling energy consumption of electric furnaces. In: 7th European Electric Steelmaking Conference, vol. 1, pp. 305–314 (2002)
38. Kleimt, B., Köhle, S., Kühn, R., Zisser, S.: Application of models for electrical energy consumption to improve EAF operation and dynamic control, pp. 1–10. Betriebsforshungsinstitut, Düsseldorf (2002)
39. Stark, A., Mühlbauer, A., Kramer, C.: Handbook of Thermo-processing Technologies, 807 p. Vukan Verlag (2005). ISBN 3-8027-2933-1

40. Pfeifer, H., Kirschen, M.: Thermodynamic analysis of EAF energy efficiency and comparison with a statistical model of electric energy demand. In: 7th European Electric Steelmaking Conference, 2002. Venice, May 26–29, 2002
41. Zuliani, D.J., Scipolo, V., Born, C.: Opportunities for Increasing Productivity, Lowering Operating Costs and Reducing Greenhouse Gas Emissions in EAF and BOF Steelmaking, pp. 35–42. Millenium Steel India (2010)
42. Nicola, G., Noferi, P.L., Papini, P.: L'alimentazione dei forni elettrici ad arco e valutazione del disturbo (flicker) da essi introdotto in rete, Ente Nazionale per l'energia Elettrica, Relazione di Studio e ricerca, n. 361. Gennaio (1979) (in Italian)
43. IEC—International Electrotechnical Commission-Technical Committee n. 27: Disturbances—Flicker, 25 pp. (1978)
44. UIE—Working Group Disturbances: Connection of Fluctuating Loads, 84 pp. UIE, Paris (1988)
45. Schauder, C.: STATCOM for compensation of large electric arc furnace installations. IEEE Power Eng. Soc. SM **2**, 1109–1112 (1999)
46. Ladoux, P., Postiglione, G., Foch, H., Nunes, J.: A comparative study of AC/DC converters for high-power DC arc furnaces. IEEE Trans. Ind. Electron. **52**(3), 747–757 (2005)
47. Singh, B., Saha, R., Chandra, A., Al-Haddad, K.: Static synchronous compensators (STATCOM): a review. IET Power Electron. **2**(4), 297–324 (2009) (with 320 references)
48. Danieli SpA: Danieli Starts Up Huge Melt Shop. Industrial Heating (2010)
49. Jones, R.T., Reynolds, Q.G., Curr, T.R., Sager, D.: Some myths about DC arc furnaces. Southern African Pyrometallurgy, Johannesburg (2011)
50. Leontev, L.I., Smirnov, L.A., Žučkov, V.I., Daševskiĭ, B.Ja.: World Production of Steel and Ferroalloys, n. 2, pp. 2–9. Elektro-metallurgia (2008) (in Russian)
51. Ryss, M.A.: Production of Ferroalloys, 344 pp. Metallurgia, Moskow (1985) (in Russian)
52. Gasik, M.I.: Söderberg electrodes for ore reducing furnaces, 368 pp. Metallurgia, Moskow (1976) (in Russian)
53. Di Stasi, L.: Forni elettrici, 451 pp. Patron, Bologna (1976) (in Italian)
54. Aliferov, A., Bikeev, R., Gorieva, L., Lupi, S., Forzan, M., Barglik, D.: Arc Furnaces, 204 pp. NGTU-Novosibirsk State Technical University (2016), ISBN 978-5-7782-2813-9 (in Russian)

Chapter 4
Resistance Furnaces

Abstract Resistance furnaces are heating installations that use the heat generated by Joule effect in appropriate heating elements (resistors) located on the walls of the furnace chamber, and transmitted to the workpiece to be heated mainly by radiation and convection. The electrical energy transformed into heat in the resistors is used in part to raise the temperature of the charge and in part to heat the walls of the chamber and to compensate for the furnace heat losses. After a description of the different constructive types of furnaces (batch, continuous, for low, medium or high temperature, with protective atmosphere or in vacuum), heating cycle, criteria and materials for wall design and test methods of this type of furnace are analyzed. In the final paragraphs, we will describe the different types of resistors, characteristics of materials used for their construction and design criteria of resistors for classical radiation or convection furnaces. At the end, furnaces with high specific power resistors are briefly discussed, making also reference to the energy balance of this type of furnaces.

4.1 Introduction to Resistance Heating

Resistance heating uses the heat generated by Joule effect in a conductive body, metallic or non-metallic, in which an electrical current flows.

Depending on whether the current flows in appropriate heating elements (resistors) independent of the workpiece to be heated or directly into it, the heaters are classified as *resistance furnaces* (or indirect resistance heating installations) or *direct resistance heaters*, where the current flows directly in the workpiece.

The first group includes different types of electrical resistance furnaces, used in a very wide range of applications in many industrial sectors. In these furnaces the heat is developed in resistors located on the walls of the furnace chamber, and is transmitted to the workpiece to be heated mainly by radiation and convection.

The heaters of the second group are less widespread since, due to their specific characteristics, they are only suitable for special applications (heating of "long" workpieces of relatively small cross-section, fluids, etc.). However, they allow to

S. Lupi, *Fundamentals of Electroheat*, DOI 10.1007/978-3-319-46015-4_4

achieve high production rates and very high energetic efficiency because the heat sources are located directly within the workpiece to be heated.

They will be discussed in Chap. 5.

4.2 Resistance Furnaces

A resistance furnace is schematically constituted as shown in Fig. 4.1.

It comprises the following main elements:

- a heating chamber of appropriate dimensions in order to accommodate the given charge, designed with refractory and insulation materials able to withstand the maximum working temperature and to keep heat losses down to an acceptable value;
- heating elements of different types, usually supported on the walls or, additionally, incorporated in the floor, roof and door;
- a temperature control system;
- the supply from the mains;
- systems for fast loading and unloading the charge in batch furnaces or mechanisms for carrying the load through the furnace in continuous furnaces;
- auxiliary devices, like fans in forced convection furnaces, or those for producing the required atmosphere in furnaces with controlled atmospheres or in vacuum.

The technical function of the furnace is to provide to the charge the amount of heat required for heating and maintaining it at a pre-set temperature or even, sometimes, to remove the heat for cooling the charge within the same chamber, in a controlled manner.

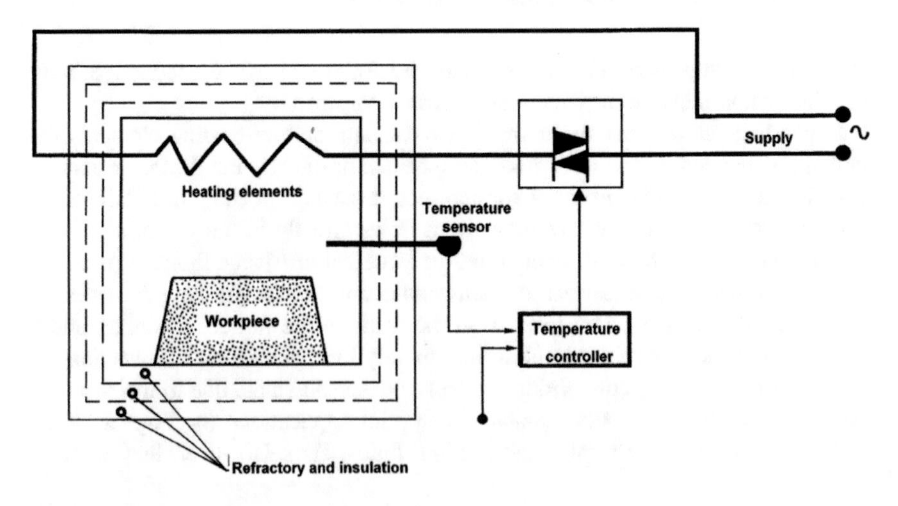

Fig. 4.1 Schematic of a resistance furnace [1]

In implementing these functions it is impossible to avoid unwanted heat losses, such as those through the chamber walls.

The electrical energy transformed into heat in the resistors is therefore used in part to raise the temperature of the charge (*useful energy*) and in part to heat the walls of the chamber and to compensate for the furnace heat losses (*energy losses*).

In thermal steady-state conditions the energy losses are totally transmitted to the surrounding environment through the thermal insulation of the heating chamber; under non-stationary conditions, i.e. when the temperatures of the furnace elements vary in time, a portion of this energy is stored in the heat capacities of the construction elements when the temperature increases, and is lost when the temperature decreases.

The analysis of operation of a furnace requires to consider the following main problems:

1. criteria of walls design, evaluation of the stored (or accumulated) heat and calculation of the losses
2. design of resistors, on the basis of the laws of the heat transfer which takes place by radiation and convection from them to the workpiece and to the chamber walls
3. analysis of the heating transient in the workpiece to be heated for the determination of optimum heating power and heating time, in order to avoid local overheating, risk of deformations and unwanted temperature differences in the workpiece.

4.3 Types and Classification of Resistance Furnaces

Resistance furnaces can be classified in different ways, according to the fields of application, mode of heat transfer, temperature levels to be reached or how the charge is introduced in the heating chamber and undergoes the heating cycle.

We can therefore distinguish:

- furnaces for melting or for heating liquids and solid charges
- radiation, convection or conduction furnaces
- furnaces with normal atmosphere (air)
- furnaces with gas or controlled atmospheres (both of batch or continuous type)
- vacuum furnaces (with vacuum tight chamber and auxiliary facilities for the production and maintenance of vacuum).

As regards temperature levels, the furnaces can be classified into three groups, which differ by their design mainly for the choice of refractory, insulating and metallic materials, the construction of the chamber walls in single or multiple layers, and the choice of resistors.

In particular, we distinguish:

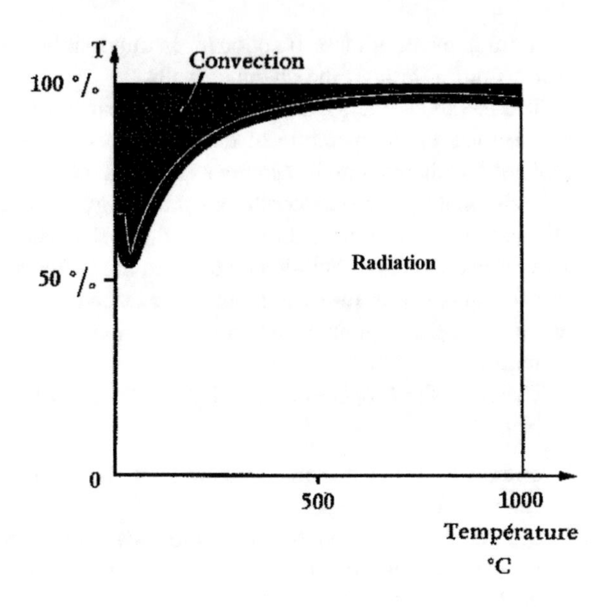

Fig. 4.2 Percentage of convection and radiation heat transfer as a function of temperature [2]

- *low temperature furnaces*, up to 600 °C, where natural or forced convection play a considerable role in the heat transmission from the resistors to the charge (as shown in Fig. 4.2)
- *medium temperature furnaces*, from 600 °C to about 1200–1300 °C, where heat transmission is mainly due to radiation
- *high temperature furnaces*, above 1200–1300 °C, that is the upper limit temperature for the use of metallic resistors.

Another classification distinguishes batch furnaces from continuous furnaces.

Batch furnaces are used in most industrial processes for heating and heat treatment of items ranging from small components to large metallic ingots, from metals to ceramics.

In these furnaces the charge is loaded and unloaded through a door of the heating chamber, and remains still during the heating cycle.

Some examples of medium/large batch units are shown in Fig. 4.3.

Since the energy efficiency of the heating process strongly depends on the furnace utilisation factor, this can be improved in some cases by using a bogie hearth on wheels, which runs outside the furnace for loading and unloading (as in Fig. 4.3b), by multiple hearths or by an elevator on which the loaded hearth is raised into the heating chamber using hydraulic rams, in the so-called elevator furnace.

4.3.1 Low Temperature Batch Furnaces

Except furnaces heated with infrared lamps, which are not considered in this book, in these furnaces the heat transfer occurs in a considerable extent by convection.

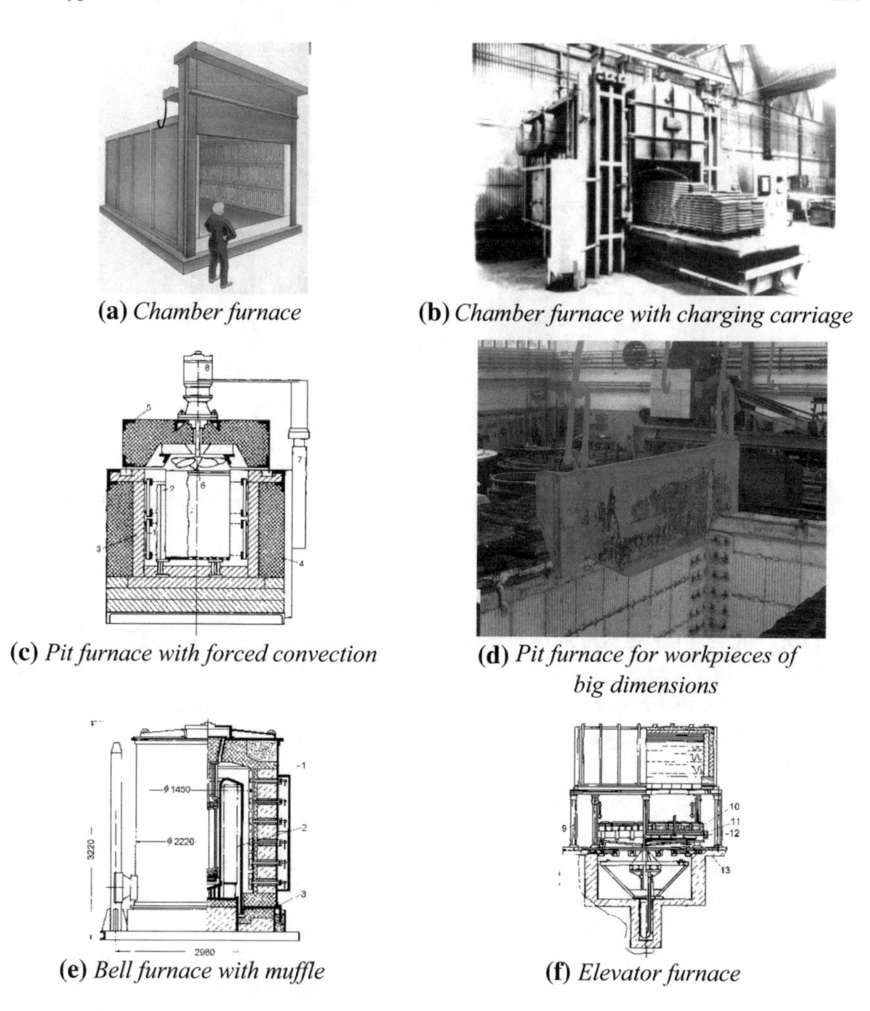

(**a**) *Chamber furnace*
(**b**) *Chamber furnace with charging carriage*
(**c**) *Pit furnace with forced convection*
(**d**) *Pit furnace for workpieces of big dimensions*
(**e**) *Bell furnace with muffle*
(**f**) *Elevator furnace*

Fig. 4.3 Different types of medium/large batch furnaces [3]

They also differ in the construction from the high temperature furnaces, in an increased use of metal parts within the heating chamber which is possible for the lower working temperature and, in some cases, the lack of refractory lining.

One of the most popular type is the so called *chamber furnace* with still charge, where the required output is obtained by successive heating cycles, preceded and followed by loading and unloading.

The simplest type is the furnace with natural air convection, shown in Fig. 4.4a.

The difficulty to achieve effective convective air flow paths and the low values of the coefficient of heat transfer (3–15 W/m °C) limit the field of application of these furnaces only to small units and low production rates.

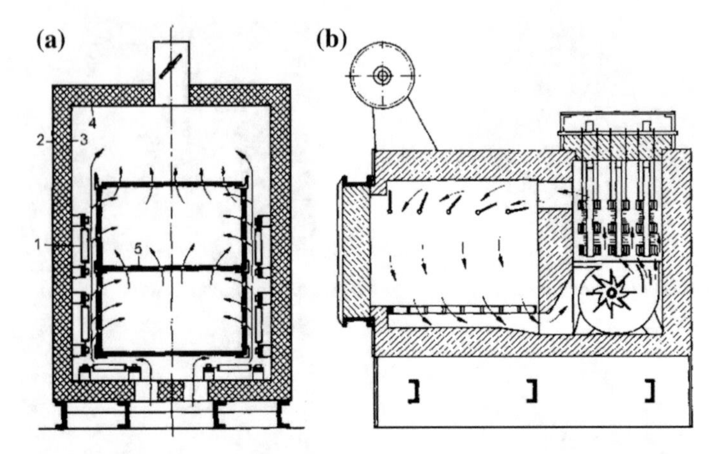

Fig. 4.4 **a** Chamber furnace with natural convection. **b** Chamber furnace with forced convection and separate air heating chamber [3]

Therefore the furnaces characteristic of this category are those with forced convection, where the heat transfer coefficient can reach values of 10–150 W/m^2 °C.

In these furnaces a fan produces a closed-circuit flow of air which, lapping the resistors and absorbing heat, rises its temperature. The hot air is then carried in contact with the charge, transferring to it the heat required for increasing its temperature.

Typical of this group are the so-called bell and pit furnaces; an example of the latter is shown in Fig. 4.3c.

In some cases, when the furnace chamber is large and high temperature uniformity is required, the resistors can be placed in an air heating chamber, separated from that where the charge is placed, as shown in Fig. 4.4b.

In other cases, in order to obtain a better control of the heating cycle, the heating chamber can be subdivided into several zones, with different installed power and independent power control of resistors of each zone.

4.3.2 Medium Temperature Batch Furnaces

The furnace walls are generally realized with refractory and insulating materials arranged in several layers, as in the example of Fig. 4.3c.

In medium temperature furnaces the walls always consists of two layers: the internal one made of refractory bricks, which often act also as support of resistors, and one of thermal insulation supported externally by the metal frame of the furnace.

In these furnaces, as in the low temperature ones, are always used metal resistors, mostly arranged on the roof, the side walls and the floor of the heating chamber, as shown in Fig. 4.5.

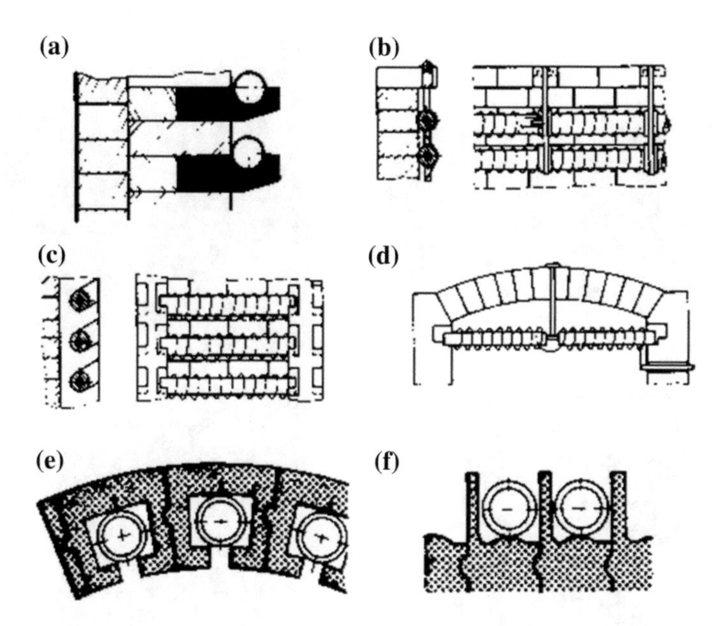

Fig. 4.5 Arrangement of resistors on the walls, roof and floor of the heating chamber [2]

The most used types of medium temperature batch furnaces are *chamber furnaces* (of Fig. 4.3a, b), *pit furnaces* (Fig. 4.3c, d) which have the charging door located on the roof of the heating chamber and *bell furnaces* (Fig. 4.3e) consisting of a heating chamber open at the bottom, which is cyclically placed on several working bases, each in turn constituting the floor of the heating chamber.

The energy efficiency depends on several factor, like furnace design, utilisation factor, method of loading and unloading the charge, time of opening the doors, dead time between subsequent heating cycles, type of atmosphere, heat recovery, etc. Therefore the losses can be considerably different in different furnaces or in the same furnace differently operated, so that they can range from 15 to 40 % of the rated furnace power.

4.3.3 Continuous Furnaces

In these furnaces the material is passed with continuous movement through the furnace chamber which is generally in the form of a tunnel.

As shown in the examples of Fig. 4.6, several types of continuous furnace are used, the main difference regarding the mechanism for carrying the load through the furnace.

As shown in Fig. 4.6a, the furnace is often divided into several zones, with different installed power and power control, so that the work-pieces can undergo, if required by the process, a predetermined heating cycle.

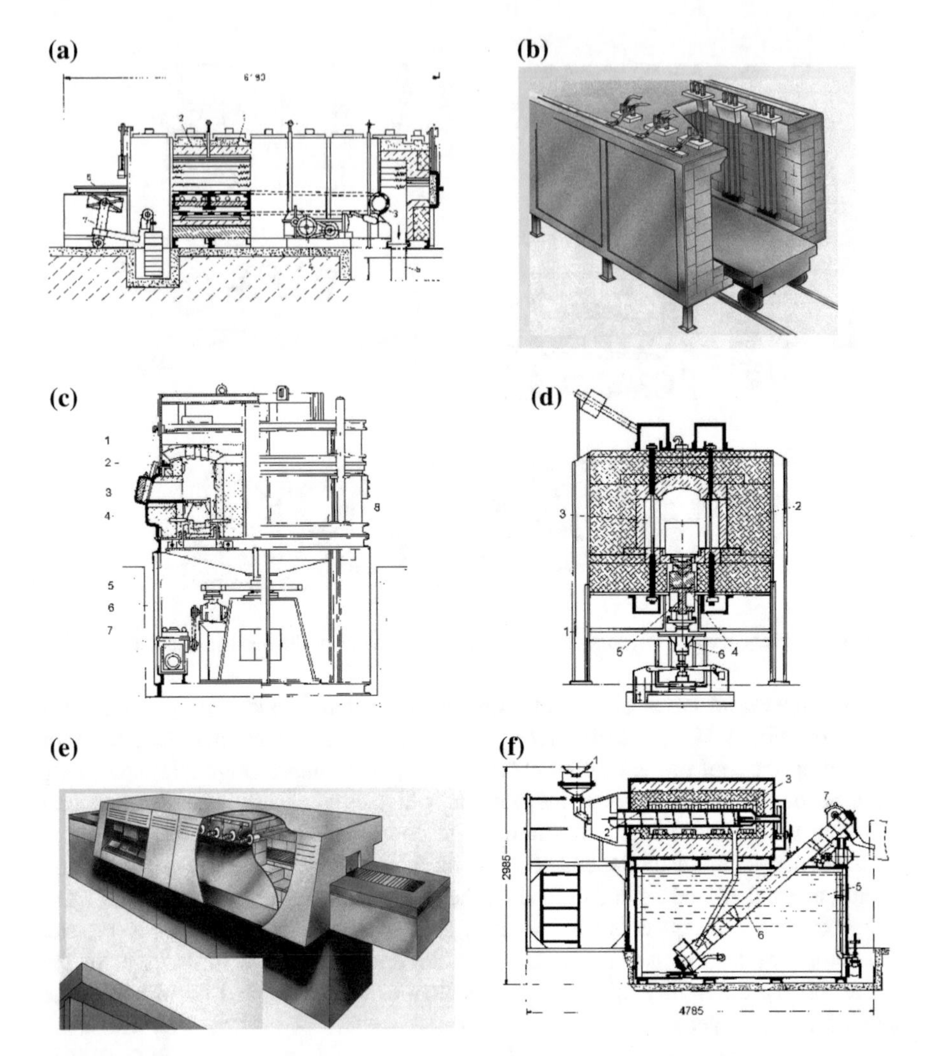

Fig. 4.6 Different types of continuous furnaces [3, 4]. **a** Conveyor belt furnace. **b** Tunnel furnace with movable hearth. **c** Carousel furnace. **d** Walking beam furnace for high temperature. **e** Furnace for thermal treatment of metallic strips. **f** Muffle furnace for hardening processes

For their ability to perform completely automated thermal treatments, these furnaces are particularly suited for high production rates and use in production chains.

Unlike batch furnace, where heating and (total or partial) cooling periods alternate, this type of furnace operates for most of the time under constant power and temperature conditions.

A major drawback is constituted by the heat losses from the openings necessary for the entry and the exit of workpieces; this necessarily reflects on the energetic efficiency of the heating process.

Moreover, this type of furnaces is less suited to operations with controlled atmospheres or in vacuum.

The following figures show some examples of the most used types of continuous furnaces.

The *conveyor belt furnaces* (Fig. 4.6a) are used mainly for small workpieces and temperatures up to 900 °C; the charge moves within a tunnel on a metal belt driven by rollers, which replaces the floor of the furnace.

Sometimes, although this may result in significant heat losses, the belt extends outside the tunnel to allow an easier loading of the charge and to ensure safe operation of the belt drive, by placing it outside the high temperature area.

The belt is made in different ways, depending on the weight of the charge.

The resistors are arranged, usually on the roof of the heating tunnel, under the conveyor belt, or even—less frequently—on the side walls.

Pusher furnaces are equipped with an electro-mechanical or hydraulic system that pushes forward the workpieces along the heating chamber on rails or rollers placed on the floor (Fig. 4.7a). The energy required for driving the pusher can be considerable at high temperature due to the friction on the rails.

In case of small items the pusher does not act directly on them but on appropriate containers (Fig. 4.7b). The containers, however, absorb a considerable part of the heat supplied resulting in considerable losses, thus decreasing efficiency. Moreover, they require a transportation system outside the furnace and have always short life due to the high working temperature.

These furnaces are mostly used for working temperatures up to about 1000 °C.

The above mentioned drawbacks, due to the severe operating conditions of the charge transport device in the high temperature zone of the heating chamber, are greatly reduced in case of workpieces of big dimensions, by the use of a lifting cam or a "pilgrim system" conveyor of the types shown in Fig. 4.8. With these systems the movement is made step by step, periodically raising the workpiece, moving it forward and then laying it down.

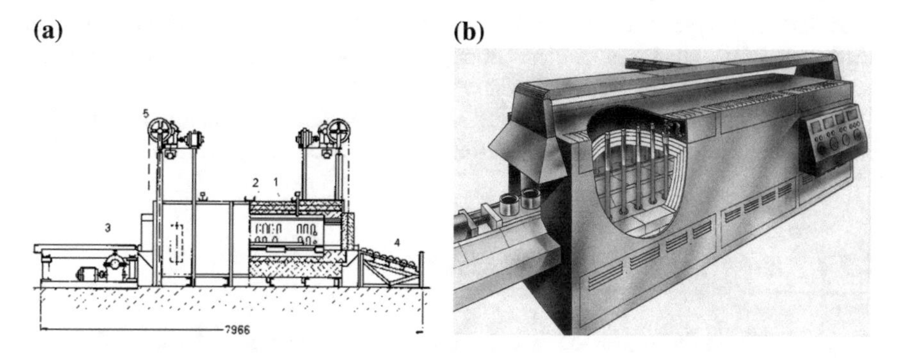

Fig. 4.7 Pusher furnaces [3]

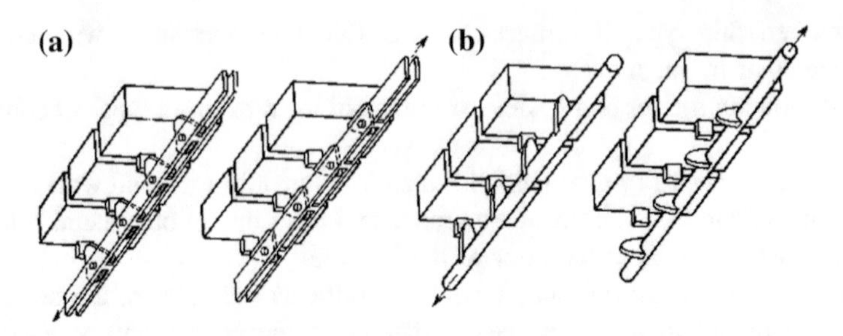

Fig. 4.8 Transport systems: **a** "pilgrim system"; **b** with lifting cams—[5]

These transport systems being housed in a longitudinal groove in the floor of the heating chamber, will work in the high temperature zone only for a relatively short period of the working cycle.

A typical furnace of this type is shown in Fig. 4.6d.

Another type of continuous furnace, with the mechanism for moving the charge located outside the high temperature zone, is the so-called *carousel furnace* of Fig. 4.6c. It consists of a heating chamber with a circular bogie which can rotate 360° around the axis of the chamber.

This type of construction is particularly suited for operation at high working temperatures.

The *muffle furnace* of Fig. 4.6f is used for heating small items up to a maximum temperature of about 900 °C, in particular for hardening heat treatments. It consists essentially of a refractory drum rotating around a horizontal axis, having inside a spiral which—by rotation—produces the advancement of workpieces.

The special feature of this furnace is that the heat transfer occurs from the resistors to the refractory drum, and then from the drum to the charge. As the figure shows, the installation may comprise also a swimming pool for the quenching process. It represents a typical example of furnace in which is performed not only the heating, but also a complete technological process.

A special design have furnaces for heating steel, non-ferrous metals strips or wires, where the charge is "pulled" through the heating chamber at high speed; an example of this type of furnace is shown in Fig. 4.9a.

To avoid very long heating chambers, with large floor occupation, the chamber can be placed vertically, forcing the band to follow a zig-zag path (as in Fig. 4.9b).

4.3.4 High Temperature Furnaces

They differ from those previously described for the choice of materials and design criteria, which are imposed by the operating temperatures.

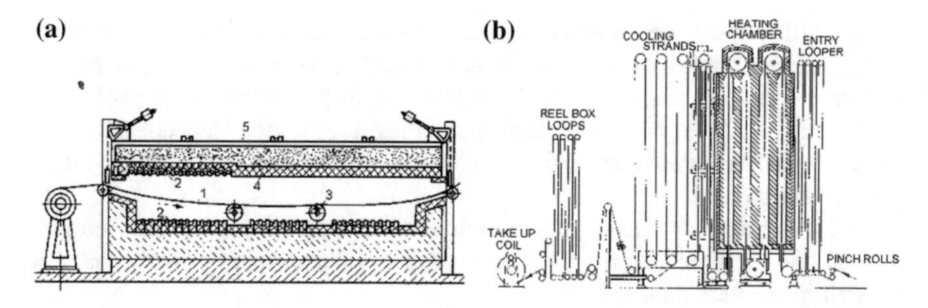

Fig. 4.9 Furnaces for strips and wires: **a** pass-through furnace; **b** furnace with vertical chamber [3, 6]

This reflects in particular on the following elements:

- *walls of the heating chamber*
  they are usually constituted by three layers, the innermost made with dense refractory material for high temperature (rich in Al_2O_3), the intermediate one of "Schamotte" with lower thermal conductivity, the third one of thermal insulating material. In furnaces with graphite resistors, the thermal insulation is made in some cases with graphite powder; in vacuum furnaces are often used to this end multiple reflecting screens.
- *building materials*
  in the high temperature areas cannot be used metal alloys, which are replaced with silicon carbide (SiC) or ceramic elements.
- *resistors*
  in this temperature range it is no longer possible to use metal resistors; they are replaced by resistors of silicon carbide (SiC) for temperatures up to 1400 °C, molybdenum disilicide ($MoSi_2$) up to 1750 °C, molybdenum up to 1650 °C (in vacuum) or to 2000 °C (in protecting atmosphere), graphite up to 2600 °C (in vacuum).

High temperature furnaces are mostly of batch type, like chamber furnaces, pit furnaces, etc. or—less frequently—of continuous type, e.g. carousel furnaces.

4.3.5 Furnaces with Protective Atmospheres

During heating and heat treatments of workpieces of steel, copper, brass and their alloys or metals such as titanium, molybdenum, zirconium and tungsten, always occurs—albeit to a different extent—surface oxidation.

Moreover, in the case of steel heated at high temperature, occurs also the decarburization of the workpiece external layer.

Given that these two phenomena have considerable influence on mechanical properties and subsequent operations (e.g. welding), it is necessary in many cases to perform the heating and the whole heat treatment in protective gas atmosphere.

Sometimes, like in surface hardening, nitriding, carburizing and carbo-nitriding, the atmosphere may have not only a protective function, but also actively participates in the process.

As already mentioned, in high temperature furnaces, special gas atmospheres can be used also for protecting the resistors or other metal parts of the heating chamber, in order to increase their life.

Different gases are used depending on the chemical behaviour required, e.g. "neutral", carburizing, decarburizing, reducing or oxidizing. These reactions depend mainly on gas composition, heated material and temperature.

Typical gases include argon, hydrogen, atmospheres of hydrogen and nitrogen obtained from cracked ammonia, gases produced by partial or total combustion of gaseous hydrocarbons, city gas, gas from coke coal and others (see Table 4.1).

Reducing atmospheres (natural gas, hydrogen, cracked ammonia, etc.) and inert gases (nitrogen, argon, vacuum) are used to remove or to avoid oxidation in processes such as bright annealing or reduce scale formation in annealing and normalising processes.

However, the use of furnaces with controlled atmosphere, along with the benefits, involves construction problems and additional costs arising from the need of the equipment for gas production and feeding, provisions for making the furnace tight, increase of heat losses, effects of certain gases on some furnace elements (like metal resistors and/or refractory lining) and danger of explosions or toxic effects of the gases themselves.

The furnaces with protective atmospheres are mostly batch, pit, bell or elevator furnaces, in many cases with a refractory gas-tight muffle for protecting the charge during cooling.

Table 4.1 Typical furnace atmospheres—[7]

Gas	Analysis % (excluding H_2O)					
	H_2	CO	CO_2	NH_3	CH_4	N_2
Cracked ammonia	75	–	–	–	–	25
Endothermic gas						
(a) from natural gas	40	20	1	–	1	Bal
(b) from propane	30	22	1	–	1	Bal
Exothermic gas						
(a) rich	17	12	4	–	1	Bal
(b) forming gas	10	–	–	–	–	90
(c) lean	2	2	10	–	–	Bal
Hydrogen	100	–	–	–	–	–

The gas consumption is in the range of 2.0–2.5 m^3/h per m^3 volume of the heating chamber, and the capacity of the gas production equipment between 5 and 250 m^3/h.

There are also realizations of continuous furnaces with gas atmosphere, but in this case, because of the difficulties of assuring gas-tightness, the gas consumption is two to three times higher than that previously indicated.

To prevent hazards caused by air infiltration into the furnace, the operation generally is done with a small over-pressure (5–15 mm of water column) in the heating chamber and the excess gas is burned in special burning "candles"; for the same purpose are also provided "curtains of fire" on the internal part of the chamber door.

4.3.6 Vacuum Furnaces

Heating processes in vacuum are used in the industry for decades, since they offer a range of benefits that allow in special applications to obtain high quality products.

Typical application areas are:

- production of special metals, semiconductors or ceramic products
- sintering of metal powders
- annealing of magnetic alloys, steels with low and high carbon content, stainless steels, etc.
- brazing without flux of steels and magnetic alloys
- hardening of special steels
- special heat treatments such as degassing titanium and zirconium, combined brazing and hardening, impregnation in vacuum, etc.

The main advantages that have led to the development of this technology can be summarised as follows:

- elimination of oxidation and contamination and possibility of degassing the metal during heating;
- way for obtaining high-purity materials and surfaces of workpieces that do not require subsequent heat treatment and further machining;
- ability to achieve high working temperatures, since the heating elements are protected from oxidation and attack of gases.

Vacuum furnaces, however, require high installation and operating costs and, therefore, are used only when they are the unique or best way to ensure a high quality product.

As regards the walls of the vacuum-tight chamber, vacuum furnaces can be subdivided into two groups called respectively "hot wall" and "cold wall" furnaces.

Hot wall furnaces are characterized by a vacuum-tight container heated externally (Fig. 4.10).

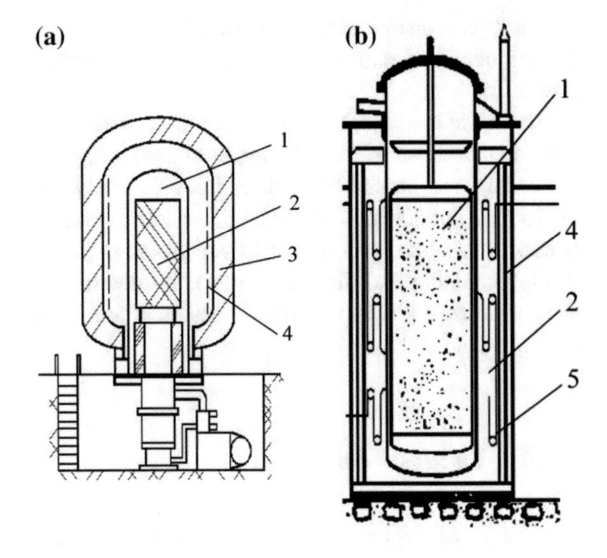

Fig. 4.10 Hot wall furnaces **a** single-vacuum bell furnace (*1* vacuum container; *2* workpiece; *3* thermal insulation; *4* resistors) **b** double vacuum pit furnace [2, 3]

Among them, a further distinction is made between *single-vacuum* and *double-vacuum* furnaces.

In the first type (Fig. 4.10a), vacuum is realized only within the container itself, while in the latter (Fig. 4.10b) a lower degree of vacuum is present also between the vacuum-tight container and the furnace walls in order to reduce, at temperature above 800–900 °C, the mechanical stress acting on the main vacuum container as a result of pressure difference between the inside and the outside of the wall.

In this way, in double vacuum furnaces, temperatures of about 1100–1200 °C can be reached.

For the same reason, in simple vacuum furnaces, the working vacuum pressure is seldom lower than 10^{-2} Torr.

However, modern vacuum furnaces are—almost without exception—*cold wall furnaces*, i.e. with the wall of the vacuum-tight container cooled with water circulation in an internal gap of the wall, while resistors and thermal insulation are placed inside the vacuum container (Fig. 4.11).

Advantages of these furnaces, in comparison with the hot wall ones, are the possibility to reach very high temperatures (up to 2000–3000 °C) and to operate at very low residual pressure without the problems due to mechanical stresses on the container at high temperature, their low thermal inertia, the absence of heat losses to the ambient and the possibility to realize large furnaces.

As shown in Fig. 4.11, the charge (1)—placed inside the container (2)—is surrounded by the resistors (3). In order to limit thermal losses, between the resistors and the cold container is interposed a thermal insulation (4) consisting of mineral fibres, graphite felts or multiple metal screens.

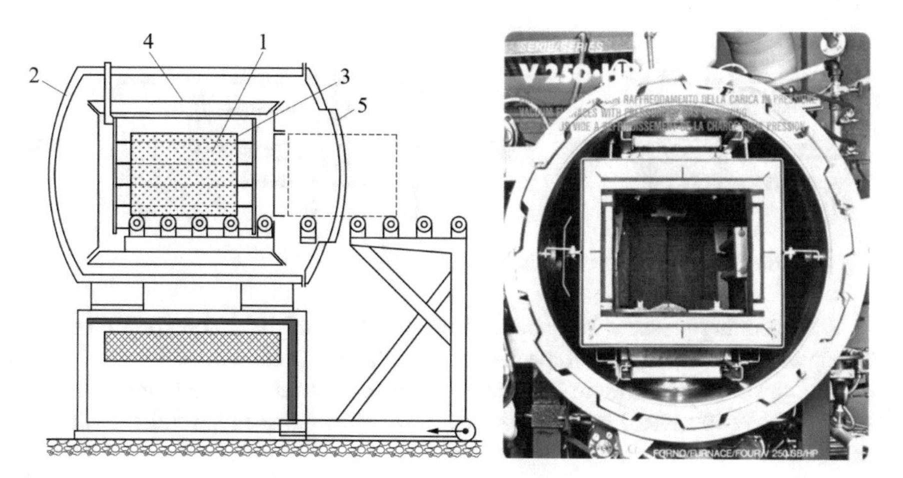

Fig. 4.11 Cold wall vacuum furnace (*1* heating chamber; *2* metallic container with water cooled walls; *3* resistors; *4* thermal insulation) [2]

The main problems in the construction of these furnaces are:

- mechanical problems and tightness of the cylindrical vacuum container, usually made of stainless steel, with rubber seals on the container flanges;
- choice of thermal insulation, which is complicated by the need of using materials which must be easily and quickly degassed and should not contaminate the charge.

The selection of the insulating material should therefore be done taking into account working pressure and temperature. The types of insulation mostly used are reflective multiple metal screens which have very low thermal inertia, graphite and graphite felts, light or very light “Schamotte”, or refractory oxides (e.g. Al_2O_3, ZrO_2).

- selection of the heating elements which, depending on temperature, residual pressure, cost of material and characteristics of the charge, are made of Ni–Cr, Mo, Graphite or Tungsten.
- problems of rapid cooling, which is necessary in order to increase the utilisation of installations. These problems are solved by the introduction of an inert gas into the container at the end of the heating period. A fan, protected during the heating by a mobile screen, conveys the gas heated by the charge to a water-cooled heat exchanger.

These furnaces are usually of batch type, with fixed charge and chamber with or without a vacuum-tight metal muffle. In some cases, always to the end of increasing productivity, they can also have additional vacuum chambers for charging and cooling the charge, separated from the heating chamber by means of diaphragms, as schematically shown in Fig. 4.12.

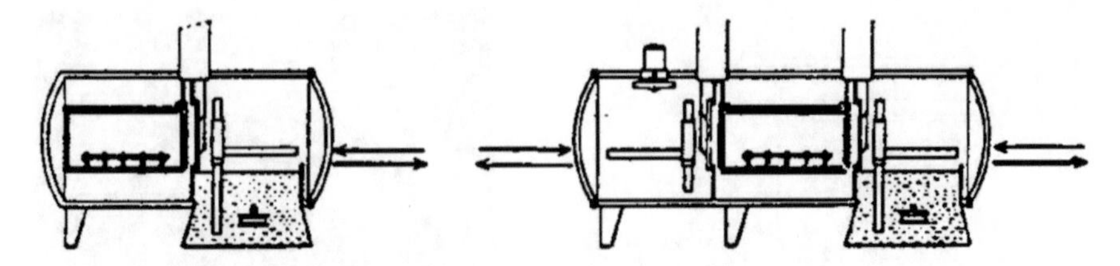

Fig. 4.12 Semi-continuous vacuum furnaces [2]

The operating pressure is generally between 10^{-1} and 10^{-5} Torr (typically 10^{-3} Torr), which is obtained with mechanical or diffusion pumps.

The power rating of these furnaces is extremely variable, with outputs ranging from few kW to several hundred kW and capacities between several tens of dm^3 and 100 m^3.

4.3.7 Special Furnaces

A particular group of furnaces is constituted by those for melting metals and heating liquids.

As schematically shown in Fig. 4.13, these furnaces are characterised by a steel crucible with an external refractory and insulating wall and heating elements placed internally or externally to the tank.

The main applications are for melting metals with low melting point such as lead, tin, zinc, aluminium or magnesium and their alloys.

An example of a melting furnace for aluminium is that of Fig. 4.14.

Other special furnaces are the *salt-bath furnaces*.

In these furnaces the charge is heated by immersion in a bath of molten salts, which generally act only as a means for heat transmission but, in some cases, have also a direct chemical function, such as in cementation and nitriding processes.

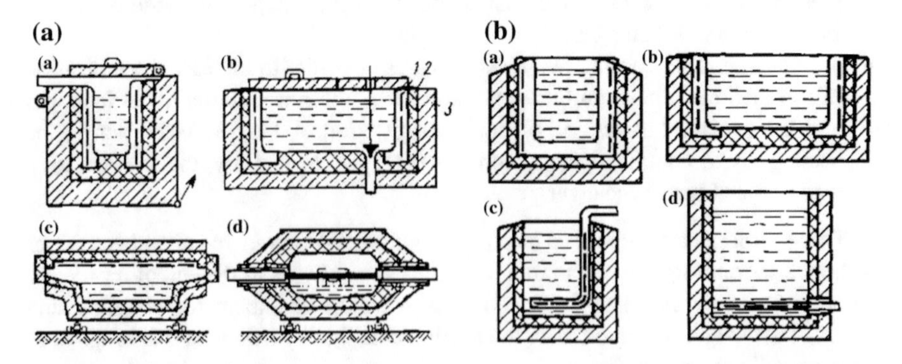

Fig. 4.13 **a** melting furnaces **b** furnaces for heating liquids (*1* heating elements; *2* refractory; *3* thermal insulation) [5]

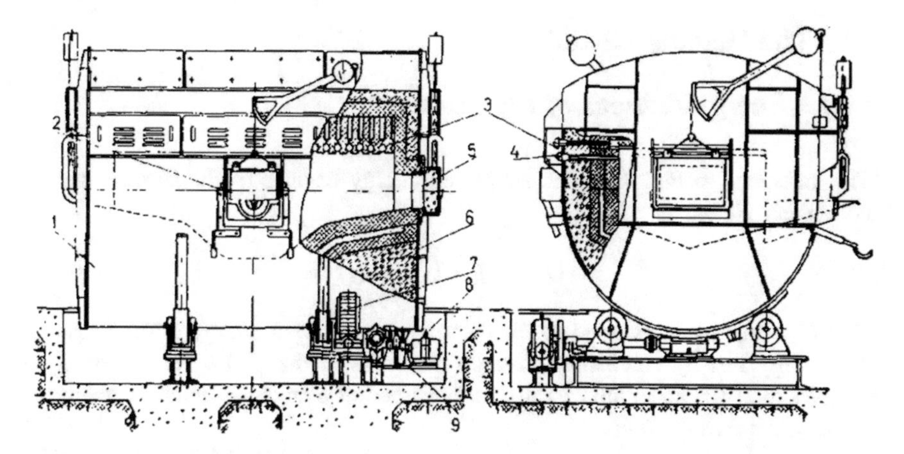

Fig. 4.14 Melting furnace for aluminium (*1* furnace shell, *2* pouring spout, *3* heating elements, *4* temperature sensors; *5* charging door, *6* refractory coating, *7*, *8*, *9* tilting devices) [3]

We can distinguish furnaces with heating elements external to the crucible containing the molten salts and the charge, from those with armored heaters immersed in the bath itself, which are constructively similar to those shown in Fig. 4.13.

Maximum working temperatures in the above two cases are about 900 and 500 °C respectively.

Table 4.2 shows the main types of salts and the temperature range of their use.

Finally, we must mention the so-called "*fluidized bed*" *furnaces*, similar to the salt-bath ones, but less dangerous for workers and environment, in which the medium transmitting heat to the charge consists of a "bed" of material with very fine granulation—e.g. alumina—kept in suspension by an upward gas stream.

The fluidized bed behaves, from a thermal point of view, as a liquid, thus assuring a high heat transfer coefficient between its constituting material and the charge.

The heating occurs in the same way described for salt-bath furnaces. The gas can actively participate in the process or it can have only function of maintaining the fluidized bed.

Maximum working temperature of these furnaces is about 1100 °C.

Table 4.2 Salts and temperature range of use

Chemical composition	Melting point (°C)	Working temperature range (°C)
28 % NaCl + 72 % $CaCl_2$	505	540–870
50 % Na_2CO_3 + 50 % KCl	560	590–815
50 % NaCl + 50 % K_2CO_3	560	590–815
35 % NaCl + 65 % Na_2CO_3	620	650–815
50 % $CaCl_2$ + 50 % $BaCl_2$	600	650–900
22 % NaCl + 78 % $BaCl_2$	630	650–900
44 % NaCl + 56 % KCl	665	700–870
20 % KCl + 80 % $BaCl_2$	750	850–1350

4.4 The Heating Chamber

4.4.1 Energy Balance of the Furnace

With reference to Fig. 4.15, the simplified energy balance of the furnace can be written as follows:

$$Q_w = Q_i - (Q_c + Q_a) \tag{4.1}$$

with: Q_w—energy transferred to the workpiece; Q_i—input energy to the resistors; Q_c—energy lost by conduction through the walls of the heating chamber; Q_a—energy stored (or accumulated) in the construction elements, which is totally or partially lost during cooling.

In order to achieve high process efficiency, the loss terms $(Q_c + Q_a)$ must be kept at the lowest possible value by design of the walls of the heating chamber and a convenient choice of refractory and insulating materials.

However, the two terms have a different impact on the efficiency, depending on the furnace heating cycle. If the unit is designed to operate at constant temperature, the walls should have a high thermal inertia in order to maintain this temperature constant under variable load conditions. On the contrary, if the operation is discontinuous, with frequent heating and cooling cycles, a high thermal capacity gives rise to a high stored energy Q_a during the heating stage which is lost in the ambient during the cooling one.

The design of the walls, which may be constituted by several layers, must also keep to a minimum the conduction losses Q_c.This can be obtained by increasing the thickness of the insulation layer(s) and by the choice of low thermal conducting materials. Finally, the selection of refractory and insulating materials must be done in relationship with the maximum temperature at which each layer constituting the wall will be operated.

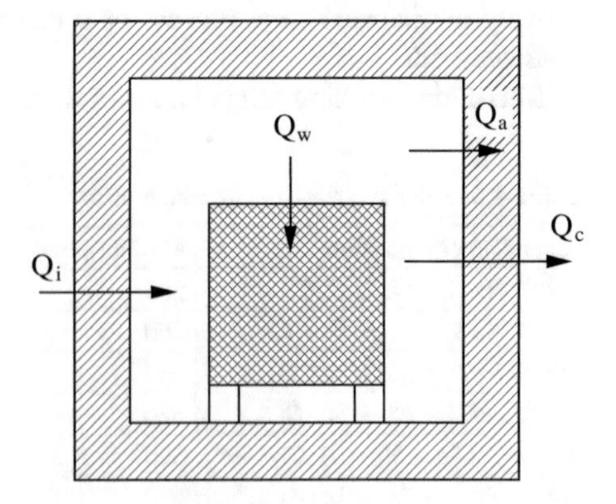

Fig. 4.15 Energy balance of a furnace [8]

4.4.2 Heating Cycle of an Empty Furnace

The knowledge of the working cycle of the furnace it is therefore of fundamental importance for designing the furnace walls and the dynamic temperature controller.

From the process control point of view, a furnace with charge and temperature sensor (e.g. thermo-couple) is made of a combination of thermal storage capacities and thermal resistances. Figure 4.16 shows the physical arrangement and an equivalent diagram [10].

The temperature sensor (T) measures a conventional temperature in the furnace chamber which depends on the influence of heating elements (H), furnace walls (W) and charge (C). However, the great diversity of the charges (form, shape, material and mass) makes almost impossible to describe in a general way the furnace together with the charge.

Fortunately, the temperature control of a furnace with charge is easier than that of an empty furnace. Therefore, choosing the temperature controller and its setting on the basis of the data of the empty furnace and applying it to a charged furnace, one is always on the safer, more stable side of its dynamic behaviour.

Moreover, the knowledge of the no-load losses in thermal steady-state conditions and the losses corresponding to the stored energy gives a quality factor of the furnace itself, independent of the characteristics of the charge.

It's therefore interesting to analyse the transient temperature and power distributions in an empty furnace when the rated voltage is applied to the heating resistors and the rated power P_N is initially taken from the supply.

As shown in Fig. 4.17, we can distinguish the following characteristic periods:

- A first period from $t = 0$ to $t = t_M$ ("*dead time*") in which from the supply is absorbed the power $P = P_N$. In this period occurs a rapid increase of the

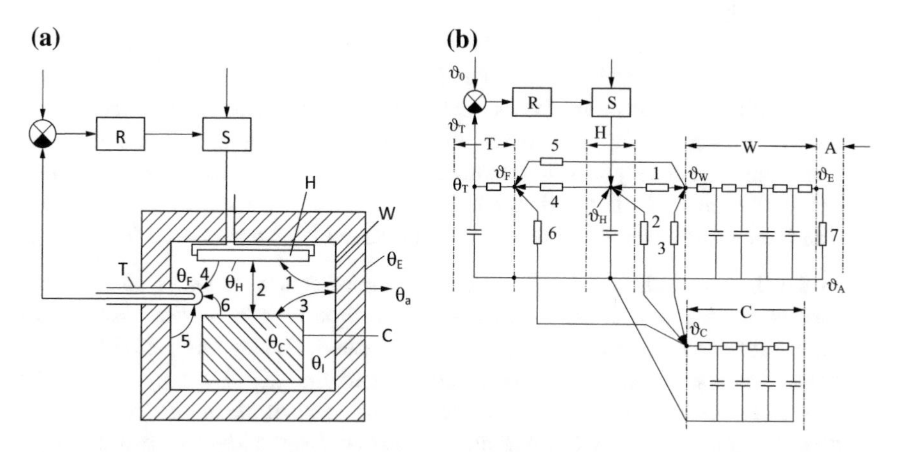

Fig. 4.16 Temperature control loop of a resistance furnace (*H* Heating elements; *W* Walls; *C* Charge; *T* Temperature sensor; *A* Ambient; *R* controller; *S* final control element; Heat transfer resistances: *1* H/W; *2* H/C; *3* W/C; *4* H/T; *5* W/T; *6* C/T; *7* W/A) [10]

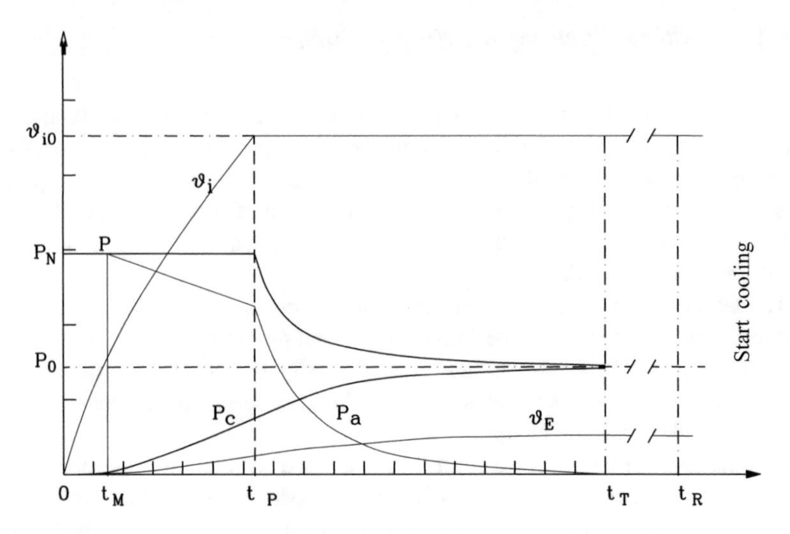

Fig. 4.17 Transient temperature and power variations in an empty furnace [1]

temperature ϑ_i of the internal surface of the walls while the temperature of the external surface remains practically at the ambient temperature, i.e. $\vartheta_E \approx \vartheta_A$ (assumed to be zero in the diagram). In this situation the heat exchange with the ambient is nil and the total energy supplied is stored in the thermal capacity of the walls.

- A second period for $t_M \leq t \leq t_P$, characterized by constant absorbed power P_N and conduction losses P_c through the wall, which are no more negligible since the temperature difference $(\vartheta_e - \vartheta_a)$ causes the transmission of heat from the wall external surface to the ambient. The time t_P, named "*practical heating up time*", is the time required to heat up the empty, dry furnace to a given temperature ϑ_{i0}, after application of the rated voltage to the resistors, i.e. the time elapsed until the temperature measured by the sensor fitted by the manufacturer reaches the set temperature for the first time. The difference between the energy absorbed from the supply and the energy lost by conduction is stored in the walls (and eventually in other furnace elements) and corresponds to the area below the line of P_a in the diagram.
  The total energy stored at the instant t_P is named "*dynamic stored heat*" Q_{AD}.
- In the third period, $t_P \leq t \leq t_T$, the temperature controller keeps constant the furnace temperature at the value ϑ_{i0} till the time t_T ("*theoretical heating up time*") at which the thermal steady state is reached. During this period both the power P_C and the power P_a, corresponding to the stored energy, vary from instant to instant. At the time t_T the thermal capacitors of the construction elements are completely charged and the power absorbed by the resistors must only compensate the steady-state conduction losses P_0. The energy Q_{AS} stored from $t = 0$ to t_T is named "*static stored heat*".

- In the period after the time t_T thermal steady state conditions are kept constant by the temperature controller, the temperatures ϑ_i and ϑ_E remain constant and the power P_0 absorbed by the resistors corresponds to the no-load losses of the furnace.
- The last period—a cooling period—starts at the time t_R when the resistors are switched-off; in this period the energy stored in the walls is transmitted (partially or totally) to the ambient and must be considered lost in the energy balance of the process.

Typical values of specific conduction losses (q_c) and stored energy losses (q_a) for different wall designs of heat treatment furnaces are given, as a function of the furnace temperature, in the diagrams of Fig. 4.18. In other types of furnaces, conduction losses can be higher, in some cases up to 800–1000 W/m^2.

4.4.3 Refractory, Insulating and Metallic Materials

As shown in previous paragraphs, the walls of the heating chamber are realized with a structure with several layers.

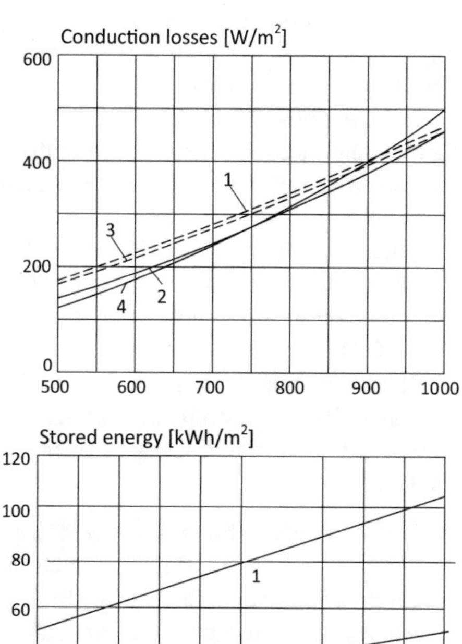

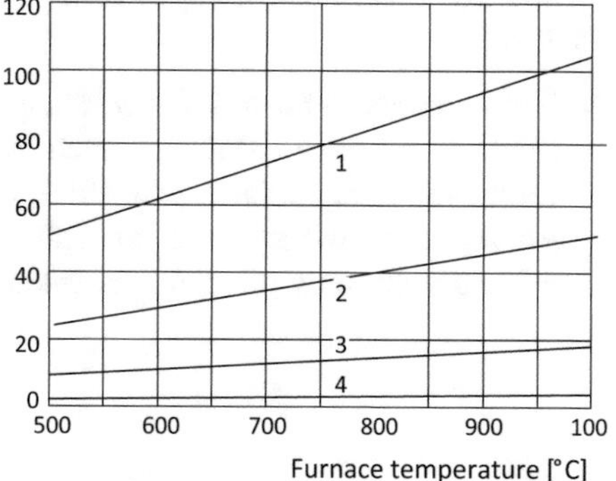

Fig. 4.18 Typical specific conduction losses q_c and stored energy losses q_a as a function of furnace temperature for different wall design (*1* 100 mm, refractory 2000 kg/m^3; 100 mm, refractory 1300 kg/m^3; 100 mm, vermiculite 700 kg/m^3; 75 mm, mineral wool 200 kg/m^3; *2* 100 mm, refractory 1300 kg/m^3; 100 mm, vermiculite 700 kg/m^3; 100 mm, mineral wool 200 kg/m^3; *3* 110 mm, refractory 500 kg/m^3; 75 mm, calcium silicate 375 kg/m^3; 50 mm, mineral wool 140 kg/m^3; *4* 260 mm, fibre 96 kg/m^3; 10 mm, fibre 64 kg/m^3; External surface temperature: 57–59 °C) [9]

Starting from inside, the heating chamber is usually constituted by a first layer of refractory bricks, a second layer of semi-insulating bricks and an external layer of insulating material.

The number of layers is reduced to two or just one in furnaces working at medium or low temperature.

The use of the layered structure is due to the possibility of using in each layer the most appropriate material in relationship to its working temperature and the mechanical, thermal and chemical requirements.

The walls of the heating chamber must meet the following requirements:

- limit heat losses to the ambient in order to ensure the achievement of the desired temperature with optimal thermal efficiency; this can be obtained by building the walls with the lowest possible values of thermal conductivity and, in batch furnaces, of heat capacity;
- constitute a "reserve" of stored heat which, in addition to the heating elements, re-radiates towards the charge; this requires, in contrast with the previous statement, a relatively high heat capacity;
- to be compatible with the atmosphere of the heating chamber during furnace operation
- ensure sufficient mechanical strength.

The main difference between refractory and insulating materials is that refractories must be able to withstand high temperatures while the insulation must exhibit low thermal conductivity.

The innermost layer, which is constituted of refractory material, must have good thermal and mechanical resistance at the operating temperature and to temperature variations, as well as low thermal conductivity and—in most cases—relatively low heat capacity.

Since the refractory bricks often directly support the heating resistors (see Fig. 4.5), they must have also high electrical resistivity and chemical stability in order to avoid any chemical reaction with these elements and short circuits between parts of them.

Refractory materials are non-metallic minerals with high melting point and "softening" point generally greater than 1500 °C. The most commonly used can be subdivided into three groups; their main characteristics are summarized in Table 4.3.

1. Silica-alumina refractories ("*acid*" *refractories*), mainly consisting of alumina (Al_2O_3) and silica (SiO_2), which include Schamotte, Sillimanite, Mullite, corundum, pure alumina, pure silica, etc.
   Among them widely used are different kinds of *Schamotte*, produced from refractory clays by pressing and baking, whose average characteristics are:

Table 4.3 Physical characteristics of main refractory materials

Material	Chemical content (%)	Melting Point (°C)	Density (Kg/dm^3)	Specific heat (Wh/KgC)	Coeff. of thermal expansion (10^{-6}/°C)	Thermal conductivity (W/m °C)	Electrical resistivity (Ω cm 10^3)
Schamotte	15–45 Al_2O_3 55–85 SiO_2	1630–1750	1.7–2.1	0.25–0.29	4.6–7.6	1.24–1.38	
Sillimanite Mullite	60–72 Al_2O_3 28–40 SiO_2	1790–18, 809	2.2–2.4	0.28	4.6	1.24–1.38	2.0
Silica	93–96 SiO_2	1700–1750	1.7–1.9	0.31	–	1.86–2.08	7.0
Corundum	80–99 Al_2O_3	1850–2000	2.5–3.2	0.31	9.4	2.20	1.0
Alumina	Al_2O_3	2050	–	–	–	–	–
Magnesium - oxide	MgO	2800	–	–	–	–	–
Magnesite	80–95 MgO, + Al_2O_3, Fe_2O_3	2000	2.6–3.1	0.34	14.0	3.06–4.44	2.0
Chromium- - Magnesite	60 MgO, + Fe_2O_3 Cr_2O_3	1920–2000	2.8–3.2	0.31	8.0	2.08	1.0
Chromerz	15–33 Al_2O_3 14–19 MgO 10–17 Fe_2O_3 30–45 Cr_2O_3	1800–1900	3.0–3.8	0.26	7.1	2.08	–
Kaolin	CaO	2200–2570	–	–	–	–	–
Dolomite	CO_3Ca CO_3Mg	2300	–	–	–	–	–
Silicon Carbide	90–95 SiC Al_2O_3	1920	2.2–2.7	0.29	4.5–5.5	9.28	5.0
Graphite	90–98 C	2300–3000	1.3–1.8	0.44	5.0	1.05–36.0	–
Zirconium	93 Zr_2O_3; 5 CaO	2677	5.9	1.57	9.4	2.3	0.03

- alumina content 15–45 %
- melting point 1630–1750 °C
- density 1700–2100 Kg/m^3
- thermal conductivity (from 600 to 1200 °C) 1.0–1.4 W/m °C
- specific heat 0.25–0.29 Wh/kg °C
- thermal expansion (from 600 to 1200 °C) 0.7–0.8 %.

   Other types of Schamotte bricks, produced with special production techniques, have lower density (from 1300–1600 to 300 Kg/m^3) and are characterized by much lower values of thermal conductivity and heat capacity.

   They are used when are not required high resistance properties to mechanical stress and temperature variations.

2. "*Basic*" *refractories* are constituted by simple or combined basic oxides such as magnesium oxide (MgO), oxides of iron or chromium (Cr_2O_3–Fe_2O_3), kaolin (CaO) and calcined dolomite (CaO–MgO).
3. *Special refractories*, used in high temperature furnaces, are silicon carbide, graphite, zirconium, or other types of oxides and carbides.

Figure 4.19 gives values of thermal conductivity as a function of temperature for some refractory materials.

Since refractory materials have mostly a relatively high thermal conductivity and therefore would produce substantial conduction heat losses, the function of thermal insulation are primarily entrusted to the outer layer of the chamber wall.

Such layer, consisting of insulating material, is thermally protected towards the inside of the chamber by the refractory layer and is also completely free from mechanical stress.

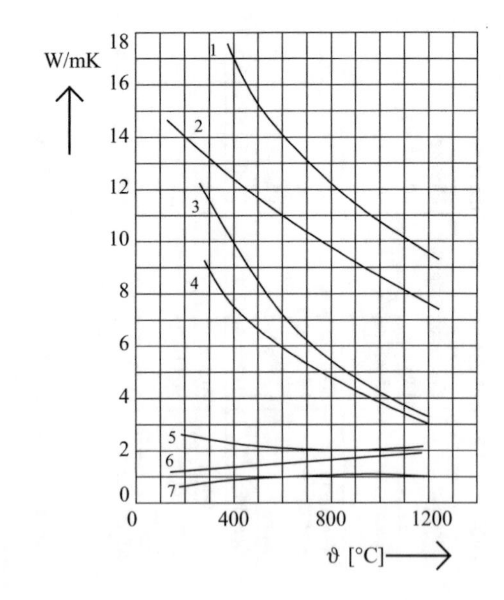

Fig. 4.19 Thermal conductivity of some refractories as a function of temperature (*1*, *2* silicon carbide; *3*, *4* magnesite; *5* corundum; *6* "silika"; *7* schamotte) [3]

All insulating materials are characterized by high porosity, natural or artificial (by convention, higher than 45 %) which contributes to reduce thermal conductivity.

The choice between different insulating materials is based on the following characteristics:

- maximum working temperature
- weight per unit volume
- thermal conductivity
- specific heat.

Among the most used materials we should mention:

- some organic substances (cork, wool, sawdust), used only at very low working temperature;
- magnesium oxide, for temperatures up to about 300 °C;
- glass wool and rock wool, up to 450–500 °C;
- mineral fibers, up to about 600 °C;
- "*Kieselguhr*", up to 900–1000 °C.

The latter, known by the trade name "*Kieselguhr*", is a fine mineral powder (diatomaceous earth), very light, consisting mainly of silica and a small percentage of alumina (about 10 %). It has the following average characteristics:

• maximum working temperature:	1000 °C
• apparent density:	350 kg/m^3
• thermal conductivity at 200 °C:	0.14 W/m °C
at 400 °C:	0.16 W/m °C
at 600 °C:	0.19 W/m °C
at 800 °C:	0.21 W/m °C
at 1000 °C:	0.23 W/m °C
• average specific heat:	0.27 Wh/Kg °C.

The diatomaceous earth insulation can be also produced in form of bricks, having a thermal conductivity slightly different from that of the bulk material.

Average characteristics of some insulating materials are given in Table 4.4, while the diagrams of Fig. 4.20 give the values of thermal conductivity as a function of temperature.

In addition to refractory and insulating materials, in the furnaces are used also metallic materials.

There are special requirements for metallic parts operating at high temperatures, depending on whether they are mechanically stressed or unstressed.

For components not mechanically stressed, in medium-temperature furnaces Cr steels are used; for the mechanically stressed ones—Ni–Cr steels.

Table 4.4 Average characteristics of insulating materials

Material	Density (Kg/m^3)	Thermal conductivity (W/m °C)
Substitutes of asbestos	5.76	0.14
Magnesium carbonate	250	0.056
Rock wool	120–250	0.03–0.06
Kieselguhr (bricks)	180–250	0.08
Cellular cement	260–900	0.06–0.20
Glass wool	220	0.056
Granulated cork	80–100	0.04
Agglomerated cork	250–300	0.06
Wool	136	0.037
Kapok	18	0.034
Felts	120	0.031
Water	1000	0.50
Dry air in motion	1.3	0.12
Dry air motionless	1.3	0.02–0.035

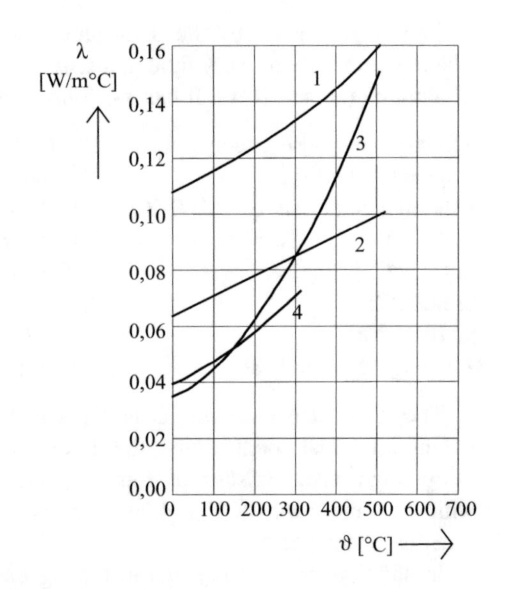

Fig. 4.20 Thermal conductivity of some insulating materials as a function of temperature (*1* Kieselgur bricks; *2* Kieselgur slabs; *3* rock wool felts; *4* bulk magnesium oxide) [3]

In fact, Chromium is the element that provides to the steel high oxidation resistance. Thus, steels with 13–15 % of Cr can be used up to temperatures of about 700–750 °C, with 25–30 % up to 1100–1150 °C.

As regards steels for parts mechanically stressed, it should be remembered that the resistance to mechanical stress decreases very rapidly with the increase of temperature. For example, a steel with high content of Ni and Cr, has a yield point of 1.5 kg/mm^2 at 800 °C, while this value reduces to 0.5 kg/mm^2 at 900 °C, and only to 0.25 kg/mm^2 at 950 °C.

Other steels, sometimes used at high temperature because of their good heat resistance, are Cr–Al steels, but their use is limited by a certain fragility and susceptibility to cracking.

4.4.4 Wall Design

The design of the walls of the heating chamber, i.e. the choice of number of layers, their thickness and types of refractory and insulating materials, are closely dependent on the furnace operating conditions (heating cycle, temperature, atmosphere) and the corresponding losses.

In particular, the design of the walls is fundamentally different for continuous or batch furnaces.

In fact, in the first case are involved on one hand technical considerations on the limitation of conduction losses and the static of construction, on the other, the economical ones related to construction and operating costs. In the second case it must be taken into account not only the above mentioned factors, but also the cost of the losses corresponding to the heat stored in the chamber walls.

With reference to the thermal cycle of Fig. 4.17 and denoting with:

E, Q_c—values of energy absorbed by the heating elements and heat transferred by conduction to the environment, from the instant of switching-on till a generic instant *t* of the thermal steady-state regime, the energy balance of the furnace can be described by the following relations:

$$\left.\begin{aligned} E &= Q_{AS} + Q_C = P_N \cdot t_p + \int_{t_p}^{t_T} P(t) \cdot dt + P_0 \cdot (t - t_T) \\ Q_{AS} &= P_N \cdot t_p + \int_{t_p}^{t_T} P(t) \cdot dt - \int_0^{t_T} P_c(t) \cdot dt \\ Q_c &= \int_0^{t_T} P_c(t) \cdot dt + P_0 \cdot (t - t_T) \end{aligned}\right\} \tag{4.2}$$

In continuous furnaces the initial phase till time t_T—during which the power supplied must also compensate for the accumulated losses—has a relatively short duration compared to the total length of the heating cycle and consequently the term corresponding to the stored heat has a low weight on the whole energy balance.

On the contrary, the amount of stored heat and the corresponding losses have a decisive role in batch furnaces where, due to the variable temperature regime, this energy must be provided at each heating cycle.

In conclusion, if the furnace is for continuous operation it is convenient to minimize by wall design the steady-state conduction losses P_o, compatibly with the economics and static of the construction.

In many cases, the thickness of the wall is determined for statics reasons by the thickness of the refractory layer. To the end of reducing the losses P_o it would be preferable to keep it as thin as possible, to the minimum value able to assure at the surface in contact with the insulating layer, the maximal admissible temperature of the insulating material.

But, in order to guarantee the static of the construction, its thickness cannot be reduced below certain values, which depend on the size of the heating chamber.

As regards the thickness of the insulating layer, on one hand it must be as high as possible in order to minimize the losses P_o; on the other hand it is limited in practice to 2–4 times the thickness of the refractory layer, depending on the operating temperature of the furnace, since—above these values—the decrease of the losses does not justify the higher costs of the construction.

In fact, as it is illustrated by the curves of Fig. 4.21, there is a range of optimal thickness that minimizes the sum of operating and construction costs.

With the increase of the thickness, increases the cost of construction (curve 2) but become lower the losses and, finally, the operating costs (curve 1).

The curve 3—sum of the previous two—shows a minimum, which corresponds to the most cost effective wall thicknesses.

The steady-state conduction losses P_o can be calculated remembering that, in analogy with Ohm's law, the heat transmitted by conduction through the wall is directly proportional to the temperature difference $(\vartheta_i - \vartheta_e)$ between the inner and outer surfaces and inversely proportional to the wall's thermal resistance.

In case of walls with several layers, the overall thermal resistance is equal to the sum of the thermal resistances of each layer.

Fig. 4.21 Range of optimum wall thickness (*1* conduction losses/operating costs; *2* stored energy/construction costs; *3* total losses/total costs; *4* optimal thickness)

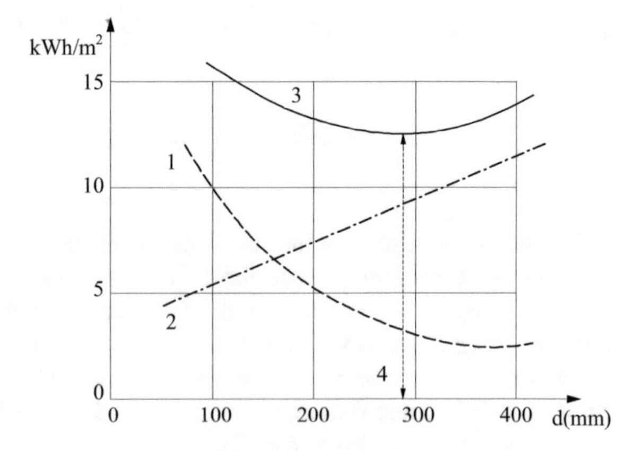

Then we can write:

$$P_0 = \frac{\vartheta_i - \vartheta_e}{\frac{s_1}{\lambda_1 A_{1m}} + \frac{s_2}{\lambda_2 A_{2m}} + \cdots + \frac{s_n}{\lambda_n A_{nm}}} = \frac{\vartheta_i - \vartheta_a}{\frac{s_1}{\lambda_1 A_{1m}} + \frac{s_2}{\lambda_2 A_{2m}} + \cdots + \frac{1}{\alpha_E A_E}} \tag{4.3}$$

with:

P_0	steady-state conduction losses (W)
$s_1, s_2, \ldots, s_n$	thicknesses of layers 1, 2, …, n (m)
$\lambda_1, \lambda_2, \ldots, \lambda_n$	thermal conductivity of layers (W/m °C) at their average temperature
α_E	heat transmission coefficient between external surface and ambient (W/m^2 °C)
A_E	area of the external surface (m^2)
$A_1, A_2, \ldots, A_n$	areas of the internal surfaces of layers 1, 2, …, n (m^2)
$A_{1m}, A_{2m}, \ldots, A_{nm}$	average areas of the layers (m^2), given by the relationship: $A_{nm} = \sqrt{A_n \cdot A_{n+1}}$.

Equation (4.3) can be used either directly, for the determination of the losses P_0 on the basis of given values of thicknesses and temperatures of the layers, or—conversely—for the calculation of these quantities corresponding to set values of the losses.

In order to evaluate the temperature of the surfaces of each layer, we can again use the analogy with Ohm's law remembering that in an electrical circuit the voltage drop is proportional to the electrical resistance. In the same way the temperature decrease in the wall will be proportional to the thermal resistance.

Then, taking into account that the same amount of heat flows through all layers and denoting with $A_i = A_1$ the inner surface of the chamber and by p_c the specific conduction losses (W/m^2), Eq. (4.3) can be rewritten in the form:

$$p_c = \frac{P_0}{A_i} = \frac{\vartheta_1 - \vartheta_2}{\frac{s_1 A_i}{\lambda_1 A_{1m}}} = \frac{\vartheta_2 - \vartheta_3}{\frac{s_2 A_i}{\lambda_2 A_{2m}}} = \cdots \tag{4.4}$$

From Eq. (4.4), the temperature distribution in the different layers will be:

$$\vartheta_{n+1} = \vartheta_n - \frac{s_n A_i}{\lambda_n A_{nm}} \cdot p_c, \quad (n = 1, 2, 3, \ldots). \tag{4.5}$$

For batch furnaces, which undergo periodical thermal cycles with alternate heating and cooling periods, the wall thickness must be so designed that the new achievement of steady-state thermal conditions, after a cooling period, does not require an excessive amount of heat.

This can be obtained by limiting the heat capacity of the walls, what ultimately means to select materials with appropriate properties and to limit the thickness of the insulating layer so as to minimize—for given power supply conditions—the sum $(Q_A + Q_C)$ of the energies corresponding to accumulated heat and conduction losses.

As seen previously, the value of this sum depends on power supplied to the furnace, wall thickness, thermal characteristics λ, c and γ of used materials and the coefficient α_E of heat transfer from the external furnace surface to the ambient.

The influence of the wall thickness on the total heat losses can be qualitatively illustrated by the diagram of Fig. 4.21, where the curve 1 represents the reduction of conduction losses with the increase of wall thickness, curve 2 the corresponding increase of the losses due to the energy stored in the wall during heating, and curve 3 the resulting losses which have a minimum at the optimal wall thickness.

The influence of resistors rated power on the heating cycle is qualitatively illustrated by the diagrams of Fig. 4.22.

The increase of the power P_N produces a decrease of the practical time t_P and an increase of the time interval $(t_T - t_P)$; as a result this produces always a reduction of the total amount of energy $(Q_A + Q_C)$.

However, the evaluation of the minimum of the quantity $(Q_A + Q_C)$ as a function of wall thickness and installed power is very complex if it is carried out either analytically with simplifying hypotheses, or with similarity laws and tests on experimental prototypes.

On the contrary, the problem can be solved relatively easily by the use of electrical analogical models that allow to study the thermal transients in plane (or cylindrical) layers with applied external heat sources ("Beuken" model).

These models are constituted by resistances and capacitors, supplied with appropriate voltages or currents, and are based on the analogy between thermal and electrical quantities.

As shown by way of example in the diagram of Fig. 4.23, which represents the model of a flat layered wall, to the thermal resistances and heat capacities of the layers (real or fictitious) of thicknesses s_1, s_2, …, s_n in which the wall is subdivided,

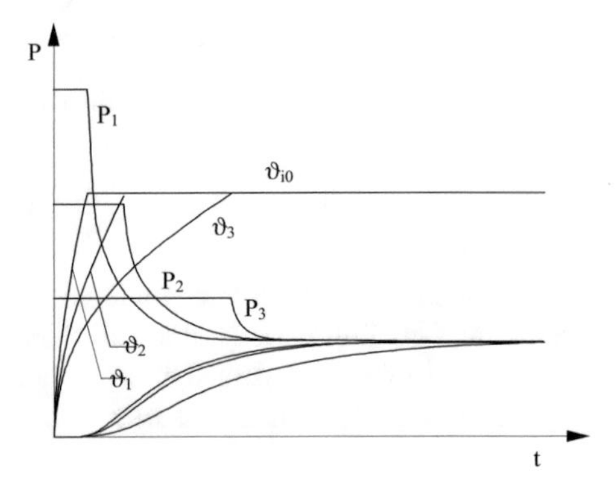

Fig. 4.22 Influence of installed power on heating cycle

correspond in the model electrical resistances and capacitors $R_1, R_2, \ldots, R_n$ and $C_1, C_2, \ldots, C_n$ respectively. To the application of a heat source at constant temperature on the surface A_1, corresponds in the model the application of a DC constant voltage V (conversely to the heating with constant power corresponds the supply with constant current); to the transient temperatures of the layers as a function of time, correspond the time variation of voltages $V_1, V_2, \ldots, V_n$ across the capacitors.

It's therefore easy to evaluate the wall temperature distributions in various instants of the heating transient (as shown by way of example in Fig. 4.24) and, from their values, to calculate the energies Q_A and Q_C.

In many cases, it is sufficient, in the model, to subdivide the wall in a very small number of fictitious layers to determine with a quite good approximation the minimum of the function $(Q_A + Q_C)$ and the optimum control programme of the power supplied to the furnace.

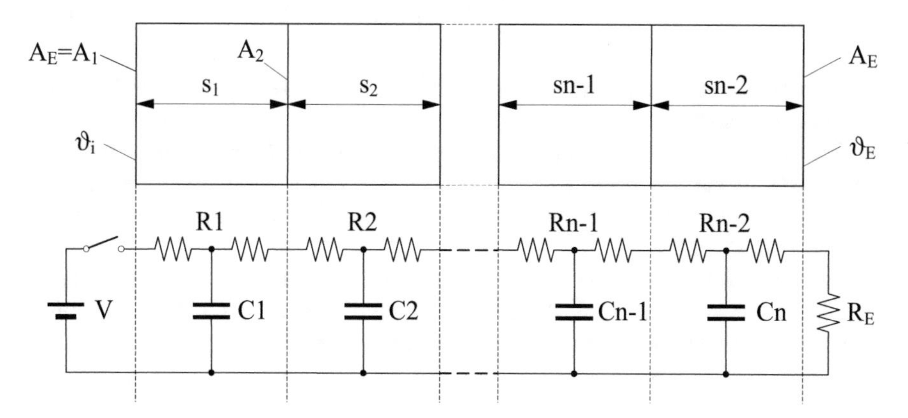

Fig. 4.23 Analogical model for analysis of the wall heating transient

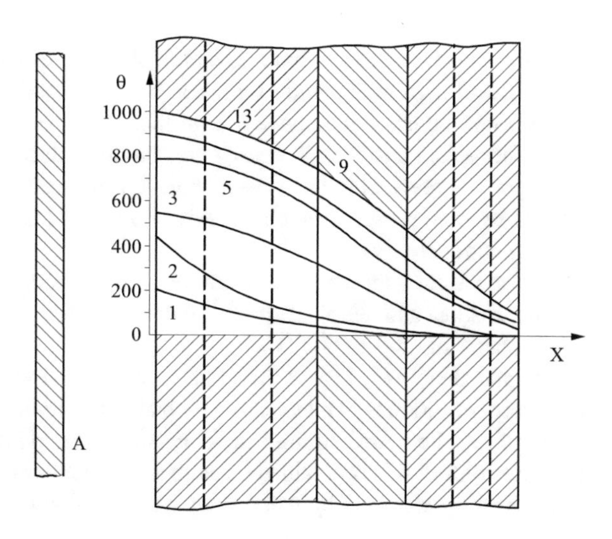

Fig. 4.24 Temperature distributions in the wall during heating (*A* heating element; *1, 2, 3…* heating times in hours) [9]

However, for this last purpose, the model must be completed including the heating elements and the charge.

4.4.5 Thermal Inertia and Cooling Speed

How shown previously, under thermal steady-state conditions the energy accumulated in the walls depends on the final temperature ϑ_{i0}.

In many cases it is useful to make reference to the concept of furnace thermal inertia I, which is defined as the amount of energy required to increase of 1 °C the furnace temperature:

$$I = \frac{Q_{AS}}{\vartheta_{i0}}, \quad \left[\frac{\mathrm{kWh}}{{}^{\circ}\mathrm{C}}\right] \tag{4.6}$$

With good approximation this amount of energy can be considered independent on the temperature.

Even more useful in the practice is the knowledge of the ratio P/I between the total losses of the furnace at a certain temperature ϑ_i and its thermal inertia.

This ratio represents the cooling speed of the furnace at the temperature ϑ_i, expressed in (°C/h).

Given that the losses P are variable with the temperature ϑ_i, while the inertia I is practically independent on it, the cooling speed also varies with the temperature and, more precisely, it is lower as the latter decreases.

4.4.6 Furnaces with Low Thermal Inertia

In traditional furnaces the inner layers of the chamber walls are made, at least in medium and high temperature furnaces, by high-density materials.

As a consequence they have a high thermal capacity that causes, especially in batch furnaces, considerable energy consumption due to the stored heat.

Major improvements have been achieved in recent years with the use of light fibrous ceramic refractories, characterized by very low values of bulk density and thermal conductivity.

The most common ceramic fibers, used at temperatures up to 1400 °C, consist mainly of silica and alumina; special fibers with high content of alumina or zirconia allow operation up to 1600 °C.

Despite their cost, higher than that of traditional materials but justified by the operating economies which can be obtained, ceramic fibers have found wide application where are required rapid heating and cooling cycles and a fine temperature control; this is obtained through a continuous control of the power supplied to the heating elements by using thyristor control systems.

Table 4.5 gives the average characteristics of these materials and, for comparison, those of some conventional refractories.

Example 4.1 Preliminary wall design and evaluation of losses and installed power in a chamber furnace

Specifications:

- Furnace for annealing containers made of steel: .
- Maximum temperature of annealing: 950 °C
- Minimum stay of containers in the chamber: 6 min
- Hour production: 100 kg/h
- Heating chamber dimensions: 1.80 × 0.85 × 0.65 m.

(1) ***Preliminary wall design***

With reference to the sketch of Fig. 4.25, we make the following assumptions:

- layer 1 made of Schamotte
- layer 2 made of Kieselguhr
- temperature of wall internal surface $\vartheta_i = \vartheta_1 = 950\,°C$

Table 4.5 Characteristics of fibrous refractories and comparison with conventional ones [2]

Material	Maximum working temperature (°C)	Density (Kg/m^3)	Thermal conductivity (W/m °C)
Alumino-silicate fibers			
• (95 % Al_2O_3)	1600	160–190	0.22[a]
• (61 % Al_2O_3)	1450	160–190	0.22[a]
• (48 % Al_2O_3)	1260	160–190	0.22[a]
Mineral wool	750	144	0.01[b]
Equivalent to asbestos fibers	500	195	0.07[c]

Material	Thermalconductivity (W/m °C)	Density (kg/dm^3)	Thickness (m)[a]	Weigth (t/m^2)[a]	Comparison of costs[a]
High density brick	1.40	2.30	1.60	3.68	100
Insulating brick	0.28	0.80	0.32	0.13	36
Insulating concrete	0.27	1.25	0.30	0.48	70
Fibrous material	0.19	0.10	0.21	0.02	50

Note [a]at 820 °C
[b]At 340 °C
[c]At 350 °C
Note [a]for walls with equal insulating power and internal surface temperature 1000 °C

Fig. 4.25 .

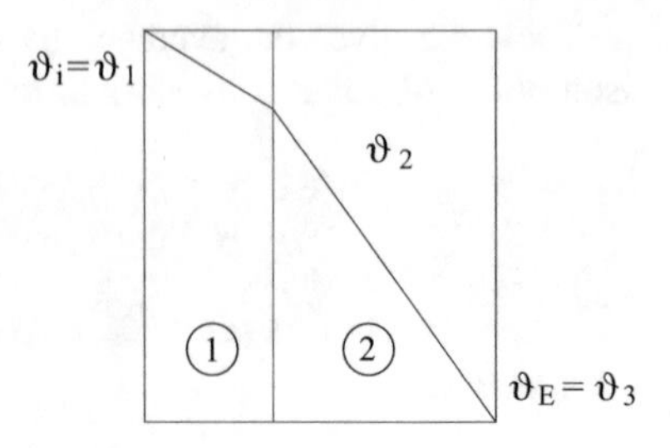

- temperature of surface between layers $\vartheta_2 = 850\,°C$
- temperature of wall external surface $\vartheta_E = \vartheta_3 = 50\,°C$
- ambient temperature $\vartheta_A = 20\,°C$
- specific conduction power losses $p_C = 800–1000\ W/m^2$

Assuming layers of equal surface area, it is:

average temperature of layer 1: $\vartheta_{1m} = 900\,°C$
thermal conductivity of layer 1: $\lambda_1 = 1\,W/m\,°C$
thickness of layer 1: $s_1 = (\vartheta_1 - \vartheta_2)\lambda_1/p_c$

$$= (950 - 850) \cdot 1/800 = 0.125\,m$$

average temperature of layer 2: $\vartheta_{2m} = 450\,°C$
thermal conductivity of layer 2: $\lambda_2 = 0.15\,W/m\,°C$
thickness of layer 2: $s_1 = (\vartheta_2 - \vartheta_3)\lambda_2/p_c$

$$= (850 - 50) \cdot 0.15/800 = 0.15\,m$$

(2) ***Check of specific losses***

- *Calculation of actual areas of surfaces:*

$$A_i = A_1 : (0.85 + 0.65) \times 1.80 \times 2 + (0.85 \times 0.65) \times 2 = 6.51\,m^2$$
$$A_2 : (1.10 + 0.90) \times 2.05 \times 2 + (1.10 \times 0.90) \times 2 = 10.18\,m^2$$
$$A_E = A_3 : (1.40 + 1.20) \times 2.35 \times 2 + (1.40 \times 1.20) \times 2 = 15.58\,m^2$$

$$A_{1m} = \sqrt{A_1 A_2} = \sqrt{6.51 \times 10.18} = 8.14\ m^2$$
$$A_{2m} = \sqrt{A_2\ A_3} = \sqrt{10.18 \times 15.58} = 12.59\ m^2$$

- *Thermal resistances of layers*

$$R_1 = s_1/(\lambda_1 A_{1m}) = 0.125/(1. \times 8.14) = 0.0154\,°C/W$$
$$R_2 = s_2/(\lambda_2 A_{2m}) = 0.15/(0.15 \times 12.59) = 0.0794\,°C/W$$
$$R_E = 1/(\alpha_E A_E) = 1./(10. \times 15.58) = 0.0064\,°C/W$$

(with $\alpha_E = 10\,W/m^2\,°C$)

- *Steady-state conduction losses [Eqs. (4.3), (4.4)]*

$$P_o = (\vartheta_i - \vartheta_a)/(R_1 + R_2 + R_E)$$
$$= (950 - 20)/(0.0154 + 0.0794 + 0.0064) = 9190\,W$$
$$p_c = P_o/A_i = 9190/6.51 = 1412\,W/m^2$$

In order to reduce specific losses p_c to about 800–1000 W/m^2, it is necessary to increase the thermal resistance R_2, i.e. the thickness of insulating layer. To obtain this result it should be:

$P_o = p_C \cdot A_i$	= 800 × 6.51	= 5200 W
$R_1 + R_2 + R_E = (\vartheta_i - \vartheta_a)/P_0$	= 930/5200	= 0.1788 °C/W
$R_2 = (R_1 + R_2 + R_E) - (R_1 + R_E)$	= 0.1788–(0.0154 + 0.0064)	= 0.1570 °C/W
$s_2 = \lambda_2 A_{2m} R_2$	= 0.15 × 12.59 × 0.1570	= 0.296 m.

Assuming s_2 = 0.3 m, the heating chamber has the dimensions of Fig. 4.26, from which results:

$$A_E = A_3 = (1.70 + 1.50) \times 2.65 \times 2 + (1.70 \times 1.50) \times 2 = 22.06\,m^2$$
$$A_{2m} = 10.18 \times 22.06 = 14.98\,m^2$$

$$R_2 = s_2/(\lambda_2 A_{2m}) = 0.3/(0.15 \times 14.98) = 0.134\,°C/W$$
$$R_E = 1/(\alpha_E A_E) = 1/(10 \times 22.06) = 0.0045\,°C/W.$$

Therefore it is:

$$P_o = 930/(0.0154 + 0.134 + 0.0045) = 6043\,W$$
$$p_C = 6043/6.51 = 928\,W/m^2$$

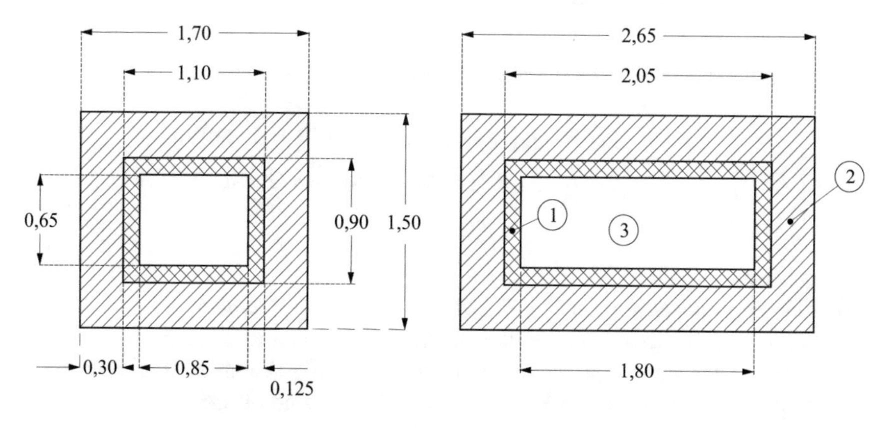

Fig. 4.26 Dimensions of the heating chamber

and from Eq. (4.5) we obtain:

$$\vartheta_2 = 950 - (928 \times 0.125 \times 6.51)/(1 \times 8.14) = 857\,^\circ\mathrm{C}$$
$$\vartheta_E = 857 - (928 \times 0.30 \times 6.51)/(0.15 \times 14.98) = 50\,^\circ\mathrm{C}$$

(3) *Stored heat in steady-state conditions*

- Volume of layers:

$$v_1 = [(0.90 \times 1.10) - (0.65 \times 0.85)] \times 1.80 + (0.90 \times 1.10) \times 2 \times 0.125 = 1.04\,\mathrm{m}^3$$
$$v_2 = (1.70 \times 1.50) \times 2.65 - (0.90 \times 1.10) \times 2.05 = 4.73\,\mathrm{m}^3$$

- Average temperature of layers:

$$\vartheta_{1m} = (950 + 857)/2 = 903\,^\circ\mathrm{C}$$
$$\vartheta_{2m} = (857 + 50)/2 = 453\,^\circ\mathrm{C}$$

- Thermal capacities of layers:

$$C_1 = \gamma_1 \cdot v_1 \cdot c_1 = 1700 \times 1.04 \times 0.25 = 442\,\mathrm{Wh}/^\circ\mathrm{C}$$
$$C_2 = \gamma_2 \cdot v_2 \cdot c_2 = 350 \times 4.73 \times 0.25 = 413\,\mathrm{Wh}/^\circ\mathrm{C}$$

- Time constants:

$$R_1 \cdot C_1 = 0.0154 \times 442 = 6.8\,\mathrm{h}$$
$$R_2 \cdot C_2 = 0.134 \times 413 = 55.3\,\mathrm{h}$$

- Stored heat in the layers in steady-state conditions:

$$Q_{A1} = C_1(\vartheta_{1m} - \vartheta_A) = 1700 \times 1.04 \times 0.25 \times (903 - 20) = 390\,\mathrm{kWh}$$
$$Q_{A2} = C_2(\vartheta_{2m} - \vartheta_A) = 350 \times 4.73 \times 0.25 \times (453 - 20) = 179\,\mathrm{kWh}$$

- Total stored heat in the wall in steady-state conditions:

$$Q_{AS} = Q_{A1} + Q_{A2} = 390 + 179 = 569\,\mathrm{kWh}$$
$$q_{AS} = Q_{AS}/A_i = 569/6.51 = 87\,\mathrm{kWh/m}^2$$

(4) *Energy required for heating the air in the heating chamber:*
It can be estimated assuming that at each loading and unloading operation it is renewed a volume of air equal to that of the heating chamber. Therefore, every hour it must be heated from 20 to 950 °C a volume of air:

$$v_A = 10 \times (0.85 \times 0.65 \times 1.80) = 9.95\,\mathrm{m}^3$$

Assuming for the air the following data:

$$\text{specific heat}: c_A = 0.20\,\mathrm{Wh/Kg\,^\circ C};\ \text{density}: \gamma_A = 1.29\,\mathrm{Kg/m}^3,$$

the energy required is:

$$Q_A = c_A \gamma_A v_A (\vartheta_i - \vartheta_A) = 0.20 \times 1.29 \times 9.95 \times 930 = 2387\,\mathrm{Wh}$$

(5) *Radiation losses from the door of the heating chamber during loading and unloading*
Assuming that each loading and unloading operation requires 30 s, the total time of radiation is 300 s/h, i.e. 0.0833 h.
The amount of heat lost by radiation is calculated, as known, with the equation:

$$Q_i = \eta \cdot F_{12} \cdot A_1 \cdot (T_1^4 - T_2^4) \cdot t$$

where:

$\eta = 5.67 \times 10^{-8}$	Stefan-Boltzmann constant, $\mathrm{W/m^2K^4}$
$F_{12} = 1/[1/\varepsilon_1 + (A_1/A_2)(1/\varepsilon_2 - 1)]$	view factor between the surfaces 1 and 2, expressing the fraction of the total energy radiated by the surface 1 that reaches the surface 2
A_1, A_2	areas of surfaces 1 and 2, m^2
$\varepsilon_1, \varepsilon_2$	emissivity and absorption factors of surfaces 1 and 2
T_1, T_2	absolute temperatures of surfaces 1 e 2, K
t	radiation time of the surfaces, h.

Assuming

$$A_1 = 2 \times 0.85 \times 0.65 = 1.11\,\mathrm{m}^2, A_2 = 10\,A_1, \varepsilon_1 = 0.87, \varepsilon_2 = 0.71,$$

it results:

$$F_{12} = 1/[1.149 + 0.1(1.408 - 1)] = 0.84$$

$$Q_i = 5.67 \cdot 10^{-8} \times 0.84 \times 1.11 \times (1223^4 - 293^4) \times 0.0833 = 9925\,\mathrm{Wh}$$

(6) *Heat absorbed by the grid supporting the load*

Taking into account that due to the particular nature of the objects to be annealed they require a grid support during the heating phase, it must be evaluated also the heat absorbed by this grid.

As regards thermal efficiency, the grid should be as light as possible, while maintaining the highest resistance to mechanical stresses.

The weight of the grid does not exceed, in general, the weight of the objects to be heated. In addition, during loading and unloading usually the grid does not cool down completely, but it is assumed here that at the re-introduction in the furnace it has a temperature of about 200 °C.

Assuming for the grid the average specific heat value of 0.19 Wh/kg °C and a weight of 10 kg, the losses corresponding to the absorbed heat in 1 h (10 charges per hour) are:

$$Q_G = c_G \cdot p_G \cdot (\vartheta_i - \vartheta_f) = 0.19 \times 100 \times 750 = 14300\,\text{Wh}$$

(7) *Total energy lost (neglecting accumulated heat)*

- without grid:

$$Q_{E1} = Q_o + Q_A + Q_i = 6043 + 2387 + 9925 = 18355\,\text{Wh}$$

- with grid:

$$Q_{E2} = Q_o + Q_A + Q_i + Q_G = 18355 + 14300 = 32655\,\text{Wh}$$

(8) *Useful energy*

$$Q_U = c \cdot p \cdot (\vartheta_i - \vartheta_A) = 0.19 \times 100 \times 930 = 17700\,\text{Wh}$$

(9) *Total energy consumption, specific energy consumption and minimum installed power*

- without grid:

$$Q_{t1} = Q_{E1} + Q_U = 18355 + 17700 = 36.055\,\text{Wh}$$

specific energy consumption: 0.36 kWh/Kg
minimum installed power: 36 kW

- with grid:

$$Q_{t2} = Q_{E2} + Q_U = 32655 + 17700 = 50355\,\text{Wh}$$

specific consumption: 0.50 kWh/kg
minimum installed power: 50 kW

It is evident the convenience of avoiding, whenever possible, the use of removable grids.

In practice the installed power should be greater than the above determined minimum power, in order to obtain a fast temperature control, a more rapid recovery of the set temperature after loading, and to compensate for the decrease of power that occurs for the aging of the heating elements.

The experience suggests to increase the calculated minimum power at least of 20–30 %. Therefore, the installed power of this furnace should be approximately 65–66 kW.

4.5 Test Methods of Batch Resistor Furnaces

Tests of batch resistor furnaces are done in order to check the following characteristics:

1. furnace internal and external geometrical dimensions, mm
2. total weight, kg
3. furnace rated voltage, V
4. power from the supply grid, kVA
5. furnace rated power P_N, kW
6. nominal temperature ϑ_{i0}, °C
7. no-load power losses P_0, kW
8. practical heating-up time t_P, h
9. stored heat in thermal steady state Q_{AS}, at the nominal temperature, kWh.

While the control of quantities from 1.0 to 6.0 is relatively simple, more problematical is the evaluation of the remaining ones and, in particular, the static stored heat.

The static stored heat Q_{AS} is the heat stored in the furnace construction elements (heating elements, thermal insulation, installed fixtures, transport mechanisms, walls), when the furnace is in thermal steady state at the nominal temperature ϑ_{i0}.

Among the quantities which must be determined in the tests, it is not included the efficiency, since it depends not only on the construction of the furnace, but also on the type of charge and the thermal working cycle.

When the working cycle is known, the comparison of different types of furnaces can be made, as shown in the preceding paragraphs, on the basis of the no-load losses and the heat stored during heating.

The static accumulated heat Q_{AS} at rated temperature and thermal steady-state conditions can be determined using two different methods, which are known as *direct method I* (or "classical method") and indirect method (or *method "Beuken-I"*) respectively [11–14].

Because of its complexity, the first method is used only as a laboratory method, while the second one—more simple—is the one normally used in industrial tests.

Similarly, for the evaluation of the variations of stored heat during the heating transient, can be used either the *direct method II* or the indirect method (or *method "Beuken-II"*).

4.5.1 Direct (or "Classic") Method

With this method the accumulated heat Q_A is determined during the heating of the empty (and dry) furnace, as the difference between the electrical energy Q_E supplied to the heating elements and the energy Q_C, corresponding to the losses through the walls, transferred from the furnace to the surrounding ambient in the same period of time:

$$Q_A(t) = Q_E(t) - Q_C(t) = \int_0^t P(t)dt - \int_0^t P_C(t)dt \tag{4.7}$$

The value of the accumulated heat Q_{AS} in steady-state conditions at nominal temperature is obtained by extending the integrals until the thermal steady-state is reached, i.e. up to the theoretical heating up time t_T (see Fig. 4.17). The first integral is evaluated by measuring the energy Q_{E0} corresponding to the total kWh supplied to the resistors till the achievement of the thermal steady-state.

With reference to Fig. 4.27 it is:

$$Q_{AS} = Q_{E0}(\text{Area}\,///) - Q_{C0}(\text{Area}\,\text{xxx})$$

For the evaluation of Q_{C0}, it must be measured during heating the temperature of the external surface of the furnace at regular time intervals Δt_k, up to the attainment of the thermal steady state conditions. The heat transmission coefficient can be evaluated from the no-load losses and temperature rise of the external surface of the furnace.

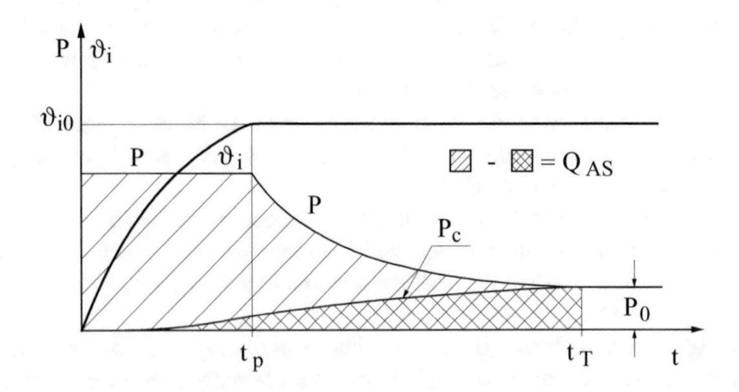

Fig. 4.27 Evaluation of static stored heat Q_{AS}

It follows that Q_C is given by:

$$Q_C = \frac{P_{C0}}{(\vartheta_{E0} - \vartheta_A)} \sum_{k=1}^{n} \left[\frac{\vartheta_{e(k-1)} + \vartheta_{ek}}{2} - \vartheta_A \right] \Delta t_k \qquad (4.8)$$

with:

P_{C0} no-load losses in thermal steady state conditions
ϑ_{e0} temperature of the furnace external surface in steady state conditions
ϑ_{ek} temperature of the furnace external surface at the end of the interval Δt_k
ϑ_A ambient temperature.

If, in addition to the steady state value, it is required also the evolution with time of the stored heat, it is necessary to measure the electrical energy $Q_E(t)$ as a function of time, to evaluate by Eq. (4.8) the energy $Q_{EC}(t)$ lost by conduction, and finally to calculate their difference at each time instant *t*.

The main difficulties of this method are: to measure the heat losses with sufficient accuracy, the evaluation of the effective (average) temperature of the external surface and the determination of the time at which the steady state thermal regime is reached.

4.5.2 *Method Beuken-I (Indirect Method)*

With this method the stored heat Q_{AS} is determined as the total amount of heat transmitted from the furnace walls to the environment during the cooling, starting from the thermal steady state at nominal temperature ϑ_{i0}.

Given the difficulty of evaluating the amount of heat transferred to the environment on the basis of temperature measurements on the external surface of the furnace, the stored heat Q_{AS} is determined indirectly by the measurement during the cooling transient of the temperature of the inner surface of the chamber wall.

Assuming that the system is linear, the cooling curve $\vartheta_i(t)$ can be represented in a normalized form, as a series of exponential functions:

$$\frac{\vartheta_i(t) - \vartheta_A}{\vartheta_{i0} - \vartheta_A} = A_1 e^{-t/\tau_1} + A_2 e^{-t/\tau_2} + \cdots \qquad (4.9)$$

where:

$\vartheta_{i0}, \vartheta_i(t)$ temperature of the inner surface of the wall at the start of the cooling transient and at a generic instant *t*
$A_1, A_2, \ldots$ numerical coefficients (with $\sum A_i = 1$)
$\tau_1, \tau_2, \ldots$ time constants.

The method Beuken-I assumes that at any point in the wall thickness the temperature varies with a similar law, when the heating power is switched-off.

In the case of a linear system, where heat transfer coefficient, thermal resistances and specific heat of the furnace components can be considered independent on temperature, the stored heat Q_{AS} can be calculated with the equation:

$$Q_{AS} = P_{C0}(\tau_1 + \tau_2 + \cdots). \tag{4.10}$$

Taking into account only the exponential with the highest time constant, Eq. (4.10) can be further approximated as:

$$Q_{AS} \approx P_{C0}\frac{\tau_1}{A_1} \tag{4.10a}$$

The quantities $\vartheta_A, \vartheta_{i0}$ and P_{C0} are measured directly at the end of the heating period, when steady-state conditions of temperature are reached.

The values $(\tau_1, \tau_2, \ldots)$ or (τ_1, A_1) are evaluated by a graphical construction or an appropriate numerical program, starting from the measured values of the temperature during the natural cooling of the furnace, given by the sensor installed by the furnace manufacturer.

Typical diagrams of the external and internal wall temperatures measured during cooling, are represented in semi-logarithmic coordinates in Fig. 4.28.

They show the following:

- the curve $[\vartheta_i(t) - \vartheta_A]$ can be generally approximated with two exponentials with time constants τ_1 and τ_2 which allow at their intersection the evaluation of the "dead time" t_M

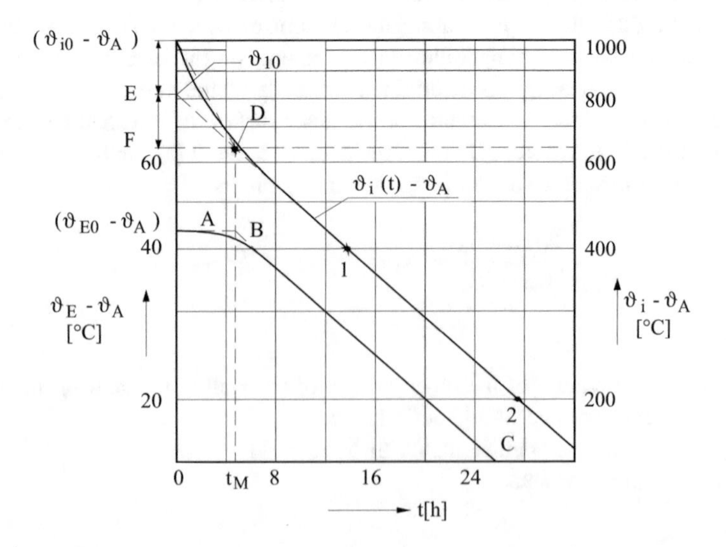

Fig. 4.28 Diagrams of external and internal temperatures of the wall during cooling

- till the instant t_M the temperature of the external surface of the chamber wall remains nearly constant at the value ϑ_{E0}, so that the conduction losses remain also constant and equal to P_{C0}
- for $t \rangle t_M$ the curves $[\vartheta_i(t) - \vartheta_A]$, $[\vartheta_E(t) - \vartheta_A]$ and $P_C(t)$ are represented, in semi-logarithmic coordinates, by parallel lines.

According to these trends it is possible to replace the actual cooling curve $[\vartheta_E(t) - \vartheta_A]$ with segments of straight lines connecting the points ABC, so that instead of Eq. (4.9) we can write:

$$\left.\begin{aligned} \frac{\vartheta_E(t)-\vartheta_A}{\vartheta_{E0}-\vartheta_A} &= 1, && \text{for } 0 \le t \le t_M \\ \frac{\vartheta_E(t)-\vartheta_A}{\vartheta_{E0}-\vartheta_A} &= e^{-\frac{(t-t_M)}{\tau_1}}, && \text{for } t_M \le t \le \infty \end{aligned}\right\} \tag{4.11}$$

The period beginning after t_M is called regular cooling phase.

Denoting with:

$P_{C0} = A_E \alpha_E (\vartheta_{E0} - \vartheta_A)$ no-load losses in thermal steady conditions
A_E area of the external surface of the furnace
α_E heat transmission coefficient between external surface and ambient,

the total stored heat, which is dissipated during the cooling stage, can be expressed in the form:

$$\begin{aligned} Q_{AS} &= A_E \int_0^{\infty} \alpha_E [\vartheta_E(t) - \vartheta_A] dt \\ &= A_E \left[\int_0^{t_M} \alpha_E (\vartheta_{E0} - \vartheta_A) dt + \int_{t_M}^{\infty} \alpha_E (\vartheta_{E0} - \vartheta_A) e^{-\frac{(t-t_M)}{\tau_1}} dt \right] \\ &= P_{C0}(t_M + \tau_1) \end{aligned} \tag{4.12}$$

The determination of Q_{AS} is thus made through the measure of the power P_{C0} lost in thermal steady-state conditions, which is equal to the electrical power supplied to the heating elements in these conditions and, according to the different formulas used, to the evaluation of the quantities $(\tau_1, \tau_2, \ldots)$, (τ_1, A_1) or (τ_1, t_M).

The time constants τ and the value of A_1 are determined, in the case of the graphical construction, first drawing in semi-logarithmic scale the dimensionless ratio

$$\Theta_i(t) = [\vartheta_i(t) - \vartheta_A]/[\vartheta_{i0} - \vartheta_A]$$

calculated from the measurement of the temperature $\vartheta_i(t)$ (see Fig. 4.29) and extrapolating to the time t = 0 (start of cooling) the straight line section of the curve.

The point of intersection of the straight line G_1 with the axis Θ gives the value of A_1, while the time constant τ_1 is obtained as the abscissa corresponding to the point

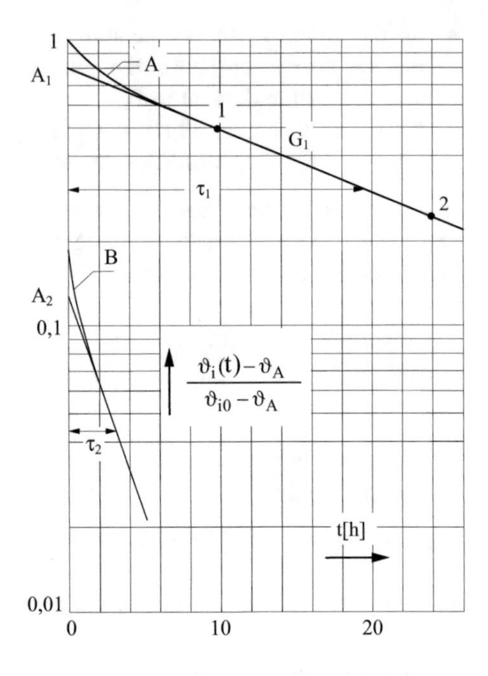

Fig. 4.29 Graphical determination of A_i and τ_i

of intersection of the straight line G_1 with the value $(A_1/e) = 0.37 \cdot A_1$ or by the values of time and temperature of any couple points 1 and 2 of the straight line G_1, through the relationship:

$$\tau_1 = \frac{(t_2 - t_1)}{(ln\,\Theta_1 - ln\,\Theta_2)}.$$

Taking into account that for evaluating τ_1 must be already negligible the exponential with time constant τ_2, it is suggested to assume as points 1 and 2 those corresponding to the temperatures $0.5 \cdot \vartheta_{i0}$ and $0.25 \cdot \vartheta_{i0}$.

Then tracing the curve B, corresponding to the difference between curve A and the straight line G_1, and extrapolating again to $t = 0$ the straight line section, A_2 can be obtained as intersection with the axis Θ, and τ_2 as intersection of the straight line with $(A_2/e) = 0.37 \cdot A_2$.

If necessary, a similar procedure can be used for the evaluation of A_3 and τ_3.

As regards the dead time t_M, it can be demonstrated that it can be approximately obtained by the graphical construction shown in Fig. 4.28, constructing the triangle D–E–F with the side DE on the line representing the exponential with time constant τ_1 and the side EF, equal to $(1-A_1)$, on the vertical axis.

Equation (4.12) has been obtained in the hypothesis of physical characteristics of materials constituting the walls and the heat transfer coefficient α_E independent on temperature.

However, the variations with temperature can be taken into account by introducing a correction coefficient k, which is a function of ϑ_{E0}, modifying Eq. (4.12) as follows:

$$Q_{AS} = P_{C0}(t_M + k \cdot \tau_1) \tag{4.13}$$

In practice, it can be assumed $k \approx 0.85$, value corresponding to $\vartheta_{E0} \approx 70\,°C$.

4.5.3 Test Method Beuken-II

It is used for evaluating the variations during heating of the heat stored in the furnace.

With reference to the diagram of Fig. 4.17, the furnace under test is heated with constant nominal power P_N until the nominal temperature ϑ_{i0} is reached. The duration of this interval corresponds to the practical heating-up time t_P. Then the power supplied to the resistors is controlled so as to maintain constant the temperature ϑ_{i0}.

At the time t_T, when the furnace has reached steady state thermal conditions, the electrical power consumption corresponding to the conduction losses P_{C0} is measured. Then, switching-off the supply, the measurement of the temperature $\vartheta_i(t)$ is done during the natural cooling, as in the method Beuken-I.

As shown in Fig. 4.30, the time variation of the "accumulated" power $P_A(t)$ corresponding to the stored energy, can be subdivided into three time intervals:

- in the first interval, between $t = 0$ and $t = t_M$, the power $P_A(t)$ is equal to P_N, since in practice the conduction losses are negligible and the total input energy is stored in the wall;

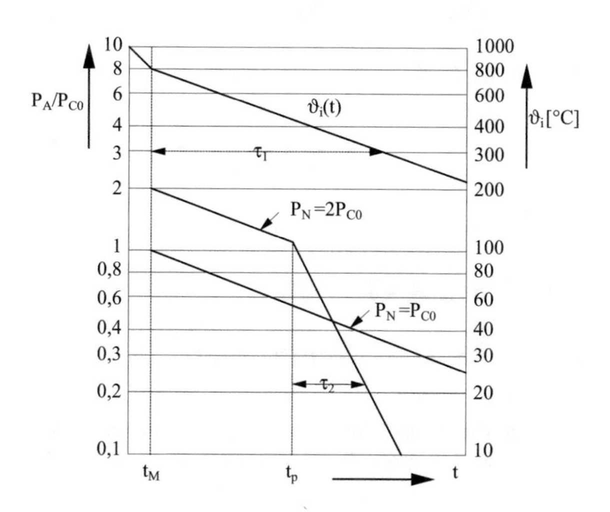

Fig. 4.30 Variation of "accumulated" power during heating

- in the second interval, between t_M and t_P, $P_A(t)$ varies exponentially with the same time constant τ_1 of the function $\vartheta_i(t)$ of the cooling period;
- in the third interval, for $t > t_P$, $P_A(t)$ varies according to an exponential function with time constant τ_2^*.

In semi-logarithmic scale, the function $P_A(t)$ therefore is that represented in Fig. 4.30. In the same figure it is also drawn the curve of $\vartheta_i(t)$, approximated with two exponentials.

The curves of $P_A(t)$ can be therefore described by the equations:

$$\left.\begin{array}{lll} P_A(t) = P_N, & \text{for} & 0 \le t \le t_M \\ P_A(t) = P_N e^{-(t-t_M)/\tau_1}, & \text{for} & t_M \le t \le t_P \\ P_A(t) = P_A(t_P) e^{-(t-t_P)/\tau_2^*}, & \text{for} & t_P \le t \le \infty \end{array}\right\} \quad (4.14)$$

For the evaluation the variation of the stored heat during heating, it's therefore sufficient to integrate Eq. (4.14) between t = 0 and a generic instant t, i.e.:

$$Q_A(t) = \int_0^t P_A(t)dt \quad (4.15)$$

The quantities in Eq. (4.14), required for evaluation of Eq. (4.15), can be determined as follows:

P_N, t_P, P_{C0}	during heating of the empty furnace;
τ_1, t_M	from the curve of $\theta_i(t)$, during cooling, as in method Beuken-I;
τ_2^*	from the equality between the stored heat Q_{AS} at the end of the heating stage and the heat transmitted to the ambient during cooling, evaluated as in method Beuken-I.

Since this equality can be written as follows:

$$Q_{AS} = P_N t_M + [P_N - P_A(t_P)]\tau_1 + P_A(t_P)\tau_2^* = P_{C0}(t_M + \tau_1) \quad (4.16)$$

after some manipulation, we obtain:

$$\tau_2^* = \tau_1 - (t_M + \tau_1)\left[1 - \frac{P_{C0}}{P_N}\right] e^{(t_P - t_M)/\tau_1}. \quad (4.17)$$

4.6 Heating Elements

4.6.1 Material Requirements

One of the main components of a resistance furnace are the heating elements ("*resistors*"), where electrical energy is converted into heat.

Given that they generally work at the maximum temperature allowed by their constituent material, it is of fundamental importance for economy of construction and safety of furnace operation a careful material selection and an accurate design.

As an example, an increase of the working temperature above the maximum admissible of only 30–50 °C can lead to a dramatic reduction (even up to one half) of the resistor life.

The materials used for the heating elements must have several special properties. In particular:

- *High operating temperature*: it's required in order to obtain an adequate transfer of energy from the resistor to the charge and the walls of the furnace chamber; it is limited by the softening and melting temperatures of the material.
- *High electrical resistivity*: given the value of the required resistance, by decreasing the resistivity increases the length and/or decreases the diameter of resistor's wire, resulting in difficulty of installation of the heating element in the furnace chamber and reduction of his life.
- *Temperature coefficient of resistivity low and positive*: the first feature allows to limit the variation of the resistance from cold to hot working conditions and, as a consequence, of power and current from the supply. The second one produces a self-regulating effect of power absorbed by the furnace.

It must be noted that a resistivity coefficient of 4 ‰ per °C, typical of pure metals, produces a variation of resistance from 1 to 4 when the temperature varies from 0 to 1000 °C. For this reason metal alloys with lower temperature coefficients are generally preferred to pure metals.

- *Electrical properties and size constant in time*: some materials undergo an "*aging*" process which can lead to increase of resistivity or lengthening of the conductor. These two phenomena may pose construction problems, which are usually solved with the use of adjustable voltage supply transformers and adopting design solutions that allow a free linear expansion of resistor.
- *High mechanical strength* in the whole working temperature range.
- *Ease of machining and welding.*
- *Low chemical reactivity* in hot conditions to oxidation, corrosion and with gases that may be used in the furnace chamber.

However, the main requirement in resistor's material selection is its maximum working temperature, which is evaluated on the basis of the absorbed power and the laws of heat transfer by radiation and convection applied to the system consisting of the heating elements, the furnace walls and the charge.

This calculation thus provides useful information for the selection of the type of resistor, the evaluation of the maximum power absorbed by the resistors without risk of overheating and the assessment of the production capacity of the furnace.

The most common materials used for the heating elements and their maximum operating temperatures are shown in Fig. 4.31. They can be subdivided in three main groups: (1) Metallic alloys for temperatures up to 1200–1400 °C, (2) Silicon

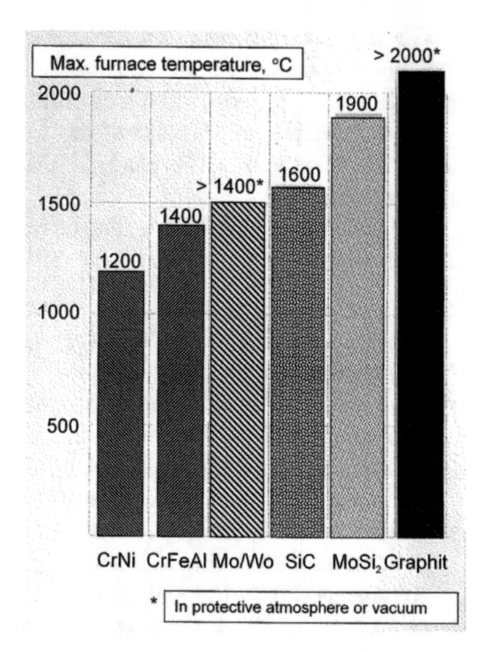

Fig. 4.31 Maximum operating temperatures of heating elements [15]

carbide (SiC) and Molibdenum disilicide ($MoSi_2$) up to 1600–1900 °C and (3) Graphite, Tungsten and Tantalum for temperatures >2000 °C, if used in protective atmospheres or in vacuum.

4.6.2 *Metallic Alloys*

For temperatures up to 1200–1400 °C in dry air, are used metal alloys with two or three components, the main of which are listed in Table 4.6.

A first group, for temperatures up to 1150–1250 °C, is constituted by *Ni–Cr alloys* with low Fe content (0.5–2.0 %), which are characterized by very good mechanical and electrical properties.

In particular, as pointed out by the diagram of Fig. 4.32, the electrical resistivity undergoes very small variations in the whole temperature working range; this allows in most cases to avoid special devices for power control during the first heating stage.

When used in oxygen rich atmospheres, on the resistor's surface forms a layer of Cr_2O_3 with melting point higher than that of the alloy which, having a coefficient of expansion very close to that of the alloy, resists very well to periodic temperature variations.

This is particularly important in furnaces where the temperature regulation is of the "*on-off*" type.

Table 4.6 Characteristics of metal alloy resistors

Alloy	Ni (%)	Cr (%)	Fe (%)	Al (%)	Melting temperature (°C)	Maximum working temperature in air (°C)	Mass density (gr/cm^3)	Specific heat (Wh/kg K)	Electrical resistivity (μΩ cm)[a]	Thermal conductivity (W/m K)[a]	Coefficient of linear expansion (1 · e−6/ °C)
Ni–Cr											
(1)	80	20	–	–	1400	1200	8.3	0.12[a]	109	14.7	15–18
	70	30	–	–	1400	1250	–	–	–	–	–
Fe–Ni–Cr											
(2)	60	15	25	–	1390	1150	8.2	0.13[a]	111	13.4	14–17
(3)	35	20	45	–	1390	1100	7.9	0.14[a]	104	–	16–19
	30	20	50	–	1390	1100	7.9	0.15[b]	–	13.0	–
	20	25	55	–	1380	1050	7.8	0.14[a]	95	13.0	16–19
Fe–Cr–Al											
	–	15	80.7	4.3	~1500	1050	7.28	0.13[a]	125	–	–
	–	20	75	5.0	1500	1200	7.2	0.17[b]	–	12.6	–
(4)	–	22	72.2	5.8	1500	1400	7.1	0.13[a]	145	–	11–15
(5)	–	22	72.7	5.3	1500	1400	7.15	0.13[a]	139	–	11–15
(6)	–	22	73.2	4.8	~1500	1300	7.25	0.13[a]	135	–	11–15
	–	25	70	5.0	1500	1350	7.1	0.17[b]	144	12.6	–

N.B. [a]at 20 °C
[b]Average value in the range 0–1000 °C

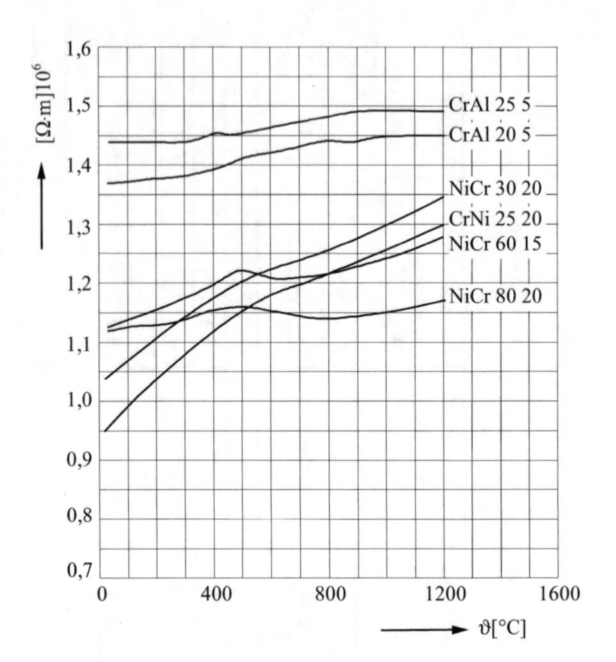

Fig. 4.32 Resistivity of metal alloys as a function of temperature [3]

Typical of this group are the alloys Ni–Cr 80/20 (80 % Ni, 20 % Cr) and 70/30 (70 % Ni, 30 % Cr).

Increasing the content of Cr, increases the material resistance to oxidation at high temperature, but at the same time reduces its workability.

Moreover, these alloys must be used with great precaution when they work in aggressive atmospheres (like SO_2, H_2S, SO_3, water vapour, rich exothermic gases, etc.) since in this case may occur material corrosion (the so-called "*green rot*") leading to a rapid destruction of the resistor.

Therefore the maximum working temperature in different furnace atmospheres must be chosen as indicated in Table 4.7.

To improve material workability, can be used three components *Fe–Ni–Cr alloys*, which are less expensive than Ni–Cr.

The resistance to oxidation of these alloys at high temperature is also ensured by a surface layer of Cr oxide, but—due to the formation of Fe oxides which are not completely impermeable to oxygen—at the same working temperature oxidation can progress more rapidly, in comparison with Ni–Cr alloys.

This limits the maximum operating temperature of these alloys in oxidizing atmosphere to 1050–1150 °C. On the contrary they can be conveniently used in reducing atmospheres, but using silica and alumina refractories, since basic refractories easily react with Cr.

Are often used also the *alloys Fe–Cr–Al*, with chemical composition 15–30 % Cr, 2–6 % Al and Fe balance (sometimes with small additions of Co for increasing the resistance to corrosion).

Table 4.7 Maximum working temperature (°C) in different atmospheres

Alloy	Ni–Cr	Fe–Ni–Cr		Fe–Cr–Al		
Atmosphere	[a](1)	[a](2)	[a](3)	[a](4)	[a](5)	[a](6)
Oxidant						
• Dry air	1200	1150	1100	1400	1400	1300
• Humid air	1150	1100	1050	1200	1200	1200
Neutral						
• Nitrogen, (N_2)	1250	1200	1150	1200–1050	1250–1100	1150–1100
• Exothermic (10CO, $15H_2$, $5CO_2$, $70N_2$)	900–1000[b]	1100	1100	1150	1150	1100
Reducing						
• Endothermic (20CO, $40H_2$, $40N_2$)	900	1100	1100	1050	1050	1000
• Hydrogen, (H_2)	1250	1200	1150	1400	1400	1300
• Cracked ammonia, (75 H_2, 25 N_2)	1250	1200	1150	1200	1200	1100
Vacuum						
• $>10^{-5}$ Torr	–	–	–	1100	1100	1050

N.B. [a](1) … (6): materials of Table 4.V
[b]"Green rot" danger

In this case the resistance to oxidation at high temperature is due to a surface layer of oxides of Cr and Al, which ensures a good resistance to corrosion in oxidizing atmospheres up to 1300–1400 °C.

The main characteristics of Fe–Cr–Al alloys are: electrical resistivity higher than that of Ni–Cr alloys ($\sim 1.4 \cdot 10^{-6}$ Ωm, see Fig. 4.32), relatively low mechanical strength, a certain material's fragility, possibility of chemical attack by the refractory lining of the furnace walls for reactions occurring at high temperature between Fe and SiO_2 and, finally, their relatively high cost.

At temperatures above 1400 °C can be used *metals and alloys with high melting point* such as nickel, chromium, platinum, platinum-rhodium, rhodium, molybdenum, tungsten, tantalum, niobium, etc.

Given their high cost, these materials are used only in special furnaces.

Their main characteristics are given in Table 4.8.

In particular, depending on the furnace atmosphere, *Molybdenum* can be used up to 1700 °C in vacuum furnaces (10^{-2}–10^{-4} Torr) or inert or reducing gases. It becomes brittle above 1000 °C and it is characterized by strong variations of resistivity with temperature.

Tungsten can be used up to 2700–2800 °C in neutral atmospheres or in vacuum (10^{-4}–10^{-6} Torr).

Tantalum up to 2400 °C in vacuum (up to 10^{-6} Torr).

Table 4.8 Characteristics of metallic materials with high melting temperature

Material	Pt	Pt–Rh 10 %	Pt–Rh 20 %	Rh	Mo	W	Ta
Density, Kg/dm^3	21.5	20	18.75	12.48	9.6–10.28	19.32	16.65
Melting temperature, °C	1769	1850	1884	1985	2610	3410	3000
Vaporization temperature, °C	–	–	–	–	4800	5930	4100
Maximum working temperature, °C	1400	1500	1600–1700	1850–1900	1500–1700	2200–2800	2400
Resistivity at 0 °C, μΩ cm	9.81	18.4	20.4	4.3	5.17	5.5	12.4
Temperature resistivity coefficient, 10^{-3} °C	31.6	14.22	12.10	5.96	5.5	5.0	3
Maximum specific power, W/cm^2	2	2	2	2	10–20	10–20	10–20

4.6.3 Non-metallic Resistors

Non-metallic resistors are made with metal-ceramic materials (also called "*cermets*") such as molybdenum disilicide and some metal oxides (e.g. lanthanum chromite, zirconium oxide) or non-metallic materials such as silicon carbide, graphite and amorphous carbon.

Silicon carbide (SiC) is the most widely used material of this group; it is produced in arc furnaces by melting together silicon and carbon at temperatures above 2200 °C. Given its very high resistivity (about 0.1 Ωcm at 1200 °C), it is extensively used in resistors having form of rods or tubes, with diameters between 10 and 50 mm and metallized terminals which ensure good "cold" contacts.

As illustrated by the curves of Fig. 4.33a, the resistivity is strongly variable with temperature: due to the presence of carbon, it firstly decreases rapidly till about 800 °C, while above this temperature it is characterized by a moderate increase up to the maximum working temperature. The curves may have significant variability in the low temperature range for the presence of material impurities.

The protection to oxidation at high temperature is provided by a layer of silicon oxide, which allows, in different atmospheres, the maximum working temperatures given in Table 4.9.

The material is also characterized by relatively low mechanical strength and shows considerable variations of resistivity with aging (which can lead even to doubling the initial value after few thousand hours of operation).

Molybdenum disilicide ($MoSi_2$) is characterized by resistivity values highly variable with temperature (from 30 μΩ cm in "cold" conditions, up to 400 μΩ cm at 1700 °C) (see Fig. 4.33b).

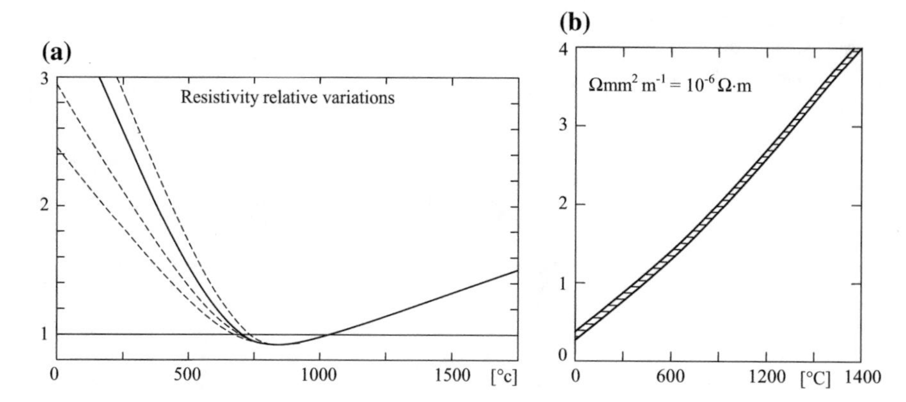

Fig. 4.33 Variations of resistivity with temperature (**a** silicon carbide; **b** molibdenum disilicide) [2, 3, 9]

Table 4.9 Maximum admissible temperature in different atmospheres

Atmosphere	Maximum temperature, °C	
	Silicon carbide	Molybdenum disilicide
Air	1575–1625	1700–1800
Nitrogen	1400	1600–1700
Argon/Helium/Neon	1575–1625	1600–1700
Hydrogen	1200	1350–1400
Exo-Gas	1250–1400	1600–1700
Endo-Gas	1250–1400	1500–1550
Vacuum	1000–1200	–
Carbon dioxide	1575–1625	1600
Hydrocarbons	1250	1350
Cracked ammonia	–	1400–1450

This implies a certain automatic power regulation when the temperature increases, which allows to reduce heating times and risks of resistor overheating.

Unlike what happens with silicon carbide, the material does not show any phenomenon of "aging", what makes it easier to replace series connected damaged resistors.

The resistance to high temperature oxidation is due to the formation of a surface layer of silica glass which allows to achieve in air maximum temperature of 1700–1800 °C and consequent high specific loads.

Table 4.8 shows the maximum admissible temperatures in some typical atmospheres.

Given the low mechanical strength of the material above 1400 °C, the resistor is generally shaped at "U" or "W" (as shown in Fig. 4.34) and hung on the walls or the ceiling of the heating chamber.

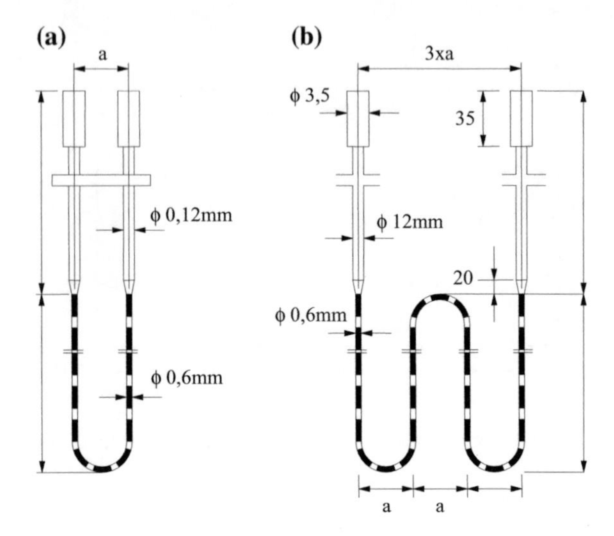

Fig. 4.34 Shape of resistors of molybdenum disilicide (**a** "U" shape; **b** "W" shape) —[9]

Graphite and *amorphous carbon* have high resistivity values, with a strong negative temperature coefficient up to about 600 °C (Fig. 4.35).

They oxidize in air at 400–500 °C and are therefore used—in particular graphite —in vacuum furnaces or in non oxidizing atmospheres up to 2200–3000 °C.

Table 4.10 gives other data of amorphous carbon and graphite.

Other special materials, sometimes used in laboratory furnaces, are some metal oxides such as *lanthanum chromite* CrO_3La (consisting of chromium Cr_2O_3 and lanthanum La_2O_3 oxides) and *zirconium oxide* (ZrO_2), both characterized by very high resistivity values in "cold" conditions. In case of zirconium, this requires even an external warm-up of resistors before switching-on.

They can be used in air up to temperatures of about 1700–1850 °C for lanthanum chromite and 2000–2200 °C for zirconium.

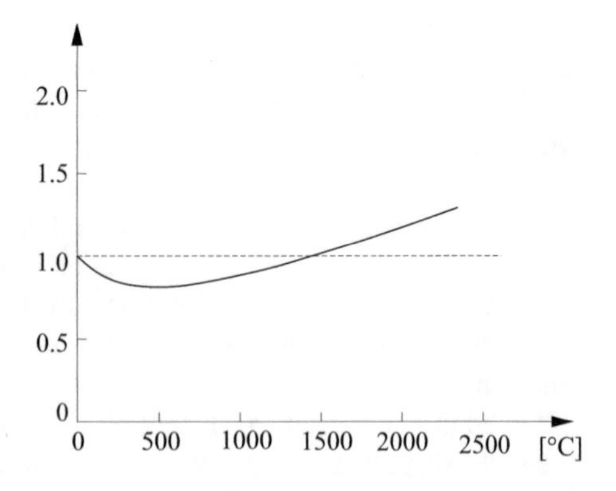

Fig. 4.35 Graphite: relative variation of resistivity as a function of temperature

Table 4.10 Characteristics of graphite and amorphous carbon [9]

Characteristics	Amorphous carbon	Graphite
Density, kg/dm^3	1.8	2.25
Apparent mass density, kg/dm^3	1.1	1.56
Specific heat, Wh/kg °C	0.20–0.23[a]	0.20–0.23[a]
	0.33–0.55[b]	0.33–0.55
Resistivity, μΩ cm	5000–8000	800–1300
Beginning of oxidation temperature, °C	400	500
Distillation temperature, °C	2200–2800	>2700
Boiling point, °C	3600	3600
Max. working temperature, °C	2200–2300	2600[c] 2800[d] 3000[e]

N.B. [a]at 20 °C
[b]In the range 1000–2000 °C
[c]In vacuum
[e]In Helium
[d]In argon

4.6.4 Resistors Life

The life of resistors mainly depends on their maximum working temperature, thermal cycles, design, furnace atmosphere and chemical attacks by solids, liquids or vapours in the heating chamber.

The influence of the maximum working temperature is due to the gradual oxidation of the external layer of the heating element that occurs at high temperatures (or to evaporation for resistors working in vacuum furnaces).

The rate of diffusion of oxidation increases slowly until a certain temperature, called the maximum admissible temperature, but very rapidly above it (a similar trend has the process of evaporation of resistors in vacuum).

As a consequence, if the operating point is near this temperature, even small variations in comparison with the design value can drastically change the life duration of the resistor, as shown as an example in Table 4.11 for some materials specified in Table 4.6.

Table 4.11 Relative values of life of some metallic resistors

Material →	(1) (%)	(2) (%)	(3) (%)	(4) (%)	(5) (%)	(6) (%)
Temperature ↓ (°C)						
1100	120	95	40	340	465	250
1200	25	25	15	100	120	75
1300	–	–	–	30	30	25

N.B. (1) … (6)—materials of Table 4.6

The resistor life can be estimated by the following empirical formula:

$$D = k\sqrt{ln\frac{A}{\vartheta_R}}$$

where is: D—resistor's life, years; ϑ_R—resistor working temperature; k, A—constants, characteristic of the resistor. (e.g., for Ni–Cr resistors made with relatively thick wire conductor: k ≈ 20, A ≈ 1250).

When the oxidation proceeds, gradually increases the thickness of the external oxide layer and, correspondingly, increases the resistor's electrical resistance and decrease heating power and mechanical strength.

When this phenomenon is localized, e.g. for imperfections in the wire cross section, it gives rise to local overheating which leads to a rapid increase of oxidation and then the "burning" of the resistor.

In order to take into account the occurrence of local overheating phenomena, the maximum working temperature is always lower than the maximum admissible temperature. The difference between them is normally assumed of 50–100 °C for metallic alloys and about 150 °C for SiC or $MoSi_2$ resistors.

For the same reason it should be avoid the use of wire conductors with too small cross-sections and it is necessary to take special care of the resistors fixing, in order to avoid additional causes of localized overheating.

In intermittent batch furnaces, the oxidation also depends on the thermal cycle of resistors and proceeds more rapidly the more frequent are the temperature variations. In fact, as a result of different linear expansion coefficients of the external oxide layer and the core material, during subsequent heating and cooling, in the surface layer will form "cracks" through which oxygen can penetrate, thus increasing the thickness of the oxidized layer.

In this case, for assuring a longer life of resistors, it must be assumed a difference between the maximum admissible temperature and the maximum working one further increased by 50 °C above the previous mentioned values.

4.6.5 Types of Resistors

Resistors made of metallic alloys are mostly manufactured with bare wires or strips with of spiral or zig-zag shape, as shown in Fig. 4.36.

For assuring life duration and mechanical strength it is recommend to use wires with diameters not lower than 3 mm (at temperatures below 1000 °C) or 5 mm (at higher temperatures) and strip thickness, not lower than 1 and 2 mm respectively.

In fact, lower diameters and thicknesses would result in a rapid "aging" of resistor, since the oxidized surface layer would reduce the useful cross-section and lead to a rapid increase of the electrical resistance.

As regards maximal dimensions, it is not cost-effective to use conductors with diameters higher than 15 mm for wires and 3 mm for strip thickness, in order to

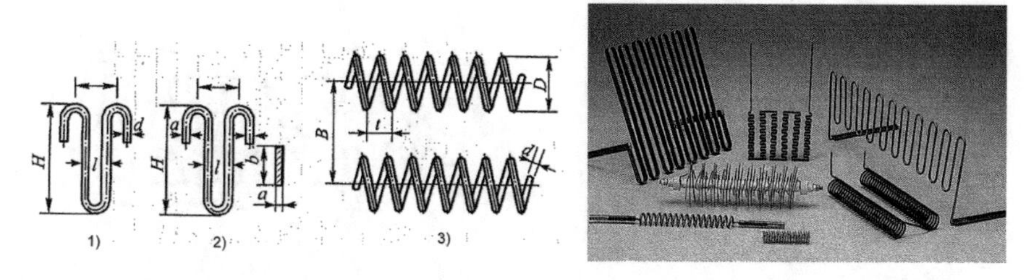

Fig. 4.36 Typical shapes of metallic heating elements (*1*, *2* zig-zag wire and strip elements; *3* spiral elements; *4* various metallic elements)

avoid low ratios of the external surface to the conductor cross-section, which would lead to an excessive use of material.

As already said, the most common types of construction in radiation furnaces, both for wire and strip conductors, are spiral and zig-zag shapes. In both cases particular care must be taken in the selection of the most appropriate geometrical dimensions of the resistor and its hanging supports, in order to allow the best radiation conditions towards the heating chamber, thus keeping to the lowest value resistor overheating, since its strength and life decrease rapidly with the increase of the working temperature.

Spiral resistors are mostly made of wire, wrapping the wire in spiral form, usually with the geometrical ratios (D/d) of spiral diameter to conductor diameter specified in Table 4.12, and between pitch of the spiral and conductor diameter (p/d) > 2.5–3.0

Spiral resistors are generally hung on the chamber walls or freely supported on wall brackets or suspended on ceramic refractory "candles" (Fig. 4.37a–d).

The supporting material must be chosen carefully to avoid chemical attack of resistors.

Sometimes the spirals are located into slots on the roof and the bottom of the heating chamber, in order to protect them from mechanical impacts during loading and unloading operations (Fig. 4.37e, f).

In case of spirals suspended on ceramic tubes, to the end of facilitating free radiation, it's suggested to adopt a ratio of inner spiral diameter to the outer diameter of the supporting tube of about 1.1–1.2 and the distance between the axes of tubes equal to 2–2.5 times the diameter of the spiral.

Table 4.12 Suggested geometrical ratios of wire spiral resistors

Max. temperature (°C)	Ratio D/d for alloys	
	Ni–Cr	Fe–Cr–Al
800	10	8
900	9	7
1000	8	6
1100	–	5

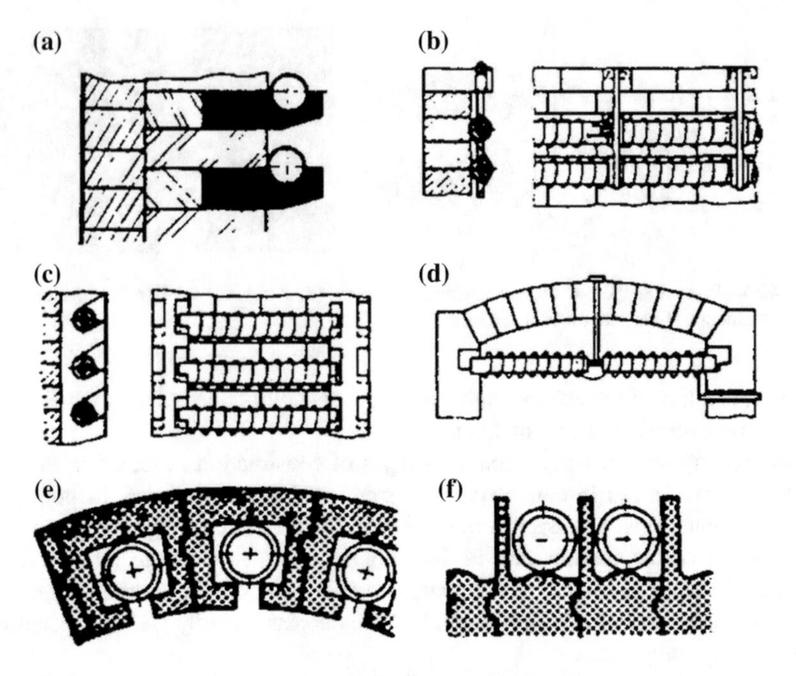

Fig. 4.37 Arrangement of spiral resistors in the heating chamber [2]

Serpentine or zig-zag resistors are mostly made either with wire of relatively big cross-section or strips. They have good mechanical strength and a very favourable ratio of radiating surface to conductor's cross-section.

Convenient geometrical dimensions must be also selected in order to limit mutual radiation between adjacent resistor sections.

For wire resistors (Fig. 4.38a) the coil pitch must be not lower than 2.5–3 times the diameter of the wire, while the height should not exceed 250–300 mm.

For strip elements, the pitch h of zig-zag should never be lower than the size b of the largest side of the strip cross-section, the ratio of strip width b to its thickness a is generally between 5 and 15, while the height B varies in the range 150–600 mm depending on material, working temperature and resistors lay-out in the heating chamber. Spacers are required at least every 200 mm.

The values most commonly used are:

$$(\mathrm{h}/\mathrm{b}) = 1.8; \quad (\mathrm{b}/\mathrm{a}) = 10; \quad (\mathrm{B}/\mathrm{b}) = 7.8.$$

Figure 4.38b shows some examples of mounting of strip elements. The hanging supports and spacers, which must be designed with particular care to avoid localized overheating, are made of refractory material (Schamotte, Al oxide, etc.) or, more frequently, with high-temperature steels or Ni–Cr.

The *choice between wire or strip resistors* depends on many factors. Among them, preference should be given to strip heating elements, which give higher

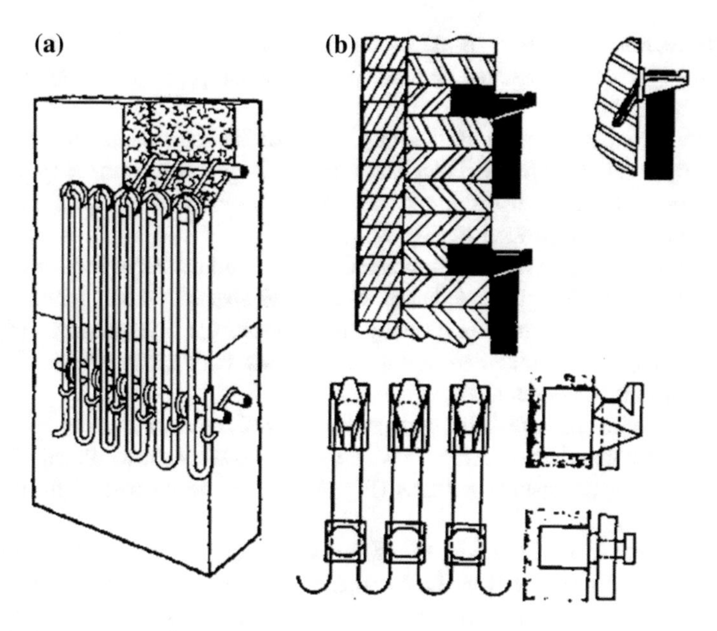

Fig. 4.38 Arrangements of zig-zag resistors in the heating chamber [2]

mechanical strength and higher installed power per square meter of the chamber wall.

However, it must be remembered that bending of conductor during manufacture of zig-zag resistors may be difficult in case of hard materials or even impossible for the brittle ones.

Moreover, for the same radiating surface, the strip conductor has a weight significantly lower than that of the wire. In fact, for having with a wire of diameter *d* the same radiating surface of a strip of dimensions *a* and *b* (with $m = b/a$), the ratio of the volumes of the two conductors must be:

$$k = \frac{\pi d^2}{4ab} = \frac{(1+m)^2}{\pi m}.$$

For example, for m = 5–10–15, the values of k are respectively 2.29–3.85–5.43 and the corresponding ratios d/a become 3.82–7.00–10.18.

This fact has not important consequences on costs of small cross-sections conductors, but for bigger cross-sections the use of strip conductors can result in significant economies. However, these economies cannot be always practically achieved, since the cost of the strip is higher than that of the wire, particularly for small cross-sections, and the manufacture and installation of wire resistors and their supports is easier and cheaper.

Where technical reasons do not suggest otherwise, the choice between the two systems is thus determined by a cost analysis.

In convection furnaces or furnaces for high temperature with protective atmosphere, the type of construction depends on the gas temperature, its aggressiveness and the possibility of explosion.

In these furnaces are often used arrangements of zig-zag resistors made either by wire or strip conductors (in the latter case with the largest side of the strip parallel to the direction of gas flow, in order to obtain low hydraulic resistance) or spiral freely blown resistors, as illustrated in Fig. 4.39.

The main problem in these arrangements is the overheating of the resistor at the contact points with hanging insulators, hooks and spacers which, for this reason, must be accurately designed (e.g. using metal hooks for high temperature of the same diameter of the wire and spiral connections with cross-section 3–4 times greater than that of conductor).

In case of wire spiral resistors arranged in several layers, the spiral coils can be wound on ceramic tubes placed crosswise the gas flow, as shown in Fig. 4.40a, or freely suspended on insulators tubes (Fig. 4.40b) or placed inside the insulators (Fig. 4.40c).

The first type of construction, allows to achieve high surface temperatures but gives relatively low values of the heat transfer coefficient; the third one, mostly used in gas furnaces with temperatures up to 700 °C, allows to concentrate high power in small volumes, but leads to strong overheating of the coil within the insulator.

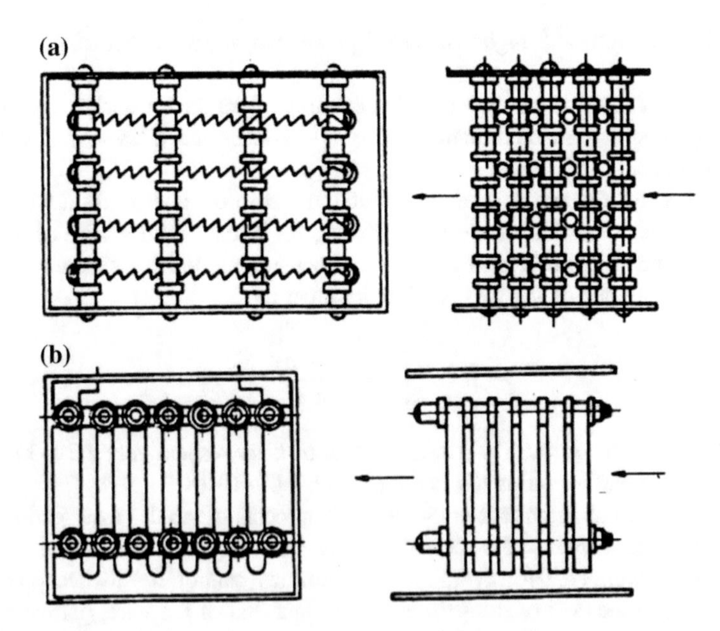

Fig. 4.39 Arrangement of resistors in forced convection furnaces (**a** with wire conductors; **b** strip conductors) [2, 3]

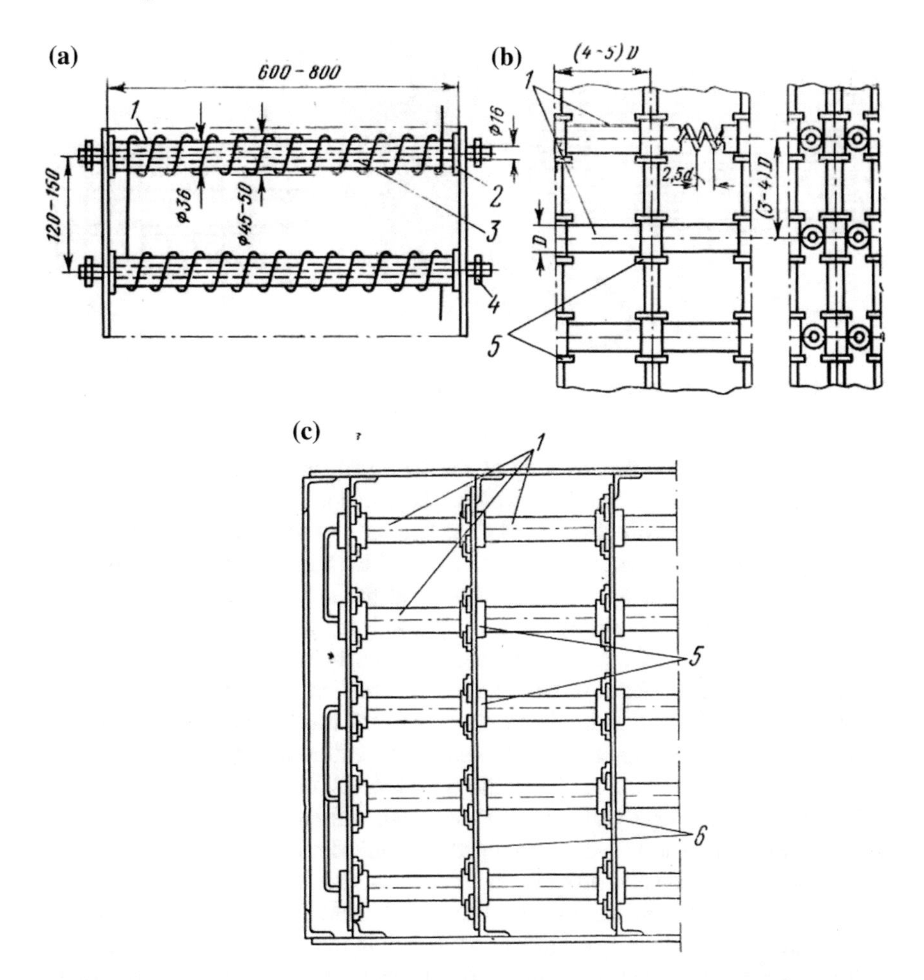

Fig. 4.40 Arrangement of resistors in forced convection furnaces (**a** wire spiral resistors on ceramic tube: **b** resistors freely suspended on insulators tubes; **c** resistors inside ceramic tubes; 1a-spiral; 2a, 5c-ceramic washer; 3a, 1c-ceramic tube; 4a-refractory rod; 5b-cylindrical insulators; 6c-metallic plate) [16]

Another particular example of construction, used for heating air up to 1000 °C, is shown in Fig. 4.41a. The air to be heated flows inside tubular resistors placed within ceramic blocks.

In some cases, when the resistor length is too high, it can be used the arrangement shown in Fig. 4.41b, which is characterized by low heat losses from the external walls, but an intense heat exchange between resistors and air and a compact lay-out.

For other special construction arrangements, the reader should refer to the bibliography [16].

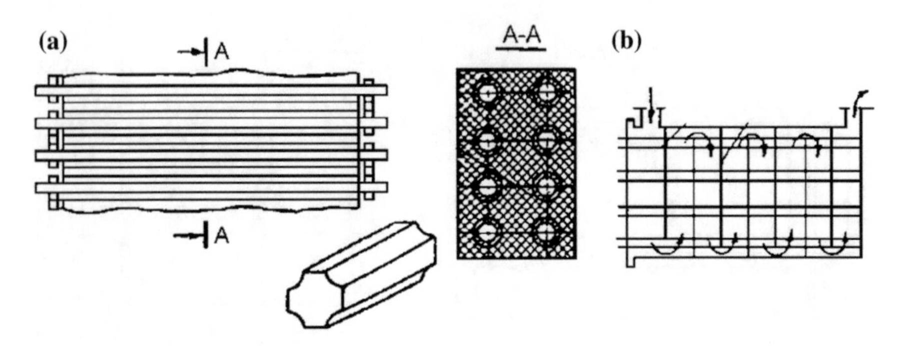

Fig. 4.41 Special arrangements for heating air [16]

Special mention should be made of the so-called *armoured heating elements*, consisting of a metallic tube within which it is placed a Ni–Cr spiral with electrical connections at the tube ends (Fig. 4.42). The space between spiral and tube is generally filled with magnesium oxide (MgO), which has good electrical insulating properties and high thermal conductivity. Good thermal conductivity is essential, since the heat must be transmitted by thermal conduction trough the insulation from the heating element to the external surface of the heater.

The choice of the tube metallic material depends on the application: copper is normally chosen for immersion heaters used to heat water, but if the resistor must be used in a corrosive surrounding, then materials with resistance appropriate to the expected type of attacks, like nickel alloys and steels, must be used.

This type of resistor is particularly widespread, with power ranging from several hundred watts to some kW, for its ruggedness and life duration, since the heating conductor is well protected from oxidation and mechanical damage.

Due to their high resistivity, the heating elements made of silicon carbide, molybdenum disilicide or graphite are produced as rods or tubes (Fig. 4.43).

They are usually fixed horizontally or vertically on the walls, the roof or the bottom of the heating chamber, as illustrated in Fig. 4.44.

Because of their susceptibility to deformation at high temperature, resistors of molybdenum disilicide cannot be fixed horizontally without a suitable support at

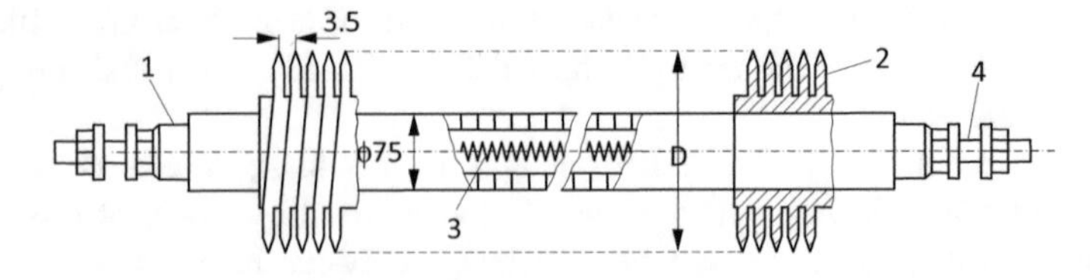

Fig. 4.42 Armored resistor (*1*, *4* end connections; *2* heat radiator; *3* spiral resistor)

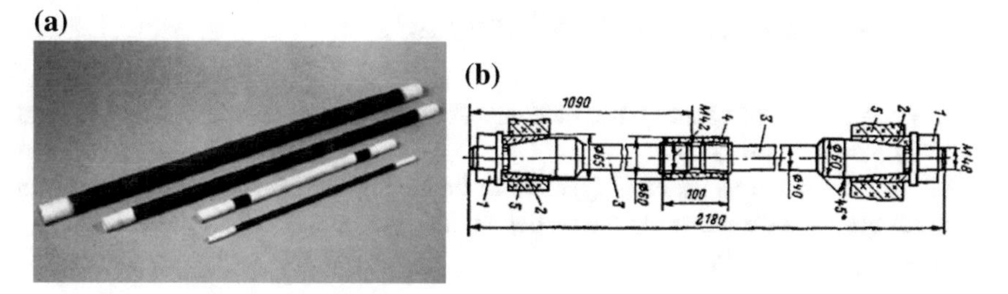

Fig. 4.43 Rod heating elements (**a** silicon carbide; **b** graphite)

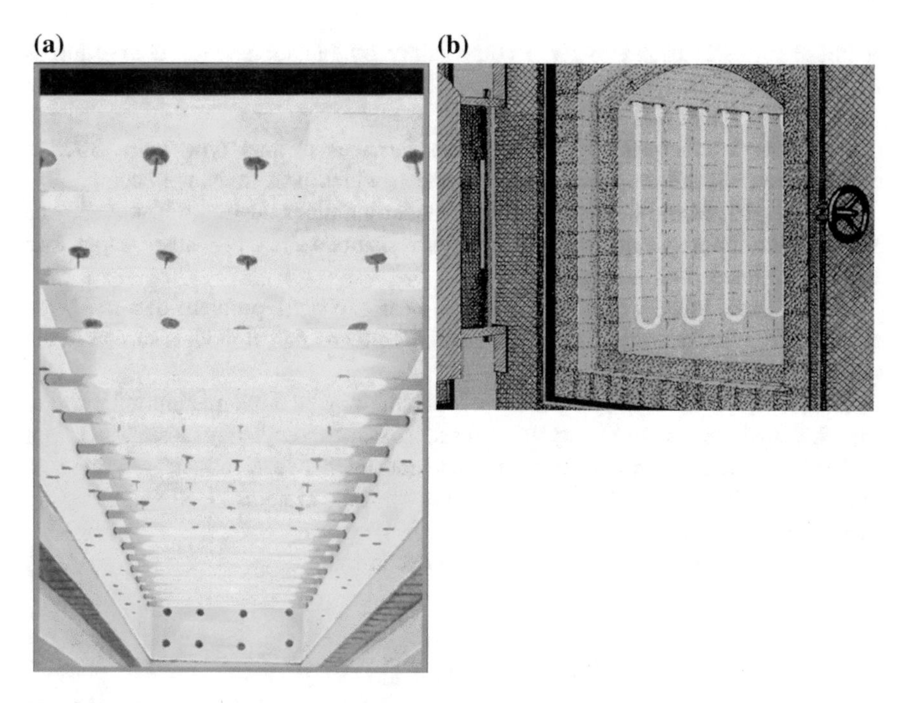

Fig. 4.44 **a** Resistors of silicon carbide installed on the roof of the heating chamber; **b** Resistors of molybdenum disilicide installed on the wall

temperature above 1400 °C. For this reason they are usually hung vertically on the walls and the roof of the heating chamber and designed with “U”or “W” shape, as show in Fig. 4.34.

Due to brittleness and the lack of mechanical resistance to heat of graphite, the terminal connections must allow a free expansion of the rod and must be so designed as to limit the overheating occurring at the contact areas, with appropriate selection of geometry and materials.

4.6.6 Design of Classical Metallic Resistors

In this paragraph only some basic elements for preliminary calculation of classical metallic resistors are given.

Given the numerous different types of resistors and the many factors involved in their design, for a deeper study the reader should refer to the extensive bibliography on this subject [2, 3, 9, 17].

4.6.6.1 Radiation Furnaces

In these furnaces heat transfer occurs mainly by radiation while convection is relatively small. This group comprises furnaces without forced air flow, operating at temperatures above 600–700 °C.

Figure 4.45 schematically shows three furnaces of this type with different resistors layouts and the main elements among which heat exchange occurs.

In general it would be necessary to consider simultaneously the heat exchange between the three systems resistors/charge, resistors/walls and walls/ charge, and the mutual interaction among them.

In this case, however, the study of the heat exchange problem becomes very complex and can be carried out, qualitatively, with analogical models similar to the one schematised in Fig. 4.46.

However, a simplified preliminary calculation may be done making reference to Fig. 4.45a where, due to the layout of resistors which completely cover walls, roof and sole of the heating chamber, the heat transfer between refractory lining and charge is negligible, and the heat exchanges between systems (1–2) and (1–3) are independent from each other.

The heat transfer equations in this case become:

$$\left.\begin{aligned} P_u &= C_{12} \cdot (T_1^4 - T_2^4) \cdot S_2 \\ P_{loss} &= C_{13} \cdot (T_1^4 - T_3^4) \cdot S_1 \end{aligned}\right\} \qquad (4.18)$$

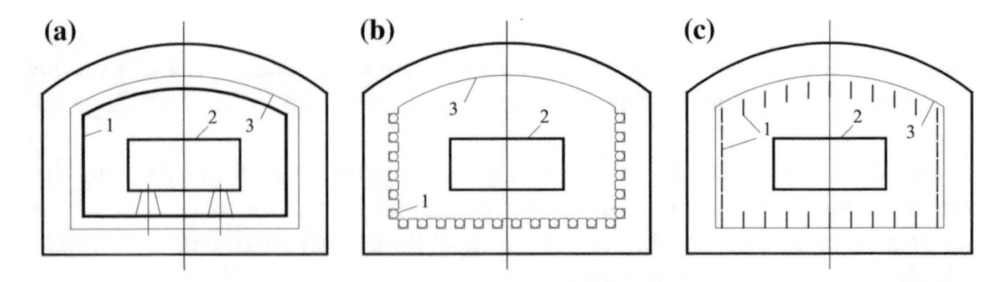

Fig. 4.45 *1* heating elements; *2* charge; *3* chamber walls

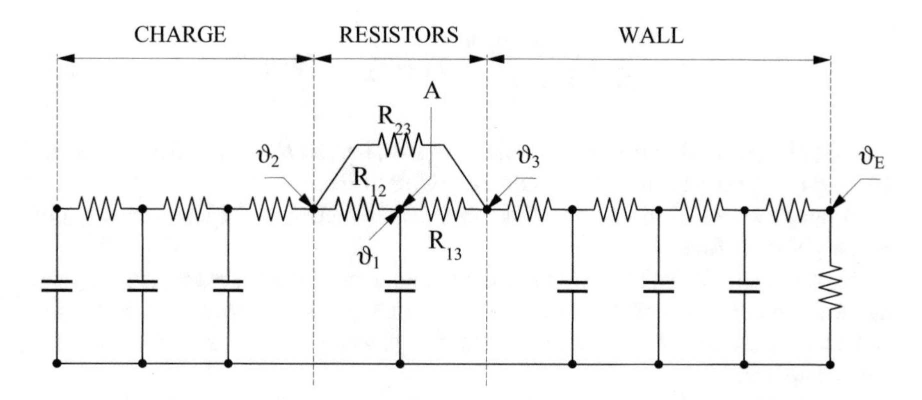

Fig. 4.46 Analogical model of heat exchanges among elements 1–2–3 of Fig. 4.44 (*A* resistor's supply; $\vartheta_1, \vartheta_2, \vartheta_3$ temperatures of resistors, surface of charge and internal wall surface; R_{12}, R_{13}, R_{23} thermal resistances between elements 1–2, 1–3, 2–3)

with:

P_u, P_{loss}	useful power in the charge and power lost in the wall lining, (W)
C_{12}, C_{13}	radiation coefficients of systems (1–2), (1–3), (W/K^4 m^2)
T_1, T_2, T_3	absolute temperature of elements 1, 2, 3, (K)
S_1, S_2, S_3	surface area of elements 1, 2, 3, (m^2).

Assuming that the inner element has the lower surface and can be considered "convex", the coefficients C_{12} and C_{13} are given by the well known relationships:

$$\left.\begin{aligned} C_{12} &= \frac{5.67 \cdot 10^{-8}}{\frac{1}{\varepsilon_1} + \frac{S_1}{S_2}\left(\frac{1}{\varepsilon_2} - 1\right)} \\ C_{13} &= \frac{5.67 \cdot 10^{-8}}{\frac{1}{\varepsilon_1} + \frac{S_1}{S_3}\left(\frac{1}{\varepsilon_3} - 1\right)} \end{aligned}\right\} \tag{4.19}$$

Equations (4.18) and (4.19) allow the determination of temperatures ϑ_1 and ϑ_3 of the heating elements and the wall internal surface:

$$\left.\begin{aligned} \vartheta_1 &= \sqrt[4]{\frac{P_u}{C_{12}S_2} + T_2^4} - 273, \quad &{}^\circ\mathrm{C} \\ \vartheta_3 &= \sqrt[4]{\frac{P_u}{C_{12}S_2} - \frac{P_{loss}}{C_{13}S_1} + T_2^4} - 273, \quad &{}^\circ\mathrm{C} \end{aligned}\right\} \tag{4.20}$$

In particular, for an ideal furnace characterized by zero losses ($P_p = 0$) and S_1 equal to S_2, from Eqs. (4.18) and (4.19) we can write:

$$p_i = \frac{P_u}{S_1} = \frac{5.67 \cdot 10^{-8}}{\frac{1}{\varepsilon_1} + \frac{1}{\varepsilon_2} - 1}\left(T_1^4 - T_2^4\right), \quad (W/m^2) \qquad (4.21)$$

where p_i is the specific power of an ideal resistor, i.e. the amount of heat which the heating element can transfer per unit area in unit time.

Average values of the emission factors of some materials, for the calculation of p_i are given in Table 4.13.

In particular, Eq. (4.21) shows the strong influence of the emission factors ε_1 and ε_2 on the specific power that the resistor can radiate per unit area. In fact, for the same resistor, we have very different values of p_i when heating loads with different characteristics.

This is illustrated, as an example, by the diagrams of Fig. 4.47a, b, which give the values of p_i as a function of the temperature of the charge for different resistors temperatures, in the case of a furnace with ideal metallic resistors (e.g. of Ni–Cr or similar) heating steel and aluminium workpieces, respectively.

In the general case, where simultaneously heat exchanges occur among resistors/charge, resistor/walls and walls/charge (see Fig. 4.45b, c), the curves of the ideal heating elements may still be used by introducing appropriate correction coefficients.

These coefficients take into account that Eq. (4.21), which links the resistor temperature $T_1 = T_R$ to the temperature $T_2 = T_C$ of the charge (considered for simplicity equal to that of the furnace chamber in thermal steady state conditions), was derived assuming that all points of the resistor surface radiate freely in all directions.

In particular, it was neglected the mutual radiation between adjacent heating elements, between different parts of a single resistor and between the latter and the chamber walls.

This would lead to a resistor temperature greater than the one occurring in case of free radiation.

This is taken into account in the calculation by introducing the *form factor F* (smaller than unity), expressing the percentage of the total power radiated by the resistor that is transmitted to the charge, and calculating the resistor with a fictitious specific power:

Table 4.13 Emission factors —[2]

Material	ε
Steel, Ni–Cr e Carborundum (o.)	0.8
Copper (o.)	0.7
Brass (o.)	0.6
Steel (u.)	0.45
Aluminium (o.), Copper and Brass (u.)	0.3
Molybdenum and Tungsten (u.)	0.2

[**o.**—oxidized; **u.**—unoxidized]

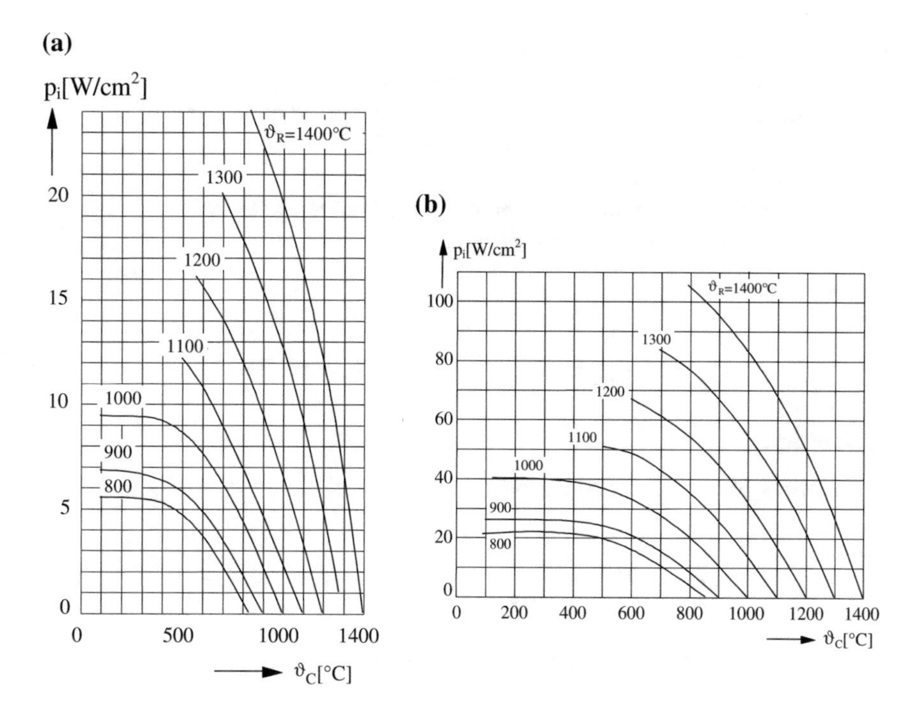

Fig. 4.47 Surface specific power p_i (W/cm^2) of ideal resistors when heating in air steel (**a**) or aluminium (**b**) charges (ϑ_R, ϑ_C—resistor and charge temperature, °C) [1, 3]

$$p = \frac{p_i}{F} = \frac{5.67 \cdot 10^{-8}}{\frac{1}{\varepsilon_1} + \frac{1}{\varepsilon_2} - 1} \cdot \frac{\left(T_1^4 - T_2^4\right)}{F} \tag{4.22}$$

The factor *F* takes into account the resistor shape (spiral, zig-zag, rod, wire, strip, etc.), its geometrical dimensions ratios, its orientation with respect to the charge and the filling factor of the walls of the heating chamber.

One must carefully evaluate the influence of these parameters, since they can significantly modify the value of *F*, which can vary in a wide range (e.g. 0.3–0.7).

On the basis of these values and the analysis of the curves of Fig. 4.47, it can be easily understood why in furnaces with metal resistors the installed power is mostly limited to about 10–20 kW/m^2.

After determination of the design specific surface power according to Eq. (4.22), can be selected the connection diagram of heating elements, the number of phases and of parallel resistors, power and voltage, and size and weight of the wire needed for resistors manufacturing.

In fact, denoting with:

P resistor power (or power of one phase)
V resistor voltage

R resistance of the heating element
S, l, s total surface, length and cross section of resistor conductor
d, a, b conductor diameter (for circular cross-section) or dimensions of sides of a strip conductor of rectangular cross-section
G conductor weight
ρ resistivity of resistor material at the working temperature
γ specific weight of resistor material,

from equations:

$$P = \frac{V^2}{R} = p \cdot S; \quad R = \rho \frac{\ell}{s}; \quad G = \gamma \cdot \ell \cdot s$$

one obtains:

- for wires of circular cross-section:

$$d = 0.740 \cdot \sqrt[3]{\frac{\rho \cdot P^2}{p \cdot V^2}}; \quad \ell = \frac{P}{\pi \cdot d \cdot p}; \quad G = \gamma \frac{P \cdot d}{4p} \tag{4.23}$$

- for strip conductors (with side ratio m = b/a):

$$a = 0.794 \cdot \sqrt[3]{\frac{\rho \cdot P^2}{m \cdot (1+m) \cdot p V^2}}; \quad \ell = \frac{P}{2(1+m) \cdot a p}; \quad G = \gamma \frac{P \cdot a \cdot m}{2(1+m) \cdot p} \tag{4.23a}$$

Example 4.2 Preliminary calculation of resistors for the furnace of example 4.1.

Data

- Total installed power: P′ = 66 kw
- N. 6 resistors in two parallel groups delta connected
- Voltage supply of each resistor: V = 220 V
- Power of each resistor: P = 11 kW
- Spiral resistors made with wire of Ni–Cr 80/20 (length of spiral L = 1.8 m, ratio spiral pitch/wire diameter d = 2)
- Form factor: F = 0.65
- Temperature of the charge: $\vartheta_2 = 950\,°C$ ($T_2 = 1223\,°K$)
- Current: $I = \frac{P}{V} = \frac{11000}{220} = 50\,A$
- Resistance: $R = \frac{V}{I} = \frac{220}{50} = 4.40\,\Omega$
- Specific power of the ideal resistor (assuming θ_R = 1050 °C):

$$p_i = \frac{5.67 \cdot 10^{-8}}{\frac{1}{\varepsilon_1} + \frac{1}{\varepsilon_2} - 1} \cdot \left(T_1^4 - T_2^4\right) = \frac{5.67 \cdot 10^{-8}}{\frac{1}{0.8} + \frac{1}{0.8} - 1} \cdot \left(1323^4 - 1223^4\right)$$
$$= 3.17 \cdot 10\,\mathrm{W/m^2}$$

- Fictitious specific power for calculation of the real resistor:

$$p = \frac{p_i}{F} = \frac{3.17 \cdot 10^4}{0.65} = 4.88 \cdot 10^4\,\mathrm{W/m^2}$$

- Resistor temperature (from curves of Fig. 4.47a), $\vartheta_C = 950\,°C$; P = 4.88 W/cm^2):

$$\vartheta_R \approx 1130\,°C$$

- Wire diameter:

$$d = 0.740\sqrt[3]{\frac{\rho P^2}{p V^2}} = 0.740\sqrt[3]{\frac{116 \cdot 10^{-8}(11000)^2}{4.88 \cdot 10^4(220)^2}} \approx 2.9 \cdot 10^{-3}\,\mathrm{m} = 2.9\,\mathrm{mm}$$

Note The optimum diameter is usually in the range 3–4 mm. With diameter lower than 2 mm the resistor is more brittle and more susceptible to aging, with diameters greater than 5 mm the resistor manufacture is more difficult.

- Length of wire:

$$\ell = \frac{P}{\pi d p} = \frac{11000}{\pi\, 2.9 \cdot 10^{-3} 4.88 \cdot 10^4} = 24.7\,\mathrm{m}$$

- Number of turns:

$$N = \frac{L}{2d} = \frac{1800}{2 \cdot 2.9} = 310$$

- Spiral diameter:

$$\ell = \pi(D + d)N \quad \Rightarrow \quad D = \frac{\ell}{\pi N} - d = \frac{24700}{\pi \cdot 310} - 2.9 = 22.5\,\mathrm{mm}$$

- Check of spiral diameter:

$$5 \cdot d \leq D \leq 12 \cdot d \quad \Rightarrow \quad 14.5 \leq 22.5 \leq 38.8$$

4.6.6.2 Convection Furnaces

In this paragraph it is considered the calculation of resistors for furnaces with forced air flow, i.e. furnaces where the heat transfer occurs mainly by convection and the working temperature of resistors is between 200 and 800 °C (working temperature of the furnace between 100 and 700 °C).

Assuming negligible the heat transfer by radiation from the resistor surface, the various construction types are analyzed in analogy to the case of circular tubes or wires, single or in bundles, invested by an air flow perpendicular to their axis.

In thermal steady state conditions the power transformed into heat in the conductor is completely transferred to the air and, with reference to Fig. 4.48, the following relation holds:

$$p = \alpha \cdot (\vartheta_R - \vartheta_A) = \frac{4\rho I^2}{\pi^2 d^3} \tag{4.24}$$

where:

p	resistor surface specific power, W/m^2
ϑ_R, ϑ_A	temperatures of resistor surface and air, °C
h_c	convection heat transmission coefficient, W/m^2 °C
I	electrical current flowing in the conductor, A
ρ	conductor material resistivity, Ωm
d	conductor diameter, m.

Therefore we can write:

$$d = \sqrt[3]{\frac{4\rho I^2}{\pi^2 h_c (\vartheta_R - \vartheta_A)}} \tag{4.25}$$

The main problem in the use of Eq. (4.25) is the evaluation of the coefficient h_c, which can be done through empirical relationships of the form:

$$\mathrm{Nu} = C\, Re^n \tag{4.26}$$

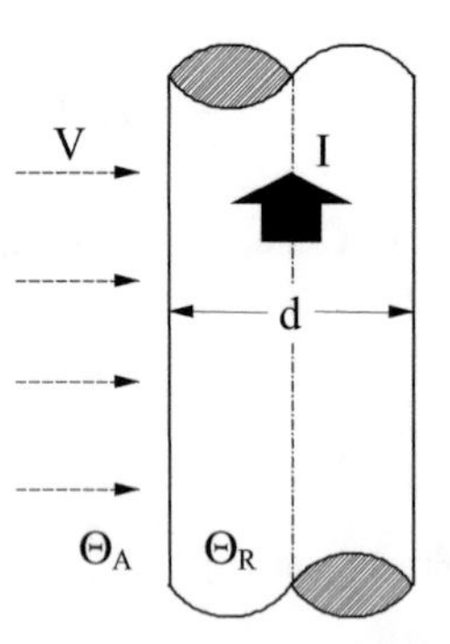

Fig. 4.48 Air flow perpendicular to the axis of a wire where the current I flows

where:

$Nu = (\alpha d)/\lambda$	Nusselt number,
$Re = (v d)/\mu$	Reynolds number,
λ, μ	air thermal conductivity (W/m °C) and kinematic viscosity (m^2/s)
v	air velocity (m/s),
C, n	experimental coefficients, which depend on resistor temperature ϑ_R and its construction (wire, tape, pipe, zig-zag, free spiral, spiral on ceramic tube, etc.).

In particular:

- *cylindrical conductors freely blown in air:* can be calculated with equations of Sect. 1.3.3.
- *wire zig-zag resistors*: for the constants *C* and *n* the following values are suggested:

$$C = 0.238 \quad n = 0.60, \quad \text{for} \quad Re > 1000$$
$$C = 0.625 \quad n = 0.46, \quad \text{for} \quad 80 < Re > 1000$$

  and the coefficient h_c can be obtained from the diagrams of Fig. 4.49 [3].
- *Strip zig-zag resistors,* with sides ratio b/a = 10, can be calculated as the previous ones by considering a wire of circular cross-section with diameter d_e, equivalent in terms of heat transfer:

$$d_e = \frac{\text{perimeter}}{1.5\pi} \quad ; \quad d_e = \frac{\text{perimeter}}{\pi}$$

  with a different value depending on whether the longer side *b* or the shorter side *a* of the strip is in the direction of the air flow, as schematically shown above. Moreover, for zigzag resistors, the coefficient h_c is practically independent on the pitch if this is greater than $(2.5 \cdot d_e)$.
- *spiral wire freely hanging resistors*: h_c is practically the same as for zig-zag strip resistors.
  Data for *Fe–Cr–Al wire resistors* with pitch greater than (2d) are given in Fig. 4.50 and Table 4.14 [18].
- *spiral wire resistors wound on ceramic tubes:* (e.g. of Schamotte) it can be assumed a heat transfer coefficient approximately equal to 60 % of that of a spiral wire freely hanging resistor.

Once the value of h_c is known, Eq. (4.24) allows to calculate also the temperature difference $(\Delta\vartheta = \vartheta_R - \vartheta_A)$ between resistor and air as a function of air velocity, for given values of specific power *p* and wire diameter *d*.

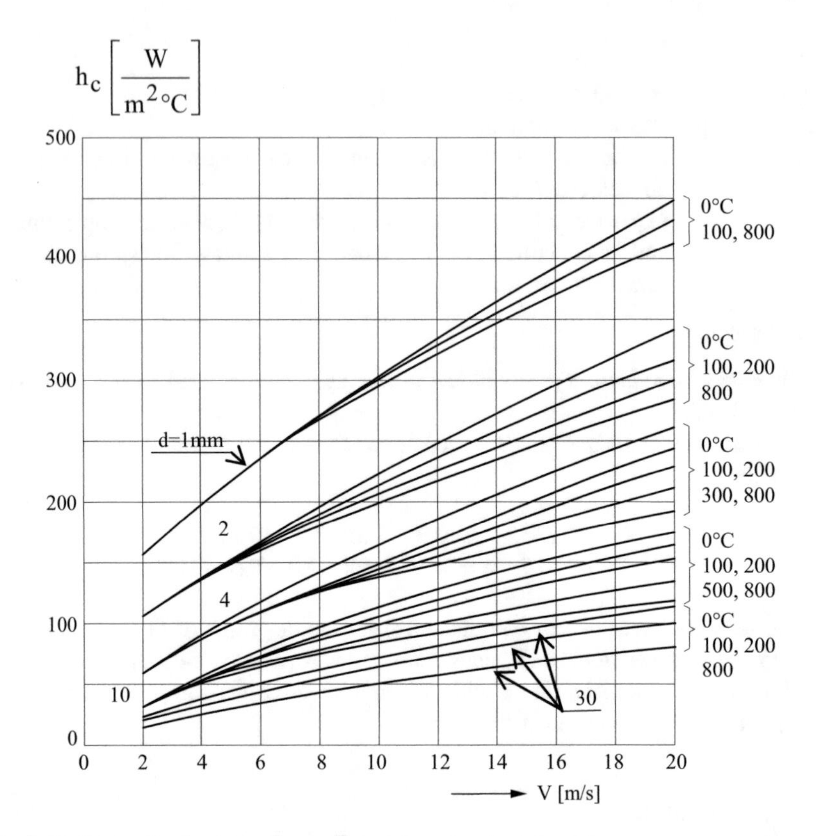

Fig. 4.49 Convection heat transfer coefficient for wire zig-zag resistors

Fig. 4.50 Nusselt number versus Reynolds number for Fe–Cr–Al spiral resistors

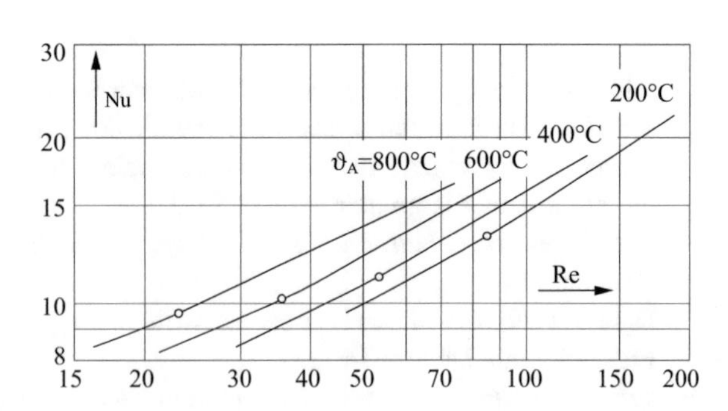

This is shown, as an example, for a wire of diameter d = 4 mm, by the diagram of Fig. 4.51 which has been calculated with the values of h_c of Fig. 4.49.

When defining the design temperature ϑ_R of resistor, it must be kept in mind that in this type of furnaces substantial temperature differences (up to 100–200 °C) may occur between different points of resistor conductor (surface points more or less

Table 4.14 Coefficients for calculation of Fe–Cr–Al wire resistors with Eq. (4.26)

ϑ_R	$Re < Re_c$		$Re > Re_c$		
°C	C	n	C	n	Re_c
200	0.393	0.527	0.222	0.627	835
400	0.570	0.479	0.304	0.579	537
600	0.827	0.432	0.461	0.532	347
800	0.227	0.384	0.699	0.484	223

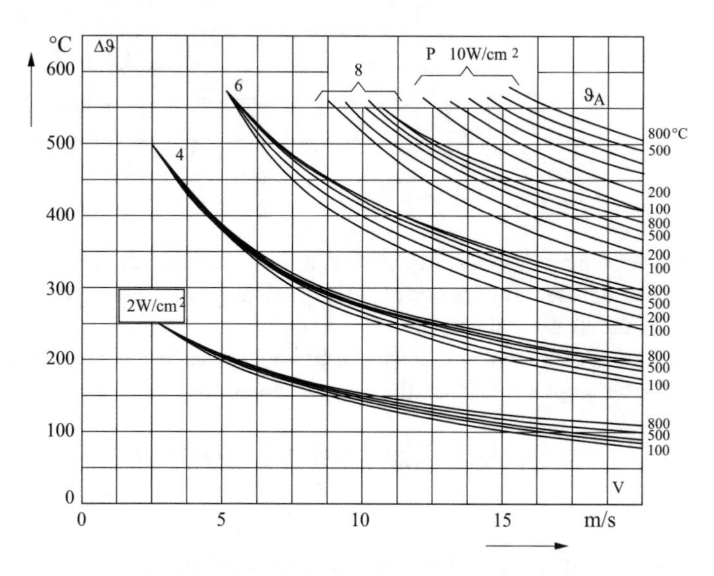

Fig. 4.51 Over-temperature above air of a wire resistor (diameter 4 mm) as a function of air velocity and temperature ϑ_A [3]

exposed to the air flow, hanging points, etc.) and that these differences are generally higher at the lower temperatures of air.

Guideline values are:

ϑ_A (°C)	ϑ_R (°C)	$\vartheta_{R\,max.}$ (°C)
300	500	700
400	500–600	–
500	600	–
700	800	–

Finally, when are given the temperatures ϑ_R and ϑ_A, the diameter d of the wire conductor and the velocity v of air, Eq. (4.24) or (4.25) allow to determine the coefficient h_c and then the specific power p, as illustrated by the example of Fig. 4.52.

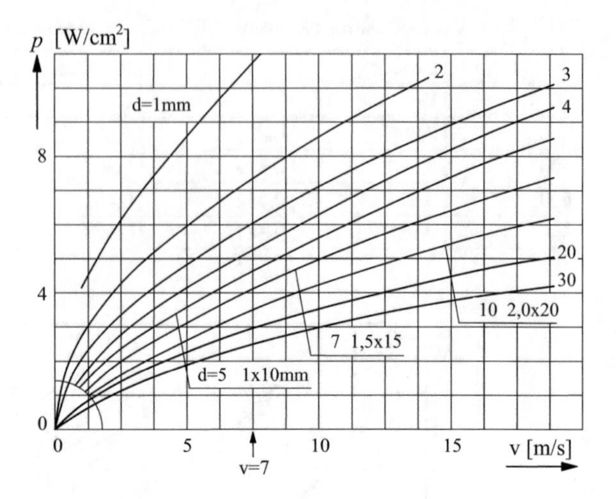

Fig. 4.52 Specific power p as a function of air velocity v for wires of various diameters d (ϑ_R = 500 °C; ϑ_A = 100 °C; p_i = 0.7–1.0 W/cm²)

However, for a more accurate calculation, it must be reminded that in addition to convective heat transfer, always occurs, to a lesser or greater extent, radiation from the surface of the heating element.

Therefore the specific power *p* is not linked only to the air velocity *v* through the coefficient h_c, but depends also on the resistor emissivity, i.e. it is:

$$p = \left[h_c \cdot (T_r - T_f) + \sigma \cdot \varepsilon_R \cdot (T_r^4 - T_f^4)\right] \cdot 10^{-4} \tag{4.27}$$

with:

T_r, T_f absolute temperatures of resistor and furnace chamber, K
ε_r resultant emissivity of resistor surface
σ Stefan-Boltzmann constant, $\sigma = 5.67 \cdot 10^{-8}$, W/(m²K⁴)
h_c heat transmission coefficient by convection, W/(m² °C).

The value of the radiated specific power p_i (second term of the right hand side of Eq. 4.27) must therefore be added to the one due to convection.

This allows, in spite of the difficulty of determining exactly the value of h_c, to obtain the curves of Fig. 4.53, which give the comparison of resistor overheating $\Delta\vartheta$ with air velocity of 6 and 12 m/s, with and without convection respectively.

In conclusion, summarizing the design procedure, there two are parameters which must be determined for the resistor design: specific power *p* and conductor geometrical dimensions.

Since these two quantities, as it is shown by Eq. (4.24), are correlated each other, they must be determined by a try and error procedure.

Defining the temperature of air on the basis of the required temperature of the charge, the velocity *v* according to the hour production and the temperature difference $(\vartheta_R - \vartheta_A)$ considering material characteristics and resistor construction criteria, are defined—on the basis of conductor geometrical dimensions—the values

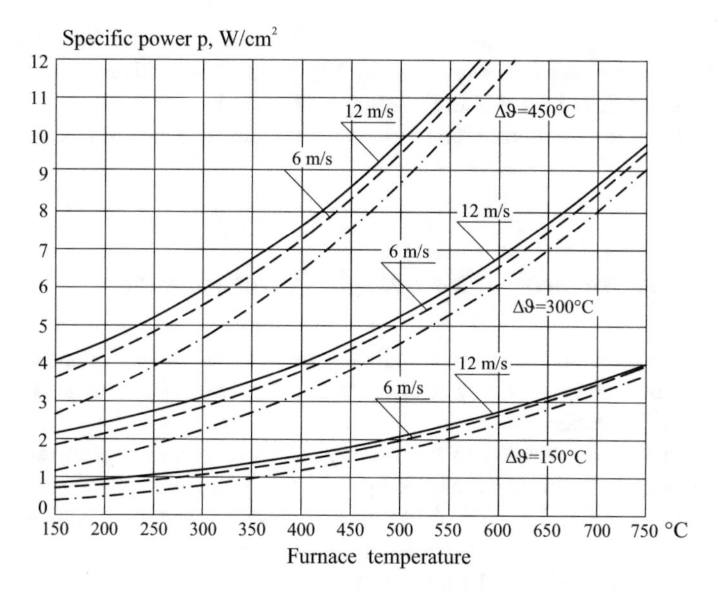

Fig. 4.53 Resistance overheating $\Delta\vartheta$ as a function of specific power p and furnace temperature ϑ_A, with emissivity $\varepsilon_R = 0.81$ [9]

of the heat transfer coefficient h_c and the specific power p (curves of Figs. 4.49 and 4.52).

However, between specific power and geometrical dimensions there is another link resulting from the requirement of developing a given total power P with a given supply voltage V. It is therefore necessary, as shown in Fig. 4.54, to draw the

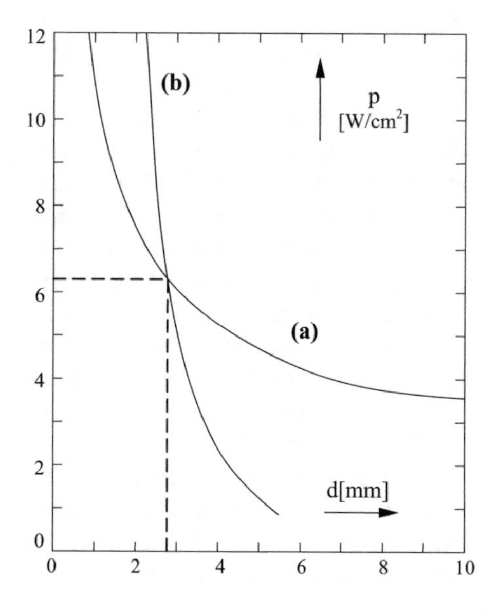

Fig. 4.54 Graphical evaluation of values of p and d (P = 11 kW; V = 220 V; ρ = 140 μΩ cm; v = 7 m/s; curve **a** from data of Fig. 4.52; curve **b** from Eq. 4.23)

curve (*a*) of variation of p as a function of diameter d starting from the data of Fig. 4.52, and the curve *(b)* calculated by Eq. (4.23), which gives as a function of p the values of diameters d corresponding to the given voltage.

The point of intersection of the two curves provides the required design values of p and d.

4.6.7 Furnaces with High Specific Power Resistors

As shown in previous paragraphs, radiation furnaces of classical construction are characterized by relatively low resistor installed power, usually between 10 and 20 kW/m^2 of chamber wall.

These values are adapted to the heating needs of most applications, since in many cases a too rapid increase of temperature may cause damages (deformation, cracks, etc.) to the heated workpieces; on the contrary they are too low—even in comparison with fuel furnaces—in some special applications in which the heating cycle requires high rates of the temperature rise.

As it is illustrated by the diagrams of Fig. 4.47, the above limitations do not depend on the possibility of heat transfer between resistor and charge, at least until the charge reaches high temperature values, but rather depend on the installed power and the way in which it is done the power control of the furnace. This control, in most cases, is based on measuring the temperature of the heating chamber in a zone between resistors and charge and the use of a contactor to switch-on or switch-off the power supplied to the resistors ("on/off" control).

This technique leads to limit drastically the installed power in order to avoid to overstress the contactor and to maintain a safety margin between the maximum temperature measured by the temperature sensor (which it is not coincident with resistor temperature) and the maximum admissible working temperature of the resistor.

The resistor specific power can be increased by using a power control system which no longer depends upon the measure of the temperature of the chamber, but on the temperatures of resistors and charge.

This is highlighted in Fig. 4.55, which refers to a furnace with resistors with maximum working temperature $\vartheta_{R\,max} = 1300\,°C$ and specific power calculated with Eq. (4.22), $\varepsilon_1 = \varepsilon_2 = 0.8$ and F = 0.6.

In the figure, curve (a) provides—on one hand—the maximum specific power p that can be exchanged between resistor and charge, as a function of the temperature ϑ_c of the charge; curves (b) and (c) show, on the other hand, the temperature values ϑ_R reached by the resistors with radiated specific power of 20 and 80 kW/m^2 respectively.

It can be observed that, limiting installed power to 20 kW/m^2 (line A–B), the resistor temperature increases according to curve b) and reaches the temperature of 1300 °C only when the temperature of the charge ϑ_c is about 1240 °C, while for lower values of ϑ_c, ϑ_R remains much below 1300 °C.

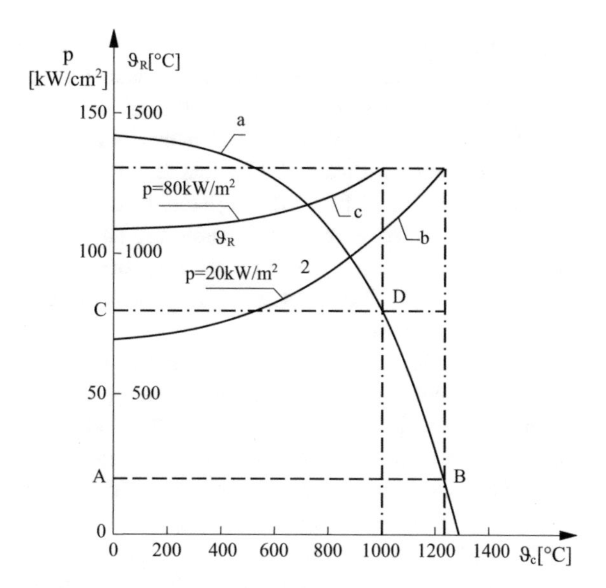

Fig. 4.55 Instantaneous power exchanged between resistors and charge and resistor temperature as a function of charge temperature

On the contrary, increasing the installed power to 80 kW/m^2 (curve c), the resistor temperature is always relatively high and reaches 1300 °C only when the charge is at 1000 °C.

In this case it is possible to make the power control following the curve C–D–B:

- in section C–D, where it is measured the temperature ϑ_R, the maximum installed power is transferred while the resistor temperature increases up to 1300 °C according to (curve c);
- in section D–B, the power control system reduces the specific power *p* in such a way as to maintain constant ϑ_R at its maximum value (control based on the temperature ϑ_R);
- finally, when the charge has reached its set final temperature, the control is made keeping constant the temperature of the charge.

This operation mode leads to a better use of the resistor material, but requires a very accurate measure with low thermal inertia of the temperatures ϑ_R and ϑ_c, and the use of a continuous power control thyristor system.

4.7 Energy Balance of Resistance Furnaces

The determination of the thermal balance of the furnace has the purpose of evaluating the energy specific consumption per unit of product (kWh/t or kWh/workpiece).

This requires the most precise possible knowledge of all flows into and out of all materials, energies, etc., involved in the process and a good interpretation of the thermal phenomena occurring in the furnace.

As mentioned previously for the efficiency, also the energy balance depends on a number of factors, such as type of furnace (continuous or batch), its different operating conditions, type and material of the charge, production rate, etc. It is therefore difficult to give values of general validity. In this paragraph we will give only some general rules and a typical example.

After definition of the type of operation for which the balance must be studied, it should be analyzed every quantity entering and sorting from the furnace (materials, emissions, absorptions of gas, vapors, energy, etc.) and every phenomenon occurring inside, like change of state, cracking, transformation of mechanical energy into thermal energy.

The analysis should refer to thermal steady state conditions and stable production rate for continuous furnaces, and to a typical thermal cycle for the batch ones. In the last case it is sometimes useful to subdivide the total cycle in sub-periods where it can be assumed that temperatures and losses do not vary too much.

Table 4.15 shows an example of the (calculated) energy balance of a batch bell furnace, with specifications of characteristics of the furnace in the first processing period and of load in the subsequent heat treatment period. Since the furnace temperature does not change too much during one cycle, reference is made to the average furnace temperature, which is lower than the maximum temperature of the charge.

Table 4.15 Energy balance of a batch bell furnace

Furnace and process data					
Opening surface	4.40 m^2				
Duration of a displacement	90 s				
Dead weights to be heated	650 kg				
First process period	$\vartheta_1 = 950$ °C; $P_1 = 20$ kW				
Second process period	$\vartheta_2 = 850$ °C; $P_2 = 15$ kW				
Furnace average temperature, °C	850	850	850	875	875
Charge					
Weight, kg	1500	1200	1200	1500	1200
Time of one cycle, h	8	8	6	8	8
Temperature, °C	875	875	875	900	900
Production rate, kg/h	188	150	200	188	150
Energy in the charge, kWh	280	224	224	291	233
Dead weights (muffle, base) kg	91	91	91	95	95
Losses in the walls, kWh	120	120	90	129	129
Bell displacement losses, kWh	9	9	9	10	10
Other losses, kWh	3	2	2	3	2
Consumption/Cycle, kWh	503	447	417	528	469
Specific Consumption, Kwh/t	336	373	348	352	391

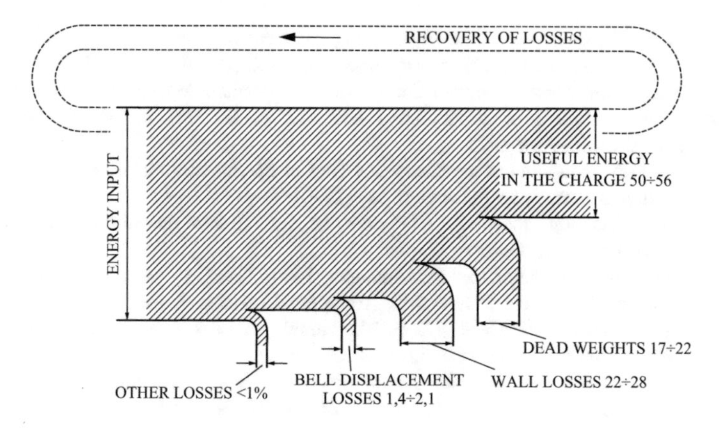

Fig. 4.56 Percent energy balance of the batch bell furnace of Table 4.15 [9]

With the same data it has been drawn the energy flow scheme of Fig. 4.56, which shows that the percent of useful energy in the different operating conditions is in the range of 50–56 %.

Finally, it must be emphasized once again that these values can be quite different for other types of furnaces and operating conditions.

References

1. Lupi S.: Appunti di Elettrotermia. Teaching Notes, 457 p. Libreria Progetto, Padova (Italy) (in Italian) (2005)
2. Orfeuil M.: Electrothermie industrielle, 803 p. Dunod, Paris. ISBN 2-04-012179-X (in French) (1981)
3. U.I.E. (International Union for Electroheat): Elektrowärme—Theorie und Praxis. Chapt.IV.1, 902 p. Verlag W. Girardet, Essen. ISBN 3-7736-0355-X (in German) (1974)
4. Nacke B., Baake E., Lupi S., Dughiero F., Forzan M., et al.: Theoretical Background an Aspects of Electrotechnologies—Physical Principles and Realization, 356 p. Intensive Course Basic I, Publishing House of ETU, St. Petersburg. ISBN 978-5-7629-1237-2 (2012)
5. Lauster F.: Manuel d'Électrothermie industrielle, 315 p. Dunod, Paris (1968)
6. Paschkis, V., Persson, J.: Industrial Electric Furnaces and Appliances. Interscience Publishers Ltd., London (1960)
7. Barber H.: Electroheat, 308 p. Granada, London (1983)
8. Nacke B.: Teaching material, ETP—Institut für Elektrothermische Prozesstechnik, University of Hannover (Germany)
9. E.D.F.: Les fours industriels à résistances électriques. Collection Electra, Dopee85, Paris. ISBN 2-86995-011-X (1989)
10. U.I.E. (International Union for Electroheat): Handbook of characteristic values of resistance furnaces, 48 p (1980)
11. Beuken, L., de Boer, J., Smeets, L.: Die Bestimmung des zeitlichen Verlaufs der Speicherwärme und Verlustwärme beim Haufheizen von Industrieöfen. Elektrowärme **22** (1), 4–11 (1964). (in German)

12. Van Jahoda, K.: Eine theoretisce Betrachtung der Beuken-I-Methode zur Bestimmung der Speicherwärme von Öfen. Elektrowärme international **28**(12), 691–695 (1970). (in German)
13. CEI—International Electrotechnical Commission: Supplement to tests methods for resistance furnaces. Measurement of accumulated heat, 12 p. CT 27: Industrial Electroheating Equipment, C.O., 22 Aug 1975
14. Report of Study Committee: Characteristic Values of Electro-heat Plants. VIII UIE Congress, Liege, Section II b, n. 9, 15 p. (in German) (1976)
15. Nacke B.: Teaching material. ETP—Institut für Elektro-thermische Prozesstechnik, University of Hannover, Germany
16. Smolenskij L.A.: Convection electrical furnaces, 168 p. Energhia, Mosca (in Russian) (1972)
17. Cherednichenko V.S., Borodachev A.S., Artemjev V.D.: Resistance Furnaces—Design and operation of resistance furnaces, 572 p. Monograph series, T.2, NSTU Publishing House, Novosibirsk. ISBN 5-7782-0674-7 (in Russian) (2006)
18. Chojnacki Z.P.: Beitrag zur Temperaturberechnung vonHeizwendeln bei erzwungener Konvektion. UIE VII Congress, Warsaw, N.423 (in German) (1972)

Chapter 5
Direct Resistance Heating

Abstract This chapter deals with Direct Resistance Heating (DRH) installations used for heating electrical conductive materials, metallic or non-metallic, by means of an electrical current flowing directly in the workpiece to be heated. According to the process technological requirements, the installation can be with still or moving workpieces and DC or AC supply at convenient frequency (mostly 50 Hz). The chapter is subdivided into four paragraphs. The first paragraph deals with installations with DC supply: the basic equations are first given, then the influence of variations material characteristics with temperature and the heating transient in the workpiece are analyzed. In the second paragraph installations with AC supply are considered, analyzing current and power density distributions in the workpiece, heating transients in non-magnetic and magnetic materials, efficiency and energy consumption. The third paragraph deals with the equivalent circuit of DRH installations and the calculation through numerical models of the influence of the installation design on the transient temperature distribution in steel bars. In the fourth paragraph data for DRH of tubes, bars with rectangular or square cross-section and continuous heaters for metal wires and strips are presented.

5.1 Introduction

Direct Resistance Heating (DRH) deals with the heating of conducting materials, metallic or non-metallic, in which an electrical current (DC or AC) flows directly in the workpiece to be heated [1–4].

According to Joule law, the electrical energy supplied is converted into heat in the material, producing the increase of its temperature up to a value defined by material thermal capacity, convection and radiation losses.

This heating technique is applied in several industrial processes, for example in:

- Metal reheating for hot metal working or heat treatment
- Glass melting, holding, refining and vitrification of scoriae, slags, cinders

S. Lupi, *Fundamentals of Electroheat*, DOI 10.1007/978-3-319-46015-4_5

- Food processing
- Metal powder sintering
- Welding
- Thermal treatments of concrete, coal and graphite
- Steam production in electrode boilers

In all applications a DRH installation comprises a contact system, constituted by electrodes of convenient shape, for supplying directly the current to the workpiece to be heated.

In some special cases the contact can be obtained by means of an intermediate liquid or salt bath. Installations for heating steel workpieces in aqueous solutions of salt ("*cathode heating*") or those for heating fluids that at the same time undergo a process of electrolysis ("*thermal electrolysis*") are some examples of this type.

The characteristic of developing heat directly into the material to be heated has some inherent advantages, among which we can mention:

- immediate starting of installations
- short heating times as a consequence of the distribution of the heating sources inside the workpiece
- very high efficiency (usually between 80 and 95 %) and consequent energy savings
- low oxidation of workpieces
- good conditions of working environment.

Drawbacks are the applicability of the process only to workpieces of convenient geometrical shape (large ratio of workpiece length to cross-section dimension and uniform cross-section in the workpiece length) and the need of contacts maintenance.

According to technological requirements, the installations can be with still or moving workpieces (the last ones for heating wires, small rods and tubes or thin narrow strips and plates) and DC or AC voltage supply at a convenient frequency (Fig. 5.1). In case of DC supply, the transformer is substituted by an appropriate rectifier system.

In stationary-type installations it is usually constant the voltage applied to the contacts while the current is variable during the heating process, due to the variations with temperature of material properties (resistivity, specific heat, magnetic permeability).

On the contrary, in installations with continuous charge advancement, after an initial heating transient occurring after switching-on the heater, the temperature distribution along the workpiece does not change with time, and everything happens as in the heating of a still workpiece with constant current supply.

The heating current is injected into the workpiece through contacts, which are generally made of low resistivity copper alloys (Cu–Cr, Cu–Cr–Zr) or sintered

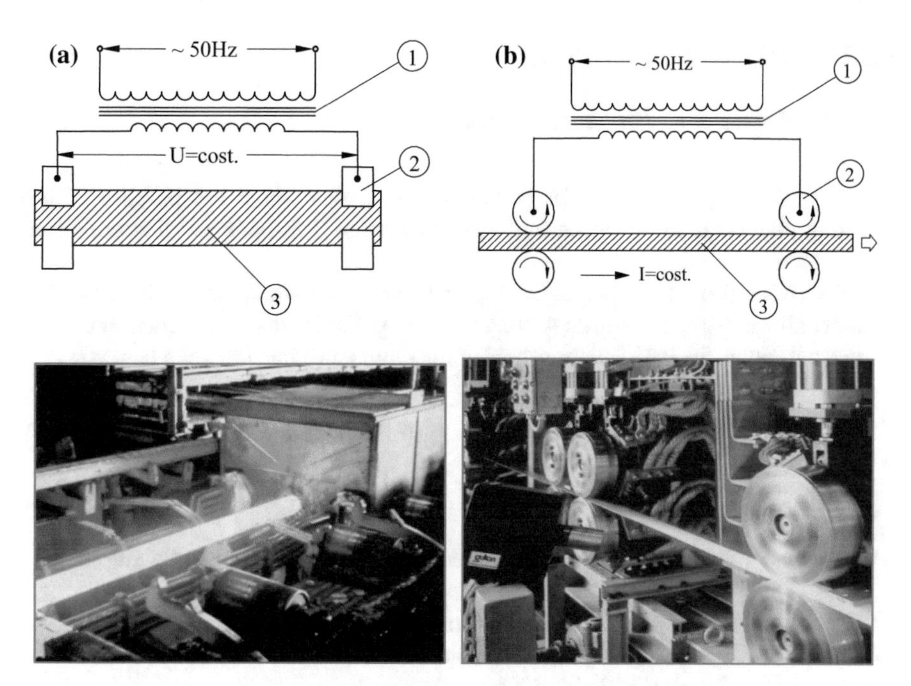

Fig. 5.1 Schematic of DRH installations **a** with still workpiece; **b** with progressive workpiece (*1* AC supply transformer, substituted by a rectifier system in DC heaters; *2* contact system; *3* workpiece)

materials (W–Cu). In addition to the heat generated by the current flowing through them, the contacts are also heated by radiation from the hot workpiece. Therefore, in order to increase life duration, it is convenient to limit their working temperature by providing the contact system with an efficient water cooling.

As concerns the choice between AC or DC supply, it mainly depends on the geometry of the body to be heated and the required final temperature distribution.

Most commonly it is used the AC supply at industrial frequency (50/60 Hz), which in many applications allows to obtain a nearly uniform final temperature distribution in the workpiece cross-section, a typical requirement in through heating processes for subsequent hot working of metals.

On the contrary, the DC supply is used for reaching an even greater temperature uniformity in the workpiece, particularly in the heating of metal strips.

In special cases medium or high frequency AC supply can also be used, when the purpose of the heating process is to produce uneven temperature distributions in different parts of the cross-section (such as, for example, in surface hardening heat treatments).

5.2 DRH with DC Supply

5.2.1 *Basic DC Electrical Equations*

This type of supply is used for particular geometries where the purpose of the heating process is to obtain an uniform temperature distribution in the workpiece cross-section.

With reference to the schematic of Fig. 5.1a, where the workpiece is a long metallic bar of uniform cross-section, neglecting the uneven distribution of current near the end contact regions, the basic relationships of the Ohm and Joule laws can be used:

$$I = \frac{V}{R_{dc}} = J \cdot S; \quad P = R_{dc} I^2 = \frac{V^2}{R_{dc}} = w \cdot S \cdot \ell \tag{5.1}$$

with:

V, I	voltage applied to the contacts, V, and current flowing in the workpiece, A
$R_{dc} = \rho\ell/S$	DC resistance of the workpiece, Ω
ℓ, S	workpiece length, m, and area of workpiece cross-section, m^2
ρ	electrical resistivity, Ωm
J	current density in the workpiece, A/m^2,
$E = \rho J$	electric field intensity, V/m
$w = \rho J^2$	volume power density, W/m^3

Note: Since vector quantities **E** and **J** have only one spatial component, in the following we will use the scalar quantities E and J.

Neglecting the losses from the surface of the workpiece to the environment and those in the power supply systems and contacts, the energy balance of a unit volume in the workpiece is:

$$w\,\Delta t = c\,\gamma \Delta\vartheta_m \tag{5.2}$$

with:

$\Delta\vartheta_m$	temperature increase, °C, in the time interval Δt, s
c, γ	specific heat, J/kg °C, and density, kg/m^3, of workpiece material

From Eqs. (5.1), (5.2) the integral and field quantities can be rewritten as follows:

$$\begin{aligned} w &= c\,\gamma \frac{\Delta\vartheta_m}{\Delta t} & P &= S\ell c\,\gamma \frac{\Delta\vartheta_m}{\Delta t} \\ J &= \sqrt{\frac{c\,\gamma}{\rho}\frac{\Delta\vartheta_m}{\Delta t}} & I &= S\sqrt{\frac{c\,\gamma}{\rho}\frac{\Delta\vartheta_m}{\Delta t}} \\ E &= \sqrt{\rho\, c\,\gamma \frac{\Delta\vartheta_m}{\Delta t}} & V &= \ell\sqrt{\rho\, c\,\gamma \frac{\Delta\vartheta_m}{\Delta t}} \end{aligned} \tag{5.3}$$

Equations (5.3) show that the specific quantities E, J and w depend only on the material characteristics and the rate of increase of temperature $\Delta\vartheta_m/\Delta t$.

Example 5.1 DC static heating (according to Fig. 5.1a) of cylindrical steel bars with the following specifications:

$$\Delta\vartheta_m = 1250\,^\circ\text{C};\ \Delta t = 60\,\text{s};\ \ell = 6\,\text{m};\ D = 20\ \text{and}\ 100\,\text{mm};$$
$$c = 0.71\,\text{kJ/kg K};\ \gamma = 7.8 \cdot 10^3\,\text{kg/m}^3;\ \rho = 0.75 \cdot 10^{-6}\,\Omega\text{m}.$$

From Eq. (5.3) we obtain:

D = 20 mm R_{dc} = 14.3 mΩ		**D = 100 mm** R_{dc} = 0.573 mΩ	
w = 115 w/cm^3	P = 217 kW	w = 115 w/cm^3	P = 5.419 kW
J = 12.4 A/mm^2	I = 3.890 A	J = 12.4 A/mm^2	I = 97.250 A
E = 9.28 V/m	V = 55.7 V	E = 9.28 V/m	V = 55.7 V

Remarks

1. These results underline the problems arising when heating workpieces of "large" cross-section in short heating times. In fact, the current required becomes so high that it would be difficult to inject it into the workpiece, taking into account that the maximum admissible current for each contact point is in the range 5–10 kA. The same applies in the case of very "short" workpieces, where the applied voltage V becomes comparable with the voltage drop in the external circuit and the contacts.
2. The temperature distribution is strongly affected by the variations of bar cross-section, like the one schematically illustrated in Fig. 5.2.

Denoting with the subscripts 1 and 2 the quantities in regions 1 and 2 of the figure, and writing the ratio of radii as $r_2/r_1 = (1 + \varepsilon)$, it is:

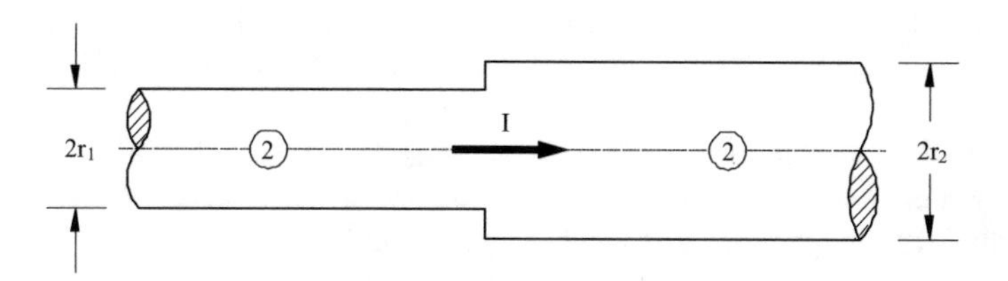

Fig. 5.2 DRH of a bar of uneven cross-section

$$\left.\begin{aligned} S_2 &= S_1(1+\varepsilon)^2 \\ J_2 &= J_1/(1+\varepsilon)^2 \\ w_2 &= w_1/(1+\varepsilon)^4 \\ \Delta\vartheta_{m2} &= \Delta\vartheta_{m1}/(1+\varepsilon)^4 \end{aligned}\right\}. \tag{5.4}$$

If ε is small compared to unity, it is:

$$\Delta\vartheta_{m2} \approx \Delta\vartheta_{m1}(1-4\varepsilon) \tag{5.4a}$$

Therefore strong temperature differences in the bar length can arise even for very small variations in the cross section dimensions; this generally restricts the field of application of this heating technique only to workpieces with very uniform cross sections along the length.

Example 5.2 Heating in 60 s, with current I = 3890 A (as in Example 5.1) a bar with the following dimensions (see Fig. 5.2): r_1 = 10 mm, r_2 = 10.3 mm.

From Eq. (5.4) it results:

$$\begin{aligned} \varepsilon &= 0.03 \\ S_1 &= 12.4 \cdot 10^6\ \mathrm{A/m^2} = 12.4\ \mathrm{A/mm^2} \\ S_2 &= 11.7 \cdot 10^6\ \mathrm{A/m^2} = 11.7\ \mathrm{A/mm^2} \\ w_1 &= 115.0 \cdot 10^6\ \mathrm{W/m^3} = 115.0\ \mathrm{W/cm^3} \\ w_2 &= 102.6 \cdot 10^6\ \mathrm{W/m^3} = 102.6\ \mathrm{W/cm^3} \\ \vartheta_{m1} &= 1250\ ^\circ\mathrm{C} \\ \vartheta_{m2} &= 1111\ ^\circ\mathrm{C} \end{aligned}$$

5.2.2 *Influence of Variations with Temperature of Material Characteristics*

The previous calculations have been done assuming constant "average" values of material characteristics during the whole heating process. However, in practice the quantities ρ, c and γ vary with temperature, as shown in Appendix in Table A.6 for various metals.

As it will be seen in the following paragraphs, in the case of steel, in addition to the after mentioned variations, it is very important the change of magnetic permeability with temperature and local magnetic field intensity.

For analysing the influence of these variations on the heating transient, the calculation must be done iteratively updating the material characteristics with temperature.

The analysis of material characteristics shows that within small temperature intervals it is admissible to assume constant average values of material parameters, corresponding to the average temperature in the interval.

Using this criterion, the heating transient is subdivided into several temperature intervals, assuming in each of them the average value of material characteristics. The accuracy of this calculation obviously increases with the number of subdivision intervals.

Applying iteratively Eq. (5.3), it must be paid attention that in static heating applications the applied voltage V across the load is constant and, as a consequence, also the electric field intensity E remains constant. On the contrary, in continuous heating installations, the electric field intensity E is variable along the workpiece length and the quantities which are constant—for a given voltage V applied between contacts—are the current I and the current density J.

Repeating in this way the calculation of Example 5.1 with D = 20 mm, E = const. = 9.28 V/m and the material characteristics of steel given in the first columns of Table 5.1, it is possible to evaluate the values of Δt, w, P, J and I for each temperature interval $\Delta\vartheta_m$, using Eq. (5.3) as shown in the following example.

Example 5.3 Heating of the bar considered in Example 5.1, with constant applied voltage V = 55.68 V (E = 9.28 V/m).

From Eq. (5.2) it is:

$$\Delta t = \frac{\Delta\vartheta_m}{E^2}(\rho c \gamma)$$

$$w = \frac{J^2}{\rho} \qquad P = S\ell w$$

$$J = \frac{E}{\rho} \qquad I = JS$$

Assuming $\gamma = 7.8 \cdot 10^3$ kg/m^3, and for ρ and c the values corresponding to the average temperature in the interval $\Delta\vartheta_m$, the data shown in Table 5.1 are obtained.

From these values the curves of Fig. 5.3 have been drawn. They show that the iterative method does not greatly modifies the total heating time (see Example 5.1), while—on the contrary—the rate of temperature increase, the volume specific power and the current density undergo considerable variations. In particular, they are much higher in the first part of the heating cycle of those calculated with constant parameters.

In addition it should be noted that, since radiation losses increase with temperature, the relatively longer period of the heating transient at high temperature would result in lower thermal efficiency, unless the workpiece is provided with suitable insulation.

Table 5.1 Calculation of heating transient parameters in different temperature ranges

$\Delta\vartheta_m$ (°C)	$\rho \cdot 10^{-6}$ (Ωm)	$c \cdot 10^{-3}$ (J/kg °C)	$\rho c \gamma$	Δt (s)	$t = \sum \Delta t$ (s)	$w \cdot 10^{-6}$ (W/m²)	P (W)	$S \cdot 10^{-6}$ (A/m²)	I (A)
0–50	0.20	0.477	0.74	0.43	0.43	431	812	46.4	14.580
50–150	0.25	0.489	0.95	1.10	1.53	344	648	37.1	11.660
150–250	0.30	0.518	1.21	1.41	2.94	287	541	30.9	9.710
250–350	0.40	0.552	1.72	2.00	4.94	215	405	23.2	7.290
350–450	0.50	0.598	2.33	2.71	7.65	172	324	18.6	5.840
450–550	0.62	0.656	3.17	3.68	11.33	139	262	15.0	4.710
550–650	0.77	0.740	4.44	5.16	16.49	112	211	12.1	3.800
650–750	0.92	0.878	6.30	7.32	23.81	93.6	176	10.1	3.170
750–850	1.10	0.698	5.99	6.96	30.77	78.3	148	8.44	2.650
850–950	1.15	0.702	6.30	7.32	38.09	74.9	141	8.07	2.540
950–1050	1.20	0.706	6.61	7.68	45.77	71.8	135	7.73	2.430
1050–1150	1.22	0.719	6.84	7.94	53.71	70.6	133	6.61	2.390
1150–1250	1.25	0.732	7.14	8.29	62.00	68.9	130	7.42	2.330

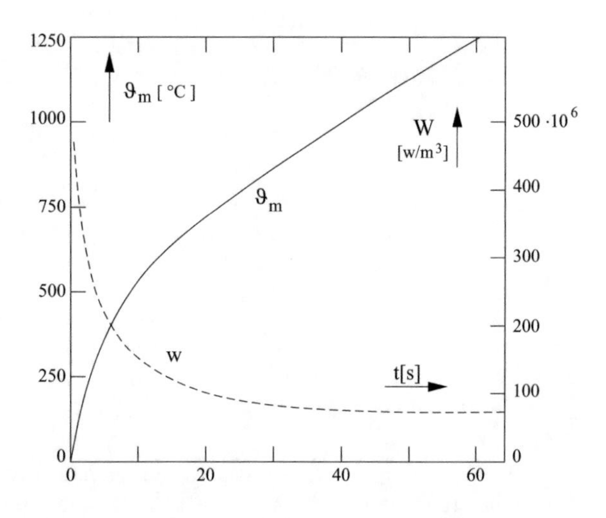

Fig. 5.3 Average temperature and volume specific power as a function of heating time (D = 20 mm; E = const. = 9.28 V/m) [1, 2]

5.2.3 *Influence of Radiation Losses on Heating Transient*

In order to analyse this phenomenon, we consider here the heating of a "long" cylindrical workpiece with constant volume power density *w*, constant material parameters and losses from the surface dependent on temperature.

The evaluation of the transient temperature distribution is obtained as solution of the differential equation:

$$\frac{\partial \vartheta}{\partial t} = k\left(\frac{\partial^2 \vartheta}{\partial r^2} + \frac{1}{r}\frac{\partial \vartheta}{\partial r}\right) + \frac{w}{c\,\gamma} \tag{5.5}$$

with the conditions:

$$\left.\begin{array}{ll} w = w_0 = \text{const.} & \\ \vartheta(r) = 0, & \text{for } t = 0 \\ (\partial\vartheta/\partial r) + h\vartheta = 0, & \text{for } r = r_e \text{ and } t > 0 \end{array}\right\} \tag{5.5a}$$

and:

r_e, r	external and generic internal radius of the cylinder, m
$k = \lambda/(c\,\gamma)$	thermal diffusivity, m^2/s
$h = \alpha/\lambda$	ratio of the heat transmission coefficient α, (W/m^2 °C) from the workpiece surface to the ambient, to the thermal conductivity λ, (W/m °C) of the heated material, m^{-1}

Introducing the notation:

$p_0 = P/(2\pi r_e \ell)$—power density referred to the external surface of the cylinder, W/m^2, and the dimensionless quantities:

$$A = h\,r_e;\ \xi = \frac{r}{r_e};\ \tau = \frac{kt}{r_e^2};\ \Theta = \frac{\lambda\vartheta}{p_0 r_e} \tag{5.6}$$

the differential equation and boundary conditions (5.5), (5.5a) become:

$$\frac{\partial \Theta}{\partial \tau} = \frac{\partial^2 \Theta}{\partial \xi^2} + \frac{1}{\xi}\frac{\partial \Theta}{\partial \xi} + 2 \tag{5.7}$$

$$\left.\begin{array}{ll} \Theta(\xi) = 0, & \text{for } \tau = 0 \\ \frac{\partial \Theta}{\partial \xi} + A\Theta = 0, & \text{for } \tau > 0 \text{ and } \xi = 1 \end{array}\right\} \tag{5.7a}$$

It should exist a time-independent partial solution of Eq. (5.7), which corresponds to the equilibrium of the heat supplied to the workpiece and the heat transmitted to the environment, which occurs for $\tau \to \infty$.

Since in these conditions it is $(\partial\Theta/\partial\tau) = 0$, according to Eq. (5.7) the partial solution must satisfy the DE:

$$\frac{\partial^2 \Theta}{\partial \xi^2} + \frac{1}{\xi}\frac{\partial \Theta}{\partial \xi} + 2 = 0.$$

The partial solution therefore is:

$$\Theta^*(\xi) = C_1 \ln \xi - \frac{1}{2}\xi^2 + C_2$$

Taking into account the boundary conditions:

$$\left.\begin{array}{ll}\Theta^* \neq \infty, & \text{for } \xi = 0 \\ \Theta^* = 1/A, & \text{for } \xi = 1\end{array}\right\}$$

(the second one corresponding to the condition $p_0 = \alpha\vartheta^*$, i.e. to the fact that for $t \to \infty$ and $r = r_e$ the total power supplied to the body is transmitted from the surface to the environment), we obtain:

$$C_1 = 0 \quad \text{and} \quad C_2 = \frac{1}{A} + \frac{1}{2}$$

and finally:

$$\Theta^*(\xi) = \frac{1}{A} + \frac{1}{2}(1 - \xi^2)$$

To this solution we must add a term which becomes zero for $\tau \to \infty$, and satisfies the initial and boundary conditions.

Taking into account the general methods of solution of thermal transients (see Sect. 1.2.5), this term takes the form [5]:

$$\sum_{n=1}^{\infty} A_n J_{0B}(\beta_n \xi)\, e^{-\beta_n^2 \tau}$$

with:

A_n	unknowns constants
$\beta_n = \alpha_n r_e$	roots of the equation $\beta \cdot J_{1B}(\beta) - A \cdot J_{0B}(\beta) = 0$, whose values as a function of A are given in appendix (Table A.2)
J_{0B}, J_{1B}	Bessel functions of first kind of order zero and one respectively.

The solution therefore is the following [6]:

$$\Theta(\xi, \tau) = \frac{1}{A} + \frac{1}{2}(1 - \xi^2) - \sum_{n=1}^{\infty} A_n J_{0B}(\beta_n \xi)\, e^{-\beta_n^2 \tau} \tag{5.8}$$

Since from the condition $\Theta = 0$ for $\tau = 0$ we can write:

$$\Theta(\xi, 0) = \frac{1}{A} + \frac{1}{2}(1 - \xi^2) - \sum_{n=1}^{\infty} A_n J_{0B}(\beta_n \xi) = 0,$$

it follows that is

$$f(\xi) = \frac{1}{A} + \frac{1}{2}(1 - \xi^2) = \sum_{n=1}^{\infty} A_n J_0(\beta_n \xi) \tag{5.9}$$

The determination of the constants A_n is done, as usual, by multiplying both sides of Eq. (5.9) by $\xi J_{0B}(\beta_n \xi)$ and integrating between 0 and 1:

$$\int_0^1 f(\xi) \cdot \xi J_{0B}(\beta_n \xi)\, d\xi = \int_0^1 \sum_{n=1}^{\infty} [A_n J_{0B}(\beta_n \xi)] \cdot [\xi J_{0B}(\beta_n \xi)]\, d\xi$$

From Eqs. (A.3.1), (A.3.5) of Table A.3 of appendix it is:

$$\int_0^1 f(\xi) \cdot \xi J_{0B}(\beta_n \xi) d\xi = A_n \frac{A^2 + \beta_n^2}{2\beta_n^2} J_{0B}^2(\beta_n),$$

therefore we obtain:

$$A_n = \frac{2\beta_n^2}{(A^2 + \beta_n^2) J_{0B}^2(\beta_n)} \int_0^1 \xi f(\xi) J_{0B}(\beta_n \xi)\, d\xi \tag{5.10}$$

According to Eq. (5.9), the integral in Eq. (5.10) can be re-written as:

$$\int_0^1 \xi \left(\frac{1}{A} + \frac{1}{2}\right) J_{0B}(\beta_n \xi)\, d\xi - \frac{1}{2} \int_0^1 \xi^3 J_{0B}(\beta_n \xi)\, d\xi$$

Moreover, it is: [6]

$$\int_0^1 \xi J_{0B}(\beta_n \xi)\, d\xi = \frac{1}{\beta_n} J_{1B}(\beta_n)$$

$$\int_0^1 \xi^3 J_{0B}(\beta_n \xi)\, d\xi = \frac{1}{\beta_n} J_{1B}(\beta_n)\left[1 - \frac{4}{\beta_n^2}\right] + \frac{2}{\beta_n^2} J_{0B}(\beta_n).$$

By using the relationship $\beta_n J_{1B}(\beta_n) - A J_{0B}(\beta_n) = 0$, finally we obtain:

$$A_n = \frac{4A}{\beta_n^2 (A^2 + \beta_n^2) J_{0B}(\beta_n)} \tag{5.11}$$

and

$$\Theta(\xi, \tau) = \frac{1}{A} + \frac{1}{2}(1 - \xi^2) - 4A \sum_{n=1}^{\infty} \frac{J_0(\beta_n \xi) e^{-\beta_n^2 \tau}}{\beta_n^2 (A^2 + \beta_n^2) J_0(\beta_n)} \tag{5.12}$$

Equation (5.12), allows to obtain also the time evolution of the average temperature $\Theta_m(\tau)$ in the bar cross-section by evaluating the integral:

$$\Theta_m(\tau) = 2\int_0^1 \Theta(\xi,\tau)\xi\, d\xi$$

It results:

$$\Theta_m(\tau) = \frac{1}{A} + \frac{1}{4} - 8A^2 \sum_{n=1}^{\infty} \frac{e^{-\beta_n^2 \tau}}{\beta_n^4 (A^2 + \beta_n^2)} \tag{5.13}$$

Figure 5.4a, b show some examples of transient temperature distributions in steel bars heated up to 1250 °C, calculated with Eqs. (5.12) and (5.13), where ϑ_a, ϑ_s and ϑ_m are respectively the axis, surface and average temperatures of bar cross section.

The curves show that, since heat sources are uniformly distributed in the cross-section and are present surface losses, the temperature of the axis is always higher than that of the surface and their difference is maximum at the end of the heating cycle.

This situation must be taken into account particularly in the through heating of some metals (e.g. aluminium) which have a hot working temperature very close to the melting one.

In some cases it may be useful to estimate approximately the heating time corresponding to Eq. (5.13) by using the diagrams of Fig. 5.5a, b.

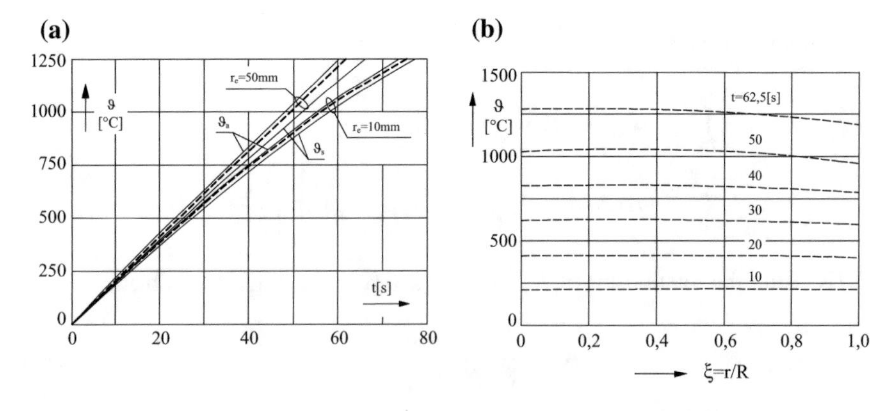

Fig. 5.4 Effect of surface losses on transient temperature distributions in DRH of steel bars with DC current (λ = 29 W/m °C; c γ = 5.54 · 10^6 J/m °C; α = 175 W/m^2 °C; w = 115 · 10^6 W/m^3; **a** r_e = 10–50 mm, ϑ_s, ϑ_a, ϑ_m—surface, axis and average temperature (- - -); **b** r_e = 50 mm

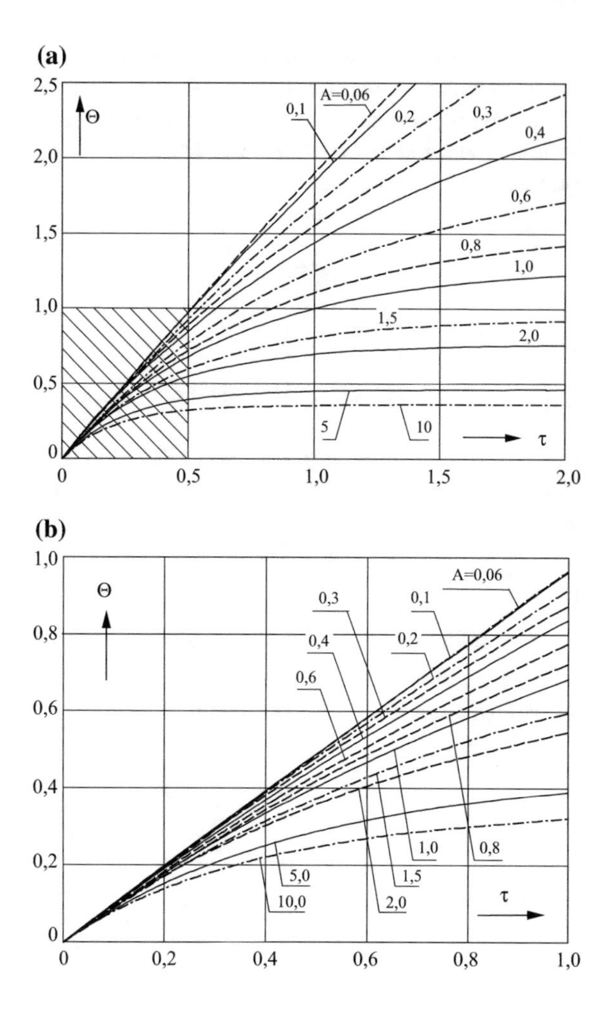

Fig. 5.5 Dimensionless graphs of the average temperature as a function of time, for different values of parameter A (τ, Θ, A—given by Eq. (4.6); **b** magnification of dotted region of fig. **a**)

5.3 DRH With AC Supply

5.3.1 *Distributions of Current and Power Density*

As known, when using AC currents the skin effect occurs in the body to be heated; it produces, particularly in ferromagnetic bodies or workpieces of large cross sections, uneven current density and volume specific power distributions.

These uneven distributions and the thermal losses from the body surface will produce different temperature values from point to point in the workpiece cross-section during the heating transient. It is therefore necessary to analyse the coupled electrical and thermal phenomena occurring during the heating transient, in order to evaluate the required power and the corresponding process heating time.

For analysis of electromagnetic phenomena, we will consider a cylindrical "long" cylindrical workpiece with AC current flowing in axial direction.

Neglecting displacement currents (due to the range of frequencies involved) and assuming sinusoidal quantities, the Maxwell's equations are written in the form (see Sect. 2.2):

$$\mathrm{rot}\,\dot{\bar{H}} = \frac{\dot{\bar{E}}}{\rho}; \quad \mathrm{rot}\,\dot{\bar{E}} = -j\,\omega\mu\mu_0\dot{\bar{H}} \tag{5.14}$$

with:

$\dot{\bar{H}}, \dot{\bar{E}}$	complex magnetic field and electric field intensities, A/m and V/m,
ρ, μ	resistivity, Ωm, and relative permeability of workpiece material
$\mu_0 = 4\pi \cdot 10^{-7}$	magnetic permeability of vacuum, H/m
$\omega = 2\pi f$	angular frequency, s^{-1}
f	frequency, Hz
$j = \sqrt{-1}$	.

With reference to a system of cylindrical coordinates (*r, φ, z*), only the components $\dot{H}_\varphi$ of $\dot{\bar{H}}$ and $\dot{E}_z$ of $\dot{\bar{E}}$ are non-zero. Therefore, omitting in the following the subscripts, from Eq. (5.14) we obtain:

$$\left.\begin{aligned} \frac{d\dot{H}}{dr} + \frac{1}{r}\dot{H} = \frac{\dot{E}}{\rho} \\ \frac{d\dot{E}}{dr} = j\,\omega\mu\mu_0\dot{H} \end{aligned}\right\} \tag{5.15}$$

From the second of Eq. (5.15) it is:

$$\frac{d\dot{H}}{dr} = \frac{1}{j\,\omega\mu\mu_0}\frac{d^2\dot{E}}{dr^2}.$$

By substitution in the first one, remembering that it is $\dot{E} = \rho\dot{J}$, we obtain:

$$\frac{d^2\dot{J}}{dr^2} + \frac{1}{r}\frac{d\dot{J}}{dr} - \frac{j\,\omega\mu\mu_0}{\rho}\dot{J} = 0 \tag{5.16}$$

Introducing the notations:

$$\left.\begin{aligned} &\xi = \frac{r}{r_e};\ \beta^2 = -\frac{j\,\omega\mu\mu_0}{\rho}r_e^2 = -jm^2;\ m = \frac{\sqrt{2}r_e}{\delta} \\ &\delta = \sqrt{\frac{2\rho}{\omega\mu\mu_0}} - \text{penetration depth, m} \end{aligned}\right\} \tag{5.17}$$

Eq. (5.16) can be written in the form:

$$\frac{d^2\dot{J}}{d\xi^2} + \frac{1}{\xi}\frac{d\dot{J}}{d\xi} + \beta^2\dot{J} = 0 \tag{5.18}$$

As known, the solution of this DE is:

$$\dot{J} = \dot{C}_1 J_0(\beta\xi) + \dot{C}_2 Y_0(\beta\xi) \tag{5.19}$$

with: J_0, Y_0—Bessel functions of order zero of first and second kind.

The constants $\dot{C}_1$ and $\dot{C}_2$ can be determined with the conditions:

$$\dot{J} \neq \infty, \quad \text{for } \xi = 0; \quad \dot{J} = \dot{J}_e, \quad \text{for } \xi = 1.$$

Taking into account that is $Y_0(0) = -\infty$, from the first condition it is $\dot{C}_2 = 0$, from the second one $\dot{C}_1 = \dot{J}_e / J_0(\beta)$.

By substitution in Eq. (5.19) it results:

$$\dot{J} = \dot{J}_e \frac{J_0(\beta\xi)}{J_0(\beta)} = \dot{J}_e \frac{J_0(\sqrt{-j}m\xi)}{J_0(\sqrt{-j}m)} \tag{5.20}$$

Sometimes it may be convenient to separate real and imaginary parts of the function $J_0(\sqrt{-j}x)$, using the relationship:

$$J_0(\sqrt{-j}x) = \text{ber}\, x + j\, \text{bei}\, x.$$

Thus Eq. (5.20) becomes:

$$\dot{J} = \dot{J}_e \frac{\text{ber}(m\xi) + j\,\text{bei}(m\xi)}{\text{ber}(m) + j\,\text{bei}(m)} \tag{5.21}$$

and

$$\left|\frac{\dot{J}}{\dot{J}_e}\right| = \sqrt{\frac{\text{ber}^2(m\xi) + \text{bei}^2(m\xi)}{\text{ber}^2(m) + \text{bei}^2(m)}} \tag{5.21a}$$

In literature Eq. (5.21) is sometimes written in the form:

$$\dot{J} = \dot{J}_e \frac{M(m\xi)\, e^{j\phi(m\xi)}}{M(m)\, e^{j\phi(m)}}$$

with:

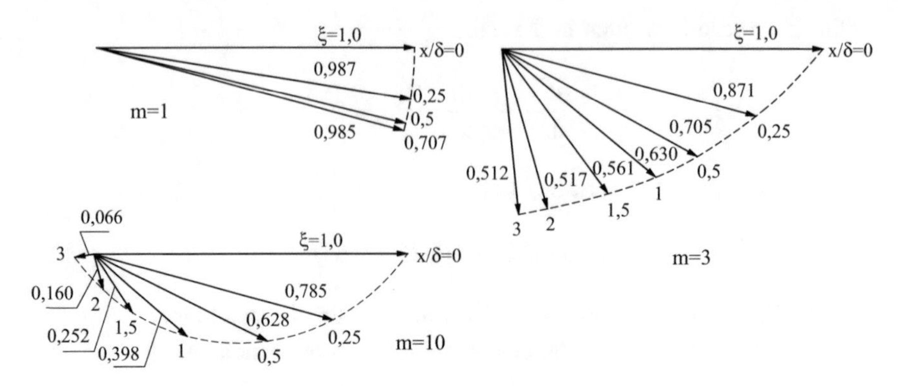

Fig. 5.6 Phasors of current density at different distances $x = r_e(1-\xi)$ from the surface, for m = 1–3–10 ($m = \sqrt{2}r_e/\delta$)

$$M(u) = \left[ber^2(u) + bei^2(u)\right]^{1/2};\ \phi(u) = \arctan[bei(u)/ber(u)].$$

Equation (5.21) show that the current density varies in amplitude and phase moving from the surface towards the axis of the workpiece, as illustrated by the diagrams of Figs. 5.6 and 5.7.

The diagram 5.7 also shows that the current density has non-uniform distribution in the workpiece cross-section for m > 1, while below this values the distributions become nearly uniform as in the case of DC current.

The value of $\dot{J}_e$ can be determined as a function of the total current flowing in the bar by applying Ampere's law at the surface of the cylinder:

$$\dot{I} = 2\pi r_e\, \dot{H}_e. \tag{5.22}$$

Since it is:

$$J_0'(u) = \frac{dJ_0(u)}{du} = -J_1(u)$$

from the second of Eq. (5.15) we obtain:

$$\dot{H} = j\sqrt{-j}\frac{\dot{J}_e}{\sqrt{2}}\delta\frac{J_1(\sqrt{-j}m\xi)}{J_0(\sqrt{-j}m)} \tag{5.23}$$

and from Eq. (5.20):

$$\dot{J}_e = \sqrt{-j}\,\frac{m}{2}\left(\frac{\dot{I}}{\pi r_e^2}\right)\frac{J_0(\sqrt{-j}m)}{J_1(\sqrt{-j}m)} \tag{5.24}$$

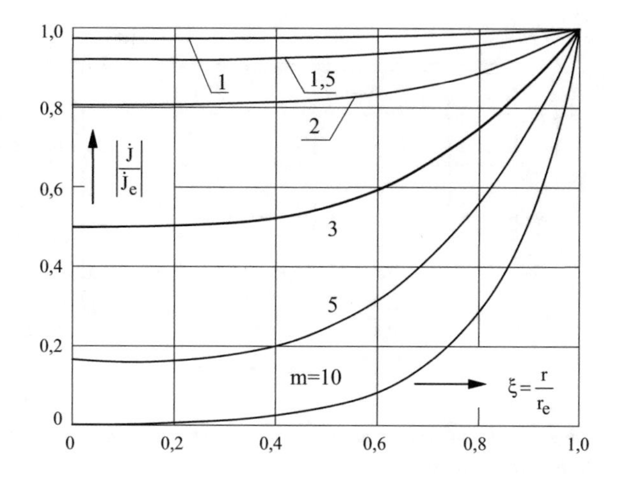

Fig. 5.7 Radial distributions of module of the current density for different values of m (r_e, r—external and generic internal radius of the workpiece; $\xi = r/r_e$; J_e—value of J for $r = r_e$; $m = \sqrt{2}\, r_e/\delta$; δ—penetration depth)

Finally, Eq. (5.20) can be written as:

$$\dot{J} = \sqrt{-j}\,\frac{m}{2}\left(\frac{\dot{I}}{\pi r_e^2}\right)\frac{J_0(\sqrt{-j}m\xi)}{J_1(\sqrt{-j}m)} \tag{5.25}$$

The specific power per unit volume w can be obtained from the relationship $w = \rho|\dot{J}^2|$, by introducing the expression of $\dot{J}$ given by Eq. (5.21). Indicating with w_e the value of w for $\xi = 1$, we have:

$$w = w_e\left(\frac{|\dot{J}|}{|\dot{J}_e|}\right)^2 = w_e\frac{\mathrm{ber}^2(m\xi) + \mathrm{bei}^2(m\xi)}{\mathrm{ber}^2(m) + \mathrm{bei}^2(m)}. \tag{5.26}$$

The corresponding radial distributions of w are given in Fig. 5.8 as a function of ξ.

The diagrams show that for values of $m \leq 1$ the heat sources are nearly uniformly distributed in the workpiece cross-section as with DC current.

Denoting with $w_{dc} = \rho(I/\pi r_e^2)^2$ the power per unit volume with DC current, from Eq. (5.26) we obtain:

$$\frac{w_e}{w_{dc}} = \frac{m^2}{4}\frac{\mathrm{ber}^2(m) + \mathrm{bei}^2(m)}{\mathrm{ber}'^2(m) + \mathrm{bei}'^2(m)} = w^*(m). \tag{5.27}$$

The values of $w^*(m)$, given in Table 5.2, show that with the same total current flowing in the bar, the value of w_e increases with m.

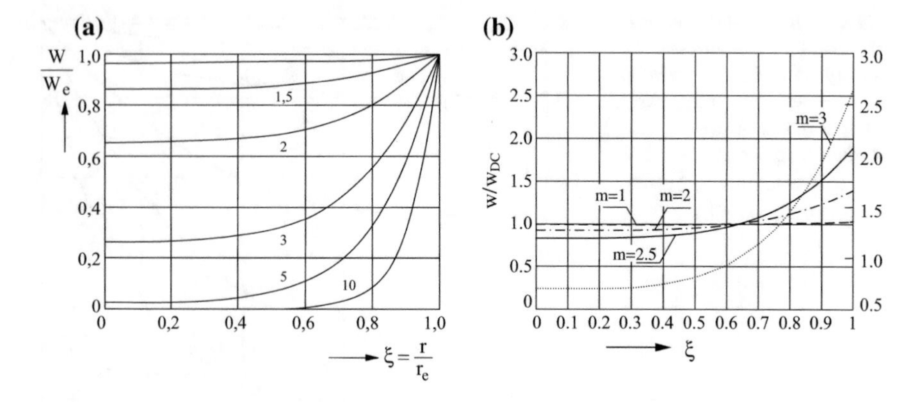

Fig. 5.8 Radial distributions of volume specific power w (r_e, r—external and generic internal radius of workpiece; w_e—value of w for r = r_e; $m = \sqrt{2}r_e/\delta$; δ—penetration depth); **a** relative values referred to the surface value w_e; **b** relative values referred to the same total current I

Table 5.2 Values of w*(m) and k_r as a function of m

$m = \sqrt{2}R/\delta$	$w*(m) = w_e/w_{dc}$	k_r
1.0	1.026	1.004
1.5	1.129	1.025
2.0	1.394	1.079
3.0	2.641	1.318
5.0	7.191	2.042
10.0	26.821	3.809

The "*internal impedance*" per unit length of the workpiece can be evaluated by the relationship:

$$\dot{Z}_{iu} = R_{acu} + j\, X_{iu} = \frac{\dot{E}_e}{\dot{I}} \tag{5.28}$$

where R_{acu} and X_{iu} are respectively the AC resistance and the "*internal reactance*" of the bar per unit length.

From Eq. (5.24) and $\dot{E}_e = \rho\dot{J}_e$, it is:

$$\dot{Z}_{iu} = \sqrt{-j}\frac{m}{2}\left(\frac{\rho}{\pi r_e^2}\right)\frac{J_0(\sqrt{-j}m)}{J_1(\sqrt{-j}m)} \tag{5.28a}$$

Denoting with $R_{dcu} = \rho/(\pi r_e^2)$ the DC resistance for unit length, the relative values of the internal impedance can be written in the form:

$$\frac{\dot{Z}_{iu}}{R_{dcu}} = \frac{R_{acu}}{R_{dcu}} + j\frac{X_{iu}}{R_{dcu}} = k_r + jk_x = \sqrt{-j}\frac{m}{2}\frac{J_0(\sqrt{-j}m)}{J_1(\sqrt{-j}m)}. \tag{5.29}$$

By separation of real and imaginary parts of Bessel functions, Eq. (5.29) gives:

$$\left.\begin{aligned} k_r = \frac{R_{acu}}{R_{dcu}} = \frac{m}{2}\frac{\mathrm{ber}(m)\mathrm{bei}'(m) - \mathrm{bei}(m)\mathrm{ber}'(m)}{\mathrm{ber}'^2(m) + \mathrm{bei}'^2(m)} \\ k_x = \frac{X_{iu}}{R_{dcu}} = \frac{m}{2}\frac{\mathrm{ber}(m)\mathrm{ber}'(m) + \mathrm{bei}(m)\mathrm{bei}'(m)}{\mathrm{ber}'^2(m) + \mathrm{bei}'^2(m)} \end{aligned}\right\} \tag{5.30}$$

Sometimes it can be convenient to use the following relationships obtained through approximation of Bessel functions:

- for $m \ll 1$:

$$k_r \approx 1 + (m^4/192);\ k_x \approx (m^2/8)[1 - (m^4/384)]$$

- for $m > 3.5$:

$$k_r \approx \frac{m}{2\sqrt{2}} + \frac{1}{4} + \frac{3\sqrt{2}}{32\,m};\ k_x \approx \frac{m}{2\sqrt{2}} - \frac{3}{64}\left(\frac{2\sqrt{2}}{m}\right) + \frac{3}{16\,m^2}.$$

The diagram of Fig. 5.9 gives the values of k_r, k_x and $\cos\varphi = 1\Big/\sqrt{1 + (k_x/k_r)^2}$ as a function of m.

They show that for $m \ll 1$, $R_{acu} \approx R_{dcu}$ and $X_{iu} \to 0$, while, for high values of *m,* it is $R_{acu} \approx X_{iu}$ and the phase displacement between applied voltage and current is about 45°.

From Eqs. (5.29) and (5.30), the active power per unit length can be written as:

$$P_u = R_{dcu} \cdot k_r \cdot I^2 = \frac{\rho}{\pi r_e^2} \cdot k_r \cdot I^2 \tag{5.31}$$

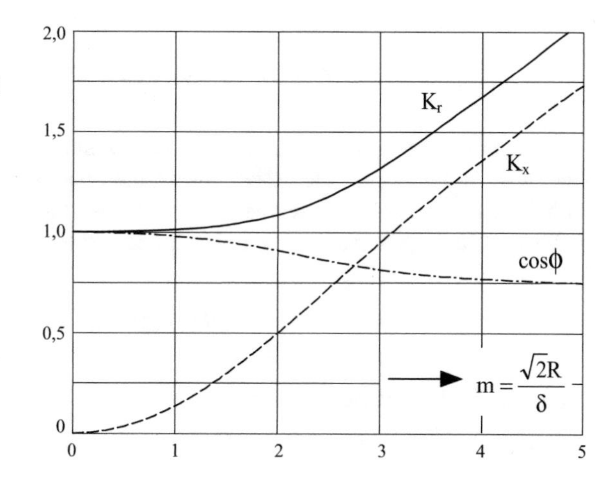

Fig. 5.9 Values of skin-effect coefficients k_r, k_x of the AC resistance R_{acu} and "internal" reactance X_{iu} and power factor $\cos\varphi = 1/(\sqrt{1 + (k_x/k_r)^2})$ as a function of $m = \sqrt{2}r_e/\delta$ in cylindrical workpieces

5.3.2 Transient Temperature Distribution in Cylindrical Body with AC Current Supply

As illustrated by the diagrams of Fig. 5.8, when AC current flows in the workpiece, uneven distributions of power density *w* will occur for values of *m* greater than 1.

In particular, when heating magnetic steel workpieces at power frequency, up to Curie point the magnetic permeability is relatively high and the penetration depth is consequently low. In this case it is generally $m > 1$ and the power sources *w* are unevenly distributed along the radius.

Moreover, in the initial stage of the heating transient the surface temperature is still relatively low and the losses from the surface to the environment can be assumed to be negligible.

In such conditions, taking into account Eq. (5.26), the differential equation (5.5), takes the form:

$$c\gamma\frac{\partial\vartheta}{\partial t} = \lambda\left(\frac{\partial^2\vartheta}{\partial r^2} + \frac{1}{r}\frac{\partial\vartheta}{\partial r}\right) + w_e\frac{\mathrm{ber}^2(\sqrt{2}r/\delta) + \mathrm{bei}^2(\sqrt{2}r/\delta)}{\mathrm{ber}^2(m) + \mathrm{bei}^2(m)} \tag{5.32}$$

and can be solved with the following boundary and initial conditions:

$$\vartheta = 0, \quad \text{for } t = 0; \quad \partial\vartheta/\partial r = 0, \quad \text{for } t > 0 \text{ and } r = r_e; \tag{5.32a}$$

Since an analytical solution of Eq. (5.32) valid for all instants of the heating transient cannot be obtained easily, and most applications of DRH are in the field of through heating, it is reasonable to derive a more simple solution that provides the temperature distribution in the workpiece when—after an initial transient period—the temperatures of all points in the cross-section increase linearly in time at the same rate.

In this situation, the energy balance for a generic cylinder of radius *r* and unit length is:

$$2\pi r\,\lambda\frac{\partial\vartheta}{\partial r} + \int_0^r 2\pi r\,w(r)\,dr = \pi r^2 c\gamma\frac{\partial\vartheta}{\partial t}. \tag{5.33}$$

It expresses the equality of the power transmitted through the external surface of the cylinder plus the power transformed into heat within the same cylinder, with the power needed for increasing the temperature at the rate $\partial\vartheta/\partial t$.

Introducing again the dimensionless parameters:

$$\xi = \frac{r}{r_e}; \quad \Theta = \frac{2\pi\lambda}{P_u}\vartheta; \quad \tau = \frac{kt}{r_e^2},$$

from Eq. (5.33) it can be written:

$$\frac{\partial \Theta}{\partial \xi} = \frac{\xi}{2}\frac{\partial \Theta}{\partial \tau} - \frac{2\pi r_e^2}{P_u \xi}\int_0^{\xi} \xi w \, d\xi. \tag{5.34}$$

The constant rate of the temperature increase can be determined with reference to the cylinder of radius r_e and unit length and the total power P_u converted into heat in it,

$$P_u = c\gamma\, \pi r_e^2 \frac{\partial \vartheta}{\partial t},$$

which in dimensionless units becomes:

$$\frac{\partial \Theta}{\partial \tau} = 2. \tag{5.35}$$

Taking into account Eqs. (5.26), (5.35) and (5.34) can be written as follows:

$$\begin{aligned}\frac{\partial \Theta}{\partial \xi} &= \xi - \frac{1}{\xi}\frac{\int_0^{\xi} \xi w \, d\xi}{\int_0^1 \xi w \, d\xi} = \xi - \frac{1}{\xi}\frac{\int_0^{\xi} \xi[\mathrm{ber}^2(m\xi) + \mathrm{bei}^2(m\xi)]\, d\xi}{\int_0^1 \xi[\mathrm{ber}^2(m\xi) + \mathrm{bei}^2(m\xi)]\, d\xi} \\ &= \xi - \frac{\mathrm{ber}(m\xi)\mathrm{bei}'(m\xi) - \mathrm{bei}(m\xi)\mathrm{ber}'(m\xi)}{\mathrm{ber}(m)\mathrm{bei}'(m) - \mathrm{bei}(m)\mathrm{ber}'(m)}\end{aligned} \tag{5.36}$$

By a further integration we obtain:

$$\Theta(\xi) = \frac{\xi^2}{2} - \frac{\int_0^{\xi} [\mathrm{ber}(m\xi)\mathrm{bei}'(m\xi) - \mathrm{bei}(m\xi)\mathrm{ber}'(m\xi)]\, d\xi}{\mathrm{ber}(m)\mathrm{bei}'(m) - \mathrm{bei}(m)\mathrm{ber}'(m)} + C$$

Indicating with:

$$\Theta = \Theta_a \quad \text{for } \xi = 0,$$

finally we can write:

$$\Theta(\xi) - \Theta_a = \frac{\xi^2}{2} - \frac{\int_0^{\xi} [\mathrm{ber}(m\xi)\mathrm{bei}'(m\xi) - \mathrm{bei}(m\xi)\mathrm{ber}'(m\xi)]\, d\xi}{\mathrm{ber}(m)\mathrm{bei}'(m) - \mathrm{bei}(m)\mathrm{ber}'(m)} \tag{5.37}$$

Equation (5.37) provides the radial temperature distributions in the cylinder above the temperature of the axis, in the stage of linear temperature increase.

These distributions are given, for different values of *m*, by the diagrams of Fig. 5.10, which were obtained from Eq. (5.37) by numerical integration.

For $m \to \infty$ the distribution tends to a parabolic curve, i.e. the same obtained for the heating transient with constant surface power density at paragraph 1.2.5.4.

In particular, the above distributions allow us evaluate the temperature difference between surface and axis for each value of *m*:

$$\Delta\Theta = \Theta_s - \Theta_a = \frac{1}{2}F(m) \tag{5.38}$$

or between the average temperature in the cross-section and the temperature of the axis:

$$\Theta_m - \Theta_a = F'(m) \tag{5.39}$$

The values of $F(m)$ and $F'(m)$ are given, as a function of m, by the diagrams of Fig. 5.11. As in Fig. 5.10, for $m \to \infty$, $F(m) \to 1$ and $\Delta\Theta = 1/2$.

Given the hypothesis of negligible surface losses, in analogy with Eq. (5.35), it is:

$$\Theta_m = 2\tau.$$

Denoting with:

$$\varepsilon = \frac{\Theta_s - \Theta_a}{\Theta_s} = \frac{\vartheta_s - \theta_a}{\vartheta_s} = \frac{F(m)}{2\Theta_s} \tag{5.40}$$

from Eqs. (5.38)–(5.40) we can write:

$$\tau = \frac{F(m)}{4\varepsilon}[1 - \varepsilon F''(m)] \tag{5.41}$$

with:

$$F''(m) = 1 - 2\frac{F'(m)}{F(m)}. \tag{5.42}$$

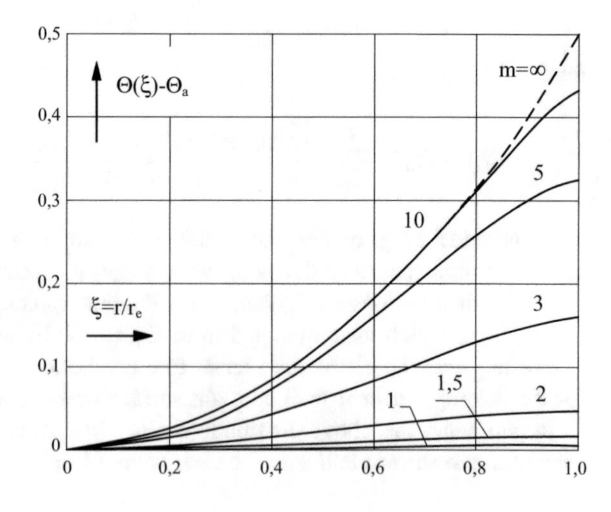

Fig. 5.10 Radial dimensionless temperature distributions above the temperature of axis, in the stage of increase with constant rate (r_e, r—external and generic radius; $\xi = r/r_e$; $m = \sqrt{2}r_e/\delta$)

Fig. 5.11 Diagrams of F(m), F′(m) and F″ (m), as a function of m

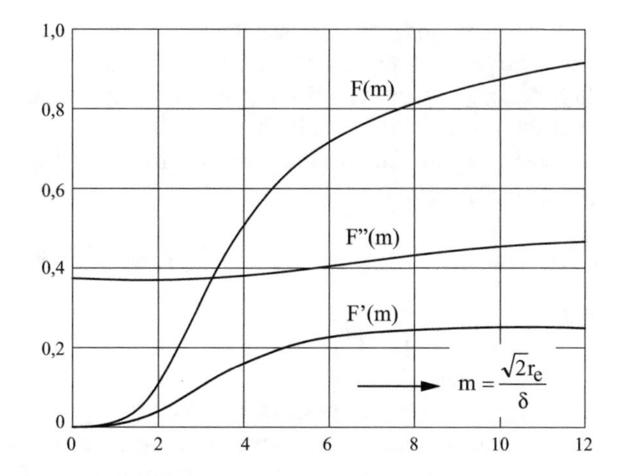

The values of $F''(m)$ are given in Fig. 5.11.

Equations (5.38) and (5.42) allow the evaluation of the heating time t_0 and the power P_u needed to heat the cylinder surface at a final temperature ϑ_s with a given differential $\Delta\vartheta = \vartheta_s - \vartheta_a$ between surface and axis, by the relationships:

$$\left.\begin{aligned} t_0 &= \frac{r_e^2}{4k\varepsilon} F(m)[1 - \varepsilon F''(m)] \\ P_u &= \frac{4\pi\lambda}{F(m)}(\vartheta_s - \vartheta_a) = 4\pi\lambda \frac{\varepsilon}{F(m)} \vartheta_s \end{aligned}\right\} \tag{5.43}$$

Equations (5.43) underline the influence of the values of ε and *m* on the heating time and the required power.

In particular, considering the trend of the curve of F(*m*) for "low" values of *m*, it appears possible, by a suitable frequency selection, to achieve the same final percent differential ε by increasing power P_u and correspondingly decreasing heating time.

The values of the function F″(m) also show that, for relatively low values of ε like those required in trough heating processes (ε = 0.05–0.10), in the first Eq. (5.43) the term [ε F″(m)] is always very small in comparison with unity.

On the other hand, the second equation shows that—all other things being equal—the differential $(\vartheta_s - \vartheta_a)$, or the percent difference ε, can be reduced by decreasing the specific power P_u.

5.3.3 *Heating Transient in Ferromagnetic Steel Bars*

As shown previously, the variations of material properties (electrical resistivity, specific heat, thermal conductivity, etc.) during heating have a considerable influence on the transient temperature distribution in the workpiece.

But, in ferromagnetic steels, still more important are the variations of the relative magnetic permeability μ with the temperature and the local magnetic field intensity. In particular, a dramatic influence has the abrupt variation of magnetic permeability near the Curie point (at ≈770 °C).

Several approximated formulas are used for the evaluation of the function $\mu = \mu(H, \vartheta)$, e.g.:

$$\mu = 1 + [\mu_{20}(H) - 1]\varphi(\vartheta) \tag{5.44}$$

with:

$\mu_{20} = 8130 \cdot H^{-0.894}$,
H—in A/cm,
$\varphi(\vartheta)$ given by the diagram of Fig. 5.12 [7, 8].

As a consequence of the above variations, the heating transients in the DRH of ferromagnetic steel bars are of the type illustrated in Fig. 5.13.

The shape of the curves can be explained as follows:

- in the first heating stage, below the Curie point, the material is characterised by relatively low resistivity (average value $\rho \approx$ 50–55 μΩ cm) and high relative permeability. In these conditions it is $m > 1$, the radial distribution of heat sources is uneven (see Fig. 5.8), the surface radiation losses are negligible or very low and—as a result—the surface temperature is higher than the axial one.
- above Curie point, the average resistivity increases to about 100–110 μΩ cm, while the relative permeability drops to one. As a consequence it is usually $m \leq 1$, the heating sources are nearly uniformly distributed along the radius and the radiation losses become high. This situation is analogous to the one illustrated at Sect. 5.2 for a DC heating transient, where the radiation losses produce an axial temperature higher than the one of the surface.

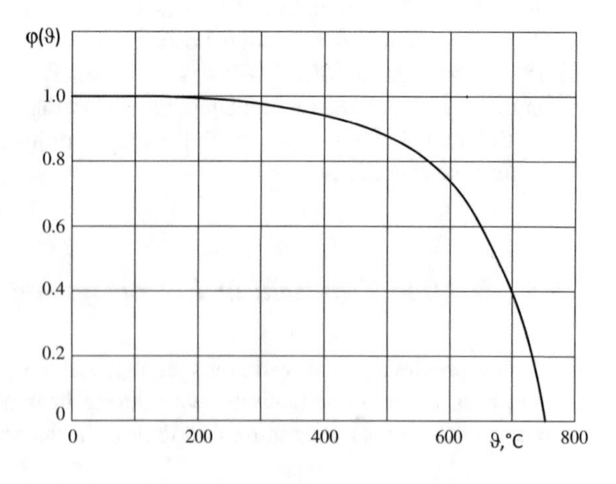

Fig. 5.12 Function $\varphi(\vartheta)$ as a function of the temperature ϑ

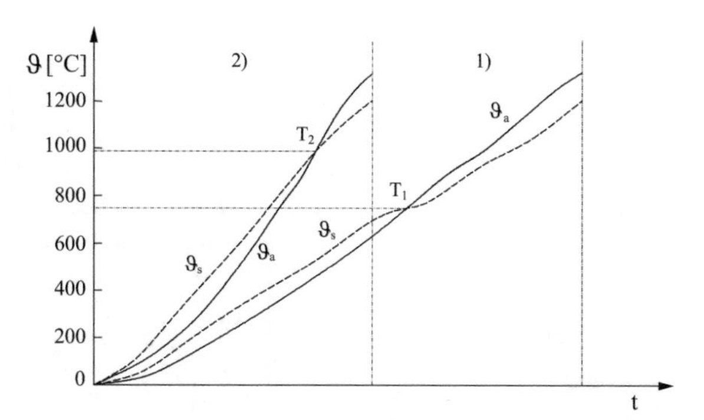

Fig. 5.13 Heating transients in ferromagnetic steel bars (ϑ_s; ϑ_a—surface and axial temperature) [19]

- the transition between these two heating stages occurs through an equalization process which produces the intersection of the curves of surface and axis temperatures at a point T which depends on material thermal conductivity and heating rate.

Since this equalization process always requires a certain time, in case of high heating rates the intersection point moves towards higher temperatures $(T_2 \rangle T_1)$.

With very high heating rates and short heating times, the curves intersection may even not occur, if the time available above Curie point is not sufficient for realization of the equalization process.

5.3.4 Installations for Bar Heating

An installation for AC direct resistance heating of bars corresponds to the schematic of Fig. 5.14, where the following typical elements are shown:

(A) three-phase supply transformer;
(B) adjustable phase balancing unit (Steinmetz scheme);
(C) control unit;
(D) single-phase transformer with primary taps and high current secondary windings;
(E) secondary circuit comprising connecting bus-bars, contacts and workpiece to be heated.

In the following some peculiarities of the installation and its equivalent circuit are described.

5.3.4.1 Equivalent Circuit

Neglecting the magnetization impedance of the single phase transformer (D), the equivalent circuit of the installation is that of Fig. 5.15a, where the following symbols are used:

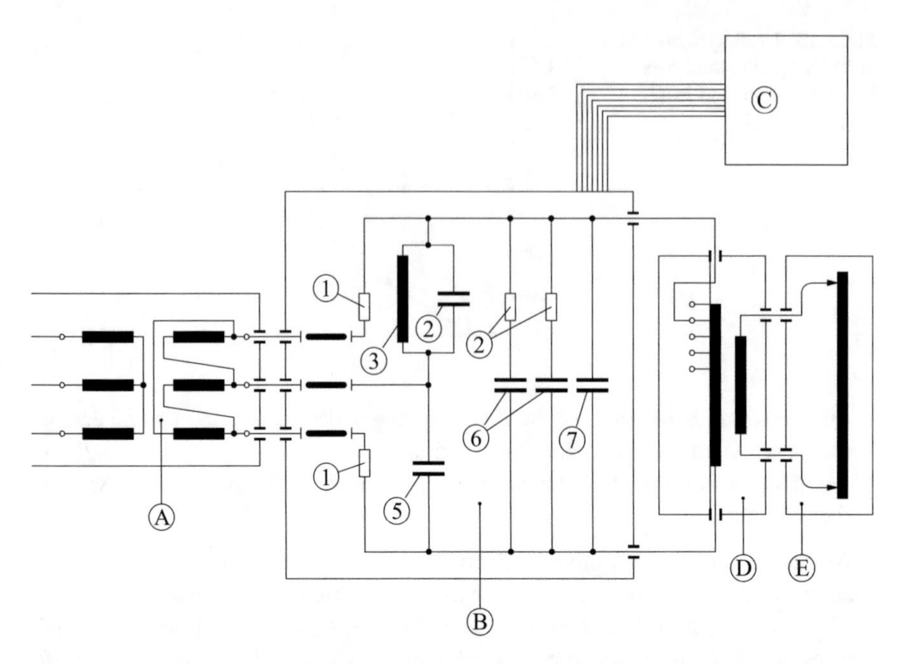

Fig. 5.14 Schematic of an AC-DRH installation (*1* on/off insertion devices according to cycles programmed from the control unit C; *2* elements for reactive power compensation; *3* reactance for load balancing; *4*, *5* capacitors for load balancing; *6* capacitors for programmed power factor compensation; *7* fixed capacitors for power factor compensation) [1]

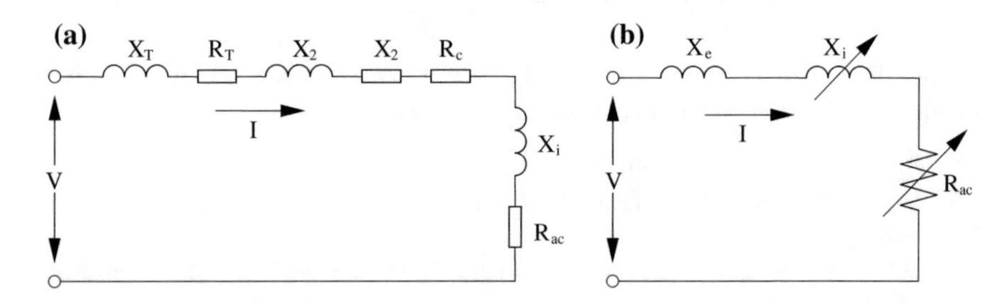

Fig. 5.15 **a** Equivalent circuit of the single-phase transformer (D) and the secondary circuit (E); **b** Simplified equivalent circuit

V, I	secondary voltage of single-phase transformer and current of secondary high current circuit;
R_T, X_T	equivalent resistance and reactance of single-phase transformer referred to the secondary side;
R_2, X_2	resistance and reactance of high current circuit connections;
R_C	resistance of contacts;
R_{ac}, X_i	AC resistance and "internal" reactance of the workpiece

$$X_e = X_T + X_2 \quad .$$

The calculation of the short-circuit impedance of the high current supply circuit can be done on the basis of the constructive design of the installation.

This circuit consists of the bus-bars (or cables) made of copper or aluminum and the contacts system (where bus-bars are parallel connected).

In the calculation of R_2 and X_2 the proximity and skin effects in the bus-bars must be considered.

The influence of the skin effect on R_2 can be taken into account by using the coefficients k_r, given in Fig. 5.9.

The evaluation of the external reactance $X_2 \approx X_e$ may be done, in a first approximation, by the diagrams of Fig. 5.16a, b [9, 10]. The diagram of figure (a) gives the value of X_2 as a function of the area covered by the circuit comprising

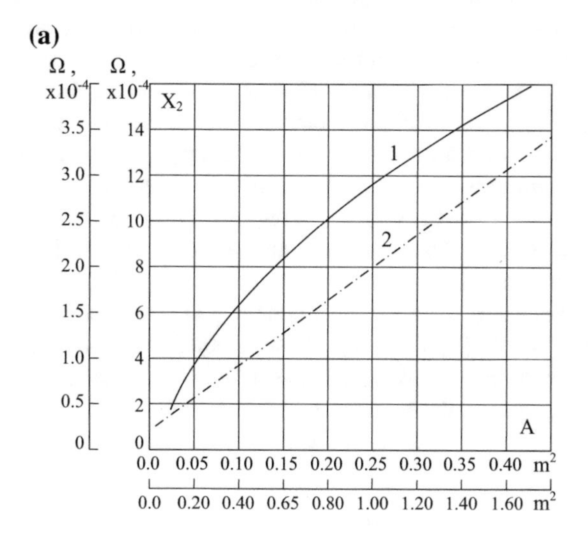

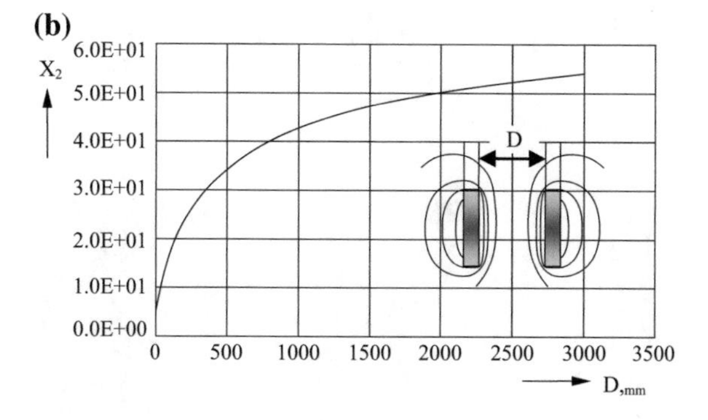

Fig. 5.16 **a** Values of reactance of a circuit as a function of its area A: (1) $A < 0.4\ m^2$; (2) $A = 0.1–1.6\ m^2$ [9]. **b** External reactance X_2 of 1 m length of a system of two parallel rectangular non-magnetic conductors of cross-section (20 × 120 mm), as a function of the distance D between them [10]

the transformer, the high-current circuit and the workpiece; the graph of figure (b)—the reactance of 1 m length of two parallel bifilar nonmagnetic conductors of rectangular cross-section (20 × 120 mm).

5.3.4.2 Contact System

Each contact "head" usually consists of a water cooled copper block with several contacts made of special alloys (e.g. Cu–Cr, Cu–Cr–Zr, Cu–Be, Cu–W) with high electrical and thermal conductivity, high surface hardness and high melting point. They have the function to inject into the load high intensity currents, even when the surface of the workpiece is irregular or strongly oxidized.

Taking into account that for large bar diameters the current can have values up to 120–140 kA and that—in order to avoid rapid wear of the contact—the maximum admissible current per point of contact should not exceed 5–10 kA, the contact system is designed with a convenient number of contacts as sketched in Fig. 5.17, according to the workpiece diameter and the current intensity.

For example, are used four contacts on the lateral side and one on the bar front (Fig. 5.17e) for currents up to 50 kA and diameters up to about 90 mm; eight lateral and four front contacts (Fig. 5.17f) for currents up to 100 kA and diameters up to 150 mm; 12 lateral and 4 front contacts for currents up to 130 kA and diameters up to 170 mm.

The contact resistance R_k of a copper contact point on steel workpieces is in the range $0.4–0.8 \cdot 10^{-4}\ \Omega$ [9]. For contacts with other materials, shapes and workpieces, data can be found in Refs. [9, 11] as a function of the force with which the contact is pressed against the workpiece.

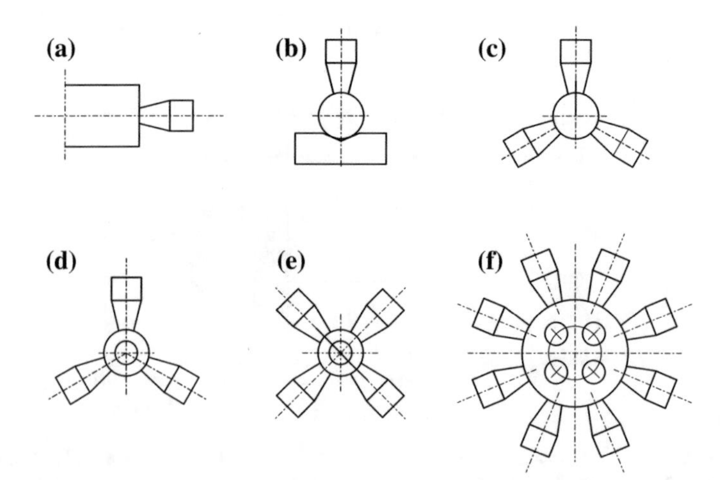

Fig. 5.17 Contact system with several lateral and front contacts according bar diameter and total current [20, 21]

The losses in a well design contact system are in the range of 1–2 kW per contact point.

Moreover, the contact system must be able by its design to offset the effects of the workpiece lengthening during heating and—sometimes—to control the contact resistance by adjusting the pressure applied to the contacts. It is worth to remember that, during the heating from 0 to 1250 °C the lengthening of steel workpieces is about 15 mm/m of length.

Finally, taking into account the heavy working conditions, the contact system must allow by design easy access for maintenance and replacement of contacts.

5.3.4.3 Load Balancing and Power Factor Correction

As seen in Sect. 5.3.4.1, the equivalent circuit of a direct resistance heater is characterised by a resistor and a reactance which change during the heating process. As a consequence also the power and the power factor undergo corresponding particularly strong variations.

Given the high values of power involved, the circuit is normally supplied from a three-phase network.

It is well known that a single-phase load in which a current I_e flows, connected to two phases of a three-phase grid, absorbs from the line a positive sequence and a negative sequence triplets of currents $\left(I_d = I_i = I_e/\sqrt{3}\right)$, as shown in Fig. 5.18.

The currents of negative sequence and the corresponding voltage drops give rise to additional losses in the supply line, for which the international standards give specific limits that cannot be exceeded at the point of installation. These limits depend on the ratio between the power of the load and the short circuit power of the grid at the point of installation.

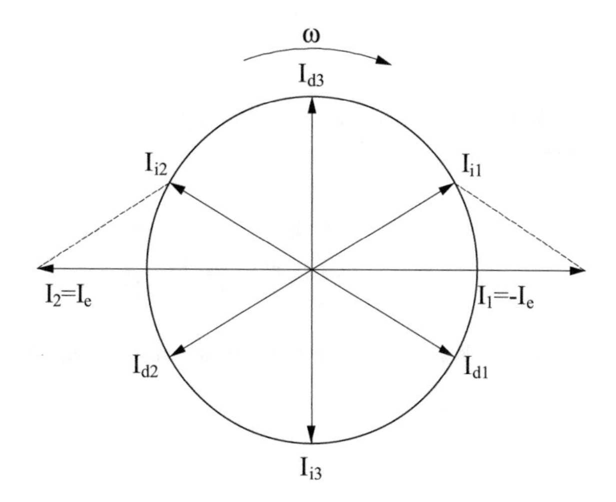

Fig. 5.18 Positive and negative sequence triplets of currents produced by a single-phase load connected to a three-phase system

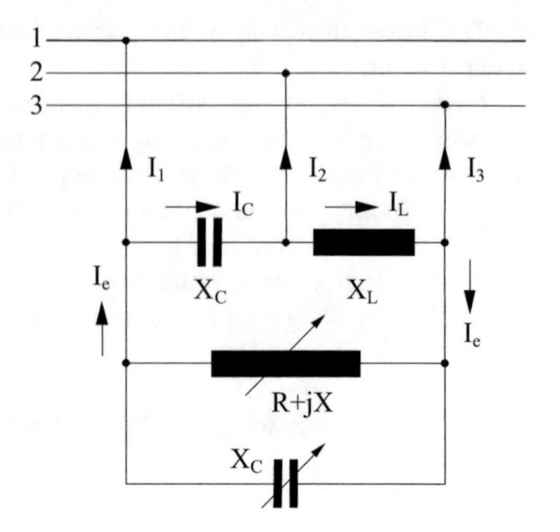

Fig. 5.19 Steinmetz circuit for power factor compensation and static load balancing of a single-phase load connected to a three-phase line

Since the above limits are often exceeded, high power installations are usually provided with systems for power factor compensation and static load balancing of the single-phase variable load.

However, since in DRH reactive power and power factor varies very rapidly during heating, a continuous and complete balancing of the load is not cost effective, and in most cases it is used the so-called Steinmetz circuit, comprising a reactor and an additional capacitor of suitable values (Fig. 5.19).

In this circuit, the values of reactor and additional capacitor giving symmetrical line currents, can be calculated from the resistance R and the reactance X of the single-phase load by the relationships:

$$(\text{for } 0 \leq X \leq R/\sqrt{3}) \quad \left. \begin{aligned} X_L &= \frac{\sqrt{3}(R^2 + X^2)}{R + \sqrt{3}X} \\ X_C &= \frac{\sqrt{3}(R^2 + X^2)}{R - \sqrt{3}X} \end{aligned} \right\} \tag{5.45}$$

With these values the line currents I_1, I_2, I_3 and the system's power factor become.

$$\left. \begin{aligned} I_1 = I_2 = I_3 &= \frac{V}{R^2 + X^2} \sqrt{\frac{R^2}{3} + 3X^2} \\ \cos \varphi_s &= \frac{1}{\sqrt{1 + 9(X/R)^2}} = \frac{\cos \varphi}{\sqrt{9 - 8\cos^2 \varphi}} \end{aligned} \right\} \tag{5.46}$$

where cos φ is the power factor of the single-phase load.

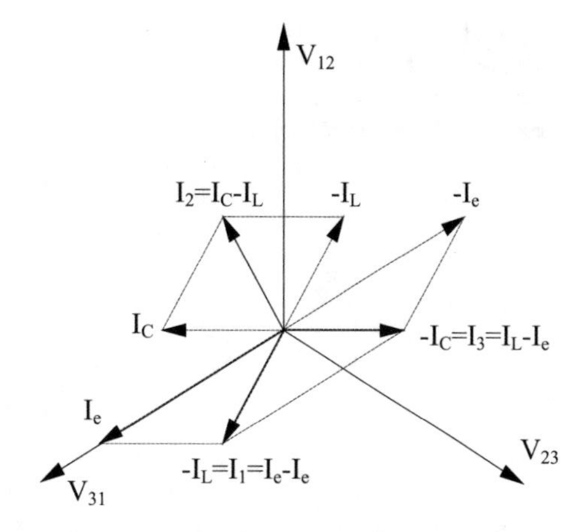

Fig. 5.20 Vector diagram of power factor compensation and static load balancing of a resistive single-phase load

If in parallel to the single-phase load (R, X) it is connected a capacitive reactance $X'_C = (R^2 + X^2)/X$, the equivalent impedance becomes purely resistive and equal to $R' = (R^2 + X^2)/R$, the power factor becomes unity and the symmetrizing inductive and capacitive reactance from Eq. (5.45) will be:

$$X_L = X_C = \sqrt{3}\frac{R^2 + X^2}{R} \tag{5.47}$$

In particular, in case of a purely resistive load (i.e. $X = 0$ and $\cos\varphi = 1$), from Eq. (5.45) one obtains:

$$X_L = X_C = \sqrt{3}R. \tag{5.48}$$

This case is represented by the vector diagram of Fig. 5.20.

As illustrated by the diagram of Fig. 5.21, in case of a pure resistive load the reversal of the phase sequence of the power supply or the exchange of the two elements of the balancing circuit, gives rise again to an unbalanced system with currents $I_2 = I_3 = \sqrt{7}I_1$.

5.3.5 *Efficiency and Energy Consumption*

5.3.5.1 Efficiency

The total efficiency of the heating process is given by the product $\eta = \eta_e \cdot \eta_t$ of the electrical efficiency η_e and the thermal efficiency η_t.

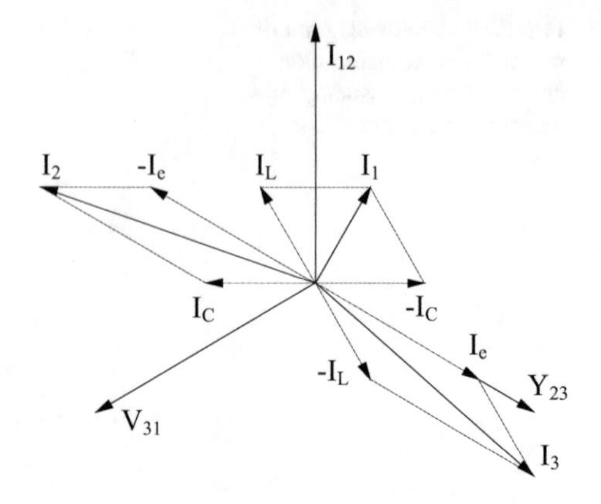

Fig. 5.21 Diagram of the same case of Fig. 5.20, but with the reversal of phase sequence of the power supply

With reference to the equivalent circuit of Fig. 5.15a, the electrical efficiency can be calculated with the relationship:

$$\eta_e = \frac{1}{1 + \frac{R_T + R_2 + R_C}{R_{ac}}}. \quad (5.49)$$

The experience has shown that in well designed installations the values of electrical efficiency η_e are in the range 0.85–0.95.

The main influence on η is therefore given by the thermal efficiency η_t, which depends on the radiation losses from the surface of the bar to the ambient.

Considering the useful energy required for increasing of $\Delta\vartheta_m$ the temperature of a workpiece of unit length in the heating time t_0:

$$P_u \eta_t t_0 = c\,\gamma\,\pi r_e^2 \Delta\vartheta_m \quad (5.50)$$

and the surface radiation losses:

$$P_\ell = p_0\, 2\pi r_e$$

with: p_0—specific losses, W/m², (see diagram A.7 in appendix), the thermal efficiency can be written as follows:

$$\eta_t = 1 - \frac{P_\ell}{P_u \eta_t} = \frac{1}{1 + \frac{2 p_0 t_0}{c\,\gamma r_e \Delta\vartheta_m}} \quad (5.51)$$

The curves of Fig. 5.22 give the values of η_t calculated with Eq. (5.51), as a function of the diameter, for steel bars heated up to 1250 °C with heating times in the range 30–240 s.

They show the need of using rapid heating cycles for achieving high thermal efficiency.

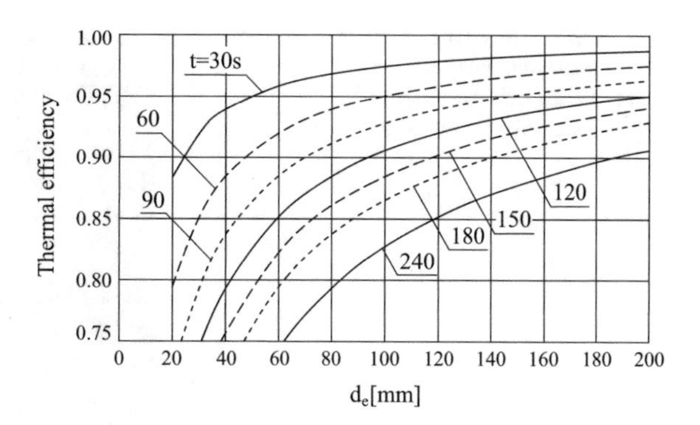

Fig. 5.22 Thermal efficiency as a function of bar diameter, in the heating of steel workpieces up to 1250 °C in 30-60-90-120-150-180-240 s

The above results also show that the efficiency is generally high and that, all other things being equal, it increases with the bar diameter.

From Eqs. (5.50)–(5.22), the following relationships for the total current in the bar and the magnetic field intensity at the workpiece surface can be obtained:

$$\begin{aligned} I &= \pi r_e^2 \sqrt{\frac{c\,\gamma\Delta\vartheta_m}{\rho k_r \eta_t t_0}}; \\ H_e &= \frac{I}{2\pi r_e} = \sqrt{\frac{c\,\gamma\, r_e^2\Delta\vartheta_m}{4\rho k_r \eta_t t_0}} \end{aligned} \tag{5.52}$$

5.3.5.2 Energy Consumption and Production Rate

The typical energy balance of Fig. 5.23 and the curves of Fig. 5.24 show that the energy consumption in the hot working of steel bars up to 1200–1250 °C is in the range 250–400 kWh/t, depending on geometrical dimensions.

The shape of the lower curves of Fig. 5.24 can be explained by fact that transformer, bus-bars and contact losses remain constant for a given value of the heating current, while the bar impedance and the power transformed into heat depend on its length and cross-section.

The diagram also shows the comparison of energy consumptions in DRH and induction heating. It confirms that the DRH process—when applicable—is the most convenient one from the energetic point of view.

The installation's production rate *M*, kg/h, can be evaluated as follows from workpiece geometrical and material characteristics and heating time t_0, taking into account the dead time t_M required for loading and unloading:

$$M = \frac{3600\ell S\,\gamma}{t_0 + t_M} \tag{5.53}$$

Fig. 5.23 Energy balance in direct resistance heating of a steel bar, 28 mm diameter, 500 mm length, up to 1250 °C, [22]

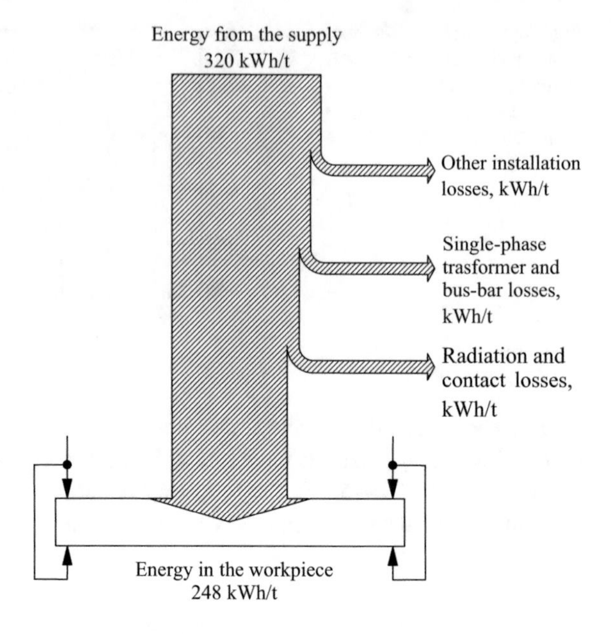

Fig. 5.24 Energy consumption in DRH heating steel bars of different diameters and lengths up to 1200 °C (*dashed* regions: corresponding values in induction heating processes) [19, 23]

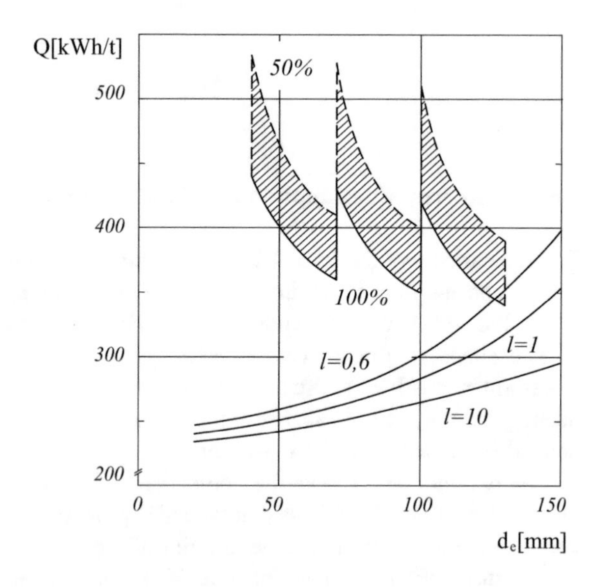

Figure 5.25 gives, as an example, the values of production rate and energy consumption of some industrial installations, corresponding to the specification data of Table 5.3.

The curves confirm the typical consumption values of 250–400 kWh/t for heating to 1250 °C steel bars of different geometrical dimensions.

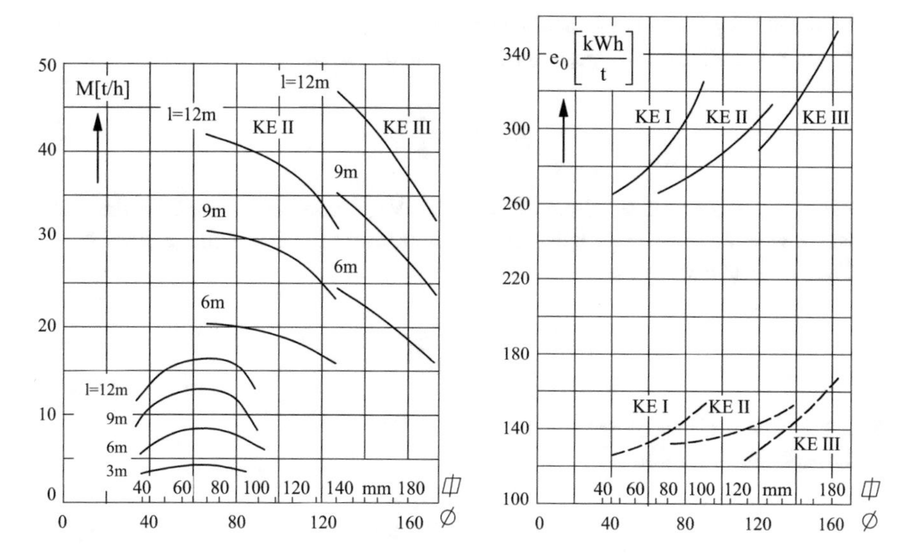

Fig. 5.25 Production rates M and average energy consumption e_0 of installations for DRH of steel bars, as a function of their geometrical dimensions (—— heating from 20 up to 1200 °C; - - - - from 700 to 1200 °C) [24]

Table 5.3 Data of some DRH installations for steel bars [29]

Installation	KE-I	KE-II	KE-III
Minimum bar dimensions (mm)			
Cylindrical bars	36	80	160
Bars of square cross-section	36 × 36	70 × 70	130 × 130
Maximum dimensions (mm)			
Cylindrical bars	100	130	180
Bars of square cross-section	90 × 90	120 × 120	170 × 170
Maximum current (kA)	50	100	130

5.3.6 Calculation of DRH Installations

As pointed out in previous paragraphs, the results of calculation of DRH systems are strongly influenced by the variations of material characteristics during heating.

However, given the high levels of power involved, this calculation is of fundamental importance because any uncertainty has the consequence of oversizing the installation components, with the result of increasing not only installation, but also running costs.

The basic method of calculating electrical and energetic parameters of a DRH heater is based on the use of its equivalent circuit, described in Fig. 5.15.

The elements of the equivalent circuit may be calculated as follows.

Impedance of the single-phase transformer.
It can be obtained from its plate data with the equation:

$$Z_T = v_{sc} \frac{V_{2n}}{I_{2n}}, \tag{5.54}$$

where is: v_{sc}—percent short-circuit voltage; V_{2n}—nominal voltage of the transformer low voltage side, V; I_{2n}—nominal current on the low voltage side, A.

Transformer resistance.

$$R_T = \frac{P_{sc}}{I_{2n}^2}, \tag{5.54a}$$

with P_{sc}—short-circuit power of transformer, W.

Transformer reactance.

$$X_T = \sqrt{Z_T^2 - R_T^2}. \tag{5.54b}$$

Resistance R_2. As already said, the *resistance R_2* can be calculated using the skin-effect coefficient k_r given in Fig. 5.9 and considering the influence of proximity effect by the method described in Refs. [9, 12, 13].

The *external reactance $X_2 \approx X_e$* can be obtained in a first approximation by using the diagrams of Fig. 5.16a, b [9, 10].

The *resistance R_k* of the contact system can be evaluated considering its design and the resistance of each contact point, which for copper/steel contact points is in the range $0.4–0.8 \cdot 10^{-4}\ \Omega$ [9]. For contacts between other materials and various shapes of contacts and workpieces, data can be found in Refs. [11, 14, 15] as a function of the force with which the contact is pressed on the workpiece.

As already pointed out, the most critical problem is to evaluate the *bar "internal" impedance* which, depending on geometrical dimensions, frequency, current flowing in it and material characteristics, generally undergoes significant changes during the heating process.

These variations, as a consequence, will produce—at constant supply voltage—large variations of absorbed current and power, which will be more or less pronounced depending on bar dimensions and the value of external impedance (in particular the reactance *Xe*), which constitutes a "ballast" impedance of the circuit.

In the following example it is shown an approximate procedure for evaluation of the variations during heating of the internal impedance and a preliminary rough calculation of a DRH installation done on the basis of the analytical formulas given in Sect. 5.3.1.

Example 5.4 AC current DRH of a cylindrical steel rod ($d = 2r_e = 40$ mm diameter, 6 m length) to the final average temperature $\vartheta_m = 1200$ °C in heating time $t = 50$ s.

With reference to the equivalent circuit of Fig. 5.15b, we assume the ratio (X_e/R_{dc}) between "external" reactance of the supply circuit and DC resistance of the bar at the start of heating (i.e. in "cold conditions") equal to 2.

It is required the evaluation of the main circuit electrical parameters.

(a) *Preliminary calculation with DC current:*

- Average required heating power (neglecting losses) in the temperature range 0–1200 °C, (Eq. 5.3):

$$P = \pi r_e^2 \ell c \gamma \frac{\Delta\vartheta_m}{\Delta t} = \pi \cdot (2 \cdot 10^{-2})^2 \cdot 6 \cdot 0.711 \cdot 7.8 \cdot 10^3 \frac{1200}{50} = 1004\,\mathrm{kW}$$

- DC resistance of the rod at beginning of heating:

$$R_{dc} = \rho \frac{\ell}{\pi r_e^2} = 18 \cdot 10^{-8} \cdot \frac{6}{\pi (2 \cdot 10^{-2})^2} = 0.859\,\mathrm{m\Omega}$$

- "External" reactance of supply circuit:

$$X_e = 2 \cdot R_{dc} = 2 \cdot 0.859 \cdot 10^{-3} = 1.718\,\mathrm{m\Omega}$$

(b) *Calculation of system parameters at the start of heating ("cold conditions").*

In most applications, the values of current required for the heating correspond to values of the surface magnetic field intensity H_e in the range 100–300 kA/m [see Eq. 5.52].

The relative magnetic permeability corresponding to a given magnetic field intensity can be deduced either from Fig. 5.26 or approximate relationships like Eq. (5.44) or, for field intensity above 100 kA/m, by the equation:

$$\mu_{20} = 1 + \frac{B_s}{H_e} \approx 1 + \frac{14{,}000}{H_e} \tag{5.55}$$

where is:

B_s saturation induction of steel, Gauss
H_e magnetic field intensity, A/cm

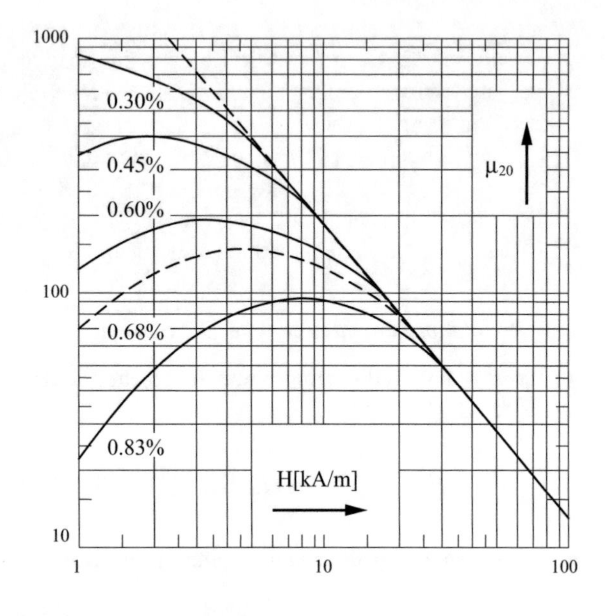

Fig. 5.26 Relative magnetic permeability as a function of magnetic field intensity for steels with different % carbon contents (*1* iron; *2* 0.3 %; *3* 0.45 %; *4* 0.6 %; *5* 0.83 %) [25]

It follows that the values of μ_{20} are mostly between 5 e 15. Thus we assume the tentative value:

$$\mu'_{20} = 10.$$

As a consequence (omitting in the following the subscript 20) we have:

- penetration depth, value of m and skin-effect coefficients k_r, k_x:

$$\delta' = 503 \cdot \sqrt{\frac{\rho}{\mu f}} = 503\sqrt{\frac{18 \cdot 10^{-8}}{10 \cdot 50}} = 0.954 \cdot 10^{-2}\,\mathrm{m}$$

$$\mathrm{m}' = \frac{\sqrt{2}\mathrm{r_e}}{\delta} = \frac{\sqrt{2}(2 \cdot 10^{-2})^2}{0.954 \cdot 10^{-2}} = 2.96$$

$$\mathrm{k}'_\mathrm{r} \approx 1.32; \mathrm{k}'_\mathrm{x} \approx 0.93 \quad \text{(from figure 5.9)}$$

- AC resistance and internal reactance of the bar [Eq. (5.29)]:

$$\mathrm{R}'_\mathrm{ac} = \mathrm{R_{dc}} \cdot \mathrm{k}'_\mathrm{r} = 0.859 \cdot 10^{-3} \cdot 1.32 = 1.13\,\mathrm{m\Omega}$$

$$\mathrm{X}'_\mathrm{i} = \mathrm{R_{dc}} \cdot \mathrm{k}'_\mathrm{x} = 0.859 \cdot 10^{-3} \cdot 0.93 = 0.799\,\mathrm{m\Omega}$$

- total current and surface magnetic field intensity:

$$I' = \sqrt{\frac{P}{R'_{ac}}} = \sqrt{\frac{1004}{1.13 \cdot 10^{-3}}} \approx 29.8\,\text{kA}$$

$$H'_e = \frac{I'}{2\pi r_e} = \frac{29.8}{2\pi \cdot 210^{-2}} \approx 237\,\text{kA/m}$$

Iterating this procedure we obtain:

- at the second iteration:

$$\mu'' = 1 + \frac{14{,}000}{2370} \approx 6.90$$

$$\delta'' = 503\sqrt{\frac{18 \cdot 10^{-8}}{6.90 \cdot 50}} = 1.15 \cdot 10^{-2}\ \text{m}$$

$$m'' = \frac{\sqrt{2} \cdot (2 \cdot 10^{-2})^2}{1.15 \cdot 10^{-2}} = 2.46$$

$$k''_r = 1.16;\ k''_x = 0.87 \quad \text{(from figure 4.9)}$$

$$R''_{ac} = 0.859 \cdot 10^{-3} \cdot 1.16 = 0.996\ \text{m}\Omega$$

$$X''_i = 0.859 \cdot 10^{-3} \cdot 0.87 = 0.747\ \text{m}\Omega$$

$$I'' = \sqrt{\frac{1004}{0.996 \cdot 10^{-3}}} \approx 31.75\ \text{kA}$$

$$H''_e = \frac{31.75}{2\pi \cdot 210^{-2}} \approx 252.7\ \text{kA/m}$$

- at the third iteration:

$$\mu''' = 1 + \frac{14{,}000}{2527} \approx 6.54$$

$$\delta''' = 503\sqrt{\frac{18 \cdot 10^{-8}}{6.54 \cdot 50}} = 1.18 \cdot 10^{-2}\,\text{m}$$

$$m''' = \frac{\sqrt{2}(2 \cdot 10^{-2})^2}{1.18 \cdot 10^{-2}} = 2.40$$

$$k'''_r = 1.14;\ k'''_x = 0.67 \quad \text{(from figure 4.9)}$$

$$R'''_{ac} = 0.859 \cdot 10^{-3} \cdot 1.14 = 0.979\,\text{m}\Omega$$

$$X'''_i = 0.859 \cdot 10^{-3} \cdot 0.67 = 0.576\,\text{m}\Omega$$

$$I''' = \sqrt{\frac{1004}{0.979 \cdot 10^{-3}}} \approx 32.02\,\text{kA}$$

$$H'''_e = \frac{32.02}{2\pi \cdot 2\,10^{-2}} \approx 254.8\,\text{kA/m}$$

Since at the next iteration we obtain $\mu = 6.49$ and $\delta = 1.18 \cdot 10^{-2}$ m, i.e. values which are practically the same of the previous iteration, the procedure can be stopped.

With these values, the following parameters can be calculated for the circuit in "cold conditions":

- equivalent impedance of the system:

$$Z_e = \sqrt{R_{ac}^2 + (X_i + X_e)^2} = \sqrt{(0.979)^2 + (0.576 + 1.718)^2} = 2.494\,\text{m}\Omega$$

- power factor:

$$\cos\varphi = \frac{R_{ac}}{Z_e} = \frac{0.979}{2.494} \approx 0.393$$

- supply voltage:

$$V = Z_e \cdot I = 2.494 \cdot 10^{-3} \cdot 32.02 \cdot 10^3 = 79.9\,(\text{V})$$

- apparent power:

$$P_a = V \cdot I = 79.9 \cdot 32.02 \cdot 10^3 = 2560\,(\text{kVA})$$

(c) *Calculation of average values of the system parameters in the "intermediate regime" of heating cycle (below Curie point).*

The procedure used in (b) is repeated with supply voltage V = 79.9 V, assuming for the material characteristics average values in the temperature range 0–750 °C.

- DC resistance:

$$R_{dc} = \rho \frac{\ell}{\pi r_e^2} = 60 \cdot 10^{-8} \cdot \frac{6}{\pi \cdot (2 \cdot 10^{-2})^2} = 2.865\,\text{m}\Omega$$

Evaluation of AC parameters

- First iteration:

$\mu'_{20} = 6.5$ (final value obtained in b)
$\mu' \approx 1 + (\mu'_{20} - 1)\,\Phi(\vartheta) \approx 1 + (6.5 - 1) \cdot 0.8 \approx 5.4$
with $\Phi(\vartheta)$ obtained from Fig. 5.12.

$$\delta' = 503 \cdot \sqrt{\frac{\rho}{\mu f}} = 503\sqrt{\frac{60 \cdot 10^{-8}}{5.4 \cdot 50}} = 2.37 \cdot 10^{-2}\,\text{m}$$

$$m' = \frac{\sqrt{2}r_e}{\delta'} = \frac{\sqrt{2}(2 \cdot 10^{-2})^2}{2.37 \cdot 10^{-2}} = 1.19$$

$$k_r' \approx 1.01; k_x' \approx 0.20 \quad \text{(from figure 5.9)}$$

$$R_{ac}' = Rdc \cdot k_r' = 2.865 \cdot 10^{-3} \cdot 1.01 = 2.894\,\text{m}\Omega$$

$$X_i' = Rdc \times k_x' = 2.865 \cdot 10^{-3} \cdot 0.20 = 0.573\,\text{m}\Omega$$

$$Z_e' = \sqrt{R_{ac}'^2 + (X_i' + X_e)^2} = \sqrt{(2.894)^2 + (0.573 + 1.718)^2} = 3.691\,\text{m}\Omega$$

$$I' = \frac{V}{Z_e'} = \frac{79.9}{3.691 \cdot 10^{-3}} \approx 21.65\,\text{kA}$$

$$H_e' = \frac{I'}{2\pi r_e} = \frac{21.65}{2\pi \cdot 2 \cdot 10^{-2}} \approx 172.4\,\text{kA/m}$$

- Second iteration:

$$\mu_{20}'' = 1 + \frac{14{,}000}{1724} \approx 9.12$$

$$\mu'' \approx 1 + (\mu_{20}'' - 1)\Phi(\vartheta) \approx 1 + (9.12 - 1) \cdot 0.8 \approx 7.50$$

$$\delta'' = 503 \cdot \sqrt{\frac{\rho}{\mu f}} = 503\sqrt{\frac{60 \cdot 10^{-8}}{7.50 \cdot 50}} = 2.01 \cdot 10^{-2}\,\text{m}$$

$$m'' = \frac{\sqrt{2}r_e}{\delta} = \frac{\sqrt{2}(2 \cdot 10^{-2})^2}{2.01 \cdot 10^{-2}} = 1.41$$

$$k_r'' \approx 1.02; k_x'' \approx 0.256 \quad \text{(from figure 5.9)}$$

$$R_{ac}'' = 2.865 \cdot 10^{-3} \cdot 1.02 = 2.922\,\text{m}\Omega$$

$$X_i'' = 2.865 \cdot 10^{-3} \cdot 0.256 = 0.732\,\text{m}\Omega$$

$$Z_e'' = \sqrt{(2.922)^2 + (0.732 + 1.718)^2} = 3.813\,\text{m}\Omega$$

$$I'' = \frac{79.9}{3.813 \cdot 10^{-3}} \approx 20.95\,\text{kA}$$

$$H_e'' = \frac{20.95}{2\pi \cdot 2\,10^{-2}} \approx 166.8\,\text{kA/m}$$

- third iteration:

$$\mu_{20}''' = 1 + \frac{14{,}000}{1668} \approx 9.39$$

$$\mu''' \approx 1 + (\mu_{20}''' - 1) \cdot \Phi(\vartheta) \approx 1 + (9.39 - 1) \cdot 0.8 \approx 7.71$$

$$\delta''' = 503\sqrt{\frac{60 \cdot 10^{-8}}{7.71 \cdot 50}} = 1.98 \cdot 10^{-2}\,\text{m}$$

Since it is $\delta''' \approx \delta''$ the iterative procedure can be stopped.

- With these values we obtain the following data for the "intermediate regime".

$$R_{ac} = 2.922\,m\Omega$$
$$Z_e = 3.813\,m\Omega$$
$$\cos\varphi = \frac{R_{ac}}{Z_e} = \frac{2.922}{3.813} \approx 0.766$$
$$I = 20.95\,kA$$
$$P = 2.922 \cdot 10^{-3} \cdot (20.95 \cdot 10^3)^2 = 1282\,kW$$
$$P_a = 79.9 \cdot 20.95 \cdot 10^3 = 1674\,kVA.$$

(d) *Evaluation of average values of the system in "hot conditions" (above Curie point)*

- Bar DC resistance:

$$R_{dc} = \rho\frac{\ell}{\pi r_e^2} = 115 \cdot 10^{-8} \cdot \frac{6}{\pi \cdot (2\,10^{-2})^2} = 5.491\,m\Omega$$
$$\delta = 503 \cdot \sqrt{\frac{\rho}{\mu f}} = 503\sqrt{\frac{115 \cdot 10^{-8}}{1 \cdot 50}} = 7.63 \cdot 10^{-2}\,m$$
$$m = \frac{\sqrt{2}r_e}{\delta} = \frac{\sqrt{2} \cdot (2 \cdot 10^{-2})^2}{7.63 \cdot 10^{-2}} = 0.371$$
$$k_r \approx 1.00; k_x \approx 0.02 \quad \text{(from figure 5.9)}$$
$$R_{ac} = R_{dc} \cdot k_r = 5.491 \cdot 10^{-3} \cdot 1.00 = 5.491\,m\Omega$$
$$X_i = R_{dc} \cdot k_x = 5.491 \cdot 10^{-3} \cdot 0.02 = 0.110\,m\Omega$$
$$Z_e = \sqrt{(5.491)^2 + (0.110 + 1.718)^2} = 5.787\,m\Omega$$
$$\cos\varphi = \frac{R_{ac}}{Z_e} = \frac{5.491}{5.787} \approx 0.949$$
$$I = \frac{79.9}{5.87 \cdot 10^{-3}} \approx 13.81\,kA$$
$$P = R_{ac} \cdot I^2 = 5.491 \cdot 10^{-3} \cdot (13.81 \cdot 10^3)^2 = 1047\,kW$$
$$P_a = V \cdot I = 79.9 \cdot 13.81 \cdot 10^3 = 1103\,kVA$$

Final remarks

The values obtained with the above procedure are only of first approximation since the variations of the material parameters with temperature and magnetic field intensity have been considered only in a very rough way.

For this reason, since in the next paragraph is presented a more accurate numerical solution, these values are not further re-adjusted here, although it is evident that the calculated power, below and above the Curie point, is higher than that required for performing the heating in 50 s, and that the corresponding heating time would therefore result shorter.

It would be necessary to adjust the supply voltage in order to obtain the given heating time and production rate.

5.3.6.1 Numerical Modelling

The analytical solutions considered so far, are very useful for a deep understanding of fundamental phenomena, but are insufficient for providing an accurate evaluation of the quantities involved, particularly in the heating of magnetic steel workpieces.

Today, however, powerful numerical methods are available which can provide accurate solutions of the coupled thermal and electromagnetic field problems and allow to calculate the main parameters needed for the design of a DRH heater.

Given the scope of this book, in order only to illustrate the possibilities existing today, here will be illustrated a simple Finite Difference 1D procedure which can be easily implemented on a PC by the students [30].

The procedure allows the analysis of the transient heating process in "long" cylindrical steel bars (neglecting the phenomena that occur near the contacts), through the solution of two coupled problems:

(a) *the electromagnetic problem*—it allows the determination of the total power converted into heat and its distribution within the bar during heating, taking into account the electrical and geometrical characteristics (such as voltage at the secondary of the supply transformer, total impedance of the secondary circuit, frequency, local values in the bar cross section of resistivity and permeability of steel, bar dimensions, etc.)
(b) *the thermal problem*—which provides the transient temperature pattern in the cross section taking into account the distribution of heat sources determined as solution of the electromagnetic problem, the material thermal parameters corresponding to the local temperature (specific heat, thermal conductivity and diffusivity) and the heat exchange conditions with the environment.

The two problems are mutually dependent, because the variations of material electrical characteristics influence the distribution of heat sources in the cross section, which—in turn—lead to a temperature increase different from point to point, thus resulting in further changes of local electrical and thermal parameters.

However, taking into account the very different values of the electromagnetic and thermal time constants, the two problems can be solved independently and alternately at predetermined time intervals, updating at each interval the values parameters as a function of the local temperature and magnetic field intensity [16–18].

(a) ***Electromagnetic problem***

The solution of this problem is illustrated through a simple finite difference method which, avoiding matrix formulation, allows high precision even with very limited requirements of computation time and memory capacity.

In the most general case, the bar is subdivided into N concentric elements with uneven thickness s_i, increasing gradually from the surface towards the axis, as shown in Fig. 5.27.

As shown in the figure, assuming that the function representative of the magnetic field intensity $\dot{H}$ is constant in each element, while the one representative of the current density $\dot{J}$ varies stepwise at the centre of each layer, it is:

$$\dot{H}_i = \text{cost.,} \quad \text{for } r_i - \frac{s_i}{2} \leq r \leq r_i + \frac{s_i}{2}$$

$$\dot{J}_i = \text{cost.,} \quad \text{for } r_{i-1} \leq r \leq r_i$$

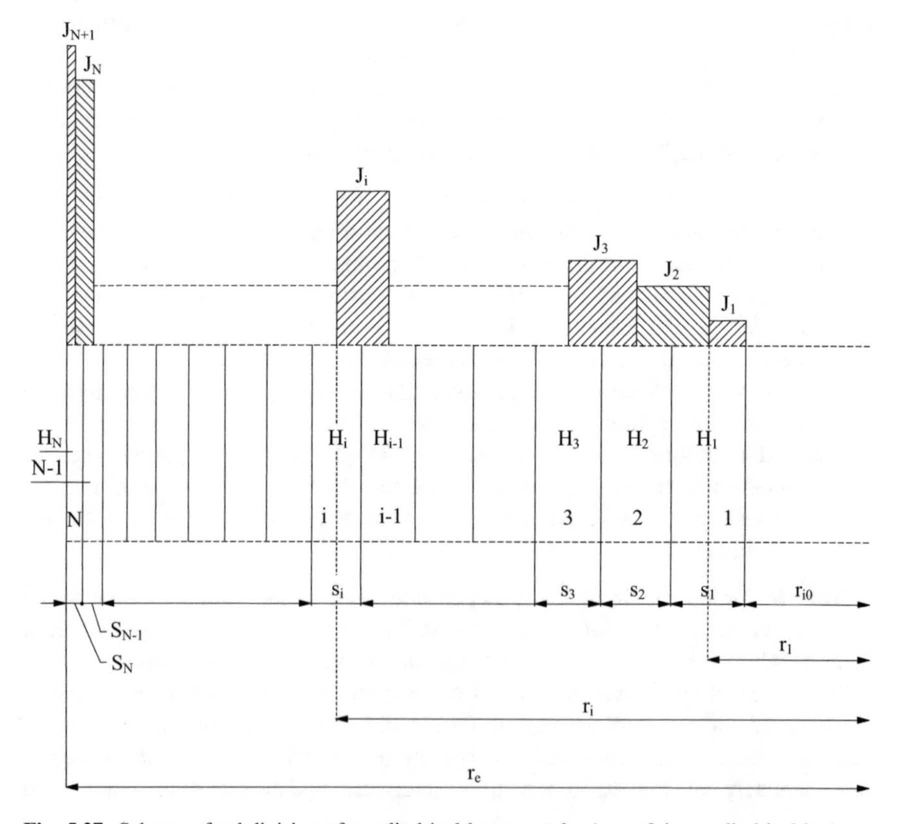

Fig. 5.27 Scheme of subdivision of a cylindrical bar or a tube ($r_{i0} = 0$ in a cylindrical bar)

The quantities of the various layers can be calculated in sequence one at a time, starting from the innermost one for which the value J_1 is fixed arbitrarily and proceeding toward the surface, as follows:

- Current $\dot{I}_1$ in the internal layer between radii r_{i0} and r_1 ($r_{i0} = 0$ for a solid cylinder):

$$\dot{I}_1 = \pi(r_1^2 - r_{i0}^2)\dot{J}_1 \cdot \tag{5.56}$$

- Magnetic field intensity $\dot{H}_1$ at radius r_1:

$$\dot{H}_1 = \frac{\dot{I}_1}{2\pi r_1} \tag{5.57}$$

- Relative permeability μ_1 of layer 1, evaluated from the curves of Fig. 5.26 or by approximate equations (e.g. 5.44), considering the permeability a scalar, all quantities sinusoidal and negligible the hysteresis losses:

$$\mu_1 = f(H_1) \tag{5.58}$$

- Current density $\dot{J}_2$, obtained from previous data and Eq. (5.15):

$$\dot{J}_2 = \dot{J}_1 + j\frac{\omega\mu_1\mu_0}{\rho_1}s_1\dot{H}_1 \tag{5.59}$$

- Iteration of the previous steps as in Eqs. (5.56)–(5.59), with i = 2, 3, …, N using the relationships:

$$\left.\begin{aligned} \dot{I}_i &= \pi(r_i^2 - r_{i-1}^2)\dot{J}_i \\ \dot{H}_i &= \frac{1}{2\pi r_i}\sum_{k=1}^{i}\dot{I}_k \\ \mu_i &= f(H_i) \\ \dot{J}_{i+1} &= \dot{J}_i + j\frac{\omega\mu_i\mu_0}{\rho_i}s_i\dot{H}_i \end{aligned}\right\} \tag{5.60}$$

with:

$\dot{I}_i$	current in the layer between radii r_{i-1} and r_i
$\dot{H}_i$	magnetic field intensity at radius r_i
ρ_i, μ_i	resistivity and relative permeability of the i-th layer
$\dot{J}_{i+1}$	current density between the layers r_i and r_{i+1}.

- Current $\dot{I}_{N+1}$ in the layer between the radii r_N and r_e:

$$I_{N+1} = \pi(r_e^2 - r_N^2) \cdot J_{N+1} \tag{5.61}$$

- Total current $\dot{I}$ in the bar:

$$\dot{I} = \sum_{i=1}^{N} \dot{I}_i \tag{5.62}$$

- Magnetic field intensity $\dot{H}_e$ at the bar surface:

$$\dot{H}_e = \frac{\dot{I}}{2\,\pi r_e} \tag{5.63}$$

Given the arbitrary choice of J_1, all calculated values will not correspond, in general, to those occurring with a given value I_0 of the total current.

In case of non-magnetic bars, the solution is obtained by multiplying all the calculated field quantities by the ratio (I_0/I) between I_0 and the calculated total current.

On the contrary, in case of magnetic steels the solution must be obtained by iterative procedure, repeating the calculation several times with initial values of J_1 appropriately updated at each iteration, until the assigned value I_0 (or H_e) is reached.

As shown by the equations, the procedure allows to consider values of ρ_i and μ_i different from layer to layer, which can be updated according to the temperature of each element during the calculation of the heating transient.

When the current density values J_i in the various layers are known, it is immediate the calculation of the corresponding specific power densities $w_i = \rho_i J_i^2$, which are the heating sources for the solution of the thermal problem.

(b) ***Thermal problem***

It consists in the numerical solution of the differential equation (5.5) at each time interval Δt of subdivision of the heating transient, with known distributions of heat sources w_i (determined as indicated in a) and temperature at the beginning of the interval Δt (calculated at the end of the previous time interval), with specified conditions of heat exchange with the environment.

For simplicity, we will refer again to a classical finite difference solution, applied to the same scheme of subdivision of the bar of Fig. 5.27.

Denoting with:

$\vartheta_{i-1}, \vartheta_i, \vartheta_{i+1}$ temperature of layers *i*−1, *i*, *i* + 1 at the instant t, start of the generic interval Δt

$\vartheta'_{i-1}, \vartheta'_i, \vartheta'_{i+1}$ temperature in the same layers at the instant $t + \Delta t$ (end of the time interval Δt)

$$a_i = \frac{s_i + s_{i+1}}{2}; \quad b_i = \frac{s_i + s_{i+1}}{2},$$

for the determination of the heat balance of the i-th layer in the time interval Δt, for a bar of unit axial length, calculating heat fluxes with the temperature at time *t*, we must consider the following terms:

- amount of heat through the surface $2\pi\,(r_i + {}^{s_i}/_2)$:

$$Q_{ei} = \lambda_i 2\pi (r_i + \frac{s_i}{2}) \frac{\vartheta_{i+1} - \vartheta_i}{a_i} \Delta t \tag{5.64}$$

- amount of heat through the surface $2\pi\,(r_i - {}^{s_i}/_2)$:

$$Q_{ui} = -\lambda_i\, 2\pi (r_i - \frac{s_i}{2}) \frac{\vartheta_i - \vartheta_{i-1}}{b_i} \Delta t \tag{5.65}$$

- amount of heat developed inside the layer by the heat sources w_i:

$$Q_{wi} = 2\pi r_i s_i w_i \Delta t \tag{5.66}$$

- amount of heat required for increasing the temperature of the layer from ϑ_i to ϑ_i':

$$Q_i = 2\pi r_i s_i c_i \gamma (\vartheta_i' - \vartheta_i). \tag{5.67}$$

Writing the heat balance as follows:

$$Q_{ei} + Q_{ui} + Q_{wi} = Q_i \tag{5.68}$$

and introducing the Eqs. (5.64)–(5.67), for i = 2, 3, …, N − 1 we obtain:

$$\begin{aligned} \vartheta_i' = \vartheta_i + \frac{k_i \Delta t}{s_i^2} \Big\{ \frac{s_i}{a_i}(1 + \frac{s_i}{2r_i})\vartheta_{i+1} - [\frac{s_i}{a_i}(1 + \frac{s_i}{2r_i}) \\ + \frac{s_i}{b_i}(1 - \frac{s_i}{2r_i})]\vartheta_i + \frac{s_i}{b_i}(1 - \frac{s_i}{2r_i})\vartheta_{i-1} \Big\} + \frac{w_i \Delta t}{c_i \gamma} \end{aligned} \tag{5.69}$$

This equation allows to evaluate in explicit form the temperature of the *i*-th layer at time t + Δ t, starting from the knowledge of the temperature in the same layer and the adjacent layers at time t.

The solution must be completed considering separately the heat balance (5.68) for the layers 1 and N, taking into account the relevant boundary conditions.

For example, in case of a solid cylinder, by applying to the layer 1 the condition $Q_{u1} = 0$, it is:

$$\vartheta_1' = \vartheta_1 + \frac{2k_1\Delta t}{s_1^2(1+s_2/s_1)}(1+\frac{s_1}{2r_1})(\vartheta_2 - \vartheta_1) + \frac{w_1\Delta t}{c_1\gamma} \tag{5.69a}$$

For the layer N, the heat balance depends on the conditions of heat exchange with the environment; for example, in case of free radiation, Eq. (5.68) must be evaluated using, instead of Eq. (5.64), the relationship:

$$Q_{eN} = -\alpha\, 2\,\pi(r_N + \frac{s_N}{2})(\vartheta_e - \vartheta_a)\Delta t \tag{5.69b}$$

with:

$\alpha = h_c + \alpha_i$	.
h_c	heat transmission coefficient by convection
$\alpha_i = \varepsilon\,\sigma\frac{T_e^4 - T_a^4}{\vartheta_e - \vartheta_a}$	.
$\vartheta_e = \vartheta_N + \frac{\vartheta_N - \vartheta_{N-1}}{b_N}\frac{s_N}{2}$	temperature of the bar external surface, °C
ϑ_a	ambient temperature, °C
T_e, T_a	temperatures ϑ_e, ϑ_a in K
ε	emissivity of the surface
$\sigma = 5.67 \cdot 10^{-8}$	Stefan–Boltzmann constant, W/m^2K^4

In particular, in case of a subdivision of the solid cylinder with layers of equal thickness *s*, assuming $\vartheta_e \approx \vartheta_N$ and $\vartheta_a = 0$, Eq. (5.69) become:

$$\left.\begin{aligned}
\vartheta_1' &= \vartheta_1 + \frac{2k_1\Delta t}{s^2}(\vartheta_2 - \vartheta_1) + \frac{w_1\Delta t}{c_1\gamma}\\
\vartheta_i' &= \vartheta_i + \frac{k_i\Delta t}{s^2}(1+\frac{s}{2r_i})\vartheta_{i+1} - (1-\frac{s}{2r_i})\vartheta_{i-1} + \frac{w_i\Delta t}{c_i\gamma}\\
\vartheta_N' &= \vartheta_N + \frac{k_N\Delta t}{s^2}\Big\{-[(1-\frac{s}{2r_N}) - \frac{\alpha s}{\lambda_N}(1+\frac{s}{2r_N})]\vartheta_N\\
&\quad + (1-\frac{s}{2r_N})\vartheta_{N-1}\Big\} + \frac{w_N\Delta t}{c_N\gamma}
\end{aligned}\right\} \tag{5.70}$$

As known, the explicit formulation is particularly simple since it does not require the simultaneous solution of a system of equations; however, for the stability of the calculation, it must be satisfied the condition:

$$\frac{k_i\Delta t}{s_i^2} \le \frac{1}{2}. \tag{5.71}$$

This means that, when the thickness of layers is given, the time step must be chosen such as to fulfil the condition:

$$\Delta t \leq \frac{s_i^2}{2k_i} \tag{5.71a}$$

In practice, this may results in a large number of calculation steps, as it is pointed out by the following example.

Example 5.5 Heating in 120 s a cylindrical steel bar of radius r_e = 50 mm up to 1250 °C. The calculation of the heating transient is done with uniform subdivision (s = 1 mm) and constant time steps.

Assuming the maximum value of the thermal diffusivity k_i equal to 0.1 cm^2/s, (see Table A.6 in appendix), from Eq. (5.71a) it is:

$$\Delta t \leq \frac{0.1^2}{2 \cdot 0.1} = 0.05 \text{ s}$$

Therefore, in the case here considered, at least 2400 time steps Δt are needed. However, even if the number of calculations may appear very high, by using modern PCs the calculation time of the whole heating transient is always of the order of few seconds.

On the contrary, if the heat fluxes in Eqs. (5.64) and (5.65) are evaluated on the basis of the temperatures at time $(t + \Delta t)$, equations similar to those previously derived, corresponding to the so-called implicit scheme, will be obtained, which at each time step Δt requires the simultaneous solution of the whole system of equations.

The method may appear more complex of the explicit one but, since the system matrix is tri-diagonal, simplified algorithms can be used that allow a significant reduction of calculation time. It also has the advantage of being inherently stable for any value of the time step, which can be increased in comparison with the explicit method, thus reducing the number of time steps needed.

Since the heat transfer coefficient α between workpiece surface and environment depends on the temperature of the surface, in the implicit method the system becomes non-linear and must be solved by an iterative procedure.

May be therefore convenient to use improved implicit methods, like Crank - Nicolson, Hopscotch, etc.

In addition to the above, it must be remembered that a well-defined amount of energy $E_t \approx 17.7$ kWh/t is required for the transition of the steel from magnetic to non-magnetic conditions, at the Curie temperature ($\vartheta_C \approx 770\,°C$).

This should be taken into account when implementing the calculation program, since in the *i*-th layer, when ϑ_C is reached during heating, the temperature cannot increase further unless to the layer is given the amount of energy ($2\pi\, r_i\, s_i\, \gamma\, E_t$).

The same amount of heat is “released” in the layer when the temperature ϑ_C is reached during cooling.

5.3.7 *Influence of Electrical Installation Design on the Heating of Steel Bars*

The use of numerical methods allows to evaluate with good precision the temperature transient in the workpiece and to analyse the influence on it of the installation design parameters.

Among these parameters particular importance have the "external" impedance of the supply circuit and the secondary voltage of the power transformer.

The voltage is defined by the need to circulate in the secondary circuit a current sufficient to obtain the required output.

This circuit, characterized by very high current values, includes the workpiece, the contacts, the fixed bus bars system and the flexible conductors that allow to adapt the supply circuit to bars of different lengths.

The workpiece "internal" impedance depends on frequency, workpiece geometrical dimensions, intensity of the current flowing in it and characteristics of the workpiece material which, generally, undergoes significant changes during heating.

Consequently, at constant supply voltage, a large variability of the absorbed current and power can be experienced, more or less pronounced depending on the workpiece dimensions and the value of the "external" impedance of the supply circuit, in particular the reactance X_e, which constitutes a "ballast" impedance in the circuit.

With reference to the simplified equivalent circuit of Fig. 5.15b and to the heating of cylindrical steel bars, this particularly complex situation will be illustrated by the examples given in the following diagrams.

In particular:

- Figure 5.28 shows the radial distributions of relative magnetic permeability at the beginning of heating, in bars of different diameters, with surface magnetic field intensity 100 kA/m;
- Figure 5.29a, b give the relative distributions along radius, at beginning of heating, of the specific power per unit volume in bars of different diameters, with surface magnetic field intensity of 100 and 300 kA/m respectively.

It should be noted that they differ significantly from the analogous specific power distributions calculated with the classical analytical solutions with constant permeability in the bar cross section (see Fig. 5.8).

To the distributions of Fig. 5.29 correspond, for different bar diameters and values of H_e, the power transformed into heat per unit surface of the bar given by the curves of Fig. 5.30.

Fig. 5.28 Radial distributions of the relative magnetic permeability μ, at the beginning of heating (i.e. material in “cold conditions”) (Steel: UNI-C45; H_0 = 100 kA/m)

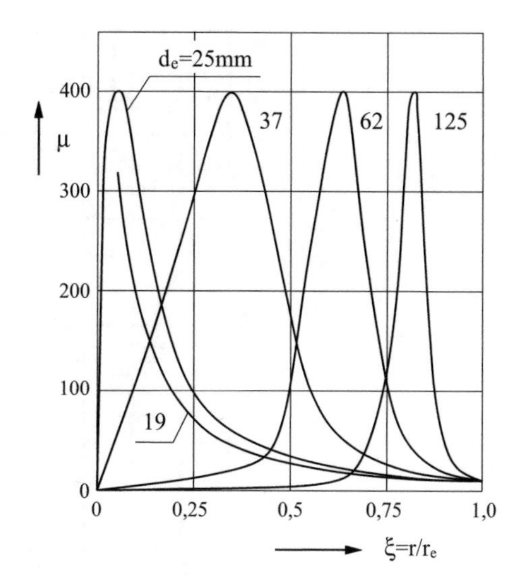

- Figure 5.31: radial distributions of specific power per unit volume at various instants of the heating transient.

The diagrams show, that in bars of “small” diameter the distributions of specific power are nearly uniform at the beginning of heating, while with “big” diameters or low current intensity this occurs only in the temperature range above Curie point, where the distributions are influenced only by the resistivity variations with temperature.

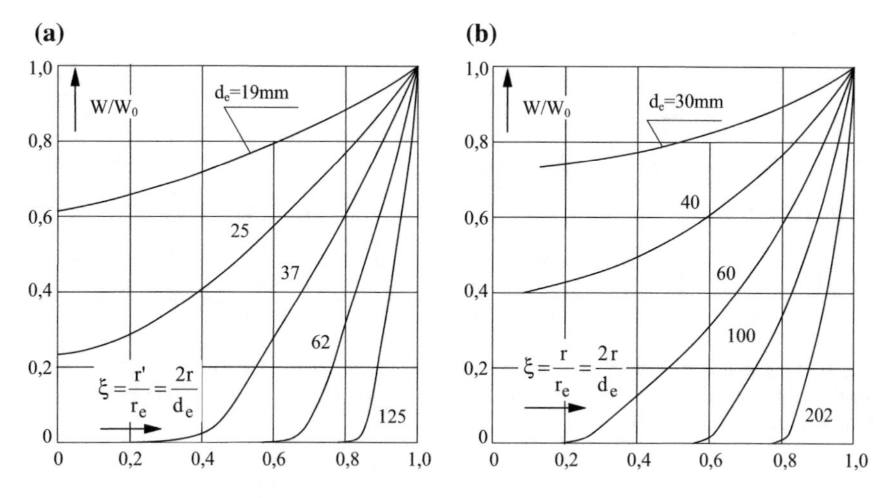

Fig. 5.29 Relative radial distributions of specific power per unit volume at beginning of heating (Steel UNI-C45; **a** H_e = 100 kA/m; **b** H_e = 300 kA/m)

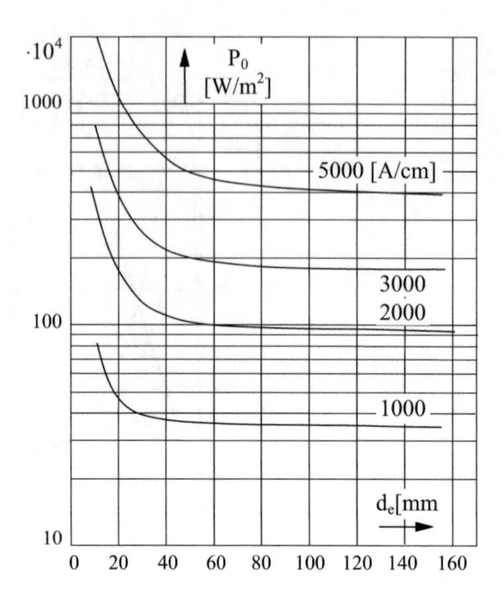

Fig. 5.30 Specific power per unit surface of the workpiece at the beginning of the heating transient, as a function of bar diameter $d_e = 2r_e$, for different values of the surface magnetic field intensity H_e, A/cm (Steel UNI-C45)

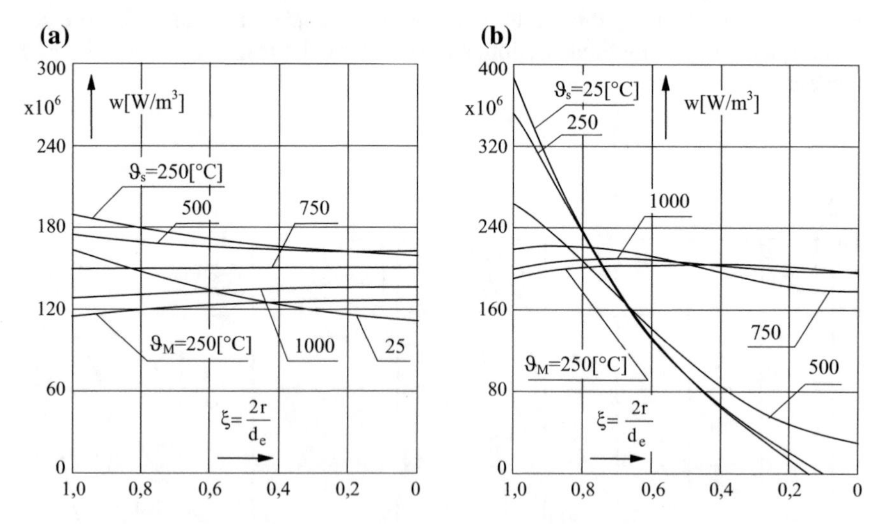

Fig. 5.31 Radial distributions of specific power per unit volume w at different instants of the heating transient, when the temperature reached at the surface is ϑ_s (**a** d_e = 20 mm, V = 80 V; **b** d_e = 60 mm, V = 100 V)

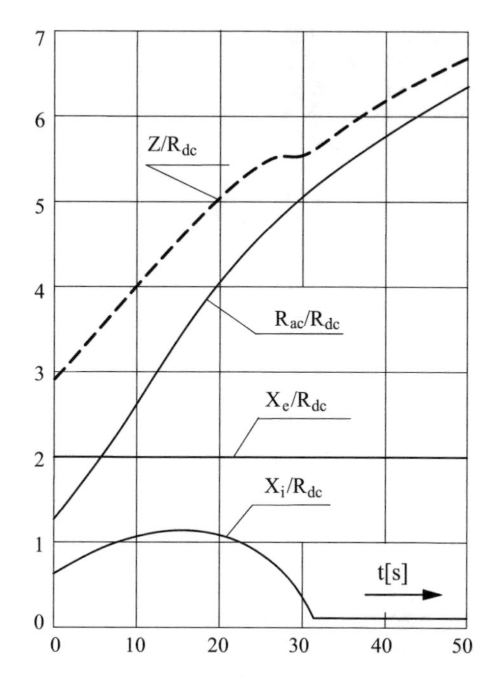

Fig. 5.32 Variations during heating of bar resistance, "internal" reactance and total impedance of secondary circuit (Steel UNI-C45; $\ell = 6\,m$; d_e = 40 mm; X_e/R_{dc} = 2; V = 80 V)

The influence of resistivity variations is particularly evident near the workpiece surface, where temperatures and resistivity values are lower due to the heat radiation losses to the environment.

- Figure 5.32: Variations during the heating transient of the AC resistance and "internal" reactance of the bar and the total impedance of the secondary circuit.

As a consequence, depending on the above parameters, the distributions of specific power in the bar cross section and with time will be very different from case to case, and therefore also the corresponding temperature transients have different shapes, as shown in Fig. 5.33.

The curves (a) correspond to relatively uniform distributions of specific power in the first stage of the heating transient, similar to those shown in Fig. 5.31a, so that below the Curie point there is not an appreciable temperature differential between surface and axis. On the contrary the differential becomes significant in the high temperature stage due to the more pronounced radiation losses from the workpiece surface.

The transient temperature distributions of Fig. 5.31b, on the contrary, are a consequence of uneven radial distributions of the specific power also in the initial stage, which will produce temperatures on the axis lower than those at the surface. The contrary occurs, due to surface losses, in the high temperature range.

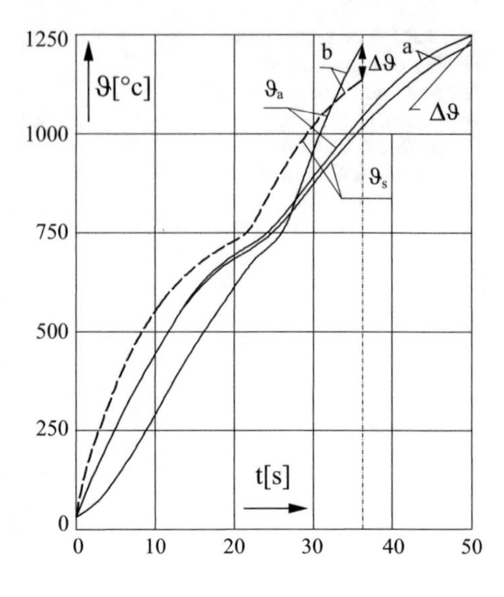

Fig. 5.33 Variations of temperature of surface (- - -) and axis (—) during heating transient (Steel UNI-C45; $\ell = 6\,\text{m}$; $X_e/R_{dc} = 2$; **a** $d_e = 20$ mm, V = 80 V, $\Delta\vartheta = 46\,°C$; **b** $d_e = 60$ mm, V = 100 V, $\Delta\vartheta = 114\,°C$)

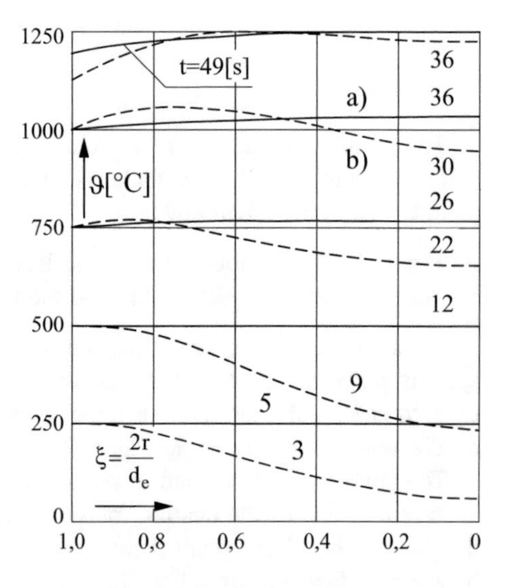

Fig. 5.34 Radial temperature distributions at different instants of heating transient (parameters as in Fig. 5.33; values on the curves: time in seconds)

For the cases (a) and (b), the radial temperature distributions in the bar cross-section at different instants of the heating transient have the shapes shown in Fig. 5.34.

Also power and power factor vary during heating in different ways from one diameter to another; these variations depend, to a considerable extent, on the ratio of the reactance of the supply system to the impedance of the workpiece.

This is confirmed by the following example, which refers to the heating with constant voltage of a bar, 40 mm diameter, 6 m length, within heaters characterised by supply circuits with different external reactance X_e, having values

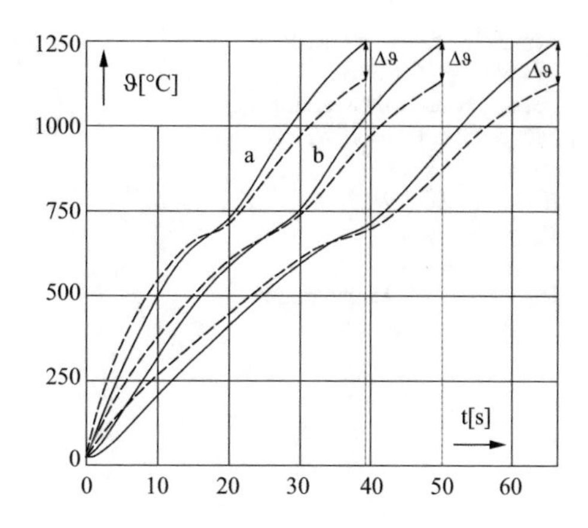

Fig. 5.35 Transient temperature distributions at surface (- - -) and axis (—) (Steel UNI-C45; ℓ = 6 m; d_e = 40 mm; V = 80 V; **a** X_e/R_{dc} = 1, $\Delta\vartheta$ = 115 °C; **b** X_e/R_{dc} = 2, $\Delta\vartheta$ = 117 °C; **c** X_e/R_{dc} = 3, $\Delta\vartheta$ = 119 °C)

corresponding to 1, 2 or 3 times the resistance R_{dc} of the workpiece in “cold” conditions.

The heating transients, shown in Fig. 5.35, are characterized by heating times which obviously, at constant supply voltage, increase when the external reactance increases but final temperature differentials $\Delta\vartheta = \vartheta_s - \vartheta_a$ which remain practically unchanged in all cases.

The relative variations of apparent power and power factor during these heating transients are given in the Fig. 5.36.

It can be observed that in case of magnetic steel workpieces the power factor has very low values, 0.4–0.6, at the start of heating, while it increases to about 0.85–0.95 above Curie temperature.

On the contrary, in case of non-magnetic materials the power factor has rather high values (0.8–0.9) during the whole heating transient.

The figure also shows the influence of the reactance X_e on the power factor and the variations of current during the heating cycle.

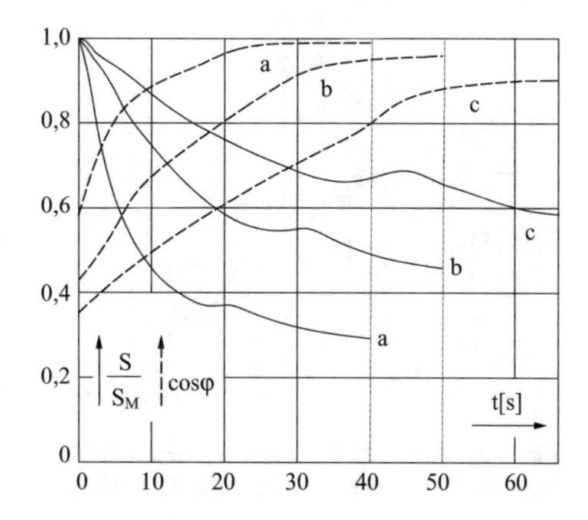

Fig. 5.36 Relative variations of apparent power (or current) and power factor during heating (circuit data as in Fig. 5.35; **a** S_M = 3472 kVA, cos φ_m = 0.911; **b** S_M = 2190 kVA, cos φ_m = 0.801; **c** S_M = 1628 kVA, cos φ_m = 0.700)

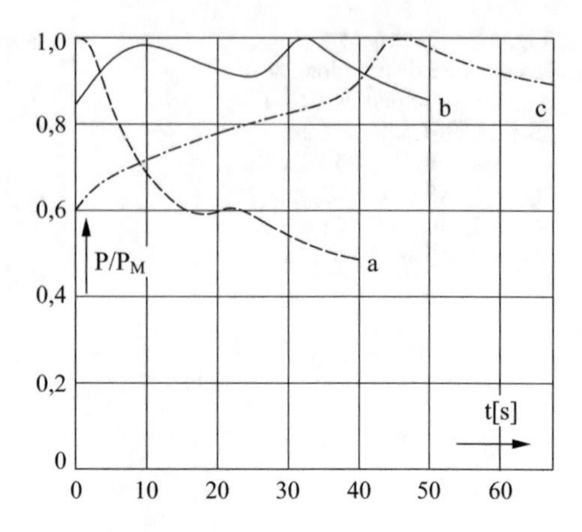

Fig. 5.37 Relative variations of active power during heating (circuit data as in Fig. 5.35; **a** $P_M = 2073$ kW, $P_m = 1337$ kW; **b** $P_M = 1123$ kW, $P_m = 1045$ kW; **c** $P_M = 960$ kW, $P_m = 801$ kW) M-maximum; m-average values

Figure 5.37 shows the variations of active power during the previous heating transients. The curves suggest that a convenient choice of the ratio X_e/R_{dc} at the design stage can be useful in order to minimize such variations during the heating cycle.

To this end may be used the curves of Fig. 5.38a, b, which provide the ratios (P'/P'') of the active power at the beginning and the end of heating, as a function of X_e/R_{dc}, for different diameters and values of H'_e (at beginning of heating) of 100 and 300 kA/m, respectively.

They show that convenient values of the ratio P'/P'' are in the range 2.5–3.5.

Finally, once it is defined the value of the installation's reactance X_e, it is possible to achieve the required heating time and production rate by adjusting the value of supply voltage V.

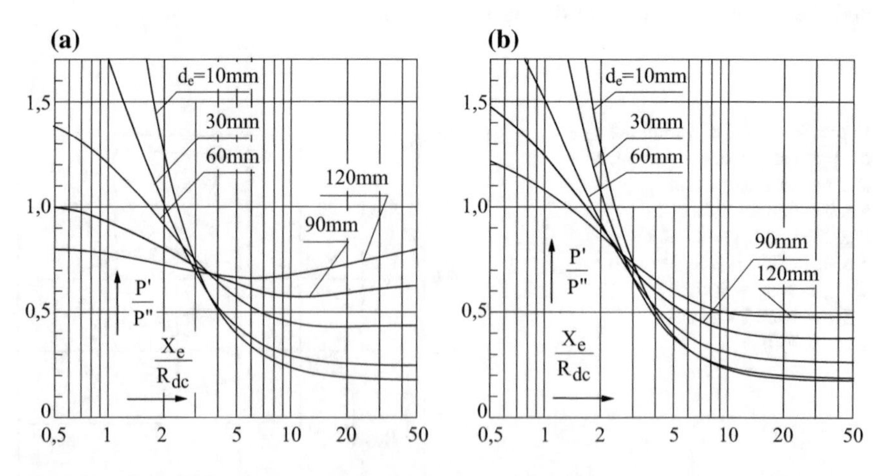

Fig. 5.38 Ratio (P′/P″) of active power at beginning and end of heating transient as a function of the ratio X_e/R_{dc} (Steel UNI-C45; **a** $H_e' = 100$ kA/m; **b** $H_e' = 300$ kA/m)

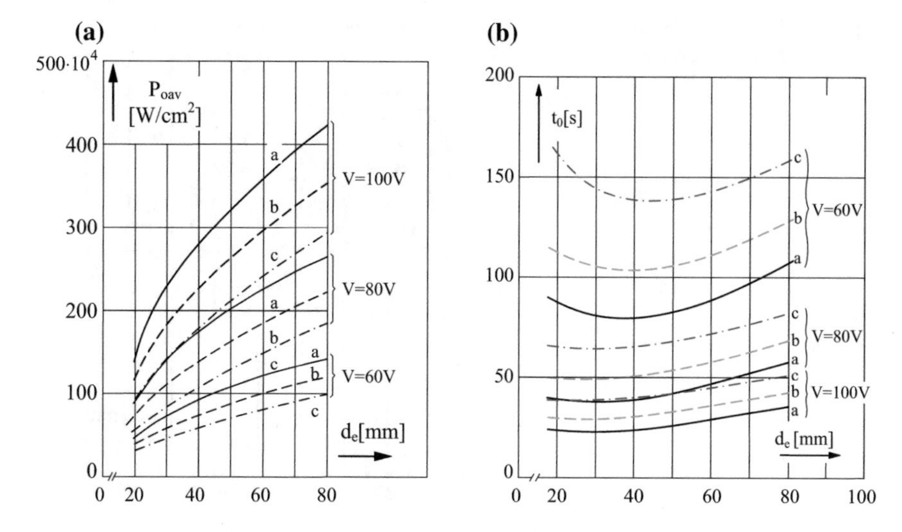

Fig. 5.39 Average specific power in the workpiece during the heating cycle as a function of diameter (on the *left*), and the corresponding heating times (on the *right*) (Steel UNI-C45; ℓ = 6 m; **a** X_e/R_{dc} = 1; **b** X_e/R_{dc} = 2; **c** X_e/R_{dc} = 3)

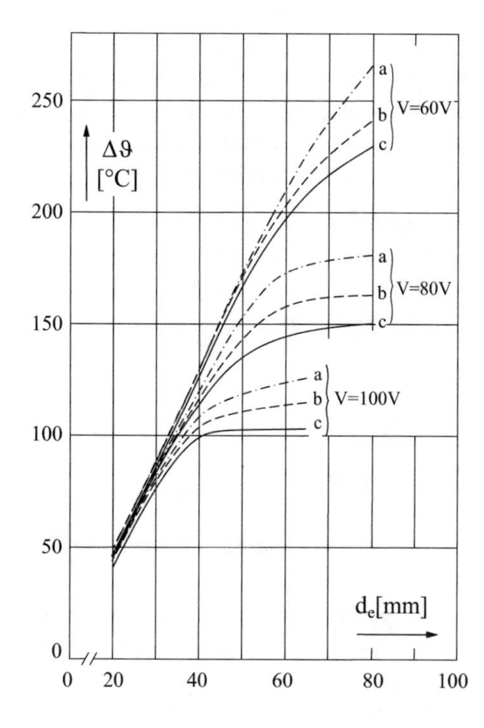

Fig. 5.40 Temperature differential $\Delta\theta$ between surface and axis at the end of heating (data as in Fig. 5.39)

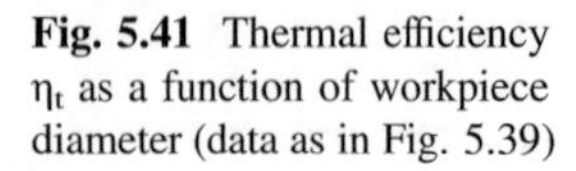

Fig. 5.41 Thermal efficiency η_t as a function of workpiece diameter (data as in Fig. 5.39)

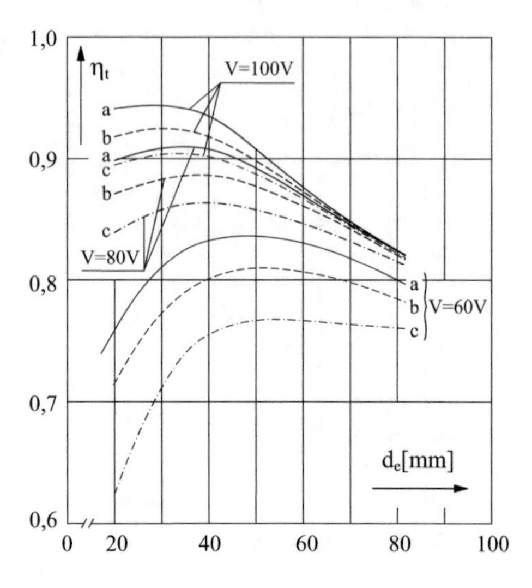

To this end, the supply power transformer is generally equipped with additional taps for stepwise regulation of the secondary voltage. This voltage control can be also used for adapting the heater to workpieces of different diameters and lengths.

In the last case, it is necessary to evaluate the heater performance for the different diameters (and lengths) of the production program; this may be done by diagrams similar to those of Figs. 5.39, 5.40 and 5.41, which refer to the heating of steel bars UNI-C45, 6 m length, to surface temperature 1250 °C.

The diagrams, which refer to different values of the ratio X_e/R_{dc} and different supply voltages, provide the following quantities:

- Figure 5.39a, b: average power transformed into heat during heating, referred to the unit surface of the bar, and corresponding heating times;
- Figure 5.40: temperature differential $\Delta\vartheta = \vartheta_s - \vartheta_a$ between surface and axis at the end of heating;
- Figure 5.41: thermal efficiency of the heating process.

5.4 Miscellanea

5.4.1 DRH Tubes Heating with DC Supply

In this case Eq. (5.3) still apply and the specific quantities J, *E* and *w* depend only on the material characteristics (ρ, c, γ) and the rate of temperature increase $(\Delta\vartheta_m/\Delta t)$.

On the contrary, due to its smaller cross section in comparison with a solid cylindrical workpiece with the same external radius r_e, the integral values of current and power are considerably lower.

In fact, with reference to a tube of external radius r_e and internal r_i, using the notation $\alpha = r_i/r_e$, if the current density is the same, the following ratios apply:

$$\frac{R_{dcbar}}{R_{dctube}} = \frac{I_{tube}}{I_{bar}} = \frac{P_{tube}}{P_{bar}} = 1 - \alpha^2 \tag{5.72}$$

As a result, for the same value of r_e, it is easier to achieve the values of current and power required for a certain production rate, as it is illustrated by the following example.

Example 5.6 Heating with DC current, as in the Example 5.1, a tube with external diameter $d_e = 100$ mm ($r_e = 50$ mm), thickness 5 mm ($r_i = 45$ mm) and length 6 m.

Assuming the same values of Example 5.1 for *w, J, E* and *V*, will change as follows the values of R_{dc}, *P* and *I*.

$$\alpha = \frac{r_i}{r_e} = \frac{45}{50} = 0.9;$$

$$1 - \alpha^2 = 1 - 0.81 = 0.19$$

$$R_{dctube} = \frac{0.573 \cdot 10^{-3}}{0.19} = 3.016 \cdot 10^{-3}\,\Omega$$

$$P = 5419 \cdot 0.19 = 1.030\,\text{kw}$$

$$I = 97.250 \cdot 0.19 = 18.478\,\text{A}$$

5.4.2 DRH Tube Heating with AC Supply

Also in this case the distributions of current and power density in the tube thickness are non-uniform.

In the hypothesis of constant resistivity and permeability of the tube material, these distributions can be obtained again from Eq. (5.19), rewritten here in the form:

$$\dot{J} = \dot{C}_1 J_0(\sqrt{-j}m\xi) + \dot{C}_2 Y_0(\sqrt{-j}m\xi) \tag{5.73}$$

and the constants $\dot{C}_1$ and $\dot{C}_2$ by the conditions:

$$\left.\begin{aligned} \dot{J} &= \dot{J}_e && \text{for } \xi = 1 \\ \oint \dot{H}\,ds &= 0 && \text{for } \xi = \alpha \end{aligned}\right\} \tag{5.74}$$

Since to the second of Eq. (5.74) corresponds the condition [18]:

$$\frac{d\dot{J}}{d\xi} = 0 \quad \text{for } \xi = \alpha \tag{5.75}$$

but it is:

$$\frac{d\dot{J}}{d\xi} = -\sqrt{-jm}[\dot{C}_1 J_1(\sqrt{-jm}\xi) + \dot{C}_2 Y_1(\sqrt{-jm}\xi)$$

from Eqs. (5.74) and (5.75) we can write:

$$\left.\begin{aligned} \dot{J}_e &= \dot{C}_1 J_0(\sqrt{-jm}) + \dot{C}_2 Y_0(\sqrt{-jm}) \\ 0 &= \dot{C}_1 J_1(\sqrt{-jm}\alpha) + \dot{C}_2 Y_1(\sqrt{-jm}\alpha) \end{aligned}\right\} \tag{5.76}$$

After determination of the constants $\dot{C}_1$ and $\dot{C}_2$, we finally obtain:

$$\frac{\dot{J}}{\dot{J}_e} = \frac{J_0(\sqrt{-jm}\xi)Y_1(\sqrt{-jm}\alpha) - Y_0(\sqrt{-jm}\xi)\, J_1(\sqrt{-jm}\alpha)}{J_0(\sqrt{-jm})Y_1(\sqrt{-jm}\alpha) - Y_0(\sqrt{-jm})\, J_1(\sqrt{-jm}\alpha)} \tag{5.77}$$

Since from Eq. (4.5.15) it is:

$$H = \frac{1}{j\omega\mu_0\mu}\frac{d\dot{E}}{dr} = -j\frac{r_e}{m^2}\frac{d\dot{J}}{d\xi}$$

by applying the relationship $\dot{I} = 2\,\pi r_e\,\dot{H}_e$, we obtain:

$$\dot{J}_e = \frac{\dot{I}}{j\sqrt{-j}\pi r_e^2}\frac{m\,[J_0(\sqrt{-jm})Y_1(\sqrt{-jm}\alpha) - Y_0(\sqrt{-jm})J_1(\sqrt{-jm}\alpha)]}{2[J_1(\sqrt{-jm})Y_1(\sqrt{-jm}\alpha) - Y_1(\sqrt{-jm})J_1(\sqrt{-jm}\alpha)]} \tag{5.78}$$

Introducing Eq. (5.78) into (5.77) it results:

$$\begin{aligned} &\frac{\dot{J}}{\frac{\dot{I}}{\pi r_e^2(1-\alpha^2)}} = \\ &= \sqrt{-j}(1-\alpha^2)\frac{m[J_0(\sqrt{-jm}\xi)Y_1(\sqrt{-jm}\alpha) - Y_0(\sqrt{-jm}\xi)\, J_1(\sqrt{-jm}\alpha)]}{2[J_1(\sqrt{-jm})Y_1(\sqrt{-jm}\alpha) - Y_1(\sqrt{-jm})J_1(\sqrt{-jm}\alpha)]} \end{aligned} \tag{5.79}$$

where the term $[\dot{I}/\pi r_e^2(1-\alpha^2)]$ represents the current density occurring if the total current $\dot{I}$ is uniformly distributed in the tube cross section.

Taking into account that it is: $\dot{Z}_i = R_{ac} + j\,X_i = \dot{E}_0/\dot{I}$, from Eq. (5.79) we obtain the internal impedance per unit length:

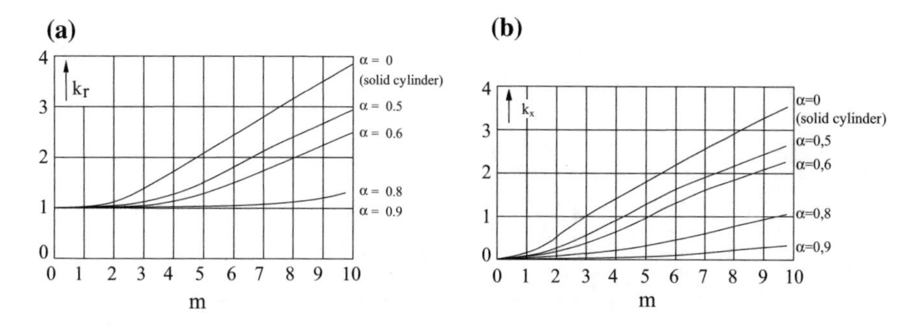

Fig. 5.42 Values of skin-effect coefficients as a function of m for tubes of various thicknesses: **a** $k_r = R_{acu}/R_{dcu}$; **b** $k_x = X_i/R_{dcu}$ [26]

$$\frac{\dot{Z}_i}{R_{dcu}} = \sqrt{-j}(1-\alpha^2)\frac{m\left[J_0(\sqrt{-j}m)Y_1(\sqrt{-j}m\alpha) - Y_0(\sqrt{-j}m)J_1(\sqrt{-j}m\alpha)\right]}{2\left[J_1(\sqrt{-j}m)\,Y_1(\sqrt{-j}m\alpha) - Y_1(\sqrt{-j}m)J_1(\sqrt{-j}m\alpha)\right]} \tag{5.80}$$

where: $R_{dcu} = [\rho/\pi r_e^2(1-\alpha^2)]$—is the DC resistance of the tube per unit length.

By separation of real and imaginary parts of Eq. (5.80), in analogy to what was done for the solid cylindrical workpiece, we can determine the coefficients $k_r = R_{acu}/R_{dcu}$ and $k_x = X_i/R_{dcu}$.

The values of these coefficients for tubes of different thickness are given, as a function of *m*, in the diagrams of Fig. 5.42a, b.

For tubes of magnetic materials, the values of k_x and k_r have been determined by numerical methods, taking into account the variations of permeability from point to point with local magnetic field intensity (and temperature).

Results obtained for tubes of the steel UNI-C30 at room temperature (i.e. at the beginning of heating), are given in Fig. 5.43a, b.

In the diagrams are also shown for comparison the values calculated with relative permeability equal to one. In particular, it is worth to observe that, due to the lower current for heating tubes, the values of the surface magnetic field intensity H_e are much smaller than in solid bars of the same external diameter, resulting in higher values of magnetic permeability and, therefore, higher values of *m*.

5.4.3 AC-DRH of Bars with Rectangular or Square Cross Section

With reference to non-magnetic bars of rectangular cross-section of sides *a* and $b = q \cdot a$ (with $b > a$), the skin effect coefficient $k_r = R_{ac}/R_{dc}$ can be evaluated by the diagram of Fig. 5.44.

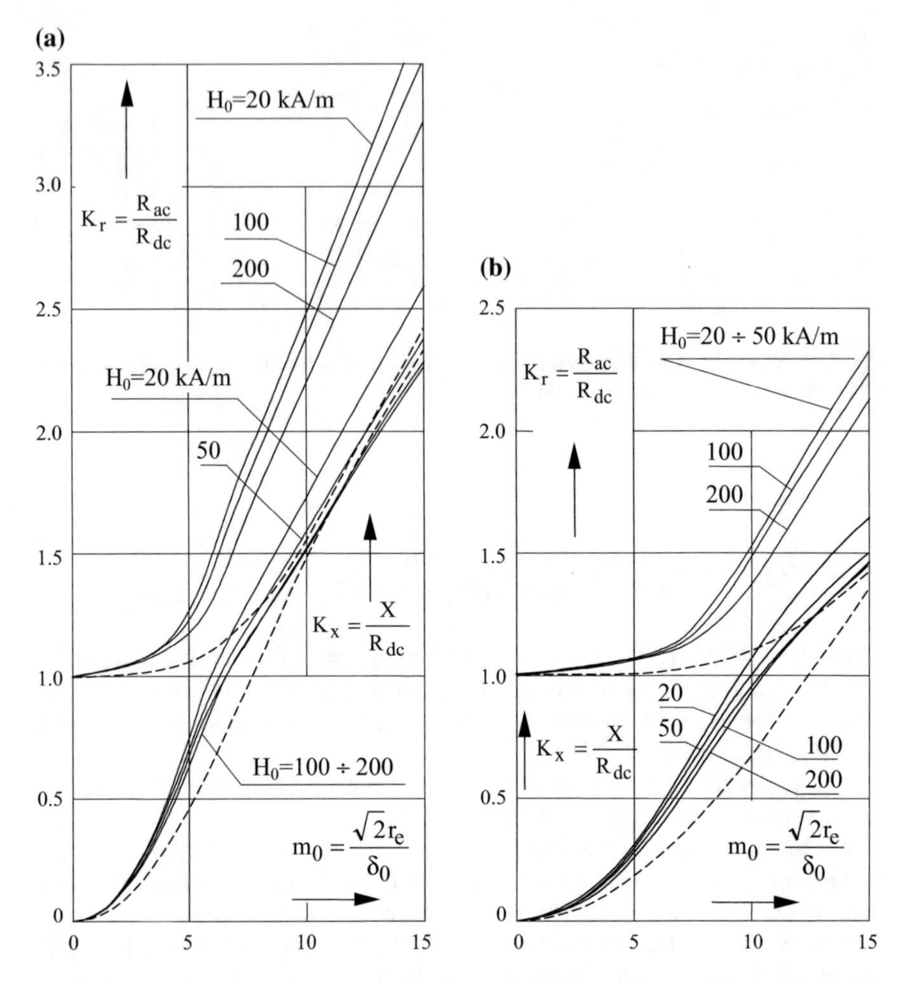

Fig. 5.43 Skin effect coefficients k_r and k_x of steel UNI-C30 tubes at room temperature, as a function of $m_0 = \sqrt{2}r_e/\delta_0$ and different values of surface magnetic field intensity H_e (**a** $\alpha = 0.75$; **b** $\alpha = 0.85$; r_i/r_e;——numerical solution with μ variable; - - - analytical solution with $\mu = 1$; δ_0: penetration depth with μ corresponding to magnetic field intensity H_e) [1]

The curves give the values of k_r for different values of the ratio q between the sides of the cross-section, as a function of the parameter:

$$p = \frac{\sqrt{2}a}{\delta}\sqrt{\frac{q}{\pi}} = \frac{\sqrt{2}r_{eq}}{\delta}$$

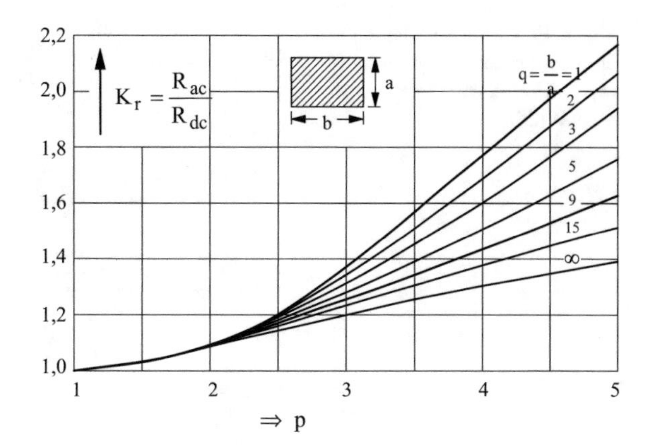

Fig. 5.44 Skin effect coefficient k_r of rectangular conductors as a function of parameter p, for various ratios q between sides of bar cross-section [27]

with:

$r_{eq} = a\sqrt{\frac{q}{\pi}}$ radius of a circle of area equivalent to the area of the cross section of rectangular bar

In particular, the curve for q = 1 give the values for bars with square cross-section.

5.4.4 DRH of Continuous Heaters for Metal Wires and Strips

Some metallurgical processes (such as lamination, tempering, annealing) are made with continuous feeding of the material to be heated ("progressive" heating).

In this case the heated workpieces have generally the form of wires, rods or thin flexible strips of small cross section, where the current density is nearly uniformly distributed.

As shown in Fig. 5.1b, during heating the material passes with constant velocity (v) through the space between contacts, which in this case are constituted by rollers or sliding contacts (Fig. 5.45).

After an initial transient at the start of heating, the temperature distribution and thus the material properties (resistivity, permeability, specific heat) between contacts do not vary in time, so that the total impedance (with AC supply) or the resistance (with DC) remain constant. Then also the supply current (*I*) remains constant during the heating.

Making reference to the average values of material characteristics in the temperature range considered and the symbols and units used in Sect. 5.2.1, neglecting the losses it is:

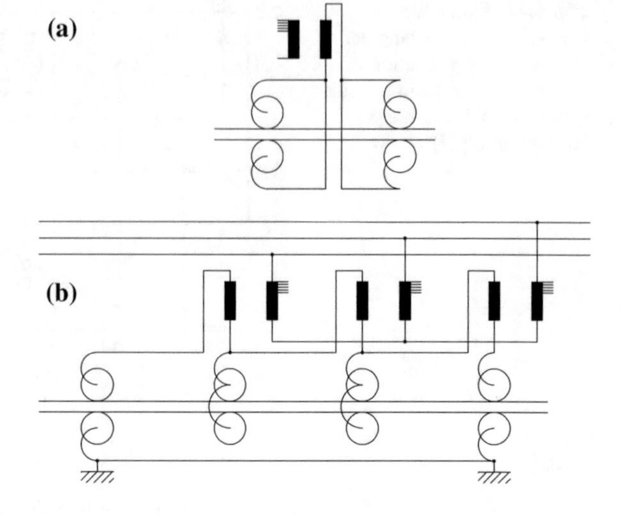

Fig. 5.45 Contact systems in progressive heating (**a** with two rollers; **b** with four rollers) [28]

- Production rate (kg/h)

$$M = 3600\,\gamma S v \tag{5.82}$$

- Power transformed into heat in the workpiece for increasing its temperature of $\Delta\vartheta_m$

$$P = S\,c\,\gamma\,\ell\frac{\Delta\,\vartheta_m}{\Delta t} = S\,c\,\gamma\,v\,\Delta\vartheta_m = \frac{M}{3600}c\,\Delta\vartheta_m \tag{5.83}$$

N.B.—independent of ρ *and* ℓ

- Current flowing through the workpiece:

$$I = \sqrt{\frac{P}{R_{dc}}} = S\sqrt{\frac{c\,\gamma\,v\,\Delta\vartheta_m}{\rho\,\ell}} = \sqrt{\frac{M\,S\,c\,\Delta\vartheta_m}{3600\,\ell\,\rho}} \tag{5.84}$$

- Voltage between contacts:

$$V = R_{dc}I = \sqrt{c\,\gamma\,\rho\,\ell\,v\,\Delta\vartheta_m} = \sqrt{\frac{M\,c\,\rho\,\ell\,\Delta\vartheta_m}{3600\,S}} \tag{5.85}$$

Considering that in order to reduce contact losses—especially in case of sliding contacts—it is convenient to limit the current, from the previous equations it can be seen that the only parameter on which the designer can act in order to reduce the current is the distance ℓ between contacts. Therefore this distance should be taken as high as possible, consistently with other practical requirements, like the need of supporting thin-wires (or strips) moving at high-temperature with high speed.

In practice, since the workpiece to be heated is usually in coils of finite length, at the beginning and the end of each coil there is a period with variable speed, which at constant voltage supply gives rise to variations of the current intensity.

It is therefore necessary to provide an adequate voltage regulation in order to obtain the desired output temperature during these variable speed periods.

References

1. Lupi, S., Nunes, M.F.: Riscaldamento dei metalli mediante conduzione diretta di corrente, 86 p. CLEUP, Padova, Italy (1990) (in Italian)
2. Aliferov, A., Lupi, S.: Direct Resistance Heating of Metals. Novosibirsk State Tecnical University Publishing House, 223 p. ISBN 5-7782-0475-2 (2004) (in Russian)
3. Aliferov, A., Lupi, S.: Induction and Direct Resistance Heating of Metals, 411 p. Novosibirsk State Tecnical University Publishing House. ISBN 978-5-7782-1622-8 (in Russian)
4. Lupi, S., Forzan, M., Aliferov, A.: Induction and Direct Resistance Heating—Theory and Numerical Modeling, 370 p. Springer International Publishing, Switzerland. ISBN 978-3-319-03479-9 (2015)
5. Carslaw, H.S., Jaeger, J.C.: Conduction of Heat in Solids. Clarendon Press, 2nd ed. Oxford (1959)
6. Geisel, H.: Grundlagen der unmittelbaren Widerstandserwärmung langgesttreckter Werksktücke, 116 p. Vulkan-Verlag Dr.W. Classen, Essen (1967) (in German)
7. Hegewald, F.: Induktives Oberflächenhärten, pp. 434–456. BBC Nachrichten, July/August (1961) (in German)
8. Nemkov, V.S., Demidovich, V.B.: Theory and calculation of installations of induction heating, p. 280. Energoatomizdat, Leningrado (1988) (in Russian)
9. Romanov, D.I.: Direct Resistance Heating of Metals, 280 p. Mashinostroenie, Moscow (1981) (in Russian)
10. Lupi, S., Forzan, M., Aliferov, A.: Characteristics of installations for direct resistance heating of ferromagnatic bars of square cross-section. In: Proceedings of International Science Collection "Modelling for Electromagnetic Processing", Hannover (Germany), pp. 43–49. ISBN 978-3-00-026003-2, 27–28 Oct 2008
11. Kovriev, G.S.: Direct Resistance Heating for Hot Working of Non-ferrous Metals, 312 p. Metallurgia (1975) (in Russian)
12. Strunskiĭ, B.M.: Short Circuit Network of Electrical Furnaces, 335 p. Metallurgizdat, Moscow (1962) (in Russian)
13. Mukoseev, Y.P.: AC Current Distribution in Conductors. Gosenergoizdat, Moscow (1959) (in Russian)
14. Neimark, B.E.: Physical Properties of Steels and Alloys Used in the Energy Field, 240 p. Energia, Moscow (1967) (in Russian)
15. Kazanciev, E.A.: Industrial Furnaces: A Reference Manual for Analysis and Design, 312 p. Metallurgia, Moscow (1975) (in Russian)
16. Nemkov, V., Polevodov, B.: Mathematical Modeling on Computer of Devices for RF Heating, 48 p. Mashinstroenie, Leningrad (1980) (in Russian)
17. Nemkov, V., Polevodov, B., Gurevich, S.: Mathematical Modelling of high-frequency heating installations, 60 p. Politecnika, Leningrad (1991) (in Russian)
18. Langman, R.D.: Worked Examples in Electroheat, 180 p. Teaching Monograph, The Electricity Council, Aston/Cambidge (1987)

19. Annen, W.: Le chauffage électrique direct des billettes par résistance, no. 11/12, pp. 667–672. Revue Brown Boveri, Tome 48 (1961) (in French)
20. Trelle Kh.: Eigenschaften und Probleme der konduktiven Erwärmung von Sthal. BBC-Nachrichten, pp. 425–433. Sept. 1967 (in German)
21. Von Jurgens, H.: Widerstandserwärmung im Walzwerksbetrieb. Elektrowärme international, 30 (1972), B1, Februar, B34–B40 (in German)
22. Mühlbauer, A. (Hrsg.): Industrielle Elektrowärme-technik, 400 p. Vulkan-Verlag Essen. ISBN 3-8027-2903-X (1992) (in German)
23. Nacke, B., Baake, E., Lupi, S., Forzan, M., et al.: Theoretical Background and Aspects of Electro-Technologies, 356 p. Course Basic I, St. Petersburg (Russia), Publishing House of ETU. ISBN 978-5-7629-1237-2 (2012)
24. Hegewaldt, F.: Erwärmung im direkten Stromdurchgang. Elektrowärme Int., 34, B4-August, B202–B211 (1976)
25. Kuvaldin, A.B.: Low Temperature Induction Heating of Steel, 111 p. Energija, Moskow (1976) (in Russian)
26. Davies, E.J.: Conduction and Induction Heating, 385 p. Peter Peregrinus Ltd., London (UK) (1990) ISBN 0 86341 174 6
27. Silvester, P.: Skin effect Coefficients for Rectangular Conductors. IEEE Trans. Power App. Syst. **PAS-86**(6), 770–774 (1967)
28. Trelle, K.: Direkte Widerstandserwärmung von Walzknüppeln. Elektrowärme Int. **Bd.26**(1–2), 54–60 (1968) (in German)
29. Jurgens, H.: Widerstanderwärmung im Walzwerks-betrieb. Elektrowärme Int., Bd. 30, B1-Feb., B34–B40 (1972)
30. Lavers, J.D.: An efficient method of calculating parameters for induction and resistance heating installations with magnetic loads. IEEE Trans. Ind. Appl. **IA-14**(5), 427–432 (1978)

Chapter 6
Induction Heating

Abstract Induction heating uses the heat produced by currents induced within a conducting body exposed to the alternating magnetic field produced by AC current flowing in an inductor coil. The main advantages of this process are transmission of electromagnetic energy from the inductor to the workpiece without direct contact, as well as fast and selective heating in defined regions of the workpiece. The first part of the chapter deals with the distributions of induced current and power density within a cylindrical body in longitudinal magnetic field, the equivalent impedance, the electrical efficiency and the quality and power factors of the inductor-load system. In the second paragraph, we will study the transient temperature pattern and the influence of variations of material characteristics during the heating of magnetic or non-magnetic workpieces. The third part deals with the calculation of inductors with approximate, analytical and numerical methods. Numerical examples are included to illustrate the calculation procedures. In the last part of the chapter the main industrial applications of induction heating are presented: in particular induction crucible furnaces, channel induction furnaces, mass through heating prior hot working of metals and heat treatments for induction surface hardening.

6.1 Introduction

In the large majority of cases induction heating is used to heat electrically conducting materials, in most cases metallic materials. Its specific features make it the best solution in many industrial heating processes.

The main factors of success which have contributed to the wide acceptance of this technology are:

- generation of heat sources inside the workpiece to be heated in very short times
- possibility of concentrating them in specified areas of the workpiece, according to the needs of the application
- rational use of electrical energy for heat generation, also in some processes where it substitutes fuel heating

S. Lupi, *Fundamentals of Electroheat*, DOI 10.1007/978-3-319-46015-4_6

- repeatability of the heating characteristics and consequent constant features of final products, with reduction of rejected workpieces
- high production rates, due to the possibility of using high power densities and to realize very short heating times, thus leading to reduction of labor costs
- possibility of achieving specific results which cannot be achieved with other technologies
- rapid thermal start-up of installations
- energy savings
- low shape and size distortion of the heated components
- great reduction of heat losses towards the surrounding environment, which leads to high efficiency and improvement of operators working conditions
- full theoretical understanding of physical phenomena involved in the process and availability of numerical calculation packages, which have led to the design of new high efficiency inductors
- utilization of new solid state components which allowed the development of high power, high efficiency frequency converters in the whole frequency range of interest for industrial applications
- full automation of installations through the use of micro-processors control systems, and possibility of integrating heaters in continuous production lines
- reliability of the installations with reduction of interruptions of production due to malfunctions or maintenance
- safety and flexibility of installations
- improvement of workplace operating environment.

The main fields of application in terms of number of installations and unit installed power are:

- Melting of metals
- Heating of metals prior hot-working operations like forging, extrusion, rolling
- Heat treatments, through and surface hardening
- Welding, soldering and brazing.

Nowadays many other non-traditional, innovative applications are being developed and used in the chemical, glass, textile, food, paper industries as well in advanced metallurgy and steel production.

The laws on which induction heating is based are (see Chaps. 1 and 2):

1. *law of electromagnetic induction (Maxwell equations)*, which states that induced currents (or "*Foucault's currents*") are generated within a conducting body whenever it is exposed to an exciting alternating magnetic field
2. *Joule effect,* under which the induced currents produce power losses directly inside the workpiece to be heated, which are the heat sources used to raise its temperature.
3. *skin, proximity and ring effects,* which produce non-uniform distributions of induced currents and heat sources

4. *laws of heat conduction (Fourier equation) and heat exchange by radiation and convection* from the workpiece surface to the surroundings, which determine the increase and distribution of temperature within the workpiece.

Peculiar characteristics of induction heating are:

- transmission of the electromagnetic energy from the induction coil, which creates the exciting magnetic field, to the workpiece, without direct contact;
- distribution of induced currents and, in turn, of heat sources inside the workpiece, which decrease from the surface towards the interior of the body;
- possibility of concentrating heat sources into well-defined regions of the workpiece or distributing them over the whole body cross-section, according to the requirements of the thermal process, by means of an adequate selection of frequency, power values and coil geometry;
- direct heat generation within the body to be heated which allows to achieve controlled heating transients and much shorter heating times in comparison with surface heating processes, like those in fuel or resistance furnaces. These results are obtained through the control of the distribution of the internal heat sources as well as the use of high specific power.

The exciting magnetic field is produced by the flow of current of suitable intensity and frequency in the exciting coil (or "*inductor*"). The inductor, mostly made with water-cooled copper tube, is usually placed in front of the workpiece to be heated ("*load*") or encircles it, as in Fig. 6.1.

According to the workpiece geometry and the required heating process, many other types of inductors may be used (see Fig. 6.2).

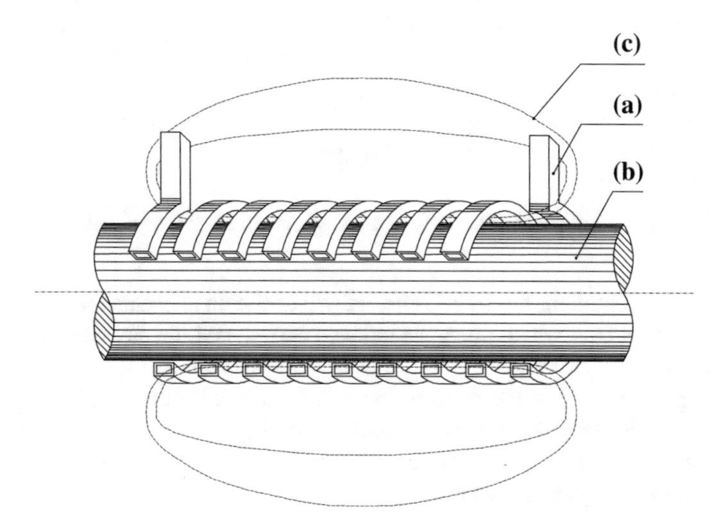

Fig. 6.1 Typical arrangement of induction coil (**a**) and load (**b**) and distribution of magnetic field lines (**c**)

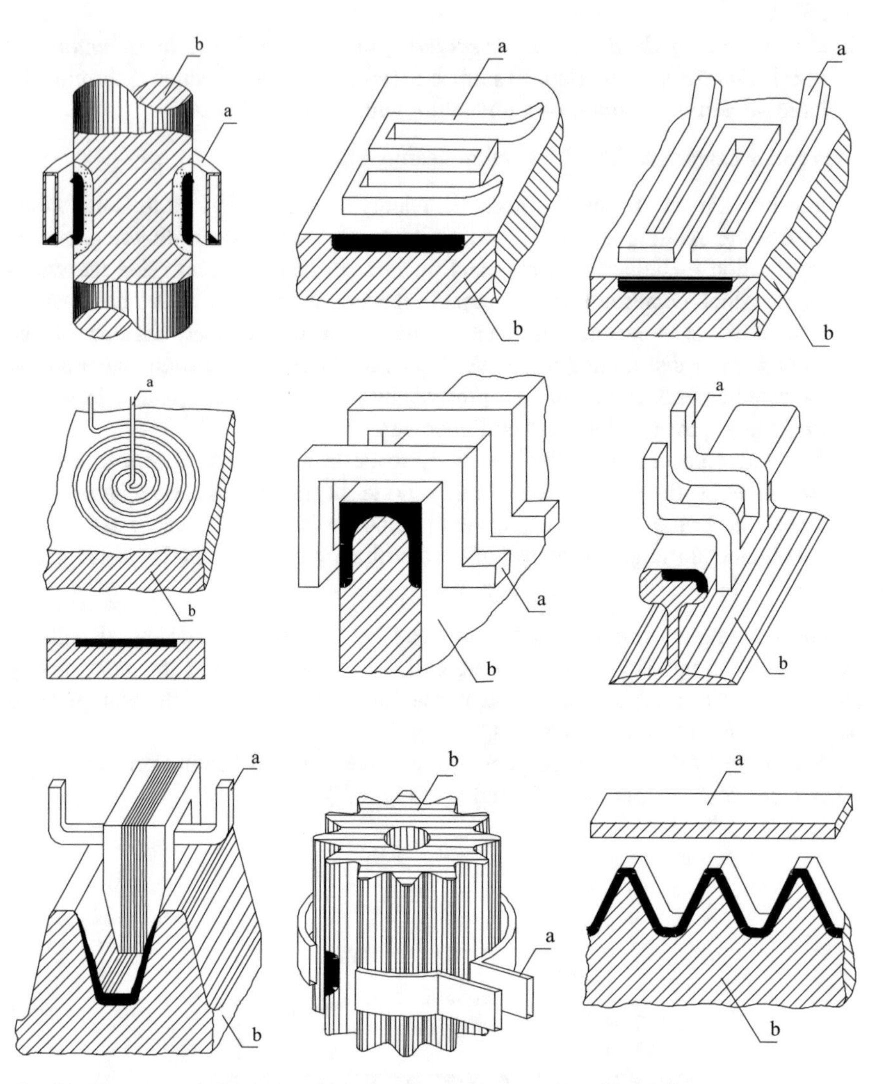

Fig. 6.2 Examples of different induction coil-load systems (**a** load; **b** workpiece)

In fact, any conductor fed by AC current will excite an alternate magnetic field in its surrounding space, which produces induced currents within any conducting body placed in its close proximity.

Keeping in mind that the distribution of the induced currents depends on the frequency of the exciting magnetic field, according to the requirements of the application, the induction coil is fed with current either at industrial frequency (50/60 Hz) or at higher frequency.

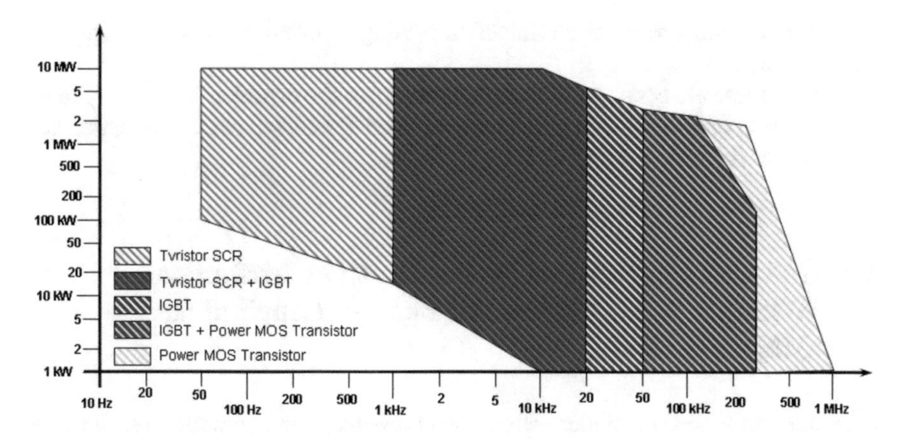

Fig. 6.3 Frequency and power ratings of commercial solid state generators

Nowadays, the required frequencies are generated by solid state converters using thyristors, IGBT, MOSFET transistors. Frequency and power range of presently available solid state generators are shown in Fig. 6.3.

Other types of frequency converters such as static magnetic frequency multipliers (for frequencies 0.15, 0.25, 0.45 kHz) or rotating machine converters (for frequencies from 1 to 10 kHz), which were used in the past, are no longer used.

Also vacuum tube oscillators, which were the only RF induction heating power source prior to 1985, have been progressively replaced in most applications by solid state units for their greater efficiency, safety, smaller size and lower cost.

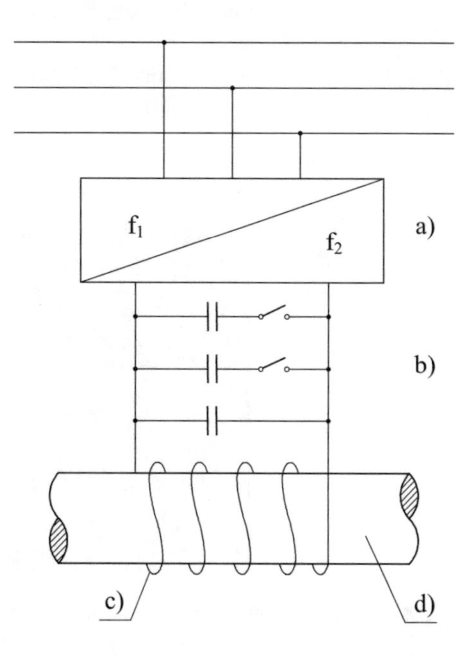

Fig. 6.4 Main elements of an induction heating installation (**a** frequency converter; **b** capacitor bank; **c** inductor coil; **d** body to be heated)

The basic components of an induction heating installation are shown schematically in Fig. 6.4.

Among these elements, particular importance has the capacitor bank required for compensation of power factor of the inductor-load system which, in most cases, can be very low (in the range 0.1–0.3).

6.2 Induced Current and Power Density Distributions Within a Solid, Cylindrical Body in Longitudinal Magnetic Field

Consider a solid metal cylinder with constant resistivity and permeability, shown in Fig. 6.5, which is encircled by an induction circular coil fed with sinusoidal current.

With reference to a cylindrical coordinate system (r, φ, z), the following notations are used:

r_e, r_i	external radius of the cylinder and inner radius of the inductor coil (m)
$\xi = \frac{r}{r_e}; \alpha = \frac{r_i}{r_e}$	dimensionless parameters
ℓ	axial length of induction coil and cylinder (m)
N	number of turns of induction coil
I	r.m.s. value of current flowing in the coil (A)
$\dot{\bar{H}}_e, \dot{\bar{H}}$	vectors of magnetic field intensity in the gap between inductor and load and inside the body (A/m)
$\dot{\bar{E}}_e, \dot{\bar{E}}$	vectors of electric field intensity at the surface and inside the body (V/m)
$\omega = 2\pi f$	angular frequency of the current I (s^{-1})
f	frequency (Hz)

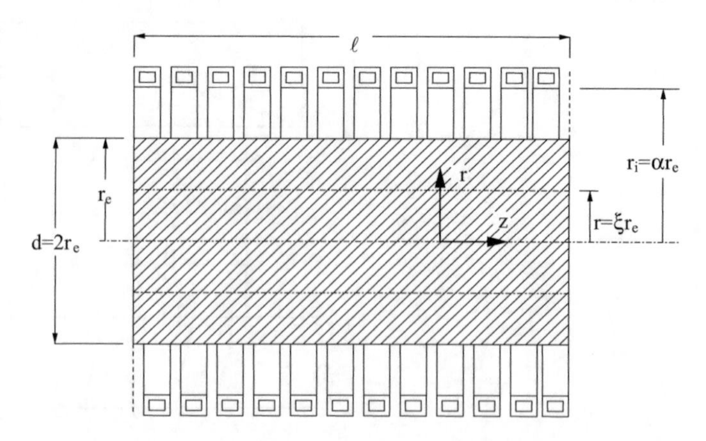

Fig. 6.5 Cylindrical inductor-load system

ρ, μ resistivity (Ω m) and relative magnetic permeability of the workpiece material
$\mu_0 = 4\pi 10^{-7}$ permeability of vacuum (H/m)

Under the assumption that the section of length ℓ of the inductor-load system is a portion of a cylindrical configuration of infinite axial length, the magnetic field between inductor and cylinder is directed along z and its intensity is equal to the one produced by the empty coil without load:

$$\dot{H}_e = \frac{N\dot{I}}{\ell} \tag{6.1}$$

Neglecting displacement currents (given the range of frequencies normally used) and under the assumption of sinusoidal wave forms, using complex notations, Maxwell equations can be written as follows:

$$\left.\begin{aligned} \text{rot}\,\bar{\dot{H}} &= \frac{\bar{\dot{E}}}{\rho} \\ \text{rot}\,\bar{\dot{E}} &= -j\omega\mu\mu_0\bar{\dot{H}} \end{aligned}\right\} \tag{6.2}$$

with: $j = \sqrt{-1}$.

Since for the considered inductor-load geometry, in cylindrical coordinate system are non-zero only the components $\dot{H}_z$ of $\bar{\dot{H}}$ and $\dot{E}_\varphi$ of $\bar{\dot{E}}$, Eq. (6.2) can be written as follows:

$$\left.\begin{aligned} \text{rot}_\varphi\bar{\dot{H}} &= -\frac{\partial \dot{H}_z}{\partial r} = \frac{\dot{E}_\varphi}{\rho} \\ \text{rot}_z\bar{\dot{E}} &= \frac{\partial \dot{E}_\varphi}{\partial r} + \frac{\dot{E}_\varphi}{r} = -j\omega\mu_0\mu\dot{H} \end{aligned}\right\} \tag{6.3}$$

Hence, neglecting (hereafter) the indexes φ and z, it results:

$$\dot{E} = -\rho\frac{\partial \dot{H}}{\partial r} = -\frac{\rho}{r_e}\cdot\frac{\partial \dot{H}}{\partial \xi} \tag{6.4}$$

and

$$\frac{\partial^2 \dot{H}}{\partial r^2} + \frac{1}{r}\frac{\partial \dot{H}}{\partial r} - j\frac{\omega\mu_0\mu\dot{H}}{\rho} = 0 \tag{6.5a}$$

or:

$$\frac{\partial^2 \dot{H}}{\partial \xi^2} + \frac{1}{\xi}\cdot\frac{\partial \dot{H}}{\partial \xi} + \beta^2\dot{H} = 0 \tag{6.5b}$$

with:

$$\beta^2 = -j\frac{\omega\mu_0\mu}{\rho}r_e^2 = -j\,m^2; \quad m = \frac{\sqrt{2}r_e}{\delta}; \quad \xi = \frac{r}{r_e};$$

$\delta = \sqrt{\frac{2\rho}{\omega\mu_0\mu}} \approx 503\sqrt{\frac{\rho}{\mu f}}$—penetration depth (m).

This equation is the Bessel differential equation of order zero whose general solution can be written in the form:

$$\dot{H} = \dot{C}_1 \cdot J_0\left(\sqrt{-j}m\xi\right) + \dot{C}_2 \cdot Y_0\left(\sqrt{-j}m\xi\right), \tag{6.6}$$

with: $J_0(u)$ and $Y_0(u)$—Bessel functions of order zero, of first and second kind respectively, of complex argument *u*; $\dot{C}_1, \dot{C}_2$—integration constants.

The constants $\dot{C}_1$ and $\dot{C}_2$ can be determined with the boundary conditions:

$$\dot{H} \neq \infty \quad \text{for} \quad \xi = 0; \quad \dot{H} = \dot{H}_e \quad \text{for} \quad \xi = 1.$$

From the first condition, being $Y_0(0) = -\infty$, it results $\dot{C}_2 = 0$, from the second one we have $\dot{C}_1 = \dot{H}_e / J_0(\sqrt{-j}m)$.

Substituting these values in Eq. (6.6), for the modulus of vector of magnetic field intensity we have:

$$\dot{H} = \dot{H}_e \frac{J_0(\sqrt{-j}m\xi)}{J_0(\sqrt{-j}m)} \tag{6.7}$$

It can be useful sometimes to calculate separately real and imaginary parts of the function $J_0(\sqrt{-j}u)$, using the relationship:

$$J_0(\sqrt{-j}u) = \mathrm{ber}(u) + j\,\mathrm{bei}(u)$$

Equation (6.7) then becomes:

$$\left.\begin{aligned} \dot{H} &= \dot{H}_e \frac{\mathrm{ber}(m\xi) + j\,\mathrm{bei}(m\xi)}{\mathrm{ber}(m) + j\,\mathrm{bei}(m)} \\ \left|\frac{\dot{H}}{\dot{H}_e}\right| &= \sqrt{\frac{\mathrm{ber}^2(m\xi) + \mathrm{bei}^2(m\xi)}{\mathrm{ber}^2(m) + \mathrm{bei}^2(m)}} \end{aligned}\right\} \tag{6.7a}$$

Equation (6.7) show that the moduli of the complex vector of magnetic field intensity $\dot{H}$ depend not only on the dimensionless radius $\xi = r/r_e$, but also on $m = \sqrt{2}r_e/\delta$.

They also point out that moving from the surface to the axis of the cylinder (i.e. varying ξ) both the modulus and the phase angle of the field intensity change in comparison to the values at the surface.

Table 6.1 Values of H/H_e as a function of ξ for different values of m

m\ξ	0	0.2	0.4	0.6	0.8	0.9	0.95	1.00
1	0.984	0.984	0.984	0.986	0.992	0.996	0.998	1.00
2.5	0.662	0.663	0.672	0.713	0.813	0.895	0.944	1.00
3.5	0.388	0.390	0.411	0.495	0.681	0.821	0.906	1.00
5	0.160	0.163	0.197	0.313	0.550	0.741	0.870	1.00
7.5	0.0378	0.0363	0.0658	0.156	0.388	0.621	0.787	1.00
10	0.0066	0.0082	0.0228	0.0764	0.272	0.519	0.721	1.00
15	–	0.0004	0.0027	0.0186	0.134	0.363	0.604	1.00
20	–	–	–	0.0045	0.0661	0.256	0.503	1.00

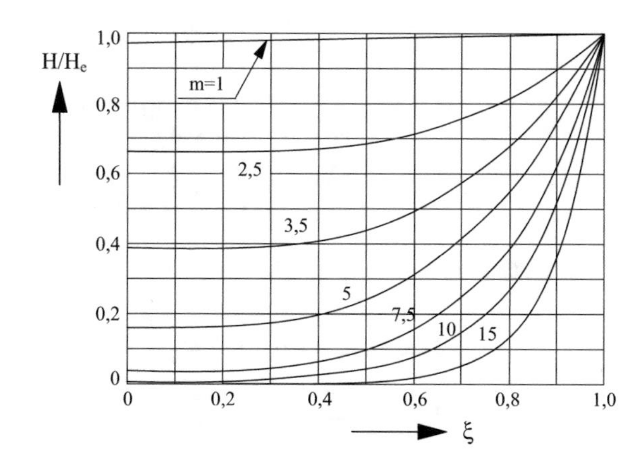

Fig. 6.6 Distributions of magnetic field intensity H along radius, referred to the value H_e at the surface ($\xi = r/r_e$; $m = \sqrt{2}\, r_e/\delta$)

Values and distributions of $\left|\dot{H}/\dot{H}_e\right|$ as a function of ξ are given in Table 6.1 and Fig. 6.6 for different values of *m*.

The curves show that the induced currents tend to expel the magnetic field from the core of the cylinder and to concentrate it in the surface layers. This phenomenon is more pronounced the higher is the value of *m*.

The moduli of complex vectors of electric field intensity $\dot{E}$ and induced current density $\dot{J}$ inside the cylinder can be derived from Eqs. (6.4) and (6.7).

Remembering that is $J_0'(u) = -J_1(u)$, we obtain:

$$\dot{E} = \sqrt{-j} \cdot \dot{H}_e \cdot \frac{\rho}{\delta} \cdot \sqrt{2} \cdot \frac{J_1(\sqrt{-j}m\xi)}{J_0(\sqrt{-j}m)}, \tag{6.8}$$

$$\dot{J} = \frac{\dot{E}}{\rho} = \sqrt{-j} \cdot \dot{H}_e \cdot \frac{\sqrt{2}}{\delta} \cdot \frac{J_1(\sqrt{-j}m\xi)}{J_0(\sqrt{-j}m)}. \tag{6.9}$$

Having denoted with $\dot{E}_e$ and $\dot{J}_e$ the values of $\dot{E}$ and $\dot{J}$ at the surface of the cylinder ($\xi = 1$), the following relationship holds for their ratios:

$$\frac{\dot{E}}{\dot{E}_e} = \frac{\dot{J}}{\dot{J}_e} = \frac{J_1(\sqrt{-j}m\xi)}{J_1(\sqrt{-j}m)} \tag{6.10}$$

Separating real and imaginary parts of the function $J_1(\sqrt{-j}u)$, by the relationship:

$$-\sqrt{-j}J_1(\sqrt{-j}u) = ber'(u) + j\,bei'(u),$$

the current density distributions can be conveniently calculated by the following relationships, obtained from Eqs. (6.9) and (6.10):

$$\dot{J} = \frac{\dot{E}}{\rho} = -\frac{\dot{H}_e}{r_e}\,m\,\frac{ber'm\xi + j\,bei'm\xi}{ber\ m + j bei\ m} \tag{6.9a}$$

$$\left.\begin{aligned} \frac{\dot{E}}{\dot{E}_e} &= \frac{\dot{J}}{\dot{J}_e} = \frac{ber'm\xi + j\,bei'm\xi}{ber'm + j\,bei'm}\rho \\ \left|\frac{\dot{E}}{\dot{E}_e}\right| &= \left|\frac{\dot{J}}{\dot{J}_e}\right| = \sqrt{\frac{ber'^2m\xi + bei'^2m\xi}{ber'^2m + bei'^2m}} \end{aligned}\right\} \tag{6.10a}$$

In particular, from Eqs. (6.9) and (6.9a) it is:

$$\left.\begin{aligned} \dot{J}_e &= \frac{\dot{J}_e}{\rho} = \sqrt{-j}\dot{H}_e\frac{\sqrt{2}}{\delta}\frac{J_1(\sqrt{-j}m)}{J_0(\sqrt{-j}m)} = -\frac{\dot{H}_e}{r_e}m(P + jQ) \\ |\dot{J}_e| &= \frac{|\dot{H}_e|}{r_e}m\sqrt{P^2 + Q^2} \end{aligned}\right\} \tag{6.11}$$

with:

$$\begin{aligned} P + jQ &= -\sqrt{-j}\frac{J_1(\sqrt{-j}m)}{J_0(\sqrt{-j}m)} = \frac{ber'm + j\,bei'm}{ber\ m + j\,beim} \\ &= \frac{ber\ m\,ber'm + bei\ m\,bei'm}{ber^2m + bei^2m} + j\frac{ber\ m\,bei'm - bei\ m\,ber'm}{ber^2m + bei^2m} \end{aligned} \tag{6.12}$$

The values of coefficients *P* and *Q* are given, as a function of *m,* in Table 6.2 and the curves of Fig. 6.7.

From the previous relationships the following conclusions can be drawn:

- the induced currents have different phase angles along the radius of the workpiece and their amplitudes decrease from the maximum value at the surface towards the core (see Table 6.3 and Fig. 6.8);
- the current density is always zero on the axis of the cylinder and, as *m* increases, it concentrates in a layer thinner and thinner below the surface;
- if r_e and H_e are kept constant, the absolute value of J_e will increase with higher values of *m*. In fact, in Eq. (6.11) the term $(m\sqrt{P^2 + Q^2})$ has, as a function of *m*, the values given in Table 6.4.

Table 6.2 Values of coefficients P and Q as a function of m

m	P	Q	m	P	Q	m	P	Q	m	P	Q
			2.1	0.3748	0.7799	**4.2**	0.5885	0.7122	**7.8**	0.6417	0.7088
0.2	0.0005	0.1000	**2.2**	0.4026	0.7827	**4.4**	0.5925	0.7110	**8.0**	0.6433	0.7087
0.3	0.0017	0.1500	**2.3**	0.4283	0.7829	**4.6**	0.5964	0.7104			
0.4	0.0040	0.1999	**2.4**	0.4513	0.7808	**4.8**	0.6006	0.7102	**8.2**	0.6450	0.7086
0.5	0.0078	0.2497	**2.5**	0.4718	0.77700	**5.0**	0.6040	0.7102	**8.4**	0.6464	0.7085
0.6	0.0135	0.2992	**2.6**	0.4898	0.7721				**8.6**	0.6478	0.7085
0.7	0.0213	0.3482	**2.7**	0.5019	0.7666	**5.2**	0.6077	0.7102	**8.8**	0.6492	0.7084
0.8	0.0316	0.3966	**2.8**	0.5188	0.7605	**5.4**	0.6113	0.7102	**9.0**	0.65.5	0.7084
0.9	0.0447	0.4440	**2.9**	0.5303	0.7542	**5.6**	0.6147	0.7102			
1.0	0.0608	0.4899	**3.0**	0.5400	0.7486	**5.8**	0.6181	0.7102	**9.2**	0.6518	0.7083
						6.0	0.6211	0.7101	**9.4**	0.6529	0.7083
1.1	0.0798	0.5339	**3.1**	0.5481	0.7429				**9.6**	0.6541	0.7082
1.2	0.1020	0.5756	**3.2**	0.5550	0.7379	**6.2**	0.6239	0.7100	**9.8**	0.6552	0.7081
1.3	0.1269	0.6144	**3.3**	0.5608	0.7296	**6.4**	0.6268	0.7099			
1.4	0.1546	0.6497	**3.4**	0.5656	0.7287	**6.6**	0.6293	0.7097	**10.0**	0.6563	0.7081
1.5	0.1844	0.6812	**3.5**	0.5699	0.7251	**6.8**	0.6317	0.7096	**12.0**	0.6648	0.7078
1.6	0.2158	0.7085	**3.6**	0.5733	0.7221	**7.0**	0.6339	0.7095	**15.0**	0.6734	0.7075
1.7	0.2483	0.7315	**3.7**	0.5764	0.7197				**20.0**	0.6819	0.7073
1.8	0.2810	0.7498	**3.8**	0.5792	0.7175	**7.2**	0.6360	0.7093			
1.9	0.3135	0.7638	**3.9**	0.5818	0.7155	**7.4**	0.6380	0.7091			
2.0	0.3449	0.7738	**4.0**	0.5843	0.7143	**7.6**	0.6398	0.7090			

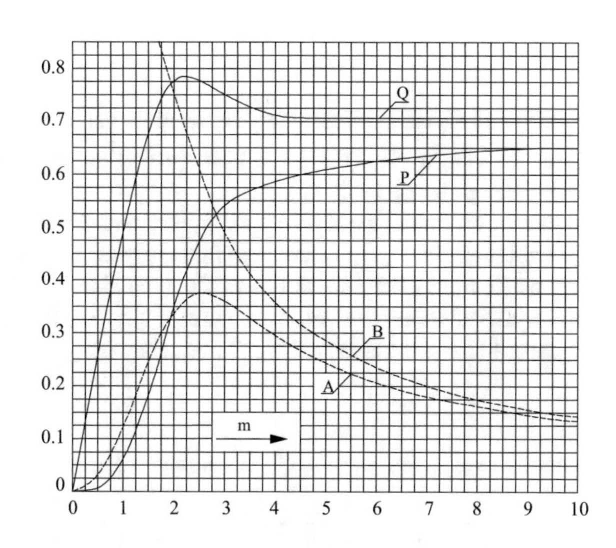

Fig. 6.7 Values of coefficients P and Q for calculation of the induced current density J_e and the active and reactive power as a function of m ($A = 2P/m; B = 2Q/m$)

Table 6.3 Values along radius of ratios J/J_e, for different values of m

m\ξ	0	0.2	0.4	0.6	0.8	0.9	0.95	1.0
1	0	0.2	0.398	0.599	0.802	0.899	0.950	1.00
2.5	0	0.182	0.365	0.553	0.758	0.873	0.934	1.00
3.5	0	0.147	0.298	0.464	0.680	0.822	0.906	1.00
5	0	0.086	0.179	0.310	0.561	0.744	0.857	1.00
7.5	0	0.027	0.064	0.151	0.384	0.651	0.825	1.00
10	0	0.0071	0.0276	0.075	0.269	0.473	0.719	1.00
15	0	–	–	0.0186	0.134	0.363	0.603	1.00
20	0	–	–	0.0045	0.0661	0.256	0.503	1.00

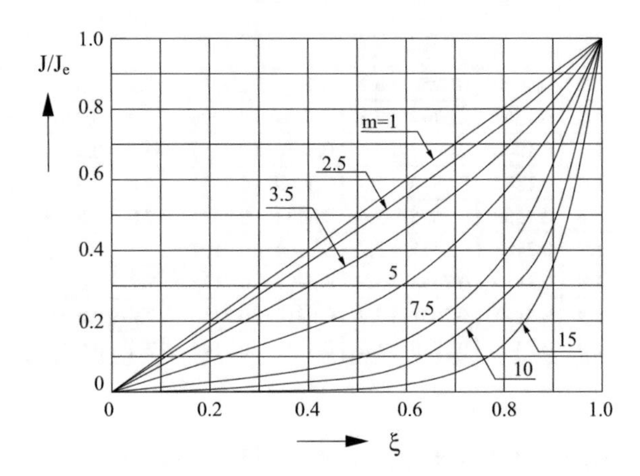

Fig. 6.8 Relative distributions of current density J along radius, for different values of m, referred to the surface values J_e

- if r_e and H_e are kept constant, the absolute value of J_e will increase by increasing *m*. In fact, in Eq. (6.11) the term $(m\sqrt{P^2+Q^2})$ has, as a function of *m,* the values given in Table 6.4.
- it follows that in reality, in a workpiece excited with the same surface magnetic field intensity H_e, the change of frequency will produce different current density distributions along the radius of the type shown in Fig. 6.9.
- in particular, at relatively "low" frequency (i.e. for $m \leq 1$), the distributions are nearly linear, whereas at "high" frequency ($m \geq 12$) they tend to an exponential.

Table 6.4 Values of $(m\ \sqrt{P^2+Q^2})$ as a function of m

m	1	2.5	3.5	5	7.5	10	15	20
$m\sqrt{P^2+Q^2}$	0.494	2.273	3.228	4.661	7.158	9.655	14.651	19.650
$m^2\,(P^2+Q^2)$	0.244	5.165	10.42	21.73	51.24	93.21	214.7	386.1

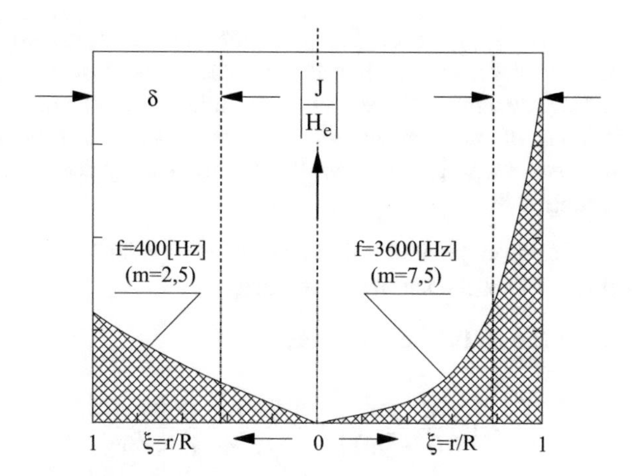

Fig. 6.9 Induced current density distributions in a cylinder heated at different frequencies with the same surface magnetic field intensity ($r_e = 10\,\text{mm}$; $\rho = 5 \cdot 10^{-8}\,\Omega\,\text{m}$; $\mu = 1$)

In fact, using the approximations of Bessel functions we have:

- for values of $m \leq 1$,

$$\text{ber}^2 u + \text{bei}^2 u \approx 1; \quad \text{ber}'^2 u + \text{bei}'^2 u \approx u^2/4$$

and Eqs. (6.7a) and (6.10a) give:

$$\left|\frac{\dot{H}}{\dot{H}_e}\right| \approx 1; \quad \left|\frac{\dot{J}}{\dot{J}_e}\right| \approx \xi \tag{6.13a}$$

- for "high" values of m,

$$\text{ber}^2 u + \text{bei}^2 u \approx \text{ber}'^2 u + \text{bei}'^2 u \approx e^{\sqrt{2}u}/(2\pi u)$$

and from the same equations we obtain:

$$\left|\frac{\dot{H}}{\dot{H}_e}\right| \approx \left|\frac{\dot{J}}{\dot{J}_e}\right| \approx \frac{e^{-m(1-\xi)/\sqrt{2}}}{\sqrt{\xi}} \approx \frac{e^{-x/\delta}}{\sqrt{1-(x/r_e)}} \tag{6.13b}$$

with: $x = r_e - r$—distance from the surface.

In conclusion, the only parameters on which the designer can act in order to modify the values and distributions of the induced current density are the intensity of the exciting magnetic field H_e and the frequency *f*.

However, these parameters cannot be chosen arbitrarily because numerous other factors must be taken into account, such as the current density distribution most convenient to the process, the efficiency of the energy transmission from the inductor to the workpiece, the cooling of the inductor the cost of frequency converter, and the electromagnetic forces acting on inductor and load.

The distributions of the *induced specific power per unit volume* along the workpiece radius are easily derived from the current density distributions.

In fact, denoting with:

$w = \rho J^2$ specific power per unit volume at radius r (W/m^3)
$w_e = \rho J_e^2$ values of w at the surface of the cylinder,

from Eqs. (6.9) and (6.10) we can write:

$$\frac{w}{w_e} = \left|\frac{J_1(\sqrt{-j}m\xi)}{J_1(\sqrt{-j}m)}\right|^2 = \frac{\text{ber}'^2\, m\xi + \text{bei}'^2 m\xi}{\text{ber}'^2 m + \text{bei}'^2 m} \tag{6.14}$$

$$w_e = \rho\left(\frac{H_e^2}{r_e^2}\right) \cdot m^2 \cdot (P^2 + Q^2) \tag{6.14a}$$

From the above equations the diagrams of Fig. 6.10 can be drawn, which show that the specific power is always unevenly distributed inside the workpiece, with a steep slope from the surface to the axis.

Moreover, w_e strongly depends on m, as shown by the values of the function $[m^2(P^2 + Q^2)]$ given in Table 6.4.

In conclusion, induction heating for its inherent characteristics gives always rise to non-uniform heating patterns; in particular, increasing the frequency, for high values of m, the heating concentrates near the surface.

In this case, Eq. (6.14) can be approximated as follows:

$$\frac{w}{w_e} \approx \frac{e^{-2x/\delta}}{1 - (x/r_e)} \tag{6.15}$$

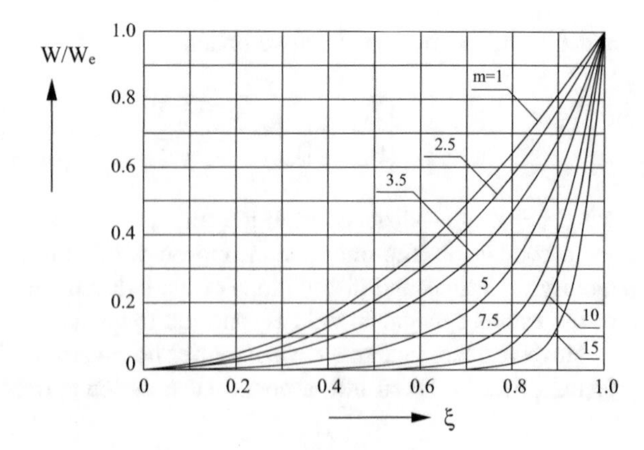

Fig. 6.10 Relative distribution of specific power per unit volume w along workpiece radius, referred to the surface value w_e, for different values of m

6.3 Power Induced in a Cylindrical Workpiece and Equivalent Impedance of the Inductor-Load System

The active and reactive power induced in the workpiece by the electromagnetic field can be determined as the real and imaginary parts of the outward flux of the complex Poynting vector through the external surface of the cylinder (of radius r_e and length ℓ)

$$\dot{\bar{S}} = \dot{\bar{E}} \cdot \dot{\bar{H}}^* \tag{6.16}$$

with $\dot{\bar{H}}^*$—complex conjugate of vector $\dot{\bar{H}}$.

Assuming as reference the phase of the magnetic field $\dot{H}_e$ and taking into account that $\dot{E}_e$ and $\dot{H}_e$ are perpendicular to each other, denoting with:

P_w, Q_w—induced active power (W) and reactive power (VAR) in the volume of the workpiece,

it results:

$$\dot{S}_e = P_w + jQ_w = -\dot{E}_e \cdot \dot{H}_e \cdot 2\pi r_e \ell. \tag{6.16a}$$

Introducing in Eq. (6.16a) the expression (6.8) evaluated for $\xi = 1$, we obtain

$$\begin{aligned} P_w + jQ_w &= H_e^2 \frac{\rho}{\delta} \sqrt{2}(P + jQ) 2\pi r_e \ell \\ &= \frac{X_{i0}}{\alpha^2} \mu (A + jB) I^2 \end{aligned} \tag{6.17}$$

with:

$P + jQ$	coefficients of active and reactive power in the cylinder given by Eq. (6.12), diagrams of Fig. 6.7 and Table 6.2,
$(A + jB) = \frac{2}{m}(P + jQ)$	a different form of coefficients of active and reactive power often used in the literature,
$X_{i0} = \omega N^2 \frac{\mu_0 \pi r_e^2 \alpha^2}{\ell}$	reactance (Ω) of empty inductor coil, considered as a portion of length ℓ of an infinite length system

The values of *A* and *B* are given as a function of *m* by the curves of Fig. 6.7. Sometimes are useful the following approximations of coefficients P, Q and A, B:

m	Active power	Reactive power	Error (%)
≤ 1	$A \approx m^2/8$	$B \approx 1$	≈1
≥ 4	$P \approx \frac{1}{\sqrt{2}} - \frac{1}{2m}$	$Q \approx \frac{1}{\sqrt{2}}$	≈4
≥ 15	$P \approx \frac{1}{\sqrt{2}}$	$Q \approx \frac{1}{\sqrt{2}}$	≈5

A parameter of fundamental importance in the design of the inductor is the *equivalent impedance of the inductor-load system.*

Under the assumption of an *ideal inductor coil with negligible own resistance*, constituted by a current layer of negligible radial thickness placed at radius r_i, for the evaluation of the total impedance it must be taken into account the reactive power Q_a in the air gap between cylinder and coil, where the magnetic field intensity $\dot{H}_e$ is constant.

By using Eq. (6.1), we can write:

$$Q_a = H_e^2 \cdot \omega\mu_0\pi(r_i^2 - r_e^2) \cdot \ell = \frac{X_{i0}}{\alpha^2}(\alpha^2 - 1) \cdot I^2 \tag{6.18}$$

By adding the expressions (6.17) and (6.18), *the total active and reactive power* of the ideal inductor-load system is:

$$P_w + j(Q_a + Q_w) = \frac{X_{i0}}{\alpha^2}\{\mu A + j[\alpha^2 - (1 - \mu B)]\} \cdot I^2 \tag{6.19}$$

The equivalent impedance $\dot{Z}_{e0}$ "seen" at the terminals of the ideal inductor can be therefore written as follows:

$$\dot{Z}_{e0} = R'_w + j(X_a + X'_w) = R'_w + j(X_{i0} - \Delta X) \tag{6.20}$$

with:

$R'_w = \frac{X_{i0}}{\alpha^2}\mu A = N^2\rho\frac{2\pi r_e}{\ell\delta}\sqrt{2}P$	resistance of the load at the inductor clamps,
$X'_w = \frac{X_{i0}}{\alpha^2}\mu B$	reactance of the load at the inductor clamps
$X_a = \frac{X_{i0}}{\alpha^2}(\alpha^2 - 1)$	reactance corresponding to the reactive power Q_a,
X_{i0}	reactance of inductor without load (see Eq. 6.17),
$\Delta X = \frac{X_{i0}}{\alpha^2}(1 - \mu B)$	variation of inductor reactance from loaded to unloaded conditions.

6.3.1 *Specific Surface Power Density*

From Eq. (6.17) and the definition of δ it follows that increasing the frequency at constant exciting magnetic field intensity H_e, the active power induced in the load increases, while at the same time the thickness of the surface layer in which this power concentrates becomes lower.

Hence, the use of high frequencies allows to induce high specific power densities, like the ones needed, for example, in surface induction hardening.

Thus, considering that increasing the frequency the heating will becomes more and more superficial, instead of referring to the volume power density *w* (W/m^3), frequently reference is made to an equivalent power density p_e (W/m^2), "induced" on the external surface of the cylinder.

On the basis of Eq. (6.17), it results:

$$p_e = H_e^2 \frac{\rho}{\delta} \sqrt{2} P \ (W/m^2). \tag{6.21}$$

Example 6.1 Heating of a non-magnetic steel cylinder (diameter $d_e = 2r_e =$ 40 mm, axial length $\ell = 700$ mm, resistivity $\rho = 1 \cdot 10^{-6}\ \Omega$ m) with an inductor of the same length and diameter $d_i = 2r_i = 70$ mm. The inductor has N = 32 turns and is excited by a current I = 1000 A at frequency f = 2 kHz.

Note: *End effects and own resistance of the inductor are neglected.*

- intensity of magnetic field between inductor and load

$$H_e = \frac{NI}{\ell} = \frac{32 \cdot 1000}{0.7} = 45.7\,kA/m = 457\,A/cm$$

- penetration depth δ, value of *m* and parameters dependent on *m*

$$\delta = 503\sqrt{\frac{\rho}{\mu f}} = 503\sqrt{\frac{1 \cdot 10^6}{2000}} = 1.125 \cdot 10^{-2}\,m = 1.125\ cm$$

$$m = \frac{\sqrt{2} r_e}{\delta} = \frac{\sqrt{2}20 \cdot 10^{-3}}{1.125 \cdot 10^{-2}} = 2.51 \approx 2.5$$

$$P + jQ = 0.4718 + j\,0.7770;\ A + jB = 0.3774 + j0.6216$$

$$m\sqrt{P^2 + Q^2} = 2.273;\ m^2(P^2 + Q^2) = 5.167$$

- specific surface values

$$J_e = \frac{H_e}{r_e} m\sqrt{P^2 + Q^2} = \frac{45{,}700}{20 \cdot 10^{-3}} 2.273 = 5194\ kA/m^2 = 519.4\,A/cm^2$$

$$E_e = \rho J_e = 1 \cdot 10^{-6} \cdot 5.194 \cdot 10^6 = 5.194\,V/m$$

$$w_e = \rho J_e^2 = 1 \cdot 10^{-6} \cdot \left(5.194 \cdot 10^6\right)^2 = 26.98 \cdot 10^6\,W/m^3 \approx 27.0\,W/cm^3$$

- reactance X_{i0}

$$\omega = 2\pi f = 1.257 \cdot 10^4; \quad \alpha = r_i/r_e = 1.75$$

$$X_{i0} = \omega N^2 \frac{\mu_0 \pi r_e^2 \alpha^2}{\ell} = 1.257 \cdot 10^4 \cdot 32^2 \frac{4\pi \cdot 10^{-7} \pi (20 \cdot 10^{-3} 1.75)^2}{700 \cdot 10^{-3}}$$
$$= 8.843 \cdot 10^{-2}\ \Omega$$

- equivalent impedance Z_{e0}

$$R'_w = \frac{X_{i0}}{\alpha^2} \mu A = \frac{8.843 \cdot 10^{-2}}{1.75^2} 0.3774 = 1.090 \cdot 10^{-2}\ \Omega$$

$$\Delta X = \frac{X_{i0}}{\alpha^2}(1 - \mu B) = \frac{8.843 \cdot 10^{-2}}{1.75^2}(1 - 0.6216)\ \Omega$$

$$\dot{Z}_{e0} = R'_w + j(X_{i0} - \Delta X)[1.090 + j(8.843 - 1.093)] \cdot 10^{-2}$$
$$= (1.090 + j\, 7.750) \cdot 10^{-2}\ \Omega$$

$$Z_{e0} = 7.826 \cdot 10^{-2}\ \Omega$$

- Supply voltage and power factor of the inductor

$$V_0 = Z_{e0} I = 7.826 \cdot 10^{-2} \cdot 1000 = 78.3\ \text{V}$$
$$\cos \varphi_0 = \frac{R'_w}{Z_{e0}} = 0.139$$

- active and reactive power

$$P_w + j(Q_a + Q_w) = \dot{Z}_{e0} I^2 = 10{,}900 + j\, 77{,}500,\ \text{W/VAR}$$
$$p_0 = H_e^2 \frac{\rho}{\delta} \sqrt{2} P = 45{,}700^2 \frac{1 \cdot 10^{-6}}{1.125 \cdot 10^{-2}} \sqrt{2} \cdot 0.4718$$
$$= 1.239 \cdot 10^5\ \text{W/m}^2 \approx 12.4\ \text{W/cm}^2$$

6.3.2 Total Flux and Impedance Ze_0

The equivalent impedance of the inductor-load system can also be derived by evaluating the total flux linked with the inductor coil.

Under the assumption that the coil has negligible own resistance and radial thickness, denoting with:

$\dot{V}_0$ voltage at inductor clamps (V)
$\dot{Z}_{e0}$ equivalent impedance of the inductor-load system (Ω)
$\dot{\Phi} = \dot{\Phi}_w + \dot{\Phi}_a$ total flux linked with the inductor coil (Wb)
$\dot{\Phi}_w$ total flux linked with the workpiece (Wb)
$\dot{\Phi}_a$ flux in air gap between cylinder and inductor coil (Wb),

it results:

$$\dot{Z}_{e0} = \frac{\dot{V}_0}{\dot{I}} = \frac{j\omega N\dot{\Phi}}{\dot{I}} = \frac{j\omega N\left(\dot{\Phi}_w + \dot{\Phi}_a\right)}{\dot{I}} \tag{6.22}$$

Taking into account Eq. (6.7), the flux $\dot{\Phi}_w$ can be calculated by the relationship:

$$\dot{\Phi}_w = \mu_0\mu \int_0^{r_e} \dot{H}2\pi r\, dr = \frac{\mu_0\mu\dot{H}_e 2\pi r_e^2}{J_0(\sqrt{-j}m)} \int_0^1 \xi J_0(\sqrt{-j}m\xi)d\xi$$

Remembering that:

$$\int_0^1 \xi J_0(\beta\xi)d\xi = \frac{J_1(\beta)}{\beta}$$

we obtain:

$$\begin{aligned}\dot{\Phi}_w &= \mu_0\mu\dot{H}_e 2\pi r_e^2 \frac{J_1(\sqrt{-j}m)}{\sqrt{-j}mJ_0(\sqrt{-j}m)} \\ &= \mu_0\mu\dot{H}_e\pi\delta^2 m(Q - jP)\end{aligned} \tag{6.23}$$

with *P* and *Q*—coefficients defined by Eq. (6.12).

The flux $\dot{\Phi}_a$ is:

$$\dot{\Phi}_a = \mu_0\dot{H}_e\pi(r_i^2 - r_e^2) = \mu_0\dot{H}_e\pi r_e^2(\alpha^2 - 1) \tag{6.24}$$

Introducing in Eq. (6.22) the expressions (6.23) and (6.24), after some steps one obtains again Eq. (6.20):

$$\dot{Z}_{e0} = \frac{X_{i0}}{\alpha^2}\{\mu A + j[\alpha^2 - (1 - \mu B)]\} \tag{6.25}$$

6.3.3 Equations of the Semi-infinite Plane

It has been previously shown that at very high values of m the electromagnetic field concentrates only in the surface layers of a cylindrical workpiece, where the values of ξ are close to unity. In this situation Eqs. (6.13b) and (6.15) can be simplified as follows:

$$\left.\begin{aligned} \frac{H}{H_e} = \frac{J}{J_e} \approx e^{-x/\delta} \\ \frac{W}{W_e} \approx e^{-2x/\delta} \end{aligned}\right\} \tag{6.26}$$

These equations are the same obtained (at Sect. 2.3) for the diffusion of the electromagnetic wave in an infinite semi-plane, excited at the surface by the magnetic field intensity $\dot{H}_e$.

This means that in the cylindrical case, for very high values of m, the effect of the curvature of the load surface is practically negligible.

The values given by Eq. (6.26), graphically displayed in the curves of Fig. 6.11, point out that at a distance x from the surface greater than (2–3)δ, the current density J reduces to some percent of its surface value, while the volume power density is practically zero.

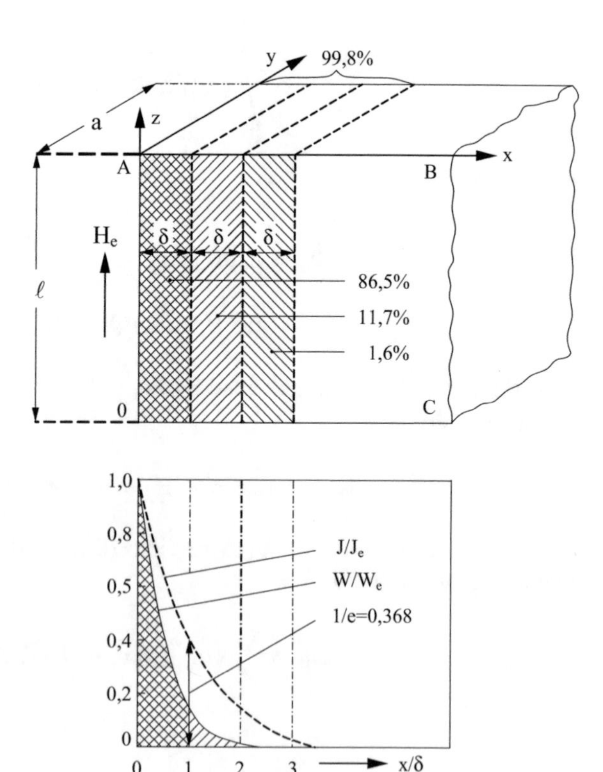

Fig. 6.11 Relative distributions of current density J and volume specific power w in a semi-infinite plane body

In particular, as indicated in the same figure, the power transformed into heat within the first three layers of thickness δ below the surface is respectively the 86.5, 11.7 and 1.6 % of the induced power, i.e. more than 99 % of the total induced power.

Under these conditions and with reference to the figure, it is:

- surface current density J_e and surface volume power density w_e [Eqs. (6.11) and (6.14) for $m \rightarrow \infty$]

$$\left.\begin{aligned} \dot{J}_e &= -\dot{H}_e \tfrac{1+j}{\delta};\, J_e = H_e \tfrac{\sqrt{2}}{\delta} \\ w_e &= \rho J_e^2 = H_e^2 \tfrac{2\rho}{\delta^2} \end{aligned}\right\} \tag{6.27}$$

- total current I_ℓ induced in an element of the semi-infinite plane of length ℓ along z

By applying the Ampere's law to the line integral along the closed path OABC and introducing the magnetic field intensity H_e, according to Eq. (6.30) we can write:

$$I_\ell = \oint_{OABC} H\,ds = H_e \ell = \ell\,\delta \frac{J_e}{\sqrt{2}} \tag{6.28}$$

We can therefore conclude that the total induced current I_ℓ can be thought as being uniformly distributed in the thickness δ, with density $J_e/\sqrt{2}$ (as already shown at Sect. 2.4.2).

- Active power P_{wx} in the surface layer of thickness x, length ℓ (along z), and depth a (along y)

$$P_{wx} = \ell a \int_0^x w\,dx = \ell a w_e \int_0^x e^{-2x/\delta} dx = \ell a H_e^2 \frac{\rho}{\delta} (1 - e^{-2x/\delta}) \tag{6.29}$$

- Total active power within an element of the semi-infinite plane of length ℓ (along z) and depth a (along y)

$$P_{w\infty} = \ell a \int_0^\infty w\,dx = \ell a\, w_e \int_0^\infty e^{-2x\delta} dx = \ell a H_e^2 \frac{\rho}{\delta} \tag{6.30}$$

- Power ratio $P_{wx}/P_{w\infty}$

$$\frac{P_{wx}}{P_{w\infty}} = (1 - e^{-2x/\delta}) \tag{6.31}$$

- Surface power density $p_{0\infty}$ [Eqs. (5.21) and (5.30)]

$$p_{0\infty} = \frac{P_{w\infty}}{\ell a} = H_e^2 \frac{\rho}{\delta} \tag{6.32}$$

- Total active and reactive power within an element of the semi-infinite body of length ℓ (along *z*) and depth *a* (along *y*):

$$\begin{aligned} P_{w\infty} + jQ_{w\infty} &= -\dot{E}_e \dot{H}_e \ell a = -\rho \dot{J}_e \dot{H}_e \ell a \\ &= H_e^2 \frac{\rho}{\delta} \ell a (1+j) = N^2 \frac{\rho}{\delta} \frac{a}{\ell} (1+j) I^2 \end{aligned} \tag{6.33}$$

- Equivalent impedance of the element of the body of length ℓ (along *z*) and depth *a* (along *y*):

$$\left. \begin{aligned} \dot{Z}_{e\infty} &= \frac{P_{w\infty} + jQ_{w\infty}}{I^2} = N^2 \frac{\rho}{\delta} \frac{a}{\ell} (1+j) \\ \left| \dot{Z}_{e\infty} \right| &= \sqrt{2}\, N^2 \frac{\rho}{\delta} \frac{a}{\ell} \end{aligned} \right\} \tag{6.34}$$

In particular, for $N = \ell = a = 1$, the impedance of a unit square surface is:

$$\dot{Z}_{e\infty u} = \frac{\rho}{\delta}(1+j). \tag{6.34a}$$

6.4 Heating of Hollow Cylindrical Workpieces in Axial Magnetic Field

6.4.1 General Solution

In this paragraph we consider the heating of the hollow, cylindrical body of infinite axial length (in the following also called "*tube*") shown in Fig. 6.12, excited by an axial magnetic field, having intensities $\dot{H}_i$ and $\dot{H}_e$ respectively at the internal and external surfaces.

This is, for example, the case of a hollow cylindrical workpiece heated with two inductors, one external and one internal to it.

The calculation of the electromagnetic quantities is developed as done at Sect. 6.1 for a solid cylindrical workpiece, determining first the radial distributions of magnetic field intensity and current density within the thickness of the tube.

In addition to the symbols used previously, the following quantities are defined:

r, r_e, r_i current, external and internal radii of the tube (m)
$s = r_e - r_i$ tube thickness (m)

$$m_r = \sqrt{2} r/\delta; \; m_e = \sqrt{2} r_e/\delta; \; m_i = \sqrt{2} r_i/\delta.$$

Fig. 6.12 Hollow cylinder in axial magnetic field

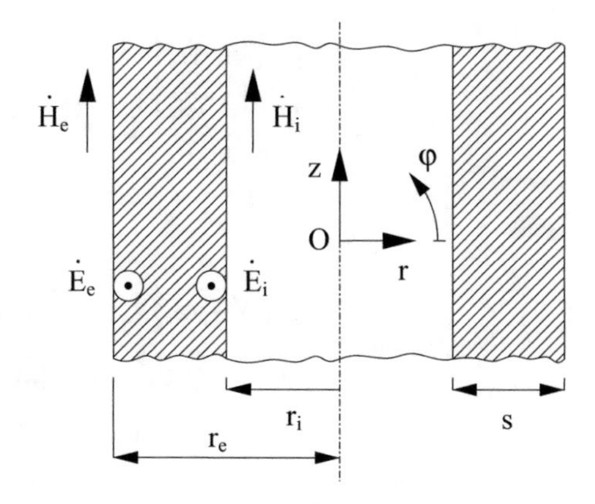

Equations (6.5a) and (6.6) still apply and are rewritten here with the new symbols:

$$\frac{d^2\dot{H}}{dr^2} + \frac{1}{r}\frac{d\dot{H}}{dr} - j\frac{\omega\mu_0\mu}{\rho}\dot{H} = 0 \tag{6.5a}$$

$$\dot{H} = \dot{C}_1 J_0(\sqrt{-j}m_r) + \dot{C}_2 Y_0(\sqrt{-j}m_r) \tag{6.6a}$$

Applying the boundary conditions:

- $$\dot{H} = \dot{H}_e \quad \text{for} \quad r = r_e$$

- $$\dot{H} = \dot{H}_i \quad \text{for} \quad r = r_i$$

it results:

$$\left.\begin{aligned} \dot{H}_e &= \dot{C}_1\, Jo(\sqrt{-j}\, m_e) + \dot{C}_2\, Yo(\sqrt{-j}\, m_e) \\ \dot{H}_i &= \dot{C}_1\, Jo(\sqrt{-j}\, m_i) + \dot{C}_2\, Yo(\sqrt{-j}\, m_i) \end{aligned}\right\} \tag{6.35}$$

and

$$\dot{C}_1 = \frac{\begin{vmatrix} \dot{H}_e & Y_0(\sqrt{-j}m_e) \\ \dot{H}_i & Y_0(\sqrt{-j}m_i) \end{vmatrix}}{F_{oo}(m_i, m_e)} = \frac{\dot{H}_e \cdot Y_0(\sqrt{-j}m_i) - \dot{H}_i \cdot Y_0(\sqrt{-j}m_e)}{F_{oo}(m_i, m_e)}$$

$$\dot{C}_2 = \frac{\begin{vmatrix} J_0(\sqrt{-j}m_e)\dot{H}_e \\ J_0(\sqrt{-j}m_i)\dot{H}_i \end{vmatrix}}{F_{oo}(m_i, m_e)} = -\frac{\dot{H}_e \cdot J_0(\sqrt{-j}m_i) - \dot{H}_i \cdot J_0(\sqrt{-j}m_e)}{F_{oo}(m_i, m_e)}$$

with:

$$\begin{aligned} F_{oo}(m_i, m_e) &= \begin{vmatrix} J_0(\sqrt{-j}m_e) Y_0(\sqrt{-j}m_e) \\ J_0(\sqrt{-j}m_i) Y_0(\sqrt{-j}m_i) \end{vmatrix} \\ &= J_0(\sqrt{-j}m_e)Y_0(\sqrt{-j}m_i) - Y_0(\sqrt{-j}m_e)J_0(\sqrt{-j}m_i) \end{aligned} \tag{6.36}$$

By substitution of the constants $\dot{C}_1$ and $\dot{C}_2$ into Eq. (6.6a), we finally obtain:

$$\dot{H} = \dot{H}_e \frac{F_{oo}(m_i, m_r)}{F_{oo}(m_i, m_e)} - \dot{H}_i \frac{F_{oo}(m_e, m_r)}{F_{oo}(m_i, m_e)} \tag{6.37}$$

where the function $F_{oo}(x, y)$ is given by the relationship:

$$F_{oo}(x, y) = J_0(\sqrt{-j}y)Y_0(\sqrt{-j}x) - Y_0(\sqrt{-j}y)J_0(\sqrt{-j}x). \tag{6.38}$$

According to Eq. (6.4), taking the derivative of Eq. (6.37), for the electric field intensity we have:

$$\dot{E} = -\rho \frac{d\dot{H}}{dr} = \sqrt{-2j}\frac{\rho}{\delta}\left[\dot{H}_e \frac{F_{o1}(m_i, m_r)}{F_{oo}(m_i, m_e)} - \dot{H}_i \frac{F_{o1}(m_e, m_r)}{F_{oo}(m_i, m_e)}\right] \tag{6.39}$$

with:

$$F_{o1}(x, y) = J_1(\sqrt{-j}y)Y_0(\sqrt{-j}x) - Y_1(\sqrt{-j}y)J_0(\sqrt{-j}x). \tag{6.40}$$

6.4.2 *Heating of Tubes with External Inductors*

In this case, in Eqs. (6.37) and (6.39) the magnetic field intensity $\dot{H}_e$ on the tube external surface is produced by the inductor, whereas the magnetic field $\dot{H}_i$ inside the tube internal cavity is constant.

The unknown value of $\dot{H}_i$ can be determined considering that the *e.m.f.* induced in an ideal circular filament conductor placed at radius $r = r_i$ must be equal to the derivative of the flux linked with the same circuit.

Therefore, one can write:

$$\oint \dot{E} d\ell = 2\pi r_i \dot{E}_i = -j\omega\dot{\Phi}_i = -j\omega\mu_0 \dot{H}_i \pi r_i^2$$

or, for Eq. (6.39):

$$\begin{aligned}\dot{E}_i &= -j\frac{\omega\mu_0 r_i}{2}\dot{H}_i \\ &= \sqrt{-2j}\frac{\rho}{\delta}\left[\dot{H}_e\frac{F_{o1}(m_i, m_i)}{F_{oo}(m_i, m_e)} - \dot{H}_i\frac{F_{o1}(m_e, m_i)}{F_{oo}(m_i, m_e)}\right]\end{aligned} \tag{6.41}$$

Remembering that it is: $F_{o1}(x, x) = \frac{2}{\sqrt{-j}\pi x}$, from the previous equation, after some steps one obtains:

$$\frac{\dot{H}_e}{\dot{H}_i} = -j\frac{\pi m_i^2}{2\mu}[F_{oo}(m_i, m_e) + F_{o1}(m_e, m_i)] \tag{6.42}$$

Hence, Eqs. (6.37), (6.39) and (6.42) allow us to calculate the relative distributions along the thickness of tube wall of the magnetic field intensity $\dot{H}$ and the current density $\dot{J} = \dot{E}/\rho$. Some qualitative distributions are illustrated in Fig. 6.13.

The curves show that at relative low frequency (*curve* f_1), H remains practically constant over the whole thickness of tube wall while J varies linearly, whereas at high frequency (*curve* f_4) the diagrams tend to exponentials.

Calculating, as in Sect. 6.3, the flux of Poynting's vector through the external surface of the tube, for the active and reactive power we obtain the relationship:

$$P_w + jQ_w = -\dot{E}_e\dot{H}_e 2\pi r_e \ell = H_e^2\frac{\rho}{\delta}\sqrt{2}(P_{te} + jQ_{te})2\pi r_e \ell \tag{6.43}$$

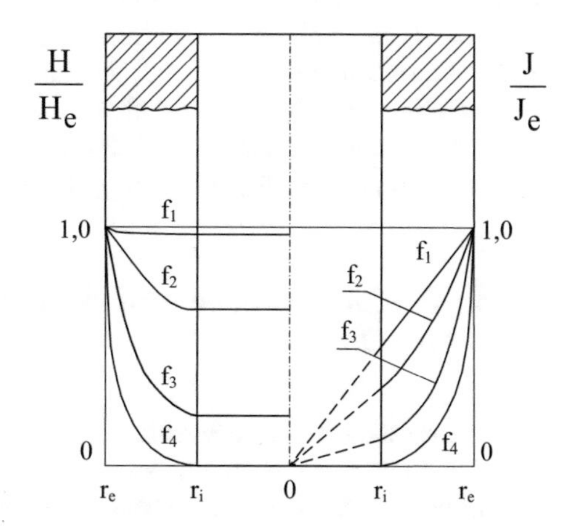

Fig. 6.13 Relative distributions at different frequencies of magnetic field intensity and current density in tubes heated with external inductors

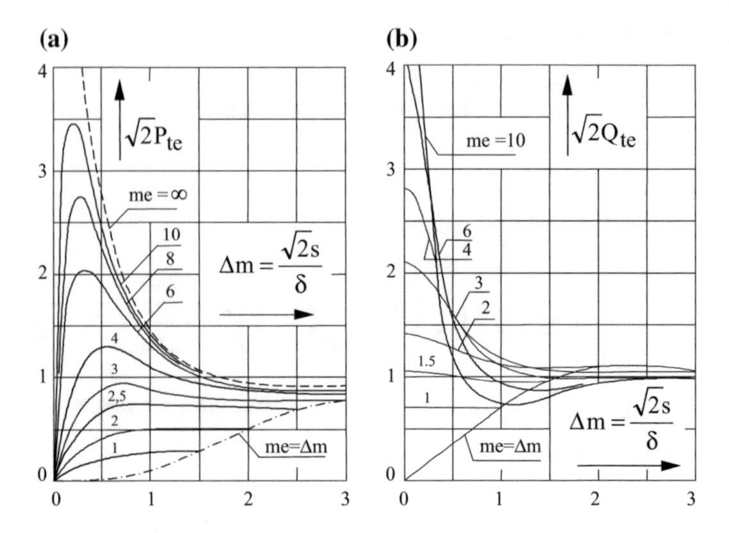

Fig. 6.14 Coefficients of active power (**a**) and reactive power (**b**) in tubes heated with external inductors$(m_e = \sqrt{2}r_e/\delta)$

with:

$$P_{te} + jQ_{te} = -\sqrt{-j}\left[\frac{F_{o1}(m_i, m_e)}{F_{oo}(m_i, m_e)} - \frac{\dot{H}_i}{\dot{H}_e}\frac{F_{o1}(m_e, m_e)}{F_{oo}(m_i, m_e)}\right] \tag{6.43a}$$

It should be noticed that Q_w comprises the reactive power produced by the constant field H_i in the internal cavity of the tube.

The diagrams of Fig. 6.14 give the values of $\sqrt{2}P_{te}$ and $\sqrt{2}Q_{te}$ for non-magnetic materials (μ = 1) as a function of $m_e = \sqrt{2}r_e/\delta$ and $\Delta m = m_e - m_i = \sqrt{2}s/\delta$.

They emphasize the following:

- for low values of Δm, the influence of induced currents is negligible, and the power (and so P_{te}) increases quasi linearly until a maximum;
- for intermediate values of Δm, the demagnetizing action of induced currents is prevailing, and P_{te} decreases after the maximum;
- for high values of Δm, the distribution of induced currents within the tube wall becomes exponential and P_{te} and Q_{te}. tend to the value $1/\sqrt{2}$ (as in a solid cylinder).

6.4.3 Hollow Cylinder with Internal Exciting Magnetic Field

In this case the exciting magnetic field H_i is applied to the internal surface of the cylinder whereas at the external surface is $H_e = 0$. This latter condition occurs in a tubular body of finite wall thickness belonging to a system of infinite axial length.

Under these conditions, Eqs. (6.37) and (6.39) can be written in the form:

$$\left.\begin{aligned}\frac{\dot{H}}{\dot{H}_i} &= \frac{F_{00}(m_r,m_e)}{F_{00}(m_i,m_e)} \\ &= \frac{J_0(\sqrt{-}m_e)Y_0(\sqrt{-j}m_r)-Y_0(\sqrt{-j}m_e)J_0(\sqrt{-j}m_r)}{J_0(\sqrt{-j}m_e)Y_0(\sqrt{-j}m_i)-Y_0(\sqrt{-j}m_e)J_0(\sqrt{-j}m_i)} \\ \frac{\dot{E}}{\dot{H}_i} &= -\sqrt{-2j}\,\frac{\rho}{\delta}\,\frac{J_1(\sqrt{-j}m_r)Y_0(\sqrt{-j}m_e)-Y_1(\sqrt{-j}m_r)J_0(\sqrt{-j}m_e)}{J_0(\sqrt{-j}m_e)Y_0(\sqrt{-j}m_i)-Y_0(\sqrt{-j}m_e)J_0(\sqrt{-j}m_i)}\end{aligned}\right\} \tag{6.44}$$

taking into account that the functions $F_{oo}(x, y)$ and $F_{o1}(x, y)$ are still given by Eqs. (6.38), (6.40) and that it is $F_{oo}(x, y) = -F_{oo}(y, x)$.

In this case the active and reactive power in a section of the cylinder of axial length ℓ can be calculated by evaluating the flux of the complex Poynting's vector flowing outwards through the surface of the cylinder of radius r_i.

Taking into account the perpendicularity of E_i and H_i, it results:

$$P_w + jQ_w = \dot{E}_i\dot{H}_i 2\pi r_i \ell = H_i^2 \frac{\rho}{\delta}\sqrt{2}(P_{ti} + jQ_{ti})2\pi r_i \ell \tag{6.45}$$

with:

$$P_{ti} + jQ_{ti} = -\sqrt{-j}\left[\frac{F_{o1}(m_i, m_e)}{F_{oo}(m_i, me)}\right] \tag{6.45a}$$

Figure 6.15 gives the values of coefficients $(\sqrt{2}P_{ti})$ and $(\sqrt{2}Q_{ti})$, for different values of m_i, as a function of $\Delta m = m_e - m_i = \sqrt{2}s/\delta$, with: $s = r_e - r_i$—thickness of the tube wall.

In particular, the curves of P_{ti} are similar to those valid for a plane wall ($m_i \rightarrow \infty$), but with values which are higher the greater is the effect of load curvature. The values

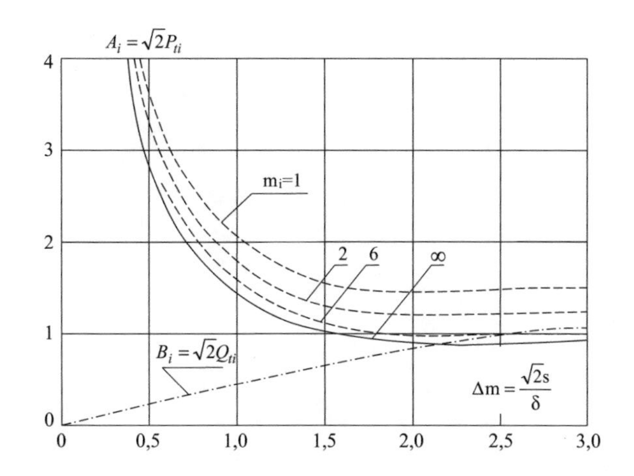

Fig. 6.15 Coefficients of active and reactive power in tubes with internal exciting magnetic field. ($m_i = \sqrt{2}r_i\delta$; $\Delta m = \sqrt{2}s/\delta$)

of Q_{ti} practically do not depend on m_i and grow linearly with the thickness s till $\Delta m \approx 2.5$. Above this value, it can be assumed $Q_{ti} \approx 1/\sqrt{2}$.

For $\Delta m \leq 1$ (thin walled tubes), can be used the approximate relations:

$$P_{ti} \approx \frac{1}{\Delta m} + \frac{1}{2m_i}; \quad Q_{ti} \approx \frac{\Delta m}{3}$$

6.5 Electrical Parameters of Inductors

6.5.1 *Resistance and Internal Reactance of the Inductor*

It can be approximately determined by assimilating the inductor coil of length ℓ_i (shown in Fig. 6.16a) to a hollow cylinder, assuming that it belongs to a system of infinite axial length and it is subdivided in N turns of rectangular cross section, which are close to each other, as shown in Fig. 6.16b.

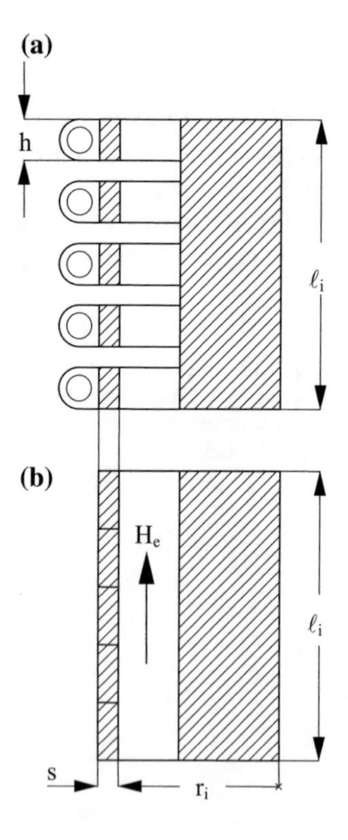

Fig. 6.16 Schematization of the inductor coil for calculation of its own resistance R_i and internal reactance X_{ii}

This configuration is the same (except for the subdivision into N windings) examined in the previous paragraph.

Therefore, Eqs. (6.44), (6.45) and the diagrams of Fig. 6.15 can be applied with the following meaning of symbols:

$\dot{H}_e = \frac{N\dot{I}}{\ell_i}$ intensity of exciting magnetic field

s radial thickness of the coil conductor

r_i internal radius of inductor coil

$r_e = r_i + s$

ρ_i resistivity of material of coil conductor

$\delta_i = \sqrt{\frac{2\rho_i}{\omega\mu_0}}$ penetration depth in the coil conductor;

$m_r = \frac{\sqrt{2}r}{\delta_i}$; $m_i = \frac{\sqrt{2}r_i}{\delta_i}$; $m_e = \frac{\sqrt{2}r_e}{\delta_i}$; $\Delta m = m_e - m_i = \frac{\sqrt{2}s}{\delta_i}$.

In particular, the active power P_i and the reactive power Q_i in the volume of the inductor, according to Eq. (6.45), can be expressed as:

$$P_i + jQ_i = H_e^2 \frac{\rho_i}{\delta_i}(A_i + jB_i) 2\pi r_i \ell_i \quad (6.46)$$

with:

$$A_i + jB_i = \sqrt{2}(P_{ti} + jQ_{ti}); \; [P_{ti} + jQ_{ti} - \text{see, Eq.(6.45a)}]$$

Therefore, denoting with:

R_i, X_{ii} AC resistance and internal reactance of the inductor coil;

k_i coefficient (>1) which takes account the axial separation from each other of the coil turns [$k_i = (\ell_i/N \cdot h)$—for conductors of rectangular cross-section with axial dimension h; $k_i = (4\ell_i/N \cdot \pi d)$—for conductors of circular cross section of diameter d]

it results:

$$\left.\begin{aligned} R_i &= \frac{P_i}{I^2} = N^2 \rho_i \frac{2\pi r_i}{\ell_i \delta_i} A_i k_i = X_{i0} \frac{\delta_i}{r_i} A_i k_i \\ X_{ii} &= \frac{Q_i}{I^2} = N^2 \rho_i \frac{2\pi r_i}{\ell_i \delta_i} B_i = X_{i0} \frac{\delta_i}{r_i} B_i \end{aligned}\right\} \quad (6.47)$$

Given that, as shown in Fig. 6.16a, the inductor coils are always made with copper water cooled tubes (with resistivity $\rho_i \approx 2 \cdot 10^{-8}\,\Omega\text{m}$ at about 50–60 °C, which is the maximum admissible temperature in order to prevent lime-scale), the arguments of Bessel functions are generally so high that it's acceptable to approximate them with exponentials.

Table 6.5 Values of coefficients A_i, B_i as a function of s/δ_i

s/δ_i	0.25	0.50	0.75	1.00	1.25	1.50
$\mathbf{A_i}$	4.00	2.01	1.37	1.086	0.959	0.920
$\mathbf{B_i}$	0.167	0.333	0.490	0.632	0.781	0.893
s/δ_i	1.57	1.75	2.00	3.00	5.00	10.0
$\mathbf{A_i}$	0.918	0.925	0.950	0.999	1.00	1.00
$\mathbf{B_i}$	0.918	0.965	1.004	1.006	1.00	1.00

In fact, with the above value of ρ_i, it is:

$$m_i = \frac{\sqrt{2}r_i\sqrt{2\pi f 4\pi \cdot 10^{-7}}}{\sqrt{2 \cdot 2 \cdot 10^{-8}}} \approx 20\, r_i\sqrt{f},$$

which, for example, with $r_i = 25 \cdot 10^{-3}$ m and $f = 1$ kHz, gives $m_i > 15$.

With such approximations, the coefficients A_i and B_i of Eqs. (6.46) and (6.47) can be expressed as follows [1, 2]:

$$\begin{aligned} A_i + jB_i &= (1+j)\frac{\cosh(\sqrt{2j}\, s/\delta_i)}{\sinh(\sqrt{2j}s/\delta_i)} \\ &= \frac{\sinh(2s/\delta_i) + \sin(2\,s/\delta_i)}{\cosh(2s/\delta_i) - \cos(2\,s/\delta_i)} + j\frac{\sinh(2s/\delta_i) - \sin(2\,s/\delta_i)}{\cosh(2s/\delta_i) - \cos(2s/\delta_i)} \end{aligned} \tag{6.48}$$

The values of A_i and B_i are given, as a function of s/δ_i, in Fig. 6.15 by the curves for $m_i \to \infty$ or in the following Table 6.5

From these values we can conclude as follows:

- all other conditions being equal, the losses in the inductor coil are minimal for $s/\delta_i = \pi/2 = 1.57$ and tend to a constant value, slightly above the minimum, when this ratio increases;
- in order to reduce inductor losses, the thickness *s* should be chosen—in relation to the frequency—so that it is $s/\delta_i \geq 1.57$, while the axial dimension *h* of the turns should give values of k_i close to unit ($k_i \approx 1.1{-}1.25$);
- for high values of s/δ_i, the reactance of the inductor coil is equal to that of a cylindrical coil of negligible radial thickness, placed at the radius $r_i + \delta_i\backslash 2$.

6.5.2 *Electrical Efficiency of the Inductor-Load System*

The electrical efficiency η_e of the "loaded" inductor is defined as the ratio between the power transformed into heat within the workpiece and the total active power absorbed by the inductor.

With the previous notations it is:

$$\eta_e = \frac{R'_w}{R_i + R'_w} = \frac{1}{1 + \frac{R_i}{R'_w}} \quad (6.49)$$

Introducing the expressions of R'_w and R_i given by Eqs. (6.20) and (6.47), one obtains:

$$\eta_e = \frac{1}{1 + \frac{r_i \ell \delta A_i k_i}{r_e \ell_i \delta_i \sqrt{2} P}} = \frac{1}{1 + \alpha \frac{\ell}{\ell_i} \sqrt{\frac{\rho_i}{\rho \mu}} \frac{A_i k_i}{\sqrt{2} P}} \quad (6.50)$$

Therefore the electrical efficiency depends on the following parameters:

- the coupling ratio α between the internal diameter of the inductor and the external diameter of the workpiece as well as on the ratio (ℓ_i/ℓ) between their axial dimensions
- the resistivity ρ_i of material of the inductor coil
- the resistivity ρ and permeability μ of the workpiece material
- the parameter $A_i k_i$ characteristic of the inductor coil
- the coefficient P, which is a function of the ratio m between the geometrical dimensions of the workpiece and the penetration depth of induced currents.

With a good inductor design and high values of m, the maximum of electrical efficiency (for $\ell_i = \ell$) is:

$$\eta_{emax} = \frac{1}{1 + \alpha \sqrt{\frac{\rho_i}{\rho_\mu}}} \quad (6.51)$$

For the workpiece materials specified in the caption and values of α between 1.4 and 2.2, from Eq. (6.50) one can draw the curves of η_e given in Fig. 6.17 as a function of m.

The analysis of the curves indicates that for $m > 2.5$ the electrical efficiency of the inductor is practically independent on the frequency, whereas for lower values of m the heating will occur with low or very low efficiency.

Moreover, the efficiency decreases considerably with the increase of the ratio α, especially when heating non-magnetic materials or magnetic materials above Curie Point (about 750–770 °C for steels). Therefore it should be avoided to use the same inductor for heating workpieces with very different diameters.

Finally it can be observed that the electrical efficiency is always relatively good (0.7–0.9) in the heating of magnetic steels or materials with high resistivity, whereas it has always low values (0.4–0.6) when heating materials with low resistivity.

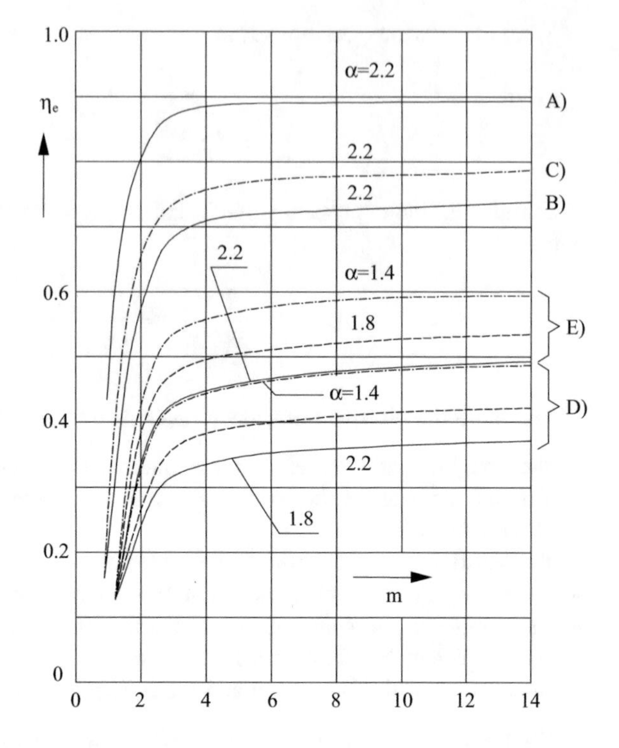

Fig. 6.17 Electrical efficiency of the inductor-load system as a function of m (*A* steel heated from 0 to 800 °C; *B* steel from 800 to 1200 °C; *C* steel from 0 to 1200 °C; *D* Aluminium and its alloys heated from 0 to 500 °C and copper to 800 °C; *E* Brass from 0 to 800 °C)

6.5.3 Quality and Power Factors

The Q-factor of the "loaded" inductor, defined as the ratio between reactive and active power at the inductor clamps, is calculated (neglecting the reactance X_{ii}) using the Eqs. (6.20), (6.47) and the inductor electrical efficiency η_e. The following relationship is obtained:

$$\begin{aligned} Q_0 &= \frac{X_{i0} - \Delta x}{R_i + R'_w} = \frac{X_{i0} - X_{i0}\frac{(1-\mu B)}{\alpha^2}}{R'_w}\eta_e \\ &= \frac{\alpha^2 - (1-\mu B)}{\mu A}\eta_e = Q'_0\,\eta_e \end{aligned} \tag{6.52}$$

where Q'_0 indicates the Q-factor of the system with ideal inductor of negligible own resistance.

For loads consisting of non-magnetic materials and for the most commonly used values of α, Fig. 6.18 shows the values of Q'_0 as a function of m.

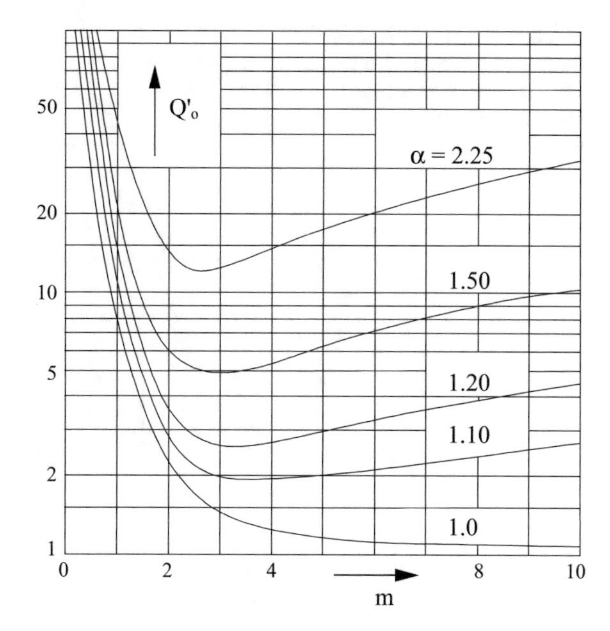

Fig. 6.18 Quality factor Q_0' as a function of m for non-magnetic materials and inductor-load system with $R_i = 0$

Similarly, for the power factor one obtains:

$$\cos \varphi = \frac{R_i + R_w'}{\sqrt{(R_i + R_w')^2 + (X_{i0} - \Delta x)^2}} = \frac{1}{\sqrt{1 + (Q_0' \eta_e)^2}} = \frac{1}{\sqrt{1 + [\frac{\alpha^2 - (1 - \mu B)}{\mu A} \eta_e]^2}} \tag{6.53}$$

The curves of Fig. 6.19 give the values of the power factor as a function of *m*, for the same values of α of Fig. 6.18 and non-magnetic loads heated with the ideal inductor of negligible resistance.

We can observe that the power factor of the inductor-load system has always relatively low values, mostly in the range 0.1–0.3. It is therefore necessary to compensate its value by connecting a bank of capacitors in parallel to the induction coil, as shown for example in Fig. 6.4.

Figure 6.20 shows other schemes of connection of the capacitor bank, which can be used in order to improve the operating conditions of the frequency converter, which can be affected by the strong variations of the impedance of the inductor-load system during heating.

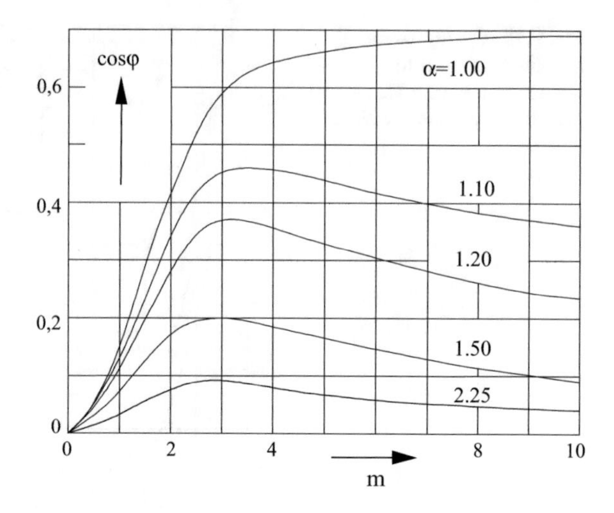

Fig. 6.19 Values of power factor as a function of m for non-magnetic materials and inductor-load system with $R_i = 0$

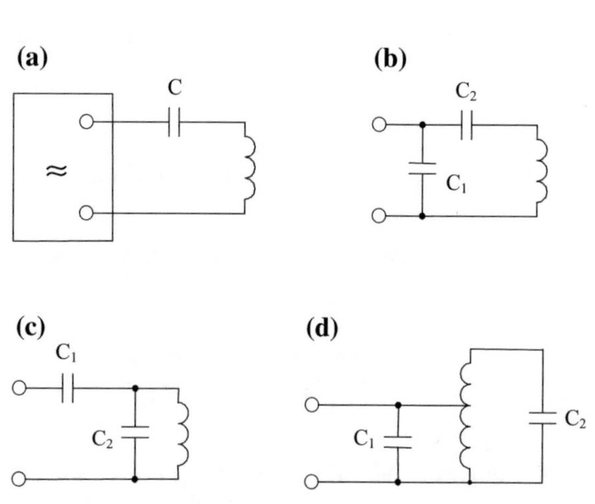

Fig. 6.20 Schemes of connection of capacitors banks for power factor correction

Example 6.2 Calculation of the inductor of Example 6.1 (N = 32 turns; r_i = 35 mm; ℓ_i = 700 mm; f = 2000 Hz; I = 1000 A).

The inductor is made of rectangular copper tube 18 × 10 × 2.5 mm.

$$\delta_i = \sqrt{\frac{2 \cdot 2 \cdot 10^{-8}}{2\pi \cdot 2000 \cdot 4\pi \cdot 10^{-7}}} = 1.59 \cdot 10^{-3}\,\mathrm{m} = 1.59\,\mathrm{mm}$$

$s/\delta_i = 2.5/1.59 = 1.57$ and from Table 6.5: $A_i = B_i = 0.918$

$$k_i = \ell_i/(Nh) = 700/(3218) = 1.22; A_i k_i = 0.918 \cdot 1.22 = 1.12$$

- equivalent impedance Z_e of the inductor-load system, comprising the resistance R_i and the reactance X_{ii}:
  $X_{i0} = 8.843 \cdot 10^{-2}\,\Omega$ (see Example 6.1)
  $R_i = 8.843 \cdot 10^{-2} \frac{1.59 \cdot 10^{-3}}{35 \cdot 10^{-3}} 0.918 \cdot 1.22 = 4.5 \cdot 10^{-3}\,\Omega$ (Eq. 6.47)
  $X_{ii} = 8.843 \cdot 10^{-2} \frac{1.59 \cdot 10^{-3}}{35 \cdot 10^{-3}} 0.918 = 3.69 \cdot 10^{-3}\,\Omega$ (Eq. 6.47)

$$R_t + R'_w = 4.5 \cdot 10^{-3} \cdot 1.090 \cdot 10^{-2} = 1.54 \cdot 10^{-2}\,\Omega$$

$\dot{Z}_e = (1.090 + j\,7.750) \cdot 10^{-2}\,\Omega$ (see Example 6.1)

$$\dot{Z}_e = \dot{Z}_{e0} + (R_i + j\,X_{ii}) = (1.540 + j\,8.119) \cdot 10^{-2}\,\Omega,$$

$$Z_e = 8.264 \cdot 10^{-2}\,\Omega$$

- supply voltage, electrical efficiency and power factor

$$V = Z_e I = 82.6\,V$$
$$\eta_e = 1.090 \cdot 10^{-2} / 1.540 \cdot 10^{-2} = 0.708$$
$$\cos\varphi = 1.540 \cdot 10^{-2} / 8.264 \cdot 10^{-2} = 0.186$$

6.6 Transient Temperature Distribution in Cylindrical Workpieces Heated by Induction

In the design of an induction heating installation, fundamental problems are the prediction of the heating time, the corresponding induced power and how the power can be controlled during heating in order to obtain in the workpiece the final temperature distribution required by the considered technological process.

From this point of view, the main industrial applications of induction heating can be classified according to the temperature distribution within the body at the end of heating, leading to the following two main categories:

- *localized heating*, e.g. that required in hardening, which is characterized by strong temperature differentials among different points of the workpiece cross-section;
- *uniform heating,* with final temperature distributions in the workpiece cross-section as homogeneous as possible, like the one required in through heating of billets before hot working.

Since the specific power distributions previously analyzed by their nature tend always to produce differential heating in the workpiece cross-section, it becomes fundamental to know the behavior of the thermal transient.

In cylindrical workpieces of length much greater than diameter, the heat transmission occurs only in radial direction and is governed by the Fourier equation, which—assuming constant material parameters—is written as follows in cylindrical coordinates (r, φ, z) [see Eq. (1.10b)]:

$$\frac{\partial \vartheta}{\partial t} = k\left(\frac{\partial^2 \vartheta}{\partial r^2} + \frac{1}{r}\frac{\partial \vartheta}{\partial r}\right) + \frac{w(r)}{c\,\gamma} \tag{6.54}$$

with:

ϑ	temperature (°C), at a generic radius *r* at time *t*
$k = \lambda / c\,\gamma$	diffusivity of the workpiece material (m^2/s); λ—thermal conductivity (W/m °C), *c*—specific heat (Ws/kg °C), γ—specific weight (kg/m^3)
$w(r)$	induced power distribution per unit volume within the cylinder (W/m^3)

Assuming that the cylinder is initially at temperature zero and the heating occurs without heat losses from its surface, Eq. (6.54) is solved with the following initial and boundary conditions:

$$\left.\begin{array}{ll} \vartheta(r) = 0, & \text{for} \quad t = 0 \\ \frac{\partial \vartheta}{dr} = 0, & \text{for} \quad t > 0 \quad \text{and } r = r_e \end{array}\right\} \tag{6.54a}$$

Introducing the dimensionless parameters:
$\xi = \frac{r}{r_e}$; $\tau = \frac{kt}{r_e^2}$; $\Theta = \frac{2\pi\lambda}{P_u}\vartheta$
with:
$P_u = 2\pi \int_{r=0}^{r=r_e} r\,w(r)dr = 2\pi r_e^2 \int_{\xi=0}^{\xi=1} \xi w(\xi)d\xi$—power induced in the workpiece per unit axial length (W/m),
and denoting with:

$$\psi(\xi) = \frac{2\pi\,r_e^2}{P_u} w(r) = \frac{w(\xi)}{\int_0^1 \xi w(\xi)d\xi},$$

Fourier equation and initial and boundary conditions can be rewritten in the following form:

$$\frac{\partial \Theta}{\partial \tau} = \frac{\partial^2 \Theta}{\partial \xi^2} + \frac{1}{\xi}\frac{\partial \Theta}{\partial \xi} + \psi(\xi) \quad \text{for} \quad 0 \le \xi \le 1 \tag{6.55}$$

$$\left.\begin{array}{ll} \Theta(\xi) = 0, & \text{for } \tau = 0 \\ \frac{\partial \Theta}{d\xi} = 0, & \text{for} \quad \tau > 0 \text{ and} \quad \xi = 1 \end{array}\right\} \tag{6.55a}$$

Since from Eq. (6.14) it is:

$$\frac{w}{w_e} = \left|\frac{J_1(\sqrt{-j}m\,\xi)}{J_1(\sqrt{-j}m)}\right|^2 = \frac{ber'^2 m\,\xi + bei'^2 m\xi}{ber'^2 m + bei'^2 m}$$

it results:

$$\psi(\xi) = m\frac{ber'^2 m\xi + bei'^2 m\xi}{ber\ m\, ber' m + bei\ m\, bei' m} \tag{6.56b}$$

A general solution of Eq. (6.55) with the conditions (6.55a) and the function $\psi(\xi)$ (6.56b) is given in the bibliography in the form of an infinite series of terms obtained by applying Laplace transformation.

However, considering on one side that owing to its complexity it requires always the use of computers, on the other side that, due to the variations of material characteristics with temperature, the assumption of a function $\psi(\xi)$ invariant during the whole thermal transient represents a very rough approximation, it is more convenient to refer—for the different types of heating—to more simple solutions.

6.6.1 *Localized Surface Heating*

A qualitative analysis can be done considering two limit solutions corresponding to simplified expressions of the ratio w/w_e.

- *for* $m \rightarrow \infty$

At very high frequency the penetration depth is small in comparison to the workpiece diameter and the heat transfer occurs from the surface. Then the solution of the heating transient is the same considered at Sect. 1.2.5.4 with constant surface heat flux:

$$\Theta = 2\tau + \frac{1}{2}\xi^2 - \frac{1}{4} - 2\sum_{n=1}^{\infty}\frac{J_0(\beta_n\xi)}{\beta_n^2 J_0(\beta_n)}e^{-\beta_n^2\tau} \tag{6.57}$$

with: β_n = 3.83, 7.02, 10.17, 13.32,…—positive roots of the equations $J_1(\beta) = 0$ (Table A.2 in appendix).

For this case, the curves of Fig. 6.21a show the variations of Θ as a function of τ, at different radii of the cylinder, in the region of non-linear increase of temperature.

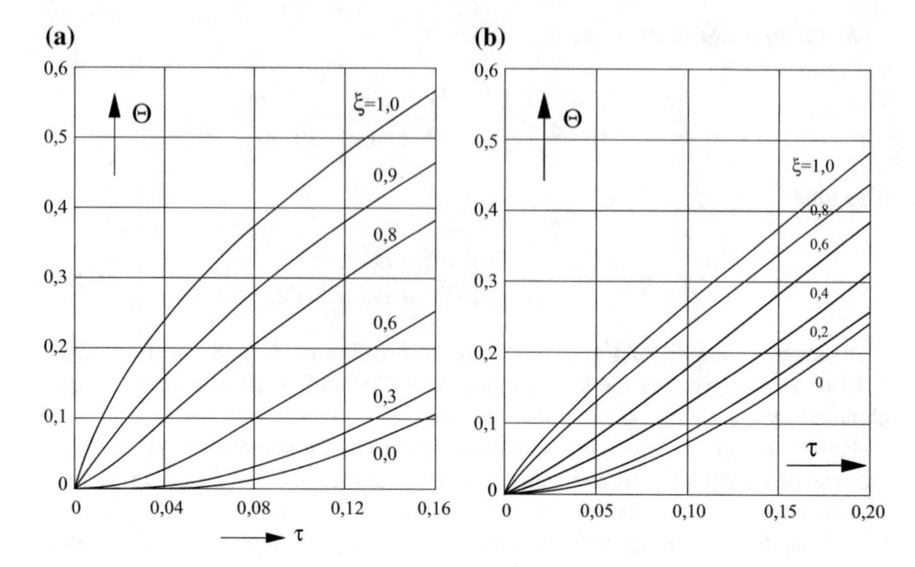

Fig. 6.21 **a** Transient temperature distribution for $m \rightarrow \infty$. **b** Transient temperature distribution for $m \leq 1$

- *for m* ≤ 1 $(w/w_e = \xi^2)$

$$\Theta = 2\tau + \frac{1}{2}(\xi^2 - \frac{1}{4}\xi^4) - \frac{1}{6} - 16\sum_{n=1}^{\infty}\frac{J_0(\beta_n\xi)}{\beta_n^4 J_0(\beta_n)}e^{-\beta_n^2\tau} \tag{6.58}$$

The corresponding diagrams of Θ as a function of τ are shown in Fig. 6.21b.

6.6.2 Quasi Uniform Heating

Relatively uniform final temperature distributions in the workpiece cross-section can be obtained only if the heating transient ends in the region of linear temperature increase.

This occurs if the end of the heating process occurs for $\tau \geq 0.25$, i.e. when in the solutions (6.57) and (6.58) all exponentials terms become negligible.

A linear increase of temperature with the same rate in all points of the cylinder cross-section can occur when the following condition is fulfilled:

$$\pi r^2 c\gamma\frac{\partial\vartheta}{\partial t} = 2\pi r\lambda\frac{\partial\vartheta}{\partial r} + \int_0^r 2\pi r w(r)dr.$$

It expresses for a generic cylinder of radius r and unit axial length the equality between the power required for raising the temperature at constant rate ($\partial\vartheta/\partial t$), and the sum of the power transmitted by conduction through its lateral surface plus the power directly induced in the cylinder.

By using the dimensionless parameters ξ, τ, Θ, the same relationship can be written as follows:

$$\frac{\partial\Theta}{\partial\tau} = \frac{4\,\pi r_e^2}{P_u \xi^2}\int_0^{\xi} \xi w(\xi) d\xi + \frac{2}{\xi}\frac{\partial\Theta}{\partial\xi}. \tag{6.59}$$

The constant rate of temperature rise can be determined with reference to the cylinder of radius r_e and unit length and the total power P_u transformed into heat inside it. In this case the relation:

$$P_u = \pi r_e^2 c\gamma \frac{\partial\vartheta}{\partial t}$$

in dimensionless terms gives:

$$\frac{\partial\Theta}{\partial\tau} = 2. \tag{6.60}$$

Equating (6.59) and (6.60), the following relation independent on τ is obtained:

$$\frac{\partial\Theta}{\partial\xi} = \xi - \frac{1}{\xi}\frac{\int_0^{\xi} \xi w(\xi) d\xi}{\int_0^{1} \xi w(\xi) d\xi}, \tag{6.61}$$

which allows to determine the distribution of Θ as a function of ξ at any instant in the range of linear temperature increase.

By substituting in Eqs. (6.61), (6.14) and calculating the integrals, one obtains:

$$\frac{\partial\Theta}{\partial\xi} = \xi - \frac{\text{ber } m\xi \, \text{ber}' m\xi + \text{bei } m\xi \, \text{bei}' m\xi}{\text{ber } m \, \text{ber}' m + \text{bei } m \, \text{bei}' m}$$

By a new integration, it results:

$$\Theta = \frac{1}{2}\xi^2 - \frac{\text{ber}^2 m\xi + \text{bei}^2 m\xi}{2m(\text{ber } m \, \text{ber}' m + \text{bei } m \, \text{bei}' m)} + C$$

Denoting with Θ_a the value of Θ for $\xi = 0$ (the dimensionless temperature at the axis of the workpiece) and determining the integration constant C with the condition:

$$\Theta = \Theta_a \quad \text{for} \quad \xi = 0$$

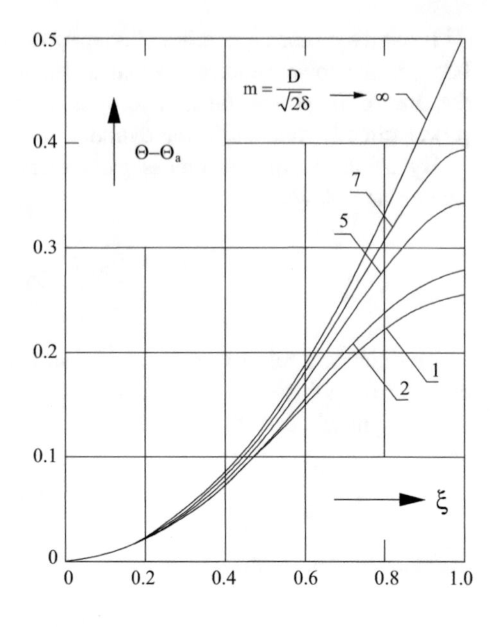

Fig. 6.22 Radial temperature distributions during the linear temperature increase

after some simple steps one obtains:

$$\Theta = \Theta_a + \frac{1}{2}[\xi^2 - \frac{\mathrm{ber}^2 m\xi + \mathrm{bei}^2 m\xi - 1}{m(\mathrm{ber}\ m\,\mathrm{ber}'m + \mathrm{bei}\ m\,\mathrm{bei}'m)}] \quad (6.62)$$

Equation (6.62) is applicable only in the region of temperature linear increase and allows the calculation of temperature distributions along radius at a generic instant, as a function of the temperature of axis.

Such distributions are given for different values of m by the curves of Fig. 6.22.

In particular, for $m \rightarrow \infty$, the temperature distribution along the radius becomes parabolic:

$$\Theta - \Theta_a = \frac{1}{2}\xi^2$$

,

and the temperature differential between workpiece surface and axis is given by:

$$\Theta_s - \Theta_a = \frac{1}{2},$$

with Θ_s—value of Θ for $\xi = 1$, i.e. dimensionless temperature of the surface.

For lower values of m such temperature difference—for the same power transformed into heat within the cylinder—is lower, and can be calculated with Eq. (6.62) for $\xi = 1$.

In this way it results:

$$\Theta_s - \Theta_a = \frac{1}{2}F(m) \tag{6.63}$$

with:

$$F(m) = 1 - \frac{ber^2 m + bei^2 m - 1}{m(ber\, m\, ber' m + bei\, m\, bei' m)} \tag{6.63a}$$

The function F(m) can be considered a correction factor to the case of surface heating, which takes into account that in induction heating the heat sources are distributed inside the workpiece.

The function F(m) is given, as a function of *m*, by the curve (a) of Fig. 6.23.

However, for the evaluation of heating times it is more convenient to make reference to the average temperature ϑ_m in the workpiece cross-section instead of the temperature ϑ_a of the axis.

Since the dimensionless average temperature is:

$$\Theta_m = \frac{2\pi\lambda}{P_u}\vartheta_m = 2\int_0^1 \Theta\xi\, d\xi,$$

taking into account Eq. (6.62) we obtain:

$$\Theta_m = \Theta_a + F'(m) \tag{6.64}$$

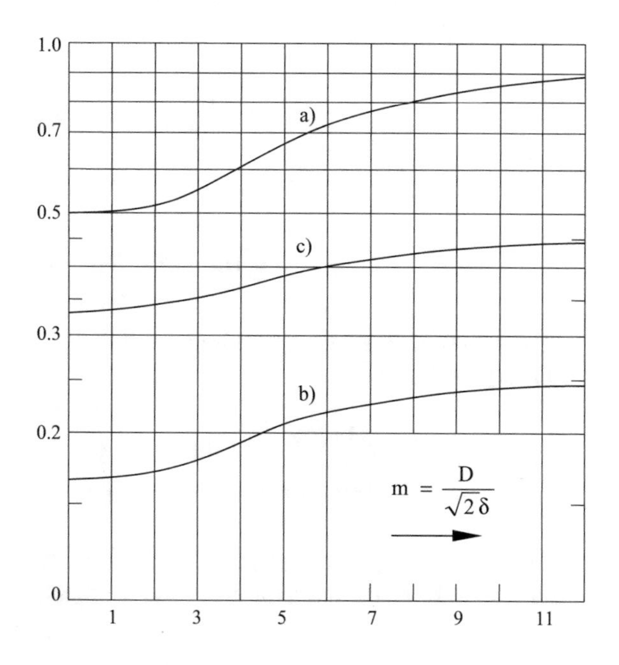

Fig. 6.23 Values of F(m), F′(m), F″(m) as a function of m [**a** F(m); **b** F′(m); **c** F″(m)]

with:

$$F'(m) = \frac{1}{4} - \frac{\text{ber } m\,\text{bei}' m - \text{bei } m\,\text{ber}' m - m/2}{m^2(\text{ber } m\,\text{ber}' m + \text{bei } m\,\text{bei}' m)} \tag{6.64a}$$

The values of $F'(m)$ are given as function of m in Fig. 6.23 (curve *b*).
For Eq. (6.60) and the assumption of negligible surface losses it is also:

$$\Theta_m = 2\,\tau, \tag{6.65}$$

thus from Eqs. (6.63) and (6.65) it is:

$$\Theta = 2\tau - F'(m) + \frac{1}{2}[\xi^2 - \frac{\text{ber}^2 m\xi + \text{bei}^2 m\xi - 1}{m(\text{ber } m\,\text{ber}' m + \text{bei } m\,\text{bei}' m)}]. \tag{6.66}$$

In particular, for $\xi = 1$:

$$\Theta_s = 2\,\tau - F'(m) + \frac{1}{2}F(m). \tag{6.67}$$

Equations (6.63) and (6.67) can be used in "*through heating*" processes for the evaluation of heating time t and power P_u required for raising the workpiece temperature to a given final surface temperature ϑ_s with a specified differential $(\vartheta_s - \vartheta_a)$ between surface and axis.

From the above equations, indicating with:

$\varepsilon = \frac{\vartheta_s - \vartheta_a}{\vartheta_s}$ relative temperature differential, referred to the surface temperature, we can write:

$$\tau = \frac{F(m)}{4\,\varepsilon}[1 - \varepsilon F''(m)] \tag{6.68}$$

with:

$$F''(m) = 1 - 2\frac{F'(m)}{F(m)} \tag{6.68a}$$

The values of $F''(m)$ are given by the curve (*c*) of Fig. 6.23.
Finally we have:

$$\left.\begin{aligned} t &= r_e^2\frac{F(m)}{4k\varepsilon}[1 - \varepsilon F''(m)] = r_e^2\frac{\pi c\gamma}{P_u}\vartheta_s[1 - \varepsilon F''(m)] \\ P_u &= \frac{4\pi\lambda}{F(m)}(\vartheta_s - \vartheta_a) = 4\pi\lambda\frac{\varepsilon}{F(m)}\vartheta_s \end{aligned}\right\} \tag{6.69}$$

These equations and the values of $F''(m)$ emphasize that for values of ε relatively small (ε = 0.05–0.10), the term $\varepsilon F''(m)$ is negligible compared to unit. In fact, it

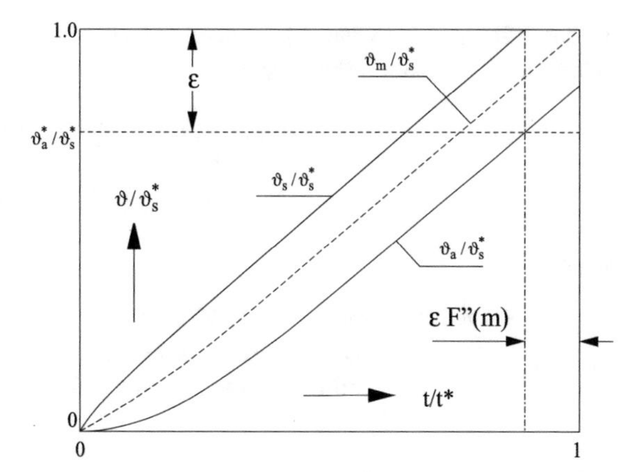

Fig. 6.24 Thermal transient in through heating of non-magnetic loads, with negligible surface losses (*—values at the end of heating)

only takes into account that the evaluation of the heating time is made with reference to the workpiece final surface temperature instead of the average one, as it is illustrated in the drawing of Fig. 6.24.

Equation (6.69) highlight the decisive influence on the heating times of the value of ε and the frequency.

In particular, they show that:

- increasing the frequency, increases the value of F(m) and, as a consequence, all other conditions being the same, reduces the power P_u and increases the heating time;
- at constant frequency, lower values of ε can be obtained reducing the specific power P_u and increasing the time for reaching the final average temperature;
- the final temperature ϑ_s (imposed by the technological process) and the heating time (defined on the basis of the required production rate) leads to the specification of the power P_u.
  Therefore, the only parameter on which the designer can act (within a limited extent) in order to reduce ε, is the frequency, which allows to modify the value of m and F(m);
- conversely, when the values of ϑ_s and ε are given (on the basis of process technological requirements), it is possible to increase the power P_u and hence to reduce within certain limits the heating time, only by adopting a lower value of F(m) through a convenient frequency selection.

In the previous analysis have been neglected the losses at the surface of the cylinder. These losses, on one side reduce the useful power and, as a consequence, give rise to longer heating times; on the other side, they produce lower temperature differentials in the cross-section, due to the decrease of temperature in proximity of the surface caused by them.

Under these conditions, an approximate evaluation of the heating time and the temperature differential can be done making reference to the useful power P_u^* (equal to the difference between the induced "electromagnetic" power P_e and the surface power losses) through the relations:

$$\left.\begin{aligned} t &= r_e^2 \frac{\pi c\gamma}{P_u^*} \vartheta_s [1 - \varepsilon F''(m)] \\ (\vartheta_s - \vartheta_a) &= \frac{P_u^*}{4\pi\lambda} F''(m) \end{aligned}\right\} \tag{6.70}$$

where:

$P_u^* = P_e - 2\pi r_e p_i$ useful power (W/m)
p_i specific surface losses (W/m^2)
$F^*(m)$ coefficient given in Fig. 5.26 as a function of m and the ratio P_u^*/P_e.

Example 6.3 For the same inductor and load of Examples 6.1 and 6.2, calculate power, heating time, supply voltage and current required in order to heat the workpiece from 0 to 1000 °C with a final temperature differential ε = 10 %.

The following average values in the considered temperature interval are used:

$$\lambda = 22.0\,W/m^\circ C;\, k = \lambda/c\gamma = 4.94 \cdot 10^{-6}\,m^2/s;\, p_i = 3.03 \cdot 10^4\,W/m^2$$

- From Figs. 6.23 and 6.25 we have:

$$\text{for } m = 2.5 \rightarrow F(m) = 0.53;\, F''(m) \approx 0.35$$

- with these values and those of Examples 6.1 and 6.2 it is: [Equation (6.69)]

$$p_u = 4\pi\lambda \frac{\varepsilon}{F(m)} \vartheta_s = \frac{4\pi \cdot 22.0 \cdot 0.1 \cdot 1000}{0.53} = 5.22 \cdot 10^4\ W/m$$

$$p_e = P_u/(2\pi r_e) = 5.22 \cdot 10^4/(2\pi \cdot 20 \cdot 10^{-3}) = 41.5 \cdot 10^4\ W/m^2$$

$$t = r_e^2 \frac{F(m)}{4k\,\varepsilon} [1 - \varepsilon F''(m)] = \frac{(20 \cdot 10^{-3})^2\, 0.53}{4 \cdot 4.94 \cdot 10^{-6} \cdot 0.1} [1 - 0.1 \cdot 0.35] = 104\,s$$

[Equation (6.21)]

$$H_e = \sqrt{\frac{p_e \delta}{\rho \sqrt{2P}}} = \sqrt{\frac{41.5 \cdot 10^4\, 1.125 \cdot 10^{-2}}{1 \cdot 10^{-6} \cdot 1.41 \cdot 0.4718}} = 8.38 \cdot 10^4\ A/m$$

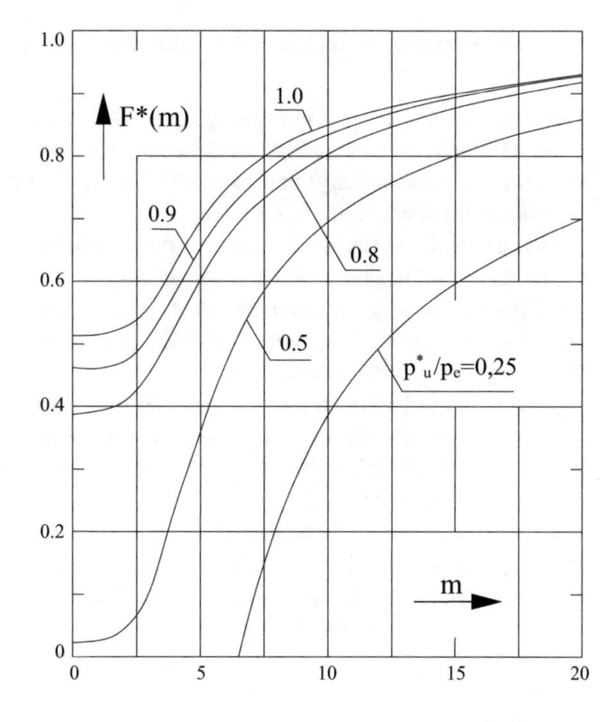

Fig. 6.25 Correction factor which takes into account surface thermal losses

[Equation (6.1)]

$$I = (H_e \ell)/N = 8.38 \cdot 10^4 \cdot 0.7/32 = 1833\,\text{A}$$
$$V = |Z_e| I = 8.264 \cdot 10^{-2} \cdot 1833 = 151.4\,\text{V}$$

- taking into consideration surface losses, the above results modify as follows:

$$P_u^* = 2\pi\, r_e (p_e - p_i) = 2\pi \cdot 20 \cdot 10^{-3} \cdot (41.5 - 3.03) \cdot 10^4 = 4.83 \cdot 10^4\,\text{W/m}$$
$$p_u^*/p_e = (pe - p_i)/p_e = 0.925 \rightarrow F^*(m) \approx 0.49$$

[Equations (6.70)]

$$t^* = t\frac{p_e}{p_u^*} = \frac{104}{0.925} \approx 112\,\text{s}$$

$$(\vartheta_s - \vartheta_a)^* = \frac{P_u^*}{4\pi\lambda} F^*(m) = \frac{4.83 \cdot 10^4}{4\pi \cdot 22.0} 0.49 \approx 85.6\,^\circ\text{C}$$

$$\varepsilon^* = (\vartheta_s - \vartheta_a)^*/\vartheta_s = 85.6/1000 \approx 8.6\,\%$$

6.7 Variation of Material Characteristics During Heating

The relationships of previous paragraphs have been derived assuming that the physical characteristics of the workpiece material (electrical resistivity, specific heat, thermal conductivity) and the conditions of thermal exchange with the environment were constant during heating.

In practice, however, the variation of these characteristics with temperature, have a remarkable influence on the heating process.

Moreover, in case of magnetic steel, in addition to the above influences, the variation of magnetic permeability with temperature and local magnetic field intensity plays a fundamental role.

Data of these parameters are given in Table A.6 of Appendix.

The analysis of these data emphasizes that the assumption of constant parameters is only admissible within limited temperature intervals, in which constant average values can be assumed.

In the following we examine in a qualitative way the consequences of such variations, making reference to a heating transient in which the average temperature rises from ϑ_1 to ϑ_2 and assuming, for simplicity, to neglect the temperature differences in the workpiece cross-section.

Furthermore, we will use the subscripts "1" and "2" to denote the values of all parameters at the above temperatures.

From equations of previous paragraphs we can draw the following conclusions:

- the penetration depth δ [Eqs. (6.5a) and (6.5b)] changes in the ratio:

$$\frac{\delta_2}{\delta_1} = \sqrt{\frac{\rho_2 \mu_1}{\rho_1 \mu_2}}$$

- the parameter m [Eqs. (6.5a) and (6.5b)] varies according to:

$$\frac{m_2}{m_1} = \frac{\delta_1}{\delta_2} = \sqrt{\frac{\rho_1 \mu_2}{\rho_2 \mu_1}}$$

  (**N.B.**—as shown in Sect. 6.6.2, for good electrical efficiency, the minimum value of m should be ≥ 2.2–2.5)
- the induced currents distribution within the body changes with m, as illustrated in Fig. 6.8
- during a heating processes with the inductor supplied with constant current, the surface induced current density J_e [Eq. (6.11)] changes in the ratio

$$\frac{J_{e2}}{J_{e1}} = \frac{m_2}{m_1} \sqrt{\frac{(P^2 + Q^2)_2}{(P^2 + Q^2)_1}}$$

- the power per unit volume w_e [Eq. (6.14a)] modifies in the ratio

$$\frac{w_{e2}}{w_{e1}} = \frac{\rho_2}{\rho_1}\left(\frac{J_{e2}}{J_{e1}}\right)^2 = \frac{\rho_2}{\rho_1}\left(\frac{m_2}{m_1}\right)^2 \frac{(P^2+Q^2)_2}{(P^2+Q^2)_1}$$

- when *heating with constant inductor current supply*, the induced active power P_w as well as the specific power p_e per unit surface [Eqs. (6.17) and (6.21)], change in the ratio

$$\frac{P_{w2}}{P_{w1}} = \frac{p_{e2}}{p_{e1}} = \frac{A(m_2)}{A(m_1)} = \frac{\rho_2}{\rho_1} \cdot \frac{m_1}{m_2} \cdot \frac{P(m_2)}{P(m_1)} = \frac{\mu_2}{\mu_1} \frac{m_1}{m_2} \frac{P(m_2)}{P(m_1)}$$

- with *constant inductor voltage supply* and ideal inductor of negligible resistance, the power P_w is given by:

$$\begin{aligned} P_w = R'_w I^2 &= \frac{V^2}{X_{i0}} \left\{ \alpha^2 \frac{\mu A}{(\mu A)^2 + [\alpha^2 - (1-\mu B)]^2} \right\} \\ &= \frac{V^2}{X_{i0}} G(\alpha, \mu, m). \end{aligned} \quad (6.71)$$

Therefore, during the heating transient, P_w modifies due to the variations of μ and m in the ratio:

$$\frac{P_{w2}}{P_{w1}} = \frac{G(\alpha, \mu_2, m_2)}{G(\alpha, \mu_1, m_1)}$$

- as shown in Sects. 6.6.2 and 6.6.3, in accordance with the variations of m will change also the values of electrical efficiency and power factor [see Eqs. (6.50) and (6.53)].

6.7.1 Non-magnetic Materials

In non-magnetic materials the relative permeability μ is equal to 1, whereas the variations of resistivity with temperature are mostly of the type

$$\rho = \rho(\vartheta) = \rho_0(1 + \alpha_0 \vartheta)$$

Typical values of $\rho = \rho(\vartheta)$ are given in in appendix, Table A.6.

As a consequence of the resistivity variations, when the temperature increases,

- increases the penetration depth δ
- decreases the value of m

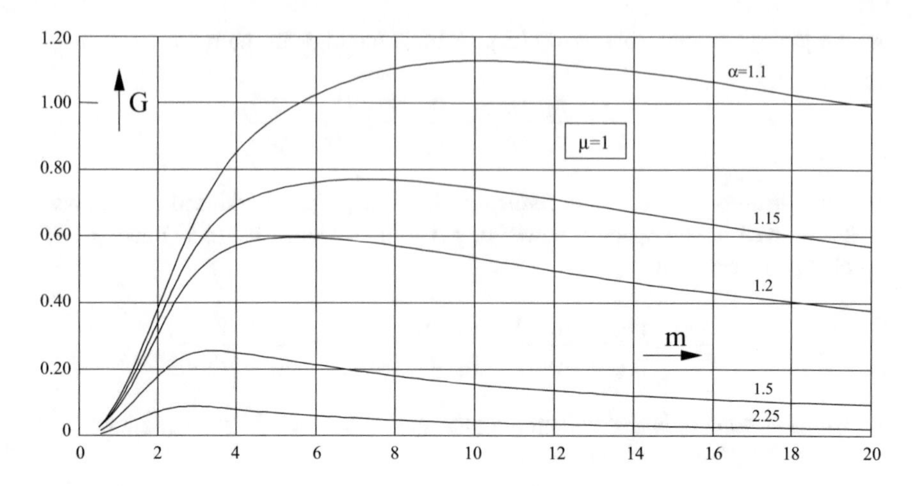

Fig. 6.26 Coefficient $G(\alpha, \mu, m)$ of active power induced into a non-magnetic cylinder at constant inductor voltage (ideal inductor of resistance $R_i' = 0$)

- the induced currents as well as the heat sources penetrate deeper into the body
- decreases the value of J_e
- *at constant inductor current*, the active power induced in the workpiece changes with *m* according to the values of the coefficient A (Fig. 6.7). This generally corresponds to an increase of P_w with temperature.
- *at constant inductor voltage*, P_w varies proportionally to the coefficient $G(\alpha, \mu, m)$, which is given as a function of *m* in the diagrams of Fig. 6.26. Thus P_w can increase or decrease during heating, depending on the initial value of *m*.

Example 6.4 Heating of a copper billet with diameter $d_e = 2r_e = 70$ mm, from ambient temperature $\vartheta_1 = 20$ °C to hot working temperature $\vartheta_2 = 800$ °C, with constant inductor current supply at f = 50 Hz.

From Table A.6 (see appendix) we have:
$\rho_1 \approx 1.72 \cdot 10^{-8}\,\Omega$ m; $\rho_2 \approx 6.55 \cdot 10^{-8}\,\Omega$ m; $\rho_2/\rho_1 = 3.81$

$\delta_1 = 503\sqrt{\frac{1.72 \cdot 10^{-8}}{50}} = 9.33 \cdot 10^{-3}\,\mathrm{m}$ $\quad\delta_2 = 503\sqrt{\frac{6.55 \cdot 10^{-8}}{50}} = 18.2 \cdot 10^{-3}\,\mathrm{m}$

$m_1 = \frac{\sqrt{2}35 \cdot 10^{-3}}{9.33 \cdot 10^{-3}} \approx 5.3$ $\quad m_2 = \frac{\sqrt{2}.35 \cdot 10^{-3}}{18.2 \cdot 10^{-3}} \approx 2.7$

$\delta_2/\delta_1 = 1.95$ $\quad m_2/m_1 = 0.51$

$m_1\sqrt{P_1^2 + Q_1^2} = 4.96$ $\quad m_2\sqrt{P_2^2 + Q_2^2} = 2.48$

$P_1 = 0.609$ $\quad P_2 = 0.502$

$J_{e2}/J_{e1} = 2.48/4.96 = 0.50$ $\quad w_{e2}/w_{e1} = 3.81 \cdot 0.5^2 = 0.953$

$$P_{w1}/P_{w2} = p_{e1}/p_{e2} = \sqrt{3.81} \cdot (0.502/0.609) = 1.61.$$

6.7.2 Magnetic Materials

In these materials (steel, iron, nickel), in addition to the change of resistivity with temperature, magnetic permeability undergoes strong variations, which depend on temperature and magnetic field intensity [3–5],

$$\mu = \mu(\mathrm{H}, \vartheta),$$

as shown as an example in Fig. 6.27.

The material most used in industrial applications is carbon steel; characteristics of steel are given in Figs. 6.28 and 6.29.

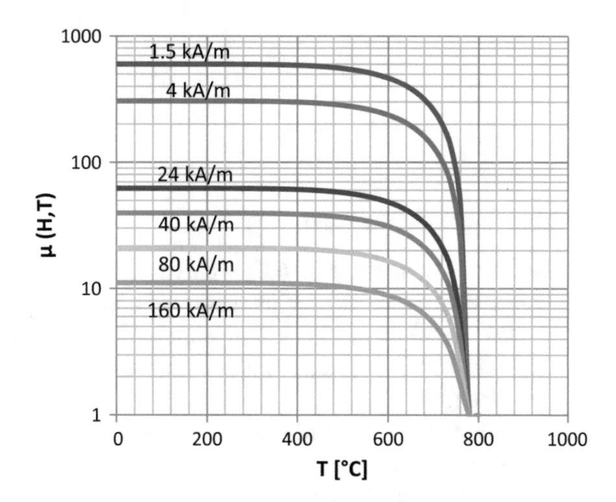

Fig. 6.27 Relative magnetic permeability as a function of temperature and magnetic field intensity [5]

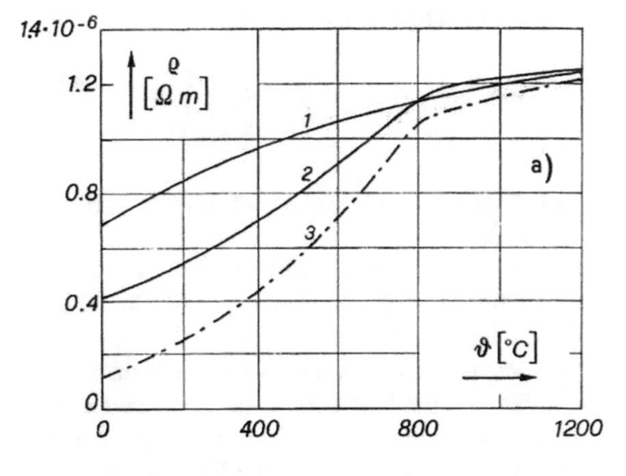

Fig. 6.28 Resistivity of different kind of steel as a function of temperature (*1* Inox 18 Cr Ni8; *2* Rapid steel; *3* Carbon Steel 0.06 %)

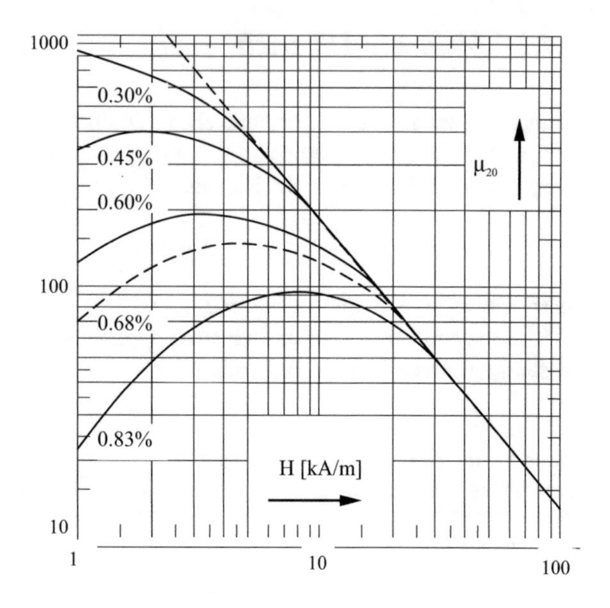

Fig. 6.29 Relative permeability of steels μ_{20} at 20 °C with various carbon content as a function of magnetic field intensity [3]

They highlight the following:

- in magnetic steels (Fig. 6.28—curves 2 and 3), the variation of resistivity is non-linear: the resistivity increases by approximately 3–5 times in the interval 0–800 °C, while at higher temperature the rate of increase is much slower. At 1200–1300 °C the value is approximately 6–7 times higher than at room temperature (curve 3);
- the value of permeability at 20 °C (μ_{20}), depends not only on magnetic field intensity, but also on the steel carbon content (Fig. 6.29). At high field intensities, which is the range of interest in applications, the value of μ_{20} can be estimated by approximate formulae proposed by different authors, which correspond to the straight line on the right in the figure.
  Formulae frequently used are:

$$\left.\begin{aligned} \mu &= 1+\frac{B_s}{H} \approx 1+\frac{14{,}000-16{,}000}{H} \\ \mu &\approx 8{,}130\,H^{-0.894} \end{aligned}\right\} \tag{6.72}$$

  with: B_S—magnetic saturation induction (Gauss); H—magnetic field intensity (A/cm).
  Note—The formulae make reference to the first time harmonics of B and H. In induction heating calculations, hysteresis losses are usually neglected, since at worst they may amount to 7 % [6].
- the magnetic permeability varies very little with temperature in the range from 20 °C to about 500–600 °C (as illustrated in Fig. 6.27), whereas it falls rapidly thereafter to the value 1, which is reached at "*Curie Point*". Its variation with temperature can be accounted for with the formula [5, 7]:

$$\mu = 1 + (\mu_{20} - 1) \cdot \varphi(\vartheta) \tag{6.73}$$

with $\varphi(\vartheta)$ given by the diagram of Fig. 6.30.

As a consequence, the heating process up to a final temperature of about 1200 °C can be schematized very roughly by two different heating stages with constant parameters: the first from room temperature to 750 °C with average resistivity $\rho_A \approx 50–60 \cdot 10^{-8}$ (Ω m), and permeability μ_A, which can be obtained from the curves of Figs. 6.29 and 6.30 or Eq. (6.72); the second one, corresponding to the heating period above Curie Point, with permeability $\mu_B = 1$ and resistivity $\rho_B \approx 100–110 \cdot 10^{-8}$ (Ω m).

With these values, and the magnetic field intensity H_e in the range most commonly used, one obtains:

$$\frac{\delta_B}{\delta_A} = \frac{m_A}{m_B} \approx 3–10 \quad \Rightarrow \quad m_A \geq (3–10) \cdot 2.5 = 7.5–25$$

However, this schematization gives only a first idea of the phenomena occurring during heating.

In reality, the value of *m* decreases progressively from a high initial value (with induced current and power distributions concentrated in "thin" surface layers) to lower and lower values. As a consequence the induced current and power density distributions penetrate progressively deeper inside the workpiece.

With reference to Eq. (6.21), rewritten in the form:

$$p_e = H_e^2 \frac{\rho}{\delta} \sqrt{2} P \approx 2.81 \cdot 10^{-3} \sqrt{\rho \mu f} P$$

and values of ρ, μ and P variable with ϑ, the power induced in the workpiece changes —during a *heating with constant inductor current*—as illustrated in Fig. 6.31.

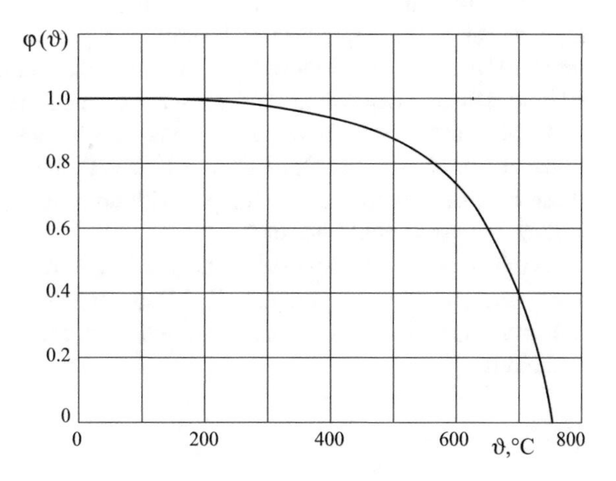

Fig. 6.30 Diagram of $\varphi(t)$ as a function of temperature

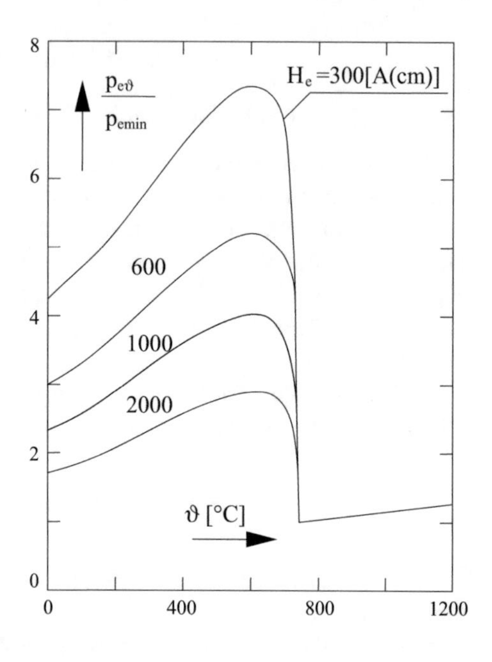

Fig. 6.31 Relative variations of equivalent surface specific power p_e with temperature during heating of magnetic steel with constant surface magnetic field intensity (p_{emin}, $p_{e\vartheta}$—minimum value of p_e during heating and at temperature ϑ; H_e—surface magnetic field intensity)

The shape of the curves is characterized by an initial slope due to the increase of resistivity, while permeability remains practically constant. In the second part, the rate of increase of resistivity is lower and the curve reaches a maximum at about 550–600 °C. Then a sudden decrease occurs due to the strong decrease of permeability, which prevails on the increase of the resistivity.

Above the Curie temperature, the power slowly starts to rise again as a consequence of the increase of resistivity. However the induced power remains at values not only much lower than the maximum, but also lower than the initial one.

In case of *heating with constant inductor voltage,* using Eq. (6.71), one obtains, for various values of m and μ, the diagrams shown in the Fig. 6.32a–d.

From the shape of the diagrams it may be concluded that, since m has always high values at the beginning of heating and progressively decreases as the temperature increases, the induced power may either increase or decrease during the heating transient depending on the value of α and the initial value of m.

In particular, for high values of α and low values of μ, in the impedance Z_{e0}, given at Eq. (6.20), it is $(X_a + X'_w) \approx X_a \gg R'_w$ and the curves of G are similar to those obtained for the case of supply with constant inductor current (i.e. constant surface magnetic field intensity).

On the contrary, for low values of α and high values of μ, the terms R'_w and X'_w vary strongly with temperature. It follows that at constant voltage supply, the inductor current and the magnetic field intensity also undergo considerable variations.

Fig. 6.32 Coefficients for evaluation of the variations of induced power in a magnetic cylinder at constant inductor voltage supply, according to Eq. (6.71) (ideal inductor with own resistance $R_i = 0$; curves: **a** $\mu = 5$; **b** $\mu = 10$; **c** $\mu = 15$; **d** $\mu = 50$ $\alpha = r_i/r_e$)

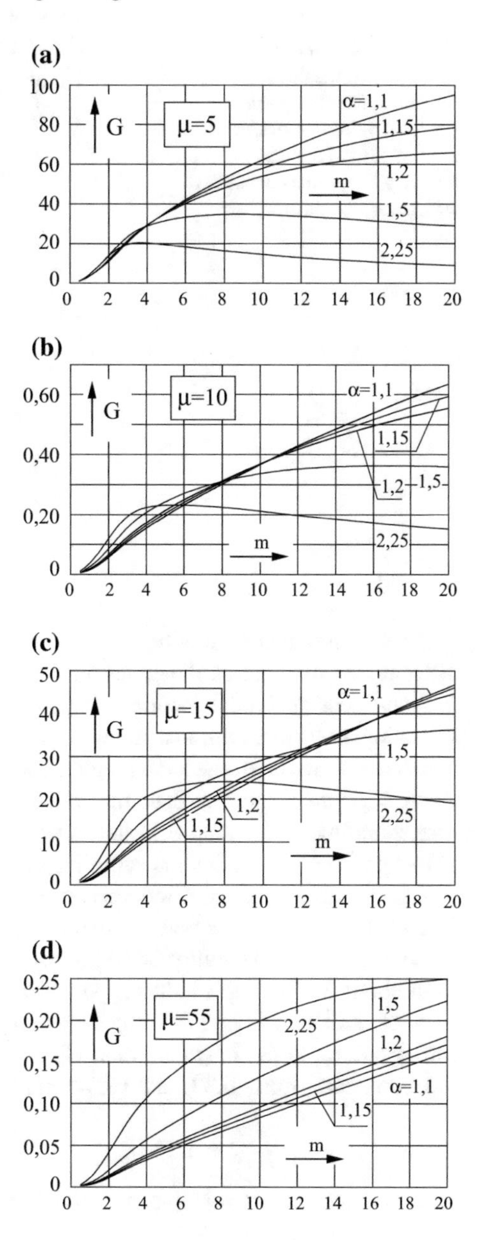

The process is further complicated by the fact that the previous description refers to a single average value of permeability, variable with temperature but constant in the cross section of the workpiece.

This does not correspond to the real phenomena which, as shown by Eq. (6.7) and Fig. 6.6, produce different values of magnetic field intensity along the radius of the workpiece, depending on H_e and *m*.

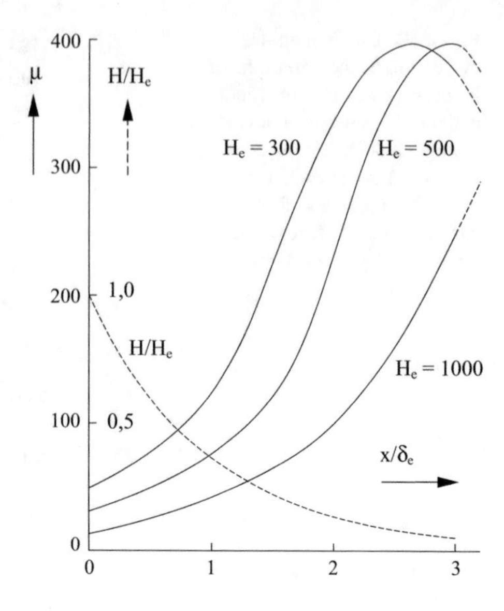

Fig. 6.33 Distribution of permeability within a ferro-magnetic body due to the local values of magnetic field intensity (ρ = cost; μ = var; x—distance from the surface; H_e (A/cm))

As a consequence, moving from the surface to the core of the workpiece, the distributions of μ are as illustrated in Fig. 6.33.

Non-linear analytical theories, developed by different authors taking into consideration such distributions, allow to state that in an infinite semi-plane with μ variable as function of the local magnetic field intensity, the induced active power is about 1.37 times greater than the value calculated with a constant value of μ, corresponding to the magnetic field intensity at the surface, whereas the reactive power practically remains the same [8, 9].

Also the distributions of induced current and power density will be modified: in fact, the electromagnetic wave penetrates to a lower depth within the body due to the increase of permeability in comparison with value at the surface. In this situation about 87 % of the induced power is transformed into heat within a surface layer of thickness $\delta_e/1.37$, with: δ_e—value of δ evaluated with the permeability corresponding to the magnetic field H_e at the surface.

In this case, Eqs. (6.33) and (6.34) must be rewritten in the form:

$$\left.\begin{array}{l} p_e = 1.37\,H_e^2\,\frac{\rho}{\delta_e} \\ \dot{Z}_{ec} = \frac{a}{\ell}\frac{\rho}{\delta_e}(1.37 + j\,0.97); Z_{ec} = 1.68\,\frac{a}{\ell}\frac{\rho}{\delta_e} \end{array}\right\} \tag{6.74}$$

Taking into consideration the high values of *m* which occur below Curie point up to 500–600 °C, the above equations can be practically applied, in this temperature range, also to the cylindrical case.

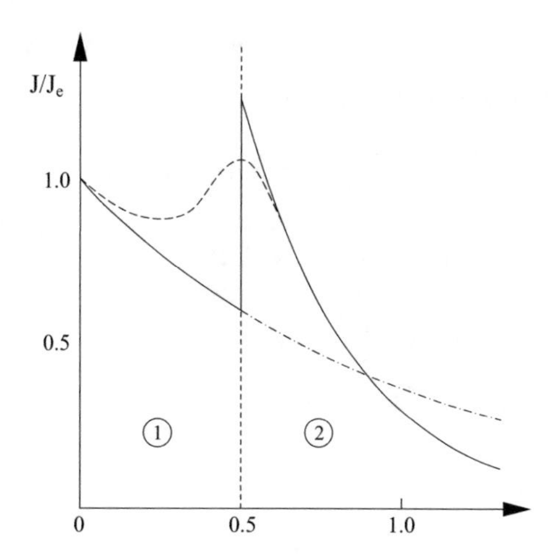

Fig. 6.34 Qualitative current density distribution in magnetic and non-magnetic layers at the transition of Curie point (*1* non-magnetic layer; *2* magnetic layer)

A second phenomenon, which modifies the theoretical distributions of induced currents and power, is due to the fact that during heating, at the transition of Curie point, a surface layer at higher temperature becomes non-magnetic, whereas the internal layers are still magnetic.

In this situation, the current density distribution can be schematized by two exponentials corresponding to the different values of δ in the magnetic and non-magnetic regions, as illustrated in Fig. 6.34.

In particular, in certain cases the value of the current density at the surface of separation of magnetic and non-magnetic regions can be even higher than the one at the surface.

However, given that in reality the values of ρ and μ change gradually from point to point, the distribution is in practice more complex, as indicated qualitatively by the dashed curve in the same figure.

Example 6.5 Heating from 0 to 1200 °C a bar of steel UNI-C45 (diameter $d_e = 40$ mm; axial length $\ell = 700$ mm) with the inductor of Example 5.1 ($N = 32$ windings; $r_i = 35$ mm; $\ell_i = 700$ mm; $f = 2000$ Hz; $I = 1000$ A).

- Considering, as in Example 6.1, a heating process with constant exciting field $H_e = NI/\ell_i = 457$ A/cm, with reference to the results of previous paragraph we assume—below and above the Curie point—the following values:

$\rho_A = 0.55 \cdot 10^{-6} \Omega\,\text{m}$ $\rho_B = 1.00 \cdot 10^{-6}\,\Omega\,\text{m}$
$\mu_A = 35$ $\mu_B = 1$

Therefore, above Curie, the same values calculated in Example 6.1 will be used here.

The results of calculation are the following:

- Penetration depth, parameter *m*, and parameters dependent on *m*

$\delta_A = 503\sqrt{\frac{0.55 \cdot 10^{-6}}{35\,2000}} = 1.41 \cdot 10^{-3}\,\mathrm{m} = 0.141\,\mathrm{cm}$ $\delta_B = 1.125\,\mathrm{cm}$

$m_A = \sqrt{2}r_e/\delta_A = \sqrt{2}20 \cdot 10^{-3}/1.41 \cdot 10^{-3} = 20.1$ $m_B \approx 2.5$

$P_A + j\,Q_A = 0.6819 + j0.7073$ $P_B + j\,Q_B = 0.4718 + j0.777$

$A_A + jB_A = (6.819 + j\,7.073) \cdot 10^{-2}$ $A_B + j\,B_B = 0.3774 + j\,0.6216$

$m_A\sqrt{P_A^2 + Q_A^2} = 19.65$ $m_B\sqrt{P_B^2 + Q_B^2} = 2.27$

$m_A^2(P_A^2 + Q_A^2) = 386.1$ $m_B^2(P_B^2 + Q_B^2) = 5.167$

- Specific surface values

$$J_{eA} = (H_e/r_e)m_A\sqrt{P_A^2 + Q_A^2} = (45{,}700/20 \cdot 10^{-3})19.65 = 44.9 \cdot 10^6\,\mathrm{A/m^2} = 4490\,\mathrm{A/cm^2}$$

$$J_{eB} = 5.194 \cdot 10^6\,\mathrm{A/m^2} = 519.4\,\mathrm{A/cm^2}$$

$E_{eA} = \rho_A J_{eA} = 0.55 \cdot 10^{-6}\,44.9 \cdot 10^6 = 24.7\,\mathrm{V/m}$ $E_{eB} = \rho_B J_{eB} = 5.194\,\mathrm{V/m}$

$w_{eA} = \rho_A J_{eA}^2 = 0.55 \cdot 10^{-6}(44.9 \cdot 10^6)^2 = 1109 \cdot 10^6\,\mathrm{W/m^3} = 1109\,\mathrm{W/cm^3}$ $w_{eB} = \rho_B J_{eB}^2 = 26.98 \cdot 10^6\,\mathrm{W/m^3} \approx 27.0\,\mathrm{W/cm^3}$

- reactance X_{io} (see Example 6.1)

$$X_{i0} = \omega N^2 \frac{\mu_0 \pi \alpha^2 r_e^2}{\ell} = 8.843 \cdot 10^{-2}\,\Omega$$

- equivalent impedance $\dot{Z}_{e0}$

(a) *Below Curie*

$$R_{wA} = \frac{X_{i0}}{\alpha^2}\mu A = \frac{8.843 \cdot 10^{-2}}{1.75^2} 356.819 \cdot 10^{-2} = 6.891 \cdot 10^{-2},\ \Omega$$

$$\Delta X = \frac{X_{i0}}{\alpha^2}(1 - \mu B) = \frac{8.843 \cdot 10^{-2}}{1.75^2}(1 - 357.073 \cdot 10^{-2}) = -4.261 \cdot 10^{-2}\,\Omega$$

$$\dot{Z}_{eA0} = R_{wA} + j(X_{i0} - \Delta X) = [6.891 + j(8.843 + 4.261)] \cdot 10^{-2} = (6.891 + j\,13.104) \cdot 10^{-2}\,\Omega$$

$$Z_{eA0} = 14.805 \cdot 10^{-2}\,\Omega$$

(b) *Above Curie*:

$R_{wB} = 1.090 \cdot 10^{-2}\,\Omega \qquad \Delta X = 1.093 \cdot 10^{-2}\,\Omega$
$\dot{Z}_{eB0} = (1.090 + j\,7.750) \cdot 10^{-2}\,\Omega \quad Z_{eB0} = 6.125 \cdot 10^{-2}\,\Omega$

- inductor voltage and power factor

$V_{0A} = Z_{e0A} I = 14.805 \cdot 10^{-2} \cdot 1000 = 148.05\ \text{V} \quad V_{0B} = 61.2\,V$
$\cos\varphi_{0A} = R_{wA}/Z_{e0A} = 0.465 \qquad \cos\varphi_{0A} = 0.178$

- active and reactive power

  (a) *Below Curie:*

$$P_{wA} + j\,(Q_a + Q_{wA}) = \dot{Z}_{e0A} I^2 = 68910 + j\,131040\ \text{W/VAr}$$

$$p_{0A} = H_0^2 \frac{\rho_A}{\delta_A} \sqrt{2} P_A = 45{,}700^2 \frac{0.55 \cdot 10^{-6}}{1.125 \cdot 10^{-2}} \sqrt{2} 0.6819$$
$$= 7.856 \cdot 10^5\ \text{W/m}^2 \approx 78.6\ \text{W/cm}^2$$

  (b) *Above Curie:*

$$P_{wB} + j(Q_a + Q_{wB}) = \dot{Z}_{e0B} I^2 = 10{,}900 + j\,77{,}500\ \text{W/VAr}$$
$$p_{0B} = 12.39 \cdot 10^4, \text{W/m}^2 \approx 12.4\,\text{W/cm}^2$$

Adding the resistance and internal reactance of the inductor coil calculated in Example 6.2 ($R_i = 4.5 \cdot 10^{-3}$ ohm; $X_{ii} = 3.69 \cdot 10^{-3}$ ohm), the values change as follows:

- equivalent impedance Z_e of the inductor-load system

  (a) *Below Curie:*

$$\dot{Z}_{eA} = \dot{Z}_{e0A} + (R_i + j\,X_{ii}) = [(6.891 + 0.45) + j\,(13.104 + 0.369)] \cdot 10^{-2}$$
$$= (7.341 + j\,13.473) \cdot 10^{-2}\,\Omega$$
$$Z_{eA} = 15.343 \cdot 10^{-2}\,\Omega$$

  (b) *Above Curie:*

$$\dot{Z}_{eB} = (1.540 + j\,8.119) \cdot 10^{-2}\,\Omega;\ Z_{eB} = 8.264 \cdot 10^{-2}\,\Omega$$

- Supply voltage, efficiency, power factor

(a) *Below Curie:*

$$V_A = Z_{eA} I = 153.4\,\text{V}$$
$$\eta_{eA} = 6.891 \cdot 10^{-2}/7.341 \cdot 10^{-2} = 0.939$$
$$\cos\varphi_A = 7.341 \cdot 10^{-2}/15.343 \cdot 10^{-2} = 0.478$$

(b) *Above Curie:*
$V_B = 82.6$ V; $\eta_{eB} = 0.708$; $\cos\varphi_B = 0.186$

6.8 Calculation of Inductors

For calculation of integral parameters of the inductor-load systems, approximate, analytical or numerical methods are available. The difference between the "long" system considered in the previous paragraphs and the real geometries is that the first were supposed to be a portion of an infinite axial length geometry, while in real systems it is not acceptable anymore to neglect the reluctance of the return path of the magnetic flux of the inductor coil.

This means that a substantial part of the *m.m.f.* produced by the coil is "used" in this part of the magnetic circuit and it is not effective in the heating process.

6.8.1 Approximate Method of the Electric Circuit Equivalent to Magnetic Reluctances

This method is based on a schematization of the magnetic circuit of the inductor-load system, obtained assuming that all inductor turns are linked with the total flux and substituting the magnetic system by its equivalent electric circuit.

For calculating magnetic reluctances, the magnetic circuit is subdivided into three parts: one constituted by the heating section (region A of Fig. 6.35) and two corresponding to the "return path" of magnetic flux (section B inside the coil and section C outside it).

In section A the total flux $\dot{\Phi}$ is subdivided in one component in air $\dot{\Phi}_a$, located in the gap between inductor and load, and one component $\dot{\Phi}_c$ within the heated workpiece.

Finally, we assume an ideal pattern of the magnetic field, with all field lines in section A parallel to the axis, whereas in sections B and C the field distribution is the same of that in the "empty" inductor (i.e. the system without load).

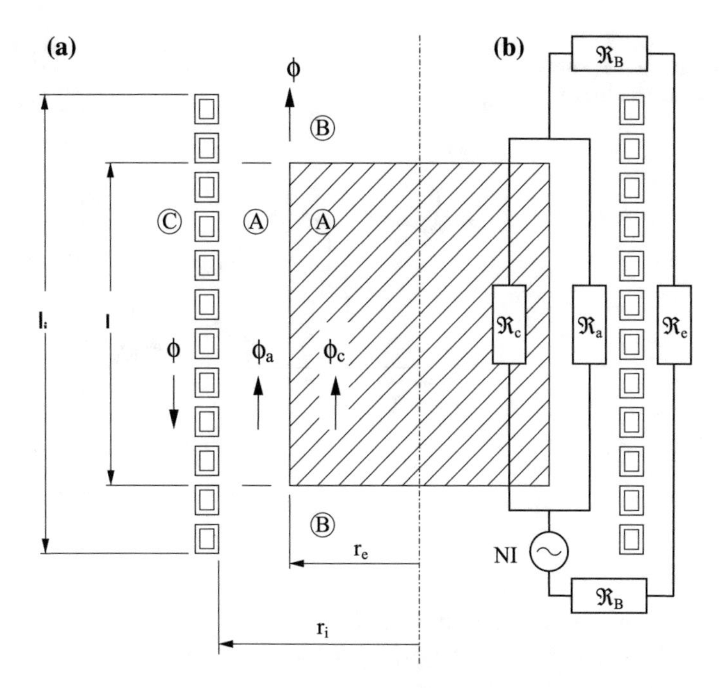

Fig. 6.35 **a** Inductor-load system; **b** equivalent magnetic circuit

With the above assumptions, to the system of Fig. 6.35a corresponds the equivalent magnetic circuit of Fig. 6.35b. It emphasizes again that the total inductor *m.m.f.* $N\dot{I}$ required for the flow of the total flux $\dot{\Phi}$ is spent partly in the reluctances $\Re_c$ and $\Re_a$ of section A and partly in the reluctance $\Re_0$ of the "return path" of the flux.

To the equivalent magnetic circuit of Fig. 6.35b corresponds the electric circuit of Fig. 6.36a and the phasor diagram of Fig. 6.36b. The equivalent electrical circuit is completed by the resistance R_i and the internal reactance X_{ii} of the inductor coil, which were not considered in the magnetic circuit.

The reactance X_0, which accounts for the return path of the flux, can be determined with reference to the total magnetic reluctance $\Re_0$ of sections B and C of the empty inductor as:

$$\Re_0 = \Re_{0B} + \Re_{0C} \tag{6.75}$$

The reluctance $\Re_{0B}$ of section B of length $(\ell_i - \ell)$ of the empty inductor, is given by the relationship:

$$\Re_{0B} = \frac{\ell_i - \ell}{\mu_0 S_i} = \Re_{i0} \frac{\ell_i - \ell}{\ell_i} \tag{6.76}$$

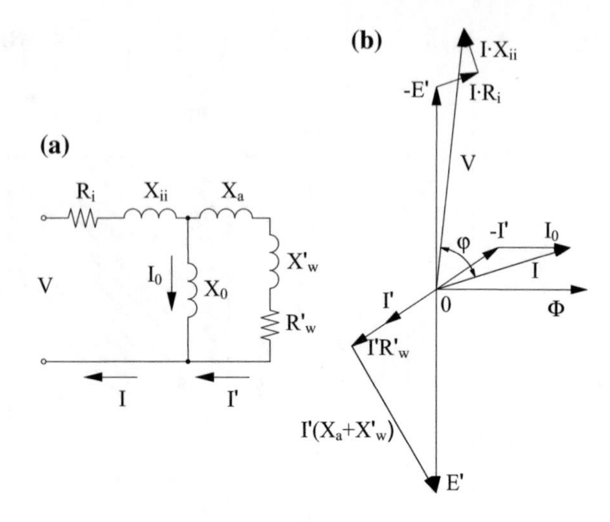

Fig. 6.36 **a** Equivalent electric circuit of magnetic scheme; **b** phasor diagram

with:

ℓ_i, ℓ	axial length of inductor and load
S_i	inductor cross-section (for cylindrical coils: $S_i = \pi r_i^2$)
μ_0	permeability of vacuum
$\Re_{i0} = \frac{\ell_i}{\mu_0 S_i}$	reluctance of the empty inductor coil considered as part of length ℓ_i of an inductor of infinite axial length

Equation (6.76) gives values of $\Re_{0B}$ lower than the real ones, because it does not take into account the distortion of flux lines near the front end surfaces of the load. However, this non-uniform flux distribution modifies the magnetic field pattern of section B only to a limited extent near the load end, leading to relatively small errors in the calculation of $\Re_{0B}$.

The reluctance $\Re_{0C}$ is evaluated by assuming it equal to the one of the external return path of an empty inductor of the same length. Therefore, it results:

$$\Re_{0C} = \Re_{i0}\left(\frac{1}{K_N} - 1\right) \tag{6.77}$$

with:

K_N—coefficient (≤ 1) for calculation of the inductance of short coils (known as coefficient of Nagaoka). The values of K_N are given in Fig. 6.37 for cylindrical coils, and Fig. 6.38 for coils of rectangular cross section.

For cylindrical coils K_N can be also calculated with the approximate formula:

$$K_N \approx \frac{2.3}{2.3 + \frac{2r_i}{\ell_i}}$$

which, for $2r_i/\ell_i \geq 0.2$, leads to errors below 3 %.

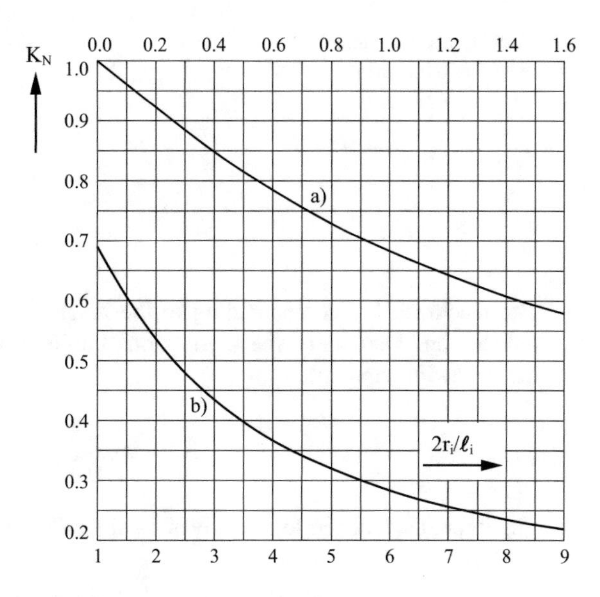

Fig. 6.37 Coefficient of Nagaoka for calculation of inductance of short cylindrical coils (curve **a** upper abscissae; curve **b** lower abscissae)

Using this formula and the symbols of Fig. 6.35, from Eq. (6.77) we can write:

$$\Re_{0C} \approx \frac{0.276}{\mu_0 R_i} = 2\Re_B + \Re_e \tag{6.78}$$

This equation highlights that the reluctance $\Re_{0C}$ is practically independent on coil length and, consequently, it can be thought as being constituted only by the reluctance $\Re_B$ of the regions close to the coil ends and not by the reluctance $\Re_e$ of the external path of length ℓ_i.

Finally, from Eqs. (6.76) and (6.77), the reactance X_0 becomes:

$$X_0 = X_{i0}\frac{K_N\ell_i}{\ell_i - K_N\ell} \tag{6.79}$$

with:

$$X_{i0} = \frac{\omega N^2}{\Re_{i0}}$$

Denoting with:

S_a—cross-section of the air gap between inductor and load,

the calculation of parameters of the equivalent electric circuit of Fig. 6.36 is completed as follows.

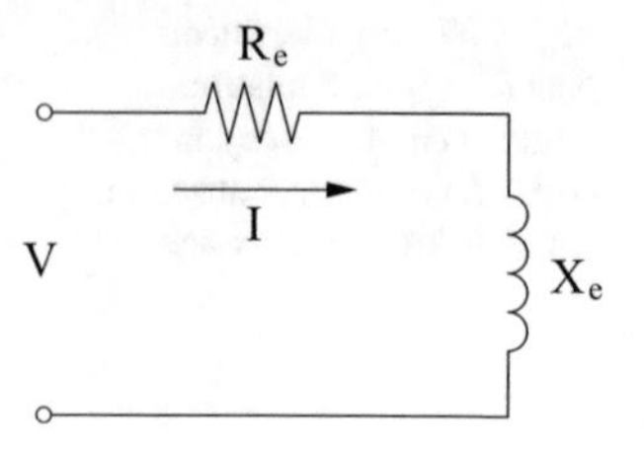

Fig. 6.38 Equivalent series circuit of the inductor-load system

- The reactance X_a, corresponding to the magnetic flux in the air gap between inductor and load, under the assumption of uniform field distribution is given by the following equation:

$$X_a = X_{i0}\frac{\ell_i S_a}{\ell S_i} \tag{6.80}$$

- The impedance of the load is given by the relationship:

$$R'_w + jX'_w = \frac{X_{i0}}{\alpha^2}\mu(A + jB)\frac{\ell_i}{\ell},$$

which is similar to Eq. (6.20), but where the values of A and B relevant for the geometry under consideration must be introduced
- The impedance of the induction coil $R_i + jX_{ii}$ can be evaluated, as shown at Sect. 6.6.1, by Eq. (6.47).

When the workpiece length exceeds that of the inductor, it is assumed $\ell_i \approx \ell$ since the induced currents are effective in the workpiece only in the zones facing the induction coil, whereas in the external part of it they decay very rapidly.

The circuit of Fig. 6.36a can be reduced to the series equivalent circuit of Fig. 6.38, whose parameters can be calculated with the relations:

$$\left.\begin{aligned} R_e &= R_i + n^2 R_w \\ X_e &= X_{ii} - n^2 X_w \end{aligned}\right\} \tag{6.81a}$$

where:

$$\left.\begin{aligned} R_w &= R'_w \\ X_w &= X_w = -\left[X_a + X'_w + \frac{(X_a + X'_w)^2 + R'^2_w}{X_0}\right] \\ n^2 &= \frac{1}{\left(\frac{R'_w}{X_0}\right)^2 + \left(1 + \frac{X_a + X'_w}{X_0}\right)^2} \end{aligned}\right\} \tag{6.81b}$$

Remarks

This method is applicable to many induction heating systems, e.g. empty inductors ($\rho \rightarrow \infty, \ell \rightarrow 0$ or $r_e \rightarrow 0$) or very high frequencies ($f \rightarrow \infty$).

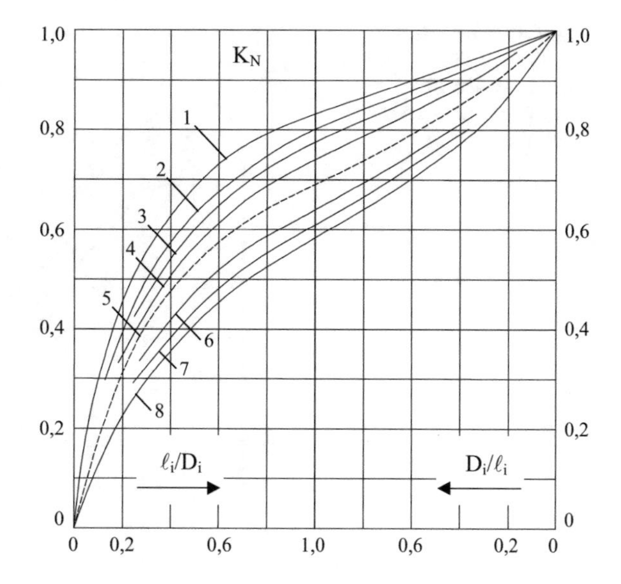

Fig. 6.39 Coefficient K_N for calculation of inductance of cylindrical solenoids with external magnetic cores (*curves 1–4*), without core (*curve 5*) or at high frequency (*curves 6–8*) (*curves 1–4*: $d_f/d_i = 1, 1.1, 1.2, 1.5$; *curve 5*: as in Fig. 5.38; curves *6–8*: $s/d_i = 0.05, 0.25, 0.6$) [10]

Moreover, it can be used for magnetic or non-magnetic loads, homogeneous or layered, with cylindrical or other shapes of the cross-section (oval, tubular, rectangular, etc.) and for systems with inductors—without internal magnetic concentrator—situated inside a cavity of the body to be heated.

However, in the following cases the calculation procedure must be applied with particular caution.

1. *Inductors with small number of turns at high frequency.* The results can be strongly influenced by the radial thickness *s* of the conductor constituting the coil turns. The coefficient K_N can be determined as a function of the ratios d_i/ℓ_i and s/ℓ_i by the curves 6–8 of Fig. 6.39.
2. *Inductors of rectangular cross-section of sides a and b.* The coefficient K_N can be determined by the diagrams of Fig. 6.40, as a function of the geometrical ratios a/ℓ_i and a/b of the cross section, whereas for the calculation of the load impedance, reference must be made to the coefficients *A* and *B* relevant in the case under consideration.
3. *Cylindrical inductors with external magnetic cores*
   The inductors considered here are those with straight magnetic concentrators without pole shoes, such as those being used in melting furnaces. The cores are mostly constituted by non-saturated lamination stacks of length ℓ_f longer than the coil $(\ell_f > \ell_i)$, uniformly distributed and conveniently spaced, along a circumference of diameter d_f outside the coil.
   The values of K_N can be determined with good approximation by the curves 1–4 of Fig. 6.39, valid for the case $\ell_f \gg \ell_i$.

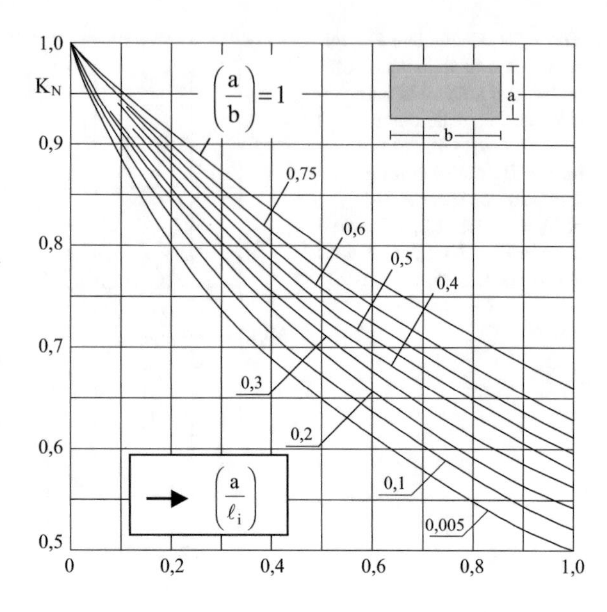

Fig. 6.40 Coefficient K_N for calculation of inductance of short solenoids with rectangular cross-section of sides a and b [10]

4. *Internal inductors without magnetic cores*
   In this case the return path of flux consists of the internal space of the inductor, and the reactance X_0 is practically equal to that of the empty inductor. Moreover, the inductor resistance can be notably lower (up to 50 %) of that given by Eq. (6.47), since in this case the current is distributed in the conductor cross section both on the external side facing the load for proximity effect and the internal one, due to ring effect [10–12].

6.8.2 Analytical Methods

The analytical methods are based on the analytical solution of electromagnetic field equations and determination of the vector potential, which allows to evaluate power distribution in the load, components of electric and magnetic field in each point of the system, induced *e.m.f.* in the inductor coil and, finally, impedance of the inductor-load system.

There are two types of solutions:

- One is based on the Fourier integral applied to the system geometry schematically shown in Fig. 6.41, which is characterized by a "short" coil and an infinitely long cylindrical load [13];
- The second one (the "*periodical fields solution*"), is developed with reference to the systems shown in Fig. 6.42, consisting of an inductor with several identical coils, equally spaced from each other along the axis, excited by the same current

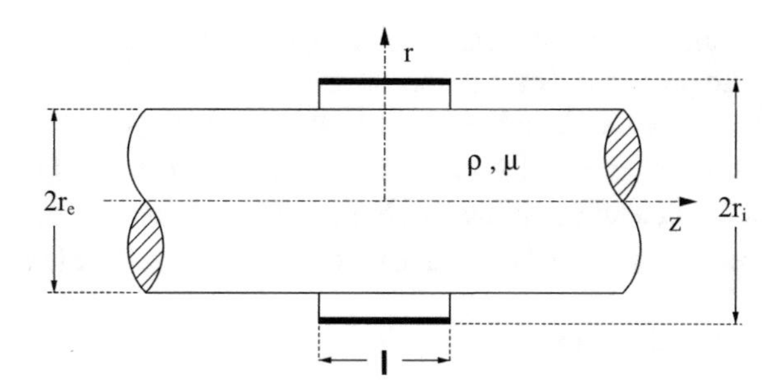

Fig. 6.41 Short inductor and cylindrical load of infinite axial length

and a "long" cylindrical load. Since the magnetic field distribution repeats itself identically along the axis, the solution is obtained by developing into Fourier series the vector potential and the field parameters derived from it [14–19].

Both methods give results with an accuracy sufficient for technical calculations; in particular, the second one has the advantage of avoiding problems of numerical integration and to be applicable to a number of inductor systems, external or internal to the load, with single or mutually coupled coils, connected in series or in opposition, and loads constituted by solid cylinders with two or more layers, tubes, tubes with internal magnetic cores, etc.

The main limitation of the method depends on the use of the superposition principle, which requires the assumption of constant material properties (resistivity and magnetic permeability).

In the following paragraph, reference is made to the configuration of Fig. 6.42, with a two-layer cylindrical load, each layer with different constant values of resistivity and permeability, and to an ideal inductor of negligible radial thickness, with an external magnetic yoke of infinite resistivity and permeability.

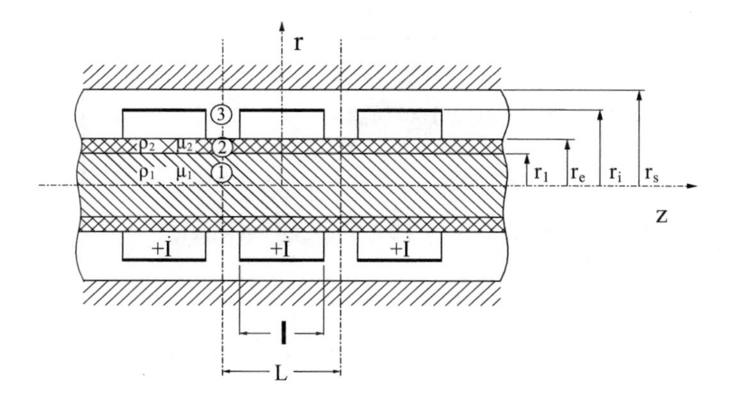

Fig. 6.42 Induction heating system with "periodical fields" inductor, a two-layer cylindrical load and an external magnetic shield

Due to the cumbersome derivation of equations, here is given only a short description of the problem, aimed to the correct use of results, whereas for a more profound study, reference should be made to the literature [18, 19].

On the basis of the knowledge of the excitation potential $\dot{A}_e$ of the empty coil, are determined the potentials $\dot{A}_1$ and $\dot{A}_2$ in the regions 1 and 2 within the cylinder as well as the reaction potential of load and yoke $\dot{A}_{3r}$ in the air region 3 comprising the inductor coils.

For region 3 therefore it is:

$$\dot{A}_3 = \dot{A}_{3e} + \dot{A}_{3r}.$$

The vector potential of the excitation field of the empty coil can be conveniently expressed by modified Bessel functions of first and second type and first order I_1 and K_1, by the equations:

$$\left.\begin{aligned} \dot{A}_e &= \frac{p_0 r}{2r_i} + \sum_{n=1,2,\ldots}^{\infty} p_n I_1(\lambda_n r) K_1(\lambda_n r_i)\cos(\lambda_n z), \quad \text{for } r \le r_i \\ \dot{A}_e &= \frac{p_0 r_i}{2r} + \sum_{n=1,2,\ldots}^{\infty} p_n I_1(\lambda_n r_i) K_1(\lambda_n r)\cos(\lambda_n z), \quad \text{for } r \ge r_i \end{aligned}\right\} \tag{6.82}$$

with:

r, φ, z	coordinate of a generic point of the system
r_i	radius of inductor coils
ℓ, L	axial length and pitch of inductor coils
μ_0	permeability of vacuum
N, I	number of turns of one coil and r.m.s. value of the current flowing in it

$\lambda_n = 2\pi n/L$

$p_0 = \mu_0 r_i NI/L$; $p_n = \frac{2\pi\mu_0 r_i NI}{\ell}\frac{\sin(\lambda_n \ell/2)}{n}$

The reaction potentials of the field, can be written as follows in a similar form:

$$\left.\begin{aligned} \dot{A}_1 &= \sum_{n=0}^{\infty} p_n F_n I_1(\beta_n r)\cos(\lambda_n z) \\ &\text{for} \quad 0 \le r \le r_1 \\ \dot{A}_2 &= \sum_{n=0}^{\infty} p_n [L_n I_1(\gamma_n r) + M_n K_1(\gamma_n r)]\cos(\lambda_n z), \\ &\text{for} \quad r_1 \le r \le r_e \\ \dot{A}_3 &= \dot{A}_{3r0} + \sum_{n=0}^{\infty} p_n [T_n I_1(\lambda_n r) + S_n K_1(\lambda_n r)]\cos(\lambda_n z), \\ &\text{for} \quad r_e \le r \le r_s \end{aligned}\right\} \tag{6.83}$$

with:

$F_n, L_n, M_n, S_n, T_n, T_0$ coefficient expressed through modified Bessel functions of real and complex arguments;

$\beta_n = \sqrt{\lambda_n^2 + j\frac{2}{\delta_1^2}}$ $\delta_1 = \sqrt{\frac{2\rho_1}{\omega\mu_0\mu_1}}$

$\gamma_n = \sqrt{\lambda_n^2 + j\frac{2}{\delta_2^2}}$ $\delta_2 = \sqrt{\frac{2\rho_2}{\omega\mu_0\mu_2}}$

$\dot{A}_{3r0} = \frac{p_0 r_i}{2r} T_0$ $j = \sqrt{-1}$

In order to calculate the impedance $\dot{Z}_e$ of one coil of the loaded system, it is sufficient to know the values of $\dot{A}_3$ at radius r_i, considering that the applied voltage $\dot{V}$ must compensate for the voltage drop in the coil own resistance R_i and the *e.m.f.* $\dot{E}_N$, sum of the *e.m.f.'s* induced in each turn.

Therefore we can write:

$$\begin{aligned} \dot{Z}_e &= R_i - \frac{\dot{E}_N}{\dot{I}} = (R_i + R'_w) + j(X_{i0} - \Delta X) \\ &= R_i + jX_i + \frac{X_{i0}}{\alpha^2}(G - jQ) \end{aligned} \tag{6.84}$$

with:

$$\dot{E}_N = -j\,\omega 2\pi r_i \frac{N}{\ell} \int_{-\ell/2}^{+\ell/2} \dot{A}_3(r_i, z)dz$$

and

R_i, R'_w own resistance of the induction coil and resistance of the load "seen" at the coil terminals

X_i, X_{i0} reactance of one coil of the unloaded system, with and without end effects

ΔX variation of reactance of one coil of the system from unloaded to loaded conditions

$\alpha = r_i/r_e$ ratio of inductor coil radius to external radius of load

G, Q dimensionless coefficients which account for the influence of load and yoke on the impedance $\dot{Z}_e$

One case of particular interest in the applications is that of a single "short" inductor coil without magnetic yoke and a load much longer than the inductor, like the one shown in Fig. 6.41.

The results for this case can be obtained through Eqs. (6.82)–(6.84), with a choice of values of L and r_s high enough, such that the mutual coupling between coils and the influence of the magnetic yoke become negligible.

As regards the mutual coupling of coils, it becomes practically negligible when the load length is greater than the sum of coil length and the double of the gap between inductor and load.

Considering that for a single coil it is $X_i = X_{i0} \cdot K_N$ (with K_N- Nagaoka coefficient), Eq. (6.84) can be rewritten in the form:

$$\dot{Z}_e = R_i + \frac{X_{i0}}{\alpha^2}\left[G + j(\alpha^2 K_N - Q)\right] \quad (6.85)$$

The coefficients G and Q, for non-magnetic, solid cylindrical loads, with $\rho_1 = \rho_2 = \rho$ and $\mu_1 = \mu_2 = 1$, depend on the dimensionless ratios:

$$m = \frac{\sqrt{2}r_e}{\delta}; \quad \alpha = \frac{r_i}{r_e}; \quad \frac{\ell}{2r_i}.$$

Their values are given the diagrams of Figs. 6.43 and 6.44, where the results for "long" inductors $(\ell/2r_i \to \infty)$ are also shown.

For loads consisting of hollow non-magnetic cylinders, the coefficients G and Q depend also on a fourth parameter, i.e. the ratio r_1/r_e of internal and external radii of the load; diagrams for this case can be found in Ref. [18].

6.8.3 *Numerical Methods*

Addressing the reader to the vide literature available on this subject [20], it is presented here a very simple 1D–FD method, analogous to that already illustrated for the direct resistance heating in Sect. 5.3.6.1 [21]. Also in this case, the electromagnetic problem for the evaluation of the power density distribution and the thermal problem for the determination of the subsequent temperature distribution at the end of one time step, are solved alternatively and separately for each time step Δt in which the thermal transient is subdivided.

With reference to a cylindrical configuration of "infinite axial length", for the solution of the electromagnetic problem the workpiece is subdivided into N concentric elements of equal radial thickness Δr, as schematized in Fig. 6.45.

As shown in the figure, it is assumed that the current density is constant inside an element and the corresponding current is concentrated in an infinitely thin current sheet situated in the middle of the element. In this way, the magnetic field intensity varies stepwise and remains constant between the middle of two adjacent elements.

Therefore it is:

- $\dot{H}_{i+1} = \cos t \quad \text{for} \quad r_i \le r \le r_i + \Delta r$
- $\dot{J}_i = \cos t \quad \text{for} \quad r_i - \frac{\Delta r}{2} \le r \le r_i + \frac{\Delta r}{2}$

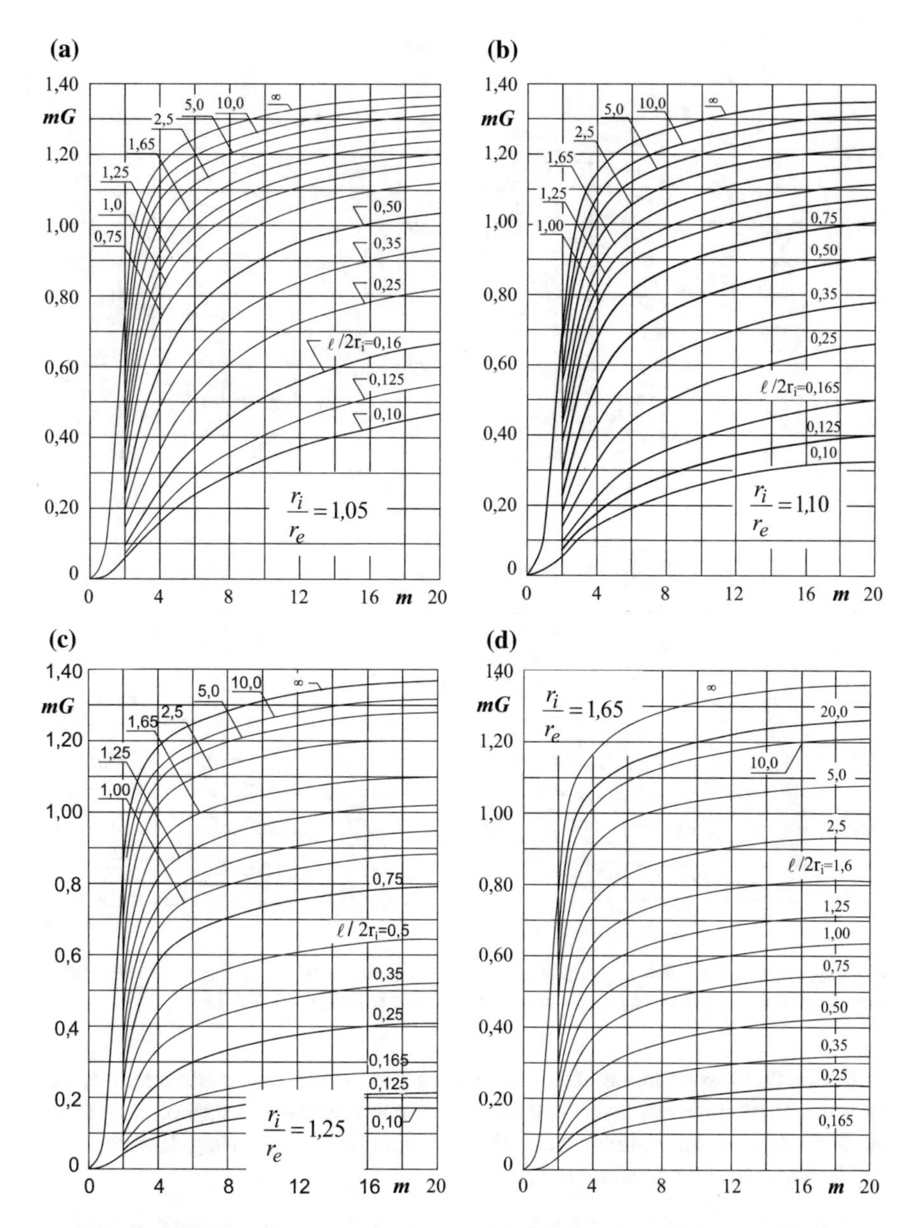

Fig. 6.43 Coefficients G as function of m, for non-magnetic loads and various ratios of length to inductor diameter (system of Fig. 6.42; **a** $r_i/r_e = 1.05$; **b** 1.10; **c** 1.25; **d** 1.65)

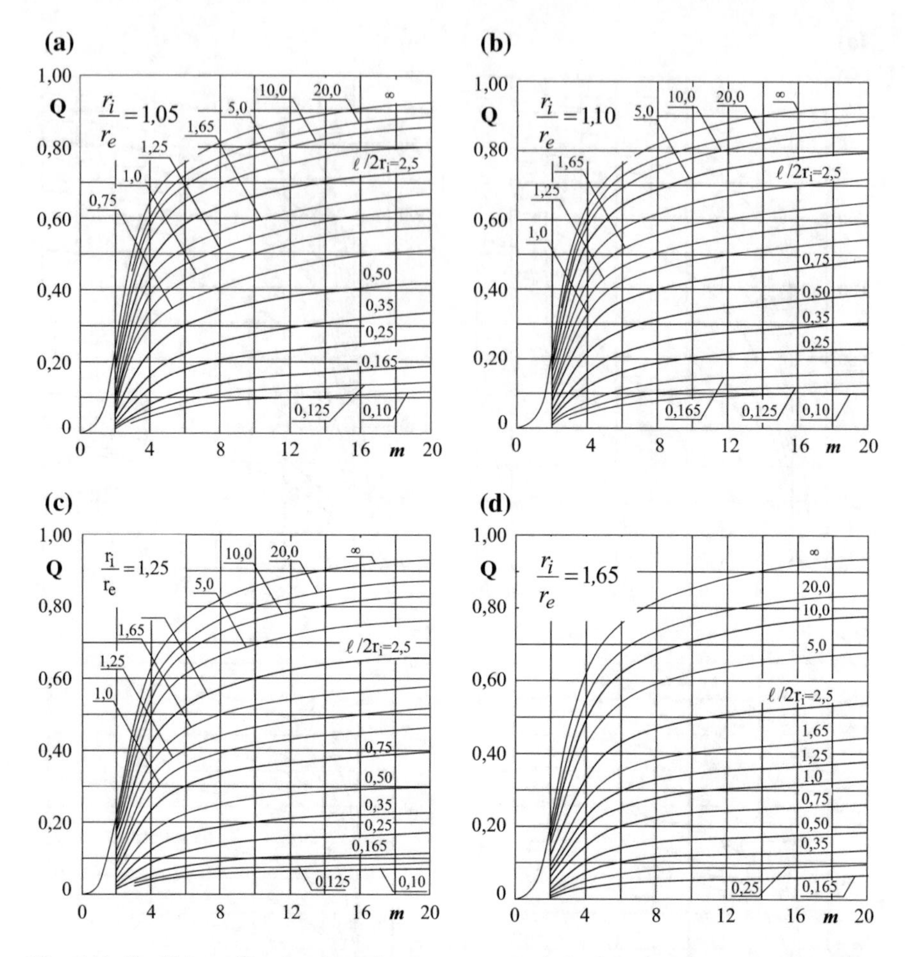

Fig. 6.44 Coefficients Q as function of m for non-magnetic loads and various ratios of length to inductor diameter (system of Fig. 6.42; **a** $r_i/r_e = 1.05$; **b** 1.10; **c** 1.25; **d** 1.65)

Under these assumptions the field values can be calculated sequentially, one after the other, for each layer, from the inner one towards the surface. The calculation starts from an arbitrary value H_1, according to the following scheme:

Calculation of field values in layer 1

- Evaluation of magnetic permeability corresponding to the assigned value H_1

$$\mu_1 = f(H_1)$$

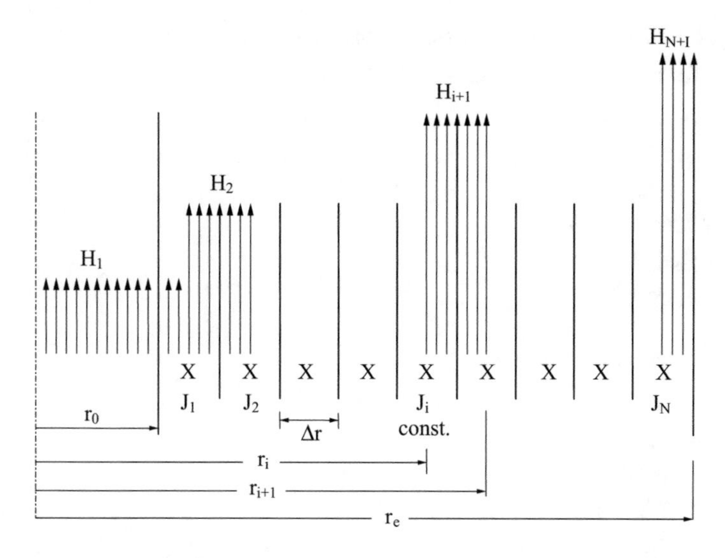

Fig. 6.45 Scheme of subdivision of a cylindrical solid or tubular load ($r_0 = 0$ for solid cylinder)

- Flux linked with the current sheet at radius r_1

$$\dot{\Phi}_1 = \pi \dot{H}_1 [\mu_0 r_0^2 + \mu_1 \Delta r (r_0 + \frac{\Delta r}{4})]$$

- *E.m.f.* induced at radius r_1

$$\dot{E}_1 = -j\,\omega \dot{\Phi}_1$$

- DC resistance of layer 1 (where uniform current density distribution is assumed)

$$R_1 = \rho_1 \frac{2\pi r_1}{\Delta r\, \ell}$$

- Current in layer 1 (supposed concentrated in the thin current sheet at radius r_1)

$$\dot{I}_1 = \frac{-j\,\omega \dot{\Phi}_1}{R_1}$$

- Current density in layer 1

$$\dot{J}_1 = \frac{-j\,\omega \dot{\Phi}_1}{\rho_1 2\,\pi r_1}$$

Calculation of field values in the second layer

Since from Eq. (6.4) it is $\frac{\partial H}{\partial r} = -J$, we have:

- $\dot{H}_2 = \dot{H}_1 - \Delta r \dot{J}_1$
- $\mu_2 = f(H_2)$
- $r_2 = r_1 + \Delta r$
- $A_2 = \pi \Delta r(2r_2 - \Delta r)$
- $\dot{\Phi}_2 = \mu_2 \dot{H}_2 A_2$
- $\dot{J}_2 = \frac{-j\omega(\dot{\Phi}_1 + \dot{\Phi}_2)}{\rho_2 2\pi r_2}$

For the subsequent layers:

- $\dot{H}_{i+1} = \dot{H}_i - \Delta r \dot{J}_i$
- $\mu_{i+1} = f(H_{i+1})$
- $r_{i+1}, = r_i + \Delta r$
- $A_{i+1} = \pi \Delta r(2r_{i+1} - \Delta r)$
- $\dot{\Phi}_{i+1} = \mu_{i+1} \dot{H}_{i+1} A_{i+1}$
- $\dot{J}_{i+1} = \frac{-j\omega(\dot{\Phi}_1 + \dot{\Phi}_2 + \cdots + \dot{\Phi}_{i+1})}{\rho_{i+1} \cdot 2\pi r_{i+1}}$

As already said, the calculation goes on until the surface layer is reached where, however, it must be taken into account that the magnetic field H_{n+1} is constant in a layer of thickness $\Delta r/2$.

Since the value H_1 was chosen arbitrarily, also the value H_{n+1} will not correspond to the given surface value $H_e = NI/\ell$, corresponding to the inductor current required for the considered heating process.

In case of a load of non-magnetic material, it is only necessary to multiply all the calculated field values by the ratio H_e/H_{N+1}.

On the contrary, in case of ferromagnetic materials, one must apply an iterative procedure, modifying at each iteration the initial value of H_1 until the convergence of H_{n+1} to the required value H_e is reached.

As the equations demonstrate, the procedure allows to take into account different values of ρ_i and μ_i from layer to layer.

When current density and resistivity values of each layer are known, it is easy to evaluate the volume power densities $w_i = \rho_i J_i^2$, which are the heat sources used for the subsequent solution of the thermal problem.

This solution, already illustrated at Sect. 5.3.6-(b), allows the calculation of temperature of any layer at the end of each elementary time interval Δt, starting from the knowledge of the temperature at the beginning of the interval.

Therefore, the method allows to update the resistivity and permeability values of layers as a function of temperature and local magnetic field intensity, before starting a new electromagnetic calculation for the subsequent interval Δt.

The solution of the thermal problem is analogous to the one given in Sect. 5.3.6.1.

6.9 Industrial Applications of Induction Heating

Induction heating is applied in a variety of industrial processes in the metallurgical, automotive, food, chemical, paper sectors.

The industrial applications of induction heating can be classified into five groups: induction melting, mass heating, heat treating, induction welding and special applications.

Since, within the limits of this book, it is not possible to describe in detail all these applications, the next paragraphs will be focused only on the most important ones for the metallurgical sector; in particular [22]:

- melting of metals.
- heating prior to hot working of metals, such as rolling, forging, and stamping
- heat treating, with particular attention to surface hardening at medium and high frequency.

For other applications or more profound knowledge, the reader should refer to the existing literature [6, 9, 23, 24].

6.9.1 Induction Furnaces

6.9.1.1 Historical Notes

The first known application of the induction heating was for melting ferrous and non-ferrous metals. It was proposed by *Sebastian Ziani de Ferranti (1864–1930)*, an English electrical engineer, who in 1887 patented the first induction furnace [25–27].

The furnace was constituted by a magnetic circuit with a coil wound around the central limb and an oval annular channel of non-conducting material, in which the metal could be melted; by using a metal through, water or other liquids could be also heated (Fig. 6.46a).

But this type of furnace became usable in industrial application only after the improvements and experiments made by *F.A. Kjellin* (Fig. 6.46b).

As shown in Fig. 6.46, the furnace was designed as a single-phase AC transformer supplied at line frequency, in which the inductor acts as primary winding and the channel containing the melt as a single secondary winding, where the induced current flows. Between 1900 and 1901 other furnaces of this type were constructed, increasing capacity from the 80 kg of the first one to 1800 kg, and

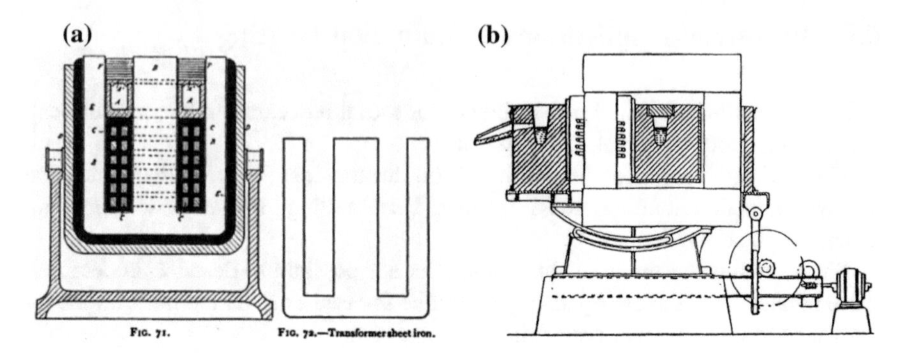

Fig. 6.46 **a** De Ferranti induction furnace (1887); **b** Kjellin furnace (1891)

reducing energy consumption from the 8,000 kWh per ton of the first furnace down to 800 kWh per ton.

In order to overcome the limitations of Kjellin furnace, due to the pinch-effect which produced break off of the liquid metal at high power levels, intensive liquid metal flow and high thermal losses, the American engineer *James R. Wyatt*, in 1915 proposed a new design with a V-shaped channel in a vertical plane below a cylindrical crucible hearth (Fig. 6.47). In this design the hydrostatic pressure of the melt, acting against the pinch-effect, gave the possibility to increase the furnace power.

Based on this design, with further modifications and improvements, this type of furnace, which is known as *Channel Induction Furnace (CIF),* became widely used and is still in use today for melting and holding ferrous and non-ferrous metals [25].

At the beginning of the 20th century, the studies on HF power sources developed for stable wireless transmissions, suggested to use them for supplying a new type of induction furnace—the coreless *Induction Crucible Furnace (ICF)*—where the use of a frequency above mains frequency allowed to increase the power induced in the charge without the need of a magnetic core.

The first patent for a furnace of this type is attributed to the *Soc. Schneider Creusot* (1905), where an "Electric induction furnace for high frequency currents", supplied with an alternative current at 100 kHz is described; but there are not further information about practical applications of this patent.

Few years later, in 1914, the Italian *Felice Jacoviello*, professor at the University of Parma obtained an Italian patent describing a coreless furnace supplied at 400 Hz (Fig. 6.48a). In this patent he also suggested to compensate the reactive power by capacitors, according to the scheme of Fig. 6.48b, which is similar to the one used later in industrial furnaces [28].

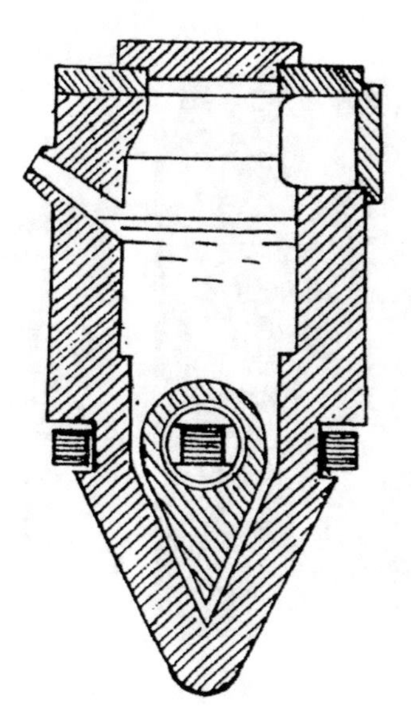

Fig. 6.47 Ajax-Wyatt furnace (1915)

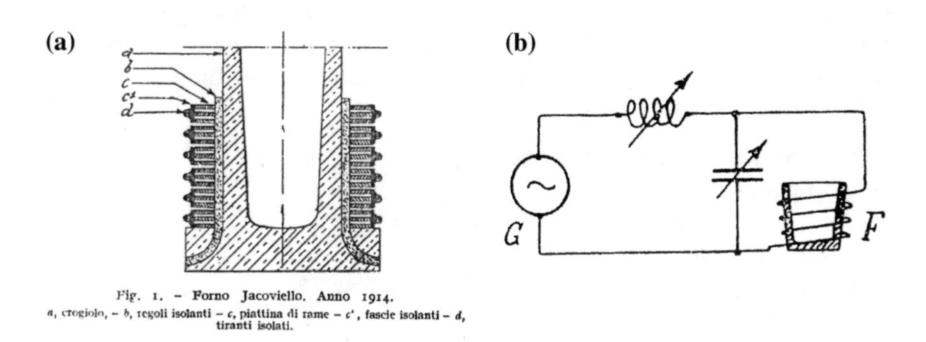

Fig. 6.48 **a** Jacoviello induction crucible furnace; **b** power factor compensation scheme proposed by Jacoviello

In 1916 *Edwin F. Northrup*, professor of physics at Princeton University (USA), designed in 1916 the first HF crucible furnace, powered by a 20 kHz spark gap generator (Fig. 6.49a, b) [29–31].

The development of Channel and Crucible Furnaces continued constantly in the years, but it received special impetus after the World War II for the increased demand of molten metal and later, around the 1980s, by the development of new high-efficiency static frequency converters and the significant improvement of refractory linings.

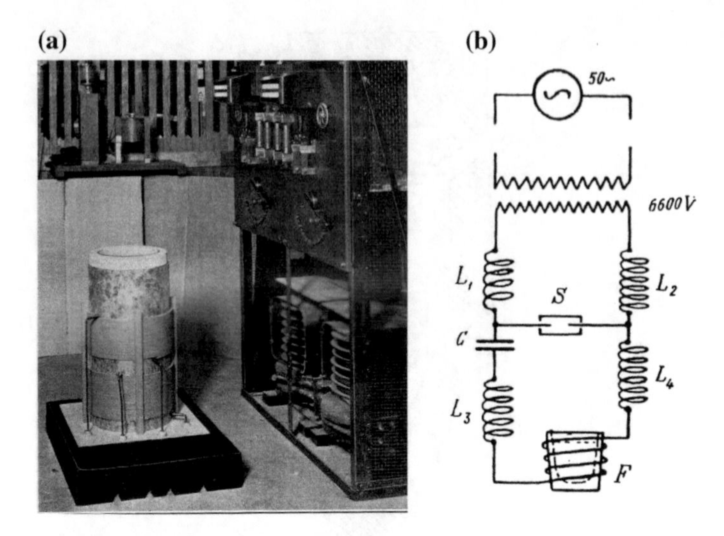

Fig. 6.49 **a** Northrup's HF furnace; **b** Spark-gap generator

6.9.1.2 Modern Induction Furnaces

Induction furnaces are widely used in foundries for their technological and economical advantages, in comparison to conventional fuel furnaces, in melting, alloying, storage, holding and casting operations of metals and alloys.

Main advantages of induction furnaces are:

- Possibility of continous production
- High power density
- Short melting times and high production rates
- Small space requirements of installations
- Flexibility and easy control of power and temperature
- High efficiency
- Low material oxidation
- High homogenity of melt due to bath stirring
- Economical operation
- Improvement of product quality
- Improvement of labor conditions and reduction of environmental pollution.

These advantages apply to crucible furnaces which are mainly used as melting furnaces, as well as to channel furnaces, which are predominantly used for holding melt at temperature for metallurgical treatments. However, recent developments of channel furnaces have made them attractive, due to their higher electrical efficiency, also as a melting unit which is used for combined melting and holding.

Table 6.6 Typical data of CIF and ICF induction furnaces [6]

Material	Capacity (t)	Power (MW)	Frequency (Hz)
CIF			
Cast Iron	10–135	0.1–3	50–60
Al, Al alloys	5–70	0.1–6	50–60
Cu, Cu alloys	4–60	0.5–10	50–60
Zn, Zn alloys	10–100	0.2–10	50–60
LF-ICF			
Steel, Cast Iron	1.3–100	0.5–21	50–60
Light Metal	0.5–15	0.2–4	50–60
Heavy Metal	1.5–40	0.5–7	50–60
MF-ICF			
Steel, Cast Iron	0.25–30	0.3–20	150–1000
Light Metal	0.1–10	0.2–4	90–1000
Heavy Metal	0.3–72	0.3–16	65–1000

Typical capacities and power ratings of today's channel induction furnaces (CIF) and induction crucible furnaces (ICF) at line frequency (LF) and medium frequency (MF) are given in Table 6.6.

In the following paragraphs are described the main design types and energetic characteristics of crucible and channel furnaces.

6.9.2 Induction Crucible Furnaces

A typical induction crucible furnace corresponds to the constructive sketch of Fig. 6.50.

Its main elements are:

- a crucible made of ceramic or conducting material (e.g. steel or graphite), which contains the melt constituting the load
- a water cooled solenoidal induction coil, surrounding the crucible
- an external metallic structure.

The inductor coil is supplied with currents at industrial frequency (LF-CIF furnaces, at 50/60 Hz) or—with the development of high power converters—at medium frequency (MF-CIF furnaces, up to 1 kHz for high capacity furnaces, up to 10 kHz for low capacity furnaces, predominantly used for melting precious metals). The current flowing in the inductor creates an alternating magnetic field that, penetrating into the load, induces currents and, in turn, ohmic heating.

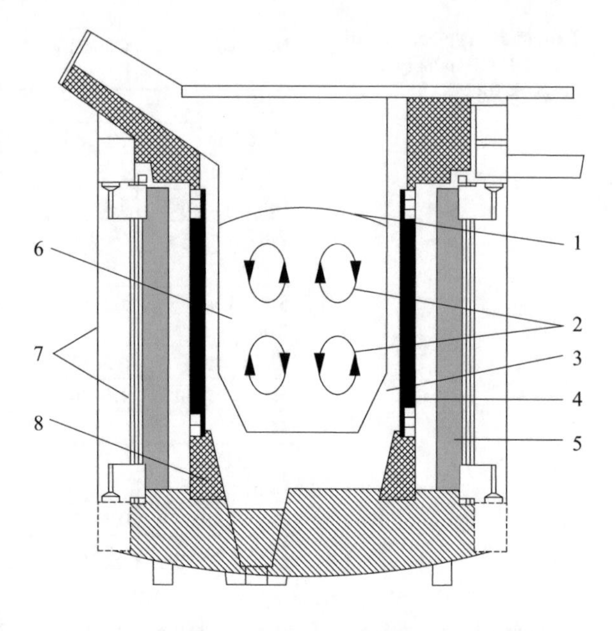

Fig. 6.50 Sketch of an induction crucible furnace (*1* meniscus; *2* melt flow; *3* crucible; *4* induction coil; *5* magnetic yoke; *6* melt; *7* steel construction; *8* construction ring) [32]

Usually, the inductor coil is surrounded by a magnetic yoke consisting of laminated stacks having the primary function, together with the external mechanical structure, of containment of mechanical stresses produced in radial direction by the coil and the loaded crucible, and the secondary task to confine the magnetic stray field, preventing uncontrolled thermal losses in the external metallic structure and exposure to the field of the operating personnel.

This last result alternatively can be obtained in low capacity furnaces by short-circuited copper shields.

6.9.2.1 Main Furnace Dimensions

Before considering technical aspects of crucible furnaces, it is useful to give an idea of their main geometrical dimensions. These dimensions are related to the capacity of the crucible by the relationship:

$$v = \frac{G}{\gamma} \cdot 10^3$$

with:

G crucible capacity (t)
v crucible useful volume (m^3)
γ specific weight of the melt (kg/ m^3)

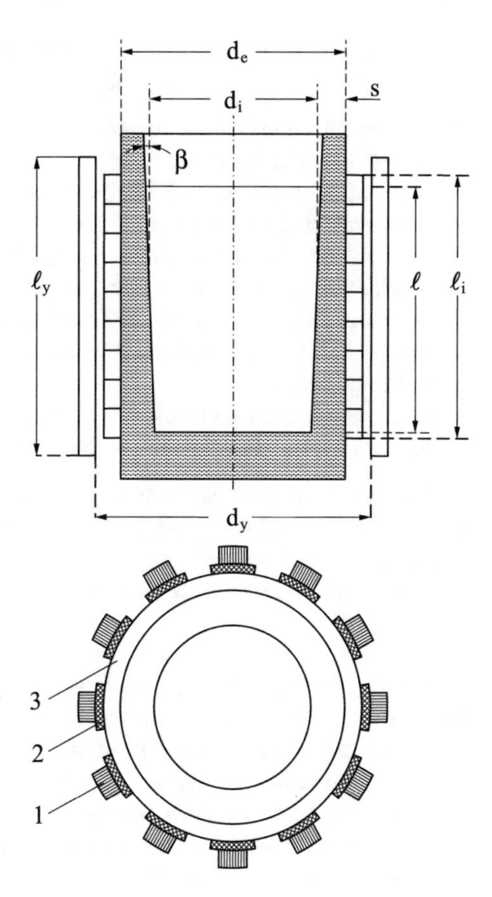

Fig. 6.51 Schematic of main furnace dimensions (*1* magnetic yokes; *2* supports of yokes; *3* inductor) [11]

In order to assure mechanical resistance to the hydrostatic pressure, the internal shape of the crucible has a conical shape, with very small values of the cone generating angle β ($\beta \approx 2°–5°$).

However, with reference to Fig. 6.51, for simplicity we assume that the crucible has cylindrical internal shape, characterized by the ratios:

$$c_1 = \frac{d_i}{\ell}; \quad c_2 = \frac{s}{d_e}; \quad c_3 = \frac{\ell_i}{\ell}$$

with:

d_e, d_i, s external and internal diameter and average radial thickness of crucible (m);

ℓ, ℓ_i, ℓ_y respectively axial height of melt, inductor and yokes (m)

The choice of optimal values of these ratios is based on technical-economic considerations:

- for minimization of thermal losses, the dimensions of diameter d_i and height ℓ of the load should be about the same.
- improvement of electrical efficiency and power factor are achieved by reducing the diameter d_i and increasing the height ℓ.
- the thickness of the crucible results from a compromise between opposing requirements: on one side reduction of investment costs and improvement of electrical efficiency and power factor, which are obtained by reducing *s*; on the other side the increase of crucible safety and life and reduction of thermal losses achieved by increasing *s*.
- moreover, it must be taken into account that the reactive power, which always results much higher than the active one, strongly depends on the thickness *s*.

As rule of thumb, it can be assumed that each annular gap layer of thickness δ/2, placed between inductor coil and load, produces a reactive power of the same magnitude of active power [24]. This reactive power must be compensated with a capacitor bank whose cost, in case of large furnaces, can even exceed the cost of the furnace itself.

Therefore, in order to reduce investment costs, the crucible should be designed with walls of minimum thickness. However, this is in contrast with other important requirements, such as safety and crucible life.

In fact, the crucible is the most stressed construction element of the furnace, since its internal wall surface is in contact with the melt which, in case of grey cast iron, can reach temperatures between 1,450 and 1,600 °C, whereas the external surface facing the water-cooled inductor coil, is at low temperature.

Another factor of stress of the crucible is due to its thermal expansion. For example, in materials of common use, thermal expansion can have values up to 1–2 %, which correspond to 20–40 mm in a crucible of 2 m diameter.

Regardless of the material used, considerable radial stresses are transmitted to the coil, the magnetic yoke and, finally, to the external metallic frame. Therefore, it is very important that the inductor coil is supported mechanically by the external metallic structure. For this reason the crucible, which is without support when it's cold, must be always preheated before being charged with scrap of large size or with molten material.

Figure 6.52 gives recommended values of the ratios c_1 and c_2 as a function of the crucible capacity G, in furnaces for melting ferrous metals or aluminum.

As concerns the ratio c_3, in order to optimize the global efficiency (product of electrical and thermal efficiency), it is recommended to use values in the range 1.1–1.3, as well as to place the inductor symmetrically in axial direction in relationship to the melt. In some cases however, to the end of reducing the meniscus height and the mixing and stirring of the bath in the upper portion of the crucible, the upper end of the inductor can even be placed below the free surface of the bath [11].

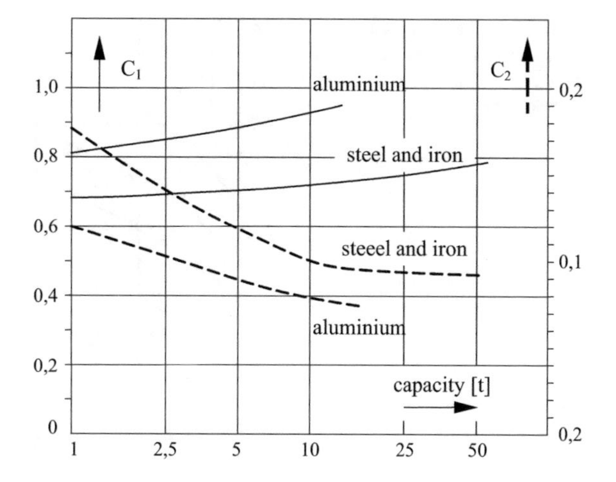

Fig. 6.52 Recommended values for ratios $c_1 e c_2$ as a function of crucible capacity [11]

In conclusion, the main dimensions of the inductor-molten metal system can be roughly estimated by the following relationships:

$$\left.\begin{array}{l} d_i = \sqrt[3]{\frac{4c_1 V}{\pi}};\ \ell = \frac{d_i}{c_1};\ s = d_i c_2; \\ \ell_i = \ell c_3;\ \alpha = \frac{d_e}{d_i} = 1 + 2c_2 \end{array}\right\} \quad (6.86)$$

Finally, it must be taken into account that the crucible height is always greater than the length ℓ of the load of about 20–40 %, in order to account for the presence of meniscus, the charging system and other technological and operational factors.

As concerns the external magnetic yokes, which shield the metallic frame from the stray magnetic field, they—ideally—should completely encircle the inductor coil. In practice, however, as shown in Fig. 6.51, they are constituted by a series of stacks of transformer crystal oriented steel laminations, oriented in radial direction.

The laminations must be accurately packed in order to prevent vibrations and tightly fixed to the external metallic structure in order to provide mechanical support to the coil and the crucible. From the electrical point of view, the cross-section of yokes, the thickness of laminations and the value of magnetic induction are chosen in relation to the frequency in order to limit the losses in the yokes to 1.5–2.0 % of the total inductor losses and the yokes overheating above the ambient to a maximum of 75 °C.

In the design of the yoke cross-section, it must be considered that only a fraction $\Phi_y = k_y \Phi$ of the total flux will be linked with laminations stacks. This fraction k_y can be determined by the graph of Fig. 6.53, as a function of the ratio (d_y/d_e) between diameters and the ratio (ℓ_y/ℓ_i) between the axial lengths of yokes and inductor. In practice, the magnetic yokes protrude out over the coil ends in order to capture the leakage field as completely as possible, and the ratio (ℓ_y/ℓ_i) is generally higher than unity, in the range 1.1–1.4.

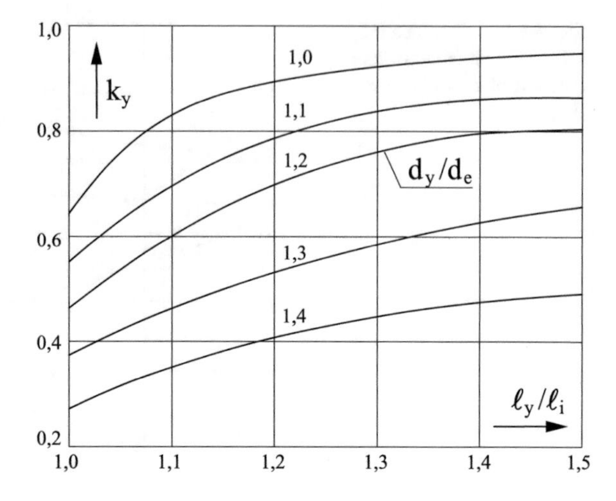

Fig. 6.53 Coefficients for determination of magnetic flux in the yokes of a crucible furnace [11]

The radial thickness of lamination stacks is approximately equal to 10 % of coil diameter and the filling factor along the circumference (ratio between circumferential length and total length of stacks) is about 50-65 %.

The geometry of the furnace corresponds to a "short" inductor-load system with external magnetic yoke of the type considered at Sect. 6.9.1.

6.9.2.2 Furnace Power

The power of the furnace can be determined starting from the theoretical heat required for heating the load from the ambient to the casting temperature and the melting time, which is determined by the production rate, accounting for the electrical and thermal efficiency of the inductor-load system and the losses in bus-bar connections to the supply as well as in compensating capacitors.

Denoting with:

E_w	theoretical heat required for heating the load from the ambient to the final temperature (kWh/t)
t_f	time, h, for heating the load capacity G, tons, up to casting temperature
P_u	useful power (kW)
P_{th}	power of thermal losses (kW)
$P_c = P_u + P_{th}$	power induced into the load (kW)
$\eta_t = P_u/P_c$	thermal efficiency
P_i	active power at inductor clamps (kW)
$\eta_e = P_c/P_i$	electrical efficiency (see Sect. 6.6.2)
P_f	furnace active power from the supply system (kW)
η_f	overall efficiency of the furnace,

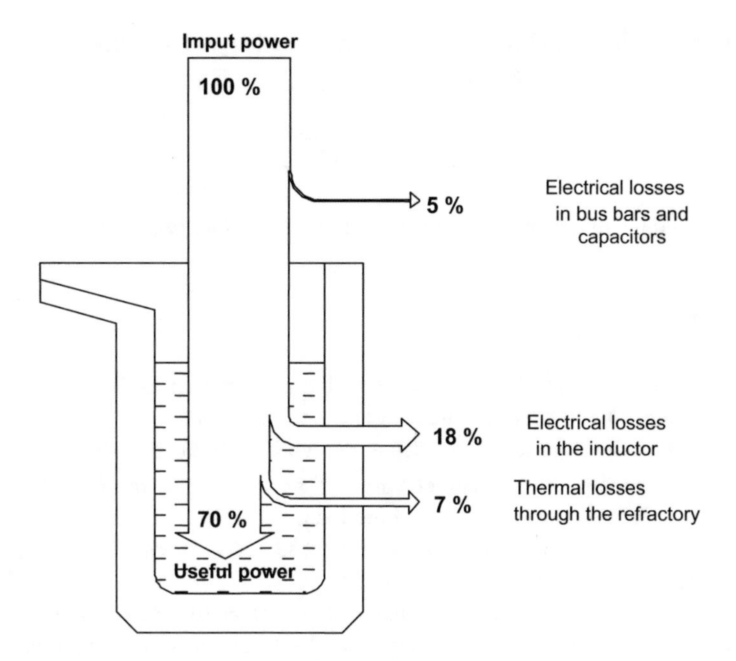

Fig. 6.54 Typical energy flow diagram in a crucible furnace

it results:

$$P_u = E_w \frac{G}{t_f}; \quad P_f \approx \frac{(1.05-1.10)}{\eta_e \, \eta_{th}} P_u \tag{6.87}$$

In Eq. (6.87), the thermal efficiency is usually in the range from 75 to 95 %, increasing with the furnace's capacity. The electrical efficiency in the melting of steels varies from 70 to 95 %, increasing with the resistivity of the metal constituting the load, whereas the coefficient (1.05–1.10) accounts for the losses in bus bars and capacitors of the supply circuit.

A typical diagram of energy flow in a coreless melting furnace is shown in Fig. 6.54 [33].

The main field of application is the melting of cast iron; but there is also a growing use in the melting of steel and aluminum, as previously shown in Table 6.6.

As regards the choice between LF or MF furnaces, with the availability of economical high power solid state frequency converters, MF furnaces have today increasing diffusion, given that their higher specific power in comparison to LF furnaces allows to design flexible melting units of smaller dimensions.

As concerns reactive power, its evaluation is the basis for the selection of the capacitor bank, which is required to bring the very low values (0.1–0.2) of power factor characteristics of this type of furnaces within contractual limits.

Its value can be estimated with good approximation, as a function of the furnace power P_f, by the relationship [24]:

$$Q \approx 1.2\, P_f \frac{\delta_i + 2s + \delta}{\delta_i + \delta} \tag{6.88}$$

with: δ_i, δ—penetration depth in the inductor and load respectively.

6.9.2.3 Frequency Selection

In principle, this selection is made following the same criteria indicated in Sect. 6.1, but accounting for the peculiar characteristics of the system.

In particular, the electrical efficiency at the working frequency should be as high as possible; therefore, a convenient value of $m = d/\sqrt{2} \cdot \delta$ must be chosen for assuring an efficient energy transfer to the load.

Figure 6.55 shows the values of electrical efficiency of induction crucible furnaces as a function of *m* for different materials.

In case of furnaces in which a considerable part of the melt (40–60 % of the nominal capacity) remains inside the crucible at the end of each casting, the diameter for the evaluation of *m* is the internal diameter d_i of the crucible, whereas for the resistivity and the magnetic permeability of the load are used the values at the temperature of the melt (i.e. relative permeability $\mu = 1$).

Moreover, it must be taken into account that for "long" systems the minimum working frequency corresponds to values of *m* equal to 2.0–2.5, while in a "short" inductor-load system like that of crucible furnaces, the efficiency does not depend only on d_e/d_i, ρ_i/ρ and s_i/δ_i, but also on the ratios ℓ/ℓ_i, ℓ_i/d_e, ℓ/d_i.

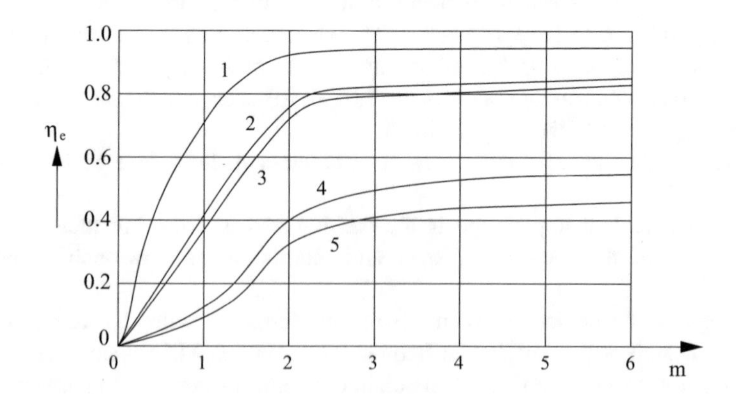

Fig. 6.55 Electrical efficiency of induction crucible furnaces as a function of the ratio $m = d/\sqrt{2} \cdot \delta$ (*1* Steel 20 °C, μ = 100, ρ = 13 μΩ cm/Steel 400 °C, μ = 30, ρ = 45 μΩ cm/ Graphite 200–1000 °C, μ = 1, ρ = 1000 μΩ cm; *2* Steel 1000 °C, μ = 1, ρ = 120 μΩ cm/Stainless steel 900 °C, μ = 1, ρ = 80 μΩ cm; *3* Stainless steel 20 °C, μ = 1, ρ = 80 μΩ cm; *4* Aluminium 100 °C, ρ = 3.8 μΩ cm; *5* Copper 100 °C, ρ = 2.2 μΩ cm) [34]

The influence of these ratios on electrical efficiency can be analyzed with the approximate or numerical methods for calculation of "short" inductors. This analysis has demonstrated that with typical geometrical ratios of crucible furnaces, the minimum admissible frequency corresponds to values of $m \geq 6$ [11].

On the contrary, when the melting starts with solid scrap, the minimum frequency should fulfil the condition $m \geq 2$ with reference to the average size of the scrap and the resistivity of material in solid state, taking into account that the relative permeability is greater than one for cold ferrous metals.

Since, as shown also by Fig. 6.55, the use of frequency values higher than the minimum one only leads to minor improvement on electrical efficiency, the value of the working frequency is finally selected on the basis of technical-economic considerations on metal stirring and investment and running costs.

6.9.2.4 Electromagnetic Forces and Melt Stirring

In induction heating applications electromagnetic forces occur due to the interaction of the induced currents with the exciting magnetic field.

Before examining the effect of these forces on the melt of crucible furnaces, we will consider the furnace as a portion of an infinitely long cylindrical system, with an inductor supplied by sinusoidal current *I*, as shown in Fig. 6.56.

With reference to this system, the distributions of magnetic field intensity $\dot{\bar{H}} = H(r) \cdot \bar{u}_z$ and induced current density in the charge $\dot{\bar{J}} = J(r) \cdot \bar{u}_\varphi$ have been given in Sect. 6.2 by the equations:

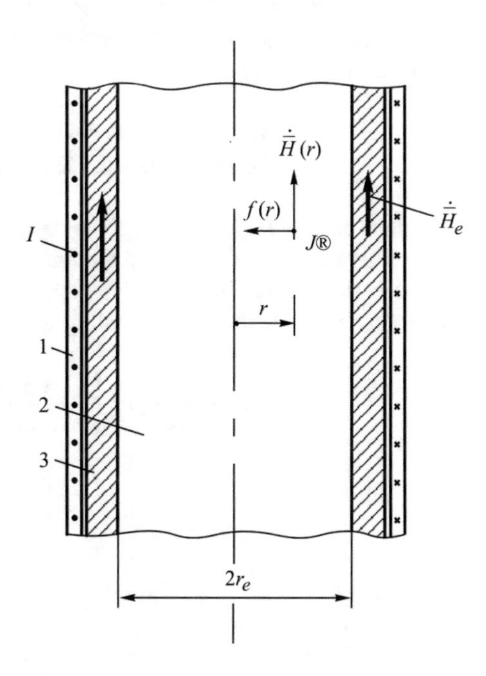

Fig. 6.56 Ideal infinitely long cylindrical induction heating system (*1* inductor; *2* charge; *3* air gap; *I* exciting current)

$$\dot{H} = \dot{H}_e \frac{J_0(\sqrt{-j}\, m\xi)}{J_0(\sqrt{-j}\, m)} = H_e \frac{\text{ber } m\xi + j \text{ bei } m\xi}{\text{ber } m + j \text{ bei } m} \tag{6.89a}$$

$$\begin{aligned} \dot{J} &= \sqrt{-j}\, \dot{H}_e \frac{\sqrt{2}}{\delta} \cdot \frac{J_1(\sqrt{-j} m\xi)}{J_0(\sqrt{-}\, jm)} \\ &= -\dot{H}_e \frac{\sqrt{2}}{\delta} \cdot \frac{\text{ber}' m\xi + j \text{ bei}' m\xi}{\text{ber } m + j \text{ bei } m} \end{aligned} \tag{6.89b}$$

From the above equations, the specific force per unit volume, in N/m^3, is:

$$f = \frac{df}{dv} = \dot{J} \cdot \dot{B} = \mu_0 \left[\dot{J} \cdot \overset{*}{H} \right] \tag{6.90}$$

Considering the orthogonality of $\dot{J}$ and $\dot{H}$, the mean value of this force at a point of radial coordinate ξ can be written in the form [35]:

$$f(\xi) = -\frac{\mu_0}{\rho} p_e \frac{\text{ber } m\xi \cdot \text{ber}' m\xi + \text{bei } m\xi \cdot \text{bei}' m\xi}{\text{ber } m \cdot \text{ber}' m + \text{bei } m \cdot \text{bei}' m} \tag{6.91}$$

with: $p_e = H_e^2 \frac{\rho}{\delta} \sqrt{2} P$—specific induced power per unit external surface of the cylinder (w/m^2); P—coefficient of active power induced in the cylinder defined in Sect. 6.1.3 (Fig. 6.7).

This force is directed radially towards the axis of the cylinder and varies from the maximum value at the surface ($\xi = 1$) to zero on the axis ($\xi = 0$); moreover, the force is proportional to the surface specific power p_e induced in the load.

From Eq. (6.91), the ratio of the force at the coordinate ξ to its maximum value at the surface becomes:

$$\frac{f(\xi)}{f(1)} = \frac{\text{ber } m\xi \cdot \text{ber}' m\xi + \text{bei } m\xi \cdot \text{bei}' m\xi}{\text{ber } m \cdot \text{ber}' m + \text{bei } m \cdot \text{bei}' m}. \tag{6.91a}$$

Values of ratio $f(\xi)/f(1)$ are given, as a function of ξ and different values of *m*, by the curves on the right in the diagram of Fig. 6.57.

The compression pressure (*pinch effect*) produced by this force increases from the surface towards the axis according to the equation:

$$p(\xi) = \frac{1}{2} \mu_0 H_e^2 \left[1 - \frac{\text{ber}^2(m\xi) + \text{bei}^2(m\xi)}{\text{ber}^2(m) + \text{bei}^2(m)} \right] \tag{6.92}$$

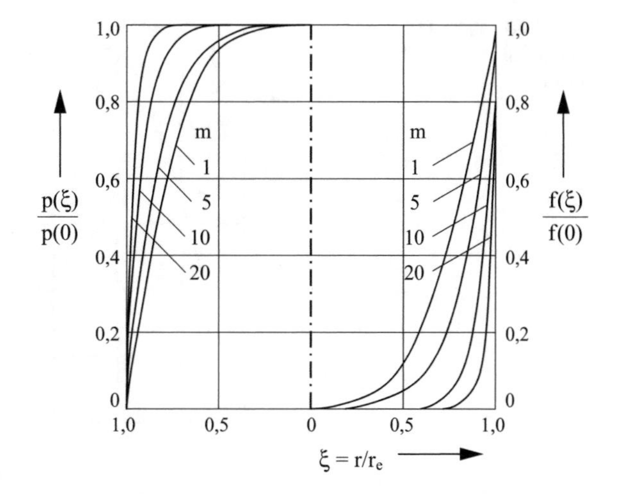

Fig. 6.57 Radial distribution of force per unit volume (on the *right*) and electrodynamic pressure (on the *left*) in an infinitely long cylinder of molten metal

It has a maximum on the axis (for $\xi = 0$), equal to:

$$\begin{aligned} p(0) &= \frac{1}{2}\mu_0 H_e^2\left[1 - \frac{1}{\mathrm{ber}^2(m) + \mathrm{bei}^2(m)}\right] \\ &= \frac{\mu_0}{\rho}\frac{\delta}{2}p_e\Psi(m) = 3.16 \cdot 10^{-4}\frac{p_e}{\sqrt{\rho f}}\Psi(m) \end{aligned} \tag{6.92a}$$

with:

$$\Psi(m) = \frac{\mathrm{ber}^2(m) + \mathrm{bei}^2(m) - 1}{\sqrt{2}[\mathrm{ber}\ m \cdot \mathrm{ber}'m + \mathrm{bei}\ m \cdot \mathrm{bei}'m]}.$$

Figure 6.58 gives the diagram of $\Psi(m)$ as function of m, while the diagrams of $p(\xi)/p(0)$ are given by the curves on the left of Fig. 6.57.

Hence, in the ideal case of a cylindrical coil and a "long" infinite load, the pressure causes a radial compression of the bath with constant intensity in different axial positions.

However, in the reality the furnace is of finite length, the exciting magnetic field is parallel to the axis only at the center of the coil and the electromagnetic forces are directed radially only in this position.

Moving from the center towards the ends of the coil, the magnetic field deforms, its axial component progressively diminishes, producing a corresponding decrease of the radial component of electromagnetic force.

Therefore, the forces of radial compression on the bath are not constant along the axis of the load and their field is not anymore free of curls, given that the integral along a given closed path is no longer equal to zero.

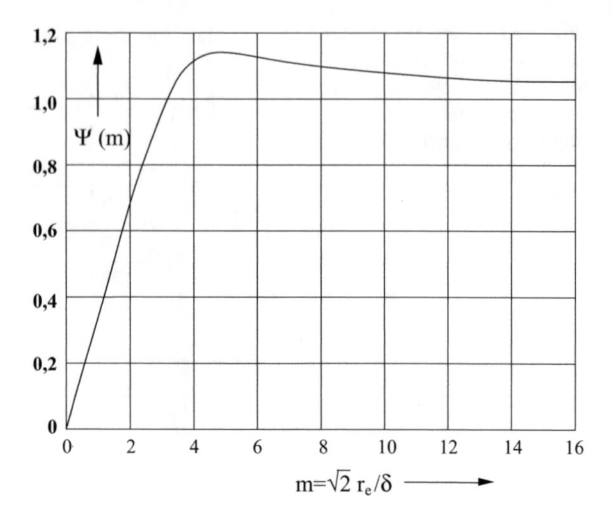

Fig. 6.58 Values of coefficient Ψ(m) as a function of m

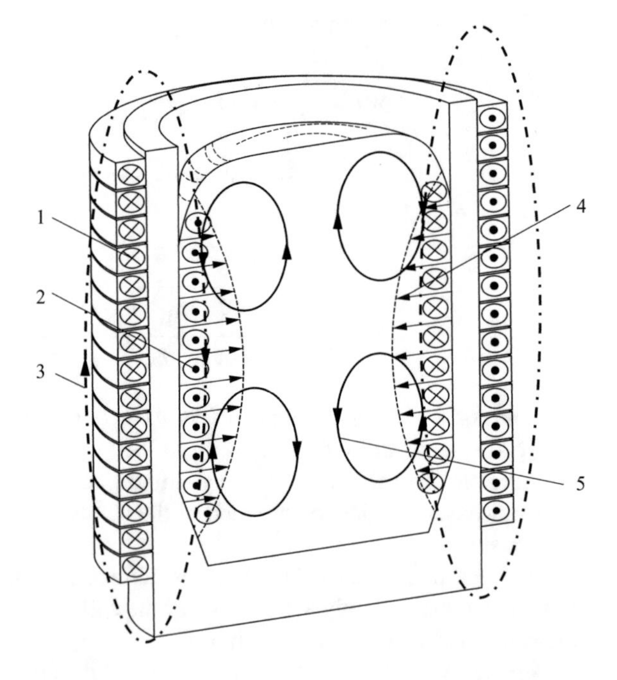

Fig. 6.59 Flow pattern and meniscus produced by the electromagnetic force density distribution (*1* inductor; *2* induced current; *3* magnetic field; *4* electromagnetic force density; *5* melt flow pattern) [32]

This curly field of force generates a flow of the melt schematically shown in Fig. 6.59, whose intensity depends on nature of curls, length of loop paths and kinematic viscosity of the melt.

It is characterized by two vortex rings in the upper part which produce the raising of the metal along the axis of the crucible above the calm bath level, and a corresponding lowering near the walls (*bath stirring and formation of meniscus*). The direction of the melt flow is opposit in the lower rings.

In geometries non-symmetric in axial direction, asymmetries between the upper and lower vortex rings will occur.

These phenomena may have positive as well as negative aspects: in fact on one hand, the stirring action shortens melting time, homogenizes temperature and chemical composition of the melt and improves the interaction between molten metal and slag. On the other hand the melt flow is limited and very low mixing occurs between the top and the bottom of the melt.

Moreover, due to the formation of meniscus, it becomes necessary to use a larger amount of slag, the area of contact between melt and slag is larger, producing a temperature decrease of the latter, which is heated only by the heat transfer from the bath surface.

The meniscus height h_m can be roughly evaluated considering only the time averaged radial component of electromagnetic forces and the equality between the pressure p(0) at the axis (Eq. 6.92a) and the hydrostatic pressure of the molten metal column.

It results:

$$\begin{aligned} h_m &\approx \frac{p(0)}{g\gamma} = \frac{3.16 \cdot 10^{-4}}{9.81} \frac{p_e}{\gamma\sqrt{\rho f}} \Psi(m) \\ &= 0.32 \cdot 10^{-4} \frac{p_e}{\gamma\sqrt{\rho f}} \Psi(m) \end{aligned} \tag{6.93}$$

with: γ—molten metal density; ρ—resistivity; f—frequency.

Usually the meniscus height does not exceed 15 % of the total height of molten metal. However, it depends on furnace geometry and filling level.

As highlighted by Eq. (6.92), the bath movement determines the maximum power for a given size of the furnace. In particular, this is true in LF furnaces for melting ferrous metals with high melting temperature.

In fact, when the temperature of the bath is above 1,550–1,580 °C, more or less strong sprinkles of liquid iron are produced by the rising of CO bubbles which penetrate the surface of the bath and the layer of slag with very high velocities. In this situation, the melt oxidizes and forms an eutectic with the SiO_2 of crucible lining, which liquefies at around 1,170 °C, producing a rapid erosion of the crucible.

Various technical solutions can be adopted in order to reduce the height of meniscus and the stirring in proximity of the bath's surface: axial asymmetric position of the inductor, subdivision of the inductor into several coils differently supplied during the melting cycle or for temperature maintenance, control of power along the height of the inductor or use of inductors with travelling magnetic field [36].

The power limitations due to the above mentioned phenomena occurring at 50 Hz, are less important in MF furnaces, given that the melt flow decreases with the increase of frequency. However, at frequencies above 50 Hz, it must be taken into account the increase of losses in the magnetic yokes. Hence, in furnaces for cast iron, the maximum specific power is currently in the range of 1 MW per ton.

6.9.2.5 Design Problems

When the main geometrical dimensions of the furnace have been defined, a first calculation to verify its electrical and thermal parameters can be made with the approximate method illustrated in Sect. 6.9.1.

In order to avoid repetitions, an example of calculation of a steel furnace is presented in the following [11].

Example 6.6 Crucible furnace for steel

Data: Capacity of crucible: 6 t; Melting time: 1.5 h; Melting start from solid scrap load: average size (diameter) of scrap: $d = 0.08$ m; Resistivity of load: (cold) $0.2 \cdot 10^{-6}\ \Omega$ m, (before melt starts) $1.2 \cdot 10^{-6}\ \Omega$ m, (during melt) $1.37 \cdot 10^{-6}\ \Omega$ m; $E_0 = 400$ kWh/t, for metal melting; $\gamma = 7200\ \text{kg/m}^3$.

1. Useful volume of the crucible:

$$v = \frac{G}{\gamma} 10^3 = \frac{6 \cdot 10^3}{7200} = 0.833\ \text{m}^3$$

2. Main geometrical dimensions: $(c_1 = 0.7; c_2 = 0.11; c_3 = 1.1)$

$$d_i = \sqrt[3]{\frac{4 c_1 v}{\pi}} = \sqrt[3]{\frac{4 \cdot 0.7 \cdot 0.833}{\pi}} \approx 0.9\,\text{m}; \quad \ell = \frac{d_i}{c_1} = \frac{0.9}{0.7} = 1.3\,\text{m};$$

$$s = d_i \cdot c_2 = 0.9 \cdot 0.11 = 0.1\,\text{m}; \ell_i = \ell \cdot c_3 = 1.3 \cdot 1.1 = 1.43\,\text{m};$$

$$\alpha = \frac{d_e}{d_i} = 1 + 2c_2 = 1 + 2 \cdot 0.11 = 1.22; d_e = 0.9 \cdot 1.22 = 1.1\,\text{m}$$

We assume an inductor symmetric to the load and the ratio $d_y/d_e = 1.1$

3. Power of furnace: (assuming $\eta_t \approx 0.93$ and $\eta_e \approx 0.85$)

$$P_u = E_0 \frac{G}{t_f} = \frac{400 \cdot 6}{1.5} = 1.600\,\text{kW}; P_c = \frac{P_u}{\eta_t} = 1.720\,\text{kW};$$

$$P_t = 120\,\text{kW}; P_i = \frac{P_c}{\eta_e} = 2.025\,\text{kW}; P_f \approx (1.05 \div 1.10) \cdot P_i \approx 2.200\,\text{kW}$$

4. Choice of frequency and feeding/supply system

- Minimum frequency—from the condition $m = d/(\sqrt{2}\delta) = 2$ referred to the average size of the scrap:

$$f_{min} \approx 2 \cdot 10^6 \frac{\rho_{max}}{d^2} = 2 \cdot 10^6 \frac{1.2 \cdot 10^{-6}}{0.08^2} = 380\,\text{Hz}$$

- Supply: thyristor frequency converter at frequency 500 Hz; voltage: 1,500 V at inductor terminals

5. Specific surface power and meniscus height:

$$p_e = \frac{P_c}{\pi D\ell} = \frac{1720 \cdot 10^3}{\pi \cdot 0.9 \cdot 1.3} = 468 \cdot 10^3\,\mathrm{W/m^2} = 46.8\,\mathrm{W/cm^2}$$

$$\delta_{melt} = 5.03 \cdot 10^3\sqrt{\frac{137 \cdot 10^{-6}}{500}} = 2.63\ \mathrm{cm} = 0.0263\ \mathrm{m};$$

$$m_{melt} = \frac{d_i}{\sqrt{2}\delta_{melt}} \approx 24.2 \Rightarrow \Psi(m) \approx 1$$

$$h_m = 0.32 \cdot 10^{-4}\frac{p_e}{\gamma\sqrt{\rho f}}\Psi(m)$$

$$= 0.32 \cdot 10^{-4}\frac{468 \cdot 10^3}{7200\sqrt{1.37 \cdot 10^{-6} \cdot 500}} \approx 0.08\,\mathrm{m}$$

The height of the meniscus is approx. 6 % of the height ℓ: acceptable.

6. Resistance and reactance of the load [Eq. (6.22) calculated for N = 1]:

$$m_{melt} = 24.2 \Rightarrow (P + jQ) \approx \frac{1}{\sqrt{2}}(1 + j)$$

$$R'_w + jX'_w = \rho\frac{\pi d_i}{\ell\delta}\sqrt{2}(P + jQ) = 1.37 \cdot 10^{-6}\frac{\pi 0.9}{1.30.0263}(1 + j)$$

$$= 0.113 \cdot 10^{-3}(1 + j)\,\Omega$$

7. Resistance and internal reactance of inductor coil [Eq. (6.47) calculated for N = 1 and $k_i = 0.9$]

$$\delta_i = 5.03 \cdot 10^3\sqrt{\frac{2 \cdot 10^{-6}}{500}} = 0.32\,\mathrm{cm} = 3.2 \cdot 10^{-3}\,\mathrm{m};$$

$$s_i = 1.57\delta_i \approx 5 \cdot 10^{-3}\,\mathrm{m} \quad \Rightarrow \quad A_i = B_i = 0.918$$

$$R_i \approx X_{ii} \approx \rho_i\frac{\pi d_e}{\ell_i\,\delta_i k_i} = 2 \cdot 10^{-8}\frac{\pi \cdot 1.1}{1.43 \cdot 3.2 \cdot 10^{-3} \cdot 0.9} = 0.0168 \cdot 10^{-3}\,\Omega$$

8. Leakeage reactance X_a [Eq. (6.80) for N = 1]:

$$X_{i0} = \omega\mu_0\frac{\pi d_e^2}{4\ell_i} = 2 \cdot \pi \cdot 500\,4 \cdot \pi \cdot 10^{-7}\frac{\pi 1.1^2}{4\,1.43} = 2.624 \cdot 10^{-3}\,\Omega$$

$$X_a = X_{i0}\frac{S_a\ell_i}{S_i\ell} = x_{i0}\frac{(d_e^2 - d_i^2)\ell_i}{d_e^2\ell} = 2.264 \cdot 10^{-3}\frac{(1.1^2 - 0.9^2)1.43}{1.1^2\ 1.3}$$

$$= 0.955 \cdot 10^{-3}\,\Omega$$

9. Reactance X_0 of the return path of flux [Eq. (6.79) calculated for N = 1]: Determining the coefficient K_N from Fig. 6.37 for the ratios $d_e/\ell_i = 0.77, d_y/d_e = 1.1$, it results $K_N \approx 0.85$ and

$$X_0 = \frac{K_N \ell_i}{\ell_i - K_N \ell} = 2.624 \cdot 10^{-3} \frac{0.85 \cdot 1.43}{1.43 - 0.85 \cdot 1.3} = 9.814 \cdot 10^{-3}\,\Omega$$

10. Parameters of the equivalent circuit [Eq. (6.81b) calculated for N = 1]:

$$n^2 = \frac{1}{\left(\frac{R'^2_w}{X_0}\right)^2 + \left(1 + \frac{X_a + X'_w}{X_0}\right)^2} = \frac{1}{\left(\frac{0.113}{9.814}\right)^2 + \left(1 + \frac{0.955+0.113}{9.814}\right)^2} = 0.813$$

$$X_w = -\left[X_a + X'_w + \frac{(X_a + X'_w)^2 + R'^2_w}{X_0}\right]$$

$$= -\left[0.955 + 0.113 + \frac{(0.955+0.113)^2 + 0.113^2}{9.814}\right] \cdot 10^{-3} = -1.186 \cdot 10^{-3}\,\Omega$$

$$R_e = R_i + n^2 R_w = (0.0168 + 0.813 \cdot 0.113) \cdot 10^{-3} = 0.1087 \cdot 10^{-3}\,\Omega$$

$$X_e = X_{ii} - n^2 X_w = (0.0168 + 0.813 \cdot 1.186) \cdot 10^{-3} = 0.9806 \cdot 10^{-3}\,\Omega$$

$$Z_e = \sqrt{R_e^2 + X_e^2} = 0.987 \cdot 10^{-3};\ \eta_e = \frac{n^2 R_w}{R_e} = 0.845;\ \cos\varphi = \frac{R_e}{Z_e} = 0.110$$

11. Total input power and losses in the inductor:
$P_i = \frac{P_c}{\eta_e} = \frac{1720}{0.845} = 2036\,\text{kW}$ (in good agreement with value at point 3)

$$P_{pi} = P_i - P_c = 2036 - 1720 = 316\,\text{kW}$$

$$P_f = (1.05 \div 1.1)P_i \approx 2200\,\text{kW}$$

12. Number of turns and current of inductor:

- Current in a single-turn coil

$$I' = \sqrt{\frac{P_i}{R_e}} = \sqrt{\frac{2036 \cdot 10^3}{0.1087 \cdot 10^{-3}}} = 136{,}859\,\text{A}$$

- Voltage at the terminals of a single-turn coil

$$V' = I' Z_e = 136{,}859 \cdot 0.987 \cdot 10^{-3} = 135.1\,\text{V}$$

- Number of turns

$$N = V/V' = 1500/135.1 \approx 11$$

- Inductor current

$$I = I'/N = 136{,}859/11 = 12{,}442\ \mathrm{A}$$

The above approximate calculation gives only a first idea of the furnace main parameters.

It must be taken into account that nowadays the final design of the furnace is always made by the use of numerical calculation programs, 2D (for simple cylindrical geometries) or 3D (for detailed analysis of specific furnace elements).

In particular, they allow:

- to calculate the electromagnetic field and the induced powers distributions in the elements of the system, in various conditions of crucible filling and with total or partial supply of different coil sections of the inductor during the melting, alloying and maintaining phases;
- to minimize the losses in inductor, magnetic yokes, short-circuited rings and external metallic structure;
- to calculate forces and velocities of the metal flow under various inductor's supply conditions, in order to limit crucible erosion, while maintaining the most convenient conditions for alloying, melt homogenizing and carburizing processes.

Typical results obtained by numerical methods are shown in Fig. 6.60a, b, which illustrate the distribution of flux lines and velocity in the melt.

In conclusion, the main problems in the design of coreless induction furnaces can be summarized as follows:

- the control of melt stirring and meniscus height—as previously stated, these phenomena are dependent on furnace power and frequency. As illustrated by the curves of Fig. 6.61, they limit the specific power at line frequency (50 Hz) to about 350 kW/t, at MF to 1 MW/t.
- limitations due to the supply grid—in some cases the maximum power of the furnace is limited by the power available from the supply network and the requirement to limit emission of harmonics into the grid.
- lifetime of crucible refractory lining—it depends on temperature and characteristics of materials constituting crucible and melt; it is also influenced by the erosion caused by the melt flow velocity.
- acoustic emissions—they are produced by mechanical vibrations due to the electromagnetic forces, which occur particularly in the inductor and the laminations stacks. They can be limited by an accurate mechanical construction and solid fixing of the furnace components to the external metallic structure.

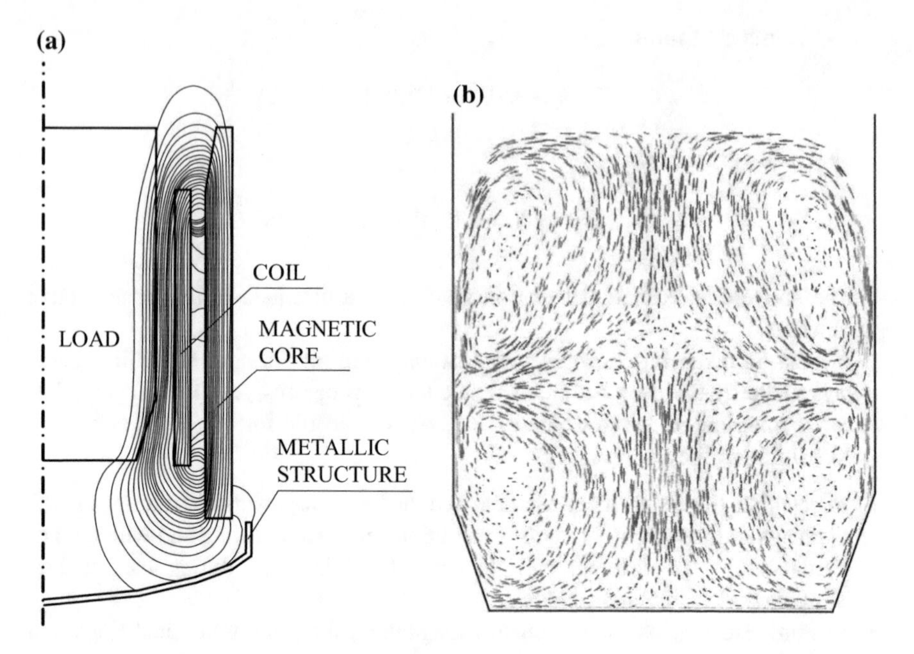

Fig. 6.60 **a** Magnetic field distribution in MF crucible furnace; **b** calculated flow pattern and meniscus [32, 33]

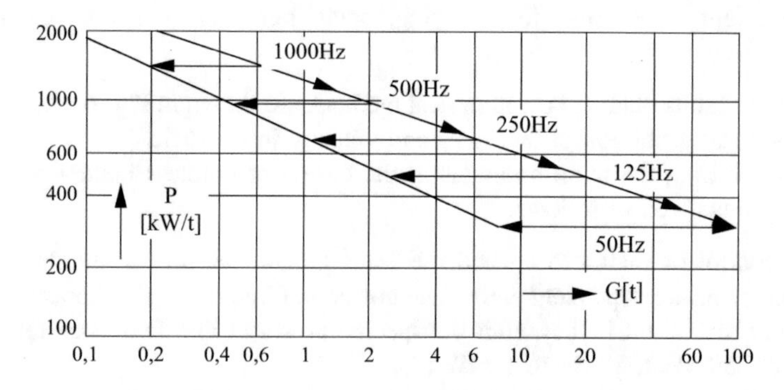

Fig. 6.61 Specific power limits of crucible furnaces at different frequencies as a function of furnace capacity [33]

6.9.2.6 Power Supply of Induction Crucible Furnaces

As shown in Table 6.6, the induction crucible furnaces can be supplied at line frequency (LF-ICFs) or at medium frequency (MF-ICFs) [32].

The schemes of connections to the three-phase supply grid of the two types of furnaces are shown in Fig. 6.62a, b.

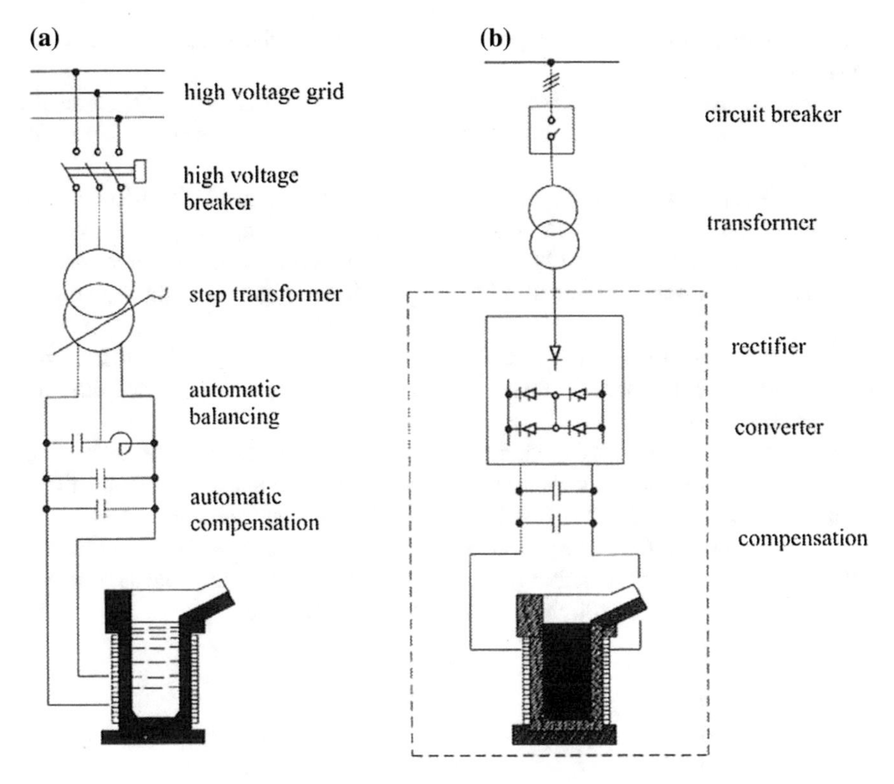

Fig. 6.62 Power supply connections to three-phase grid of induction crucible furnaces (**a** LF-ICF, **b** MF-ICF) [32]

LF-ICFs are connected to the grid by means of a balancing circuit (Fig. 6.62a) which allows to compensate power factor to cos $\varphi = 1$ with a capacitor bank connected to one phase of the delta configuration, and to balance the single phase load with capacitors and reactance of the same size connected in the other phases according to the Steinmetz scheme. (see Sect. 5.3.4.3).

Balancing in different furnace operating conditions can be achieved by adapting the compensation and balancing elements.

As shown in Fig. 6.62b, the connection to the grid of MF-ICFs is constituted by a furnace transformer, a rectifier which converts the line frequency into DC voltage or current, and a thyristor inverter that generates the medium frequency current that feeds the furnace inductor coil.

Since the 1980s, thyristor converters have completely substituted motor-generators, which were practically the only supply of coreless induction melting furnaces in the frequency range 0.25–10 kHz and the magnetic frequency multipliers (150–450 Hz) which had a certain diffusion at the beginning of the 60s.

The main advantage of thyristor converters is the possibility of constant voltage regulation, due to the self-adapting resonance frequency (between 70 to 110 % of nominal frequency) in the high power circuit between the furnace coil and the capacitor bank.

In fact the available power can be kept constant in different conditions of the charge in the crucible, generating it either at a higher or lower current from the converter.

If the melting stock is cold, magnetic or loosely filled, it has high resistance and full power can be obtained at a low current and a low frequency, but maximum voltage. On the contrary when the crucible is filled with melt, it has low resistance and the same power is obtained with lower voltage, higher current and higher frequency.

As regards the operation of the furnace plant, it is recognized that the optimization of power utilization allows to obtain the longest possible service time and the most favorable power costs.

This can be achieved in modern MF-ICFs by the constant power regulation previously described an by switching the power supply to a second furnace after a charge has been melted in the first furnace, in the so-called *tandem-system*.

As shown in Fig. 6.63, the power can be distributed between the two furnaces either by a mechanical switching, or electronically without any lag and with a power sharing from 0 to 100 %, by installing a second inverter.

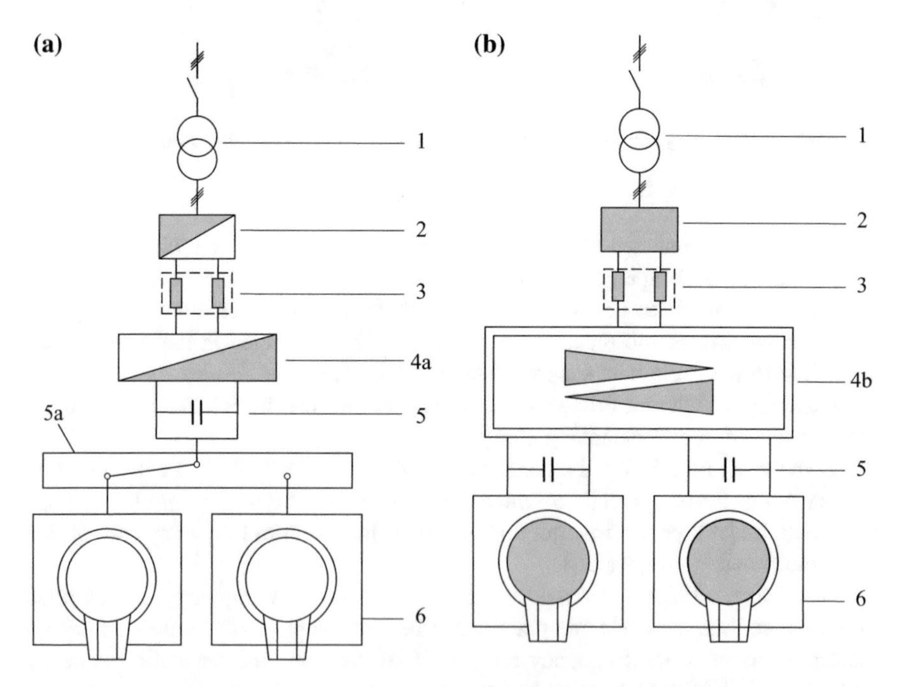

Fig. 6.63 Schematic layout of a tandem-system plant (**a** with mechanical switching; **b** with electronic power distribution; *1* transformer; *2* rectifier for 100 % capacity; *3* smoothing reactor; *4a* converter; *4b* inverter for variable power sharing 0–100 %; *5* capacitor bank; *5a* mechanical switch; *6* furnaces 1 and 2) [32]

The second scheme offers some advantageous possibilities, i.e.:

- operate one furnace in melting mode, while keeping the melt hot ready for pouring in the second one
- to share the total power input between the two furnaces in any proportion at the same time, according to the operation needs
- to replace the refractory lining of one furnace while the second is in the melting mode.

6.9.3 Channel Induction Furnaces

Channel Induction Furnaces (CIFs) are the first type of induction furnace used in the industry (see Sect. 6.10.1 and Fig. 6.47).

They correspond to the schematic of Fig. 6.64 and are characterized by the following elements:

(a) a furnace shell (or crucible), with refractory lining, designed for containing most of the melt
(b, c) a channel, obtained in the refractory material and full of molten metal, connected at its ends to the bottom of the shell
(d) one or several inductor coils, fed at industrial frequency, wound around a magnetic core
(e) a magnetic core, realized with stacks of transformer laminations.

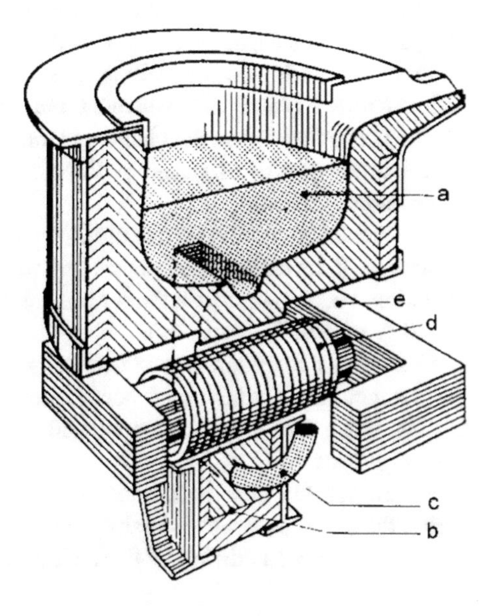

Fig. 6.64 Scheme of a channel induction furnace (**a** melt in the furnace crucible; **b** refractory lining of the channel; **c** channel; **d** inductor coil; **e** magnetic core)

The electrical operating principle of a CIF is similar to that of a transformer with iron core, whose primary and secondary windings are constituted respectively by the inductor coil (d) and the channel (c) full of molten metal.

The metal heats up inside the channel due to the Joule effect caused by the secondary currents of this transformer.

The power of the furnace corresponds to the power transformed into heat within the channel: in a channel of constant cross-section of area S_c and length ℓ_c, where the induced current I_c flows, the power is:

$$P_c = k_r \rho \frac{\ell_c}{S_c} I_c^2 \tag{6.94}$$

with: k_r—coefficient which accounts for the distribution of the AC current I_c in the cross-section of the channel [see Eq. (5.30)].

In channels of normal size, the current density is in the range of 5–15 $\mathrm{A/mm^2}$, the current I_c can be of some tens of thousands Amps and the power P_c is always relatively low. The reactive power is usually of the same order of magnitude of the active one ($\cos\varphi \approx 0.6{-}0.7$).

Example 6.7 Uniform current density in the channel cross-section: $S_c = 10\,\mathrm{A/mm^2}$; channel cross-section: 40 $\mathrm{cm^2}$; channel length: 1.8 m; molten metal: brass (resistivity $\rho = 40\,\mu\Omega\,\mathrm{cm}$.)

$$I_c = 40\,\mathrm{kA}; \quad P_c = 40 \cdot 10^{-6} \frac{180}{40} 40{,}000^2 = 288\,\mathrm{kW}$$

This type of furnace, even though characterized by relative low melting power in comparison to its capacity, has a wide diffusion for holding and casting ferrous and non-ferrous metals and for melting non-ferrous metals due its very high efficiency and economy in terms of investment and operating costs.

The heat generated in the channel is transmitted to the melt in the furnace vessel by the turbulent flow produced by the electromagnetic forces and the forces due to temperature gradients.

The necessity of the presence of molten metal in the channel creates considerable difficulties at the start-up and makes this type of furnace suitable only to continuous operation, keeping in the shell after each casting a residual quantity of melt (at least 15–20 % of crucible capacity).

Continuous operation is also necessary in order to prevent dangerous cracks of the channel lining, which may occur as a consequence of subsequent cooling and heating, when the furnace is stopped and put again into operation after a long period without power.

Moreover, this type of furnace is characterized by very low flow of the melt within the crucible, which is mainly caused by vertical temperature gradients. This results in a reduced stirring of the melt and a very calm surface of the bath.

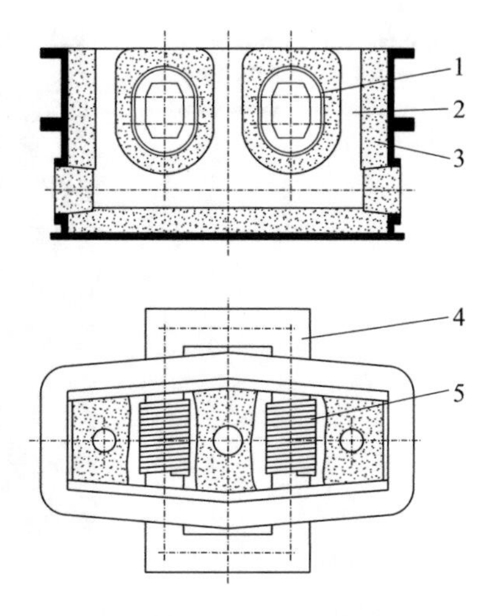

Fig. 6.65 Double- channel inductor (*1* cooling cylinder; *2* channel; *3* refractory; *4* magnetic yoke; *5* inductor)

This characteristic is beneficial in storage and temperature holding operations, where there is the need to prevent the access of oxygen from the air to the bath. For these operations may also be used tight furnaces, with charging and discharging siphons, protected by inert atmosphere of CO, which—in the case of cast iron—avoids the combustion of carbon as well as the oxidation of iron and alloying elements. These furnaces allow to maintain invariant the analysis data for long periods (e.g. even during a whole weekend).

The inert atmosphere also inhibits formation of iron and manganese oxides which would result harmful to the duration of the crucible lining.

In particular for accumulation and holding metal at casting temperature, the most convenient type is the so called rotary tilting furnace, which allows to tilt the crucible around the tapping spout, even in furnaces of capacity up to 100–200 t (see Fig. 6.65).

For special process requirements, like holding steel in vacuum, vacuum furnaces are also available.

For its high efficiency, this type of furnace is increasingly used also for melting non-ferrous metals. Since in these applications high power values are needed, double-channel inductors have been developed (Fig. 6.65) where, besides higher power, it is also achieved a better thermal exchange between channel and crucible, which produces a lower over-temperature between the melt in the channel and the temperature in the crucible.

The maximum power of single channel furnaces for grey cast iron is in the range 800–1000 kW and about 1500–2500 kW for double-loop inductors. The corresponding values are about 300–400 kW for aluminum and copper alloys and 700–800 kW for other non-ferrous metals respectively.

For higher power, more channel inductors are installed on the same crucible, as shown for example, for the rotary furnace of Fig. 6.66.

Fig. 6.66 Junker rotary furnace for copper melting with 4 inductors, each rated 450 kW

Typical capacities and power values of induction channel furnaces for melting different metals, are given in Table 6.6.

6.9.3.1 Design Problems of Channel Furnaces

The main task of the furnace design is the optimization of channel geometry and melt flow inside the channel in order to

- minimize local flow velocities
- prevent pinch phenomena
- avoid erosion or clogging in the channel
- improving intermixing of melt in the furnace.

In fact, the maximum output power of each channel is limited by the electromagnetic forces acting on the molten metal inside the channel, and the heat transfer from the molten metal in channel to the melt in the crucible.

Similarly to what occurs in crucible furnaces, a radial pressure acts on the melt inside the channel, producing a radial compression ("*pinch-effect*").

This radial pressure tends to restrict the transversal cross-section of the molten metal and in normal operation is balanced by the hydrostatic pressure of the bath above it. On the contrary, the operation of the furnace becomes instable when the electromagnetic pressure is not compensated by the hydrostatic one, thus producing a restriction of the channel or even the interruption of its electrical continuity.

The consequences of interruptions are heavy vibrations, which make it necessary to switch off the furnace and restart it at a lower power level.

The maximum pressure produced by the pinch effect, at the center of a channel of constant circular cross-section of radius r_e in which flows the current I_c, is given by the equation [37]:

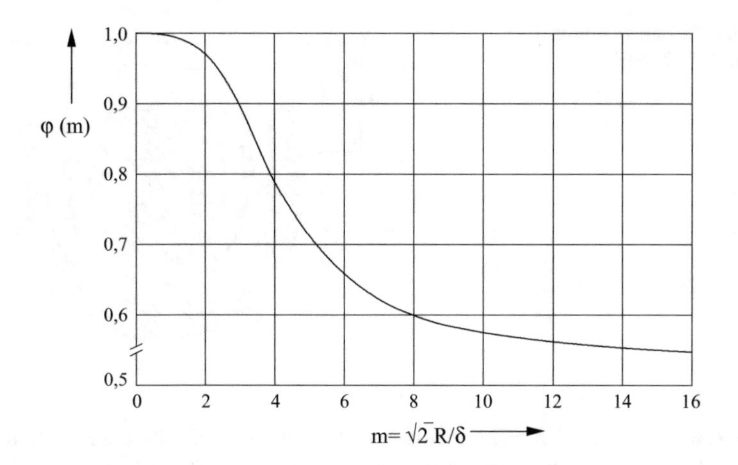

Fig. 6.67 Values of φ(m) as a function of m

$$p_{max} = 1.02 \frac{I_c^2}{\pi r_e^2} \varphi(m) \cdot 10^{-8}, \text{kp/cm}^2, (I_c, A; r_e, \text{cm}) \tag{6.95}$$

where the coefficient $\varphi(m)$ is given as a function of *m* in Fig. 6.67. In particular, if the current density is uniformly distributed in the cross-section of the channel, it is $\varphi(m) \approx 1$, whereas in the limit case $\delta \ll r_e$, $\varphi(m)$ tends to 0.5.

Introducing in Eq. (6.95) the value of power transformed into heat in the channel given by (6.94), one obtains:

$$p_{max} = 1.02 \frac{P_c}{\rho \ell_c} \frac{\varphi(m)}{k_r} \cdot 10^{-8}. \tag{6.96}$$

In particular,

(a) for $\delta \gg r_e$ $(m \ll 1) \Rightarrow k_r \approx 1; \varphi(m) \approx 1$

$$p_{max} \approx 1.02 \frac{P_c}{\rho \ell_c} \cdot 10^{-8}$$

(b) for $\delta \ll r_e$ $(m \to \infty) \Rightarrow$ $k_r \approx \frac{1}{2}\frac{r_e}{\delta}; \varphi(m) \approx \frac{1}{2}$

$$p_{max} \approx 1.02 \frac{P_c \delta}{\rho \ell_c r_e} \cdot 10^{-8}$$

In case (a), for a given load and constant cross-section of the channel, the pinch effect intensity depends on the power per unit length of the channel and the maximum power is a function of the channel's cross section; whereas, in case (b), it is proportional to the internal surface of the channel.

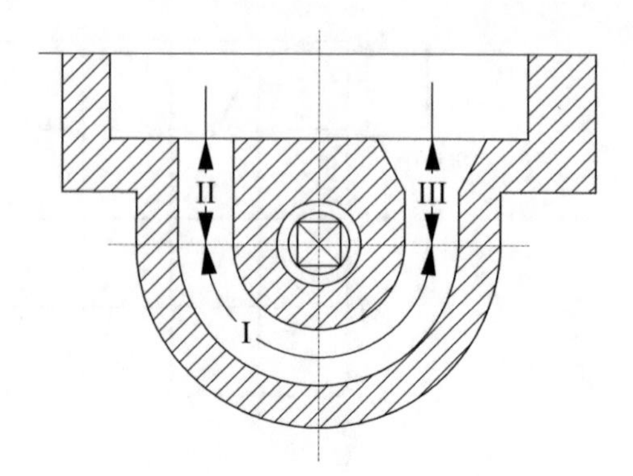

Fig. 6.68 Sketch of different zones of the channel

However, in practice the channel is not designed with constant cross-section, but schematically can be subdivided into three zones as illustrated in Fig. 6.68. A first zone (*I*) with constant cross-section, coaxial with the inductor; a second zone (*II*) with constant cross-section, non-concentric to the inductor; a third one (*III*)—in proximity of the channel ends between the inductor throat and the furnace vessel—non concentric to the inductor, having a variable cross-section along the length of the channel [36].

In the first zone, thanks to the symmetry of the system, the current density has only the component along the channel axis and the prevailing effect is the pinch effect previously described. In the second and third zone, due to the variations of transversal cross-section and hydrostatic pressure along the channel and the presence of radial components of current density in the third zone, the movement of the melt has a three-dimensional character.

In these regions the fluid flow is similar to that of a crucible furnace, and the melt moves from the walls towards the axis of the channel, then—along the axis—from the end openings of the channel towards the crucible and finally, turning back from the bath, along the channel walls (Fig. 6.69).

In reality the melt flow is further complicated for the interaction of the current flowing in the channel not only with the field produced by the current itself, but also with the leakage field of the inductor, thus resulting in a flow non-symmetric to the axis of the channel.

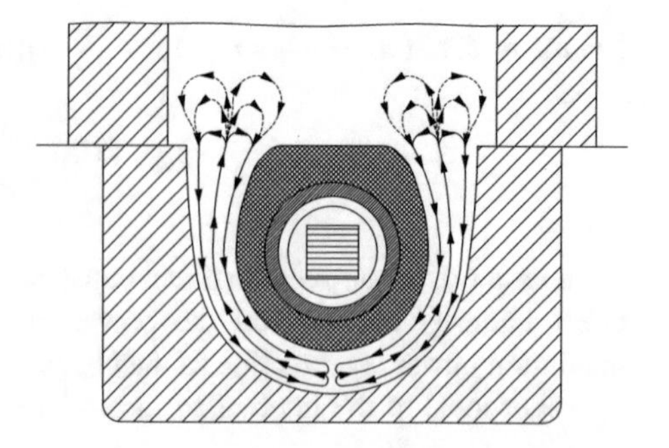

Fig. 6.69 Sketch of melt flow in the channel

Due to the ratio between size of channel cross-section and length of the current path, which is much smaller than in the crucible furnaces, the movement of the molten metal results always very weak, producing a low thermal exchange with consequent higher temperature inside the channel in comparison with the temperature in the crucible.

Moreover, results of recent numerical and experimental investigations have confirmed that the integral through flow velocity is very low, but they have also revealed that intensive transversal turbulent vortices are present in the channel, which are the main responsible of the heat transfer and mass exchange from the channel to the furnace vessel [32, 38].

As a result of the combined action of through melt flow and turbulent vortices substantial over-temperature of the melt occur in the channel above that of the furnace bath, which can reach values of 150–250 °C in single-channel furnaces.

In case of metals with very high melting points, e.g. above 1500 °C, this over-temperature limits the power of a channel of given dimensions, due to the consequent thermal stress of the refractory delimiting the channel walls.

This problem becomes even more critical in case of a periodic turn-on and turn-off of the furnace power, which may lead to sharp temperature variations producing cracks of refractory due to the fatigue.

As indicated previously, the above phenomena limit the power of a single channel to about 800–1000 kW in furnaces for melting cast iron, to about 300–400 kW for aluminum and copper alloys and to 700–800 kW for other non-ferrous metals.

A noteworthy improvement can be obtained with double-channel inductors (Fig. 6.66) which allow to control the melt flow, thus reducing the over-temperature of the channel to few tens of degrees (10–30 °C).

The flow control is realized with different connections of the inductor coils or by superposition of an external magnetic field in different positions of the channel.

For example, with series connection of the coils, the melt flow will be as illustrated in Fig. 6.70a; whereas, by the use a C-shaped electro-magnet, giving an induction field B_n normal to the surface of the drawing at the intersection of the central channel with the two lateral ones, can be produced a force directed along the axis of the intermediate channel, towards the top or the bottom according to the direction of B_n, in this way giving rise to the flow patterns schematically shown in Fig. 6.70b, c.

In particular, by varying the field intensity B_n, it is also possible to control the velocity of the flow independently from the power regulation of the furnace.

The use of the double-channel design allows the realization of inductors for maximum power up to 2000–2500 kW.

6.9.3.2 Power Supplies of Channel Induction Furnaces

Channel furnaces for melting iron are normally supplied at mains frequency via a low-voltage or high voltage transformer. As in channel furnaces, the Steinmetz

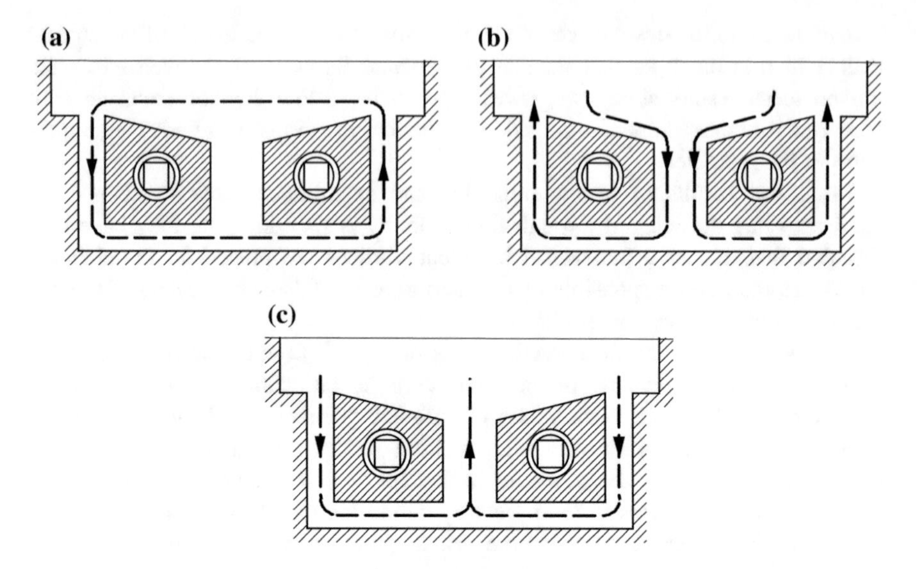

Fig. 6.70 Control of melt flow in double-channel inductors

balancing circuit, comprising a reactance and an associated capacitor, is used for loading with equal currents the three phases of the supply grid. The circuit allows also to compensate the power factor of the inductor-load system by a switchable capacitor bank parallel connected to the inductor and to adapt the compensation and balancing elements to the furnace operating conditions (Fig. 6.71).

On the contrary, the channel furnaces for melting non-ferrous metals are increasingly fed by load—commutated converters, which allow simple infinite power adjustment and the optimal supply for each operating condition, as described in Sect. 6.9.2.6 for ICFs.

6.9.4 Through Heating Prior Hot Working of Metals

6.9.4.1 Historical Background

After the theoretical works and patents of *Dr. E. Northrup* in early 20s of the XXth century and the development at the end of the 20s of powerful and reliable motor-generators in the low frequency range (1–10 kHz), this technology was ready for industrial use. However, mainly for economical reasons, the industrial application of through heating of solids workpieces started only at the end of the 30s [25–27].

In 1937, at *Ajax Electrothermic Co. (USA)* were introduced two installations for through induction heating. The first one, with power supply at 2 kHz (Fig. 6.72a),

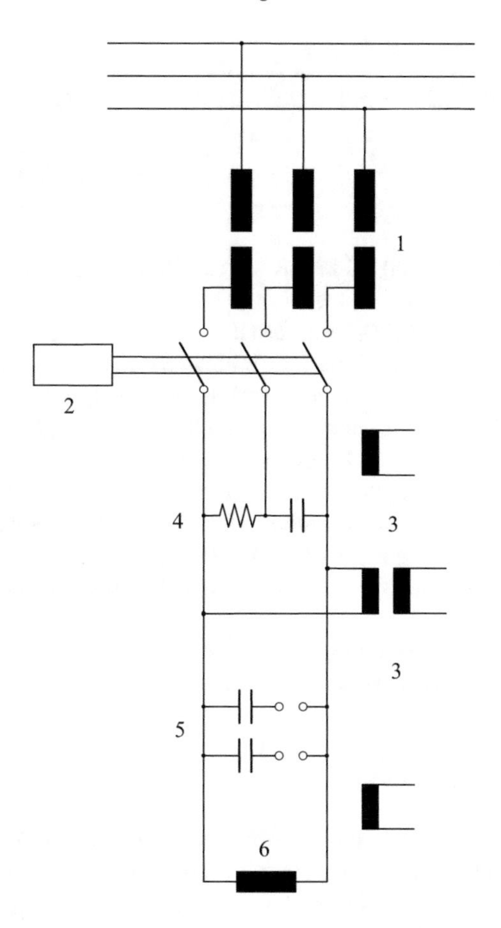

Fig. 6.71 Line frequency power supply of a CIF (*1* step-down transformer; *2* circuit breaker; *3* control units; *4* balancing system; *5* power factor compensation capacitor bank; *6* single phase inductor) [32]

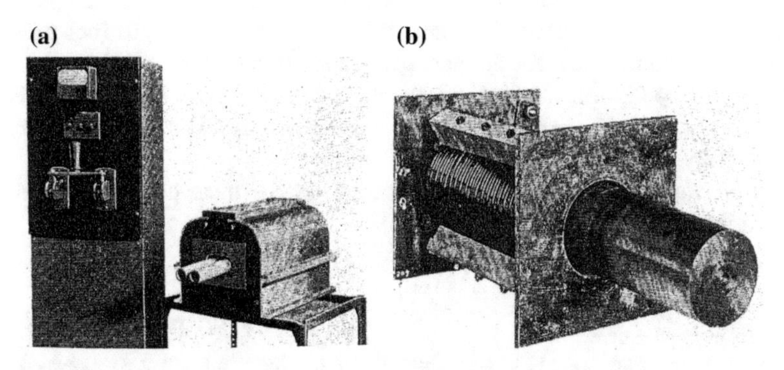

Fig. 6.72 Ajax installations for tubes through heating (1937)

Table 6.7 Hot working temperatures (°C)

Process	Carbon Steel	Stainless Steel	Copper	Aluminum	Titanium
Extrusion	1230	1300	870	480	950
Rolling	1230	1260	760–870	480–540	930
Forging	1180–1300	1200–1320	870	450	930

was used for heating to 1200 °C in less than 1 min the ends sections of steel tubes before hot forming.

The second one (Fig. 6.72b) was installed for heating the ends of large steel tubes, 125 mm diameter and 12 mm wall thickness, in a multi-turn coil, excited at 2 kHz. Heating time was 1.5 min and final uniform temperature 1100 °C.

From 1937 to 1943 several other induction heaters for forging were installed by the *Swedish company ASEA* in Europe and *General Motors Corporation* in USA.

During the WWII research and development in induction heating continued intensively with the main efforts concentrated on applications to the military needs—preheating for forging, surface hardening, etc. Special demand was for production of aluminium alloys and parts for aviation, hardening of tanks and other machine parts.

But the large diffusion of through heating in industry took place after WWII for the development of automotive and aeronautics industries.

6.9.4.2 Process Requirements

The heating of billets, bars, tubes, cables or metal sheets for subsequent hot working[1] is realized for increasing the workpiece temperature to a specified level and degree of uniformity, suitable to provide optimal technological conditions for plastic deformation of the metal during the subsequent operations of forging, stamping, rolling, extruding, etc.

The uniformity requirements may include maximum admissible temperature differentials in the workpiece, surface-to-core, end-to-end and side-to-side. However, in some cases, such as extrusion of aluminum or aluminum alloys billets, it could be required a temperature gradient along the billet length (hot nose and cooler tail) to compensate for the heat generated during the process.

The main applications regard mostly the different types of steel, but also of noteworthy importance is the heating of non-ferrous metals such as copper, aluminum, brass, and titanium.

Typical values of hot working temperature for the most important metals are listed in the Table 6.7.

[1]*Hot working operation* is the processes of plastic deformation of a metal at a temperature of or above 50 % of the melting temperature, variable depending on material and type of work.

Due its technical-economical advantages, induction heating is systematically replacing traditional processes such as heating in gas or oil furnaces.

The advantages of induction heating can be summarized as follows:

- continuous and automated heating processes with qualitative improvement of final products
- ability of in-line heating and processing
- short heating times and fine temperature control
- short start up and shutdown times of heaters
- increase of production rate and machine utilization
- reduction of scale and surface decarburization
- decrease of material consumption
- increased duration of the dies
- limited space requirements for the installation
- possibility of interrupting the work for short periods
- improvement of ambient and working conditions.

In contrast, *disadvantages* are:

- relatively high investment costs
- difficulty to fulfil uniformity requirements in workpieces with complex shape
- necessity of substituting the inductor when the heated workpiece changes, as in case of a production program with small quantities of billets of the same type, given that high electrical efficiency can be obtained only when the inductor is "adapted" to the body to be heated.

Today, it is common to realize installations with rated power of hundreds or thousands of kW and production rates up to ten thousands kg/h.

At Geneva Steel, Utah, it is installed world's largest re-heater of carbon steel slabs, with total power 42 MW and production rate of 540 t/h. The installation with the highest rated power (210 MW) constructed up to now, is the one at McLouth Steel Corporation (USA), also for heating of steel slabs, with rated power 210 MW [25].

The bodies to be heated are usually distinguished according their geometry in:

Billets—workpieces of circular or square cross-section (max. length: 1 m; diameter or side: 20–200 mm.);

blooms—bodies of circular or square cross-section (length: 1–20 m; diameter or side: 40–250 mm);

slabs—parallelepiped bodies (length: 4–10 m; height: 1–2 m; thickness: 150–300 mm);

bars, rods, wires and *tubes.*

In the following we use the term *load* for workpieces of different shapes.

As shown in Fig. 6.73, an installation for through heating basically consists of one or more inductor coils, the electrical equipment (transformers, inverter, switches and capacitor banks) for supplying the inductors with currents and voltage (s) of appropriate frequency and a set of auxiliary equipment for loading, unloading

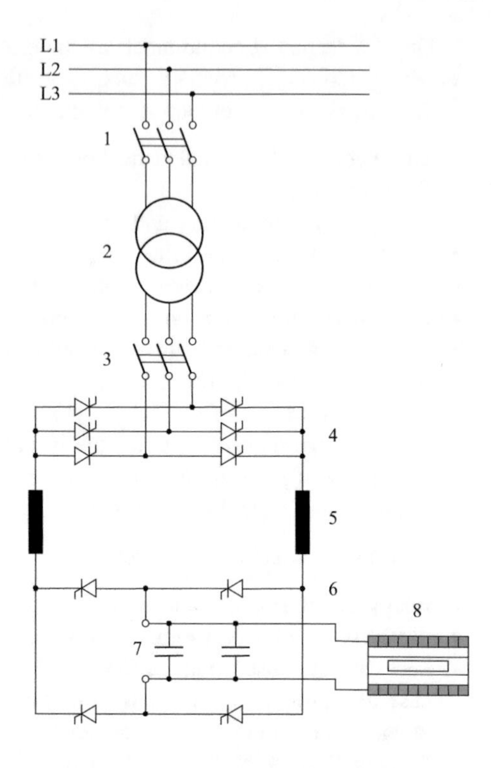

Fig. 6.73 Electrical scheme and components of a medium frequency installation for hot working of metals (*L1*, *L2*, *L3* three-phase supply line; *1*, *3* switches; *2* three-phase transformer; *4* rectifier bridge; *5* smoothing inductance; *6* inverter; *7* medium frequency capacitors for power factor compensation; *8* induction coil and load) [39]

and handling workpieces, cooling inductor, inverter, capacitors and other components for scheduled connection of capacitors, quick inductor substitution and process control.

Several design concepts are used in billet heating as regards charge handling in the inductor coil.

In particular, the heating can be classified as static, stepwise, with progressive continuous advancement or of oscillating type.

- *Heating of static type* is the one where, during the working cycle, the workpiece to be heated is introduced in the inductor coil, hold stationary during heating until the set hot working temperature is reached and then extracted from the inductor and delivered to the next hot working operation. Another cold billet is then loaded into the coil and the subsequent heating cycle starts.
  In this case, the transient temperature distribution in a steel workpiece is of the type shown in Fig. 6.74. As it will be illustrated later, time-temperature diagrams in non-magnetic billets are noticeably different.
- *Stepwise heating,* uses simultaneously 2 or 3 inductors; in each of them the workpiece remains stationary for a time equal to 1/2 or 1/3 of the whole heating cycle and then is moved into the adjacent inductor until reaching the final prescribed temperature. After the first cycle, in all inductors there are billets undergoing heating. In this way it is realized a semi-continuous production

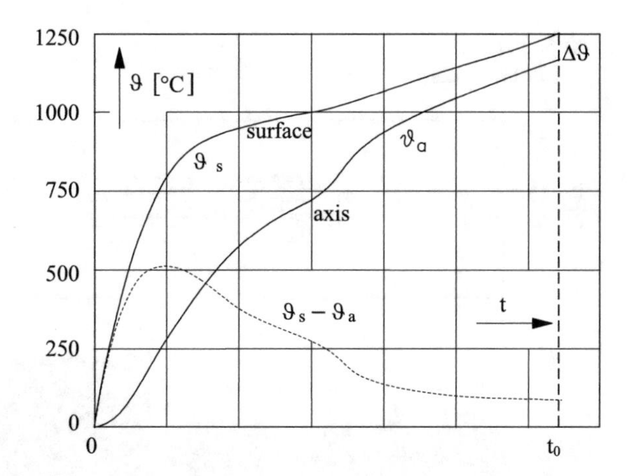

Fig. 6.74 Transient temperature distribution in a steel billet during static heating or with continuous advancement through one single inductor

process with a thermal transient of the type illustrated in Fig. 6.75. One advantage of this method is the fact that at the terminals of each inductor is seen an impedance which varies relatively little during the working cycle, unlike to what occurs during the static type heating.

- *Heating with progressive continuous advancement* is realized when the billets to be heated pass through the inductor, or a sequence of inductors, with continuous or intermittent advancement, according to the type of mechanical device used for the movement. In this case the transient temperature distribution is of the same type indicated in Fig. 6.74 or Fig. 6.75 where, assuming a constant advancement speed, the values displayed on the abscissae represent the distance from the entrance edge of the inductor.
- *Oscillating heating* is the heating mode where the workpiece is moved oscillating forwards and backwards inside the inductor during the heating process, with a convenient oscillating stroke, in order to reduce the installation's space requirements.

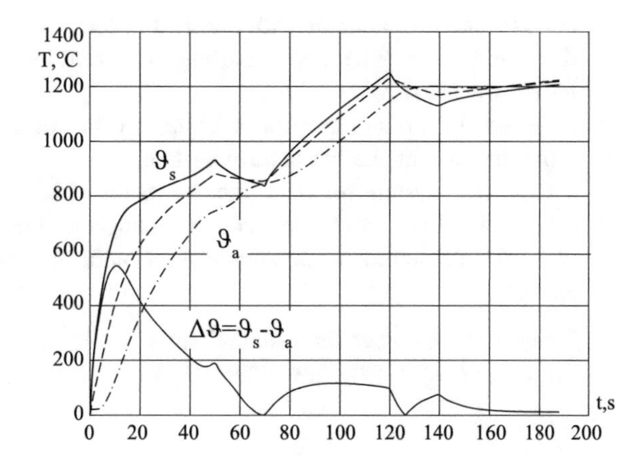

Fig. 6.75 Transient temperature distribution in steel billet during stepwise heating or with progressive continuous advancement within an inductor with three spaced coils

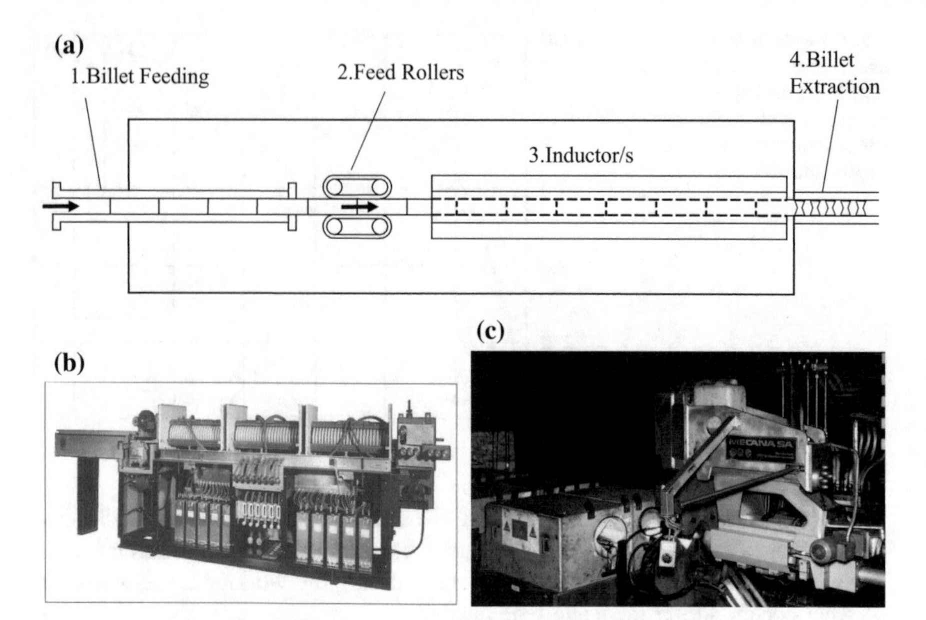

Fig. 6.76 Heating installation for heating billets with continuous advancement (**a** layout; **b** constructive view; **c** rapid extraction system)

In the progressive continuous heating, after a start-up transient, the temperature distribution along the column of heated billets remains invariant inside each inductor and, consequently, the impedance at the inductor terminals remains also constant during the working cycle.

The typical layout of an installations for billet heating with progressive continuous advancement is shown in Fig. 6.76.

It comprises the following basic elements:

1. a manual or automatic type feeding system of the workpieces to be heated;
2. a mechanical system which ensures the continuous advancement of the billets column into the inductor with constant speed;
3. the heating line, generally consisting of several inductors into which the billets move supported by rails;
4. a system for rapid extraction of heated workpieces;
5. a pyrometer temperature control system;
6. a chute or a system for rapid transportation of billets to the press;
7. a control cabinet containing compensating capacitors, start-up and switch-off devices, measurement instrumentation and, in some cases, the frequency converter.

A similar layout have the installations with continuous advancement for heating bars or tubes (Fig. 6.77); they differ only from the mechanical point of view in the

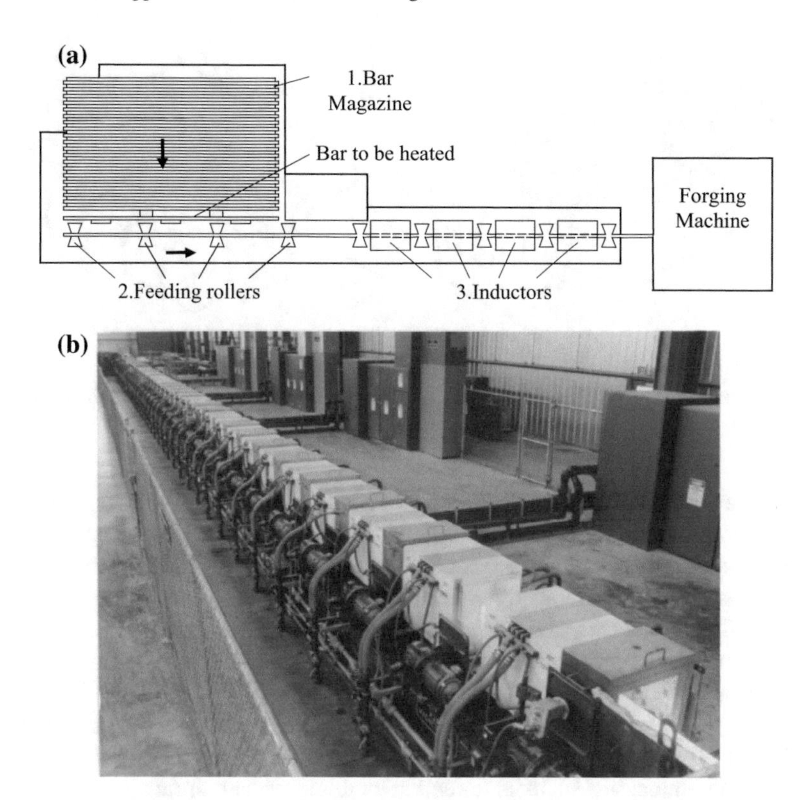

Fig. 6.77 **a** Typical layout of an installation for bar heating; **b** heater for round and square bars: diameters 100–175 mm, length 2.7–10.7 m, production rate 35 tons per hour, rated power: 12 MW (courtesy of Ajax Magnethermic)

design of the bar magazine and feeding system (1) and the bar movement system, which usually is made on rollers (2) inserted between two adjacent inductor coils (3).

The installations for heating billets and bars of circular or square cross-sections are mostly provided with inductors similar to those of Fig. 6.78. The advancement

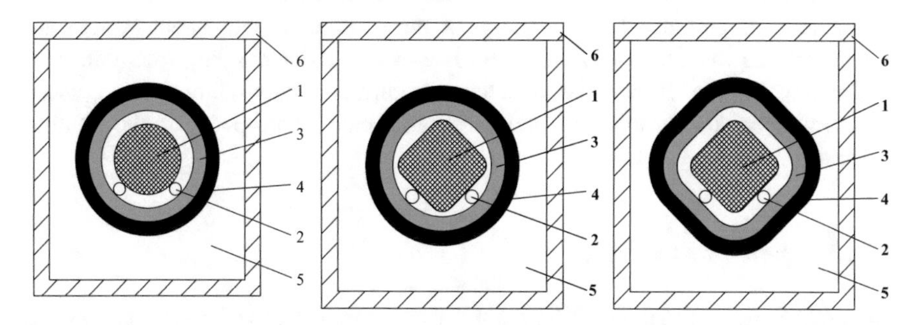

Fig. 6.78 Cross-section of typical inductors for "through" heating (*1* workpiece to be heated; *2* skid rails; *3* thermal insulation; *4* inductor coil; *5* filling material; *6* external case)

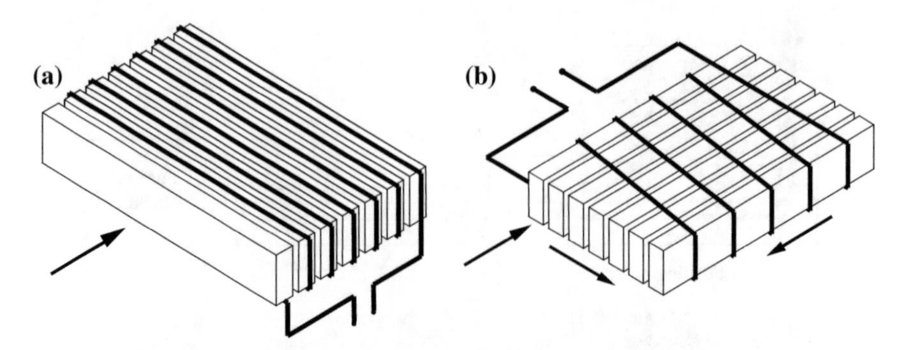

Fig. 6.79 Inductors "en nappe" (**a** with transversal advancement of bars; **b** with axial-transversal movement)

of the billets occurs in the direction of the axis of the inductor, which is provided with a thermal insulation layer (also called refractory or liner) between coil and load and self-supporting internal water cooled skid rails for support and advancement of workpieces and protection of the liner surfaces.

The rails are usually made of non-magnetic steel tubes in order to guarantee the required mechanical robustness but, at the same time, to be "quasi-transparent" to the electromagnetic field in order to avoid significant losses.

Other types of inductors are those so-called "*en nappe*" for bar heating: the movement of workpieces within them can be transversal (as in Fig. 6.79a) or with loading and unloading in axial direction and movement during heating in the transversal one (Fig. 6.79b).

Other inductors of special design are those known as *oval coil* and *channel coil* (or *slot* or *skid coil*) *inductors,* shown in Fig. 6.80, which are used for bar or rod end heating. The channel coil, however, has some drawbacks regarding reduced efficiency and non-uniformity of temperature distribution in the cross-section of the heated workpieces.

The loading and unloading system are different depending on geometry and weight of workpieces, type of inductor and degree of automation of the installation.

For these operations electrical, pneumatic or hydraulic drives are used. Of fundamental importance for the design of the continuous movement system are dimensions and weight of workpieces, length of production line, possibility of evacuating workpieces from the inductor and number of admissible rejected pieces at restarting of the installation after an unexpected interruption of the heating process.

6.9.4.3 Basic Process Parameters

The preliminary evaluation of the main process parameters is the first step of the billet heater design. This step comprises the evaluation of heating power and cycle

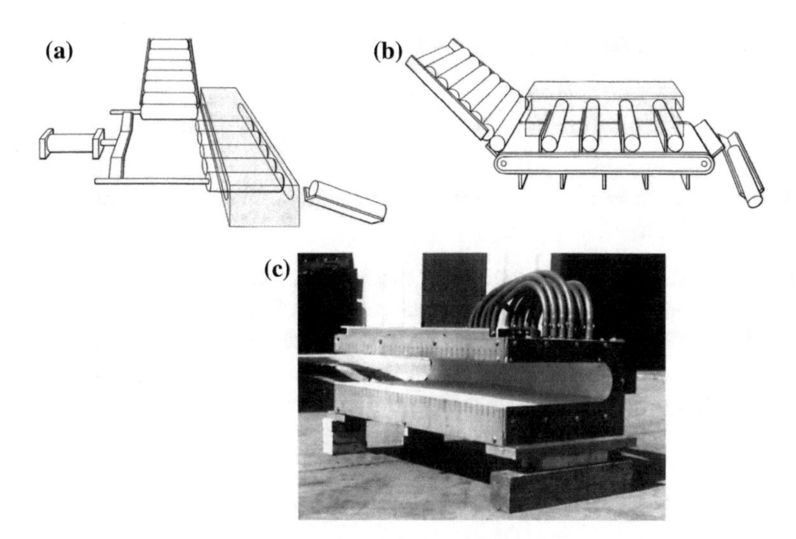

Fig. 6.80 **a** Oval coil end bar heater; **b** channel coil bar end bar heater; **c** channel coil inductor [40]

time, the selection of frequency and power control and the estimation of the heater energetic parameters like efficiency and energy consumption and other process-related factors.

- *Workpiece power*—It is the power that must be induced in the workpiece by the inductor coil(s) in order to increase its average temperature at the required production rate.

  A rough estimate of it can be obtained by the relationship:

$$P_{wav} = c \cdot m \cdot \frac{\vartheta_f - \vartheta_i}{t_h} \; (W) \tag{6.96}$$

where ϑ_f, ϑ_i are the average values of initial and final temperature in the workpiece (°C); c is the average value of specific heat in the above temperature interval (J/kg °C); m is the mass of the heated workpiece (kg) and t_h is the heating time (s).

Alternatively, the estimate of P_{wav} can be done by using diagrams giving the required kWh/t as a function of temperature (see Tables A.8 in Appendix).

It must be kept in mind that Eq. (6.96) gives the average value of power during the heating period, while the instantaneous power induced in the workpiece changes due to the variations with temperature of electrical resistivity end relative magnetic permeability of the material.

- *Power at the inductor terminals*—it is given by:

$$P_{iav} = \frac{P_{wav}}{\eta} = \frac{P_{wav}}{\eta_e \cdot \eta_{th}} \tag{6.97}$$

with: $\eta_e, \eta_{th}, \eta = \eta_e \cdot \eta_{th}$—respectively electrical, thermal and total efficiency of the inductor.

In particular, η_e includes the power dissipated in the coil copper turns and the undesirable losses in electrically conductive bodies located near the inductor, like magnetic shunts, liners, rolls, support beams, etc., while η_{th} takes into account heat losses by radiation and convection from the workpiece surface and by conduction from the billet to water cooled liners, guides and other support structures.

6.9.4.4 Frequency Selection

In any induction heating application it is important to evaluate the correct power required and to select the proper frequency.

In fact, a value of frequency too low may lead to poor coil efficiency, while a frequency too high, producing concentration of the induced power in a thin surface layer, will require long heating times in order to allow for heat transmission towards the core, which is necessary for reaching the required temperature uniformity. Therefore the frequency selection is the result of a reasonable compromise among several factors, like type of heated material, size and geometry of workpieces, required final temperature uniformity, etc.

When heating solid cylindrical workpieces, this selection is done with reference to the size of the transversal cross-section, keeping in mind that in order to reduce the heating time, with the same final temperature differential between surface and axis, the value of $m = d_e/(\sqrt{2}\delta)$ should be as low as possible, but also to fulfil the condition $m \geq 2.0–2.5$ during the whole heating cycle for assuring good electrical efficiency (see diagram of Fig. 6.17).

On the other side, high values of *m*, which could be acceptable from the point of view of electrical efficiency, will produce a distribution of heat sources concentrated near the surface which leads to an increase of heating time and thermal losses.

In practice, the optimal frequency corresponding to $m \approx 2.5$—with δ evaluated with the characteristics of the "hot" material—is not always the most convenient from an economical point of view, given that also other factors must be taken into account, like range of dimensions of workpieces in the production program, cost of converters at different frequencies, etc.

This means that in practice for each diameter exists a frequency range within which the heating can be conveniently done.

Fig. 6.81 Economical range of frequencies for heating magnetic steel billets up to 1200 °C, as a function of workpiece diameter (**a**, **b** lower and higher limits; **c** optimum range)

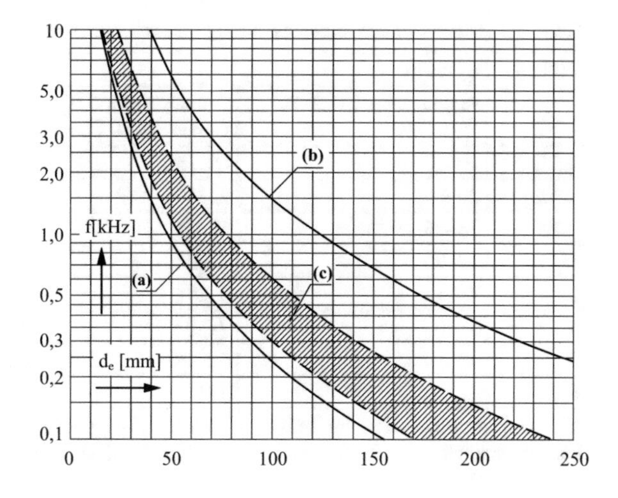

This range can be defined taking into account the following criteria [11]:

- $m < 2$ low electrical efficiency
- $m = 2.5$ optimal frequency
- $2 \leq m \leq 5$ efficiency fairly good
- $m > 7$ low thermal efficiency

Considering these criteria as well as the average values of material's characteristics in the interval from the ambient to the hot working temperature, for each frequency can be defined a range of workpiece diameters that can be heated economically.

For magnetic steels heated to 1200 °C, these frequencies are given as a function of billet diameter in the diagram of Fig. 6.81 (range between curves *a* and *b*).

The frequencies of the dashed region (*c*), which correspond to the inequality [11]:

$$\frac{3}{d_e^2} < f < \frac{6}{d_e^2} \quad (\text{with } d_e \text{ in m}), \tag{6.98}$$

give the minimum heating time.

Considering the influence of magnetic permeability below the Curie point on the values of *m*, one can use a much lower frequency for heating steel billets up to 770 °C (see first column in Table 6.8).

The same table gives also the suggested frequency for the through heating of some non-magnetic materials.

It should be also taken into account that for metals with high thermal conductivity can be used values of *m* higher than those given above, and that efficiency and costs are different at medium or industrial frequency.

When the production mix includes workpieces of different dimensions, the selection of frequency is done with reference to the lower size. Sometimes,

Table 6.8 Diameters economically through heated at various frequencies

f (kHz)	d_e (mm)			
	Magnetic steel (770 °C)	Aluminum (500 °C) Copper (870 °C)	Brass (800°C)[a]	Ti (950 °C)
0.050	>35	60–150	>70	>250
0.300	20–35	25–60	30–70	100–250
0.500	15–25	18–45	20–50	75–190
1	10–20	13–35	15–35	55–140
2	7–15	9–25	10–25	40–95
3	5–10	7–18	8–20	30–80
4	<6	6.5–16	7–15	25–70
8		5–12	5–12	20–50

[a]The frequency selection for brass and other non-magnetic alloys may be influenced by the material characteristics, which can be strongly dependent on composition

however, it is appropriate to supply the heater with frequency converters designed for working at different nominal frequencies.

In particular, in case of ferromagnetic billets, where the penetration depth δ is small in the initial heating stage but increases significantly (by a factor of ten or even more) above the Curie temperature, a dual-frequency heating can be beneficial. In this technique a low frequency is used below Curie point, when the steel is magnetic, while above this temperature a higher frequency is used more efficiently.

In the heating of *slabs, plates, blooms and rectangular bars*, as in the case of cylindrical billets, the frequency selection affects not only the temperature profile within the workpiece, but also the coil electrical efficiency.

The optimal frequency for maximum coil electrical efficiency, is obtained with ratios of the slab thickness h to the penetration depth δ greater than 2.5. Values of frequency higher than the optimal one will only slightly modify efficiency, but for $d/\delta \gg 2.5$ the total efficiency tends to decrease due to higher power losses in the inductor coil.

6.9.4.5 Heating Time and Power Density

In the design of the inductor, the same importance of the frequency selection has the definition of the heating time and the power density which give the required production rate with the shortest length of the heater.

These parameters can be evaluated preliminarily by a simplified analysis of the transient temperature distribution in the workpiece, assuming constant material parameters equal to the "integrated" values in the considered temperature interval [41].

Two technological limits must be taken into account: the first is to reach the required temperature uniformity in the workpiece cross-section at the end of heating; the second one, to avoid during the whole heating transient too high

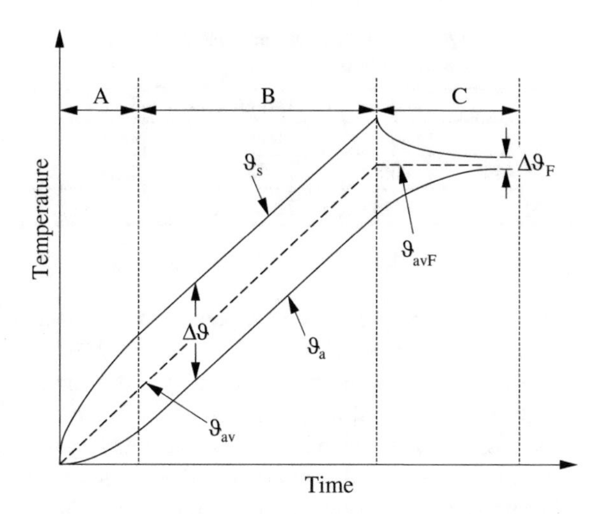

Fig. 6.82 Schematic time-temperature diagram for static heating of non-magnetic cylindrical billets (*A* transient stage; *B* steady-state stage; *C* soaking stage; *F* final values)

temperature differentials, which in some kinds of steel could produce cracks due to thermal stress.

Further requirements to be considered are low energy consumption, good electrical and thermal efficiency, and minimum oxidation of the heated material.

(a) *Non-magnetic materials*

As described in Sect. 6.1.7, the transient temperature distribution in static heating with constant power of non-magnetic billets (nonmagnetic steel, copper, aluminum, brass, etc.), has the shape schematically illustrated in Fig. 6.82. For progressive heating at constant feed rate, the same diagram applies to the temperature growth along the inductor length.

The analysis of the heating transients has shown that the heating time t_0 (s), and the power density p_e (W/m^2) or p_u $(W/m,)$ can be obtained by using Eq. (6.69), if surface losses are neglected, or Eq. (6.70) if these losses must be taken into account.

Equations (6.69) are rewritten for convenience here in the form:

$$\left.\begin{aligned} t_0 &= d_e^2 \frac{F(m)}{16k\,\varepsilon}[1-\varepsilon F''(m)] = d_e \frac{\pi c\gamma}{p_e}\vartheta_s[1-\varepsilon F''(m)] \\ p_u &= \pi d_e p_e = \frac{4\pi\lambda}{F(m)}(\vartheta_s - \vartheta_a) = 4\pi\lambda\frac{\varepsilon}{F(m)}\vartheta_s \end{aligned}\right\} \tag{6.99}$$

where: $\varepsilon = (\vartheta_s - \vartheta_a)/\vartheta_s$—final relative temperature differential; $F(m), F''(m)$—coefficients given by the diagrams of Fig. 6.23.

In hot working processes the final temperature differential ε is mostly in the range 5–7.5 %.

Taking into account that in heating processes with constant power supply the surface to core differential $\Delta\vartheta$ remains practically constant after a transient initial

Table 6.9 Power density and heating time in through heating of non-magnetic materials with $\varepsilon \approx 0.075$, m = 2.5 ($d_e$, in cm)

	Copper	Aluminum	Brass	Titanium	Stainless steel
ϑ_s (°C)	850	500	800	950	1200
p_u (W/cm)	5900	2100	2100	340	650
t_0 (s)	$0.435 \cdot d_e^2$	$0.518 \cdot d_e^2$	$1.04 \cdot d_e^2$	$5.65 \cdot d_e^2$	$6.95 \cdot d_e^2$

stage, from the previous equations can be obtained the guideline values of Table 6.9, calculated with m = 2.5 and $\varepsilon \approx 0.075$.

As shown by Eq. (6.99), the heating time is inversely proportional to ε. Hence, this parameter has a strong influence on the inductor and the heating line lengths. Its choice is therefore a compromise on one hand between the requirements of the hot working process and the duration of dies, which would suggest to adopt the lowest possible value of ε, and—on the other hand—the costs of the installation and its space requirement which would ask for the highest possible values of ε.

In inductors with continuous progressive load, when the heating time t_0 and the hour production rate M (kg/h) are given, the inductor length is determined by the relationship:

$$\ell = \frac{4 \cdot t_0}{\pi \gamma d_e^2} M \approx \frac{c}{4 \pi \lambda} \frac{F(m)}{\varepsilon} M. \tag{6.100}$$

Table 6.10 gives the values of the minimum inductor length, for a production rate of one ton/h, corresponding to the data of Table 6.9.

Example 6.8 Static heating at 50 Hz of an aluminum billet, 150 mm diameter, to the final temperature $\vartheta_s = 520°$ with $\varepsilon \approx 0.075$.

Average characteristics of aluminum:

$$\rho = 8.9\,\mu\Omega\,\text{cm}; \quad \lambda = 2.18\,\text{W/cm K}; \quad c\gamma = 2.68\,\text{W s/cm}^3\,\text{K}$$

It results:
$\vartheta_s - \vartheta_a = 39$ °C; $m \approx 5 \rightarrow F(m) = 0.67; \quad F''(m) = 0.38$

Table 6.10 Minimal inductor length for through heating of non-magnetic materials with production rate M = 1.0 t/h ($\varepsilon \approx 0.075$ and m = 2.5)

	Copper	Aluminum	Brass	Titanium	Stainless Steel
ℓ (m/t)	0.18	0.68	0.41	4.43	3.15

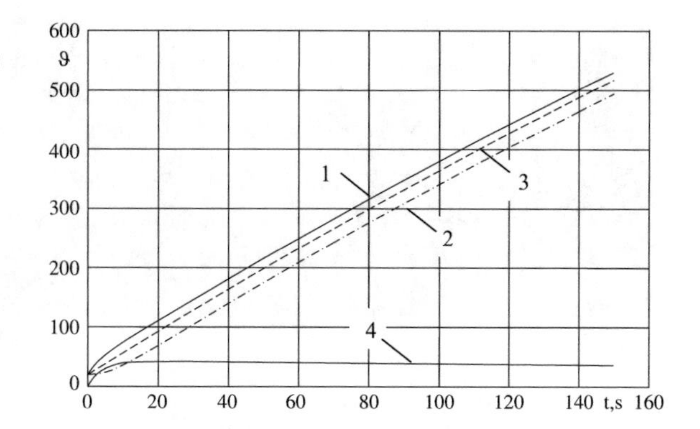

Fig. 6.83 Transient temperature distribution in the through heating of the aluminium billet of Example 6.8 (*1* ϑ_s; *2* ϑ_a; *3* ϑ_{av}; *4* $\Delta\vartheta$)

$$
\begin{aligned}
t_0 &= d_e^2 \frac{F(m)}{16k\,\varepsilon}[1 - \varepsilon F(m)] \\
&= 15^2 \frac{0.67 \cdot 2.68}{16 \cdot 2.18 \cdot 0.075}[1 - 0.075 \cdot 0.38] \approx 150\,\mathrm{s} \\
p_u &= \frac{4\,\pi\lambda}{F(m)}(J_s - J_a) = \frac{4\pi \cdot 2.18}{0.67} \cdot 39 = 1595\,\mathrm{W/cm} \\
p_e &= \frac{p_u}{\pi\, d_e} = \frac{1595}{\pi \cdot 15} = 33.8\,\mathrm{W/cm^2}.
\end{aligned}
$$

Figure 6.83 shows the transient temperature distribution calculated with the same input data by using the numerical software ELTA, which takes into account the variations of material characteristics with temperature. The agreement with the results obtained with constant material parameters is fairly good.

However, in many cases in the heating stage are used values of ε higher than those previously mentioned, reducing the final temperature differential by a subsequent soaking period, as illustrated in Fig. 6.82.

During this period the surface temperature decreases while the core temperature increases, providing the final temperature differential corresponding to the required temperature uniformity. The soaking stage can take place inside the inductor or during transportation of the billets from the exit of the inductor to the press.

From the analysis of the soaking transients, (see Sect. 1.2.5.7-(b), the time Δt_0 necessary for reducing the difference $(\vartheta_s - \vartheta_a)$ respectively to 30 and 10 % of its value at the exit of the inductor (see Figs. 1.11 and 6.82) can be determined by the relationship:

$$\Delta t_0 \approx (0.075\text{–}0.15)\frac{r_e^2}{k} \qquad (6.101)$$

Example 6.9 Heating at 50 Hz of an aluminum billet, 150 mm diameter, to the final temperature $\vartheta_s = 520$ °C with $\varepsilon \approx 0.175$, followed by a soaking period for reducing $\Delta\vartheta$ to 30 % of the value at the inductor exit ($\varepsilon = 0.075$).

- Characteristics of aluminum: as in Example 6.8
- $\vartheta_s - \vartheta_a \approx 90$ °C (at the end of heating stage);
- $\vartheta_s - \vartheta_a \approx 40$ °C (at the end of soaking period).

It results:

$$
\begin{aligned}
t_0 &= d_e^2 \frac{F(m)}{16\,k\varepsilon}[1 - \varepsilon F(m)] \\
&= 15^2 \frac{0.67 \cdot 2.68}{16 \cdot 2.18 \cdot 0.175}[1 - 0.175 \cdot 0.38] \approx 62\,\text{s} \\
\Delta t_0 &\approx 0.075 \frac{d_e^2}{4\,k} = 0.075 \frac{15^2 \cdot 2.68}{4 \cdot 2.18} \approx 5.2\,\text{s} \\
t_{tot} &= t_0 + \Delta t_0 \approx 67\,\text{s} \\
p_u &= \frac{4\,\pi\lambda}{F(m)}(\vartheta_s - \vartheta_a) = \frac{4\pi \cdot 2.18}{0.67} \cdot 90 = 3680\ \text{W/cm} \\
p_e &= \frac{p_u}{\pi\,d_e} = \frac{3680}{\pi \cdot 15} = 78\ \text{W/cm}^2
\end{aligned}
$$

The corresponding transient temperature distribution obtained with the numerical software ELTA is shown in Fig. 6.84. Also in this case, the approximate formulae lead to values in good agreement with numerical results.

(b) *Magnetic steels*

For magnetic steels, considering that above and below the Curie Point the material physical characteristics modify drastically (see Sect. 6.10), the heating

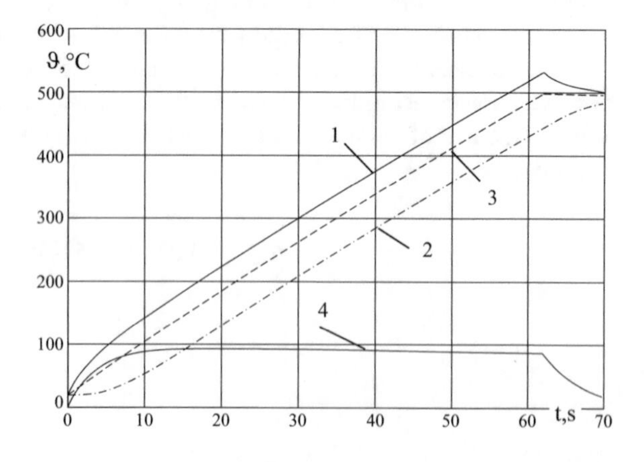

Fig. 6.84 Transient temperature distribution with final soaking period (data of Example 6.9; *1* ϑ_s; *2* ϑ_a; *3* ϑ_{av}; *4* $\Delta\vartheta$)

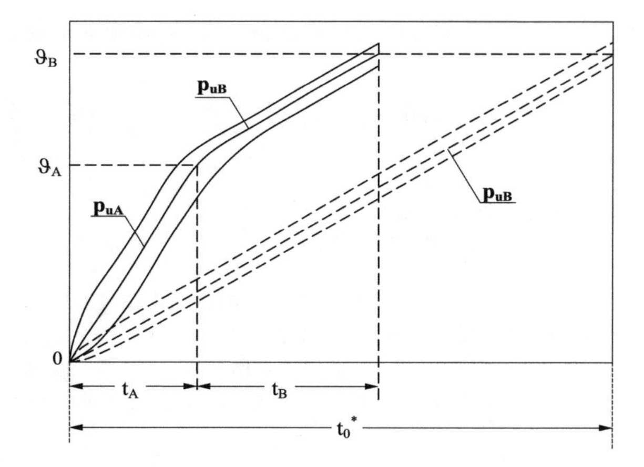

Fig. 6.85 Schematization of an equivalent transient temperature distribution in magnetic steel billets for approximate calculation of the heating time

transient can be schematically subdivided into two intervals A and B. We assume that—with constant excitation field—the power p_u remains constant in each interval.

With reference to previous symbols, we can write:

$$\left.\begin{aligned} \frac{m_B}{m_A} &= \sqrt{\frac{f_B}{f_A}}\sqrt{\frac{\rho_A}{\rho_B\,\mu_A}} \\ \frac{p_{uB}}{p_{uA}} &\approx \left(\frac{H_{0B}}{H_{0A}}\right)^2\sqrt{\frac{f_B}{f_A}}\sqrt{\frac{\rho_B}{\rho_A\,\mu_A}} \quad \left(\frac{P_B}{P_A}\right)\frac{1}{1.37} \end{aligned}\right\} \tag{6.102}$$

In this hypothesis, an approximate evaluation of the heating time can be done with reference to the average temperature of the billet, neglecting in Eq. (6.99) the term $\varepsilon F''(m)$. Thus, from the diagram of Fig. 6.85, we can write[2]:

$$t_0 = t_A + t_B \approx \pi r_e^2\left[\frac{c_A\vartheta_A}{p_{uA}} + \frac{c_B(\vartheta_B - \vartheta_A)}{p_{uB}}\right].$$

Considering that the integrated average values of c_A and c_B, in the temperature intervals A and B, are nearly the same, the calculation of time t_0 can be done by considering the heating with constant power p_{uB} of an ideal billet having the same physical characteristics of the steel above the Curie point during the whole thermal transient.

[2]N.B.—In this approximation it is also neglected the time interval corresponding to the transition of the Curie point, during which the absorbed energy, equal to 17.7 kWh/t (or 500 Ws/cm^3), does not produce increase of temperature.

Indicating with t_0^* the duration of this ideal transient, it results:

$$\frac{t_0}{t_0^*} = \frac{t_A + t_B}{t_0^*} \approx \frac{\frac{c_A \vartheta_A}{p_{uA}} + \frac{c_B(\vartheta_B - \vartheta_A)}{p_{uB}}}{\frac{c_B \vartheta_B}{p_{uB}}} \approx \frac{\vartheta_A \frac{p_{uB}}{p_{uA}} + (\vartheta_B - \vartheta_A)}{\vartheta_B}$$
$$\frac{t_A}{t_B} \approx \frac{\vartheta_A}{\vartheta_B - \vartheta_A} \frac{p_{uB}}{p_{uA}} \tag{6.103}$$

Considering the values of μ_A with the magnetic field intensity mostly used in through heating, assuming that at the end of heating transient m_B is about 2.2–2.5 and introducing for ρ_A and ρ_B the average integrated values in the temperature intervals (0–760 °C) and (760–1200 °C), one obtains:

$$\frac{m_A}{m_B} \approx 10\text{–}20; \frac{p_{uB}}{p_{uA}} \approx 0.15\text{–}0.20; \frac{t_A}{t_B} \approx 0.20\text{–}0.36; \frac{t_0}{t_0^*} \approx 0.47\text{–}0.53$$

Therefore, for inductors with uniform distributed ampere-turns it can be concluded as follows:

- the value of m is always very high below Curie point, in many cases even higher than 25;
- above the Curie point the induced power is much lower than the one in the first heating interval;
- as a consequence, in inductors with progressive movement, the Curie point always occurs in a zone close to the inductor entrance;
- a rough estimation of the heating time can be done by assuming it equal to 50 % of the time calculated for the ideal billet heated with constant power p_{uB}.

According to Eq. (6.99) it results:

$$t_0 \approx 0.5\, t_0^* = 0.5 \frac{F(m)}{16\, k\, \varepsilon} d_e^2 \approx (0.27\text{–}0.3) \frac{d_e^2}{\varepsilon} \tag{6.104}$$

with: d_e in cm and $\varepsilon = (\vartheta_s - \vartheta_a)/\vartheta_s$.

Heating times of steel billets from ambient temperature to 1200 °C with a final temperature differential $\Delta\vartheta = \vartheta_s - \vartheta_a = 75\,°C$ are given in the diagrams of Fig. 6.86.

As Eqs. (6.102) and (6.103) show, a reduction of heating time can be obtained by modifying the ratio (p_{uB}/p_{uA}) by using one or more inductors with different ampere-turns per unit length (i.e. modifying the ratio H_{0B}/H_{0A}) or different supply frequencies $f_A \neq f_B$ for the inductors below and above Curie point.

Another way of reducing the heating time is the so-called "*rapid heating*", in which—using a high specific power in the initial stage—the surface of the billet is raised as rapidly as possible to the final surface temperature with a corresponding high temperature differential $\Delta\vartheta = \vartheta_s - \vartheta_a$, and applying after this moment, a

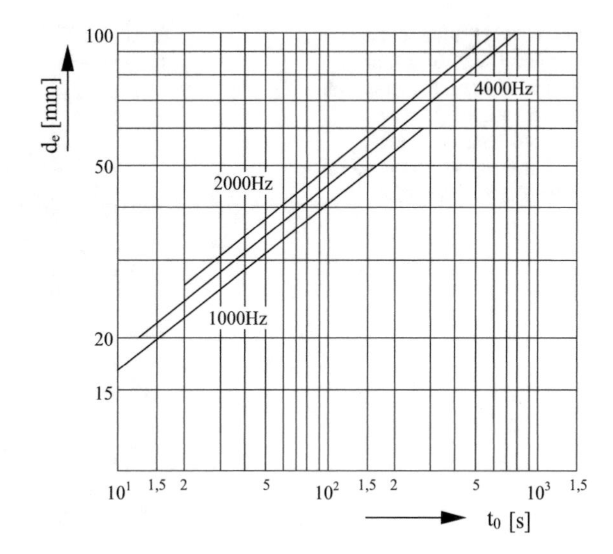

Fig. 6.86 Heating times for through heating of steel billets with normal heating, $\varepsilon = 6.25$ % [41]

conveniently low specific power in order to maintain the surface temperature as close as possible to the final one, till the reduction of the differential ε to the desired value is reached [42].

This heating technique is limited only by the maximum temperature differential admissible for the used steel grade in order to avoid radial cracks, and allows to reduce heating times up to 50 % of values given by Eq. (6.104).

The diagrams of Fig. 6.87 give the heating times which can be obtained with "rapid heating", for the same final temperature difference $\Delta\vartheta = \vartheta_s - \vartheta_a$ equal to 75 °C.

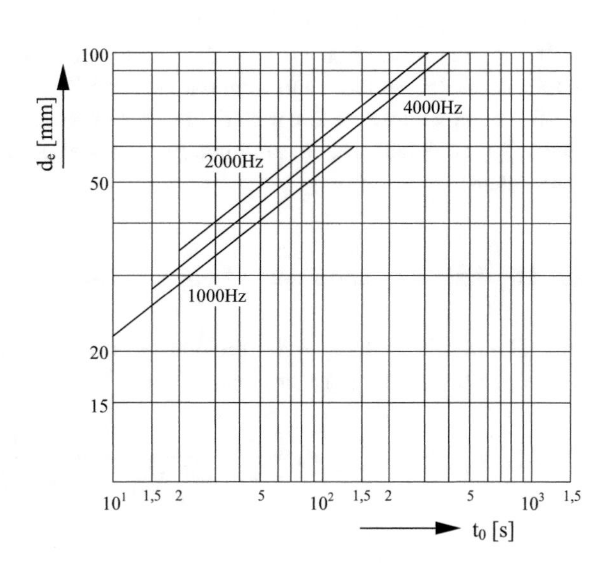

Fig. 6.87 Heating times in the "rapid" through heating of steel billets with $\varepsilon = 6.25$ % [41]

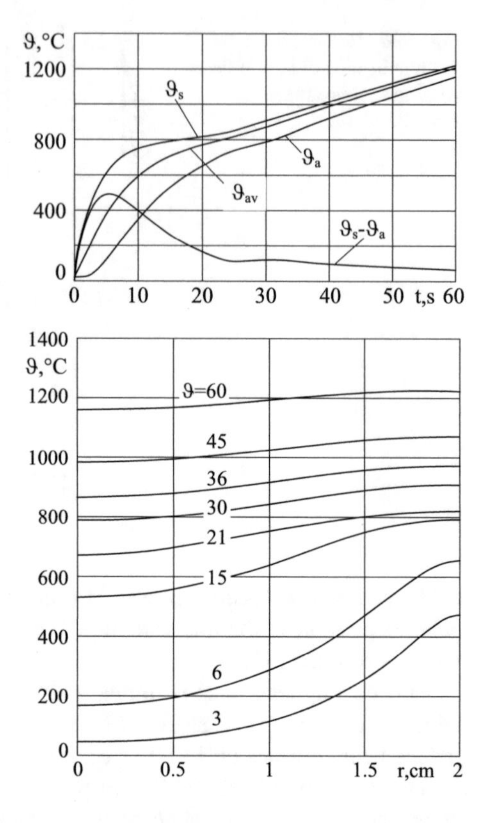

Fig. 6.88 Heating transient calculated with ELTA of a steel billet in an inductor with uniform ampere-turns (d_e = 40 mm; f = 2 kHz; I = 2.200 A; other data as in Examples 5.1 and 5.2)

The diagrams of Figs. 6.88 and 6.89 show two examples of the transient temperature distributions produced respectively by an inductor with uniformly distributed ampere-turns and with the "rapid" heating of a steel billet, 40 mm diameter, at frequency $f = 2$ kHz (see data of Examples 6.1 and 6.2).

They have been obtained with the numerical package ELTA-1D, which takes into account the variations of material characteristics with temperature.

The results allow us to conclude that the approximate calculation procedure is generally sufficient for a preliminary engineering evaluation of the process parameters.

6.9.4.6 Efficiency and Energy Consumption

(a) *Electrical efficiency*—As previously said, the electrical efficiency η_e in induction heating of *non-magnetic loads* can be calculated with the Eq. (6.50):

$$\eta_e = \frac{1}{1 + \frac{r_i \ell \delta A_i k_i}{r_e \ell_i \delta_i \sqrt{2} P}} = \frac{1}{1 + \alpha \frac{\ell}{\ell_i} \sqrt{\frac{\rho_i}{\rho \mu}} \frac{A_i k_i}{\sqrt{2} P}}$$

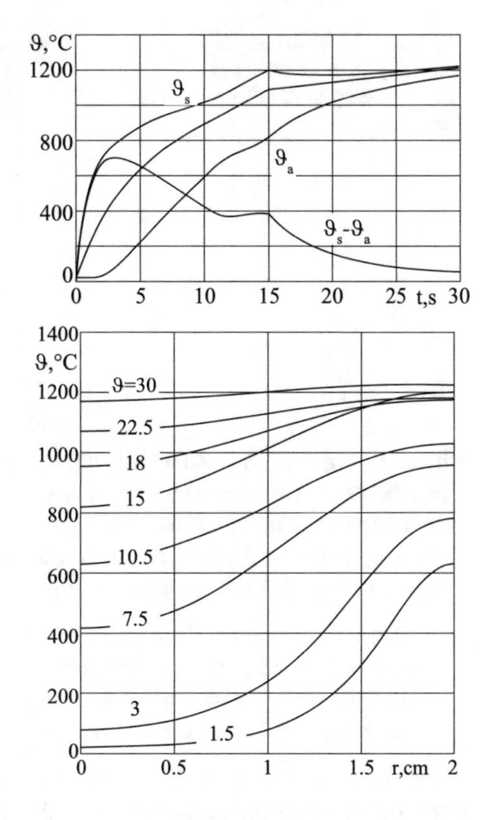

Fig. 6.89 "Rapid" heating calculated with ELTA of a steel billet in an inductor with different ampere-turns (d_e = 40 mm; f = 2 kHz; I ′ = 4.000 A; I″ = 2.000 A)

As shown in Fig. 6.17, for m > 2.5 the electrical efficiency is practically independent of frequency, whereas below this value, the heating will occur with low or very low efficiency.

Moreover, the efficiency is always relatively low when heating materials with low resistivity, and diminishes significantly with the increase of the coupling ratio α of internal diameter of the inductor to external diameter of the billet. It follows that in general it is not recommended the use of the same inductor for heating billets of very different size.

Considering the geometry of Fig. 6.78 and the space required for the thermal insulation and the skid rails, the ratio α, as a function of the billet diameter, has the indicative values of Fig. 6.90.

For *magnetic steels*, the average electrical efficiency can be evaluated by the relationship:

$$\eta_e = \frac{\eta_A t_A + \eta_B t_B}{t_A + t_B} = \frac{\eta_A \frac{t_A}{t_B} + \eta_B}{\frac{t_A}{t_B} + 1} \tag{6.105}$$

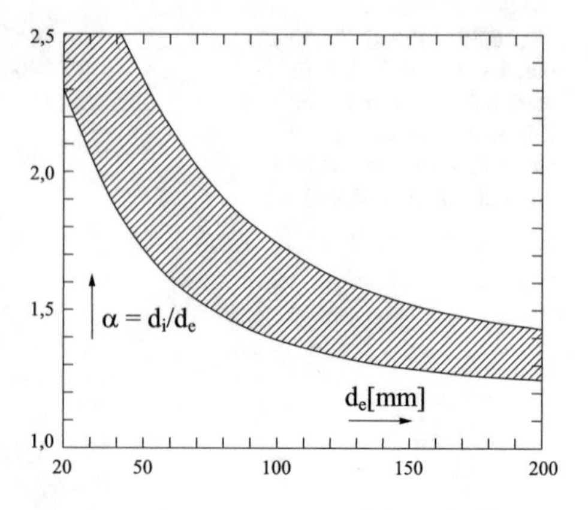

Fig. 6.90 Values of coupling ratio α between inductor and load as a function of billet diameter

where: η_A, t_A—are the values of efficiency and heating time in the temperature range between ambient temperature and Curie point; η_B, t_B—the corresponding values between Curie and the final heating temperature.

Hence, average values of electrical efficiency in the heating of magnetic billets are in the range 0.7–0.8, as already shown in Fig. 6.17.

(b) *Thermal Efficiency*—The thermal efficiency η_t accounts for the losses by convection and radiation from the surface of the billet to the inside surface of the inductor or to the ambient in the zones where thermal insulation is not provided, e.g. between inductors or from the exit of the inductor to the press.

As shown in Fig. 6.78, in order to limit losses, inside the inductor is placed a refractory layer of thickness between 10 and 20 mm, which increases with billet diameter.

The thermal losses inside the inductor coil can be evaluated, under operating conditions, considering the equilibrium of the radiation losses from the external surface of the billet to the internal surface of the refractory and the losses by conduction through the refractory layer, whose external surface is maintained at constant low temperature by the water cooling of the inductor.

Typical values of these losses during the heating process are shown in the diagrams of Fig. 6.91 with reference to the examples of Fig. 6.88. It can be observed that, given the longer period of time during which the billet surface is at high temperature, the average thermal losses are higher in case of rapid heating.

Depending on inductor design, average values of η_t are generally between 0.8 and 0.9.

(c) *Total Efficiency*—The total efficiency is defined by the ratio between the energy required for increasing the workpiece temperature to the final one and that absorbed from the supply line. Its value is influenced primarily by the losses in the frequency converter and connecting transformer, in the current-carrying components and supply bus-bars or cables (proportional to the square of current intensity) and in the skid rails (proportional to inductor length).

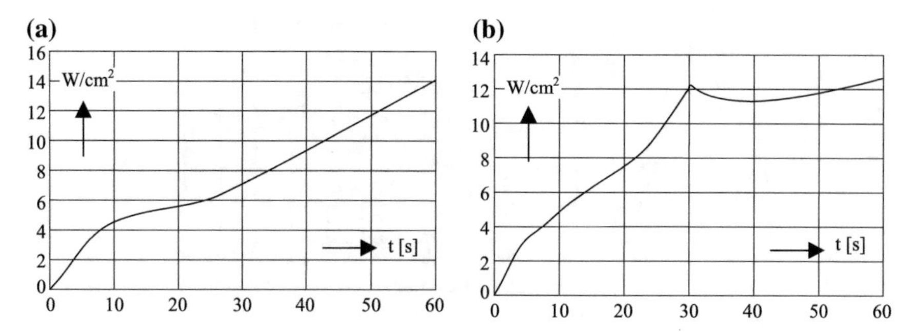

Fig. 6.91 Specific thermal losses during for the heating transients of Fig. 6.90 (**a** normal heating; **b** "rapid" heating)

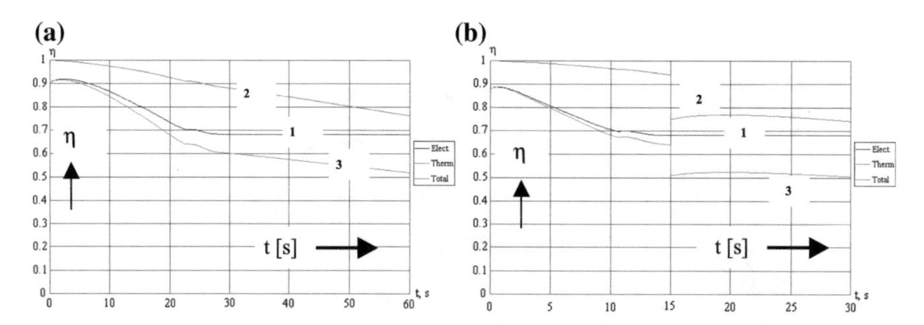

Fig. 6.92 Electrical (*1*), thermal (*2*), and total efficiency (*3*) during heating transients of Fig. 6.88 (**a** normal heating; **b** "rapid" heating)

Considering all these losses, the total efficiency of heating magnetic billets can reach values of 65–70 %.

The Fig. 6.92a, b show the values of electrical (1), thermal (2), and total (3) efficiency during the heating transients of Fig. 6.88.

(d) *Specific Energy Consumption*

Starting from the theoretical heat needed to bring the metal to the final process temperature (see Tables A.8 and A.9 in Appendix), the specific energy consumption can be evaluated by taking into account the above mentioned efficiency values.

For the hot working of the metals most used in industry, the values of Fig. 6.93 are obtained. They can be strongly influenced by the coupling α between inductor and load as well as by the utilization factor of the installation.

As indicated by the typical energy flow diagram of Fig. 6.94 for steel heated up to 1250 °C, when the coupling between the inductor and load is optimal and the installation works at full power, the specific energy consumption is about 350 kWh/t at medium frequency and about 375 kWh/t from the supply line.

These values will increase noteworthy if the same inductor is used for heating billets with diameters smaller than the optimal or the heater works at reduced load.

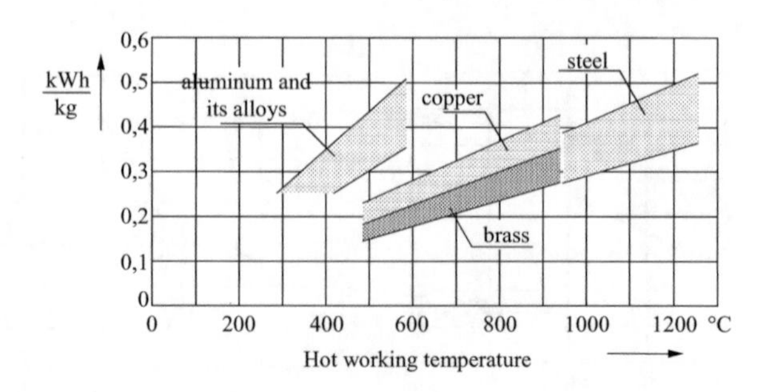

Fig. 6.93 Energy consumption in hot working of metals [23, 42, 43]

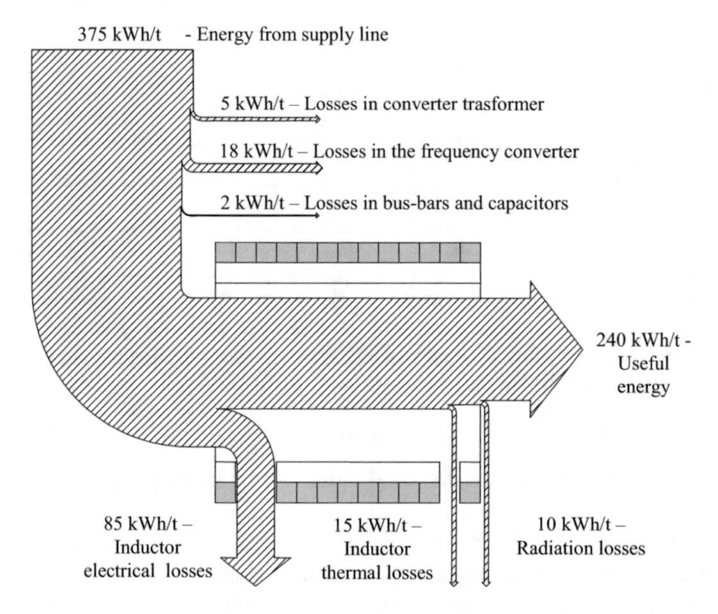

Fig. 6.94 Typical energy flow in the induction though heating of steel billets up to 1250 °C [23]

The diagrams of Fig. 6.95 show the influence of these factors in the case of a heater with rated power 1000 kW at 1 kHz with inductor designed for a billet of 100 mm diameter.

6.9.4.7 Surface Oxidation and Decarburization

In the induction through heating processes, oxidation and decarburization losses are much smaller of those occurring in gas or oil furnaces. These losses are estimated to be in the order of 0.5–1.0 % in induction heating and 2.5 % in fuel furnaces.

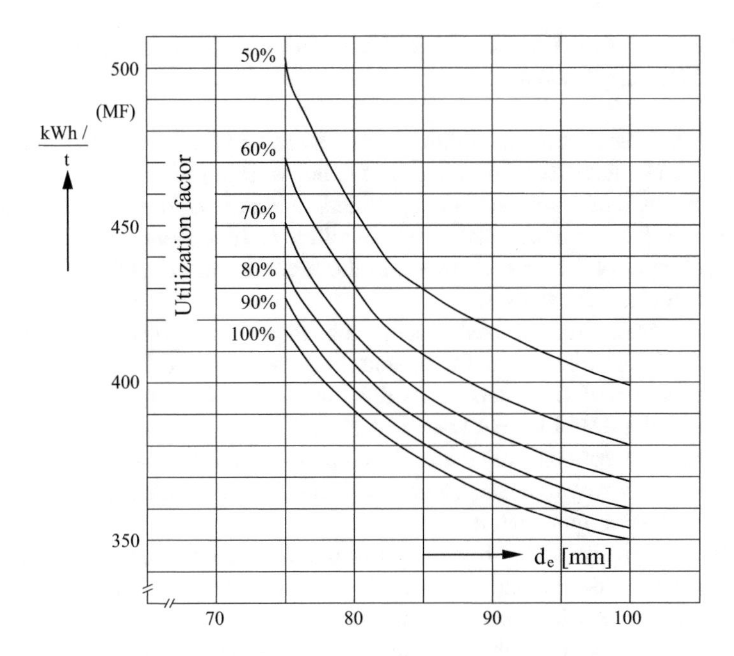

Fig. 6.95 Influence on energy consumption of coupling α and installation's utilization factor [44]

This can be explained by the considerable influence on the oxidation process (scale formation and decarburization of material) of the holding time at high temperature.

The curves of Fig. 6.96, which give some experimental data on scale formation and surface decarburization in induction heating, emphasize the decisive influence of surface temperature and heating time [45, 46].

This also explains the difference with conventional furnaces, since in the induction heating process the time of permanence of the material at high temperature is much shorter.

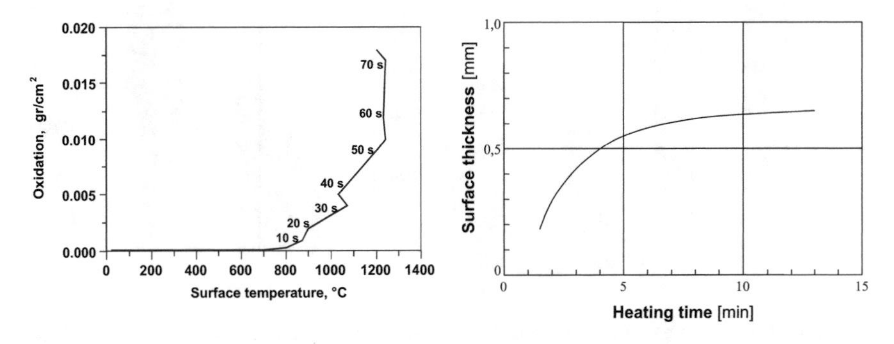

Fig. 6.96 Oxidation and surface decarburization as a function of temperature and heating time

6.9.4.8 Economical Comparison of Induction and Fuel Furnaces Heating Processes

The data available on the use of different energy sources show that nowadays many installations use fossil fuels (oil, gas) in the hot working of metals.

In many cases this situation is not justified, neither from the technological point of view nor the environmental one, but often is based only on the very low investment costs of fuel installations and the much lower cost of the fuel in comparison with electrical energy.

Data on primary and final user energy consumption, the losses due to oxidation and surface decarburization and CO_2 emissions can be found in the literature [47].

A comparison of these factors in different types of furnaces is given in the diagrams of Fig. 6.97 for steel heated to 1200 °C.

However, the economical-technical comparison of different heating processes cannot be limited only to these factors, but should include some items which cannot be easily evaluated quantitatively.

In fact, in many cases this comparison may lead to very different results depending on the design and the operating conditions of the installation, and the investment, energy, labor and materials costs.

In most cases, some of these factors can be exactly quantified, and numerical data can be found in the literature. All of them demonstrate the advantages that can be obtained by using induction heating.

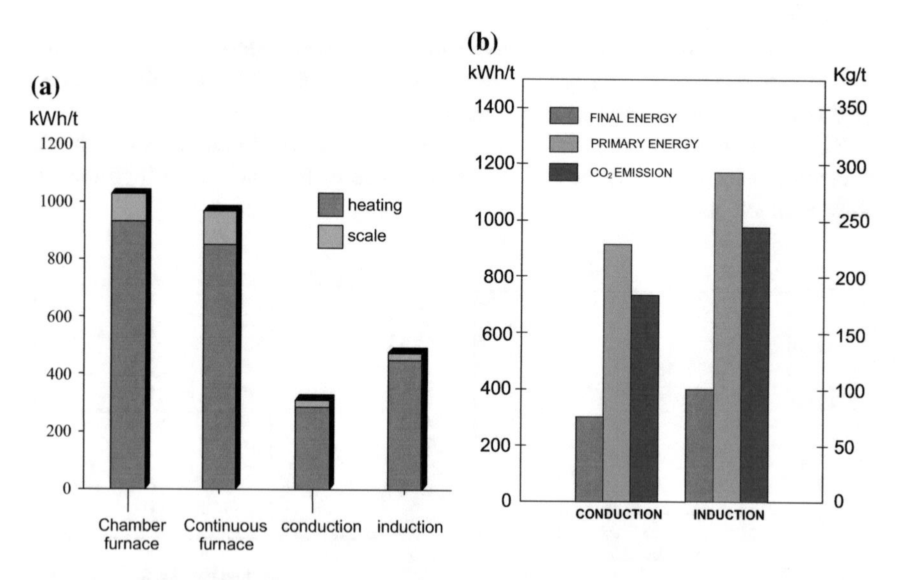

Fig. 6.97 **a** Final energy consumption and energy for oxidation and decarburization in different types of furnace, **b** final and primary energy consumption and production of CO_2 in induction and conduction heating

Table 6.11 Energetic and technical evaluation of a heating process prior to warm forging

Characteristic considered in the evaluation	Gas/oil	Induction	Ideal
Low investment costs	3	1	4
Low energy costs	3	1	4
Minimum requirement of personnel	3	3	4
Minimum space occupation	2	4	4
Low necessity of maintenance	3	3	4
Possibility of process automation	3	4	4
Flexibility	3	3	4
Possibility of integration in production line	2	4	4
High power density	2	3	4
High heating rate	1	4	4
Short time for reaching thermal state conditions	1	4	4
Minimum oxidation of metal	2	4	4
Minimum decarburization of steel	2	4	4
Rapid turn-off in case of failure	1	4	4
Possibility of preventing overheating of material	1	4	4
Independence from the geometry of workpiece	4	1	4
Minimization of smoke and powder/dust	2	3	4
Minimization of noise	2	3	4
Minimization of thermal losses	2	3	4
Sum of marks	42	60	76
Overall evaluation	0.55	0.79	1.00

On the other hand, other positive factors which can be obtained with induction heating in comparison to fuel processes, like reduced rate of scrap formation, improvement of quality of forged workpieces, lower space requirements for the installation, working flexibility, improvement of operating conditions, lower production losses due to installation's failures or maintenance and increased duration of dies, are difficult to be evaluated quantitatively and partly depend on subjective evaluations.

A comparative method which takes into account also the aspects which are difficult to be quantified, is suggested by the German norm VDI 2225.

The first step for this technical evaluation is the specification of the process characteristics which are considered to be important for the comparison, such as indicated in the first column of the example shown in Table 6.11 [45].

Then, each characteristic is valued by a mark between 0 and 4, according the following scale: 4—optimal, 3—good; 2—sufficient, 1—still acceptable, 0—unsatisfactory, where the mark 2.5 corresponds to the average value.

Finally it is calculated the ratio between the sum of all marks given in this way (column 2 and 3) to that of an installation having all ideal characteristics (column 4).

This ratio (≤ 1) represents a quality index giving an overall technical evaluation of the installation as regards the characteristics considered. A value of the ratio higher or equal to 0.8 is very good, 0.7—good, lower than 0.6—unsatisfactory.

It is evident that the result is dependent on subjective evaluations, which certainly cannot be considered to be absolute. However, although a given single mark can differ from one person to another, the experience has shown that the overall technical evaluation gives basically reliable results.

In some cases, a more accurate evaluation can be obtained by giving a different weight to each characteristic, according to its importance in the process.

6.9.5 Induction Heat Treatments

6.9.5.1 Introduction

Heat treatments are used in the industry to modify the properties of materials and improve the mechanical characteristics of workpieces, like hardness, wear resistance, tensile strength, resilience, ductility.

During heat treatment the material undergoes a thermal cycle comprising a heating phase with predetermined time-temperature pattern, a maintenance phase at a certain temperature and a cooling phase at convenient speed.

Induction heating is used in several heat treatment processes, such as surface hardening, through hardening, tempering, annealing, normalizing, stress relieving, which are applied to various metals, steel, cast iron, copper, aluminum, etc.

Among such treatments, surface hardening of steel components is the most common application, widely used in automotive, mechanical manufacturing, tooling and aeronautical sectors.

Historically, induction heating was used in the first decades of 1900 for melting metals, while the introduction of surface hardening into industry was slow because of the lack of powerful high frequency generators and insufficient knowledge of metallurgical results of the hardening process.

This situation changed in 1932, when at *Ohio Crankshaft Corporation (TOCCO),* a manufacturer of diesel engine crankshafts, started the first high production of crankshafts hardened by induction using motor generators at 1.9 and 3 kHz.

In 1933 a single-turn inductor, similar to inductors still in use today, was patented by *F.F. Deneen* and *W.C. Dunn* from TOCCO. This inductor, encircling the journal, allowed to heat up the rim zone of the crankshaft and to quench it as well (Fig. 6.98) [25].

In 1935 professor *V.P. Vologdin* was the first to apply for two Russian patents on surface heat treating, both issued in 1936 [26]. The first patent was for the induction hardening of railroad rails. This work was of extraordinary importance for USSR and found immediate application in industry.

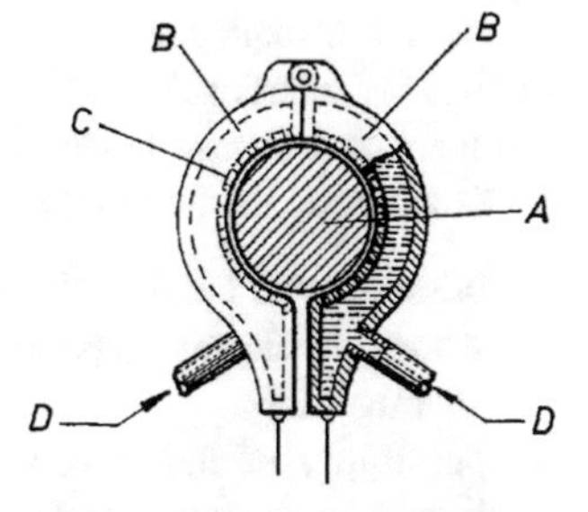

Fig. 6.98 USA, 1933: TOCCO inductor (*A* journal to be hardened; *B* half-cylindrical cheeks; *C* spraying nozzles; *D* quench)

Fig. 6.99 Prof. Vologdin in the hardening department of LETI (Leningrad—1940)

The second patent was for crankshaft hardening. This work was done according to the demands of the *Moscow automotive plant ZIS* (Fig. 6.99).

After these successful initial developments professor Vologdin continued R&D-work on surface hardening of different steel parts, including gears, camshafts, etc., mainly for the automotive, tractor and later military industry.

But the heat treatment techniques became increasingly important, in particular after the Second World War, with the development of the automotive and aeronautical industry, so that nowadays no material, no cars, no planes or aerospace engines could be built without the use of these techniques.

6.9.5.2 Induction Surface Hardening

Induction surface hardening represents the most interesting application of the induction heat treatments: with the hardening of a surface layer, different mechanical properties can be obtained in the surface layer of the workpiece in comparison to those of the core material.

The heat treatment cycle essentially consist of three stages: (1) a heating period during which the layer to be hardened is heated to a temperature higher than the "critical" one at which austenite forms, (2) a holding period at this temperature for a time sufficient for complete formation of austenite, (3) a subsequent fast quenching stage below the M_s temperature in order to "freeze" at low temperature the structure stable at high temperature or one convenient transformation of it (the *martensitic structure*).

As a consequence of such radical structural modification, it occurs a strong change of material's mechanical properties (hardness, wear resistance, etc.) in comparison to those before heat treatment.

Characteristics that make induction heating especially suited for this process are:

- possibility to localize the heat sources within a surface layer facing the inductor, whose thickness depends on frequency, material thermal conductivity and heating time;
- possibility of inducing very high volume power densities in the above mentioned layer and, therefore, to obtain, by the use of high frequencies and short heating times, very thin hardened thicknesses.

There are no general rules for defining the optimal distribution of induced currents in the workpiece to be treated, optimum heating time, time interval between end of heating and beginning of quenching, and quenching rate.

Besides material properties, these parameters depend on shape of the workpiece, presence of holes, teeth or edges, thickness of the required hardened layer, distance between inductor and workpiece surface, relative position of inductor and hardened zone, etc.

Even the results obtained with modern numerical models require always subsequent experimental validation and adjustment of the inductor in order to achieve the expected results.

Moreover, in many cases, the hardened profile is not the one desired by the designer but the one that can be obtained by the best practice of induction heating. Therefore, the design engineer when specifying the required surface hardness and hardness profile cannot use the same criteria applied in other technologies, for example carburizing.

The most important factors determining the results of the hardening process are:

- *characteristics of steel*
- *temperature distribution in the region to be hardened*
- *quenching rate*
- *type and geometry of inductor.*

6.9.5.3 Characteristics of Steel

Steels and steel alloys suitable for heat treating operations are those with carbon content less than 0.8 wt%.

The diagram of Fig. 6.100 for plain iron-carbon alloys gives, as a function of carbon content, the values of hardness and the percent of martensite which can be obtained with the hardening process (the curve at 99 % corresponds to complete hardening with total transformation of austenite into martensite). It shows that in practice only a small improvement of hardness may be achieved when the carbon

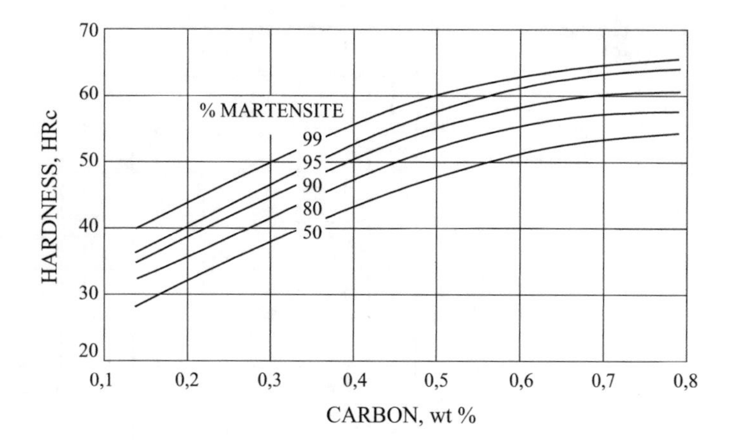

Fig. 6.100 Relationship between hardness, carbon content and amount of martensite in plain iron-carbon alloys [48]

content of the steel exceeds 0.6 %. For this reason, the steels with a carbon content in the range 0.4 –0.6 % are those mostly used.

From the above result it follows that, for the repeatability of hardening results, it is necessary to know in advance the carbon content of the steel, which must be guaranteed by the steel supplier.

The analysis of the steel treating process is based primarily on the consideration of the phase iron-carbon diagram of Fig. 6.101.

In the diagram the following critical temperatures are important:

- A_{c1}: temperature at which austenite begins to form during heating (*eutectoid temperature* A_{c1} *= 724 °C*)
- A_{c3}: temperature at which austenite formation is completed and the structure is fully austenitic.

In particular, the diagram shows that the temperature above which the formation of austenite is completed changes with the carbon content (temperature A_{c3}, curve A–B).

Also other parameters have strong influence on the value of temperature A_{c3}, like heating rate, alloying components, prior microstructure.

In particular, the temperatures of the curve A–B can diminish or increase, depending on steel grade and percent content of alloying elements, as it is shown in the example of Fig. 6.102, which shows the variation of the eutectoid temperature.

Another example is given in Fig. 6.103 which illustrates the effects of prior microstructure and heating rate on the temperature A_{c3}.

Moreover, at the high heating rates used in induction hardening, the hardening temperature must be increased above A_{c3} in order to allow for a convenient holding time above the temperature of austenite formation. Recommended values for plain carbon steels are given in Table 6.12 [48].

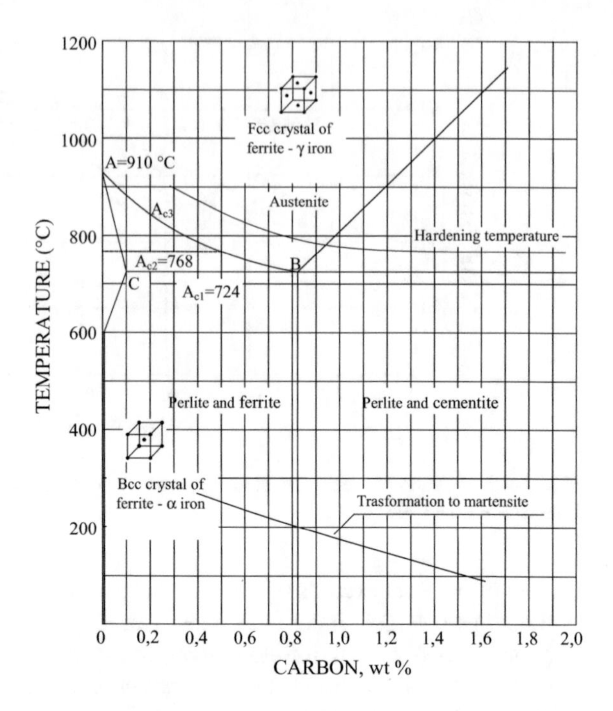

Fig. 6.101 Phase diagram of iron-carbon steel [49]

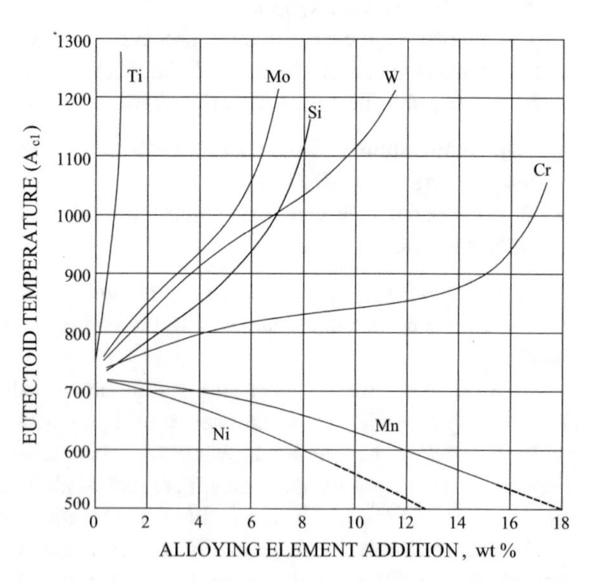

Fig. 6.102 Influence of alloying elements on eutectoid temperature [48]

These temperatures are further increased by at least 100 °C for steels containing alloying elements that promote formation of carbides, like titanium, molybdenum, vanadium or tungsten.

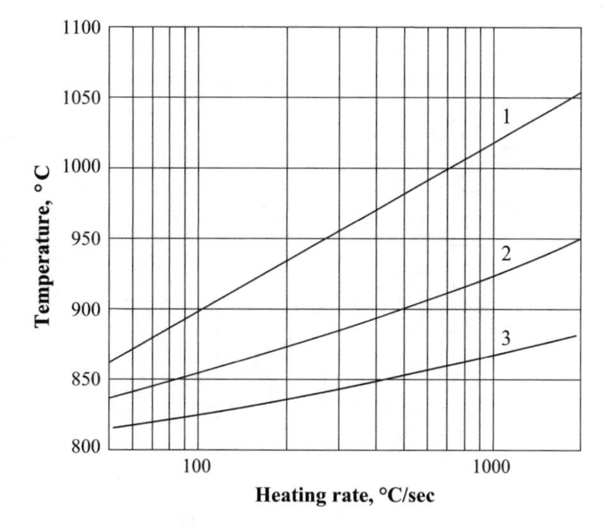

Fig. 6.103 1042 steel: effects of prior microstructure and heating rate on temperature A_{c3} (*1* annealed; *2* normalized; *3* quenched and tempered) [6]

Table 6.12 Recommended hardening temperature for plain carbon steels

C (%)	Hardening temperature (°C)	Quenchant
0.30	900–925	Water
0.35	900	Water
0.40	870–900	Water
0.45	870–900	Water
0.50	870	Water
0.60	845–870	Water/oil
>0.60	815–845	Water/oil

The above considerations implicate that in induction hardening of the most frequently used alloyed steels, there is a large variability of temperature and heating times which allows to obtain the same required thickness of the hardened layer.

6.9.5.4 Temperature Distribution in the Austenitisation Depth

Once the required hardened depth and steel grade are defined, it is necessary to heat all points of a surface layer Δ of thickness comparable with the penetration depth δ, to a temperature equal or higher to the austenitisation temperature (i.e. A_{c3} or a higher temperature, depending on heating rate), but at the same time to keep the core of the workpiece at lower temperature values. The depth of the layer Δ is usually called "*austenitisation depth*".

This temperature distribution can be achieved at high or medium frequency using high heating rates, which give rise to considerable temperature gradients in the layer to be hardened, much higher than the gradients achievable in flame hardening.

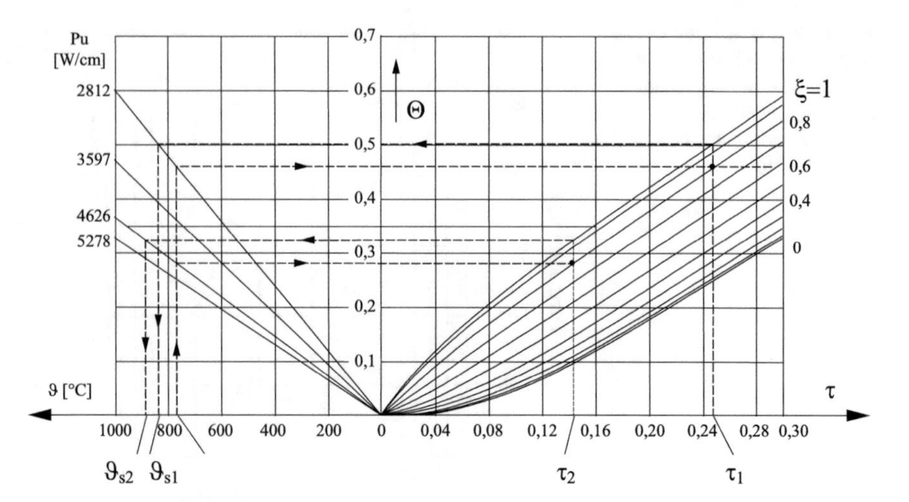

Fig. 6.104 Graphical determination of radial temperature distribution at the end of the heating stage (Steel C45; m = 3; d_e = 25 mm)

The analysis of the heating transient, illustrated in Sect. 6.6, emphasizes that the final temperature distribution in the austenitisation depth depends on heating time, specific induced power and frequency.

It was also shown that in the heating of magnetic steel, the calculation of the heating transient is a complex task—also in the simple case of a "long" cylindrical configuration—because it is strongly influenced by the variations of material characteristics (thermal conductivity, electrical resistivity and magnetic permeability) with temperature and local magnetic field intensity. Therefore, in case of complex geometries, an accurate analysis of the transient temperature distributions, can be obtained only with the use of very sophisticated 3D numerical models.

However, approximate values of the austenitizing depth as well as the corresponding specific power and heating times, can be obtained under the assumption of material characteristics constant during heating [50].

This is illustrated in the diagram of Fig. 6.104 which gives, on the right, the transient temperature evolution in a cylindrical workpiece and, on the left, the straight lines representing the relationship:

$$\Theta = \frac{2\pi\lambda}{p_u}\vartheta = \frac{\lambda}{p_e r_e}\vartheta$$

with: $\xi = \frac{r}{r_e}$, $\tau = \frac{kt}{r_e^2}$, $\Theta = \frac{2\pi\lambda}{p_u}\vartheta$—dimensionless parameters of radial position, time and temperature, and λ, p_u—constant values.

This diagram, given the required depth Δ and the austenitization temperature, allows to determine the temperature radial distribution at the end of the heating stage, the corresponding surface temperature as well as duration of heating and power required.

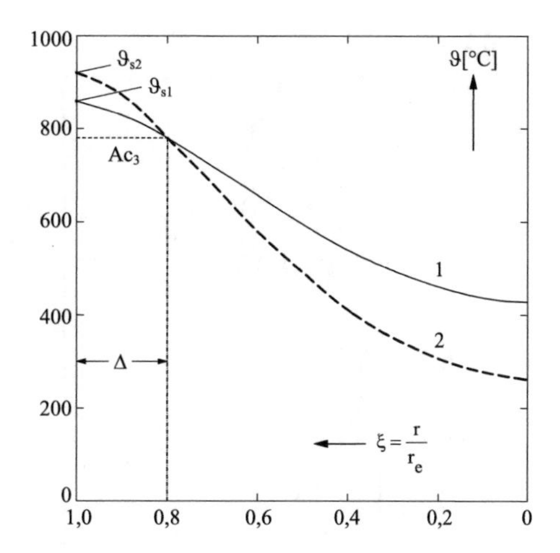

Fig. 6.105 Radial temperature distribution at end of heating stage obtained from the diagram of Fig. 6.104

In particular, as shown by the lines with arrows for $\Delta = 0.2r_e$, the same value of Δ can be obtained with different time/power recipes and correspondingly different temperature distributions along radius at the end of heating (see Fig. 6.105).

Repeating the graphical procedure with different values of $m = \sqrt{2}r_e/\delta$, one can draw diagrams similar to those given in Fig. 6.106, which refer to *m = 3* and *m = 6*, frequency values of 8 and 100 kHz and steel C45.

In reality, in the first instants of the quenching stage, the temperature gradients give rise to heat transmission towards the core of the workpiece and the depth $\mathbf{\Delta'}$ of layer above A_{c3} becomes thicker than the austenitisation depth $\mathbf{\Delta}$ evaluated at the end of the heating transient.

This phenomenon is illustrated in Fig. 6.107, where the temperature distributions along radius at different instants of the quenching stage are shown. In this example the depth $\mathbf{\Delta'}$ has its maximum value about 2 s after the beginning of quenching.

All previous remarks confirms the difficulty of an exact theoretical estimate of the hardening parameters and the consequent need of final experimental testing.

This occurs to an even bigger degree in the progressive scan hardening, where a "short" inductor moves along the axis of the workpiece to be hardened (Fig. 6.108).

In this case, the analytical calculation becomes extremely complex, even under the assumptions of constant material parameters [52]. Therefore, one must refer to experimental diagrams, analogous to those shown in Fig. 6.109 or to numerical models.

In the cases of hardening workpieces "stationary" in relationship to the inductor or of scan hardening, the power of the high frequency generator required for hardening a given area will be substantially different.

In fact, in the first case the instantaneous power varies during heating due to the resistivity and permeability variations as indicated in Fig. 6.110, where P_{av} denotes the average power and P_M the maximum power which must be supplied by the HF

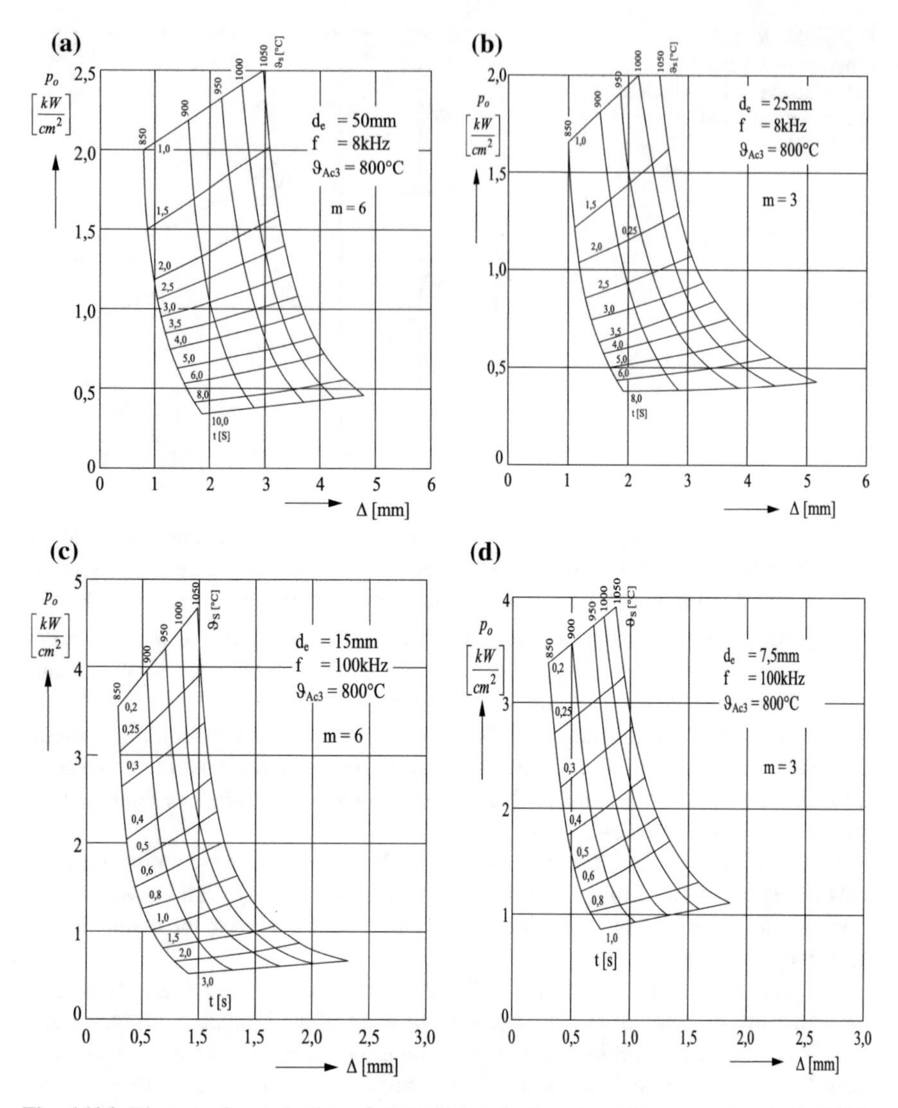

Fig. 6.106 Diagrams for evaluation of austenitizing depth and heating parameters (**a**, **b** 8 kHz; **c**, **d** 100 kHz; steel C45; $\vartheta_{Ac3} = 800$ °C; $\rho = 110$ μΩ cm; $\mu = 1$; $\lambda = 0.268$ W/cm°C; $k = 0.05$ cm^2/s) [50]

generator. The ratio P_M/P_{av} varies from case to case according to heating conditions and steel grade. Normal values are in the range 1.2–2.

In the second case (relative axial movement between inductor and workpiece), the generator is rated for the average power, which has the same value of the instantaneous power.

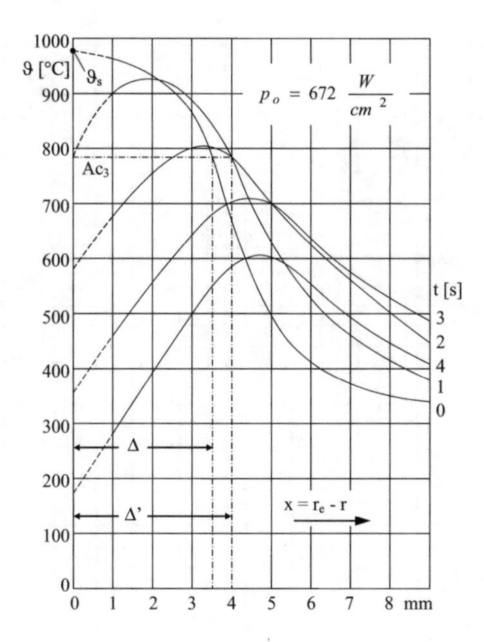

Fig. 6.107 Radial temperature distributions at different instants during quenching

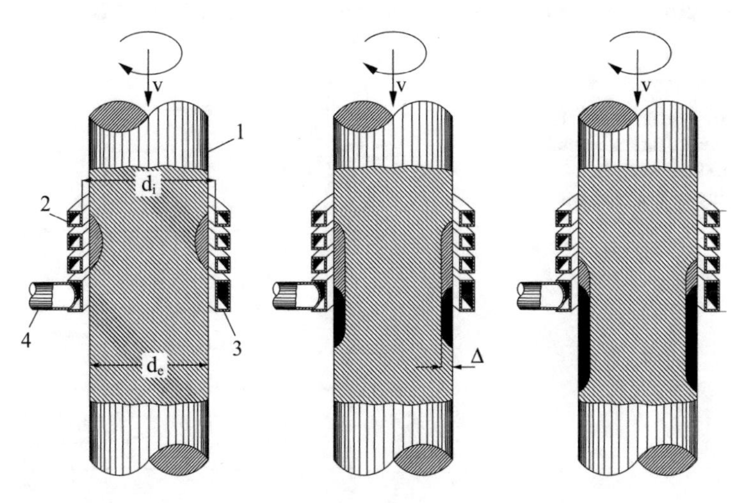

Fig. 6.108 Typical arrangement of workpiece and inductor in scan hardening (*1*, *2* inductor; *3*, *4* quenching "showers") [51]

For definition of the power of the generator, starting from the power p_u or p_e given by the previous diagrams, it is necessary to take into account the inductor efficiency which can be up to 70–80 % for inductors encircling the workpiece but drops to 20–30 % for inductors simply facing the surface to be hardened.

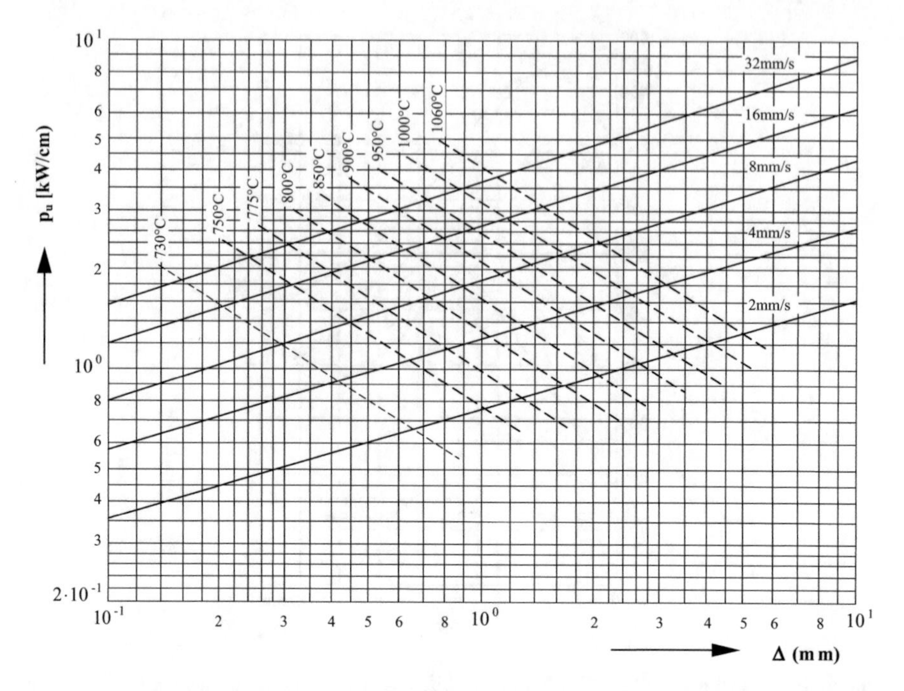

Fig. 6.109 Scan hardening: diagram for evaluation of hardened depth as a function of the maximum surface temperature of the workpiece and the scanning rate (Steel C45; f = 500 kHz) [53]

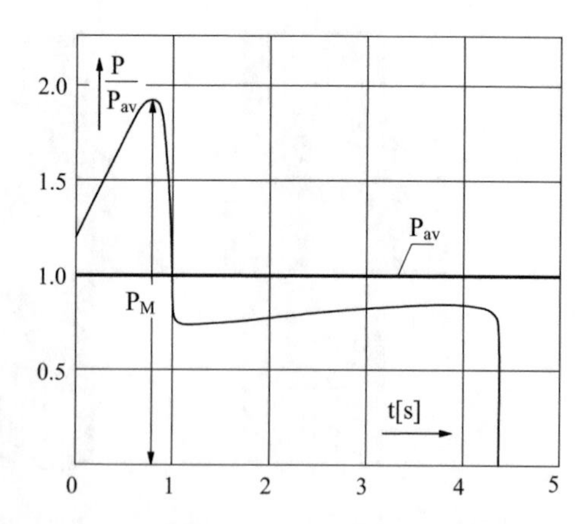

Fig. 6.110 Variation of HF power during hardening with workpiece stationary in relationship to the inductor

6.9.5.5 Rate of Quenching and Quenchants

After the austenitisation heating stage, the steel can be hardened to various degrees by controlling the cooling rate and the temperature at which the steel is held below the temperature A_{c1} for a given holding time prior to further cooling. Cooling rate and means used for quenching are therefore of fundamental importance in determining the hardening results.

The influence of the cooling stage on the depth that overcomes the temperature A_{c3} was illustrated in Fig. 6.107.

In the same way, in surface hardening of bodies of considerable mass in comparison to the heated material, the quenching velocity is strongly influenced by the "*heat sink*" effect of the unheated mass. In some special cases of very thin hardened depths and very short heating times (like in pulse hardening), sufficient quenching action able to produce the desired surface hardness can result even from the heat lost by radiation from the surface and the internal cooling effect of the unheated core *(self quenching)*.

Hardening reactions, resulting microstructures and transformation products are usually predicted by the transformation diagrams, i.e. the TTT (temperature-time-transformation) and the CCT (continuous-cooling transformation) diagrams, which can be found in bibliography for different steel grades, e.g. Ref. [54].

The TTT diagrams apply to isothermal transformations produced by holding the steel at different fixed temperatures below A_{c1}, whereas CCT diagrams give similar information obtained during continuous cooling.

The shape of TTT curves, which are also known as S-curves or Bain-curves, is shown schematically in the diagrams of Fig. 6.111a, b [55].

The curves represent the start and finish points of isothermal transformations at various temperatures. The left-hand curve connects all points representing the time required at a given temperature for starting the transformation reaction; the right-hand curve gives the time necessary to complete it.

In the lower part of the diagram there are two horizontal lines corresponding to the critical points *Ms* (*Martensite start*) and *Mf* (*Martensite finish*). At these temperatures, for a given quenching rate, respectively initiates and finishes the transformation of austenite into martensite.

Martensite is a hard and brittle constituent whose hardness is higher, the higher is the carbon content (Table 6.13).

The temperature *Ms* is characteristic of each steel and varies with its composition: in general, it diminishes with the increase of carbon content, as shown in Table 6.13 [56].

The CCT diagrams are basically similar to the isothermal ones, but are characterized by a shift towards right (and towards bottom) of the curves of start and finish of transformation. An analogous behavior is produced by a higher content of alloying elements, which reduces the critical quenching rate as illustrated in Fig. 6.112.

Finally, the transformations occurring during continuous cooling can be approximatively evaluated by superimposing the cooling curves on the CCT-diagram of the material under consideration.

In fact, the curves drawn on the diagram with different cooling rates starting from the AC_3-temperature, intersect the transformation start and finish lines in different points, leading at the end of quenching—all other conditions being the same—to different structures of the hardened material.

Fig. 6.111 TTT (Time—Temperature—Transformation) diagram (*A* Austenite; *B* Bainite; *C* Cementite; *F* Ferrite; *M* Martensite; *P* Perlite; *Ms* "Martensite start"; *Mf* finish; V_c critical quenching rate (A-Austenite: solid solution of cementite in Fe_γ; B-Bainite: microstructure consisting of ferrite and particles of cementite or complex carbides; C-Cementite: Fe_3C; F-Ferrite: solution of Fe_α or Fe_δ containing small amount of carbon; M-Martensite: supersaturated solution of carbon in Fe_α; P-Perlite: eutectic lamellae of ferrite and cementite.)

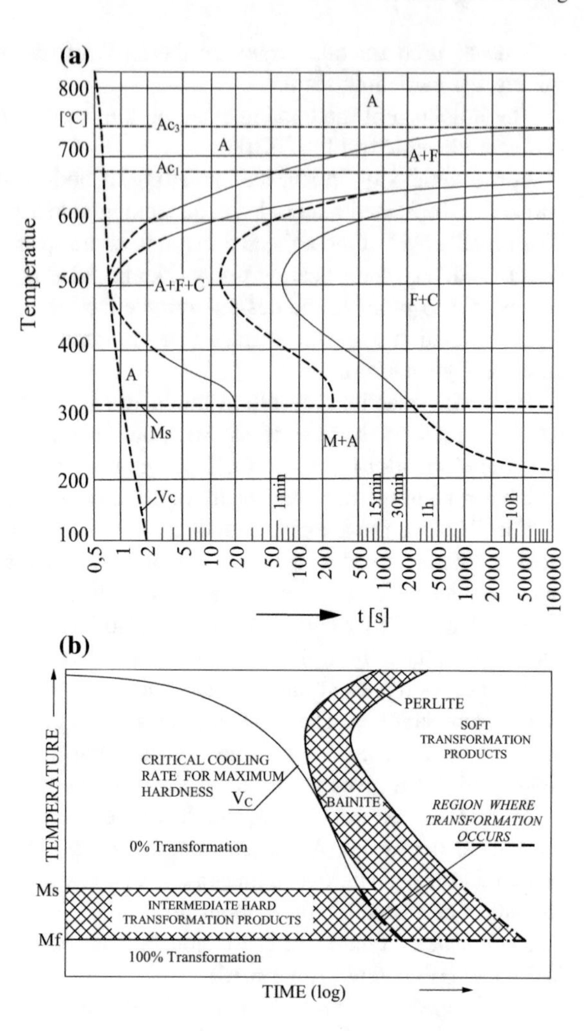

Table 6.13 Ms temperature for different types of steel

Steel	Ms (°C)	Steel	Ms (°C)
C30	400	38NiCrMo4	320
C40	360	40Cr4	310
C60	290	55Si7	270
35CrMo4	330	100Cr6	210

In particular, as schematized in Fig. 6.113, the curves corresponding to cooling rates higher or equal to the critical one (V_c), prevent intersection with the perlite and bainite transformation regions.

In this case a complete transformation of austenite into martensite occurs, with only some percentage of residual austenite if the steel is highly alloyed, thus realizing the so-called complete hardening (curve 4).

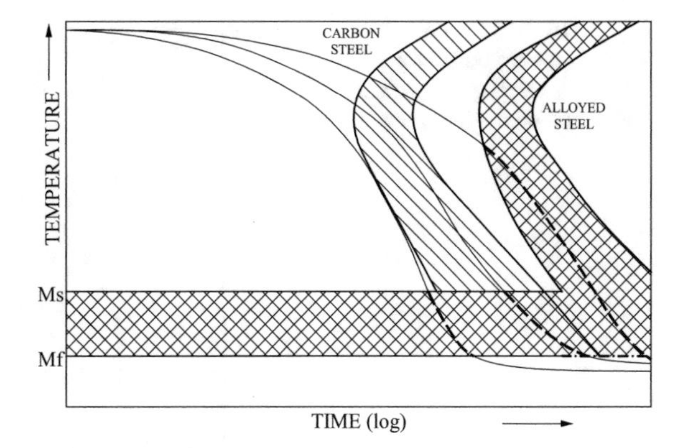

Fig. 6.112 Influence of alloying elements on CCT diagrams

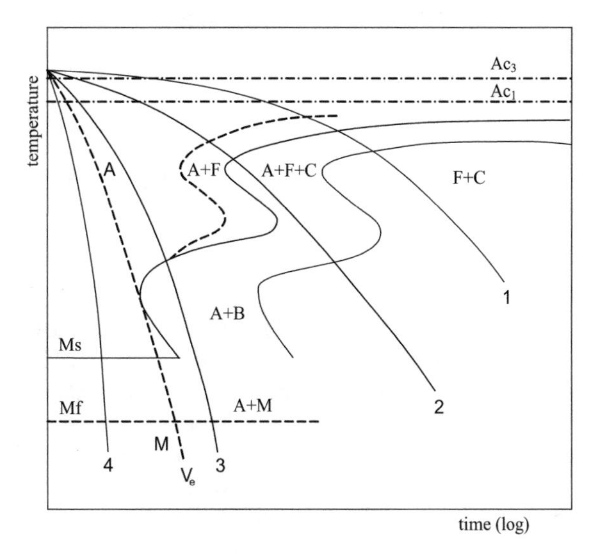

Fig. 6.113 Schematic of influence of cooling rate on final structure (symbols as in Fig. 6.111)

For lower cooling rates, various intermediate microstructures are obtained corresponding to the presence of the different constituents determined by the points of intersection between the cooling curve and the transformation curve.

In particular, for cooling rates lower than the critical one (curve 3), an incomplete hardening is obtained in which austenite transforms partially into bainite and partially into martensite. For even lower rates (curve 2) normalizing of steels with medium alloying content is realized, whereas curve 1 represents a complete annealing cycle.

The resulting hardness can be evaluated as shown in Fig. 6.114, which refers to the steel 35NC6 austenitized at 900 °C for 30 min: the solid line phase boundaries are those already shown in the TTT diagram, while the superimposed negatively-sloped curved lines correspond to different cooling rates. At the lower end of each line is indicated the resulting hardness (HV or HR).

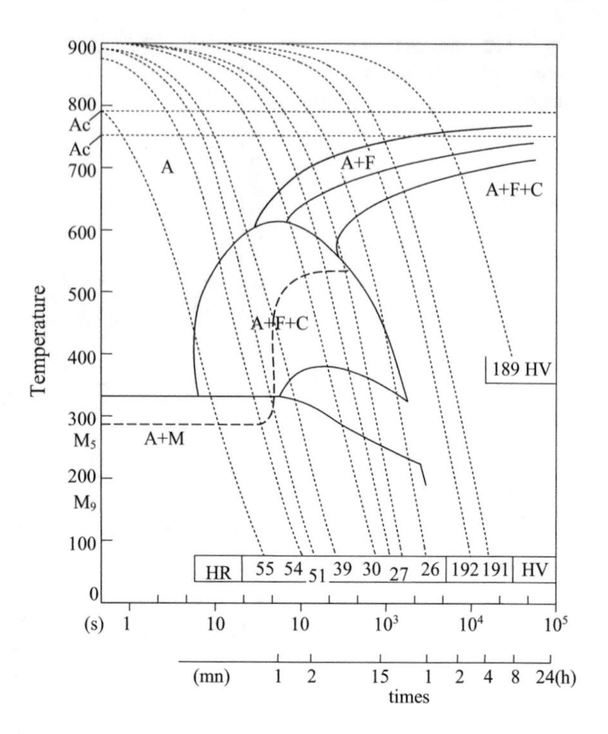

Fig. 6.114 35NC6 steel austenitised at 900 °C for 30 min: TTT diagram with superimposed cooling lines [24]

In the hardening of workpieces of considerable mass, the surface layers cool down at a rate higher than the critical one, whereas in internal layers the cooling rate progressively decreases from the surface towards the core (Fig. 6.115a).

Different cooling rates will produce different microstructures: in particular, constant hardness values in the regions where the cooling rate is higher than V_c, and hardness gradually decreasing towards the core to the values characteristic of the thermally unaffected material (Fig. 6.115b).

As the figure shows, the microstructure before heat treatment has a decisive influence on the hardened final structure and hardness distribution [24, 48].

The selection of the quenching medium has the same importance as the heating regime. This choice must be done taking into account the mechanical and metallurgical properties required, the required hardness pattern, the steel grade, size and geometry of the workpiece and the process features (e.g. automatic, manual, horizontal or vertical position of the workpiece).

The most used quenching media are: water, water with additives (sodium and calcium chloride, sodium and potassium hydroxide), hardening oils, polymer solutions, forced air.

As pointed out by the diagrams of Fig. 6.116, they allow to obtain different cooling rates in different temperature ranges and, therefore, to choose the medium most adapted to dimensions, shape and material of the workpiece to be hardened.

Moreover, it must be considered the velocity of the martensitic transformation in relation to the geometry of the workpiece to be hardened. In fact, on one side it is

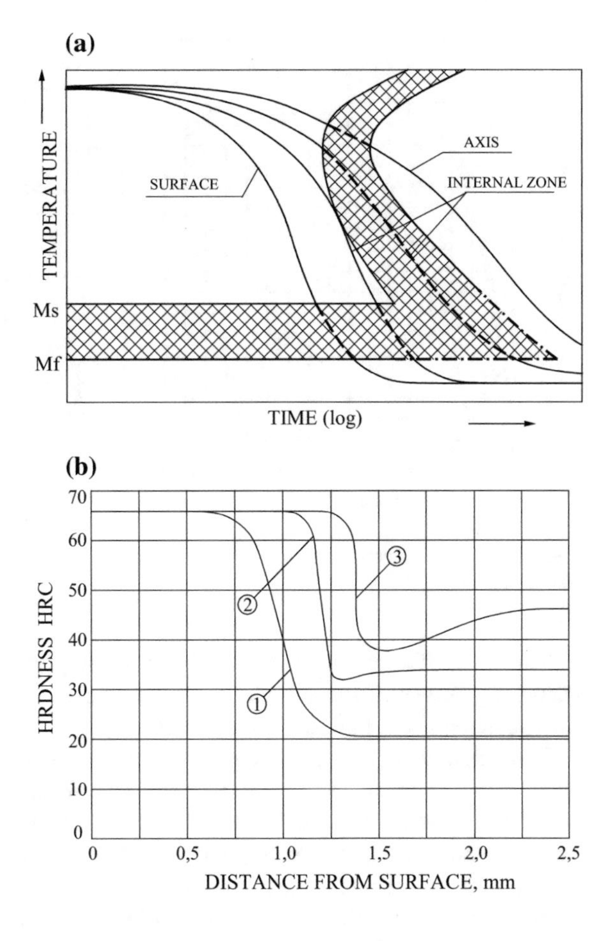

Fig. 6.115 Influence of quenching rate on microstructure and hardness profiles in the workpiece cross-section after surface hardening (material before heat treatment: *1* annealed; *2* normalized; *3* hardened and tempered)

necessary to achieve a quenching rate sufficiently high in order to prevent the perlitic "elbow", on the other side it would be useful to have a quenching velocity as low as possible below the M_s temperature.

However, it must be taken into account that the classical cooling curves of the type shown in Fig. 6.116 are usually obtained for immersion quenching. They cannot be directly applied to spray quenching where the heat transfer conditions in the initial stage of quenching, and formation, growth and removal of bubbles from the surface of the workpiece are considerably different when the part is submerged in a quench tank or is spray quenched. In fact, in the last case they depend also on the flow rate, impingement angle, rotation of the part, etc.

Since the martensitic transformation produces a volume increase, a too fast or irregular transformation induces substantial internal stresses which can cause deformations and cracks, especially in workpieces with complex geometries or in presence of surface defects.

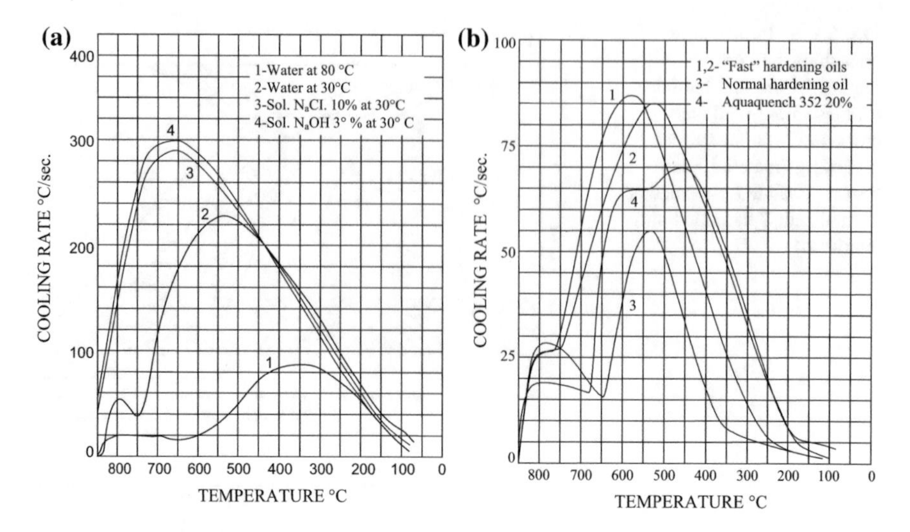

Fig. 6.116 Classical cooling curves: **a** water and aqueous solutions; **b** hardening oils and polymer solutions [57]

This is linked to the fact that the time/temperature difference between the quenching curves of the surface layers and those of the core, which cools down slower, increases at high quenching rates.

With high quenching rates (e.g. with water quenching), when the surface layer reaches the temperature M_s initiates the transformation of austenite, which is completed at M_f. But when the transformation of this layer is already finished, a large internal part of the workpiece is still austenitic and has not reached the temperature M_s. Therefore, a considerable mass of instable austenite remains "trapped" within a casing of hard and brittle martensite. Later, when the cooling of the core reaches the martensite transformation interval, it causes the volume increase of the internal mass, characteristic for this transformation. The expansion of the internal mass against the hardened surface layer causes an internal residual stress which is higher at high quenching velocity, with the likely occurrence of workpiece distortion and formation of cracks.

In case of water as quenching medium, especially with plain carbon steels good results are obtained with spray quenching, where the water is sprayed over the hot surface at high velocity, preventing the formation of a vapor film which would reduce the heat transfer between workpiece and quenchant. However, it may cause local "soft spots" or unhardened areas.

At lower quenching rates (e.g. with oil quenching), the temperature difference between internal and external layers is lower. In this case, before the surface layer has reached the point M_f causing the brittle casing of martensite, the most internal part of the workpiece is already transformed and has undergone the consequent volume expansion. The smaller is the internal mass that has not yet undergone transformation after the completion of the process in the surface layers, the lower are the above-mentioned problems.

Therefore, the right selection of the quenchant allows to use in the phase of martensitic transformation the quenching rate most suited to the steel grade and the characteristics of the part to be treated.

Sometimes, for workpieces of complex shape, air quenching (in still air, or using fans or compressed air) can be acceptable for certain grades of steel with sufficient hardenability since, due to the low quenching velocity, the transformation occurs practically at the same time in the whole mass, thus reducing to a minimum internal stress and hence risk of cracks.

In case of quenching by immersion, workpieces are usually kept in rotation and the quenching bath is maintained at controlled temperature. Due to its inflammability, hardening in oil is always done by immersion or in similar conditions.

6.9.5.6 Inductor Design and Types of Construction

The design of hardening inductors is a particularly delicate and complex task which, even with the use of sophisticated numerical programs, needs always validation by experimental tests.

The main design problems are the achievement of specified local metallurgical results on workpieces of very different shapes and dimensions, made of materials whose characteristics often are not known with sufficient accuracy, and the realization of well-defined power distributions suited to obtain the required hardening pattern.

Also when using numerical tools, it is of great interest—for reducing calculation and experimental work and inductor costs—the possibility to make a rough preliminary prediction of the effect produced by an inductor in different parts of the workpiece and, in particular, to estimate the total induced power, its distribution within the body to be treated and the inductor impedance.

In some cases, namely the hardening of the surface layer of a cylindrical body of axial length greater than that of the inductor (like in Figs. 6.41 and 6.42) or the inside wall of a cylindrical cavity, useful information can be obtained, as a function of material characteristics, geometric dimensions and applied frequency, with the so-called "*method of periodical fields*" described in Sect. 6.9.2.

With this method one can take into account in a very rough way the radial variations with temperature of electrical and magnetic material characteristics, by assuming that the cylindrical workpiece is constituted by a non-magnetic surface layer, approximately representing the layer that has exceeded temperature AC_3, and an internal magnetic core at lower temperature.

The diagrams of Fig. 6.117 refer to this schematization of the hardened workpiece and provide the coefficients G and Q for the evaluation of the inductor impedance $\dot{Z}_e$, according to the scheme of Fig. 6.42, as a function of the ratio

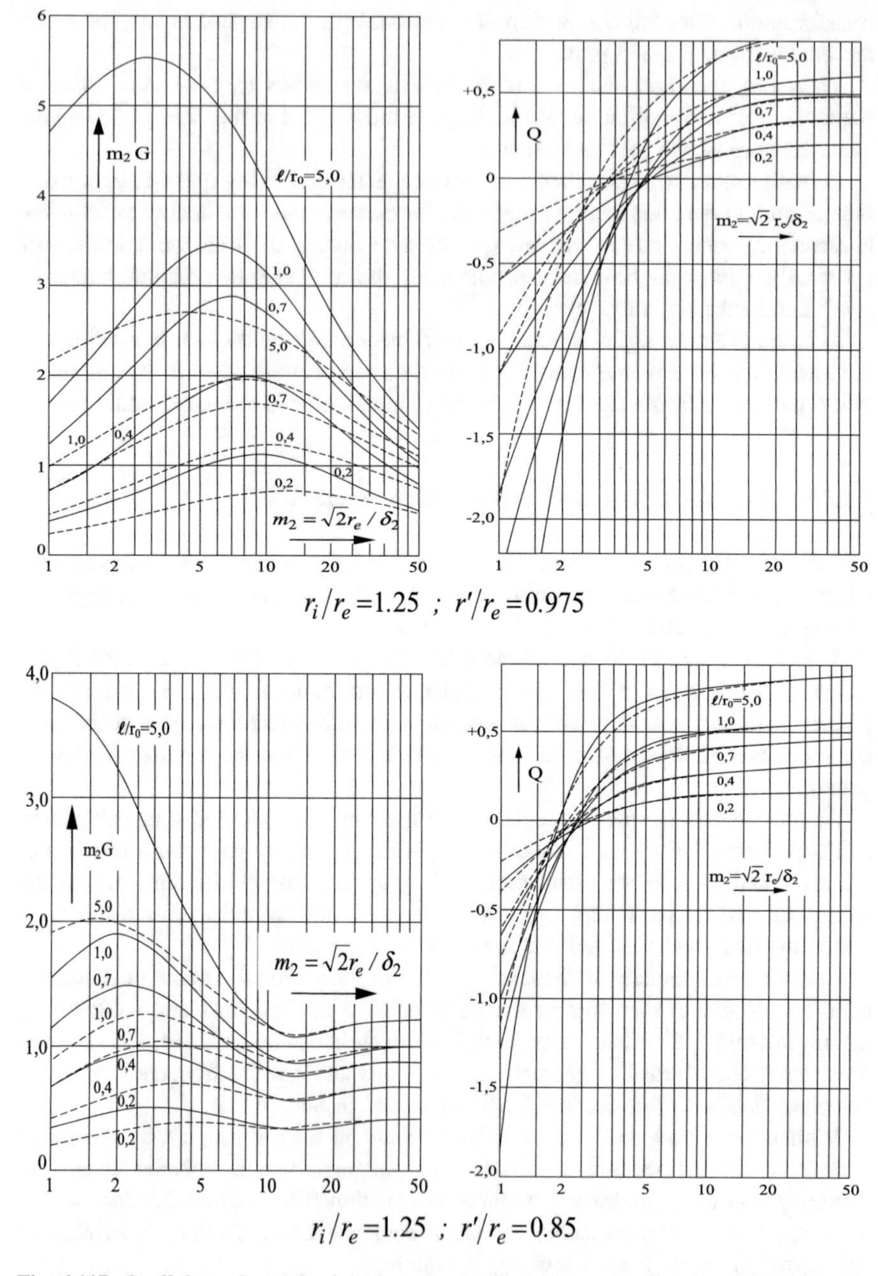

Fig. 6.117 Coefficients G and Q of the impedance of inductors for surface hardening of "long" cylindrical bodies as in Fig. 6.42 ρ_1 = 60 μΩ cm; ρ_2 = 100 μΩ cm; μ_1 = 10 (*dashed lines*); μ_1 = 50 (*solid lines*) [18]

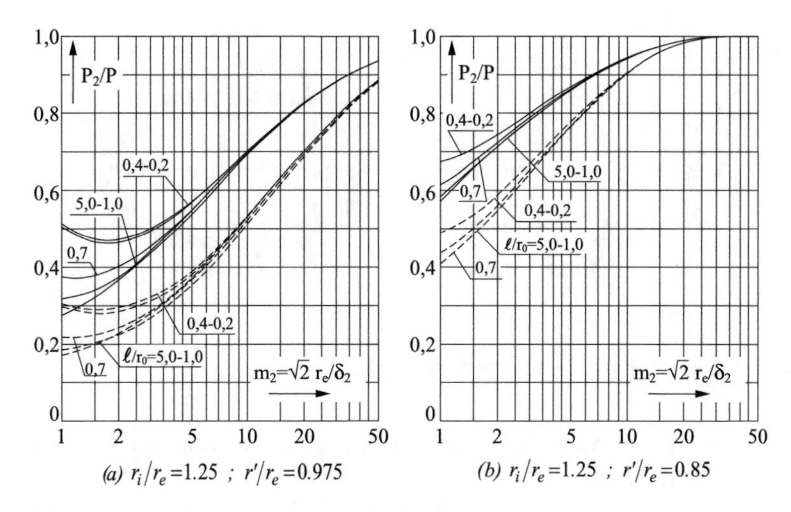

Fig. 6.118 Ratio of power P_2 in the surface layer to the total power P induced in the cylinder [18]

$m_2 = \sqrt{2}r_e/\delta_2$ (calculated with material characteristics of the surface non-magnetic layer) and different ratios of length to coil diameter:

$$\dot{\underset{e}{Z}} = r_i + \frac{x_{i0}}{\alpha^2}[G + j(\alpha^2 K_N - Q)].$$

For the same conditions, the diagrams of Fig. 6.118 give the ratios between the power P_2 in the surface layer and the total power P induced in the cylinder.

These diagrams are particularly useful for the frequency selection: in fact, assuming on the basis of the experience a convenient value of the ratio P_2/P, the value of m_2 is determined and the frequency therefore results:

$$f = \frac{\rho_2 m_2^2}{2\,\pi\mu_0 r_e^2} \approx 12.7 \cdot 10^3 \left(\frac{m_2}{r_e}\right)^2, \quad (r_e \text{ in cm}).$$

Other diagrams for the same inductor-load geometry with different values of parameters or for hardening the internal surface of cylindrical cavities are available in literature [18].

In practice, taking into consideration the generator costs and the possibility of obtaining the same hardened depth with different recipes of power and heating time, as was illustrated by the diagrams of Fig. 6.104, the frequency selection is usually made in the following ranges, depending on the required hardness depths:

- high frequency range (from 100 to 600 kHz) for hardened depths 0.3–0.75 mm and parts where wear resistance is the main requirement;
- from 6 to 50 kHz for depths 2–4 mm, in workpieces requiring both wear resistance and moderate loading;

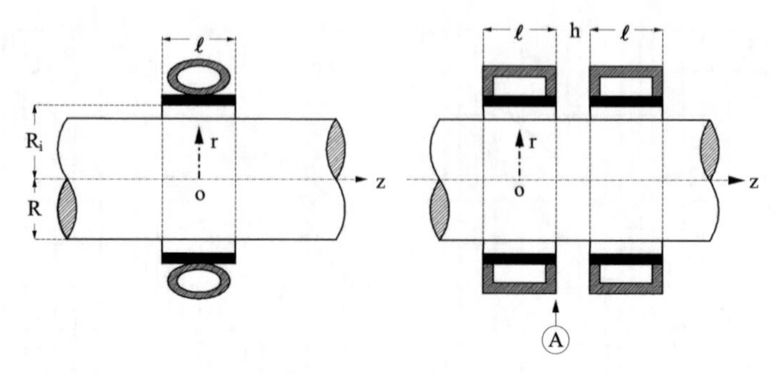

Fig. 6.119 **a** Single-turn inductor; **b** multi-turn inductor

- from 1 to 10 kHz for parts that must withstand heavy loads, requiring hardened depths between 3 and 10 mm;
- range 0.5–3 kHz for hardened depths 8–25 mm in heavy machinery components.

The method of periodical fields allows also to analyze the axial distribution of the power induced in a cylindrical body by a coaxial single-turn inductor (Fig. 6.119a) or the influence of the spacing between turns, in the case of multi-turn coils (Fig. 6.119b).

On the basis of equations of Sect. 6.9.2, the following coefficient characterizing the axial distribution of the induced power can be derived [17]:

$$k_P = \frac{p_{uz}}{p_{u0}} \approx \frac{w(r_e, z)}{w(r_e, 0)} = \left| \frac{\dot{A}(r_e, z)}{\dot{A}(r_e, 0)} \right|^2$$

where:

$A(r_e, z), A(r_e, 0)$	values of vector potential on the load surface ($r = r_e$) at the axial points of coordinates z and $z = 0$ respectively;
$w(r_e, z), w(r_e, 0)$	specific induced power (w/cm^3), on the load surface at the coordinates z and $z = 0$;
p_{uz}, p_{u0}	total induced power per unit length (w/cm), at the coordinates z and $z = 0$

An example of such distribution is given in the diagrams of Fig. 6.120 for $m = 5.1$: the curve (a) refers to a single coil inductor as in Fig. 6.119a; the remaining curves refer to the case of Fig. 6.119b and show the influence of the space between inductor turns on the axial distribution of induced power.

The method is not applicable when the axial length of the workpiece is equal or lower to the length of the inductor. In these cases, particularly at the front end edges of the load, the induced current and power density can result considerably different from those at the center of the inductor-load system.

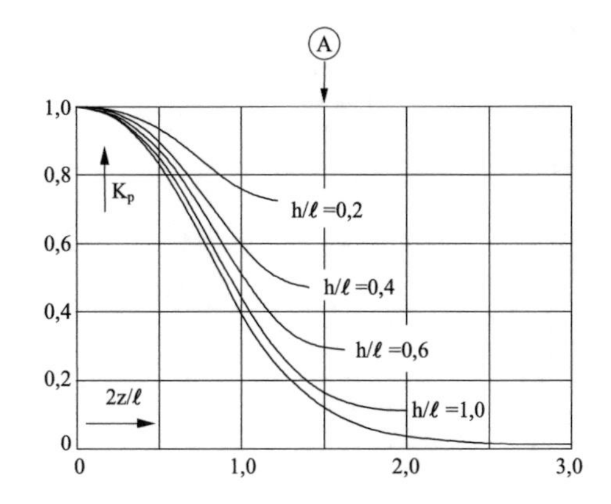

Fig. 6.120 Influence of space between inductor turns on axial distribution of induced power, characterized by the values of coefficient k_P (m = 5.1)

This "*load end effect*" can be analyzed with more complex analytical procedures [58] or, nowadays much more effectively, with numerical methods.

As it has been emphasized before, high frequency currents from few kHz to some hundreds of kHz are used in order to concentrate the induced power in a thin layer below the workpiece surface.

At these frequencies, the penetration depth is very small in comparison with the geometric dimensions of the body, since the shielding effect of induced currents is very pronounced. As an example, during a heat treatment at 450 kHz, the penetration depth in magnetic steel vary in range 0.2–0.7 mm.

Under these conditions, useful suggestions for the inductor design can be obtained from the analysis of field distribution on the surface of the workpiece made with the "*method of images*". This particularly simple method, allows the approximate analysis of various system configurations, also different from the cylindrical one [59].

The method is based on the hypothesis that the induced current is concentrated in the workpiece in a surface layer of negligible thickness facing the inductor.

In order to illustrate this method, let us consider the plane inductor-load system of Fig. 6.121, consisting of an inductor made by a single, straight, ideal filament wire conductor of infinite length, normal to the plane of the drawing, placed at distance h from the surface of the workpiece. The total exciting current I is supposed to be concentrated in this conductor.

In this system, the tangential component H_z of the magnetic field intensity at the surface of the workpiece is produced by the combined action of the inductor current and the current induced in the workpiece.

Under the previous assumptions, the same effect of the induced current is that of a current of the same magnitude and opposite direction of I, mirror on the plane of treatment of the inductor current.

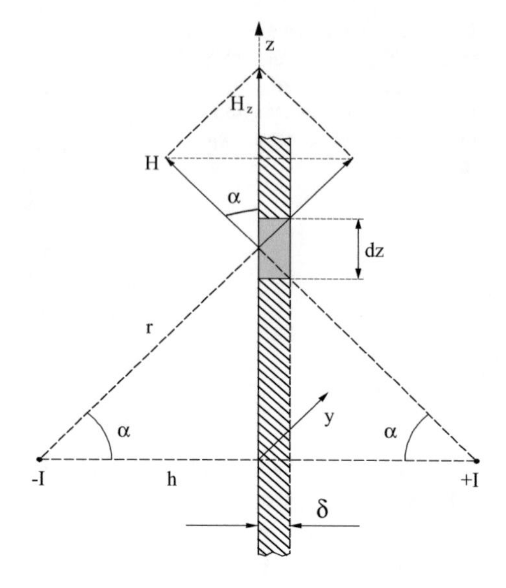

Fig. 6.121 Heating of a plane surface with a straight wire inductor of infinite length in which the current I flows

With the notations of Fig. 6.121 it is:

$$H_z = 2\frac{I}{2\pi r}\cos\alpha = \frac{I}{\pi}\left(\frac{h}{r^2}\right) = \frac{I}{\pi}\left(\frac{h}{z^2+h^2}\right) = \frac{I}{\pi h\left[1+\left(\frac{z}{h}\right)^2\right]} \tag{6.106}$$

If the conductor that represents the inductor is not a filament wire but has radius r_0 and its axis is at distance *h* from the workpiece surface, the magnetic field component H_z can be still calculated with reference to a filament wire conductor placed at distance $h' = \sqrt{h^2 - r_0^2} = h\sqrt{1-(r_0/h)^2}$ from the surface, modifying Eq. (6.106) as follows:

$$H_z = \frac{I}{\pi h}\frac{\sqrt{1-(r_0/h)^2}}{1+(z/h)^2-(r_0/h)^2}.$$

The power per unit surface induced at a point of coordinate *z* is obtained from Eq. (6.32). Using Eq. (6.106), we can write:

$$p_{0z} = \frac{\rho}{\delta}H_z^2 = \frac{\rho}{\delta}\left(\frac{I}{\pi h}\right)^2\left[\frac{1}{1+(\frac{z}{h})^2}\right]^2. \tag{6.107}$$

Indicating with p_{00} the value of p_{0z} at $z = 0$, the above distribution can be characterized by the coefficient k_p, defined by the ratio:

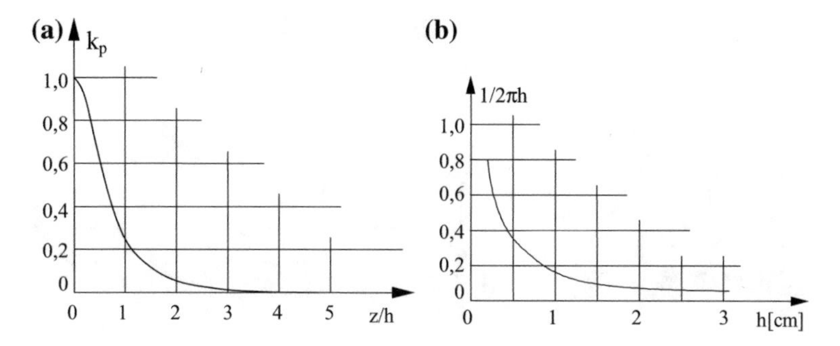

Fig. 6.122 **a** Values of coefficient k_p as a function of z/h; **b** variation of total induced power as a function of distance between inductor and workpiece surface

$$k_p = \frac{p_{0z}}{p_{00}} = [\frac{1}{1 + (\frac{z}{h})^2}]^2.$$

The diagrams of Fig. 6.122a give the values of k_p for the inductor constituted by a straight wire conductor of infinite length.

Denoting with $dP = p_{0z}dz$ the power induced in an elementary volume of cross-section $(\delta \cdot dz)$ and unit depth along *y*, the total active power $P\ (W/m)$, induced in the workpiece in a "strip" of unit depth, is:

$$P \approx \int_{-\infty}^{+\infty} dP = 2\int_{0}^{+\infty} dP = \frac{2\rho I^2}{\pi^2 h^2 \delta}\int_{0}^{\infty} \frac{dz}{[1 + (\frac{z}{h})^2]^2} = \frac{1}{2\pi h}\frac{\rho}{\delta} I^2. \qquad (6.108)$$

The diagram of Fig. 6.122b, which represents graphically Eq. (6.108), highlights that for inductors facing the workpiece, even a relatively small variation of the distance *h* can produce strong variations of the total induced power.

As already mentioned, this procedure is suited for a qualitative study of a number of configurations of practical interest. The results for some of them are shown in the tables of Fig. 6.123. For other system configurations the reader can address to Ref. [60].

In particular, for inductors placed in front of the workpiece, it can be observed that the heating is more intense in front of the conductors of inductor and the value and distribution of total induced power depend on geometric dimensions, spacing between turns and distance from the workpiece surface.

This gap is generally chosen as small as possible in order to obtain good electrical efficiency and to induce sufficient power into the workpiece. In practice, for mechanical reasons, it is never lower than 1–2 mm.

In many cases, a noteworthy advantage can be achieved, as concerns inductor efficiency and achievement of the required power distribution, by inductors with magnetic concentrators.

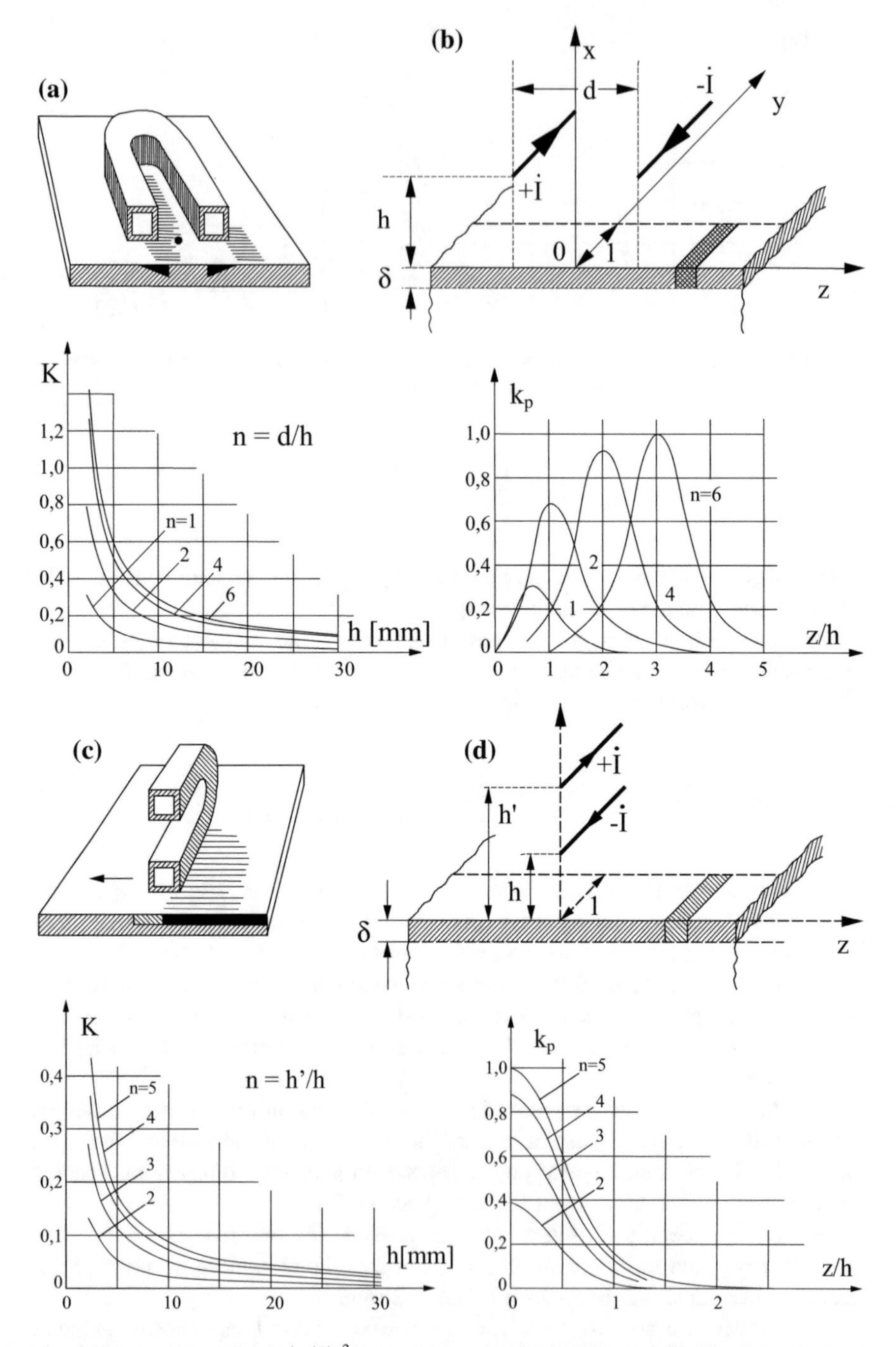

Fig. 6.123 **a**, **c** Power $P = K(\rho/\delta)I^2$ induced in a strip of unit width; **b**, **d** Distributions of induced specific power, referred to its maximum value, as a function of z/h, for different values of n

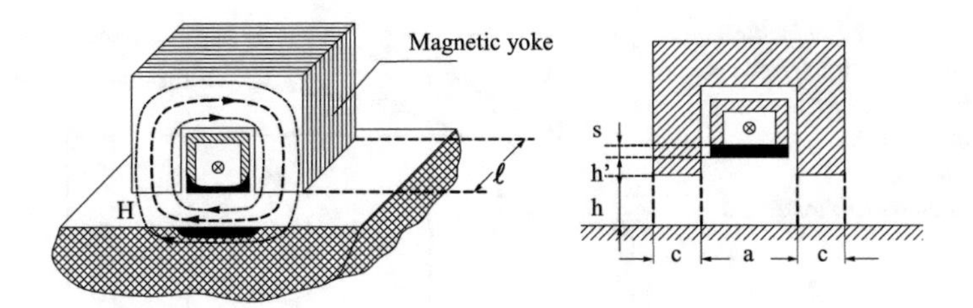

Fig. 6.124 Inductor with magnetic concentrator

Figure 6.124 shows schematically one of these inductors for the treatment of a plane surface.

The magnetic yoke can be realized with iron laminations (of thickness 0.1–0.3 mm), sintered iron (ferrites) or magneto-dielectric material which can be easily machined and is particularly suited for hardening inductors at medium and high frequency.

An example of the distributions of field lines and induced power with and without magnetic concentrator is shown in Fig. 6.125. For sake of comparison, in the figure are shown also the distributions produced by the same inductor without magnetic yoke (this last represented in the drawing only for a better understanding).

Typical characteristics of these magnetic materials are given in Fig. 6.126 [61].

In case of laminations, for thermal reasons, the induction of the magnetic yoke is always relatively low (in the range 0.2–0.8 Wb/m^2 depending on thickness of laminations and frequency). Hence its magnetic reluctance is practically negligible ($\mu \rightarrow \infty$).

The reluctance of the concentrator is always very low and may be neglected in a first approximation also with other magnetic materials.

Therefore—with reference to the equivalent scheme of Fig. 6.127, analogous to the one described in Sect. 6.9.1—the reluctance $\Re_0$ of the return path of the flux is

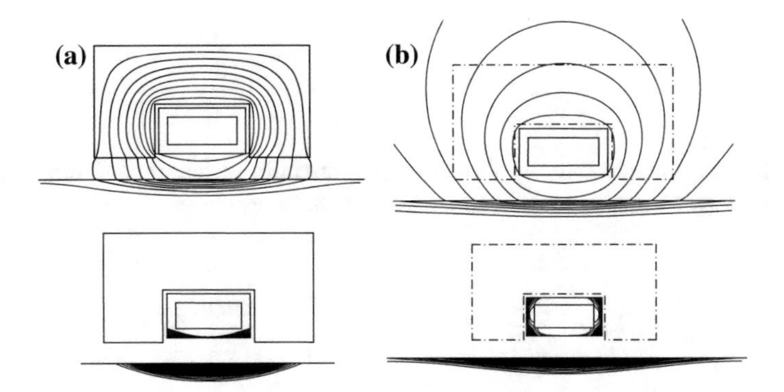

Fig. 6.125 Typical distributions of field lines and induced power with (**a**) and without (**b**) magnetic concentrator

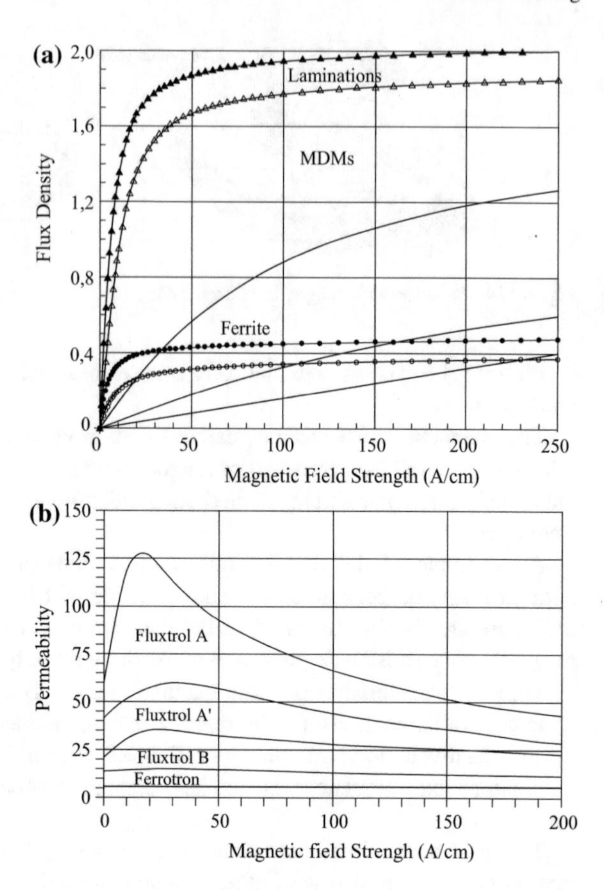

Fig. 6.126 **a** Magnetization curves of laminations, ferrites, and magneto-dielectric materials, MDMs. **b** Permeability of magneto-dielectric materials

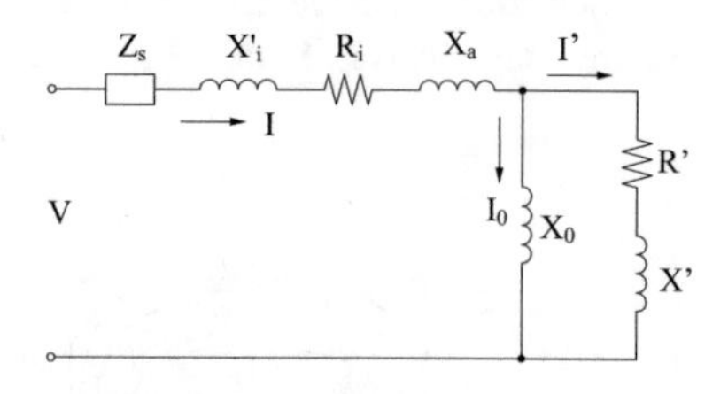

Fig. 6.127 Equivalent circuit of an inductor with magnetic concentrator

determined in practice only by the air gaps (generally of length 2–5 mm) between the pole shoes of concentrator and surface of the workpiece.

So it is:

$$\Re_0 = \frac{2h}{\mu_0 \ell c} \quad \Rightarrow \quad X_0 = \frac{\omega N^2}{\Re_0} = \frac{\omega \mu_0 \ell c}{2h} N^2 \qquad (6.109)$$

with: h—air gap between pole shoes and workpiece surface; ℓ—yoke length; *c*—width of pole shoes.

The reactance $X_s = X_a + X'_i$ can be approximately evaluated as the leakage reactance of a conductor placed in an open slot of uniform width *a* extended till the surface of the body, and taking into account that, for the presence of the magnetic yoke, the conductor current is concentrated in the region facing the surface to be hardened (see Fig. 6.125a) [11].

With these simplifying assumptions, we can write:

$$\Re_a = \frac{a}{\mu_0(h + h')\ell} \quad \Rightarrow \quad X_a = \omega\mu_0 N^2 \frac{\ell}{a}(h + h') \tag{6.110}$$

with: *a*—slot width; $h + h'$—distance of the conductor from the surface to be treated; X'_i—conductor internal reactance.

Finally, the resistance R' and the reactance X' of the surface layer of the load where the induced currents are concentrated can be calculated with reference to a width *a* equal to the width of the slot and a length ℓ equal to that of the inductor.

Taking into account that the exciting magnetic field is parallel to the surface, applying Eq. (6.34) to an element of length *a* one obtains:

$$R' + jX' = N^2 \frac{\rho}{\delta} \frac{\ell}{a}(1 + j). \tag{6.111}$$

This method gives sufficiently good results for ferromagnetic loads and air gaps $(h + h')$ relative small in comparison with the size of the slot, whereas it's precision reduces rapidly with the increase of air gap length or pole shoes size.

However, the availability today of precise commercial programs makes it advisable to calculate such systems with numerical methods.

Besides the simple cases examined until now, there are numerous complex geometries that occur in practical applications. From case to case, according to the configuration, the previous approximate or numerical methods may be used in order to obtain quick or more precise data useful for the inductor design.

6.9.5.7 Gear Contour Hardening

One typical example of a complex system geometry occurs in the simultaneous contour hardening of gears, where the purpose of the hardening process is to obtain a uniform hardened layer along the whole gear contour, namely at teeth tips, flanks, and roots.

In this case, the frequency selection is of fundamental importance not only from the electromagnetic point of view, but also the thermal one.

From the electromagnetic point of view, the frequency selection depends on the required hardened thickness, but at the same time it must fulfil the condition for good energy transmission from the inductor to the different parts of the workpiece ($m > 2$–2.5). This leads to conflicting criteria and the selection of a single frequency is necessarily a compromise between conflicting requirements.

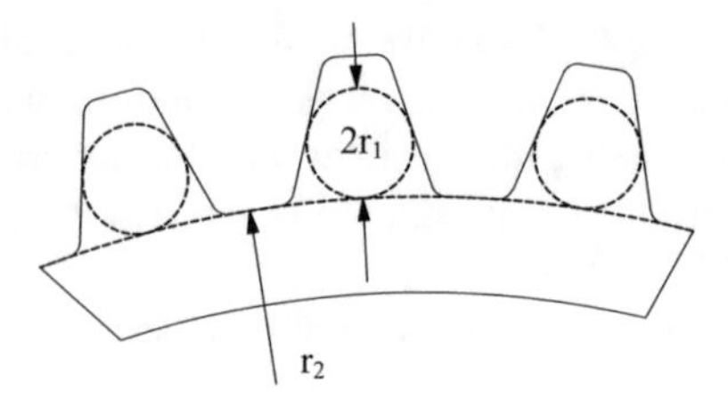

Fig. 6.128 Schematization of a gear with cylinders of different radii r_1 and r_2

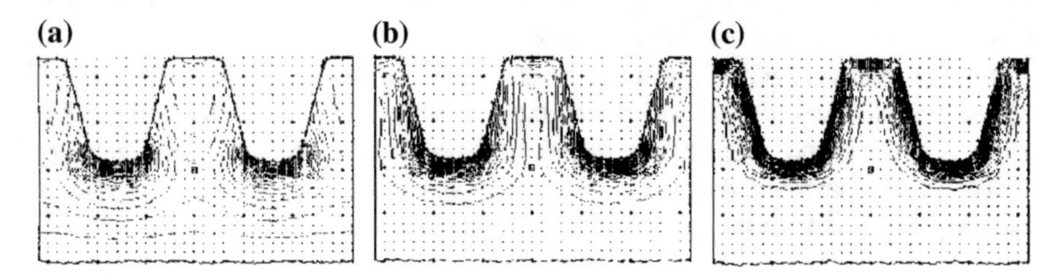

Fig. 6.129 Specific power distribution per unit volume in a gear of modulus M = 8 mm (**a** f = 1 kHz; **b** f = 4 kHz; **c** f = 16 kHz; diametral pitch: d = 1 m) [62, 63]

In fact, as schematized in Fig. 6.128, by substituting to each tooth of the gear a cylinder of radius r_1 and considering the cylinder of radius r_2 corresponding to the circumference of the root circle of teeth, it becomes evident that the fulfillment of the above conditions can be obtained only by two different frequencies.

The same conclusion can be reached by looking at the distributions of the volume specific power of Fig. 6.129.

In this example, at frequency 1 kHz (figure *a*), the heat sources are concentrated in the roots (of radius r_2) and therefore they heat predominantly the areas near the root circle of the gear. At frequency 16 kHz (figure *c*), their distribution is more uniform along the contour of the gear, which—at first sight—seems to produce more favorable heating conditions along the contour, since it affects also the cylinders of radius r_1.

However, considering that the surface of tips and flanks facing the inductor is much greater than that of the roots and the power transmitted is approximately proportional to these surfaces whereas the heat capacity of the teeth is much lower than that of the root, a different criterion must be used, namely to induce a lower specific power at the tips in comparison to that at the roots (as in Fig. 6.129b).

The influence of frequency on the temperature distribution at the end of the heating stage is further emphasized by the diagrams of Fig. 6.130 [64].

It has been proposed to select the optimal frequency f_{opt} for simultaneous contour hardening of gears with modulus M (in mm) with the relationship [65]:

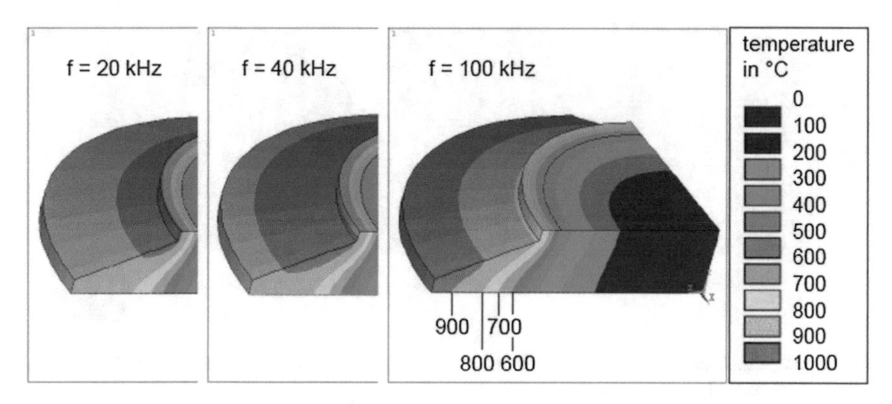

Fig. 6.130 Influence of frequency on temperature distribution in single-shot hardening of gears

$$f_{opt} \approx \frac{300{,}000 - 460{,}000}{M^2}, \tag{6.112}$$

and the corresponding process parameters with the diagram of Fig. 6.131.

The curves highlight the difficulty of the single frequency contour hardening of gears with modulus lower than 4, since it would require high frequency specific power higher than 2–3 kW/cm^2 and heating times in the range of hundreds of milliseconds. Moreover, for such low values of modulus, the optimal frequency generally produces through hardening of tooth.

The above conditions have led to the development of the dual-frequency hardening technique, where two different frequencies are applied either in sequence or simultaneously [6, 40, 64, 66, 67].

The dual frequency induction hardening process usually include a preheating stage at relatively low frequency (3–10 kHz) and a final heating stage with high-frequency power supply at 150–400 kHz.

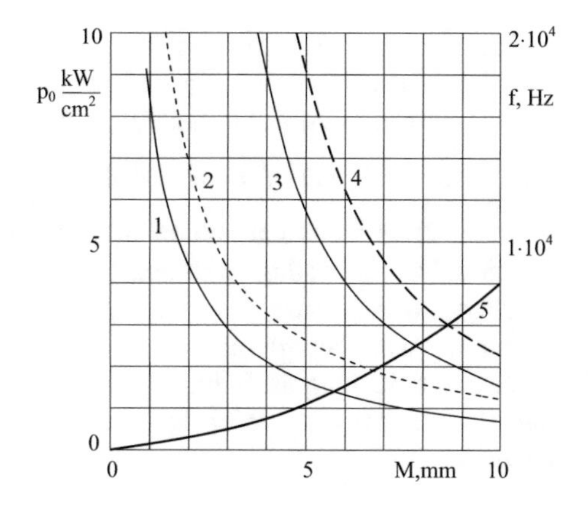

Fig. 6.131 Parameters for simultaneous contour hardening of gears with modulus M (*1* specific surface induced power; *2* generator specific power; *3*, *4* optimum frequency range; *5* heating time)

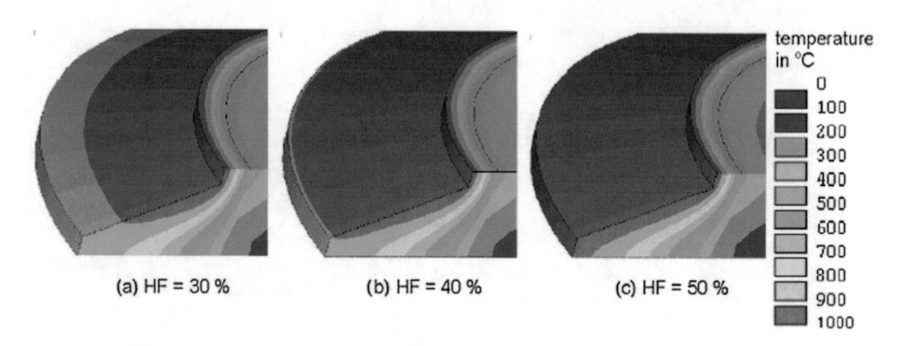

Fig. 6.132 Final temperature distributions in simultaneous double frequency hardening with different ratios of MF to HF power

The dual-frequency process, combining a deeper penetration depth during the low frequency preheat and a shallow penetration depth during the final stage, allows to produce contour hardening of the gear with maximum strength and minimum distortion. Moreover, the possibility of using different ratios of MF to HF power, gives to the process a high flexibility suitable for adapting the heating to the gear characteristics.

An example of the results obtained with this technique and different ratios of MF to HF power is shown in the diagrams of Fig. 6.132 [64].

6.9.5.8 Examples of Numerical Simulations

The complexity of electromagnetic, thermal, metallurgical and mechanical stress phenomena mentioned in the previous paragraphs, as well as the difficulty of analyzing their mutual interactions with traditional analytical methods and—at the same time—the fast development of powerful computers have led to the availability of powerful (scientific and commercial) numerical programs which nowadays allow a more complete simulation and a more profound understanding of the hardening process.

Without dealing here with numerical methods and the analysis of existing commercial programs, some typical examples of their application shall be presented in the following in order to underline their great potential.

(a) *Scan hardening of an axle shaft*

Axle shafts induction hardening represents a very widespread application for which induction heating is particularly suited since it allows to increase hardness near the surface, where it is mostly needed, and leaves the surface in compression, which improves fatigue life.

In scan hardening the inductor coil and the shaft move relative to each other, and the shaft rotates in order to obtain an even hardened pattern around the circumference.

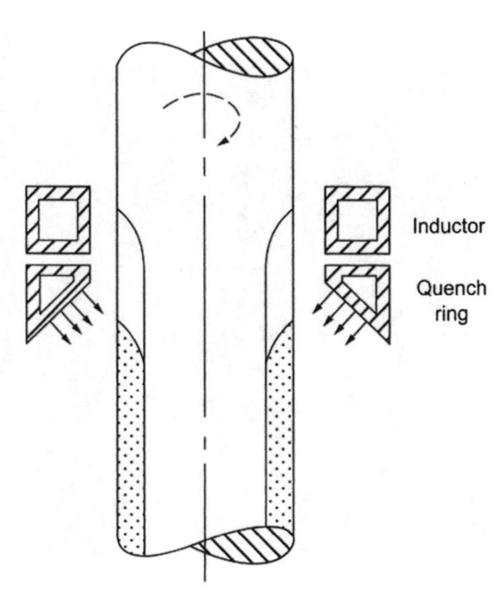

Fig. 6.133 Typical scanning induction hardening setup [68]

The scan inductor typically encircles the shaft and a quench ring is positioned next to the coil in order to spray the quenching medium on the area that has been heated (Fig. 6.133).

Some process data have been already given in Fig. 6.109, whereas Fig. 6.134 shows the temperature distributions in the heating and quenching phases of the scan hardening of a shaft for automotive applications with diameter 40.5 mm.

The simulation gives the values of power, scanning speed (7.5 mm/s), inductor axial length and frequency (3 kHz) required in order to realize the design hardness depth (4 mm).

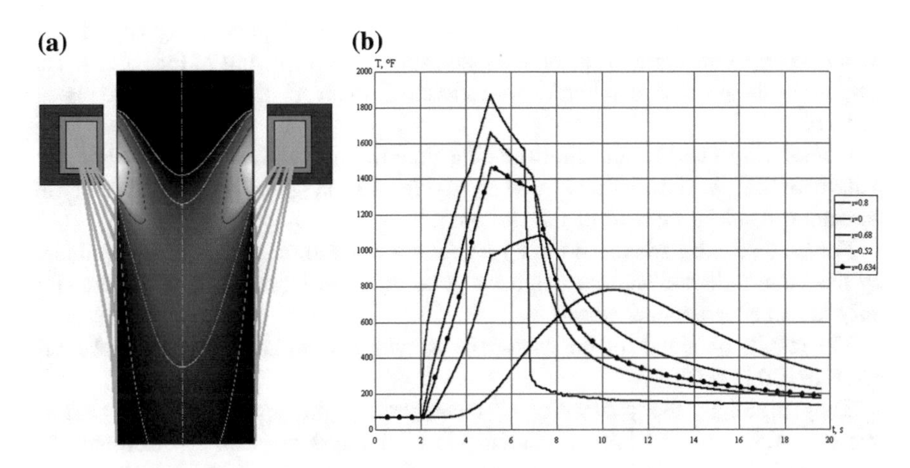

Fig. 6.134 **a** Scan hardening temperature pattern in the shaft; **b** transient temperature distributions at different radial positions (calculations with ELTA-1D) [69]

Fig. 6.135 The biggest world installation for scan hardening (axle shaft: length 10 m, diameter 0.75 m, weight 15,000 kg; frequency: 60 Hz; power: 1.5 MW; hardened depth: 75 mm). **a** Overall view; **b** closer view with tank for hardening liquid [Courtesy of Ajax Tocco Magnethermic]

This example shows that for certain applications, even a program 1D like ELTA can give a fast and reliable simulations of the process and allows to optimize a 2D system like the one here considered.

An idea of the importance of scan hardening is given by Fig. 6.135 which refers to the biggest installation of this type existing nowadays

(b) *Localized hardening of a plane surface*

This example concerns the localized heating of a steel plate using an inductor with magnetic concentrator facing the plate, according the layout of Fig. 6.124. The purpose of the process is to harden of a stretch of about 25 mm width on the plate's surface.

Considering that due the short heating time the heat diffusion is very low, the pattern of the volume induced power of Fig. 6.125a gives also an idea of the hardening result at the end of the treatment.

For sake of comparison, in the figure are shown also the distributions produced by the same inductor without magnetic yoke (this last represented in the drawing only for a better understanding)

The results obtained for the two systems, with and without magnetic yoke, are given in Table 6.14.

They highlight the possibility of concentrating the power in the region of interest, improving the integral parameters of the system (such as electrical efficiency and power factor) and reducing—for the same power within the load—the inductor losses as well as the losses in load adapter and bus-bars (which are

Table 6.14 Comparison of integral parameters of inductors of Fig. 6.125 with same induced power and same inductor current

	Inductor without core	Inductor with core	Inductor with core
Current (A)	**8130**	**8130**	4894
Supply voltage (V)	51.0	114.3	65.3
Induced power (kW)	**129.0**	356.0	**129.0**
Inductor losses (kW)	17.2	39.4	14.3
Electrical efficiency	0.88	0.90	0.90
$\cos \varphi$	0.353	0.425	0.425

proportional to the square root of the current). Calculated data were in good agreement with experimental data [61].

In reality, the improvement due to the presence of the concentrator are even bigger since, considering that the power is concentrated in a narrower zone, it is possible to reduce the total power induced in the plate, in comparison with the value given in the table, for reaching the same hardening result.

(c) *Surface hardening of the internal surface of a ball bearing*

In many hardening applications, the inductor-load system can be schematized by 1D or 2D geometries, as in the example of Fig. 6.136.

In more complex workpiece geometries this it not possible and it is necessary to resort to 3D simulations.

A 3D example is illustrated in Fig. 6.137, which refers to the surface hardening of the inside surface of the external ring of a ball bearing.

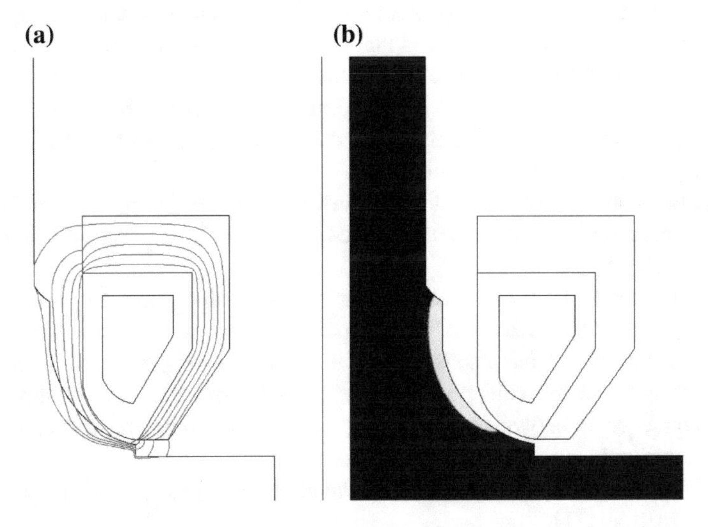

Fig. 6.136 Hardening of a ball bearing using a shaped inductor with magnetic concentrator (**a** field lines; **b** final temperature distribution)

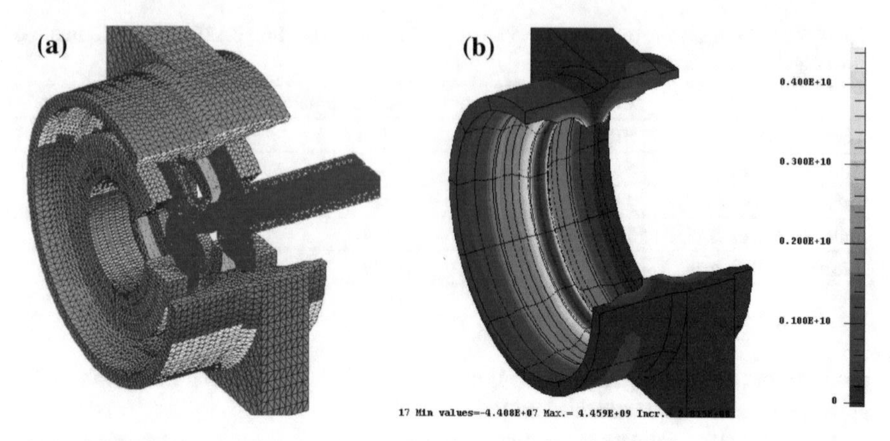

Fig. 6.137 Outer ring of a ball bearing hardened with an internal inductor provided of magnetic concentrator: **a** system geometry; **b** distribution of induced power [70]

Since the component is "short" in axial direction and the inductor has a magnetic concentrator without cylindrical symmetry, only a 3D calculation (electromagnetic and thermal) allows to obtain an accurate solution of the problem as well as a precise design of the inductor.

It is obvious that, given the complexity of the geometrical model and time and cost of calculation, a solution of this kind is justified only in case of mass production of the same workpiece.

(d) *Metallurgical and stress models*

The transient thermal distributions occurring in induction hardening processes have a significant impact not only on the metallurgical and mechanical properties of the heat treated parts, but also on residual stresses and distortion which can strongly influence the service performance of the component.

More recent studies have therefore incorporated in the computer simulation not only the electromagnetic, thermal and metallurgical phenomena, but also calculation of stress and shape deformation [71].

In the following is given an example illustrating todays computing capabilities; it refers to the hardening of a truck axle made of AISI 1541, with shaft length of about 1 m (Fig. 6.138).

The main concerns in this process are the bowing distortion and the amount of growth in length. Optimization of the hardening process allows not only to meet the hardening requirement, but also to prevent excessive heating in selected areas of the component in order to avoid possibility of cracking and excessive distortion.

The coil used is shown in Fig. 6.139. The gap between the inductor and the spray is approximately 25.4 mm.

Further details of geometry, induction heating and spray quenching process are described in Ref. [71].

The heating and spray quenching stages are summarized in Table 6.15.

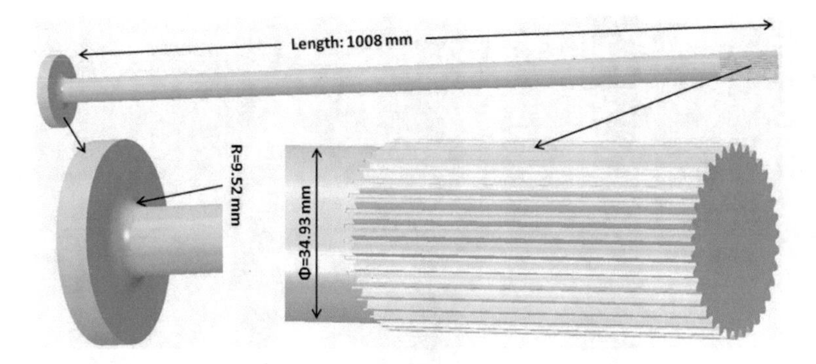

Fig. 6.138 Geometry of a full-float truck axle [71]

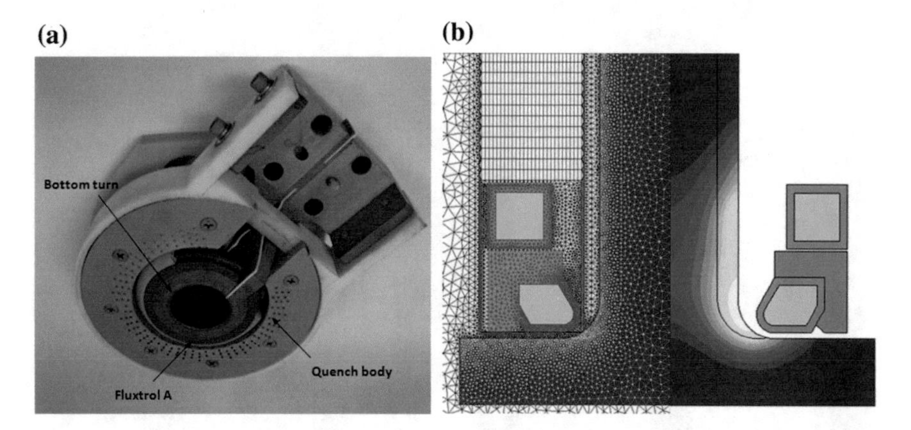

Fig. 6.139 **a** Two-turn axle scan coil with quench body; **b** fillet area of axle modelled with FLUX 2D [71]

Table 6.15 Induction heating and spray quench schedule for scan hardening of truck axle of Fig. 6.139

Step no.	Time period (s)	Inductor speed (mm/s)	Spray quench
1	9.00	0.00	No
2	1.50	15.00	No
3	6.00	8.00	Yes
4	99.00	8.00	Yes
5	14.00	8.00	Yes
6	60.00	8.00 (power off)	Yes

The spray starts at the beginning of step 3. At the end of step 3, the power distribution inside the shaft is stabilized, and step 4 covers the shaft. The inductor delivers heat to the spline during step 5, and the power is turned off at the end of

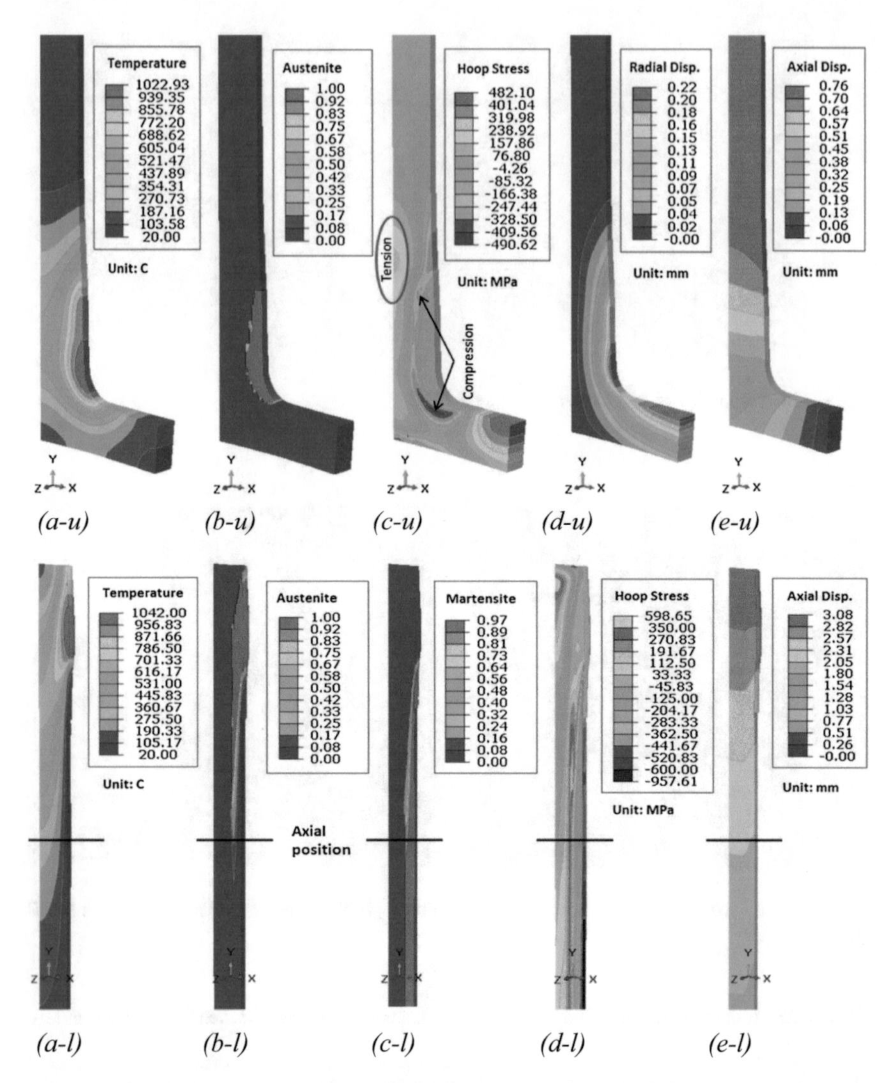

Fig. 6.140 **a** Temperature, **b** austenite phase, **c** hoop stress, **d** radial displacement, **e** axial displacement distributions (*u* upper figures: at the end of the first 9 s period; *l* lower figures: at the end of heating process, 130.15 s)

step 5 to avoid excessive heat of the edge of the spline, which is not hardened for high ductility. During the step 6, the power of the inductor is off, but the spray is continued to ensure a complete martensitic transformation (Table 6.15).

The temperature and austenite distributions at the end of the first step are shown in Fig. 6.140a-u, b-u. Because austenite has low strength at elevated temperature, the hoop stress in the austenitized zone is low.

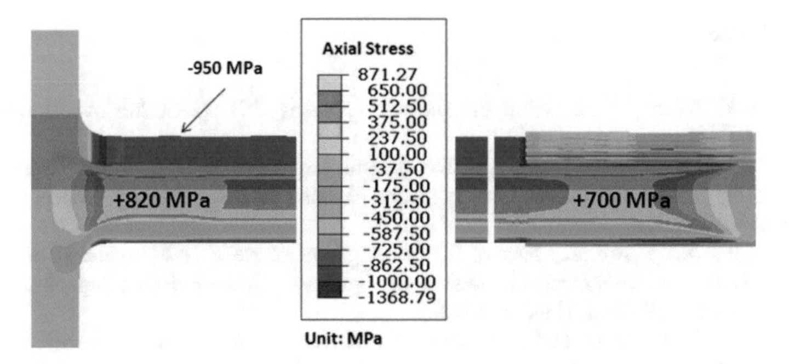

Fig. 6.141 Axial residual stress distribution at the end of hardening process

Due to thermal expansion of the austenitized region, the region below the austenitized layer is under compression, and the centre above the austenite zone is in tension as shown in figure (c-u).

At the end of the first step of 9 s, the inductor moves up with a speed of 15 mm/s for 1.5 s to avoid over heat the region above the fillet, then the speed drops down to 8 mm/s and the spray quench operates until the end of the whole process. Then power and temperature distributions of the shaft remain stable for most of the process.

Figure 6.140b-l, c-l show respectively the austenite and martensite distributions, and the phase transformation from austenite to martensite at the end of the scanning spray quench process. Figure (d-l) is a snapshot of in-process hoop stress distribution, which intuitively shows the effect of thermal gradient and phase transformation.

Figure (e-l) shows the in-process axial displacement distribution. Since at the lowest axial position in the figure the temperature drops to room temperature, the axial displacement below it represents a permanent distortion.

The axial residual stress distribution of the whole axle at the end of the hardening process is shown in Fig. 6.141.

The analysis shows that high surface compression on the surface of the shaft occurs due to the high cooling rate, while high tensile residual stresses are predicted at the case-core interface location of the as-quenched axle. The total length of the axle grows is about 2.3 mm after induction hardening, and the shaft shrinks about 4 μm radially.

This example allows us to conclude that all the induction heating process parameters including frequency, power and scan speed, have a strong influence on the temperature profile, which in turn will affect the residual stress distribution in the component. Also the cooling rate can influence significantly the residual stress distributions.

More generally it can be concluded that the metallurgical and stress models allow today to have a better understanding of the hardening results and can be used for process optimization aimed to failure prevention and service properties prediction.

References

1. Langer, E.: Theory of Induction and Dielectric Heating, 291 pp. Československà Academie, Praga (1964) (in Czechoslovak)
2. Sundberg, Y.: Induction Heating, p. 165. Västra Aros Tryckeri Aktiebolag, Västerås (1965)
3. Kuvaldin, A.B.: Low Temperature Induction Heating of Steel, 111 pp. Energija, Moskow (1976) (in Russian)
4. Zedler, T., Nikanorov, A., Nacke, B.: Investigation of Relative Magnetic Permeability as Input Data for Numerical Simulation of Induction Surface Hardening. Int. Scientific Colloquium MEP 2008, Hanover (2008)
5. Di Barba, P., Lowther, D.A., Dughiero, F., Forzan, F., et al.: A benchmark problem of induction heating. In: International Conference on HES-16, June 2016, Padua (2016)
6. Rudnev, V., Totten, G. (eds.): Induction Heating and Heat Treatment Handbook, Vol. 4C, 820 pp. ASM Heat Treating Society, ASM International (2014). ISBN-13 978-1-62708-012-5
7. Hegewaldt, F.: Induktives Oberflächenhärten. Induktionsöfen 62–84 (1961)
8. Neiman, L.R.: Skin Effect in Ferromagnetic Bodies, 188 pp. State Energetic, Leningrad (1949) (in Russian)
9. Lupi, S., Forzan, M., Aliferov, A.: Induction and Direct Resistance Heating—Theory and Numerical Modeling, 370 pp. Springer Int. Publishing, Switzerland (2015). ISBN 978-3-319-03478-2
10. Nemkov, V.S., Demidovich, V.B.: Theory and Calculation of Installations of Induction Heating, 280 pp. Energoatomizdat, Leningrado (1988) (in Russian)
11. Sluhockiĭ, A.E., Nemkov, V.S., Pavlov, N.A., Bamunier, A.B.: Induction Heating Installations, 325 pp. Energoizdat, Leningrad (1981) (in Russian)
12. Lupi, S., Forzan, M., Aliferov, A., Meleshko, A.: Resistance of inductors for induction heating of cylindrical internal surfaces. Elektrotechnika **5**, 43–47 (2010) (in Russian)
13. Mahmudov, K.M., Nemkov, V.S., Sluhockiĭ, A.E.: Methods of Calculation of Inductors, pp. 3–27, n. 114. Izvestija of Leningrad Electrotechnical Institut LETI, Leningrad (1973) (in Russian)
14. Lavers, I.D., Biringer, P.P.: An analysis of the coreless induction furnace: axial distribution of electric and magnetic fields. Elektrowaerme int. **29**(4), 232–237 (1971)
15. Lupi, S., Morini, A.: Induction heating of cylindrical rods using multiple coils. In: Proceedings of Institute of Electrical Engineering, UPEE, 69/13 (in Italian) and Elektrowaerme int. **29**(12), 663–667 (1971)
16. Lavers, I.D., Biringer, P.P.: An Improved Method of Calculating the Induction Heating Equivalent Circuit Parameters, n. 602. VII UIE Congress, Warsaw (1972)
17. Lupi, S., Nemkov, V.: The calculation of inductors with periodical fields. Elektrowaerme int. **35**(2), 103–109 (1977)
18. Lupi, S., Nemkov, V.: Riscaldamento ad induzione delle superfici esterne ed interne di cilindri bimetallici cavi mediante induttori corti. L'Elettrotecnica **66**(2), 141–147 (1979) (in Italian)
19. Lupi, S., Nemkov, V.S.: Il calcolo analitico di sistemi ad induzione cilindrici, pp. 43–47, n. 6. Elektrichestvo, Energija (1978) (in Russian)
20. Lavers, J.D.: State of the art of numerical modeling for induction processes. In: HES-07—Heating by Electromagnetic Sources, Padua, pp. 13–24, 19–22 June 2007 (2007)
21. Lavers, J.D.: An efficient method of calculating parameters for induction and resistance heating installations with magnetic loads. IEEE Trans. Indus. Appl. **IA-14**(5), 427–432 (1978)
22. Lupi, S.: Appunti di Elettrotermia. Teaching Notes, 457 pp. Libreria Progetto, Padova (2005) (in Italian)
23. Lupi, S., Mühlbauer, A., Nemkov, V., et al.: Induction Heating Industrial Applications, pp. 144. U.I.E.—International Union for Electroheat, WG Induction Heating, Paris (1992)
24. Électricité de France: Induction Conduction électrique dans l'industrie, 780 pp. Electra - Dopee85, Paris (1996). ISBN 2-86995-022-5 (in French)

25. Mühlbauer, A.: History of Induction Heating and Melting, 202 pp. Vulkan-Verlag GmbH, Essen (2008). ISBN 978-3-8027-2946-1
26. Lupi, S.: Research in the Field of Induction Heating at the University of Padua, 33 pp. Lecture given at the Dept. of Electrical Engineering, Dec. 13, 2010. SGE, Padua (2010)
27. Lupi, S.: Survey on induction heating development in Italy. In: HISTELCON 2012, Pavia (2012)
28. Brev. Ital. Reg. Gen. N. 140 341, Reg Atti N. 245 – Descrizione del trovato che ha per titolo Forno ad induzione a frequenze elevate e dispositivo per la sua connessione con gli apparecchi generatori (in Italian)
29. Northrup, E.F.: Usa Patents No. 1 286 394; No. 1 286 395, Dec. 3, 1918; No. 1 297 393, March 18, 1919; No. 1 328 336, Jan. 20, 1920; No. 1 330 133, Feb. 10, 1920
30. Northrup, E.F.: Principles of induction heating with high frequency currents. Trans Amer. Electrochem. Soc. **35** (1919)
31. Northrup, E.F.: Recent progress in high frequency inductive heating. Trans. Amer. Electrochem. Soc. **39** (1921)
32. Nacke, B., Baake, E., Lupi, S., Dughiero, F., Forzan, M., et al.: Theoretical Background and Aspects of Electrotechnologies, 356 pp. Physical Principles and Realization. Intensive Course Basic I. St. Petersburg, Publishing House of ETU (2012). ISBN 978-5-7629-1237-2
33. Mühlbauer, A: Industrielle Elektrowärme-technik, 400 pp. Vulkan-Verlag, Essen (1992). ISBN 3-8027-2903-X (in German)
34. Conrad, H., Mühlbauer, A., Thomas, R.: Elektrothermischen Verfahrenstechnik, 338pp. Vulkan-Verlag, Essen (1994). ISBN 3-8027-2911-0
35. Mühlbauer, A.: Über die elektrodynamichen Kräfte in der Schmelze von Induktionsöfen. Elektrowärme, Bd. **25**(12), 461–473 (1967)
36. Fomin, N.I., Zatulovskiĭ L.M.: Electrical Furnaces and Induction Heating Installations, 247 pp. Metallurghia, Moskow (1979) (in Russian)
37. Hegewaldt, F.: Forni ad induzione per la fonderia. La fonderia italiana (4), 105–117 (1973) (in Italian)
38. Baake, E., Jakovics, A., Kirpo, M.: Influence of the channel design on the heat and mass exchange of induction channel furnace. In: HES-10—Heating by Electromagnetic Sources, Padua, pp. 155–162, May 18–21
39. Wirtsch, D., Sonnenschein, P.: Induktives Erwärmen zum Umformen von Stahl - Gesenkschmiede, 23 pp. RWE AG-Information zentrum, Essen (1989) (in German)
40. Rudnev, V., Loveless, D., Cook, R., Black, M.: Handbook of Induction Heating, 777 pp. Marcel Dekker, Inc., New York (2003). ISBN 0-8247-0848-2
41. Fasholz, J., Orth, G.: Induktive Erwärmung – Physikalische Grundlagen un technische Anwendungen, 103 pp. RWE Energie Aktiegesellschaft, Essen (1991) (in German)
42. Geisel, H.: Konduktives oder induktives Erwärmen von Werkstücken grosser Abmessungen mit Frequenzen $f \leq 50$ Hz. Elektrowärme international **44**(B3), 107–115 (1986)
43. Lauster, F.: Manuel d'Electrothermie Industrielle, 315 pp. Dunod, Paris (1968) (in French)
44. Poncin, R.: New developments in billet heating. In: UNIPEDE—Workshop on Reheating of Metals by Induction, ENEL, Milano, pp. 29–44, March 23 1987 (1987)
45. Baake, E.: Energetische und technische Bewertung industrieller Erwärmungsverfharen. Elektrowärme International **B1**, 7–13 (1997)
46. Boergerdin, R., Baake, E., Drewek, R., Joern, K.U., Mühlbauer, A.: Reduction of scale, energy consumption and CO_2 emission of forging processes. In: International Congres EPM —Electromagnetic Processing of Materials, Paris, pp. 61–66, May 27–29 1997 (1977)
47. Nacke, B., Baake, E.: Reduction of CO_2 emissions using efficient melting and heating technologies. In: HES-10—Heating by Electromagnetic Sources, Padua, pp. 263–270, May 18–21 2010 (2010)
48. Semiatin, S.L., Stutz, D.E.: Induction Heat Treatment of Steel, 308 pp. ASM—American Society for Metals, Metals Park (1986). ISBN 0-87170-211-8
49. Davies, J., Simpson, P.: Induction Heating Handbook, 419 pp. McGraw-Hill Co. (1979). ISBN 0-07-084515-8

50. Kolbe, E., Schuster, R.: Möglichkeiten zur Vorausbestimmung der Einhärtetiefe beim Induktionshärten. Elektrowärme, Band **23**(12), 535–544 (1965)
51. Lozinskii, M.G.: Industrial Applications of Induction Heating, 672 pp. Pergamon Press, London (1969)
52. Zischka, K.A.: Beitrag zur matematischen Theorie der Induktionshärtung. Teil 1: Elektrischer Teil, Elektrowärme International, Band **28**(3), 135–142; Teil 2: Thermischer Teil, 142–149 (1970)
53. Kegel, K., Lacmann, B.: Vorausbestimmung der Einhärtungstiefe bei der induktiven Oberflächenhärtung. Elektrowärme, Band **24**(12), 419–424 (1966)
54. Atlas of Isothermal transformation and Cooling Transformation Diagrams. American Society of Metals, Metals Park (1977)
55. Brunst, W.: Die induktive Wärmebehandlung, p. 237. Springer, Berlin (1957)
56. Hougton Italia: Fast Hardening Oils. Servizio Tecnico Ricerca-Produzione, Genova
57. Hougton Italia: Principles and Hardening Techniques, 44 pp. Servizio Tecnico Ricerca-Produzione, Genova
58. Koller, L.: Einfluss der geometrischen Parameter von Induktor-Einsatz-Systemen auf die Richtung der Wirbelströme im Einsa*tz*. Elektrowarme International **40**, B5 – Oktober, B247–B252 (1982)
59. Brown, G.H., Hoyler, C.N.: Theory and Application of Radio-Frequency Heating, 370 pp. D. Van Nostrand Inc., New York (1947)
60. Merigliano, L.: Induttori per trattamenti termici ad alta frequenza, n. 706. Rendiconti dell'AEI, Roma (1957) (in Italian)
61. Ruffini, R.T., Nemkov, V., Goldstein, R.: Materials for high frequency magnetic flux control —properties and applications. In: HIS-01—Heating by Internal Sources, pp. 13–19, Padua Sept. 12–14 2001 (2001)
62. Di Pieri, C., Lupi, S., Crepaz, G.: The design of induction heating machines for single-shot hardening of large ring-gears or gear-wheels. In: UIE-11—XI International Congress on Electroheat, Malaga (Spain), Oct. 3–7 1988, n. B 7.6 (1988)
63. Di Pieri, C., Lupi, S., Crepaz, G.: Moderni impianti di riscaldamento ad induzione. L'Elettrotecnica **LXXV**(12), 1185–1197 (1988) (in Italian)
64. Nacke, B., Wrona: Design of complex induction hardening problems by the use of numerical simulation. In: REI'05—International Conference on Research in Electro-technology and Applied Informatics, pp. 159–164, August 31–Sept. 3, 2005, Katowice (2005)
65. Sluhockiĭ, A.E., Ryskiĭ, S.E.: Induction Heating Inductors, 263 pp. Energija, Leningrad (1974) (in Russian)
66. Lupi, S., Rubatto, G.: Tempra di ingranaggi mediante riscaldamento ad induzione con due frequenze. AIM—Associazione Italiana di Metallurgia, Giornata di studio su Trattamenti superficiali, Milano (Italy), 31 Marzo (1998) (in Italian)
67. Lupi, S., Forzan, M., Aliferov, A.: Induction and Direct Resistance Heating, Chap. 4.6, 370 pp. Springer International Publishing, Switzerland (2015). ISBN 978-3-319-03478-2
68. Zinn, S.: Quenching of Induction Heated Steel. ASM Handbook, Vol. 4C, Induction Heating and Heat Treatment, pp. 87–102. ASM International (2014)
69. Nemkov, V., Goldstein, R.: Computer simulation for fundamental study and practical solutions to induction heating problems. In: Proceedings of HIS-01—International Seminar on Heating by Internal Sources, Padua, pp. 435–442, 12–14 September 2001 (2001)
70. www.cedrat.com/software/flux/flux.html
71. Ferguson, L., Zhichao Li, Nemkov, V., Goldstein, R., Jackowski, J., Fett, G.: Modeling stress and distortion of full-float truck axle during induction hardening process. In: HES-13—Heating by Electromagnetic Sources, Padua, pp. 109–116, 21–24 May 2013 (2013)
72. Lavers, J.D.: An analysis of the coreless induction furnace: load end effects. Elektrowärme International **29**(7), 390–396 (1971)

Chapter 7
High Frequency and Microwave Heating

Abstract This chapter presents the industrial applications of high frequency and microwave heating, which are based on the thermal effect produced in a dielectric material exposed to a high frequency alternating electrical field. Since dielectric materials are characterized by low electrical and thermal conductivity, dielectric heating has found wide acceptance in the industry because it represents the best way to achieve fast and uniform heating in dielectric workpieces. The first paragraphs of the chapter deal with the theory of polarization processes, the equations of the power transformed into heat in the workpiece and the influence of frequency and temperature on the properties of dielectrics. In a subsequent paragraph the distribution of the electric field depending on the shape of the workpiece and the geometry of the working capacitor are studied. A paragraph deals with the calculation of the equivalent circuit of the working capacitor with load and the transient temperature distribution in the workpiece. The last theoretical paragraph deals with the peculiarities of microwave heating and the main elements of a microwave installation (magnetron, waveguide and resonant cavity). In the second part of the chapter are described the main industrial processes based on high frequency and microwave heating. In particular: gluing of wood, welding of sheets of thermoplastic materials, preheating of thermosetting resins and rubber, and applications in the textile, paper and food sectors.

7.1 Introduction

High frequency and Microwave heating are based on the thermal effect produced in a dielectric material when it is exposed to a high frequency alternating electric field.

The heating effect is due to the polarization process which takes place in the dielectric medium, and is connected with the microscopic displacements of the bound charges and the conduction current produced by the free charges moving on a macroscopic scale under the action of the electric field.

Dielectric materials are mostly characterized by low electrical conductivity and, as a consequence, by very low conduction currents. For this reason they are not

S. Lupi, *Fundamentals of Electroheat*, DOI 10.1007/978-3-319-46015-4_7

conveniently heated with induction or conduction processes. Moreover, since they have very low thermal conductivity, also surface heating (like heating in resistance furnaces) is not effective because of the long time needed for heat transfer from the surface to the core of the body to be heated.

For these reasons dielectric heating has found wide acceptance in industrial processes which require fast and uniform heating in the workpiece volume, even if it makes use of a relatively expensive form of energy.

7.2 Theoretical Background

It is well known that in a dielectric material where an electric field $\overline{E}$ (V/m) and an electric flux density $\overline{D}$ (C/m^2) are present, the specific electrostatic energy per unit volume w_E (J/m^3 in the SI system) can be expressed by the equation:

$$w_E = \frac{\overline{E} \cdot \overline{D}}{2} \tag{7.1}$$

with:

$$\overline{D} = \varepsilon_0 \overline{E} + \overline{P} = (1 + \frac{\overline{P}}{\varepsilon_0 \overline{E}}) \varepsilon_0 \overline{E} = (1 + \chi)\varepsilon_0 \overline{E} = \varepsilon_0 \varepsilon \overline{E}$$

$\overline{P} = \chi \varepsilon_0 \overline{E}$	dielectric polarization vector
$\varepsilon_0 = 8.86 \cdot 10^{-12}$	dielectric constant of vacuum (F/m)
χ	dimensionless quantity known as electric susceptibility of material.

With the exception of some dielectric materials, like ferroelectric materials, which are able to produce spontaneous polarization, the electric susceptibility doesn't depend on electric field intensity, up to values near to the dielectric strength.

In a domain where $\overline{E}$ and $\overline{D}$ are present, the first Maxwell equation can be written as follows:

$$\text{rot}\,\overline{H} = \sigma \overline{E} + \frac{\partial \overline{D}}{\partial t} \tag{7.2}$$

where:

$\overline{H}$	magnetic field intensity (A/m)
σ	electrical conductivity (S/m)
$\overline{J} = \sigma \overline{E}$	conduction current density (A/m^2)
$\frac{\partial \overline{D}}{\partial t}$	displacement current density (A/m^2)

As known, in case of time harmonic regime, expressing the sinusoidal quantities in phasor form $\overline{E} = \overline{E}\, e^{j\omega t}$, Eq. (7.2) can be re-written as follows:

$$\text{rot}\,\dot{\overline{H}} = \sigma\,\dot{\overline{E}} + j\,\omega\,\varepsilon_0\varepsilon\,\dot{\overline{E}}. \tag{7.3}$$

The right side term of Eq. (7.3) represents the total current density:

$$\dot{\overline{J}}_t = \sigma\,\dot{\overline{E}} + j\,\omega\,\varepsilon_0\varepsilon\,\dot{\overline{E}}, \tag{7.4}$$

sum of the conduction and displacement current densities.

Since, under the influence of the electric field, the polarization process of bound charges is not instantaneous, in an electric field time varying with sinusoidal law the polarization vector lags the electric field vector. In this situation we can consider χ as a complex number and the phasor $\dot{\overline{D}}$ lags the electric field $\dot{\overline{E}}$ of an angle δ_p which characterizes the active power delivered during the dielectric polarization process.

Figure 7.1 shows a phasor diagram where $\sigma_e\,\dot{\overline{E}}$ represents the component of the total current density in phase with $\dot{\overline{E}}$, which is sum of the term $\sigma\,\dot{\overline{E}}$ due to the non-null electrical conductivity of the material, and the component $\omega\,\varepsilon_0\varepsilon''\,\dot{\overline{E}}$, in phase with $\dot{\overline{E}}$, due to the displacement current density $j\omega\dot{\overline{D}}$.

Hence the total current density can be represented in two ways:

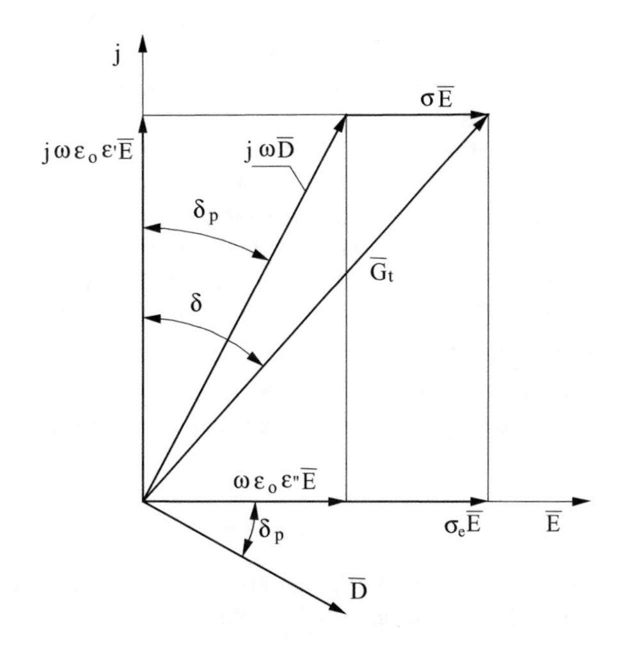

Fig. 7.1 Phase diagram of quantities characterizing a dielectric material in alternative electric field

- either as a *conduction current*

$$\dot{J}_t = (\sigma + j\,\omega\,\varepsilon_0\,\dot{\varepsilon})\dot{E} = \dot{\sigma}_t\,\dot{E} \tag{7.5}$$

with:
$\dot{\varepsilon} = \varepsilon' - j\,\varepsilon''$
$\dot{\sigma}_t = (\sigma + \omega\,\varepsilon_0\,\varepsilon'') + j\,\omega\,\varepsilon_0\,\varepsilon'$—complex equivalent electrical conductivity of dielectric

- or as a *displacement current*

$$\dot{J}_t = j\omega\varepsilon_0\,(\dot{\varepsilon} - j\frac{\sigma}{\omega\,\varepsilon_0})\dot{E} = j\omega\varepsilon_0\varepsilon'(1 - j\,\mathrm{tg}\,\delta)\dot{E} \tag{7.6}$$

with: tg δ—tangent of the dielectric loss angle.

The diagram highlights that the angle δ of the total dielectric losses is greater than the polarization loss angle δ_p, because the first one takes into account not only the thermal energy due to the polarization losses produced by the alternative electric field but also the energy developed by the conduction current.

From Eqs. (7.5), (7.6) and the diagram of Fig. 7.1 we can write:

$$\dot{\sigma}_t = j\omega\varepsilon_0\varepsilon'(1 - j\mathrm{tg}\delta) \tag{7.7}$$

$$\mathrm{tg}\,\delta = \frac{\varepsilon''}{\varepsilon'} + \frac{\sigma}{\omega\varepsilon_0\varepsilon'} = \mathrm{tg}\,\delta_p + \frac{\sigma}{\omega\varepsilon_0\varepsilon'} \tag{7.8}$$

The parameters ε' and tgδ completely define the behavior of a dielectric material submitted to a sinusoidal electric field.

To the diagram of Fig. 7.1, which represents the physical phenomena from the electromagnetic point of view, can be associated the equivalent circuit of Fig. 7.2, which describes the same phenomena from the point of view of circuit theory.

In this circuit, C represents the capacity of a capacitor constituted by the same dielectric material considered in Fig. 7.1 and supplied with the sinusoidal voltage V; R_C represents the conduction losses due to the conduction current density component ($\sigma\dot{E}$), and R_D the losses corresponding to the component of the displacement current density in phase with $\dot{E}$.

The complex power per unit volume of the dielectric can be expressed as follows:

$$w_a + j\,w_r = \dot{E}\cdot\dot{J}_t^* = \dot{\sigma}_t^*\cdot\dot{E}^2 \tag{7.9}$$

where the symbol (*) denotes the complex conjugate of a phasor quantity. From Eq. (7.7) this equation can be rewritten as follows:

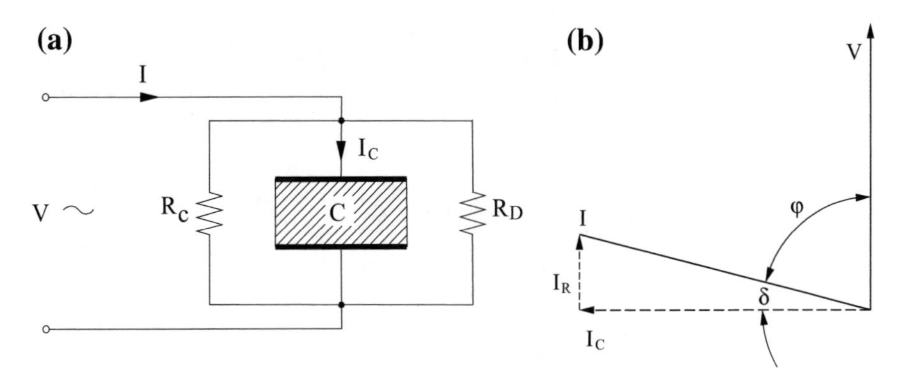

Fig. 7.2 **a** Equivalent circuit of a dielectric material in AC electric field; **b** Phasor diagram of the circuit of figure (**a**)

$$w_a + j\, w_r = (\sigma + \omega\, \varepsilon_0\, \varepsilon'' - j\, \omega\, \varepsilon_0\, \varepsilon')E^2 = \omega\, E^2\, \varepsilon_0\, \varepsilon'(\text{tg}\,\delta - j). \tag{7.9a}$$

Equations (7.9) and (7.9a) highlights again the resistive-capacitive nature of the electric circuit equivalent to the working capacitor.

The real part:

$$w_a = \omega\, E^2\, \varepsilon_0\, \varepsilon'\, \text{tg}\,\delta = 5.55 \cdot 10^{-11} \cdot f \cdot E^2 \cdot \varepsilon'\, \text{tg}\,\delta, \tag{7.10}$$

represents the total power, W/m^3, transformed into heat in the material, which depends on the frequency (f), the square of the electric field intensity (E) and the quantity ($\varepsilon'\, \text{tg}\,\delta$) which is usually named material "*loss factor*".

7.2.1 Power Transformed into Heat

In order to take into account not only the local specific quantities but the integral ones, i.e. the real dimensions of the heated workpiece, we can make reference to an ideal plane capacitor where the vectors $\overline{E}$ and $\overline{D}$ are uniformly distributed in the dielectric.

Denoting with $C = \varepsilon_0\, \varepsilon' \frac{S}{d}$ the capacitance of the plane capacitor, with reference to the phasor diagram of Fig. 7.2b and assuming that tgδ is negligible, we can write:

$$I_C = j\, \omega\, C\, V = j\, \omega\, V\, \varepsilon_0\, \varepsilon' \frac{S}{d} \tag{7.11a}$$

$$I_R = I_C\, \text{tg}\,\delta = \omega\, V\, \varepsilon_0\, \varepsilon'\, \text{tg}\,\delta \frac{S}{d} \tag{7.11b}$$

$$I = I_C + j\,I_R = j\,\omega\,V\,\varepsilon_0\frac{S}{d}(\varepsilon' - j\,\varepsilon'\,tg\,\delta) = \\ = j\,\omega\,\varepsilon_0\frac{S}{d}(\varepsilon' - j\,\varepsilon_e'') \tag{7.11c}$$

$$P + j\,Q = \dot{V}\cdot\dot{I}^* = \omega\,V^2\varepsilon_0\frac{S}{d}(\varepsilon'' tg\,\delta - j\varepsilon') = \\ = \omega\,V^2\varepsilon_0\,\varepsilon'\frac{S}{d}(tg\,\delta - j) \tag{7.12}$$

$$w_a = \frac{P}{S\,d} = \omega\,E^2\varepsilon_0\,\varepsilon_e'' = \omega\,E^2\varepsilon_0\varepsilon' tg\,\delta \tag{7.12a}$$

Equation (7.12a) is the basic equation of dielectric heating: it shows that the specific power per unit volume is proportional to the frequency, the square of the electric field intensity and the loss factor which in turn, as will be discussed in the next paragraph, is a function of frequency and temperature.

Given that the loss factor is a material characteristic and the electric field intensity cannot exceed the limit of the dielectric strength, Eq. (7.12a) can be used to evaluate the minimum working frequency below which the required power cannot be developed.

Usual values of *w* are in the range 0.5–5 W/cm^3.

7.2.2 *Parallel and Series Equivalent Circuits*

Denoting with:

R_p resistance equivalent to the parallel of R_c and R_D,
$X_p = 1/\omega C$ reactance of the capacitor C,
$P + j\,Q = (w_a + j\,w_r)\cdot vol$ active and reactive power absorbed by the circuit,
$Q_0 = Q/P = R_p/X_p$ Q factor of the circuit,

for the circuit of Fig. 7.2a Eq. (7.9a) gives:

$$Q_0 = \frac{1}{tg\,\delta};\quad R_p = \frac{1}{\omega\,C\,tg\,\delta}. \tag{7.13}$$

In many cases it is convenient to refer to the series equivalent circuit of Fig. 7.3. With the notation:

R_s, $X_s = 1/\omega\,C_s$—resistance and reactance of the equivalent series circuit, equating the impedances of the series and parallel circuits and separating real and imaginary parts we have:

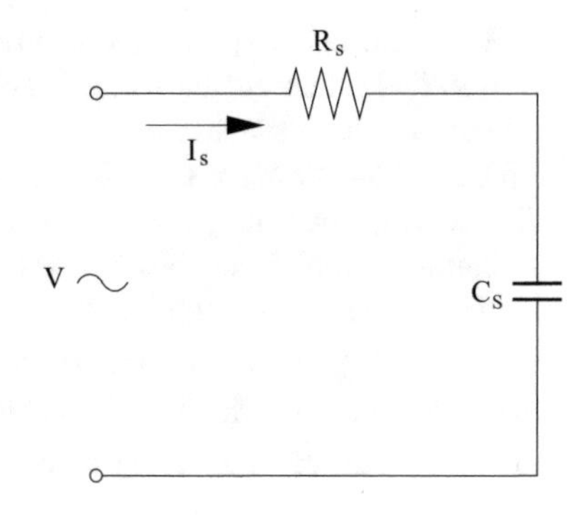

Fig. 7.3 Series equivalent circuit

$$R_s - j\,X_s = \frac{-j\,X_p\,R_p}{R_p - j\,X_p}$$

and

$$\left.\begin{aligned} R_s &= R_p\frac{X_p^2}{R_p^2 + X_p^2} = \frac{\mathrm{tg}\,\delta}{\omega C\,(1 + \mathrm{tg}^2\,\delta)} \\ X_s &= X_p\frac{R_p^2}{R_p^2 + X_p^2} = \frac{1}{\omega C(1 + \mathrm{tg}^2\,\delta)} \end{aligned}\right\} \tag{7.14}$$

7.3 Physical Properties of Dielectric Materials

7.3.1 Polarization Processes

Before discussing the characteristics of radio frequency or dielectric heating processes, we will consider the behavior of dielectric materials.

In fact the main heating effects are due to the conduction current and the polarization phenomena which, in turn, depend on frequency and material characteristics.

In the following we will describe the polarization by means of the Debye theory.

A dielectric material differs form a conductive one mainly for the presence of polarization under the effect of an electric field.

Polarization phenomena can be described by means of three different mechanisms occurring at different frequencies.

- *Electronic Polarization*—it can be described as an elastic deformation of electronic orbits when an electric field is applied to the material. When the centers of electron cloud and nucleus are no longer coincident, the atom behaves like an electric dipole oriented by the external field, with a momentum proportional to the electric field.

This is an elastic polarization process, with oscillatory character in the range 10^{14}–10^{16} Hz, that occurs when the electric field is applied to the material.

Since in dielectric heating the most used frequencies are lower than 10^9 Hz (typically 10–100 MHz for radio frequency applications, 915 and 2450 MHz for Microwaves), the semi-period of the electric field intensity is much longer than the oscillation period of the elastic polarization, and the alignment of electric dipoles is almost instantaneous at the application of the electric field.

As a consequence, the polarization vector $\dot{P}$ and the electric flux density $\dot{D}$ are in phase with the electric field $\dot{E}$. With reference to Fig. 7.1, the loss angle δ_p is practically zero and hence the term $(\omega\, \varepsilon_0 \varepsilon'' \dot{E})$ corresponding to the losses is null.

- *Ionic Polarization*—it occurs when the molecular bonds prevent the formation of dipoles in absence of the applied electric field. The behavior of the material is similar to the previous one and also in this case the polarization occurs only if an external electric field is applied to the material. The external electric field contributes to separate the linked charges, moving the positive ones in one direction and the negative in the opposite direction, thus creating a macro-dipole.

This behavior differs from the electronic polarization only for the frequency of the elastic polarization which, in this case, is in the range of 10^{11} Hz.

- *Dipolar Polarization*—it occurs in "polar" materials where, also in absence of an applied external electric field, the molecules behave like electrical dipoles. These dipoles, when the electric field is not present, are random oriented due to thermal agitation; on the contrary, if an electric field is applied, they rotate around their axes and tend to orientate in the direction of the electric field. This orientation is hindered by the thermal agitation and, as a consequence, in equilibrium conditions the dipoles are not completely oriented in the field direction, but they show a sort of "preferential" direction.

This process is known as *relaxation polarization* and, in contrast with elastic polarization, has an aperiodic character whose behavior, under the influence of an applied stepwise electric field, can be described by the equation:

$$P_r = P_{0r}(1 - e^{-t/\tau}) \tag{7.15}$$

where: P_{0r}—relaxation polarization in steady-state conditions; τ—time constant, known as *relaxation time*, characterizing the velocity of increase of relaxation polarization.

In dielectric materials with ionic or dipolar polarization the time constant τ is in the range 10^{-7}–10^{-13} s, depending on chemical structure and temperature. High values of τ modify the material frequency behavior.

- *Volume Polarization*—it is a polarization process which occurs in materials constituted by two or more components, one of them including macro-volumes of conductive materials. Typical examples are wet materials, where water inclusions are present, or foods and wood where a "cell" structure is present. The cells are separated by the dielectric material, while inside the cells a conductive liquid is present. When an electric field is applied, ions and free electrons present in the conductive inclusions move inside the inclusions, which behave like giant polarized molecules with dipole momentum. The non-uniform material structure allows the movement of charges only inside the macro-inclusions, which therefore behave like bound charges.

Also in this case the thermal agitation movements limit the charge separation inside the conductive inclusions and the polarization process becomes aperiodic, i.e. of the relaxation type described by Eq. (7.15).

The growth of polarization velocity is greater the greater is the electrical conductivity of inclusions and, in presence of conduction ions, the polarization time constant is in the range 10^{-3}–10^{-8} s.

In dielectric materials with relaxation polarization it is always present also elastic polarization. For frequencies below 1000 MHz, we can assume that elastic polarization appears instantaneously after the application of an external electric field.

The polarization occurring at the application of a stepwise electric field:

$$E(t) = E \cdot u_1(t). \tag{7.16}$$

with: $u_1(t)$—step unit function, can expressed as a function of time by the equation:

$$P(t) = P_{el} \cdot u_1(t) + P_{0r}(1 - e^{-t/\tau}). \tag{7.17}$$

Using the Laplace transform of Eqs. (7.16) and (7.17) we can define an "*operatorial dielectric constant*":

$$\varepsilon(s) = 1 + \frac{P(s)}{\varepsilon_0 E(s)} = \varepsilon_1 + \frac{\Delta\varepsilon}{1 + s \cdot t} \tag{7.18}$$

where: *s*—variable of Laplace transform; $\Delta\varepsilon$—increment of dielectric constant produced by relaxation polarization.

The introduction of the operatorial dielectric constant gives the possibility to analyze the behavior of transient polarization processes due to different time variation laws of the applied electric field E(t), in particular the sinusoidal one.

7.3.2 *Influence of Frequency on Relaxation Polarization*

If we apply a sinusoidal electric field to the dielectric material, Eq. (7.18) can be expressed as follows in complex form:

$$\dot{\varepsilon} = \varepsilon' - j\varepsilon'' = \varepsilon_1 + \frac{\Delta\varepsilon}{1 + j\omega\tau} \tag{7.19}$$

Decomposing real and imaginary parts we have:

$$\varepsilon' = \varepsilon_1 + \frac{\Delta\varepsilon}{1 + (\omega\tau)^2}; \quad \varepsilon'' = \frac{\omega\tau \cdot \Delta\varepsilon}{1 + (\omega\tau)^2} \tag{7.19a}$$

From the previous equations and Eq. (7.8) the loss angle δ_p, can be expressed as follows:

$$\operatorname{tg} \delta_p = \frac{\omega\,\tau \cdot \Delta\varepsilon}{\Delta\varepsilon + \varepsilon_1 [1 + (\omega\tau)^2]} \tag{7.20}$$

Calculating the derivative of Eq. (7.20) and equating it to zero, we obtain the frequency corresponding to the maximum value of $\tan\delta_p$:

$$\omega = \omega_{max} = \omega_0 \sqrt{1 + \frac{\Delta\varepsilon}{\varepsilon_1}} \tag{7.21}$$

with: $\omega_0 = 1/\tau$—relaxation frequency.

The maximum value of the loss angle then becomes:

$$\operatorname{tg} \delta_{p\,max} = \frac{\Delta\varepsilon}{2\sqrt{\varepsilon_1(\varepsilon_1 + \Delta\varepsilon)}}. \tag{7.22}$$

The frequency behavior of the complex vector that represents the complex permittivity $\dot{\varepsilon}$ can be described by the circular diagram of radius $(\Delta\,\varepsilon/2)$ shown in Fig. 7.4.

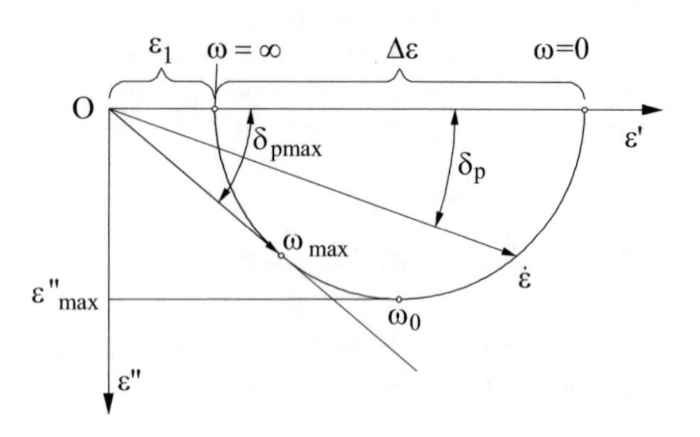

Fig. 7.4 Circular diagram of complex dielectric permittivity as a function of frequency

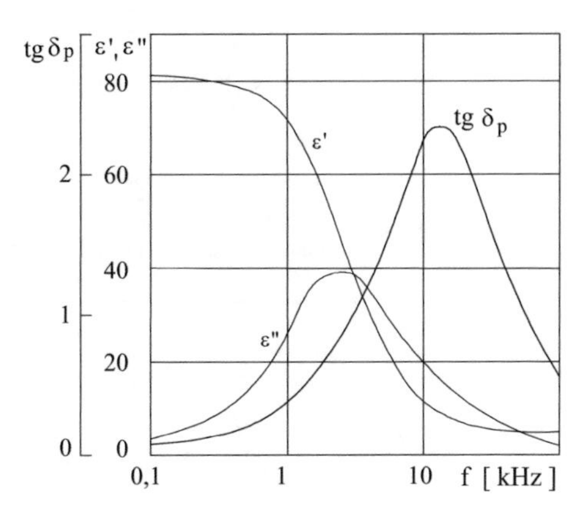

Fig. 7.5 Dielectric characteristics of ice obtained from distilled water at temperature—10 °C [1]

If the frequency increases, moving along the circumference from the point $\omega = 0$ to $\omega = \infty$, the real part ε' decreases continuously, while the imaginary part ε'' initially increases from zero to the maximum value $\Delta\varepsilon/2$ for $\omega = \omega_0$, and then decreases to zero.

A similar behavior to that of ε'' has tg δ_p, whose maximum value, which can be obtained from Eq. (7.22), is reached for $\omega_{max} > \omega_0$.

A typical example of this behavior is shown by the ice obtained from distilled water, whose characteristics at the temperature $\vartheta = -10\,°C$ are $\tau = 0.6 \cdot 10^{-4}$ s; $\varepsilon_1 = 3.5$; $\Delta\varepsilon = 78$.

To these values corresponds the diagram of Fig. 7.5, which shows the behavior of ε', ε'' and tg δ_p as a function of frequency.

In particular, it is:

for $\omega = \omega_0 \Rightarrow f_0 = \frac{\omega_0}{2\pi} = \frac{1}{2\pi\tau} = 2.7\,\text{kHz}$; $\varepsilon' = 3.5 + 39 = 42.5$; $\varepsilon'' = 39$

for $\omega = \omega_{max} \Rightarrow f_{max} = 13\,\text{kHz}$; tg $\delta_{p\,max} = 2.31$.

From a physical point of view the diagram of Fig. 7.5 can be explained as follows:

- At low frequency the semi-period of the applied electric field is greater than the time τ and the relaxation polarization can develop completely. In this situation the polarization vector $\dot{P}$ is in phase with the electric field $\dot{E}$, and hence it is $\varepsilon' = \varepsilon_1 + \Delta\varepsilon$ and tg $\delta_p = 0$.
- When the frequency increases, the polarization cannot be completed within a semi-period and therefore it lags on the applied electric field. In fact, the particles of dielectric material which are moving under the effect of the applied electric field are subjected to collisions due to thermal agitation which occur at times intervals much lower than the polarization time. These collisions hinder the preferential orientation of the particles and a sort of "molecular friction" is

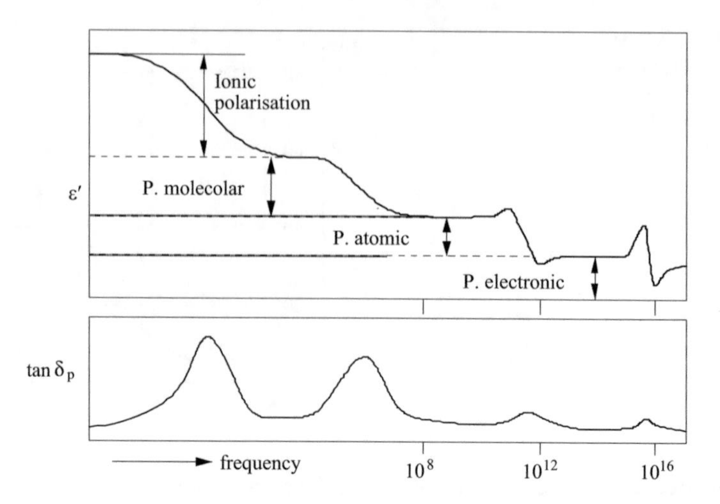

Fig. 7.6 Behavior of ε' and $\mathrm{tg}\,\delta_p$ as a function of frequency due to different polarization mechanisms

produced, which must be balanced by a certain amount of energy. This friction is responsible of a lag between the orientation of particles and the applied force.

This is the main reason of the lag between the polarization vector $\dot{P}$ and the applied electric field $\dot{E}$. This lag reaches the maximum value $\delta_{p\,max}$ for $\omega = \omega_{max}$.

- If the frequency further increases, the semi-period of the applied electric field becomes more and more shorter than τ and hence the orientation of particles is just beginning when the electric field inverts orientation. In this situation the relaxation polarization is practically negligible and it is present only the elastic polarization which occurs instantaneously. This condition is fully verified for $\omega = \omega_\infty$, where it is $\varepsilon' = \varepsilon_1$ and $\mathrm{tg}\,\delta_p = 0$.

In reality, most of the dielectric materials with relaxation polarization shows a qualitative trend of the loss factor as a function of frequency similar to that previously described, but with maximum values lower than $\Delta\varepsilon/2$.

This phenomenon can be explained taking into account that dielectric materials usually have not a single value of relaxation time, but rather a spectrum of values of τ and a probabilistic distribution of the elements of the structure that will relax in the relaxation time.

In addition, different types of polarization may occur in a dielectric at the same time, each characterized by very different relaxation times, so that in a given material, the behavior of ε' and $\mathrm{tg}\,\delta_p$ as a function of frequency can be qualitatively described by the diagram of Fig. 7.6.

7.3.3 Complex Dielectric Constant with Relaxation Polarization and Conduction

In the previous paragraphs only the effects of polarization on the dielectric constant have been considered.

However, in most cases in addition to such phenomena, the effects of the conduction current due to the material conductivity σ must be taken into account.

In this more general case the complex dielectric constant (7.19), must be rewritten in the form:

$$\dot{\varepsilon} = \varepsilon' - j\varepsilon''_e = \varepsilon_1 + \frac{\Delta\varepsilon}{1 + j\omega\tau} - j\frac{\sigma}{\omega\varepsilon_0}, \tag{7.23}$$

where ε' is still given by Eq. (7.19a), while ε''_e can be described by the equation:

$$\varepsilon''_e = \frac{\omega\tau \cdot \Delta\varepsilon}{1 + (\omega\tau)^2} + \frac{\sigma}{\omega\varepsilon_0} \tag{7.23a}$$

The second term of Eq. (7.23a) is responsible of a considerable increase of ε''_e and the total loss angle $\text{tg}\,\delta = \varepsilon''_e/\varepsilon'$ in the range of lower frequencies, where conduction current plays a role more important than relaxation in the heating of the material.

Therefore, in the industrial frequency range, the behavior of ε''_e can be described by the diagram of Fig. 7.7.

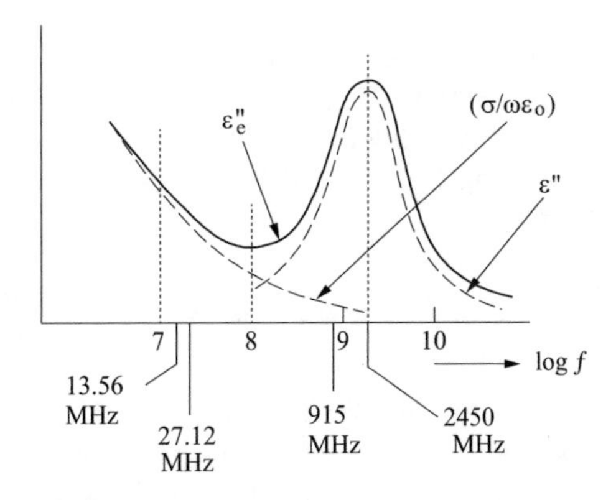

Fig. 7.7 Behavior of ε''_e in the range of industrial frequencies due to relaxation polarization and conduction

7.3.4 Influence of Temperature on Dielectric Characteristics

The influence of temperature on the characteristics of dielectric materials is due primarily to the following factors, both related to the increase of thermal motion with the temperature ϑ:

- the decrease of relaxation time τ when ϑ increases, with a law of the type:

$$\tau = a \cdot e^{b/\vartheta} \tag{7.24}$$

where a, b—are constant coefficients for a given type of relaxation polarization.

In fact, as the temperature increases, also the intensity of thermal motion increases and the material particles vary their equilibrium position most frequently, thus promoting the process of polarization.

- the decrease of the relaxation polarization $\dot{P}_r$ and the term $\Delta\varepsilon = \dot{P}_r/(\varepsilon_0\dot{E})$, which produce a corresponding decrease of permittivity.

In fact, the intensification of thermal motion hinders completion of the polarization process, since it tends to disturb the preferred orientation of bi-pole moments in direction of the electric field.

In addition to these phenomena, it must be taken into account the variation of conductivity σ which, having mostly ionic character, varies both in liquid and solid dielectrics with a law of the type:

$$\sigma = c \cdot e^{-d/\vartheta} \tag{7.25}$$

with c, d—material constants.

Equation (7.25) describes the fact that when the temperature increases, increases also the mobility of ions.

Taking into account Eqs. (7.24) and (7.25), Eq. (7.23a) can be written in the form:

$$\dot{\varepsilon} = \varepsilon' - j\varepsilon''_e = \varepsilon_l + \frac{\Delta\varepsilon}{1 + je^{\ln\omega} a e^{b/\vartheta}} - j\frac{c}{e^{\ln\omega} e^{d/\vartheta}\varepsilon_0} \tag{7.26}$$

It shows that the variations of $\dot{\varepsilon}$, ε', ε''_e, as a function of $(1/\vartheta)$, are similar to those already discussed in the previous paragraph as a function of $\omega (= e^{\ln\omega})$.

It can be concluded that the characteristics of the dielectric as a function of temperature, neglecting the variations of $\Delta\varepsilon$ are similar to those given in Fig. 7.5 as a function of frequency, but assuming the direction in which the frequency decreases from the right to the left.

In reality, taking into account that $\Delta\varepsilon$ decreases above a certain temperature $\vartheta_{\varepsilon\,\max}$, when the temperature increases ε' does not reach a constant saturation value, as it would be predicted by the curve of Fig. 7.5 looking it in direction of the frequency decrease but, after reaching a maximum at the temperature $\vartheta_{\varepsilon\,\max}$, it starts again to decrease toward the value ε_1, according to the curve shown in Fig. 7.8.

Figure 7.9 shows, as an example, some characteristics of vulcanized rubber, which highlight the influence of the simultaneous variations of temperature and frequency on the parameters of a dielectric material with relaxation polarization.

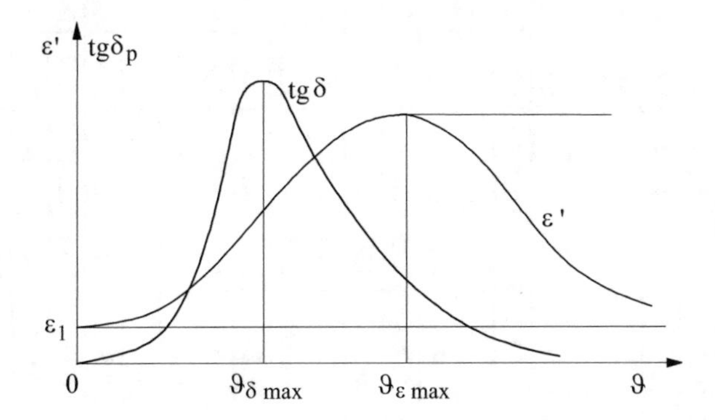

Fig. 7.8 Shape of relaxation polarization curve as function of temperature

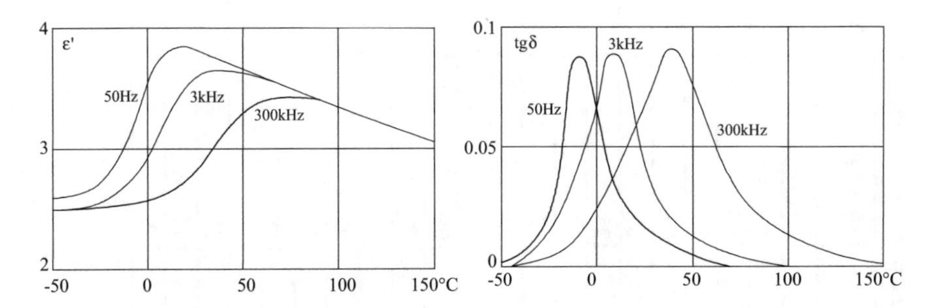

Fig. 7.9 Dielectric properties of vulcanized rubber with sulfur content 10 % as a function of temperature at different frequencies [1]

Table 7.1 HF dielectric characteristics of materials [2–4]

Material	Temperature (°C)	Moisture (%)	1 (MHz)		10 (MHz)		100 (MHz)		1.000 (MHz)		2.450–3.000 (MHz)	
			ε'	ε''_e	ε'	ε''_e	ε'	ε''_e	ε'	ε''_e	ε'	ε''_e
Ice	−12		4.25	0.50	3.7	0.067	–	–	–	–	3.2	0.003
Distilled water	1.5		87	1.65	87	0.174	87	0.609	–	–	80.5	24.96
Distilled water	15		81.7	2.53	–	–	81	–	–	–	78.8	16.15
Distilled water	25		–	–	–	–	–	–	–	–	76.7	12
Distilled water	65		64.8	5.61	–	–	64	–	–	–	64	4.89
Distilled water	95		55	7.87	–	0.725	52	0.174	–	–	52	2.44
Salted water	25		–	–	–	–	–	–	69	269	67	41.87
Porcelain	25		6.32	0.015	6.3	0.013	6.3	0.016	–	–	6.23	0.028
Nylon (610)	25		3.14	0.068	3.05	0.063	3.0	0.060	–	–	2.84	0.033
Nylon (610)	85		4.4	0.757	3.7	0.426	3.4	0.228	–	–	2.94	0.105
Nylon (FM10001)	25		–	–	3.24	0.07	–	–	3.06	0.043	3.02	0.036
PVC (QYNA)	20		2.88	0.046	2.87	0.033	2.85	0.023	–	–	2.84	0.016
PVC (QYNA)	96		3.3	0.244	2.8	0.140	2.7	0.086	–	–	2.6	–
PVC (VG5904)	25		4.3	0.602	3.7	0.407	3.3	0.22	–	–	2.94	0.10
PVC (VU1900)	25		3.3	0.29	2.95	0.165	2.8	0.087	–	–	2.65	0.035
Mylar	20		–	–	2.4	0.039	–	–	2.2	0.0088	–	–
Araldite (E134)	25		4.4	0.339	4.1	0.410	3.7	0.481	–	–	3.2	0.147
Araldite (adhesive)	25		3.71	0.111	3.46	0.121	3.27	0.111	–	–	3.14	0.072
Natural rubber	25		2.4	0.004	2.4	0.008	2.4	0.012	–	–	2.15	0.0065
India rubber	25		–	–	2.5	0.08	–	–	–	–	2.5	0.03

(continued)

Table 7.1 (continued)

Material	Temperature (°C)	Moisture (%)	1 (MHz)		10 (MHz)		100 (MHz)		1.000 (MHz)		2.450–3.000 (MHz)	
Neoprene	25		5.7	0.541	4.7	0.940	3.4	0.544	–	–	2.84	0.136
Plexiglass	27		–	–	2.71	0.027	–	–	2.66	0.017	2.6	0.015
Bakelite	25		–	–	4.3	0.18	–	–	–	–	3.7	0.15
Melamine	25		–	–	5.5	0.23	–	–	–	–	4.2	0.22
Glass (96 %SiO_2)	25		–	–	3.85	0.0023	–	–	–	–	3.84	0.0026
Glass (Borosilicate)	25		4.05	0.002	4.05	0.003	4.05	0.004	–	–	4.05	0.004
Glass Soda-Silica	25		5.4	0.07	–	–	5.1	0.051	–	–	5.05	0.066
Glass Pyrex	25		–	–	4.84	0.015	–	–	–	–	4.82	0.026
Wood (Fir)	25		1.93	0.05	1.9	0.06	1.88	0.062	–	–	1.82	0.049
Wood (Douglas)	20	0	–	–	–	–	–	–	2.05	0.04	2	0.02
Wet wood	25		–	–	2.6	0.1	–	–	–	–	2.1	0.07
Dry wood	25	0	–	–	2	0.04	–	–	–	–	1.9	0.01
Wool	25	20	–	–	1.2	0.01	–	–	–	–	–	–
Cotton	25	7	–	–	1.5	0.03	–	–	–	–	–	–
Paper	25		–	–	3.5	0.4	–	–	–	–	3.5	0.4
Paper (Royal Grey)	25		2.99	0.013*	2.86	0.163	2.77	0.183	–	–	2.7	0.151
Paper (Royal Grey)	80		3.31	0.076	3.14	0.138	3.08	0.194	–	–	2.94	0.235
Paper (78 g/m^3, E ∥ web)	22	7	–	–	3.1	0.25	–	–	–	–	3.2	0.5
Cardboard (230 g/m^3, E ∥ web)	22	5	–	–	2.8	0.3	–	–	–	–	2.7	0.3
Leather	25	0	3.2	0.089	3.1	0.093	3.1	0.118	–	–	–	–
Leather	25	15	5.6	0.784	4.9	0.49	4.5	0.45	–	–	–	–

(continued)

Table 7.1 (continued)

Material	Temperature (°C)	Moisture (%)	1 (MHz)		10 (MHz)		100 (MHz)		1.000 (MHz)		2.450–3.000 (MHz)	
Polyester	25		–	–	4.0	0.04	–	–	–	–	4.0	0.04
Polyethylene	25		–	–	2.25	0.0004	–	–	–	–	2.25	0.001
Polystirene	25		–	–	2.35	0.0005	–	–	–	–	2.55	0.0005
PTFE	25		–	–	2.1	0.0003	–	–	–	–	2.1	0.0003
Steak	25		–	–	50	1300*			50	39	40	12
Lean frozen steak			–	–	–	–	–	–	4.4	0.72	3.95	0.3

Note Different authors give considerably different values for materials denoted with (*)

7.3.5 Typical Values of Loss Factors

As discussed previously, the loss factor is not a well defined characteristic of a given material, since its value depends on frequency, temperature and moisture content.

Moreover, even small changes in chemical composition or molecular structure of the material can lead to big variations of the power transformed into heat, all other conditions being equal.

Therefore, the values given in Table 7.1 must be considered only as indicative and it is always required experimental verification in the installation prototyping.

The analysis of these data, allows us to make some interesting remarks:

- water has a very high loss factor, highly variable with temperature and physical state;
- materials with different moisture content are characterized by very different values of ε_e'' (see for example wood, leather, paper).

This can be explained by the fact that the water contained in "moist" materials is present in two different states: the so-called "free" water, which is present in material capillarities and cavities, and the water "bound" to other molecules or chemically "absorbed" and bound to the first unimolecular surface layer of the material.

The water in these two states has very different dielectric properties, as shown as a function of the frequency in Fig. 7.10, or the percent of moisture content U% in Fig. 7.11.

The last diagram shows that the behavior of ε_e'' as a function of U% is characterized by two distinct regions: in the first one (I), occurring at relatively low

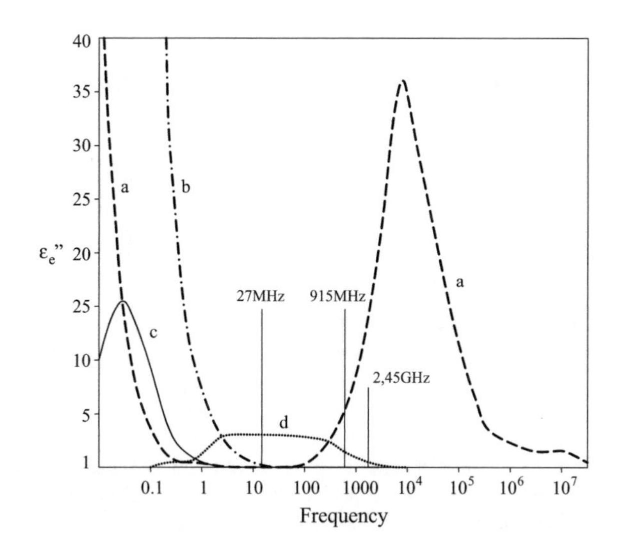

Fig. 7.10 Values of ε_e'' for water at room temperature as a function of frequency (**a** "free" water; **b** ionized; **c** crystallized; **d** "bound" water) [5]

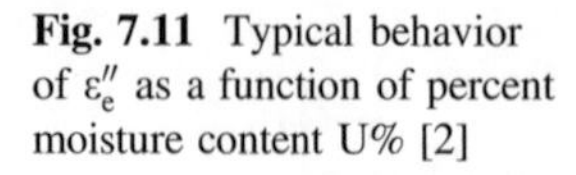

Fig. 7.11 Typical behavior of ε_e'' as a function of percent moisture content U% [2]

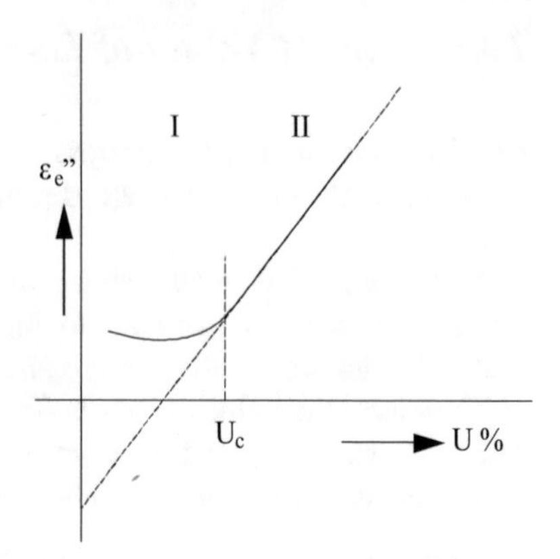

values of U%, the "bound" water plays a predominant role, while in the second one (II) the main influence is that of "free" water, which more easily follows the alternations of the applied field, thus producing more intense dielectric losses.

A typical example of the influence of the moisture content U% is shown in the diagrams of Fig. 7.12a, b, which give the loss factor of paper at two frequencies, with different orientations of the electric field with respect to the product to be heated.

The curves show that dielectric heating at 27 MHz with electric field parallel to the plane of the product, is particularly suited in order to obtain a uniform moisture content in cellulose-based products (such as paper or cardboard), because the most humid regions will absorb a higher specific power and then will be dried more quickly.

- Materials with the same commercial name but with different structure (like PVC and Araldite), may have very different loss factors.

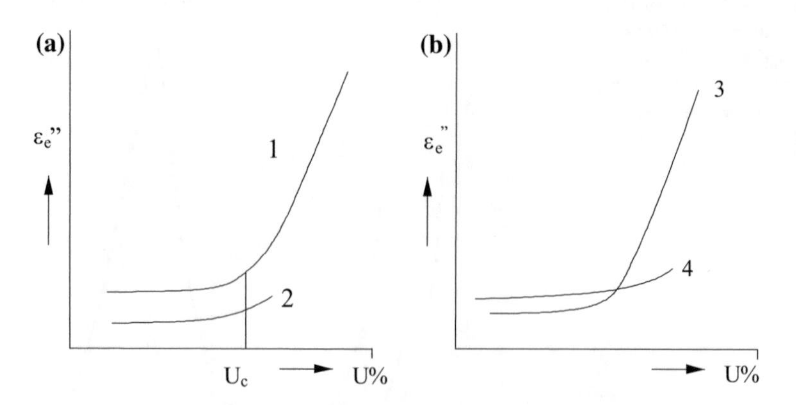

Fig. 7.12 Paper: values of ε_e'' as a function of moisture content of U% at various frequencies, with field parallel or perpendicular to the material (**a** 27.12 MHz: *1* parallel, *2* perpendicular; **b** field parallel: *3* 27.12 MHz, *4* 2450 MHz) [2]

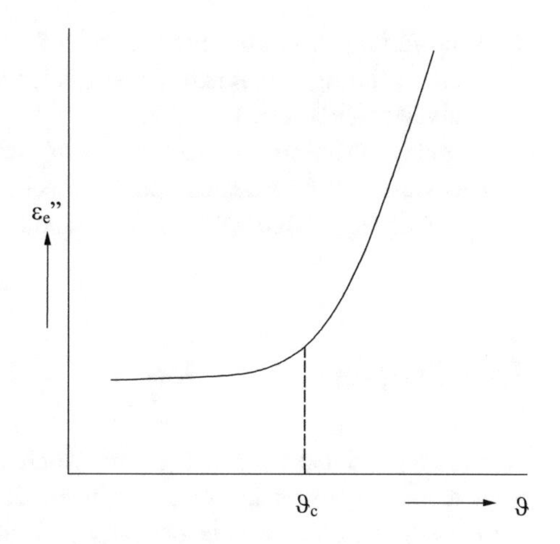

Fig. 7.13 Diagram of ε_e'' as a function of temperature [2]

- Materials with similar properties, such as rubber and neoprene, may differ greatly in the value of ε_e''. In this case the difference is due to the carbon content of neoprene. In other cases, small amounts of special additives allow to increase the loss factor in materials that otherwise would be unfit to high-frequency heating. This is the case, for example, of the addition of urea in the resins used for gluing wood laminates or NiO or C_2O_3 in refractories based on Al_2O_3.
- Particular attention must be paid to the change of ε_e'' with temperature. In fact, some materials at certain frequencies, show a behavior of the type sketched in Fig. 7.13. According to this figure, if during the heating process a portion or the mass of the product exceeds a critical temperature ϑ_c, a sudden increase of dissipated power will occur, which may "burn" or damage the heated material.

A similar problem may arise in the thawing of foodstuffs with microwaves, where at the passage of the water contained in the material from solid state (ice) to liquid one ("free" water) occurs a sharp increase of the product $(f \cdot \varepsilon_e'')$, as shown by the diagram of Fig. 7.14.

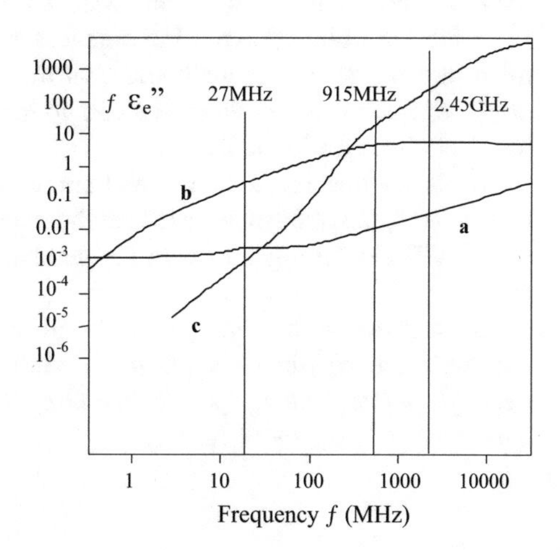

Fig. 7.14 Values of $f\,\varepsilon_e''$ as a function of frequency (**a** ice; **b** "bond" water; **c** "free" water) [5]

- The variations with temperature of ε' and ε''_e, will produce corresponding changes during the heating process of the resistance and capacitance of the equivalent circuit of Fig. 7.2.
- In many materials the decrease of ε''_e when the frequency increases does not necessarily imply a decrease of the specific power transformed into heat, since—as shows Eq. (7.10)—this is determined by the value of the product $(f \cdot \varepsilon''_e)$.

7.4 Frequency Selection

As shown previously, the parameters which can be selected by the designer in order to increase the power density in the material to be heated and reduce heating time, are the applied electric field intensity and the frequency.

The first one should be chosen as high as possible taking into account the dielectric strength of the material and the possible presence of air, moisture or vapors between the electrodes of the working capacitor.

Given that the dielectric strength of air is of the order of 30 kV/cm and that many processes are performed in presence of moisture or vapors, for safety reasons the applied field should not exceed 1–3 kV/cm.

Moreover, to avoid effluvia or sparks on electrodes, the voltage applied to them should not exceed 15 kV.

Considering these limits and the values of the product $(f \cdot \varepsilon''_e)$ in different materials, it is easy to verify that the only way to get the heating rates required in industrial applications is to use very high frequencies, in the range of tens or hundreds of MHz.

In general, however, for each material there is a large frequency range which can be used for carrying out the heating process. The final selection of the working frequency within this range must take into account several other factors, such as availability and cost of the HF source, position of the resonance peaks in the curve of the loss factor, size of the body to be heated and the frequency bands available for ISM (Industrial, Scientific and Medical) applications.

In fact, in order to avoid interferences with television broadcasts, radio communications systems for air traffic control or other electronic equipment, international regulations have allocated a certain number of frequency bands for ISM use, in which the radiation is free.

It follows that for economy and simplicity of construction of generators and installations, the frequencies used are those specified by such regulations, which are listed in Table 7.2 together with the corresponding bandwidth and wavelength in air.

Conventionally heating processes in the frequency range between 1 and 100 MHz are classified as *dielectric heating* (or *radio frequency heating*—RF heating), while heating processes using higher frequencies are designated as *microwave heating* (MW heating).

Table 7.2 Frequency bands for ISM applications

Frequency (MHz)	Bandwidth	Wavelength in air (m)
13.56	±0.05 %	22.12
27.12	±0.6 %	11.06
40.68	±0.05 %	7.37
915	±13 MHz	0.328
2.450	±50 MHz	0.122
5.800	±75 MHz	0.052
24.125	±125 MHz	0.012

This classification is based on the fact that in dielectric heating the wavelength is typically much larger than the size of the workpiece to be heated and that the two types of heating require the use of completely different frequency generators.

In practice, due to the larger bandwidth, the most used frequency for dielectric heating is 27.12 MHz, while in the microwave range, for the availability of power generators, the frequencies mostly used are 915 and 2450 MHz.

7.5 Electric Field Distribution in the Load

7.5.1 "Thermal" Penetration Depth

When the electromagnetic wave interacts with the surface of a dielectric, a part is reflected and a part is absorbed by the material, gradually attenuating inside it and heating the workpiece.

In the case of an indefinite half-plane such attenuation occurs with the exponential law:

$$E = E_s \, e^{-\gamma x}$$

with: E—electric field intensity at depth x from the surface; E_s—value of E at the surface; $\gamma = \alpha + j\,\beta$—propagation constant; α—attenuation coefficient; β—phase constant.

For values of $\mathrm{tg}\,\delta \ll 1$, a case often occurring in practice, the attenuation constant can be expressed by the approximate relationship [6]:

$$\alpha \approx \frac{\pi}{\lambda_0}\sqrt{\varepsilon'}\,\mathrm{tg}\,\delta \qquad (7.27)$$

with: λ_0—velocity of propagation in vacuum, equal to $3 \cdot 10^8$ m/s.

Assuming as conventional "*thermal penetration depth*" the distance x_T from the surface at which the power density reduces to 37 % (= 1/e) of its surface value, we obtain:

$$x_T = \frac{1}{2\,\alpha} \approx \frac{\lambda_0}{2\,\pi\sqrt{\varepsilon'}\mathrm{tg}\,\delta} = \frac{4.77 \cdot 10^7}{f\sqrt{\varepsilon'}\,\mathrm{tg}\,\delta}\,(\mathrm{m}). \tag{7.28}$$

Introducing in Eq. (7.28) the characteristic values of materials most commonly processed in dielectric heating, it can be easily verified that the thickness of the heat penetration depth is generally in the range from few meters to tens of meters. It follows that in most practical cases the heat sources can be considered uniformly distributed in the volume of the load material, if the surface applied field is uniform.

The situation is different in microwave heating where, due to the higher frequency used, the value of x_T is much lower and becomes one of the parameters that can affect the homogeneity of heating.

7.5.2 *Electric Field Intensity and Equivalent Parameters of Heterogeneous Charges*

In many cases the material inside the working capacitor is constituted by several workpieces to be heated simultaneously or by non-homogeneous materials, consisting of components with different dielectric properties.

In these cases, it is convenient to characterize the working capacitor by an equivalent dielectric constant, which takes into account the properties of the individual bodies and different components.

Often one of the above components consists of an airgap or air inclusions present, for design reasons or undesired, in addition to the material to be heated between the electrodes of the working capacitor.

The evaluation of the equivalent parameters of the working capacitor is based on the analysis of the electric field in an ellipsoid of dielectric constant ε_1 with semi-axes (*a*, *b*, *c*), placed in a dielectric medium with dielectric constant ε_2 and subjected to an uniform external electric field directed along the axis *x*, as illustrated in Fig. 7.15a, b [1].

If the direction of the field is along the axis (*a*) of the ellipsoid, the direction of the internal field E_1 is the same of E_0, while its modulus is bigger than E_0 for $\varepsilon_1 < \varepsilon_2$, or smaller for $\varepsilon_1 > \varepsilon_2$.

In this case it is:

$$E_1 = E_0 \frac{\varepsilon_2}{\varepsilon_2 + (\varepsilon_1 - \varepsilon_2)N} \tag{7.29}$$

with: N—dimensionless depolarization coefficient, which takes into account the presence of bound charges at the surface of separation of dielectrics and can be calculated with the relationship:

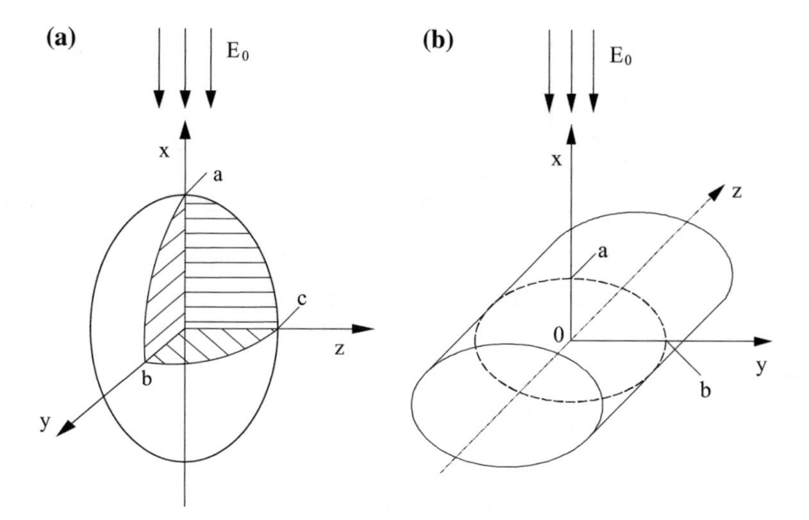

Fig. 7.15 Relative position of ellipsoid and applied electric field E_0 (**a** $a \neq b \neq c$; **b** $a \neq b$, $c \to \infty$)

$$N = \frac{a\,b\,c}{2}\int_0^{\infty}\frac{ds}{(s+a^2)\sqrt{(s+a^2)(s+b^2)(s+c^2)}}. \tag{7.29a}$$

From this general solution, the following particular cases can be obtained:

- Field inside a sphere in air ($\varepsilon_2 = 1$; $a = b = c$)

$$N = \frac{1}{3};\quad E_1 = E_0\frac{3}{\varepsilon_1 + 2} \tag{7.30}$$

- Field inside a cylinder in air, placed in a longitudinal electric field ($\varepsilon_2 = 1$; $b = c$; $a \to \infty$):

$$N = \frac{1}{2};\quad E_1 = E_0\frac{2}{\varepsilon_1 + 1} \tag{7.31}$$

- Field inside a cylinder with ellipsoidal cross-section in air, placed in a transversal electric field (as in Fig. 7.15b; $\varepsilon_2 = 1$; $c \to \infty$):

$$N = \frac{1}{2}\left(1 + \frac{b-a}{b+a}\right) \tag{7.32}$$

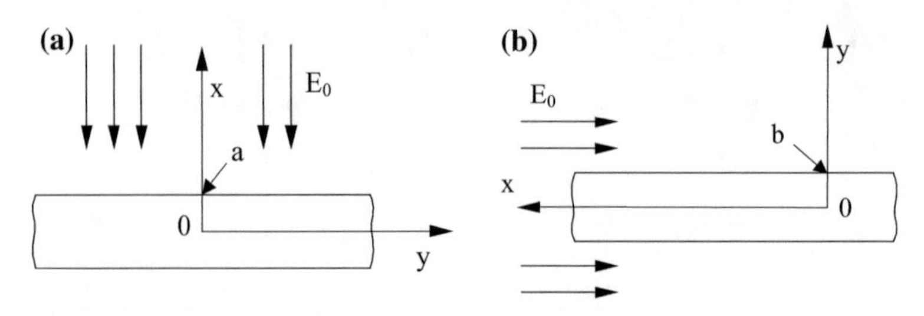

Fig. 7.16 Plane layer normal (**a**) or parallel (**b**) to the field E_0

From Eq. (7.32), we also obtain:

- In a plane layer with ($b \gg a$), normal to the applied electric field as in Fig. 7.16a:

$$N = 1; \quad E_0 = \varepsilon_1 E_1 \tag{7.32a}$$

- In a plane layer with ($a \gg b$), placed in the applied electric field as in Fig. 7.16b:

$$N = 0; \quad E_0 = E_1 \tag{7.32b}$$

Applying the previous equations, have been derived approximate formulas for the evaluation of the equivalent dielectric constant ε_e of workpieces of different shapes (spherical, cylindrical, flat, etc.), constituted by a base material of dielectric constant ε_2, which contains inclusions with dielectric constant ε_1 and volume concentration υ_1.

The volume concentration is defined as the ratio between the sum of the volumes of inclusions to the volume of the whole body.

Referring to the bibliography for the derivation of these formulas, we give in the following those of more practical interest [1].

- Workpiece with inclusions uniformly distributed in the base dielectric material:

$$\varepsilon_e = \varepsilon_2 \left[1 + \frac{(\varepsilon_1 - \varepsilon_2)\upsilon_1}{\varepsilon_2 + (\varepsilon_1 + \varepsilon_2)(1 - \upsilon_1)N} \right] \tag{7.33}$$

where N has the values previously given for the different shapes of inclusions.

Equation (7.33) can also be used for loads consisting of two or more dielectric plane layers with applied field E_0 normal or along the surfaces, as schematically illustrated in Fig. 7.17a, b.

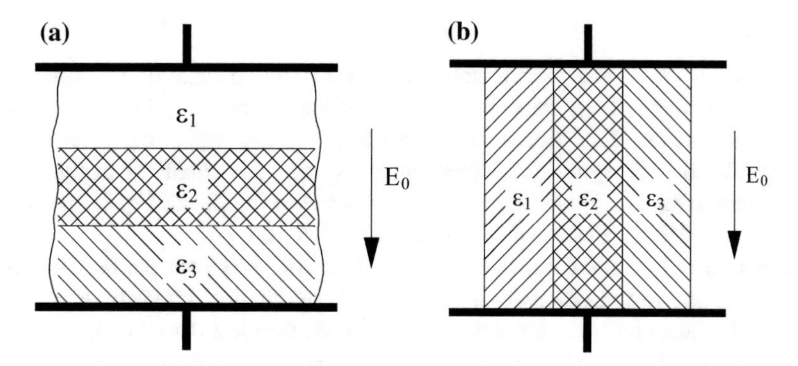

Fig. 7.17 Layered workpiece with applied field normal or along the surfaces of layers

- In case of an applied field normal to the layers, with N = 1 we obtain:
  - for two layers of thicknesses d_1 and d_2:

$$\frac{1}{\varepsilon_e} = \frac{\upsilon_1}{\varepsilon_1} + \frac{\upsilon_2}{\varepsilon_2} \quad \text{and} \quad \varepsilon_e = \frac{\varepsilon_1\,\varepsilon_2 d}{\varepsilon_1\,d_2 + \varepsilon_2\,d_1} \tag{7.34}$$

with: $\upsilon_2 = 1 - \upsilon_1$; $\upsilon_1 = d_1/d$; $\upsilon_2 = d_2/d$; $d = d_1 + d_2$.
According to the formulas for planar capacitors series connected, with obvious meaning of symbols, it is also:

$$\left.\begin{aligned} E_1 &= \frac{D_1}{\varepsilon_1} = \frac{Q}{S\,\varepsilon_1} = \frac{C_e\,V}{S\,\varepsilon_1} = \frac{\varepsilon_e\,V}{\varepsilon_1\,d} = \frac{\varepsilon_2}{\varepsilon_2\,d_1 + \varepsilon_1\,d_2}V \\ E_2 &= \frac{D_2}{\varepsilon_2} = \frac{\varepsilon_1}{\varepsilon_2\,d_1 + \varepsilon_1\,d_2} \end{aligned}\right\} \tag{7.34a}$$

  - for n layers of thickness d_k:

$$\frac{1}{\varepsilon_e} = \sum_{k=1}^{n} \frac{\upsilon_k}{\varepsilon_k} \tag{7.34b}$$

- In case of applied field parallel to the layers, from Eq. (7.33) with N = 0, it is:
  - for two layers

$$\varepsilon_e = \varepsilon_1\,\upsilon_1 + \varepsilon_2\,\upsilon_2 \tag{7.35}$$

  - for n layers:

$$\varepsilon_e = \sum_{k=1}^{n} \varepsilon_k\,\upsilon_k \tag{7.35a}$$

Remarks

1. Equations (7.34) and (7.35) show that a layered dielectric is anisotropic, since the value of ε_e depends on the direction of the applied field. This is the case for example of fibrous materials like textiles, where the fibers can be considered as cylindrical inclusions: when the field is in the fiber direction, Eq. (7.35) applies; when it is normal to the fibers, Eq. (7.34) is used with N = 1/2 for fibers with circular cross section.
2. It was already pointed out that in order to increase the power per unit volume, it is necessary to use relatively high values of electric field intensity. However, it must be taken into account that, when between the electrodes in addition to the materials to be heated are present air spaces, such as for example in continuous processes where the workpieces pass through the electrodes of the working capacitor on a conveyor belt, the field intensity in air is always greater than in the workpiece material [see Eqs. (7.30)–(7.32)], with possible occurrence of flashovers and damage of the installation or the heated product.

- Workpieces with inclusions random distributed in the base dielectric material:

$$\varepsilon_e = \frac{h + \sqrt{h^2 + 4\,\varepsilon_1\,\varepsilon_2(1 - N)N}}{2(1 - N)} \tag{7.36}$$

with:

$$h = \varepsilon_2 + (\varepsilon_1 - \varepsilon_2)\upsilon_1 - N(\varepsilon_1 + \varepsilon_2)$$

- Materials with two or more finely dispersed components with similar concentrations:

$$\text{– for two components: } \varepsilon_e = \varepsilon_1^{\upsilon_1} \cdot \varepsilon_2^{\upsilon_2} \tag{7.37}$$

$$\text{– for n components: } \varepsilon_e = \prod_{k=1}^{n} \varepsilon_k^{\upsilon_k} \tag{7.37a}$$

- Load constituted by cylindrical bodies with circular cross section (as in Fig. 7.18)

$$\varepsilon_e = \frac{\varepsilon_2\,d}{d - 2\,\pi\,m} \tag{7.38}$$

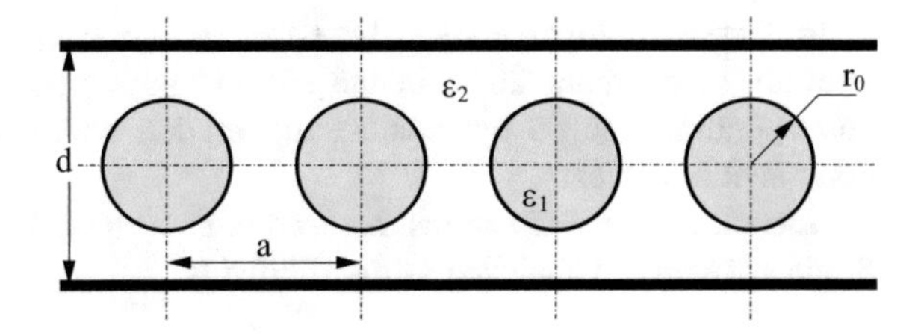

Fig. 7.18 Load constituted by cylindrical bodies of circular cross section

with:

$$m = \frac{3\,a\,r_0^2\,k_\varepsilon}{3\,a^2 + \pi^2\,r_0^2\,k_\varepsilon}; \quad k_\varepsilon = \frac{\varepsilon_1 - \varepsilon_2}{\varepsilon_1 + \varepsilon_2}$$

Example 7.1

(a) Heating at $f = 27.12$ MHz with power density $w = 3$ W/cm^3 of a dielectric workpiece with thickness $d_2 = 10$ cm, $\varepsilon' = 5$ and tg $\delta = 0.1$.

From Eq. (7.12a) we obtain:

$$E_2 = \sqrt{\frac{3}{2\pi 27.12 \cdot 10^6\, 8.86 \cdot 10^{-14}\, 0.5}} = 630.4\ \mathrm{V/cm}$$
$$V = E_2 \cdot d_2 = 630.4 \quad 10 = 6.304\ \mathrm{kV}$$

(b) Same workpiece with an air gap of thickness 2.5 cm.

In order to have the same thermal effect, the value of E_2 must be same as in (a). With reference to Eq. (7.34a) it must be:

$$E_1 = \varepsilon_2 \quad E_2 = 5 \cdot 630.4 = 3.152\ \mathrm{kV/cm}$$
$$V = E_1\, d_1 + E_2\, d_2 = 630.4 \quad 10 + 3152 \cdot 2.5 = 14.184\ \mathrm{kV}$$

Note—These values are near the limit of 10 % of dielectric strength of air and just below the limit of 15 kV suggested for the maximum voltage applied between the electrodes.

7.5.3 *Field Uniformity*

Given that the heat sources within the material vary with the square of the electric field intensity, [see Eq. (7.12a)], and that dielectric materials are generally characterized by very low thermal conductivity, the only way to achieve uniform heating is to obtain distributions as uniform as possible of the field in the load.

In fact, a difference of only 10 % in the electric field intensity leads to unevenness of about 20 % in the specific power per unit volume and hence a corresponding uneven temperature distribution within the load, if the heating is done in short times.

There are several situations that lead to uneven distributions of the electric field. Some of them are analyzed in the following.

7.5.3.1 Shape of the Load to be Heated

In case of loads with variable thickness as illustrated in the Fig. 7.19, uniform distribution of the electric field in the material can be achieved by adopting a suitable profile of the upper electrode, conveniently shaping the airgaps between the electrode and the material to be heated.

In the simplifying assumption of using also in this case Eq. (7.34a), with the notations of the figure and $\varepsilon_2 = 1$, we obtain:

$$E_{20} = \frac{1}{d_{10} + \varepsilon_1 d_{20}} V; \quad E_{2x} = \frac{1}{d_{1x} + \varepsilon_1 \, d_{2x}} V$$

Therefore, a nearly uniform distribution of the electric field intensity within the material can be obtained with:

$$d_{2x} = d_{20} + \frac{d_{10} - d_{1x}}{\varepsilon_1}. \tag{7.39}$$

However, a more accurate solution can be obtained only by the use of numerical methods which are able to take into account the actual distribution of field lines due to the edge effects of the working capacitor and the refraction phenomena occurring at the surfaces of separation between materials with different dielectric constants.

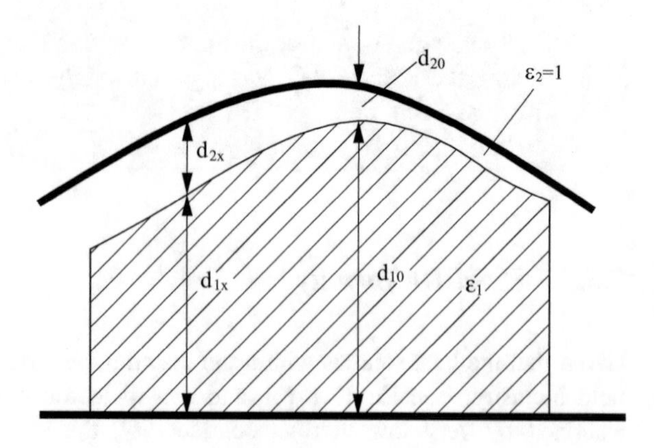

Fig. 7.19 Load with variable thickness

7.5.3.2 "Long" Electrodes

In some applications, like wood gluing, where the size of electrodes are comparable with the wavelength, the electrodes surfaces are no longer equipotential for the presence of standing waves. It is therefore necessary to evaluate the voltage distribution along the electrodes and the corresponding field strength in the material to be heated.

In many practical cases are used electrodes "long" in comparison with their width, fed at one end (or in one or more intermediate points).

These cases can be analyzed by using the equations of transmission lines with distributed parameters, assimilating the electrodes to a two-wire line of length L (as in Fig. 7.20), fed at the end (A–B) with sinusoidal voltage V of pulsation ω, and open at the other end (C–D).

Between the electrodes is placed a non-magnetic load with dielectric constant ε', which completely fills the space between them.

With reference to Fig. 7.20a, b and denoting with:

r, ℓ	resistance and inductance per unit length,
g, c	parallel conductance and capacitance per unit length,
$\dot{k} = \alpha + j\beta = \sqrt{(r + j\omega\ell)(g + j\omega c)}$	propagation constant ($\alpha > 0$; $\beta > 0$)
$\lambda_0 = 1/(f\sqrt{\ell c}) = 3 \cdot 10^8/f$	wavelength in vacuum (or in dry air),

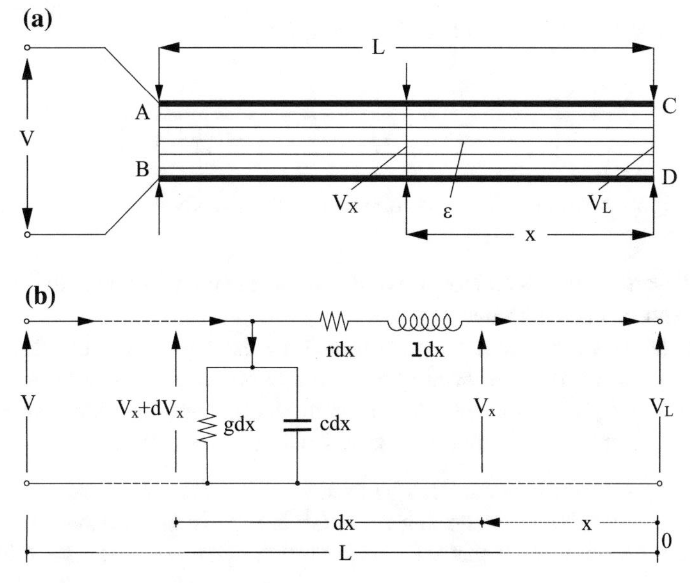

Fig. 7.20 "Long" electrodes (**a**) and electrical equivalent model (**b**)

$\lambda = 2\pi/\beta = \lambda_0/\sqrt{\varepsilon'}$ wavelength in the load material with dielectric constant ε',

$\upsilon = \omega/\beta = f \cdot \lambda$ phase velocity,

V supply voltage,

V_X voltage between electrodes at distance x from the open end,

V_L voltage at the open side,

as known it is:

$$\dot{V}_x = \dot{V}_L \cosh(\dot{k}\,x). \tag{7.40}$$

For x = L, from (7.40) we can write:

$$\dot{V}_L = \frac{\dot{V}}{\cosh(\dot{k}\,L)} \tag{7.40a}$$

Therefore an uneven distribution of the voltage occurs across the electrodes with maxima and minima at distance of λ/4. It depends on the electrodes length and the propagation constant.

In particular, for $r \ll \omega\ell$ and $g \ll \omega c$, it is:

$$\dot{k} = j\beta = j\omega\sqrt{\ell c}$$

$$\cosh(\dot{k}\,L) = \cosh(j\,\beta\,L) = \cos(\frac{2\pi}{\lambda}L) = \cos(\frac{2\pi}{\lambda_0}L\sqrt{\varepsilon'})$$

and Eqs. (7.40), (7.40a) will simplify as follows:

$$\frac{\dot{V}_x}{\dot{V}_L} = \frac{\cos(\frac{2\pi}{\lambda_0}\,x\,\sqrt{\varepsilon'})}{\cos(\frac{2\pi}{\lambda_0}\,L\,\sqrt{\varepsilon'})}; \quad \frac{\dot{V}_L}{\dot{V}} = \frac{1}{\cos(\frac{2\pi}{\lambda_0}\,L\,\sqrt{\varepsilon'})} \tag{7.41}$$

The diagrams of Fig. 7.21a, b, obtained with the above equations, show the following:

- for $L/\lambda < \pi/4$ there is an increase of the voltage from the supply side toward the open end of the electrode;
- since the specific power is proportional to the square of the electric field intensity, nearly uniform heating patterns can be obtained only by limiting the electrode length L to a few percent of the wavelength λ_0. For example, with $V_L/V < 0.9$ the temperature unevenness will be about 20 %.

The previous equations can also be used in case of electrodes supplied in their central point (Fig. 7.22a), assuming as length L the distance between the feeding point and the open end. In this way, with same frequency and the same length of

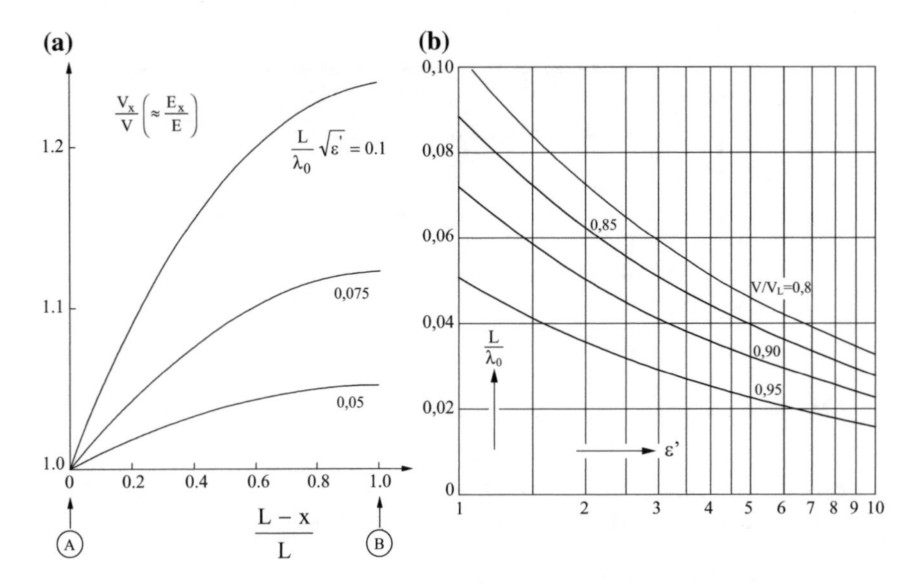

Fig. 7.21 **a** Voltage distribution along electrodes between the supplied end (*A*) and the open end (*B*); **b** values of the ratio L/λ_0 as function of ε' for prefixed values of V/V_L

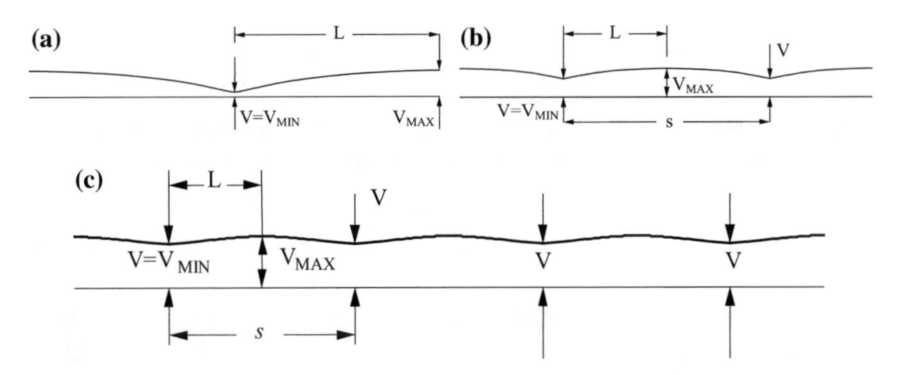

Fig. 7.22 Distributions of electric field in case of voltage supply at different intermediate points of electrodes (**a** supply in the central point; **b**, **c** supply in two and four intermediate points)

electrodes, it is possible to reach a more uniform voltage distribution, or—with the same ratio V_L/V—to use a double frequency.

This result is easily extended to the case where the same voltage is supplied to several intermediate points at distance *s*, introducing in Eq. (7.41) the length $L = s/2$. The resulting distributions of the electric field along the electrode are shown in Fig. 7.22b, c in case of supply at 2 or 4 intermediate points, respectively.

Since the supply of several intermediate points with voltages of the same amplitude and phase may be difficult, sometimes it is used the solution schematically shown in Fig. 7.23, whereby between the electrodes are connected the

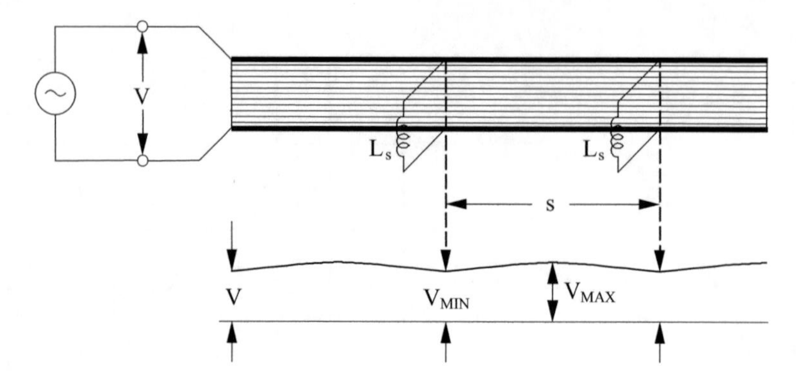

Fig. 7.23 Quasi-uniform electrode voltage distribution obtained with "tuning stubs"

inductances L_S (*"tuning stubs"*), spaced of a distance suitable to create a condition of parallel resonance, which provides at the point of connection the same voltage of the supply.

According to the Eq. (7.41) the distance *s* between two adjacent inductances can be expressed, as a function of the ratio of maximum to minimum voltage along the electrode, by the relationship:

$$s = \frac{\lambda_0}{\pi\sqrt{\varepsilon'}}\cos^{-1}\left(\frac{V_{min}}{V_{max}}\right) = \frac{3 \cdot 10^8}{\pi f \sqrt{\varepsilon'}}\cos^{-1}\left(\frac{V_{min}}{V_{max}}\right) \tag{7.42}$$

Since the set of N inductances L_S must give parallel resonance conditions with the total capacity C of the working capacitor, each of them must have the value:

$$L_S = \frac{N}{(2\pi f)^2 C} \tag{7.42a}$$

This method has the advantage of an easy experimental setup; however, it produces significant energy losses in the inductance coils, resulting in problems regarding their design and cooling.

7.5.3.3 Large Planar Electrodes

As a typical example of large plane electrodes with dimensions comparable with the wavelength, we consider a working capacitor with circular electrodes and cylindrical load, fed at its midpoint as shown in Fig. 7.24.

The distribution of the electric field intensity within the load can be obtained from Maxwell's equations, assuming sinusoidal quantities and electric field lines perpendicular to the capacitor plates.

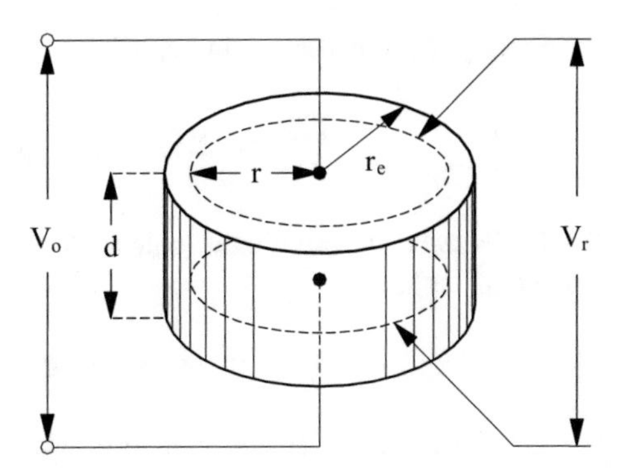

Fig. 7.24 Large plane circular electrodes with cylindrical load

In such hypotheses, with reference to a cylindrical coordinates system (*r*, *φ*, *z*), are non-null only the components $\dot{H}_\varphi$ of the magnetic field and $\dot{E}_z$ of the electric field, which for the system symmetry are functions only of the coordinate *r*. Thus in the Maxwell's equations:

$$\left.\begin{aligned} \mathrm{rot}\,\dot{\bar{H}} &= (\sigma + j\,\omega\,\varepsilon_0\,\varepsilon')\,\dot{\bar{E}} \\ \mathrm{rot}\,\dot{\bar{E}} &= -j\,\omega\,\mu_0\,\mu\,\dot{\bar{H}} \end{aligned}\right\} \tag{7.43}$$

it is:

$$\dot{\bar{H}} \equiv (0, \dot{H}_\varphi, 0); \quad \dot{\bar{E}} = (0, 0, \dot{E}_z).$$

Introducing the curl components:

$$\left[\mathrm{rot}\,\dot{\bar{H}}\right]_z = \frac{\dot{H}_\varphi}{r} + \frac{\partial\,\dot{H}_\varphi}{d\,r}; \quad \left[\mathrm{rot}\,\dot{\bar{E}}\right]_\varphi = -\frac{\partial\,\dot{E}_z}{d\,r}$$

and omitting in the following the subscripts φ of $\dot{H}_\varphi$ and z of E_z, we obtain:

$$\left.\begin{aligned} &\dot{E}(\sigma + j\,\omega\,\varepsilon_0\,\varepsilon') = \frac{\dot{H}}{r} + \frac{\partial\,\dot{H}}{d\,r} \\ &\frac{\partial\,\dot{E}}{d\,r} = j\,\omega\,\mu_0\,\mu\,\dot{H} \end{aligned}\right\}, \tag{7.44}$$

from which we can write:

$$\dot{H} = \frac{1}{j\,\omega\,\mu_0\,\mu}\frac{\partial\,\dot{E}}{d\,r}; \quad \frac{\partial\,\dot{H}}{d\,r} = \frac{1}{j\,\omega\,\mu_0\,\mu}\frac{\partial^2\dot{E}}{d\,r^2}. \tag{7.45}$$

Finally, by substitution in Eq. (7.44), we have:

$$\frac{\partial^2 \dot{E}}{d r^2} + \frac{1}{r}\frac{\partial \dot{E}}{d r} - j\,\omega\,\mu_0\,\mu(\sigma + j\,\omega\,\varepsilon_0\,\varepsilon')\dot{E} = 0. \tag{7.46}$$

By introducing the dimensionless parameter of the radial position $\xi = r/r_e$ and denoting with:

$$\beta^2 = -j\,\omega\,\mu_0\,\mu(\sigma + j\,\omega\,\varepsilon_0\,\varepsilon')r_e^2,$$

Eq. (7.46) can be rewritten as follows:

$$\frac{\partial^2 \dot{E}}{d\,\xi^2} + \frac{1}{\xi}\frac{\partial \dot{E}}{d\,\xi} + \beta^2 \dot{E} = 0. \tag{7.46a}$$

As known, the general solution of Eq. (7.46a) is:

$$\dot{E} = \dot{C}_1\, J_0(\beta\xi) + \dot{C}_2\, Y_0(\beta\xi) \tag{7.47}$$

with: J_0, Y_0—Bessel functions of order zero; $\dot{C}_1$, $\dot{C}_2$—integration constants.
From Eq. (7.45) it is also:

$$\dot{H} = \frac{1}{j\,\omega\mu_0\mu\, r_e}\frac{\partial \dot{E}}{d\,\xi} = -\frac{\beta}{j\,\omega\mu_0\mu\, r_e}\left[\dot{C}_1\, J_1(\beta\xi) + \dot{C}_2 Y_1(\beta\xi)\right]. \tag{7.48}$$

The determination of the constants $\dot{C}_1$, $\dot{C}_2$ is done by using the boundary conditions:

$$\dot{E} = \dot{E}_0 = \frac{\dot{V}_0}{d}; \quad \dot{H} = 0 \quad (\text{for}\,\xi = 0) \tag{7.48a}$$

Taking into account that $J_0(0) = 1$ and $Y_0(0) \to \infty$, from Eq. (7.47) it is $\dot{C}_2 = 0$ and $\dot{C}_1 = \dot{E}_0 = \dot{V}_0/d$.
Finally Eqs. (7.47) and (7.48) can be written as follows:

$$\left.\begin{aligned} \dot{E} &= \dot{E}_0\, J_0\,(\beta\,\xi) \\ \dot{H} &= -\frac{\beta}{j\omega\mu_0\mu\, r_e}\dot{E}_0\, J_1(\beta\xi) \end{aligned}\right\}. \tag{7.49}$$

Considering that β is a generic complex number, for the calculation of the distributions of $\dot{E}$ and $\dot{H}$ it can be convenient to simplify Eq. (7.49) by expressing them through Bessel functions of real argument, by using the following series expansions:

$$J_0(z) = 1 - \frac{z^2}{2^2} + \frac{z^4}{(2\cdot 4)^2} - \frac{z^6}{(2\cdot 4\cdot 6)^2} + \cdots$$
$$J_1(z) = \frac{z}{2} - \frac{z^3}{2^2\cdot 4} + \frac{z^5}{(2\cdot 4)^2\cdot 6} - \cdots$$

Denoting with:

$$\dot{\beta}^2 = -j\,\omega\,\mu_0\,\mu(\sigma + j\,\omega\,\varepsilon_0\,\varepsilon')r_e^2 = \omega^2\mu_0\mu\,\varepsilon_0\,\varepsilon'(1 - j\frac{\sigma}{\omega\,\varepsilon_0\varepsilon'})r_e^2 =$$
$$= \left(\frac{2\,\pi\,r_e}{\lambda}\right)^2(1 - j\,\mathrm{tg}\,\delta) = k^2(1 - j\,\Delta)$$

with:

$$k = \frac{2\,\pi\,r_e}{\lambda} = \frac{2\,\pi\,r_e}{\lambda_0}\sqrt{\varepsilon'} = \frac{2\,\pi\,r_e}{3\cdot 10^8}f\sqrt{\varepsilon'};\quad \Delta = \mathrm{tg}\,\delta = \frac{\sigma}{\omega\,\varepsilon_0\varepsilon'},$$

and neglecting the terms containing Δ^2 in comparison with those with Δ, we have:

$$(\dot{\beta}\,\xi)^2 = (k\,\xi)^2(1 - j\,\Delta);$$
$$(\dot{\beta}\,\xi)^4 \approx (k\,\xi)^4(1 - j\,2\Delta);$$
$$(\dot{\beta}\,\xi)^6 = (k\,\xi)^6(1 - j\,3\Delta);\ldots.$$

Introducing the above series expansions and separating real and imaginary terms, Eq. (7.49) can be written in the form:

$$\left.\begin{aligned} \dot{E} &= \dot{E}_0\left[J_0(k\,\xi) + j\,\Delta\tfrac{k\,\xi}{2}J_1(k\,\xi)\right] \\ \left|\frac{\dot{E}}{\dot{E}_0}\right| &= \sqrt{J_0^2(k\,\xi) + (\Delta\tfrac{k\,\xi}{2})^2 J_1^2(k\,\xi)} \end{aligned}\right\}. \tag{7.50}$$

From it we also obtain:

$$\begin{aligned} \dot{H} &= \frac{1}{j\,\omega\,\mu_0\mu\,r_e}\frac{\partial\dot{E}}{d\,\xi} = \\ &= \frac{k}{\omega\,\mu_0\mu\,r_e}\dot{E}_0\left[\Delta\frac{k\,\xi}{2}J_0(k\,\xi) + jJ_1(k\,\xi)\right]. \end{aligned} \tag{7.51}$$

The curves of Fig. 7.25 give the values of the ratio $|\dot{E}/\dot{E}_0|$ as a function of the product $(k\xi)$ in the cases $\Delta = 0$ and $\Delta = 1$.

Considering that only in very few practical applications $\Delta(= \mathrm{tg}\,\delta)$ exceeds the unity, the diagrams allow to draw the following conclusions:

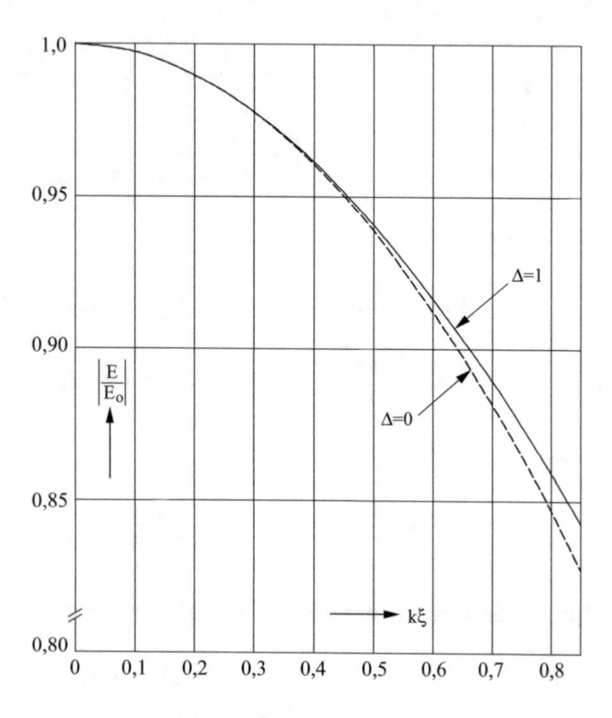

Fig. 7.25 Electric field distribution along radius of electrode as a function of kξ

- the electric field intensity decreases gradually from the supply central point ($\xi = 0$) towards the edge of the electrode ($\xi = 1$).
- sufficiently uniform electric field distributions (and corresponding uniform heating patterns) can be achieved only with relatively small values of k. For example, with k = 0.65, at the edge of the electrode (for $\xi = 1$) it is $|E/E_0| \approx 0.9$. This corresponds to unevenness of the distribution of the heat sources of approximately 20 %.
- given the definition of k and the values of ε', this means that a sufficiently uniform heating pattern can be obtained only when the value of the ratio

$$\frac{r_e}{\lambda_0} = \frac{k}{2\,\pi\sqrt{\varepsilon'}}$$

is in the range of few percent.

In the previous example (k = 0.65), with $\varepsilon' = 3$, it is $r_e/\lambda_0 \approx 0.06$ and then, at 27.12 MHz, $r_e \approx 0.66$ m.

- for $|E/E_0|$ greater than 0.85–0.9 (i.e. quasi-uniform distributions), the second term of Eq. (7.50) can be neglected in comparison with the first one, and we can write:

$$E \approx E_0 J_0(k\xi) \tag{7.50a}$$

The active (P_c) and reactive power (Q_c) in the dielectric can be calculated respectively as the real and imaginary parts of the flux of Poynting's vector through the lateral surface of the cylinder (of radius r_e and thickness d).

Calculating $\dot{E}$ and $\dot{H}^*$ by Eqs. (7.50) and (7.51) evaluated at $\xi = 1$ and taking into account their orthogonality, for the active and reactive power we obtain:

$$\begin{aligned} P_c + j Q_c &= \dot{E}_1 \dot{H}_1^* 2\pi r_e d = \\ &= 2\pi r_e d \frac{k}{\omega\mu_0\mu r_e} \dot{E}_0^2 \cdot \left[J_0(k) + j\Delta\frac{k}{2}J_1(k)\right] \cdot \left[\Delta\frac{k}{2}J_0(k) - jJ_1(k)\right] = \\ &= d\frac{2\pi k}{\omega\mu_0\mu}\dot{E}_0^2 \left\{\frac{\Delta k}{2}\left[J_0^2(k) + J_1^2(k)\right] - jJ_0(k)J_1(k)\left[1 - \left(\frac{\Delta k}{2}\right)^2\right]\right\}, \end{aligned}$$

which can be re-written in the compact form:

$$P_c + j Q_c = (d \cdot \pi r_e^2) \cdot \omega E_0^2 \varepsilon_0 \varepsilon' \cdot (\text{tg}\,\delta \cdot P - j Q) \tag{7.52}$$

with:

$$\left.\begin{aligned} P &= \left[J_0^2(k) + J_1^2(k)\right]; \\ Q &= \tfrac{2}{k}\left[1 - \left(\tfrac{\Delta k}{2}\right)^2\right] J_0(k)J_1(k) \approx \tfrac{2}{k} J_0(k)J_1(k) \end{aligned}\right\} \tag{7.52a}$$

The comparison of Eqs. (7.12) and (7.52) shows that they differ only for the coefficients P and Q, which take into account the non-uniform radial field distribution.

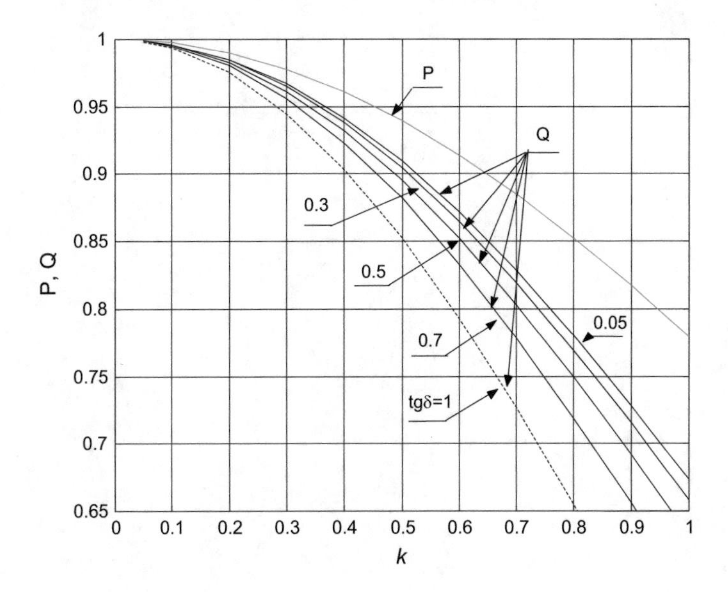

Fig. 7.26 Coefficients P and Q for calculation of active and reactive power with Eq. (7.52)

The values of the coefficients P and Q are given, as a function of k and $\Delta = \text{tg}\,\delta$, in the diagrams of Fig. 7.26.

Example 7.2 Heating at 27.12 MHz of a cylindrical load, of radius $r_e = 50$ cm and thickness d = 10 cm, with specific power w = 3 W/cm^3 and system geometry as in Fig. 7.24. Load material is the same as in Example 7.1.

(a) *Assumption of an ideal working capacitor*

In order to obtain the same thermal effect as in the Example 7.1, the applied voltage must be V = 6.304 kV.

It is therefore:

– Active Power:

$$P_c = \pi r_e^2 d w = \pi \cdot 50^2 \cdot 10 \cdot 3 = 235.62 \text{ kW}$$

– Parameters of the parallel equivalent circuit (see Sect. 7.4):

$$R_p = \frac{V^2}{P_c} = \frac{(6.304 \cdot 10^3)^2}{235.62 \cdot 10^3} = 168.66\ \Omega$$

$$C = \varepsilon_0 \varepsilon' \frac{\pi r_e^2}{d} = 8.86 \cdot 10^{-14}\, 5 \frac{\pi 50^2}{10} = 347.93 \cdot 10^{-12} \text{ F}$$

$$X_p = \frac{1}{\omega C} = \frac{1}{2\pi 27.12 \cdot 10^6\, 347.93 \cdot 10^{-12}} = 16.867\ \Omega$$

– Parameters of the series equivalent circuit [Eq. (7.14)]:

$$R_s = \frac{1}{\omega C} \frac{\text{tg}\,\delta}{1 + \text{tg}\,\delta^2} = 16.867 \frac{0.1}{1 + 0.1^2} = 1.67\ \Omega$$

$$X_s = \frac{1}{\omega C} \frac{1}{1 + \text{tg}\,\delta^2} = 16.867 \frac{1}{1 + 0.1^2} = 16.7\ \Omega$$

– Current:

$$I = \frac{V}{\sqrt{R_s^2 + X_s^2}} = \frac{6.304 \cdot 10^3}{\sqrt{1.67^2 + 16.7^2}} = 375.61 \text{ A}$$

(b) *Calculation of the same configuration taking into account the radial distribution of electric field*

$$k = \frac{2\pi r_e}{\lambda_0} \sqrt{\varepsilon'} = \frac{2\pi 50}{11.06 \cdot 10^2} \sqrt{5} = 0.635$$

$$J_0(k) = 0.90171; \quad J_1(k) = 0.30176$$
$$J_0(k)^2 + J_1(k)^2 = 0.90414; \quad J_0(k)\ J_1(k) = 0.27210$$

– Active and reactive power:

$$P_c + jQ_c =$$
$$= \frac{dk}{f\,\mu_0\mu}\dot{E}_0^2\left\{\frac{\Delta k}{2}\left[J_0^2(k) + J_1^2(k)\right] - jJ_0(k)J_1(k)\left[1 - \left(\frac{\Delta k}{2}\right)^2\right]\right\} =$$
$$= \frac{10 \cdot 0.635}{27.12 \cdot 10^6 \cdot 4\,\pi \cdot 10^{-9}}\left\{\frac{0.1 \cdot 0.635}{2} 0.90414 - j0.27210\left[1 - \left(\frac{0.1\,0.635}{2}\right)^2\right]\right\} =$$
$$= 0.21297 \cdot 10^6 - j\,2.0166 \cdot 10^6$$

– Parallel equivalent circuit:

$$R_p = \frac{V^2}{P_c} = \frac{(6.304 \cdot 10^3)^2}{212.97 \cdot 10^3} = 186.6\ \Omega$$
$$X_p = \frac{1}{\omega C} = \frac{V^2}{Q_c} = \frac{(6.304 \cdot 10^3)^2}{2.0166 \cdot 10^6} = 19.707\ \Omega$$

– Series equivalent circuit:

$$R_s = R_p \frac{X_p^2}{R_p^2 + X_p^2} = R_p \frac{1}{1 + \left(\frac{R_p}{X_p}\right)^2} = 2.058\ \Omega$$
$$X_s = X_p \frac{R_p^2}{R_p^2 + X_p^2} = X_p \frac{1}{1 + \left(\frac{X_p}{R_p}\right)^2} = 19.489\quad \Omega$$

– Current:

$$I = \frac{V}{\sqrt{R_s^2 + X_s^2}} = \frac{6.304 \cdot 10^3}{\sqrt{2.058^2 + 19.489^2}} = 321.67\ A$$

– Active power:

$$R_s\,I^2 = 2.058 \quad 321.67^2 = 212.97\ kW$$

7.6 Calculation of the Working Capacitor

The calculation of parameters of the equivalent circuit of the working capacitor, where is placed the material to be heated, requires the evaluation of the total active and reactive power.

Given the variety of geometries of electrodes and bodies to be heated, the presence of heterogeneous loads with different values of ε' and tg δ and the edge effects of the working capacitor, an accurate calculation can be made only by means of 3-D numerical methods.

However, an approximate calculation of the equivalent circuit of the loaded working capacitor, which takes into account the capacitor edge effects, can be made on the basis of an idealized schematization of the electric field, neglecting the deformation of field lines at the surface of separation between regions with different dielectric constants ("*load edge effect*") and, in particular, assuming negligible the flux of the vector D through the lateral surface of the body to be heated.

Figure 7.27a schematically shows the working capacitor and the workpiece to be heated, the real pattern of field lines (solid lines, on the left) and their idealized pattern (dashed lines, on the right). Moreover, the figure shows an ideal

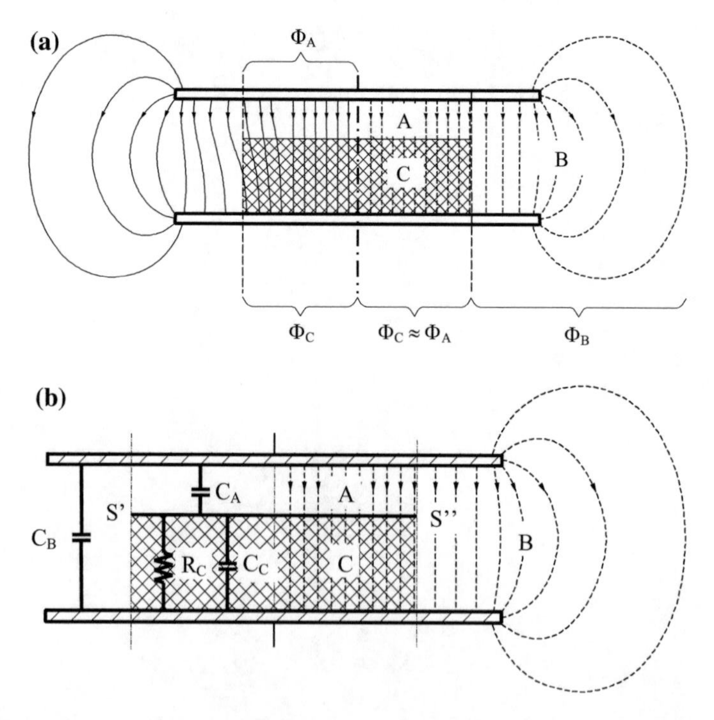

Fig. 7.27 Working capacitor: **a** real (*solid lines*) and ideal (*dashed lines*) patterns of electric field; **b** subdivision regions A, B, C and equivalent circuit

subdivision of the capacitor space into three regions A, B, C and the series and parallel connected elements of the equivalent circuit (Fig. 7.27b).

The idealized pattern allows us to make the following simplifying assumptions:

(a) equality of the fluxes $\dot{\Phi}_A$ and $\dot{\Phi}_C$ of vector D in the air region A above the load and in the load region C;
(b) the surface $S'-S''$ of the load is equipotential;
(c) to consider separately the flux $\dot{\Phi}_A \approx \dot{\Phi}_C$ through the load and the one $\dot{\Phi}_B$ in the space external to the capacitor edges;
(d) equality of the flux $\dot{\Phi}_B$ of the loaded capacitor and the one in the same capacitor without load.

The following remarks can be made:

1. the hypothesis (a) fits better with the reality the smaller is the distance between electrodes and the lower the dielectric constant of material to be heated. In fact, on one hand the deformation of the field at the edge of the dielectric modifies the local distribution of the internal heat sources but, on the other hand, it has a small influence on the integral parameters of the system, such as the resistance and the capacity of the equivalent circuit.
2. the hypothesis (d) is similar to that used in the derivation of the approximate equivalent electrical circuit corresponding to the magnetic one in induction heating systems. In that case (see Sect. 6.9.1) it was assumed that the magnetic field pattern in the regions "external" to the load was the same with or without load material.
3. at high frequency it is generally negligible the own resistance of electrodes of the working capacitor, which depends on the electrode shape and the point of connection to the supply.

The calculation of parameters of the equivalent circuit of Fig. 7.27b, in case of a working capacitor with plane electrodes, then proceeds as follows:

- Capacitance C_0 of the working capacitor without the workpiece:

$$C_0 = \varepsilon_0 \frac{S_0}{d} k_B \tag{7.53}$$

with: S_0—surface of electrodes, d—distance between electrodes; k_B—coefficient taking into account the capacitor edge effects.

The values of k_B are given in Table 7.3 for circular electrodes of radius r_e, square electrodes of side a and rectangular electrodes of sides a, b [1].

Table 7.3 Values of k_B

(a) Plane circular electrodes with radius r_e

d/r_e	0	0.1	0.4	0.6	0.8	1	1.2	1.5	2	2.5	3	5	10	20
k_B	1	1.175	1.580	1.830	2.075	2.318	2.561	2.920	3.531	4.149	4.767	7.268	13.59	26.28

(b) Plane square electrodes with side a

d/a	0	0.005	0.025	0.05	0.1	0.2	0.5	1
k_B	1	1.003	1.071	1.148	1.281	1.499	2.121	3.168

(c) Plane rectangular electrodes with sides a, b

d/a	2	1.5	1.2	1.18	1	0.667	0.5	0.376	0.333	0.3	0.25	0.2	0.196	0.167
d/b	3.53	2	2	3.53	2	2	1	0.5	1	0.5	0.5	0.333	0.59	0.333
k_B	6.03	4.62	4.27	5.46	4	3.34	2.55	1.96	2.36	1.89	1.82	1.63	1.86	1.59

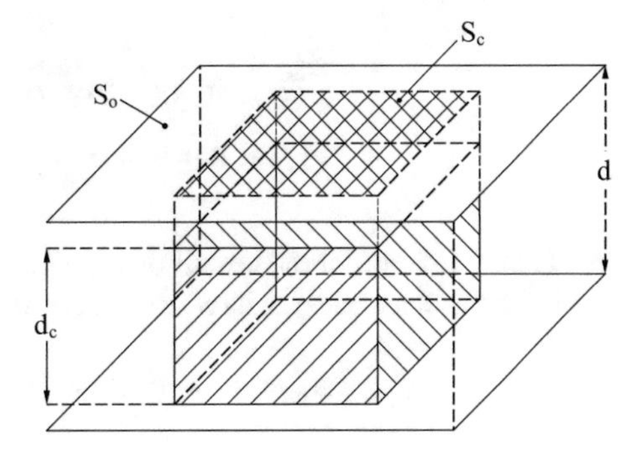

Fig. 7.28 Working capacitor with the workpiece to be heated

- Capacitance of the central part of the working electrode of surface S_C equal to upper surface of the load (Fig. 7.28):

$$C_0^* = \varepsilon_0 \frac{S_C}{d} \tag{7.54}$$

- Edge (or "external") capacitance and reactance:

$$C_B = C_0 - C_0^*; \quad X_B = 1/\omega C_B \tag{7.55}$$

- Capacitance and reactance of the central region in air of the working electrode

$$C_A = \varepsilon_0 \frac{S_C}{d_a}; \quad X_A = \frac{1}{\omega C_A} \tag{7.56}$$

- Capacitance C_C, reactance X_C and resistance R_C of the central part C inside the workpiece (Fig. 7.27b):

$$C_C = \varepsilon_0 \varepsilon' \frac{S_C}{d_c}; \quad X_C = \frac{1}{\omega C_C}; \quad R_C = \frac{1}{\omega C_c \operatorname{tg} \delta} \tag{7.57}$$

- Series reactance X_{SC} and resistance R_{SC} of the load obtained by substituting the parallel branch of Fig. 7.29a with the equivalent series one (Fig. 7.29b), using Eq. (7.14):

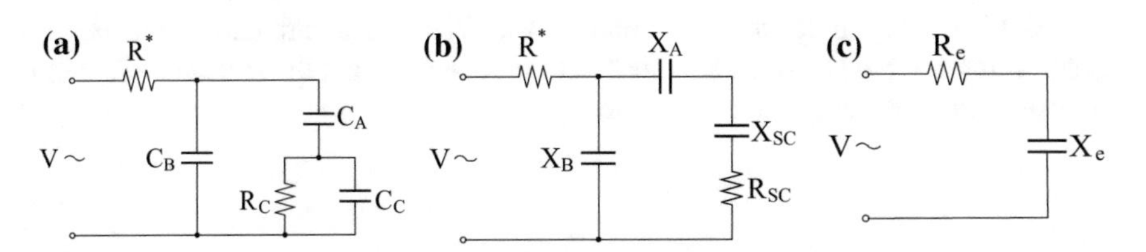

Fig. 7.29 Equivalent circuits of the working capacitor

$$X_{SC} = \frac{1}{\omega C_C(1+tg^2\delta)} = \frac{d_C}{\omega \varepsilon_0 \varepsilon' S_C(1+tg^2\delta)}$$
$$R_{SC} = X_{SC}\, tg\,\delta = \frac{tg\,\delta}{\omega C_C(1+tg^2\delta)} \tag{7.58}$$

- Transformation of the circuit of Fig. 7.29b (which comprises the own resistance R^* of the electrodes) in its equivalent series circuit of Fig. 7.29c, by the equations:

$$R_e = R^* + \Psi R_{SC} \approx \Psi R_{SC}$$
$$X_e = \Psi\left[(X_A + X_{SC}) + \frac{R_{SC}^2 + (X_A + X_{SC})^2}{X_B}\right]$$
$$\Psi = \frac{1}{\left(\frac{R_{SC}}{X_B}\right)^2 + \left(1 + \frac{X_A + X_{SC}}{X_B}\right)^2} \tag{7.59}$$

- Capacitance of the loaded working capacitor:

$$C_e = 1/\omega X_e \tag{7.60}$$

- Circuit quality factor:

$$Q_0 = X_e/R_e \tag{7.61}$$

- Absorbed current:

$$I = \frac{V}{\sqrt{R_e^2 + X_e^2}} \tag{7.62}$$

- Active power in the load and average specific power per unit volume:

$$P_a = R_e I^2; \quad w_a = \frac{P_a}{S_C d_C} \tag{7.63}$$

In conclusion, we remark the complete analogy between the approximate equivalent circuit of the inductor-load system in the induction heating (see Sect. 6.9.1 and Fig. 6.36) and the circuit of Fig. 7.29b. The circuits are practically identical if we replace the capacitances with inductances and the resistance R^* with the own coil resistance of the inductor.

Example 7.3 Calculation of a system of Fig. 7.28 with cylindrical electrodes and load, and the following data:

- Working capacitor: d = 12.5 cm; r_e = 62.5 cm; S_0 = 12,272 cm^2; f = 27.12 MHz; ω = $170.4 \cdot 10^6$ rad/s; resistance R^* negligible
- Load: d_c = 10.0 cm; r_c = 50 cm; S_c = 7854 cm^2; w = 3 W/cm^3; ε′ = 5; tg δ = 0.1.

In order to produce in the load a specific power w = 3 W/cm^3, as in Example 7.1, we should have in the material the electric field intensity E_c = 630.4 V/cm, which is obtained applying to the electrodes a voltage V = 14.184 V.

Therefore, it will result:

- Capacitance C_0 of the unloaded working capacitor: (from Table 7.3 with $d/r_e = 0.2$ is $k_B \approx 1.31$)

$$C_0 = \varepsilon_0 \frac{S_0}{d} k_B = 8.86 \cdot 10^{-14} \frac{12{,}272}{12.5} 1.31 = 113.95 \text{ pF}$$

- Capacitance of central region:

$$C_0^* = \varepsilon_0 \frac{S_C}{d} = 8.86 \cdot 10^{-14} \frac{7854}{12.5} = 55.67 \text{ pF}$$

- Edge capacitance and reactance:

$$C_B = C_0 - C_0^* = 58.28 \text{ pF};$$
$$X_B = 1/\omega C_B = 1/(170.4 \cdot 10^6 \, 58.28 \cdot 10^{-12}) = 100.70 \ \Omega$$

- Capacitance C_A and reactance X_A of the upper central portion in air:

$$C_A = \varepsilon_0 \frac{S_C}{d - d_C} = 8.86 \cdot 10^{-14} \frac{7854}{2.5} = 278.35 \text{ pF};$$
$$X_A = \frac{1}{\omega C_A} = \frac{1}{170.4 \cdot 10^6 \, 278.35 \cdot 10^{-12}} = 21.08 \ \Omega$$

- Capacitance C_C and reactance X_C of Fig. 7.29a:

$$C_C = \varepsilon_0 \, \varepsilon' \frac{S_C}{d_c} = 8.86 \cdot 10^{-14} \, 5 \, \frac{7854}{10} = 347.93 \text{ pF};$$
$$X_C = \frac{1}{\omega C_C} = \frac{1}{170.4 \cdot 10^6 \, 347.93 \cdot 10^{-12}} = 16.87 \ \Omega$$

- Reactance X_{SC} and resistance R_{SC} of Fig. 7.29b:

$$X_{SC} = \frac{1}{\omega C_C(1+tg^2\delta)} = 16.87 \frac{1}{1+0.1^2} = 16.70\ \Omega$$
$$R_{SC} = X_{SC}\quad tg\,\delta = 16.70 \quad 0.1 = 1.67\ \Omega$$

- Series equivalent circuit of Fig. 7.29c:

$$\Psi = \frac{1}{\left(\frac{R_{SC}}{X_B}\right)^2 + \left(1+\frac{X_A+X_{SC}}{X_B}\right)^2} = \frac{1}{\left(\frac{1.67}{100.70}\right)^2 + \left(1+\frac{21.08+16.70}{100.70}\right)^2} = 0.5287$$

$$R_e = R^* + \Psi R_{SC} \approx \Psi R_{SC} = 0.5287 \cdot 1.67 = 0.8830\ \Omega$$
$$X_e = \Psi\left[(X_A+X_{SC}) + \frac{R_{SC}^2+(X_A+X_{SC})^2}{X_B}\right]$$
$$= 0.5287\left[(21.08+16.70) + \frac{1.67^2+(21.08+16.70)^2}{100.70}\right] = 27.48\ \Omega$$

- Capacitance of the loaded working capacitor:

$$C_e = 1/\omega X_e = 1/(170.4 \cdot 10^6\, 27.48) = 213.56\ \text{pF}$$

- Q factor:

$$Q_0 = X_e/R_e = 27.48/0.883 = 31.12$$

- Absorbed current:

$$I = \frac{V}{\sqrt{R_e^2+X_e^2}} = \frac{14{,}184}{\sqrt{0.883^2+27.48^2}} = 515.9\ \text{A}$$

- Power and average power per unit volume in the load material:

$$P_a = R_e I^2 = 0.883 \quad 515.9^2 = 235.0\ \text{kW};$$
$$w_a = \frac{P_a}{S_C\, d_C} = \frac{235.0 \cdot 10^3}{7854\ 10} = 2.99\ \text{W/cm}^3.$$

7.7 Transient Temperature Distribution in the Load

In this paragraph we consider the heating of a plane dielectric workpiece with thickness s placed between and in contact with the electrodes of a plane working capacitor, as shown in Fig. 7.30.

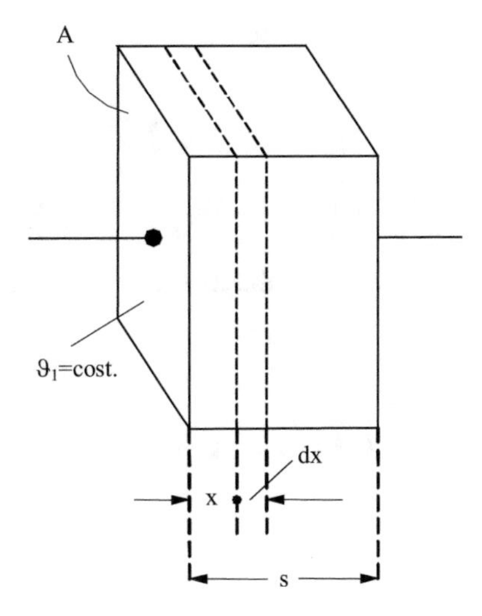

Fig. 7.30 Geometry of the plane electrodes-load system

In this system there is conduction heat transfer between electrodes and workpiece and convection and radiation losses from the lateral surface of the body toward the environment.

However, given that the maximum temperature in dielectric heating applications in most cases does not exceed 150–250 °C, it is reasonable, as a first approximation, to neglect the losses from the lateral surface of the workpiece.

With this assumption, in the following it is analyzed the temperature distribution with uniformly distributed power density *w* in a plane infinite body with constant thermal properties λ, c, γ and initial temperature ϑ_0, placed between two electrodes kept at constant temperature ϑ_1 (Fig. 7.30).

In this case the heat transfer occurs only in the direction *x* normal to the electrodes, according to the Fourier equation (see Eq. 1.10a):

$$\frac{\partial^2 \vartheta}{\partial x^2} - \frac{1}{k}\frac{\partial \vartheta}{\partial t} = -\frac{w}{\lambda} \tag{7.64}$$

with $k = \lambda / c\,\gamma$—thermal diffusivity.

Assuming the origin of the axis *x* on one electrode, the initial and boundary conditions are:

$$\left.\begin{array}{ll} \vartheta = \vartheta_0 & \text{for } t = 0 \\ \vartheta = \vartheta_1 & \text{for } t > 0;\ x = 0 \text{ and } x = s \end{array}\right\}. \tag{7.64a}$$

It can be easily verified that the expression:

$$\vartheta = \vartheta_1 + \frac{w}{2\lambda} x\,(s - x), \tag{7.65}$$

which also gives:

$$\frac{\partial \vartheta}{\partial x} = \frac{w}{2\lambda}(s - 2x); \quad \frac{\partial^2 \vartheta}{\partial x^2} = -\frac{w}{\lambda},$$

is the solution of Eq. (7.64) in thermal steady-state conditions, i.e. when it is $\partial\vartheta/\partial t = 0$.

Introducing the new variable *u* defined by the relation:

$$\vartheta = u + \vartheta_1 + \frac{w}{2\lambda} x (s - x),$$

Eq. (7.64) and the initial and boundary conditions (7.64a) become:

$$\frac{\partial^2 u}{\partial x^2} - \frac{1}{k}\frac{\partial u}{\partial t} = 0 \tag{7.66}$$

$$\left.\begin{array}{ll} u = \vartheta_0 - \vartheta_1 - \frac{w}{2\lambda} x (s - x), & \text{for } t = 0 \\ u = 0 & \text{for } t > 0;\ x = 0 \text{ and } x = s \end{array}\right\} \tag{7.66a}$$

If the initial distribution of *u* along *x* were:

$$u_0(x) = A_n \sin\frac{n\pi x}{s}$$

both the Eq. (7.66) and the conditions (7.66a) would be satisfied by the equation:

$$u = A_n \sin\frac{n\pi x}{s} e^{-\frac{n^2\pi^2 kt}{s^2}}.$$

Therefore, if the function $u_0(x)$ can be developed in a series of the form:

$$u_0(x) = \sum_{n=1}^{\infty} A_n \sin\frac{n\pi x}{s},$$

the required solution is:

$$u = \sum_{n=1}^{\infty} A_n \sin\frac{n\pi x}{s} e^{-\frac{n^2\pi^2 kt}{s^2}}, \tag{7.67}$$

and the constants A_n can be evaluated by the integrals:

$$A_n = \frac{2}{s}\int_0^s u_0(x) \sin\frac{n\pi x}{s} dx. \tag{7.68}$$

In our case, being for t = 0:

$$u_0(x) = \vartheta_0 - \vartheta_1 - \frac{w}{2\lambda} x(s - x),$$

it is:

$$A_n = \frac{2}{s} \int_0^s [\vartheta_0 - \vartheta_1 - \frac{w}{2\lambda} x(s - x)] \sin \frac{n\pi x}{s} dx \tag{7.68a}$$

Substituting the values of the integrals:

(a)
$$\int_0^s \sin \frac{n\pi x}{s} dx = \frac{s}{n\pi} (1 - \cos n\pi)$$

(b)
$$\int_0^s x \sin \frac{n\pi x}{s} dx = -\left(\frac{s}{n\pi}\right)^2 n\pi \cos n\pi$$

(c)
$$\int_0^s x^2 \sin \frac{n\pi x}{s} dx = -\left(\frac{s}{n\pi}\right)^3 \left[2(1 - \cos n\pi) - (n\pi)^2 \cos n\pi\right],$$

finally we obtain:

$$A_n = -2(\frac{1 - \cos n\pi}{n\pi}) [(\vartheta_1 - \vartheta_0) + \frac{w s^2}{2\lambda} \frac{2}{n^2\pi^2}] \tag{7.68b}$$

$$u = -2 \sum_{n=1}^{\infty} (\frac{1 - \cos n\pi}{n\pi})[(\vartheta_1 - \vartheta_0) + \frac{w s^2}{2\lambda} \cdot \frac{2}{n^2\pi^2}] \cdot \\ \cdot \sin \frac{n\pi x}{s} e^{-\frac{n^2\pi^2 k t}{s^2}} \}. \tag{7.69}$$

This expression can be written in the variable ϑ in the more compact form:

$$\vartheta = \vartheta_1 + \frac{w s^2}{2\lambda} M - (\vartheta_1 - \vartheta_0) N \tag{7.70}$$

with:

$$\left.\begin{aligned} M &= (\frac{x}{s} - \frac{x^2}{s^2}) - 4 \sum_{n=1}^{\infty} (\frac{1 - \cos n\pi}{n^3\pi^3}) \sin \frac{n\pi x}{s} e^{-\frac{n^2\pi^2 k t}{s^2}} \\ N &= 2 \sum_{n=1}^{\infty} (\frac{1 - \cos n\pi}{n\pi}) \sin \frac{n\pi x}{s} e^{-\frac{n^2\pi^2 k t}{s^2}} \end{aligned}\right\}. \tag{7.70a}$$

The values of the coefficients M and N are given, as a function of (*x/s*) and the ratio $(\mathrm{k\,t/s^2})$, in the diagrams of Fig. 7.31.

The curves of Fig. 7.31a show that for high values of time (t) the values of M, as a function of *x/s*, tend to the parabolic distribution which corresponds to the first term $\left(\frac{\mathrm{X}}{\mathrm{S}} - \frac{\mathrm{X}^2}{\mathrm{S}^2}\right)$ of Eq. (7.70a).

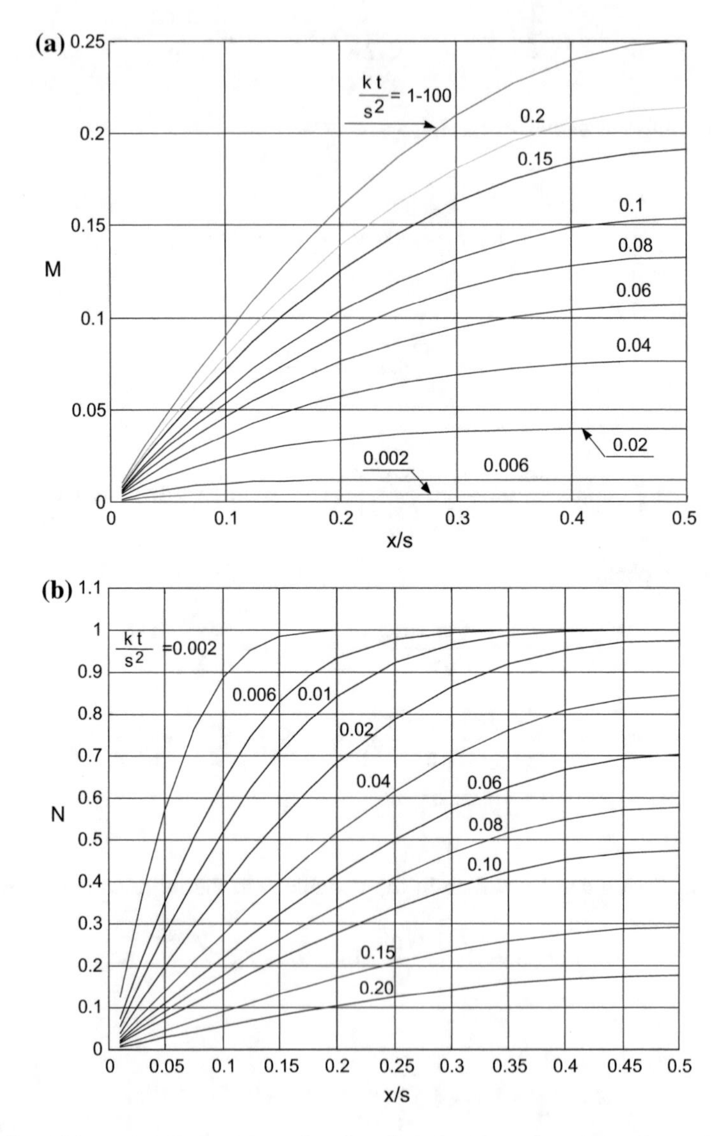

Fig. 7.31 **a** Values of coefficient M as a function of x/s; **b** Values of coefficient N as a function of x/s and various values of $\mathrm{k\,t/s^2}$

The Eq. (7.70) can be used for the analysis of some practical cases.

(a) *Dielectric heating with "cold" electrodes*

$$w \neq 0; \quad \vartheta_1 = \vartheta_0.$$

In this case we have:

$$\vartheta - \vartheta_0 = \frac{w\,s^2}{2\,\lambda} M.$$

The diagrams of M thus are proportional to the temperature distributions inside the workpiece as a function of *x/s*.

The curves show that nearly uniform distributions (with the exception of what occurs near the cold electrodes) can be obtained only with very short heating times.

This is more clearly shown in Fig. 7.32, where the curves provide some temperature distributions in per unit values, referring each curve to its maximum value, which occurs at x/s = 0.5.

(b) *Dielectric heating with "hot" electrodes*

$$w = 0; \quad \vartheta_1 > \vartheta_0\ (= 0)$$

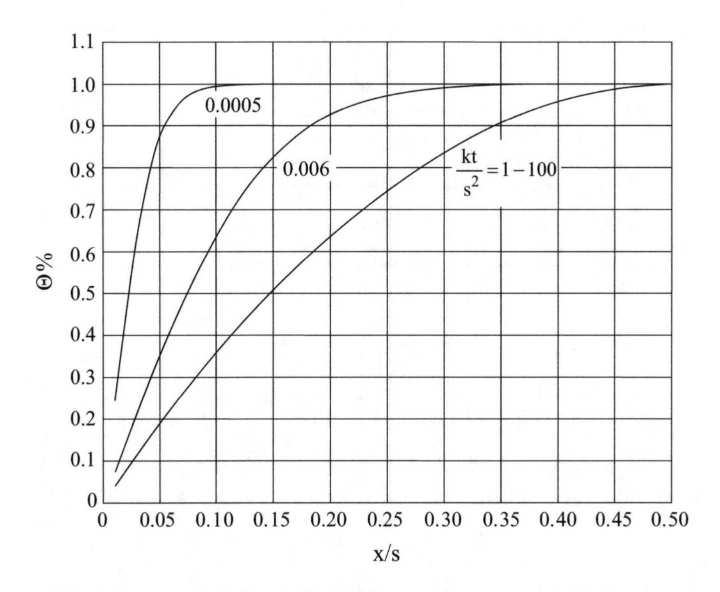

Fig. 7.32 Percent distributions of temperature along x in the heating with "cold" electrodes for different heating times

In this case Eq. (7.70) can be written as follows:

$$\vartheta = \vartheta_1 (1 - \mathrm{N}).$$

Since the heat transfer occurs only by conduction from the electrodes to the dielectric material, uniform temperature distributions can be obtained only with relatively long heating times; with short heating times, on the contrary, the heating concentrates only near the electrodes.

Apart the multiplicative factor ϑ_1, the temperature distributions in the workpiece thickness are given, as function of time, by the curves of Fig. 7.33, which represent the term $(1 - \mathrm{N})$.

In some applications (e.g. heating polymerization of resins), the uneven distribution produced in dielectric heating by the contact of the workpiece with cold electrodes is not acceptable, while the heating process which uses only "hot" electrodes requires too long heating times for achieving a sufficiently uniform temperature distribution.

In such cases, it can be convenient to use a combined process with simultaneous application of dielectric and "hot" electrodes heating, with electrodes maintained at a prefixed temperature by means of steam or electricity [7].

Selecting convenient process parameters (i.e. heating time *t* and power density *w*) the sum of the two distributions corresponding to Eq. (7.70) allows to achieve a sufficiently uniform temperature inside the workpiece, as shown, as an example, in the diagrams of Fig. 7.34.

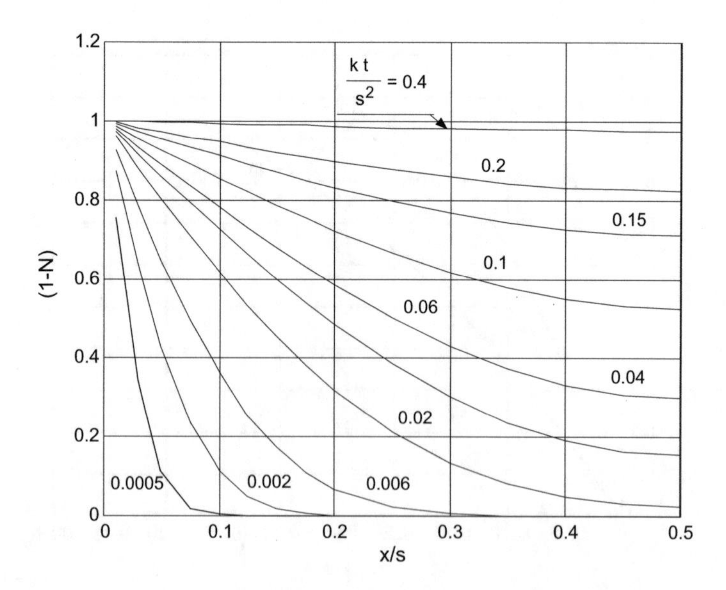

Fig. 7.33 Coefficient $(1 - \mathrm{N})$ in the equation of the heating with "hot" electrodes only

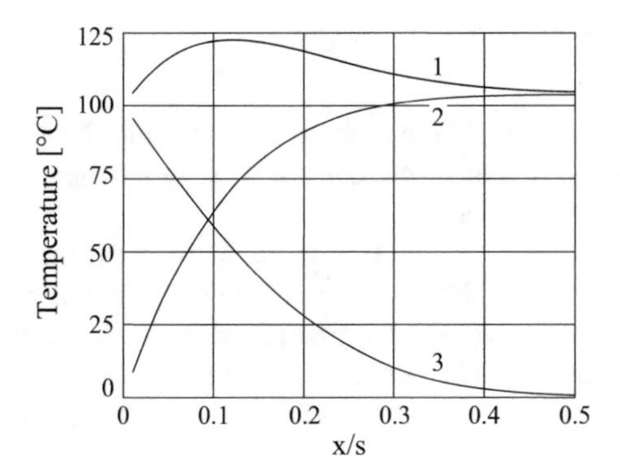

Fig. 7.34 Temperature distribution in simultaneous dielectric and "hot" electrodes heating (*1* combined heating; *2* dielectric heating; *3* "hot" electrodes heating)

In this combined process, in order to obtain a nearly uniform final temperature distribution with a given minimum temperature, the temperature of electrodes must be equal or greater than this value. Moreover, the power density must be adjusted in order to achieve approximately the same final temperature at the surface and the center of the body to be heated.

Under these conditions, as shown in Fig. 7.35, the temperature of all internal points is higher than the surface temperature and its distribution near the electrodes is characterized by a maximum, whose value and position depends on $k\,t/s^2$.

The curves of Fig. 7.34, give the distributions near the electrodes of the over-temperature $\Delta\vartheta$ above the value ϑ_1: they show that the maximum temperature unevenness is mostly lower than 20 %.

In conclusion, the use of the combined heating can lead to a considerable reduction of the heating time when a maximum temperature non-uniformity of about 20 % can be accepted, but it is less beneficial when high temperature uniformity is required.

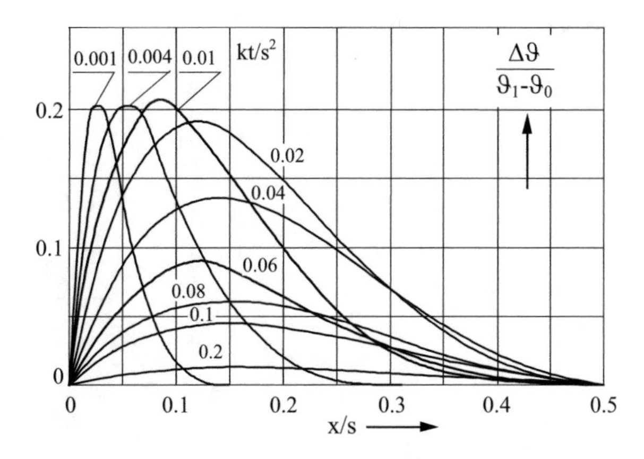

Fig. 7.35 Over-temperature near electrodes in combined heating, as a function of $k\,t/s^2$

7.8 Other Types of Applicators

Depending on the geometry of the body to be heated and the process requirements, many other configurations of electrodes are used in practice, in order to optimize the energy transfer to the load.

Among them, particular importance have those used for heating thin workpieces (like paper, textiles, fibers) with continuous movement through the electrodes. In these cases, the air gap interposed between electrodes and load is necessarily larger than the thickness of the material, and the electrode geometries previously considered do not provide sufficiently high field intensities and specific power values.

In addition, in drying processes the electrode arrangement must allow for the removal of vapors that form during heating.

Typical electrodes used in this case consist of a series of tubular metal bars, facing the material to be heated on one or both sides, which are connected to the power supply in different ways as it is illustrated in Figs. 7.36, 7.37 and 7.39.

The geometry of Fig. 7.36a produces a pulsed transverse electric field when the space L between electrodes is relatively large. Reducing this space gradually, the fields produced by two adjacent electrodes partially overlap, giving a resulting power distribution of the type shown in Fig. 7.36b, which is suitable also for static heating.

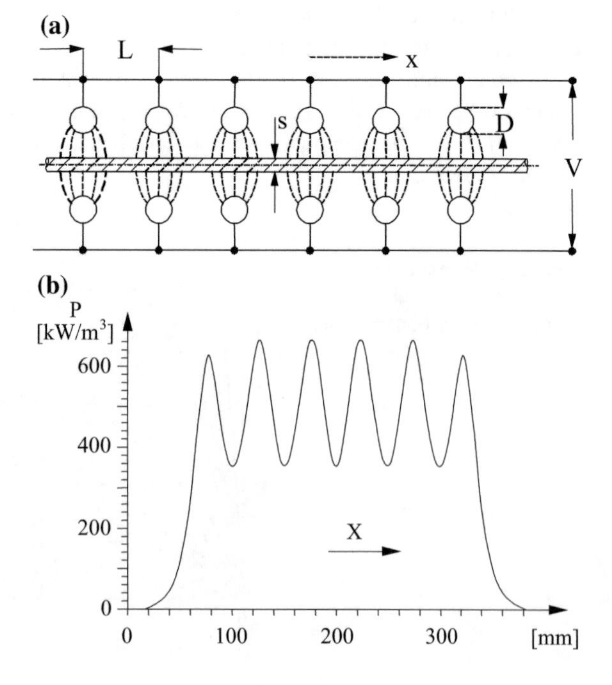

Fig. 7.36 Transverse pulsed electric field applicator: **a** geometry and field configuration at no-load; **b** example of power distribution in the load ($D = 10$ mm; $s = 4$ mm; $L = 2H = 50$ mm; $f = 13.56$ MHz; material: wood, $\varepsilon_r = 10$, $tg\delta = 0.05$) [8]

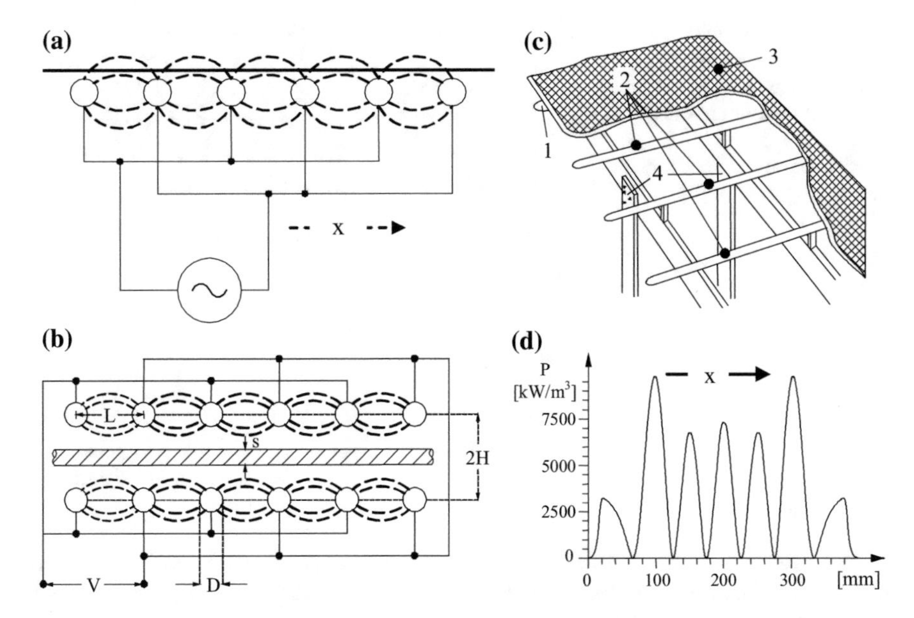

Fig. 7.37 Applicator with longitudinal alternating electric field: **a**, **b** geometry and field schematic pattern in unloaded applicator; **c** constructive arrangement (*1* electrodes, *2* tubular bars with alternating polarity, *3* conveyor belt, *4* insulating supports); **d** typical power distribution in the load (geometry and load characteristics as in Fig. 7.36) [8]

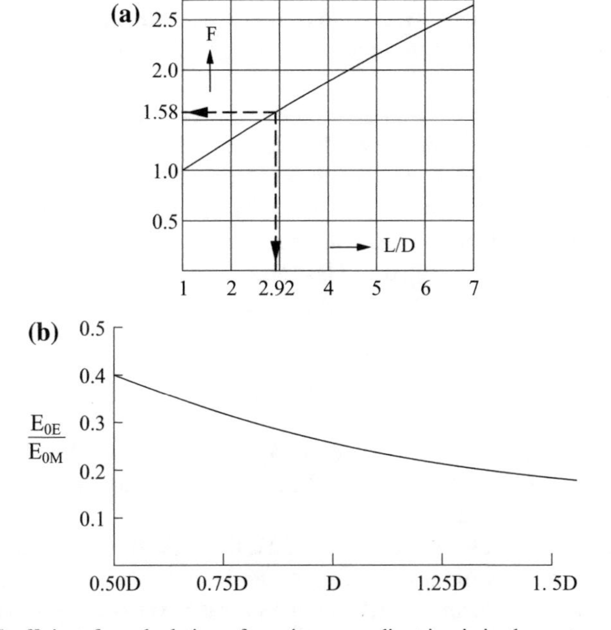

Fig. 7.38 **a** Coefficient for calculation of maximum gradient in air in the system of Fig. 7.37a; **b** Average equivalent field in the load between two electrodes, with L/D ≈ 2.92 [9]

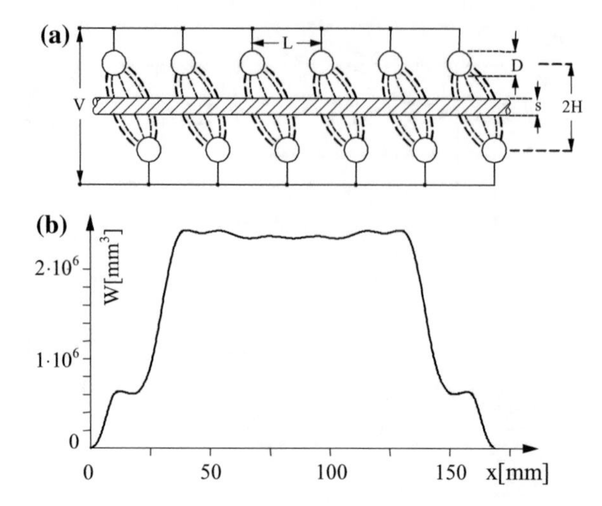

Fig. 7.39 "Staggered applicator": **a** system geometry and schematic pattern of field in unloaded applicator; **b** power distribution in the load with optimised geometry [8]

For materials having form of thin films, such as paper or cardboard, which are characterized by high loss factors in the direction of "rolling", it is more convenient the use of applicators of the type shown in Figs. 7.37 and 7.39, which produce a considerable component of the electric field in the movement direction.

The applicator of Fig. 7.37, which can be placed in front of one or both sides of the material to be heated, is characterized by a longitudinal field and is known as "*stray-field applicator*".

With the same geometrical dimensions this is the most effective type of applicator as regards transfer of power to the load but, as shown in figure (d), the power distribution is particularly non-uniform in longitudinal direction, so that it is not suitable for stationary heating.

All other conditions being equal, the active power in the load increases with the decrease of the distance H between electrodes and load, while by reducing L an optimum can be found where the power is maximum.

The lower admissible value of L is given by the maximum allowable electrical gradient E_{0M} in air, which occurs at the surface of the electrodes on the line between the centers of two adjacent electrodes.

The value of the maximum gradient is given by the relationship [9]:

$$E_{0M} = \frac{V}{L - D} F$$

with: F—function of the ratio L/D, given by the diagram of Fig. 7.38a.

Keeping constant the distance L and the voltage V, the gradient E_{0M} has a minimum for $L/D \approx 2.92$.

Figure 7.38b provides, as a function of H, the ratio of the average equivalent gradient E_{0e} to the maximum value E_{0M} in the load material during its movement between two adjacent electrodes, when it is fulfilled the condition $L/D \approx 2.92$ [9].

These values, although evaluated with the field pattern of an unloaded applicator, provide useful data for a preliminary design. In fact, the presence of load materials with relatively high values of ε tends to concentrate the field lines inside the load.

Widely used for workpieces of thickness up to 10 mm is the system of Fig. 7.39a, known in bibliography as *"staggered applicator"*. It is similar to that of Fig. 7.37a, but with the upper and lower electrodes shifted horizontally one from the other, in order to produce a high component of the field in the direction of movement. With a convenient choice of the parameters L and 2H, this system allows to obtain very uniform longitudinal power distributions in the load, as shown for example in Fig. 7.39b.

7.9 Microwave Heating

As shown in previous paragraphs, the power per unit volume produced in a dielectric material by an alternative electric field, is given by the relation:

$$w = \omega E^2 \varepsilon_0 \varepsilon \, \mathrm{tg}\delta.$$

Then, all other parameters being the same, *w* is higher the greater is the frequency. The use of high frequency therefore allows to shorten heating time or to reduce the value of electric field intensity.

This is the fundamental reason that makes attractive the use of microwaves in industrial heating applications.

As already observed, the selection of the working frequency, which theoretically should be the one giving the maximum of the product $(\varepsilon \, \mathrm{tg}\delta)$, is in practice bound by international regulations and availability of generators sufficiently powerful and safe. For these reasons the frequencies mostly used in practice are 915 and 2450 MHz.[1]

In addition, there are some fundamental differences between radio frequency and microwave heating.

In radio frequency heating the material to be heated, placed between the electrodes of the working capacitor, constitutes a capacitance that is an integral part of the generator's AF circuit. Therefore the variations of material dielectric properties during heating and the load geometry affect directly the characteristics of the circuit and the generator working conditions.

This does not occur in microwaves heating, where the electromagnetic energy produced by the generator is extracted by means of a coupled loop and then is conveyed through a waveguide towards a resonant cavity where the body to be heated is placed.

[1]Slightly different values are used in UK.

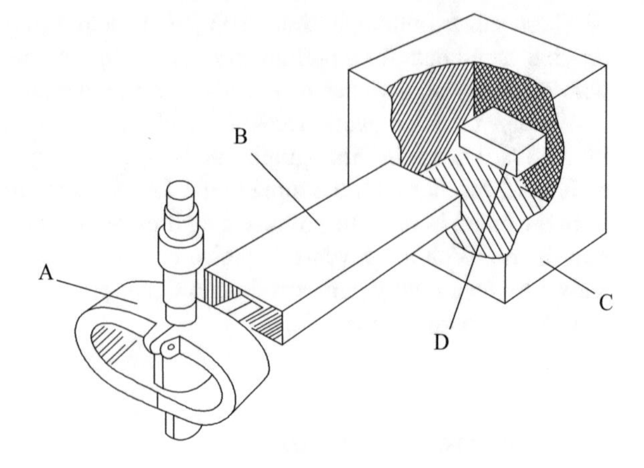

Fig. 7.40 Schematic of a microwave heating system (*A* magnetron; *B* wave guide; *C* resonant cavity; *D* material to be heated)

In this way the shape of the workpiece and its dielectric properties may vary significantly without affecting appreciably the generator working conditions.

Another feature of microwave heating is the use of short wavelengths (33 and 12 cm for the above mentioned frequencies), which can produce uniform heating patterns only through a high number of reflections of the waves on the walls of the resonant cavity and the use of mode stirrers.

Let us now examine in detail the main elements of a microwave heating system, schematically illustrated in Fig. 7.40.

7.9.1 Generators

The microwave generator most commonly used in these applications is the industrial magnetron for continuous operation.

It allows to overcome the limits met, as the frequency increases, in the realization of power tube generators, which are linked to the fact that the dimensions of constructive elements become comparable with the wavelength.

In fact, in a magnetron, generator, the oscillating and the coupling circuits constitute a single unit.

As illustrated in Fig. 7.41, a magnetron consists of three fundamental elements: a cylindrical cathode, an anode block with cavities connected with the space between cathode and anode by means of slits, and a magnet (not visible in the figure) that produces, in the above mentioned space (in the following denoted with A–C), a magnetic field directed along the cathode axis.

The high DC voltage applied between anode and cathode produces, in the space A–C, an electric field in radial direction, which intersects perpendicularly the lines of the axial magnetic field.

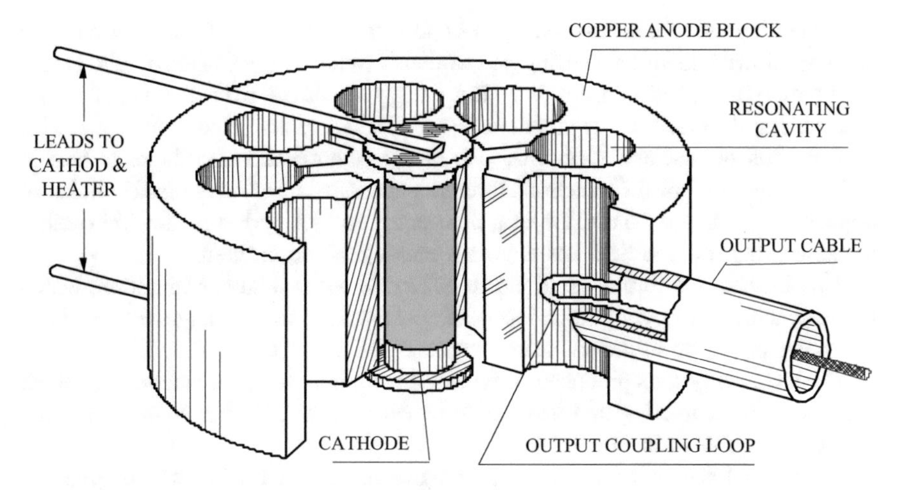

Fig. 7.41 Elements constituent a magnetron [10]

For the combined action of such fields, an electron that leaves the surface of the cathode is subjected—in absence of HF oscillations—to forces that, depending on its emission velocity, force him to follow different trajectories, like those indicated with *a*, *b*, *c*, *d* in Fig. 7.42a.

In this way a number of electrons can pass in front of the slits without falling on the anode (curve *c*).

On the contrary, when the magnetron oscillates, in front of the slits, to the continuous one produced by the anode voltage, is superimposed the alternative electric field of the resonant cavities. The paths of electrons then modify as illustrated in Fig. 7.42b.

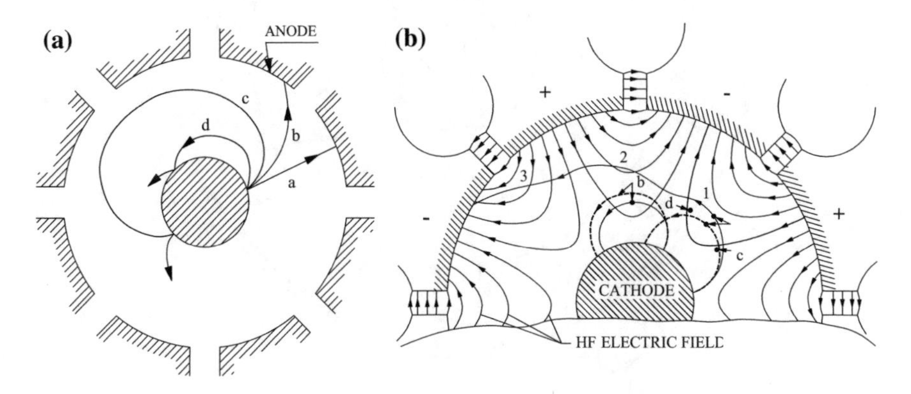

Fig. 7.42 Trajectories of electrons with different values of axial magnetic field: **a** without HF oscillations; **b** with HF oscillation [11]

If the alternative field is such as to oppose to the motion of the electrons, the latter are slowed down transferring part of their kinetic energy to the HF field before falling onto the anode. In this way the energy absorbed from the DC field for accelerating the electrons, is transferred to the HF alternative electromagnetic field.

A portion of "useless" electrons, falls on the cathode and contributes to heat it.

The transit of "useful" electrons through successive alternative electric fields of opposite sign should take place in a time equal to half period of the HF oscillations, in order that the field has always a braking effect on them.

This leads to the formation of electrons clouds with radial configuration, which rotate synchronously with the induced microwave field with an angular velocity of two poles per cycle in the so-called "*mode* π" (Fig. 7.43).

The HF energy thus produced is extracted from one cavity of magnetron, which is inductively coupled with a waveguide or with a coaxial cable, connected to the applicator as shown in Fig. 7.41.

Figure 7.44 shows an example of magnetron characteristics which give the output power and the efficiency as a function of average anode current. The power has approximately a linear variation with the current, while the efficiency decreases considerably at low current values, but generally is higher than 50 %.

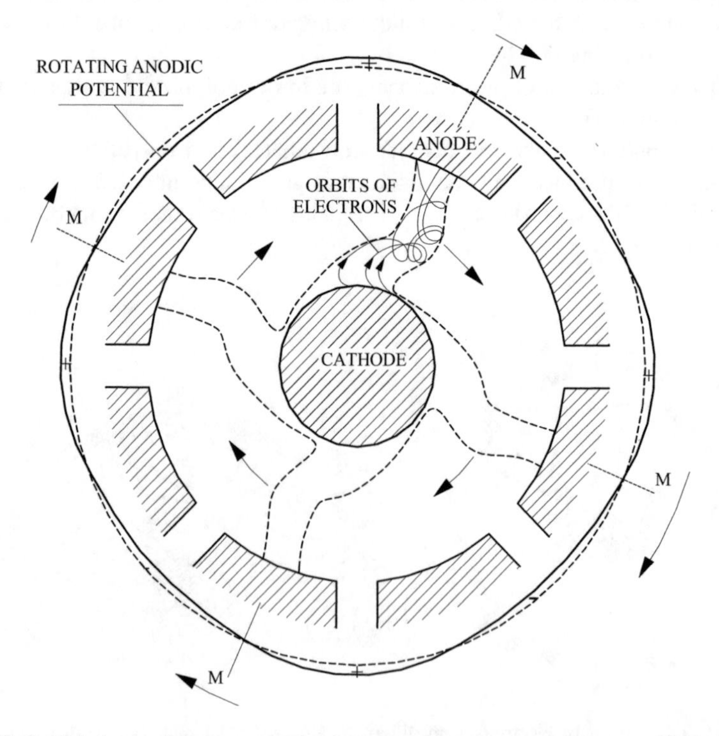

Fig. 7.43 Radial clouds of electrons rotating in the anode structure synchronous with the induced microwave field [10]

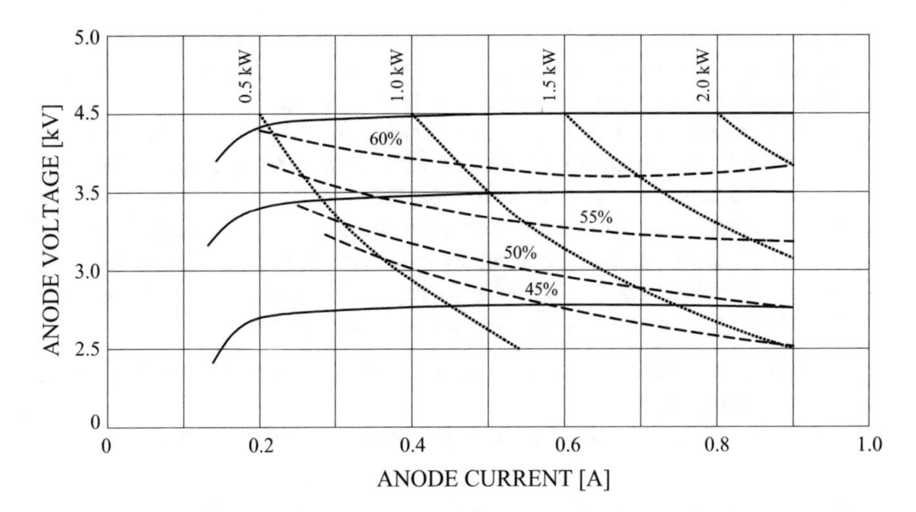

Fig. 7.44 Output power and efficiency of a magnetron as a function of the average anode current

In applications where it is needed an output power higher than some kW, are generally used several magnetrons parallel connected or, in some cases, klystrons or amplitrons.

With magnetrons parallel connected are presently supplied installations with power ratings up to several hundred kW.

7.9.2 *Waveguides*

Waveguides are used for the energy transfer from the generator to the resonant cavity where is generally placed the body to be heated. Sometimes the waveguide is used directly as applicator, for heating special loads (like sheets, thin strips, liquids, etc.).

The most common waveguide structure is constituted by a hollow conductive metal pipe, with rectangular or circular cross-section, in which the electromagnetic wave can propagate. As a rule of thumb, the width of the waveguide is of the same order of magnitude of the guided wave wavelength.

The wave is confined inside the waveguide due to total reflection on the waveguide walls, so that the propagation inside the waveguide can be described approximately as a "zigzag" between the walls. The waveguide confines the wave propagation in one direction, so that (under ideal conditions) the wave does not lose power while propagating.

The electromagnetic wave propagation and the spatial distribution of the time-varying electric and magnetic fields along the axis of the waveguide, is described by the wave equation and the boundary conditions imposed by the shape and materials of the waveguide. These boundary conditions eliminate an infinite number of solutions of the wave equation, and the remaining ones are the only possible solutions of the wave equation inside the waveguide.

Therefore the electromagnetic wave can propagate along the waveguide only in a limited number of *propagation "modes"*, each of them being characterized by a different configuration of the electric and magnetic fields.

The lowest frequency in which a certain mode can propagate is named the *cutoff frequency* of that mode. The mode with the lowest cutoff frequency is the basic mode of the waveguide, and its cutoff frequency is the waveguide cutoff frequency.

The propagation modes can be classified into two basic categories:

- TE modes, characterized by a Transverse Electric component and a longitudinal magnetic component;
- TM modes, with a Magnetic Transverse component and a longitudinal electric one.

The behavior is, in some respects, similar to that of a transmission line for the presence of waves which propagate with phase and attenuation constants, reflected waves at the end of the guide or in presence of irregularities, and stationary waves.

The most important difference is the presence of a critical wavelength, above which propagation is practically impossible.

The critical wavelength is different, in a given guide, for the different propagation modes; the mode for which it is maximum is called the "*dominant mode*".

In certain applications, the heating process is carried out inside the guide itself.

For example, if the waveguide is excited in the dominant mode TE_{10} (Fig. 7.45a), the current lines in the mid points of the broad face are parallel to the axis of the guide; it is therefore possible to make in this position a slit, without practically modifying the field distribution.

Moreover, since the electric field has direction parallel to the shorter side of the guide and has the highest value in correspondence to the slits (see Fig. 7.45b), it is possible to achieve the maximum power transfer to the load material placing it inside the slits.

This configuration is particularly used for processing planar materials (like sheets or tapes), which are fed through the slits at a suitable speed in the direction indicated in the figure.

Other guide configurations, such as the so-called "*meander applicator*" shown in Fig. 7.46, allow to obtain a high power transfer to the material by passing the tape within the slits several times.

The bends of the waveguide are so designed as to introduce minimum power losses.

To prevent dangerous reflections back towards the magnetron, which would occur in the waveguide when the power supplied by generator is not completely absorbed in the heated material (and therefore even more at no-load operation), the remainder power at the far end of the guide is usually dissipate in a water load.

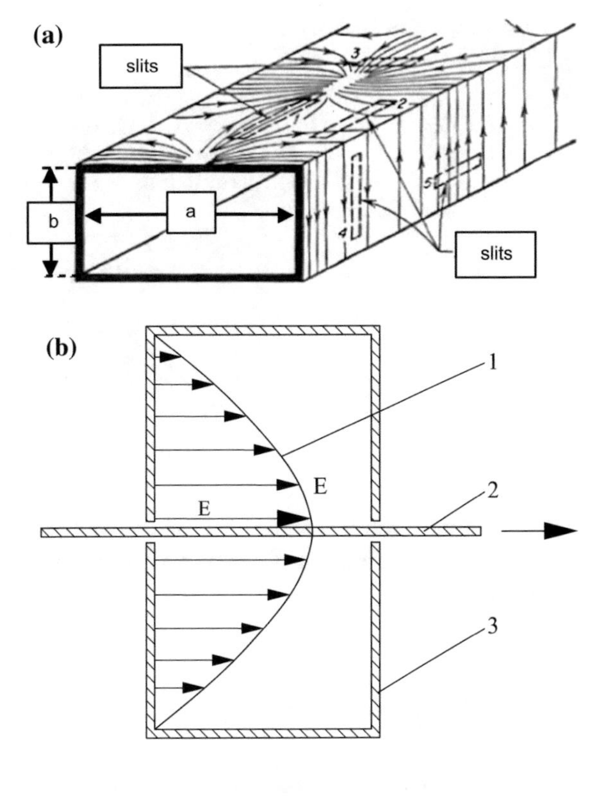

Fig. 7.45 **a** Current flow in walls of a rectangular guide in the TE10 mode (slits 1 and 4 do not modify it); **b** Electric field pattern along the longer side of the guide (1 Electric field, 2 Product, 3 Wave guide) [11]

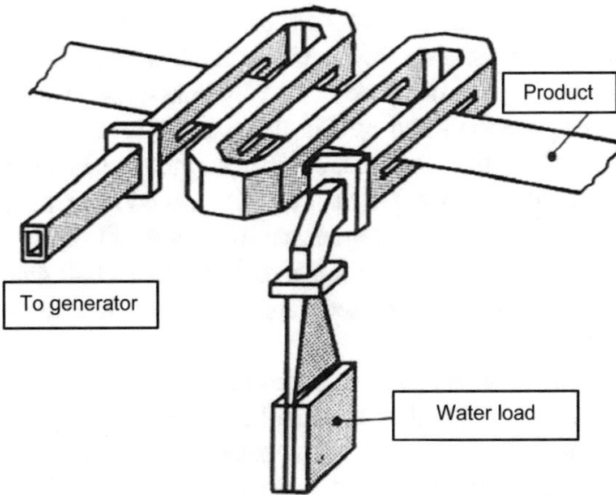

Fig. 7.46 Heating planar products in meander wave guide applicator [12]

The configuration of Fig. 7.47 is used for the heating of liquids (e.g. pasteurization of milk).

In most cases the preliminary calculation of the power absorbed by the workpiece is a complex task, due to the variations during heating of the attenuation

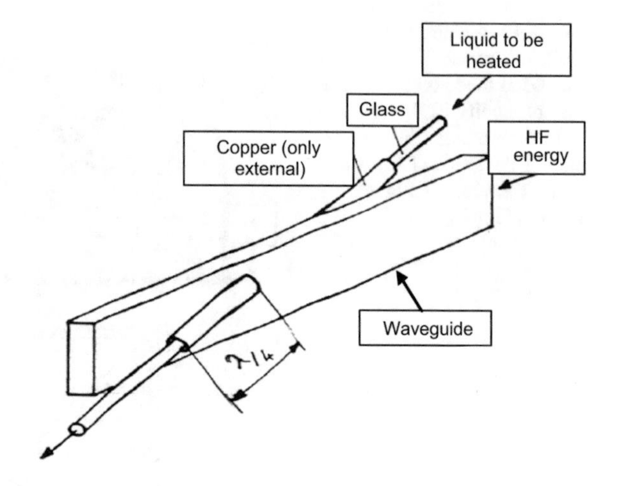

Fig. 7.47 Heating of liquids in waveguide

constant of the electromagnetic wave and the radiation phenomena produced by the presence of the workpiece. These last phenomena can be reduced by appropriate shielding.

7.9.3 The Resonant Cavity

The fundamental problem in microwave heating is the energy transfer to the load. The method most frequently used for this purpose is to use of a stationary wave in a resonant cavity where is placed or passes the material to be heated.

Any region of space closed by conductive walls constitutes a resonant cavity. The space enclosed by the walls has a resonance frequency for each configuration of the field, or "*mode*", that might exist inside this space.

As an analogy, lets us consider a section of a waveguide closed at both ends: such system may support a standing wave of frequency such that the length between the conducting end plates is a multiple of the half guide wavelength. This wave may exist inside the enclosed region without interference from, or radiation to the outside.

If an external source excites the wave, in steady-state conditions it is needed to supply only the relatively small amount of energy which is lost for the conductivity of the walls, while a large constant amount of energy is stored inside and passes back and forth between the electric and magnetic field.

Another simple analogy is based on the schematic of Fig. 7.48, which justifies intuitively the existence of a resonant frequency, considering the resonant cavity as a circuit with distributed constants, derived from a L–C circuit with lumped parameters.

Moreover, a simple form of a resonant cavity can be also derived from a rectangular or circular cross-section waveguide, closed at the two ends. This also

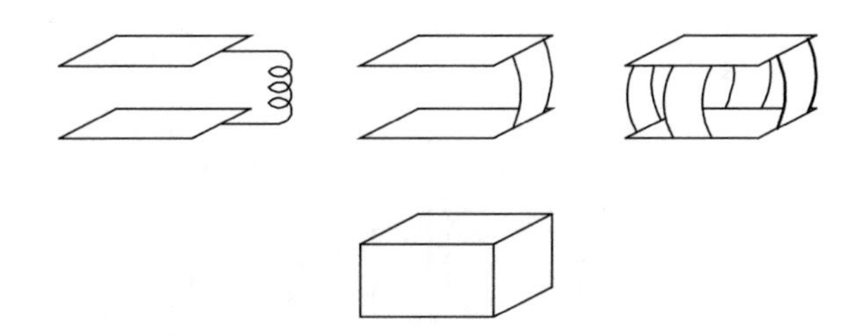

Fig. 7.48 Evolution from a resonant circuit with lumped elements to a closed cavity [13]

justifies—in an intuitive way—the presence of an infinite number of resonance frequencies.

As in the case of waveguides, the lowest resonance frequency is called the "*dominant mode*".

As illustrated in Fig. 7.49, on the metal walls of the cavity are produced multiple reflections of the electromagnetic wave which give rise, by overlapping, to a stationary distribution of the field. Since all these reflected waves interfere with each other, the intensity of the field is not uniformly distributed inside the cavity.

The most important characteristics that determine the efficient heating of the workpiece placed inside the cavity are the following:

- *Number of resonance modes supported by the cavity*

The uniformity of heating is greater the more uniform is the distribution of the electric field within the cavity. This occurs when the cavity is large in comparison to the wavelength at the operating frequency, because the resulting distribution is given by the overlap, within the relevant frequency band, of several resonant modes, thus giving an efficient coupling with the load.

The number of resonant frequencies supported in an empty rectangular cavity of dimensions (a_x, a_y, a_z), within a given frequency band, can be calculated with the equation:

$$f_r = \frac{3 \cdot 10^8}{2} \sqrt{\left(\frac{m}{a_x}\right)^2 + \left(\frac{n}{a_y}\right)^2 + \left(\frac{p}{a_z}\right)^2} \tag{7.71}$$

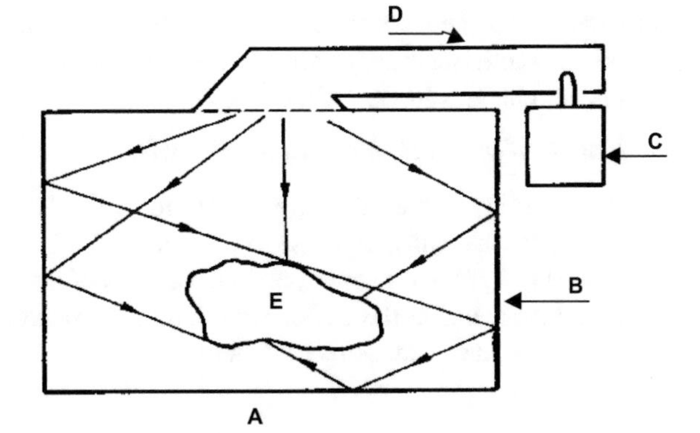

Fig. 7.49 Heating in multimode applicator (**A** resonant cavity; **B** conducting walls; **C** magnetron; **D** wave guide; **E** material to be heated) [12]

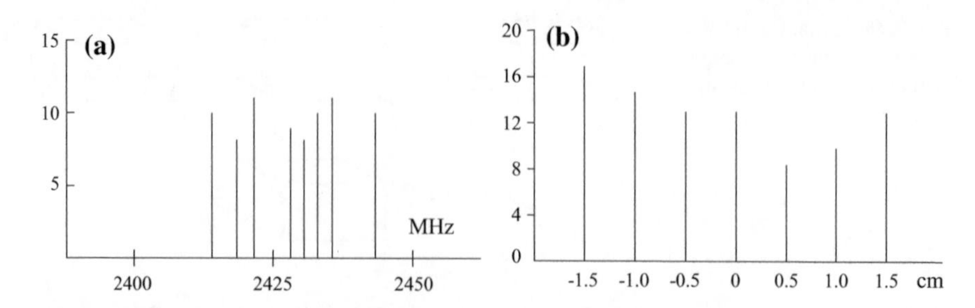

Fig. 7.50 **a** Number of resonant frequencies in a multimode cavity of size 50 × 40 × 40 cm. **b** Total number of useful resonances as a function of the variation of size a_z

with: *m*, *n*, *p*—integers, representing the number of half wavelengths in *x*, *y*, *z* directions respectively.

Given the sizes (a_x, a_y, a_z) of the cavity, all possible combinations *m*, *n*, *p* which satisfy Eq. (7.71) in the given frequency band, represent the number of resonant modes in which the cavity can oscillate.

Therefore the number of modes that can be excited in the cavity depends critically on its size, and also a variation of few percent of a single dimension can lead to a considerable variation of the number of useful resonances. This is shown, as an example, in Fig. 7.50 which refers to a cavity of a microwave oven with dimensions 50 × 40 × 40 cm, operating at nominal frequency 2.450 MHz, where the size a_z is changed of a few percent.

- *The volume ratio* K_v—It is defined by the relationship:

$$K_v = \frac{V_c}{V_m \, \varepsilon} \tag{7.72}$$

with: V_c—volume of the cavity; V_m—volume of the material to be heated; ε—average relative dielectric constant of material.

Uniform heating patterns can be obtained only with high values of K_v. Taking into account the most frequent values of ε, this means that it is of fundamental importance that the resonant cavity is much larger than the load, so that the latter—whose characteristics vary considerably during the heating—does not produce significant changes in the field configuration.

- *Motion of the workpiece inside the cavity*

The standing waves that are produced within the cavity as a result of the reflections on the walls, give rise to maxima and minima of the electric field which are spaced of $\lambda_r/4$, with consequent unevenness of the heating patterns. The motion of the material inside the cavity (translation or rotation) allows to obviate to this drawback because, for effect of motion and the variations of the resonance

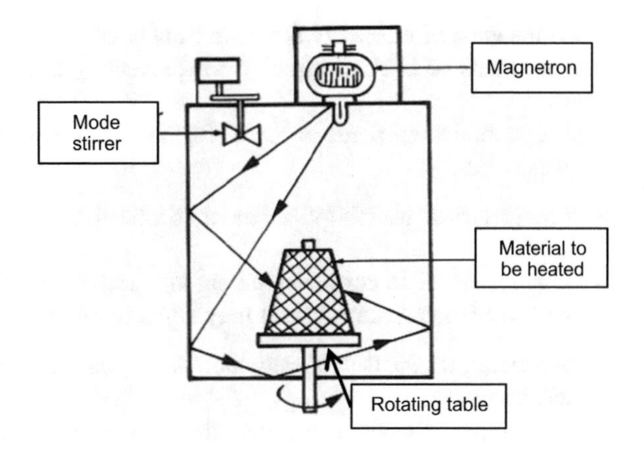

Fig. 7.51 Resonant cavity with rotating table and mode stirrer [3]

configurations produced by the change of material characteristics from point to point and from one instant to another, every point of the workpiece undergoes on average the same thermal cycle.

Another method used to improve the uniformity the heating pattern, is the introduction in the cavity of a discontinuity in the form of a "*mode stirrer*" (shown schematically in Fig. 7.51) such as a metallic multi-blade fan placed in the upper part of the cavity, which continuously perturbs the electromagnetic field produced by the microwave source by its rotation inside the resonant enclosure.

The resonant cavities used in industrial applications are much larger than those used in the domestic ovens; their volume can be up to several m^3.

In most cases the heating process is of continuous type, with advancement of the workpieces by means of a conveyor belt, like in the scheme of Fig. 7.52. In this case, the heating chamber can be constituted by several multimodal cavities, each powered by its own generator, or from a single cavity fed by several generators distributed in its length.

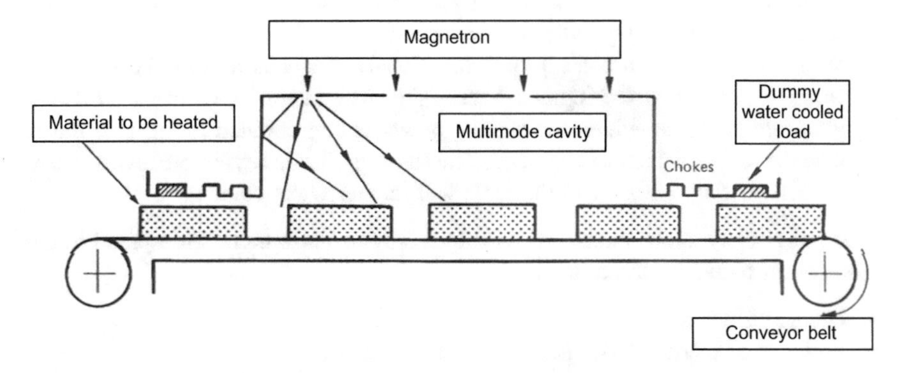

Fig. 7.52 Multimode cavity used in continuous ovens [12]

At the ends of the cavity there are usually special devices, such as water cooled dummy loads or chokes, installed for preventing stray field radiation outside the cavity.

The conducting material used for the cavity walls must meet two basic characteristics:

- low electrical resistivity, in order to constitute a low resistance path for the current;
- high resistance to corrosion, taking into account that the material of the body to be heated (many times food) may come in contact with them.

In general, for the wall construction is used copper, internally coated with Teflon or stainless steel.

Other types of cavities, e.g. cylindrical or elliptical, are also used in addition to the parallelepiped one.

7.10 Industrial Applications of High Frequency and Microwave Heating

Dielectric and microwave heating technologies are widely applied in several industrial processes because of their many technological advantages.

Their main attractive features are:

- *Heat sources produced directly inside workpieces*—This property allows to achieve uniform temperature distributions and short heating times even in bodies of considerable size and low thermal conductivity. It also helps to avoid surface overheating that can damage the material or to prevent undesired complete drying of some products.
- *Rapid heating*—the uniform distribution of heat sources allows to apply volume power densities up to 100 W/cm^3 and to reach high heating rates, up to 20 °C/s. A typical example is the welding of thermoplastic sheets, which is achieved with short heating times, typically of few seconds.
- *Selective heating*—In some processes, it's advantageous that the heat sources are concentrated in the regions with higher loss factor. A typical example is gluing of wood, in which the heating power is concentrated in the glue rather than the wooden parts to be joined. The same applies in drying processes where the HF heating is higher in the parts with higher water content.

Among many applications, we will refer only to those more widespread in the industry, in particular the following:

- Gluing of wood
- Welding of thermoplastic materials (mostly sheets)

- Pre-heating of thermosetting resins and rubber
- Reduction of moisture in textiles, paper and food
- Tempering or post-baking of foodstuffs.

Other applications, like production of medical products, ceramics, leather and others, will not be discussed for space reasons.

7.10.1 Gluing of Wood

This process is widely used in industry for bonding together wooden parts by using adhesives based on thermosetting resins.

The gluing operation is performed by clamping together the parts, with pressure of about 35 kN/m^2, and applying heating up to the achievement of the degree of polymerization of the glue that ensures a perfect adhesion of them. The polymerization is then completed without pressure thanks to the stored heat only.

The glues mostly used are based on resorcinol, melamine, urea-formaldehyde, and PVA[2] (more rarely, phenols or epoxy resins), and are characterized by a polymerization temperature of about 90–130 °C and a loss factor much higher than that of wood.

Typical values of the dielectric properties of the glue and the wood are the following [12]:

	ε'	ε''
Glue (non polymerized)	25	17.5
Glue (polymerized)	5	0.5
Wood (moisture 7–14 %)	4	0.2

It can be remarked that the loss factor of the non-polymerized glue is about 90 times greater than that of the wood and it gradually decreases during heating. When the polymerization is completed, the values of ε' and ε'' of wood and glue are nearly the same.

This allows, with suitable electrodes configurations, to concentrate the heat directly in the glue, reducing significantly the process time and the energy requirements.

The best results are obtained when the moisture content of the wood is about 10 % and is uniformly distributed. In fact, if the moisture content is lower than 7 %, the wood tends to absorb the glue; if the moisture is above 14 %, steam will be produced, leading to a lower mechanical strength of the joint.

[2]Polyvinile Acetate.

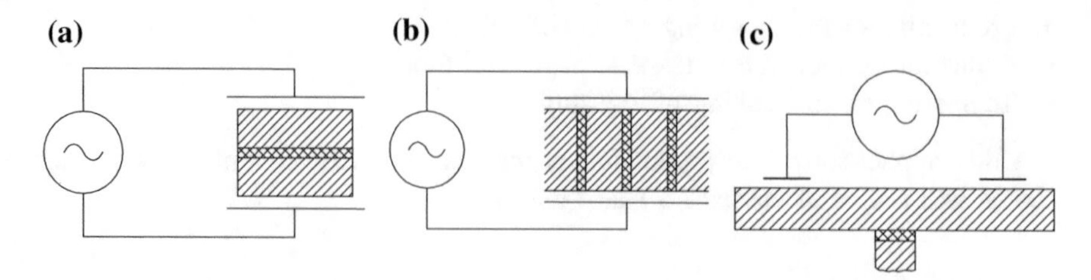

Fig. 7.53 Typical configurations used for bonding wooden parts (**a** through field heating; **b** line heating; **c** stray field heating)

In addition, a higher water content leads to increase the power dissipated in the wood and thus a higher energy consumption and to increase the risk of arc discharges.

Frequencies mostly used are 13.56 and 27.12 MHz.

The bonding is typically performed with electrodes placed over the parts to be glued, as shown Fig. 7.53.

The first configuration (Fig. 7.53a) is used, for example, in the manufacture of plywood or parquets. The layers of glue are parallel to the electrode surfaces and the field lines affect in the same way both the glue and the wood, so that also the wood is heated. Nevertheless, due to the high value of loss factor of the non-polymerized glue, most of the energy is transferred to the layer to be bonded with typical specific consumption of 0.037 kWh/kg of bonded material.

This configuration is also commonly used, with appropriate electrodes and clamps geometries, for production of bent plywood elements (for example, backs and seats of chairs) used in the furniture industry.

In the layout of Fig. 7.53b, the bonding surfaces are normal to the electrodes and specific power is distributed between glue and wood in proportion to their loss factors.

The heating is therefore localized mostly in the areas to be bonded, resulting in energy savings with consumption of about 0.26 kWh/m^2 and very short process times (less than 10 s).

To achieve uniformity of the joint and prevent the risk of electric discharge, the distance between electrodes must not exceed 90 mm and the field in the glue must not be higher than 1.5 kV/cm.

An installation of this type for bonding wooden boards is shown in Fig. 7.54.

The third scheme, shown in Fig. 7.53c, is the less effective in terms of energy consumption, but also the most suited for complex geometries or very large bonding areas, where the two previous schemes cannot be applied easily.

This configuration uses for heating the workpiece material the stray field of electrodes, resulting in a relatively high energy consumption of about 0.72 kWh/m^2.

Fig. 7.54 Bonding of boards according to the arrangement of Fig. 7.53b (AB Elphiac)

7.10.2 Welding Sheets of Thermoplastic Materials

This process is used for joining plastic sheets using dielectric heating with simultaneous application of pressure between the electrodes.

It is applied in the manufacture of a wide range of products, such as stationery articles, leather goods, shoes, toys, inflatable mattresses, inner linings of cars, etc.

The main advantages of this process are perfect finish, mechanical strength of joint and very short welding time, which is in the order of few seconds whatever is the size of the joint to be welded.

Suited to this process are thermoplastic materials with high loss factor. Among these, the material most often used is polyvinyl chloride (PVC) in sheets of thickness ranging from 0.1 to 1 mm.

Other materials suited for welding are: rigid PVC, some plasticized cellulose acetates, nylon (nylon 6 and 66) and others.

Polyethylene is not in itself suited for the process, but it can be used for welding if it is covered with a layer of weldable material.

The welding process is generally performed at 27.12 MHz, applying RF by means of shaped electrodes. Other frequencies can be used in special cases, such as 13.56 MHz for large areas of welding or 70 MHz for welding of nylon.

The more common type of welding machine is the so-called “C” type machine shown in Fig. 7.55, with pneumatic, hydraulic or by foot actuation of the press.

During operation, the sheets to be welded (M) are interposed between the welding plane (1) and the electrode (2), both made of metal, according to the arrangement of Fig. 7.56a.

The electrode has the same shape of the weld and acts on the welding plane with the pressure given by the press.

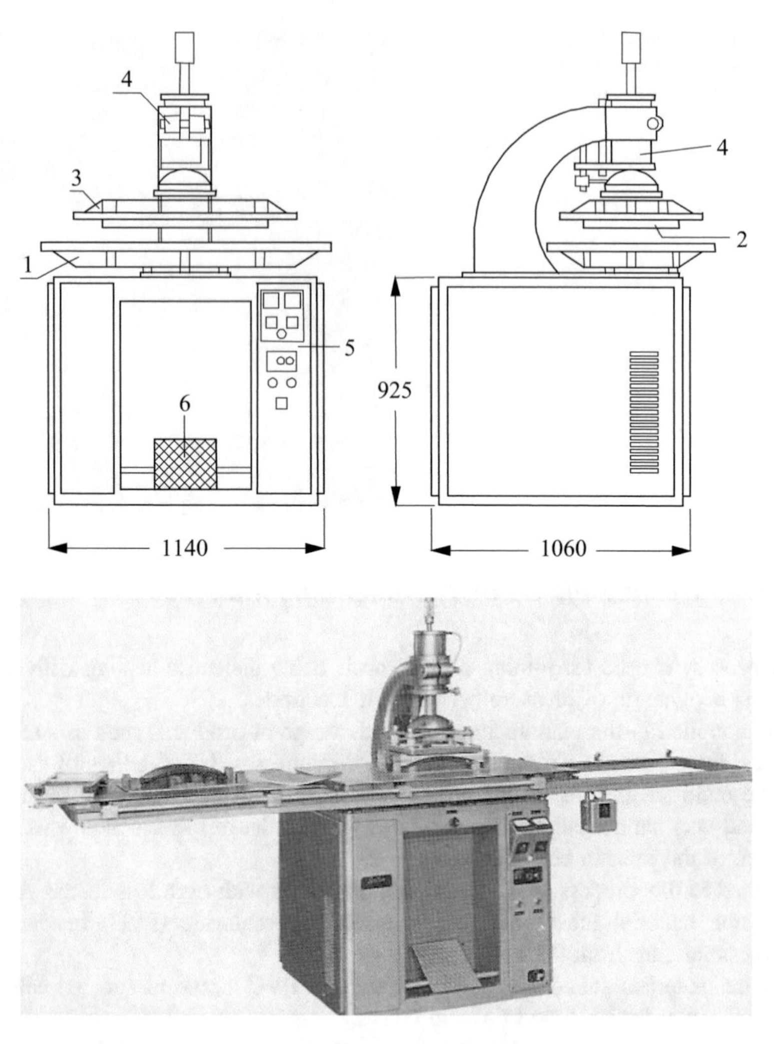

Fig. 7.55 Welding machine for sheet thermoplastic material—power: 7.5 kW; Frequency: 27 MHz (*1* welding table at ground potential; *2* mobile electrode at high voltage; *3* electrode support; *4* compressed-air cylinder; *5* instrumentation for control and regulation; *6* foot actuator of the press)

Special electrode geometries, of the type shown in Fig. 7.56b, can be used not only for welding, but also for cutting the sheets along the welding profile at the end of the process.

The HF voltage applied between the electrode and the welding table produces the required heat in the sheets to be welded, but—due to the contact with the cold welding table and the upper electrode—the temperature increases only in a very narrow region near the contact surfaces of the two facing sheets. When the material in this area becomes plastic, the HF is shut-off and the welding is completed only

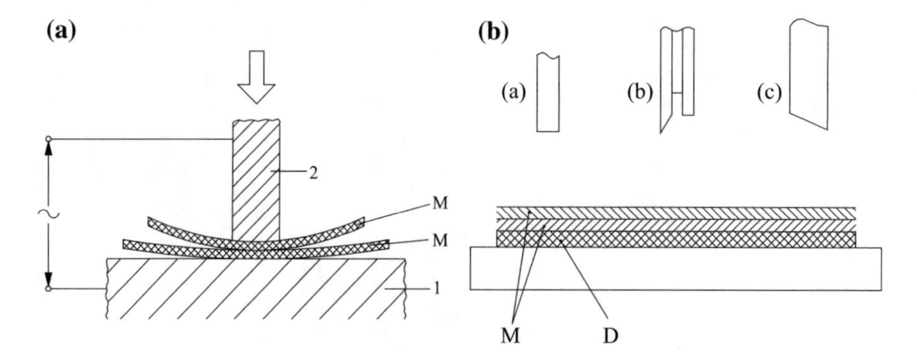

Fig. 7.56 **a** schematic diagram of welding thermoplastic sheet materials (M-material to be welded; *1* welding table, *2* electrode); **b** different geometries of electrodes and welding plane with interposition of an insulating layer (**a** electrode for straight welding; **b** electrode for welding and cutting; **c** jerk welding electrode; D-insulating material; M-material to be welded)

under pressure. Then the material rapidly solidifies thanks to the cooling effect of the electrode and the welding table.

Optimal temperature distributions can be obtained only with very short welding times and sheets of small thickness.

In order to achieve uniform welds and avoid disruptive discharges that could damage the product, the mechanical construction must be robust and precise, assuring the same working condition after several welding cycles, i.e. a constant parallelism between the electrode and the welding table.

The interposition of an insulating material (as in Fig. 7.56b) is another solution which avoids the occurrence of arcs when the distance between the electrodes decreases due to the pressure, when the material becomes plastic.

The range of pressure regulation can be relatively wide, usually between 1 and 10 kg/cm^2 of welded surface, depending on the required degree of plasticity of the material.

Depending on characteristics of the product and production rate, these welding machines may have rated power between 1 and 100 kW.

As rule of thumb, the needed HF power is around one kW for welding 20–25 cm^2.

However, the required power depends on several factors, including:

- material and thickness of the sheets to be welded
- area of the surface to be welded
- type and size of electrodes, with or without thermal insulation
- use of heated electrodes, that reduces the cooling effect on the welded material.

As an example, Fig. 7.57 gives, as a function of the HF generator power, the values of typical weldable areas of two PVC sheets, 0.5 mm thick, with electrodes of 2–3 mm width and welding time of about 3–4 s.

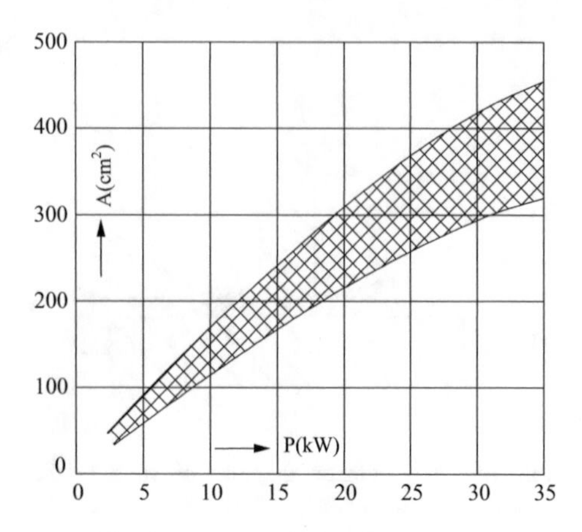

Fig. 7.57 Average weldable areas of 2 PVC sheets, 0.5 mm thick, with electrodes of 2–3 mm width (P—HF power of generator) [9]

From the electrical point of view, a key problem is the "*load matching*", which must take into account the strong variations of equivalent capacitance and resistance occurring during the heating process.

An electronic control can compensate these changes, thus achieving an almost constant power during the welding process, as indicated by curve A of Fig. 7.58.

Other matching circuits, with inductive coupling, allow to obtain power distributions of the type shown by the curve B of the same figure.

The second solution allows, for the same maximum power, to weld a smaller area, but has some advantages: (a) at the start of welding, when the material is cold and the electrodes have not a perfect contact with the plastic sheet, the voltage is lower and there is a lower risk of surface discharges; (b) at the end of the process, when the material is nearly plastic, the reduction of power makes less critical the final stage, reducing the risks of contact between electrodes.

In order to avoid arcs, which would damage both final product and electrodes, modern welding machines are equipped with arc suppressor circuits. These circuits detect the increase of the anodic current that occurs with the arc and shut down the HF power in a very short time, even lower than 100 μs.

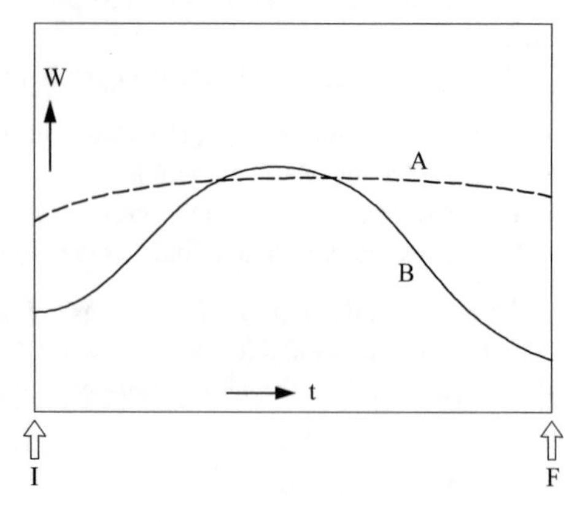

Fig. 7.58 Power variation during welding of thermoplastic material with two different systems for load matching (I, F—beginning and end of the process)

As regards operating costs, the main items to be considered are the following:

- cost of the electrical energy required by the process, which can be estimated taking into account the efficiency of the HF generator, the variation of power during the welding cycle (e.g. by the curves of Fig. 7.58) and the time of HF application during the process;
- generator maintenance costs, which are affected to a considerable extent by the cost of the triode tubes, whose life is in the range of about 5000–6000 h;
- maintenance costs of press and automation systems for transporting and positioning the material under the electrodes, which is similar to that of any industrial plant of the same type.

7.10.3 Preheating of Thermosetting Resins and Rubber

This process is used in the compression moulding of parts made of rubber or thermosetting resins.

The cold material, in the form of pellets or powder, is placed between two plane electrodes (Fig. 7.59), fed with high voltage at a frequency able to produce the power required for heating the material to about 130–165 °C in a very short time, even lower than 30 s, in order to avoid the start of material hardening.

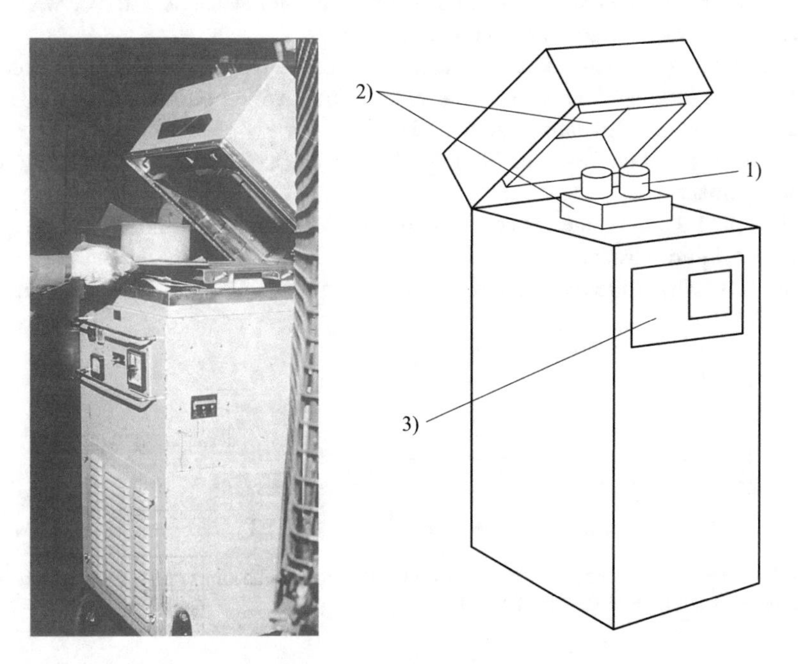

Fig. 7.59 Typical preheater for pellets of thermosetting resins with movable upper electrode for load matching (*1* pellets of thermosetting resins; *2* electrodes; *3* control panel with timer)

When the material becomes homogeneously plastic by this preheating process, it is rapidly placed under the press in a hot mold, which has the function of giving the final shape and to allow to complete the hardening phase.

The preheating at HF allows to use molds at relatively high temperature (180–205 °C) without damaging the product and, consequently, to reduce the process time of about 40–50 %.

Typical preheaters are manufactured at frequency in the range from 27 to 90 MHz, with power up to 10 kW, in order to develop high specific power even in materials with low loss factor and to keep at relatively low values the electrical gradients in the load and the air-gap.

The advantages of this process are:

- reduction of molding time
- reduction (20–40 %) of molding pressure
- increased life of molds
- improvement of the final product.

For processing thermoplastic materials and rubber are also used microwave preheaters, operating at 900 or 2450 MHz, generally provided with a rotary table for supporting the charge and a “mode stirrer” in order to achieve a more uniform heating pattern.

Microwave heating is also used in the process of *vulcanization of rubber*.

This is an irreversible chemical reaction—used for giving to the product mechanical strength and wear resistance—which occurs at about 200 °C, when the natural rubber is heated with mixtures of sulfides.

The process, after the extrusion stage, requires a rapid heating stage followed by a sufficient holding time at suitable temperature, which allows to complete the reaction without overheating the material.

As shown in Fig. 7.60, the holding phase is carried out in a hot air or infrared tunnel furnace.

The final stage of the process is constituted by a rapid cooling with spray of water or liquid nitrogen.

In this way different shapes of final extruded products can be obtained (Fig. 7.61).

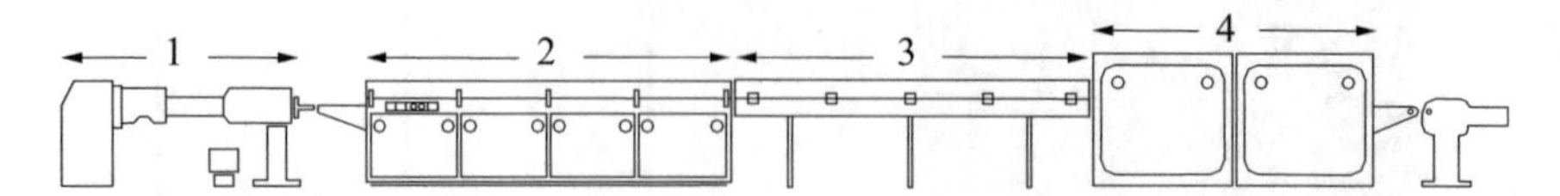

Fig. 7.60 Installation for rubber vulcanization with microwave heating (*1* extrusion; *2* microwave applicator; *3* hot air tunnel; *4* cooling) [12]

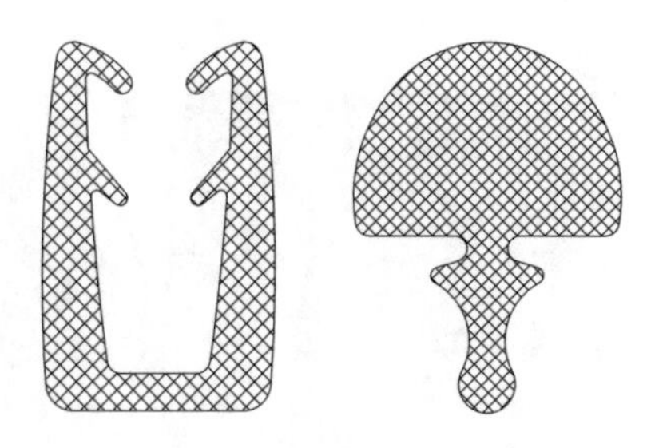

Fig. 7.61 Examples of vulcanized extruded products

7.10.4 Applications in Textile, Paper and Food Sectors

HF heating is widely used in various applications in the industry for drying textiles, paper and food.

In *drying processes* the use of HF can lead to exceptional results for the possibility, in some cases to heat the body uniformly throughout the mass, in other cases to obtain a selective heating of the water that must be evaporated without affecting the material containing it.

In this field, however, the HF is in competition or is used in conjunction with other heating methods (such as heat pumps or air with forced ventilation) for the relatively high energy consumption of the drying process.

In fact, the energy required in a drying process is constituted by the amount necessary for rising the temperature without change of the physical state and the amount required for evaporation of liquid.

For water evaporated at normal pressure, from the relationship:

$$W_H = c\,\gamma\,\Delta\vartheta + c^*\,\gamma \tag{7.73}$$

where: W_H—energy, kWh/m^3; c—specific heat content, kWh/kg °C; c^*—specific heat for the change of physical state (for evaporation of water: 0.626 kWh/kg); γ—density, kg/m^3, we can get that the evaporation of 1 kg of water requires an energy consumption of approximately 0.75 kWh.

Remembering that the efficiency of HF generators is about 50–60 %, we can conclude that the energy consumption for evaporation of 1 kg of water is about 1.5 kWh.

In practice, the use of dielectric heating is convenient only when the amount of the water to be removed does not exceed 10–20 % of the weight of the dry product.

A typical application is in the last stage of drying textile fibers, after a stage of mechanical centrifugation, where it is required the removal of about 10 % of the residual moisture.

This process is used for different types of fibers, as synthetic acrylic, polyamide, polyester, viscose, but also cotton and wool. It allows to improve the quality of the product and to increase the production rate by ensuring a final predetermined level

Fig. 7.62 Continuous tunnel dielectric dryers (**a** for cotton or linen yarns onto conical plastic bobbins, 40 kW at HF (courtesy of Stalam); **b** for not worked raw materials, 400 kg / h, humidity from 40 to 18 %, 140 kW at HF) (courtesy of RF systems)

of moisture. This is an example of a process where the HF heat is concentrated almost exclusively into the water to be evaporated and not in the material to be dried: it allows to reduce in some cases the process time from hours to a few tens or hundreds of seconds.

The products to be dried, which can be unprocessed raw material, semi-finished products like coils or skeins, or finished products, are treated in "batch" type furnaces or, more frequently, in continuous tunnel furnaces, provided with planar or "staggered" type electrodes (see Sect. 7.10) and a conveyor belt of insulating material with low losses.

As shown in Fig. 7.62a, b, the HF generator is generally situated in the upper part of the tunnel, which is also provided with a system for circulation and extraction of the hot air, having the function to avoid condensation problems and to contribute to the drying process.

Typical data of these generators are: frequency 27.12 and 13.56 MHz; rated power 6–60 kW for "batch" type furnaces, 25–300 kW for tunnel furnaces.

The main advantages of the process are:

- selective heating, concentrated mainly in the water to be evaporated. It must be taken into account that the drying of textile fibers requires uniform heating of a material which is characterized by low thermal conductivity and, as a consequence, that very long heating times are needed with classical heating methods like hot air convection;
- consequent reduction of the process time by using HF;
- uniform and well defined final moisture content, which allows to avoid excessive and useless moisture reduction, which would be rapidly re-acquired after the end of the drying process;
- reduction of 40–60 % of the cost of drying, taking into account the high time requirements of hot air processes, their lack of effectiveness in the final stage of drying and their high energy consumption, which is in the order of 3.0–3.5 kWh/kg of evaporated water.

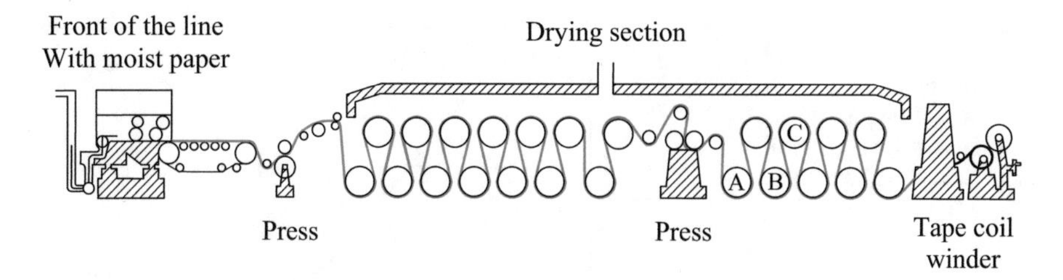

Fig. 7.63 Schematic of a machine for production of paper and possible positions (A, B, C) of HF for leveling moisture profile [14]

Another important area of application of HF is the *paper industry*, where for example it is used for obtaining a uniform moisture profile in thick products.

In fact, machines for production of paper in rolls, generally produce a non-uniform moisture content distribution in the cross section of the continuous sheet. The conventional methods of drying, such as those with contact with rollers of cast iron steam-heated or infrared, produce in the transverse cross-section a moisture profile with differences of about ±2.5 %, or may lead to excessive drying, with possible damage of product.

The use of HF, in series with heating roller as shown in Fig. 7.63, allows to reduce the lack of uniformity to ±0.5 %, and therefore to guarantee a final product with moisture content of about 1 %. Furthermore, since the HF is concentrated in the regions with higher water content, it also has an effect of self-regulation in leveling the moisture profile.

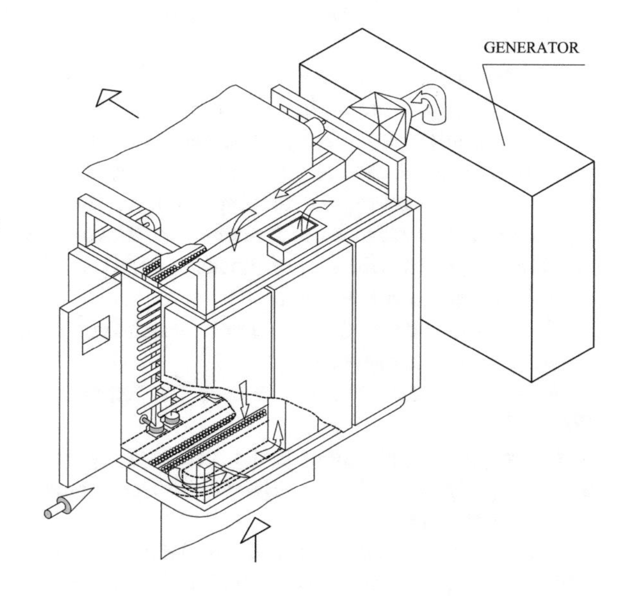

Fig. 7.64 Installation of HF paper drying (2 × 100 kW, 2 × 180 kVA; frequency: 13.56 MHz; evaporated water: 240–300 kg/h; 2 tubular electrodes normal to the direction of paper movement, 2.5 m width, 2 m length; paper-electrode distance: 5–10 mm) [12]

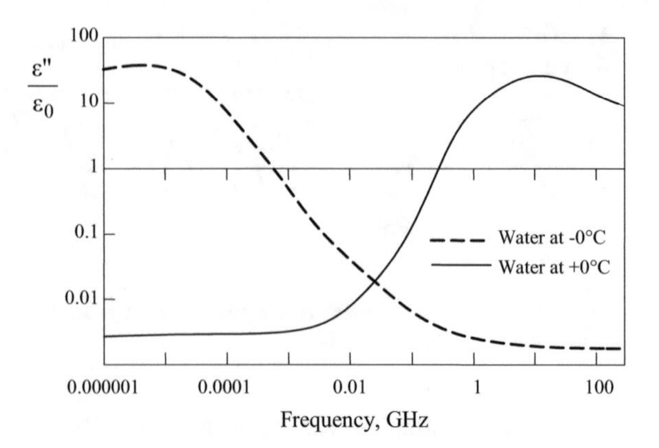

Fig. 7.65 Variations of water loss factor at transition of freezing point, as a function of frequency (continuous line: water at +0 °C; dashed line: water at −0 °C) [5]

Figure 7.64 shows an installation of this type for the production of paper 70–250 g/m^2 weight, 2400 mm width, with speed of 300 m/min.

The HF applications—including microwaves—are widely used also in the *food sector*, in processes of partial (*tempering*) or full thawing of frozen food, pasteurization and sterilization, cooking (typical "*post-baking*" of biscuits), reduction and control of moisture in food products.

The tempering process consists in bringing a product (e.g. meat, fish, butter) frozen at low temperature to a temperature just below the freezing point (−3 to −6 ° C) in order to allow to make on it subsequent working phases.

It is used, in place of total thawing, since it allows to re-freeze the product with low energy consumption, and because in the change of the phase state (ice/water) the loss factor abruptly increases significantly, as shown in the diagram of Fig. 7.65, with the possibility of producing localized overheating effects which can deteriorate the product.

The advantage is an enormous reduction of the time of the process: as an example the treatment from −18 to −4 °C of blocks of meat of 150–175 mm thickness requires several days with the traditional method which employs a temperature controlled furnace chamber, while by the use of RF or MW the process time can be reduced to 4–7 min depending on the quality of the meat.

The choice between radio frequency or microwaves depends on the characteristics of the product, the size of blocks and the danger of flashover for the presence of moisture and condensation.

The furnaces used can be either of batch type or tunnel type with continuous movement on a conveyor belt.

An example of a "batch" furnace for "tempering" meat blocks is shown in Fig. 7.66.

Other typical applications in the food industry are those regarding the final stage of cooking *("post-baking")* or the elimination of moisture to the end of reducing its content to less than 5 %.

In these processes it is generally used dielectric or microwave heating, combined with a classical heating process, such as hot air.

Fig. 7.66 MMicrowave batch furnace for tempering frozen meat. (Production: 1.5 t/h; HF power: 30 kW; Frequency: 896 MHz; Load: 27 kg of meat blocks; Process temperature: from −18 to −4°C) [12]

In fact the conventional processes, where the heat is transferred from the external surface to the interior of the workpiece, are effective only in the initial phase of cooking or drying, when the product is still very moist, but they lose effectiveness in the final stage, when the residual moisture must be extracted from the core.

Moreover, the combined process allows to use in the initial stage the kind of energy economically more convenient.

The diagrams of Fig. 7.67, which show the reduction of the moisture content in the material as a function of time, provide a qualitative representation of the

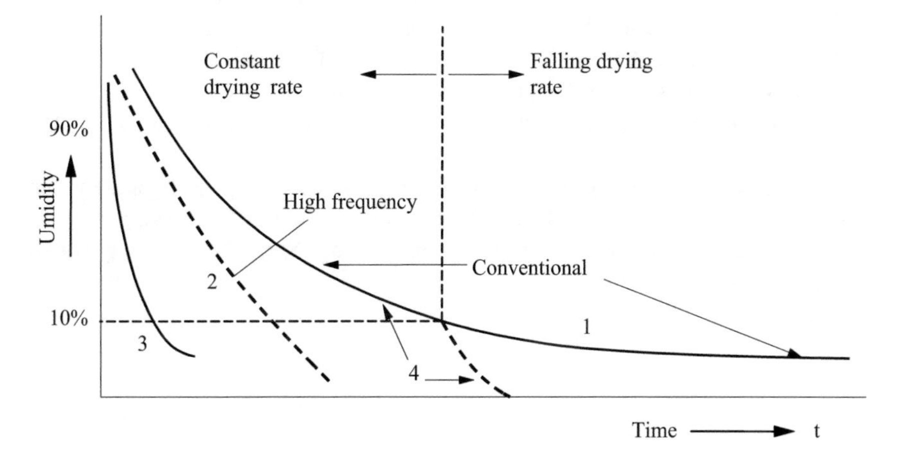

Fig. 7.67 Reduction of moisture as a function of time with different heating methods [2]

effectiveness of the drying processes with conventional, RF, MW or combined heating.

Curve 1, which refers to a conventional heating from the surface of the workpiece, shows that in the initial phase, when the material contains a large amount of water, the moisture reduction proceeds at a high constant speed. At this stage there is a balance between the movement of the moisture from the more humid internal parts towards the drier surface, and the evaporation from the surface towards the environment, which occurs at constant speed if the conditions of the environment do not change.

In a second period, when the moisture content decreases below a critical value of abut of 10 %, the rate of evaporation from the surface progressively decreases and the reduction of humidity takes place at a reduced speed. This drawback can be partially overcome by changing the conditions of heat exchange between the surface and the environment (e.g. by increasing the temperature of the air lapping the surface), bearing in mind—however—that for each material to be dried there is a limit temperature above which the surface layers of the product will be damaged.

The curve 2 refers to the heating with dielectric losses or microwaves; in this case the energy is uniformly absorbed in the volume of the wet material, and is concentrated particularly in the water, until it reaches the boiling point. The generation of steam which thus takes place, leads to an increase of internal pressure that pushes the moisture from the inside toward the surface where the evaporation occurs. In addition, the evaporation at the surface decreases the temperature of the external layers, producing a temperature gradient which contributes to the removal of moisture from the inside toward the surface. In some high-density materials, slightly porous or fragile, it is necessary to limit energy density, because an excess of internal pressure can lead to the cracking of the product.

In practice, the heating processes which use only RF or MW are seldom used for the high costs of the energy.

Curve 3 is typical for the combined heating by convective hot air and high-frequency (the so-called *ARFA*—"*Air Radio Frequency Assisted*" process).

In this case the heating by hot air provides the energy necessary to achieve the evaporation and maintain dry the surface, while the HF selectively heats the water and forces the moisture to migrate from core toward the surface.

The HF energy usually amounts to 10–15 % of the total energy required by the process.

In other cases the two types of heating are successively applied separately; the removal of moisture occurs in the first stage with convective air following curve 1, while HF is used only in the final stage for eliminating the residual humidity below the critical value, according to the curve 4.

References

1. Sluhockiĭ, A.E., Nemkov, V.S., Pavlov, N.A., Bamunier, A.B.: Induction Heating Installations, p. 325. Energoizdat, Leningrad (1981) (in Russian)
2. Metaxas, A.C.: Foundations of Electroheat—A Unified Approach, p. 500. John Wiley & Sons (1996). ISBN 0-471-95644-9
3. Orfeuil, M.: Electrothermie industrielle, p. 803. Dunod, Paris (1981). ISBN 2-04-012179-X
4. Barber, H.: Electroheat, p. 308. Granada Publishing, London (1983) (in French)
5. Nizou, P.Y.: Traitements thermiques par micro-ondes et radiofréquences, p. 125. Centre français de l'électricité, Paris (2001) (in French)
6. Mühlbauer, A.: Industrielle Elektrowärme-technik, p. 400. Vulkan-Verlag, Essen (1992). ISBN 3-8027-2903-X (in German)
7. Nelson, H.M.: Temperature distribution with simultaneous platten and dielectric heating. Br. J. Appl. Phys. 79–86 (1952)
8. Fireteanu, V., et al.: Numerical models of an RF-applicator of successive bar electrode type. HES-04 Heating by Electromagnetic Sources, Padua (Italy), 253–258 (2004)
9. Di Pieri, C.: Elektrowärme—Theorie und Praxis, cap. IV.5, 631–666. UIE-Paris, Verlag W. Girardet, Essen (1974). ISBN 3-7736-0355-X (in German)
10. Dilda, G.: Microonde, p. 335. Levrotto & Bella, Torino (Italy) (1956) (in Italian)
11. Terman, F.E.: Electronic and Radio Engineering, p. 1078. McGraw Hill, New York (1955)
12. Hulls, P., et al.: Dielectric Heating for Industrial Processes, p. 174. UIE—International Union for Electroheat, Paris (1992)
13. Ramo, S., Whinnery, J.R., Van Duzer, T.: Fields and Waves in Communication Electronics, p. 754. John Wiley & Sons, New York (1965)
14. BNCE—British National Committee for Electroheat: Dielectric heating for industrial processes, p. 30. BNCE, London (1983)

Appendix
Tables and Formulae

Appendix—Table A.1—Functions of Bi

$Bi = \alpha\delta / \lambda$	ν	ν^2	$Bi = \alpha\delta / \lambda$	ν	ν^2
0.00	0.0000	0.0000	2.2	1.1054	1.222
0.01	0.0998	0.0100	2.4	1.1300	1.277
0.02	0.1410	0.0199	2.6	1.1541	1.332
0.04	0.1987	0.0397	2.8	1.1747	1.380
0.06	0.2425	0.0584	3.0	1.1925	1.420
0.08	0.2791	0.0778	3.5	1.2330	1.520
0.10	0.3111	0.0968	4.0	1.2646	1.599
0.12	0.3397	0.1154	4.5	1.2880	1.659
0.14	0.3656	0.1337	5.0	1.3138	1.726
0.16	0.3896	0.1518	5.5	1.3340	1.780
0.18	0.4119	0.1697	6.0	1.3496	1.821
0.20	0.4328	0.1874	7.0	1.3766	1.895
0.22	0.4525	0.2048	8.0	1.3978	1.954
0.24	0.4712	0.2220	9.0	1.4149	2.002
0.26	0.4889	0.2390	10	1.4289	2.042
0.28	0.5058	0.2558	12	1.4420	2.079
0.30	0.5218	0.2723	14	1.4560	2.120
0.35	0.5590	0.3125	16	1.4700	2.161
0.40	0.5932	0.3516	18	1.4830	2.199
0.45	0.6240	0.3894	20	1.4961	2.238
0.50	0.6533	0.4264	25	1.5070	2.271
0.55	0.6800	0.4624	30	1.5200	2.310
0.60	0.7051	0.4972	35	1.5260	2.329
0.70	0.7506	0.5634	40	1.5325	2.349
0.80	0.7910	0.6257	50	1.5400	2.372
0.90	0.8274	0.6846	60	1.5451	2.387
1.00	0.8603	0.7401	70	1.5490	2.399

(continued)

S. Lupi, *Fundamentals of Electroheat*, DOI 10.1007/978-3-319-46015-4

(continued)

$Bi = \alpha\delta / \lambda$	ν	ν^2	$Bi = \alpha\delta / \lambda$	ν	ν^2
1.20	0.9171	0.8411	80	1.5514	2.407
1.40	0.9649	0.9310	90	1.5520	2.409
1.60	1.0008	1.002	100	1.5560	2.421
1.80	1.0440	1.090	$\propto$	1.5708	2.467
2.00	1.0769	1.160			

Appendix—Table A.2—First Six Roots of Equation $\beta J_1(\beta) - A J_0(\beta) = 0$

A	β_1	β_2	β_3	β_4	β_5	β_6
0	0	3.8317	7.0156	10.1735	13.3237	16.4706
0.01	0.1412	3.8343	7.0170	10.1745	13.3244	16.4712
0.02	0.1995	3.8369	7.0184	10.1754	13.3552	16.4718
0.04	0.2814	3.8421	7.0213	10.1774	13.3267	16.4731
0.06	0.3438	3.8473	7.0241	10.1794	13.3282	16.4743
0.08	0.3960	3.8525	7.0270	10.1813	13.3297	16.4755
0.1	0.4417	3.8577	7.0298	10.1833	13.3312	16.4767
0.15	0.5376	3.8706	7.0369	10.1882	13.3349	16.4797
0.2	0.6170	3.8835	7.0440	10.1931	13.3387	16.4228
0.3	0.7465	3.9091	7.0582	10.2029	13.3462	16.4888
0.4	0.8516	3.9344	7.0723	10.2127	13.3537	16.4949
0.5	0.9408	3.9594	7.0864	10.2225	13.3611	16.5010
0.6	1.0184	3.9841	7.1004	10.2322	13.3886	16.5070
0.7	1.0873	4.0085	7.1143	10.2419	13.3761	16.5131
0.8	1.1490	4.0325	7.1282	10.2516	13.3835	16.5191
0.9	1.2048	4.0562	7.1421	10.2613	13.3910	16.5251
1.0	1.2558	4.0795	7.1558	10.2710	13.3984	16.5312
1.5	1.4569	4.1902	7.2233	10.3188	13.4353	16.5612
2.0	1.5994	4.2910	7.2884	10.3658	13.4719	16.5910
3.0	1.7887	4.4634	7.4103	10.4566	13.5434	16.6499
4.0	1.9081	4.6018	7.5201	10.5423	13.6125	16.7073
5.0	1.9898	4.7131	7.6177	10.6223	13.6786	16.7630
6.0	2.0490	4.8033	7.7039	10.6964	13.7414	16.8168
7.0	2.0937	4.8772	7.7797	10.7646	13.8008	16.8684
8.0	2.1286	4.9384	7.8464	10.8271	13.8566	16.9179
9.0	2.1566	4.9897	7.9051	10.8842	13.9090	16.9650

(continued)

(continued)

A	β_1	β_2	β_3	β_4	β_5	β_6
10.0	2.1795	5.0332	7.9569	10.9363	13.9580	17.0099
15.0	2.2509	5.1773	8.1422	11.1367	14.1576	17.2008
20.0	2.2880	5.2568	8.2534	11.2677	14.2983	17.3442
30.0	2.3261	5.3410	8.3771	11.4221	14.4748	17.5348
40.0	2.3455	5.3846	8.4432	11.5081	14.5774	17.6508
50.0	2.3572	5.4112	8.4840	11.5621	14.6433	17.7272
60.0	2.3651	5.4291	8.5116	11.5990	14.6889	17.7807
80.0	2.3750	5.4516	8.5466	11.6461	14.7475	17.8502
100	2.3809	5.4652	8.5678	11.6747	14.7834	17.8931
∞	2.4048	5.5201	8.6537	11.7915	14.9309	18.0711

Appendix A.3—Integrals of Bessel Functions

$$\int_0^R r \cdot J_o(\alpha_k r) \cdot J_o(\alpha_n r) \cdot dr = 0, \quad \text{for} \quad k \neq n \tag{A3.1}$$

$$\int_0^R r \cdot J_o(\alpha_n r) \cdot dr = \frac{R}{\alpha_n} J_1(\alpha_n R), \quad \text{for any } \alpha_n \tag{A3.2}$$

$$\int_0^R r \cdot J_o^2(\alpha_n r) \cdot dr = \frac{R^2}{2} J_1^2(\alpha_n R), \quad \text{for} \quad J_o(\alpha_n R) = 0 \tag{A3.3}$$

$$\int_0^R r \cdot J_o^2(\alpha_n r) \cdot dr = \frac{R^2}{2} J_o^2(\alpha_n R), \quad \text{for} \quad J_1(\alpha_n R) = 0 \tag{A3.4}$$

$$\int_0^R r \cdot J_0^2(\alpha_n r) \cdot dr = \frac{R^2(h^2 + \alpha_n^2)}{2\,\alpha_n^2} J_0^2(\alpha_n R), \quad \text{for} \quad -\alpha_n J_1(\alpha_n R) + h\,J_0(\alpha_n R) = 0 \tag{A3.5}$$

$$\int_0^R r^3 \cdot J_o^2(\alpha_n r) \cdot dr = \frac{R^3}{\alpha_n} J_1(\alpha_n R) - \frac{2\,R^2}{\alpha_n^2} J_2(\alpha_n R) \tag{A3.6}$$

$$\int_0^x r^3 \cdot J_0(\alpha_n r) \cdot dr = (x^2 - 4) \cdot x \cdot J_1(x) + 2 \cdot x^2 \cdot J_0(x) \tag{A3.7}$$

$$Z_0'(u) = -Z_1(u); \quad Z_1'(u) = Z_0(u) - \frac{1}{u} Z_1(u) \tag{A3.8}$$

$$J_2(u) = \frac{2}{u} J_1(u) - J_0(u) \tag{A3.9}$$

$$\int_0^\xi \xi\,(\mathrm{ber}^2\, m\xi + \mathrm{bei}^2\, m\xi)\, d\xi = \frac{\xi}{m}(\mathrm{ber}\, m\xi \cdot \mathrm{bei}'\, m\xi - \mathrm{bei}\, m\xi \cdot \mathrm{ber}'\, m\xi) \tag{A3.10}$$

$$\int_0^\xi \xi\,(\mathrm{ber}'^2\, m\xi + \mathrm{bei}'^2\, m\xi)\, d\xi = \frac{\xi}{m}(\mathrm{ber}\, m\xi \cdot \mathrm{ber}'\, m\xi + \mathrm{bei}\, m\xi \cdot \mathrm{bei}'\, m\xi) \tag{A3.11}$$

$$\int (\mathrm{ber}\, m\xi \cdot \mathrm{ber}' m\xi + \mathrm{bei}\, m\xi \cdot \mathrm{bei}'\, m\xi)\, d\xi = \frac{1}{2\,m}(\mathrm{ber}^2\, m\xi + \mathrm{bei}^2\, m\xi) \tag{A3.12}$$

Rotor components in cylindrical coordinates (r,φ,z):

$$\left.\begin{aligned} (\mathrm{rot}\,\overline{A})_r &= \tfrac{1}{r} \quad \tfrac{\partial A_z}{\partial \varphi} - \tfrac{\partial A_\varphi}{\partial z} \\ (\mathrm{rot}\,\overline{A})_z &= \tfrac{1}{r}\,[\tfrac{\partial}{\partial r}(r\, A_\varphi - \tfrac{\partial A_r}{\partial \varphi}] \\ (\mathrm{rot}\,\overline{A})_\varphi &= \tfrac{\partial A_r}{\partial z} - \tfrac{\partial A_z}{\partial r} \end{aligned}\right\} \tag{A3.12}$$

Appendix A.4

Average radiation losses in the temperature range $(\vartheta_1 \div \vartheta_2)$ from a body at temperature $\vartheta_i(t)$ radiating towards an environment at temperature ϑ_a.

As known, radiation losses are calculated with the relation:

$$p_i = \frac{5.67 \cdot 10^{-12}}{\frac{1}{\varepsilon} + \frac{S}{S_a}(\frac{1}{\varepsilon_a} - 1)}[T^4 - T_a^4] \tag{A4.1}$$

where:

p_i radiated power per unit'surface, W/cm^2
ε, ε_a emission and absorption factors
S, Sa radiating and absorbing surfaces, cm^2
T, T_a absolute temperature of these surfaces, K

Assuming that it is Sa ≫ S, as occurs in many practical cases, the previous relation simplifies in the form:

$$p_i = 5.67 \cdot 10^{-12}\, \varepsilon [T^4 - T_a^4] \quad \text{(A4.2)}$$

With the further assumption of a linear increase of temperature with time, the average radiation losses in the temperature range $(\vartheta_1 - \vartheta_2)$ can be calculated with the relationship:

$$p_{im} = \frac{1}{T_2 - T_1}\int_{T_1}^{T_2} p_i \, dT \approx 5.67 \cdot 10^{-12} \cdot \frac{\varepsilon}{5} \cdot \left[\frac{T_2^5 - T_1^5}{T_2 - T_1}\right] \quad \text{(A4.3)}$$

Appendix—Table A.5—Characteristics of Dry Air at Atmospheric Pressure

ϑ	γ	$c \cdot 10^{-3}$	$\lambda \cdot 10^2$	$a \cdot 10^6$	$\mu \cdot 10^6$	$\nu \cdot 10^6$	Pr
°C	Kg/m³	kJ/(kg K)	W/(m K)	m²/s	Pa s	m²/s	
Temperature	Mass density	Specific heat content at constant pressure	Thermal conductivity	Coefficient of thermal diffusivity	Coefficient of dynamic viscosity	Coefficient of cinematic viscosity	Prandtl number
−50	1.534	1.013	2.034	13.15	1.460	9.54	0.728
−20	1.365	1.010	2.283	14.02	1.627	11.93	0.716
0	1.252	1.010	2.371	18.77	1.725	13.70	0.707
20	1.164	1.013	2.521	21.30	1.822	15.70	0.703
40	1.092	1.013	2.650	24.02	1.920	17.60	0.699
60	1.025	1.018	2.800	26.83	2.010	19.60	0.696
80	0.968	1.021	2.930	29.60	2.097	21.70	0.692
100	0.916	1.021	3.070	32.80	2.176	23.78	0.688
140	0.827	1.026	3.324	39.20	2.372	28.45	0.684
180	0.755	1.034	3.570	45.88	2.500	33.17	0.681
200	0.723	1.034	3.700	49.50	2.590	35.82	0.680
250	0.653	1.043	3.977	59.00	2.794	42.80	0.677
300	0.596	1.047	4.290	69.00	2.970	49.90	0.674
350	0.549	1.055	4.570	79.00	3.15	57.50	0.676
400	0.508	1.060	4.850	90.10	3.29	64.9	0.678
500	0.450	1.072	5.400	111.2	3.62	80.4	0.687
600	0.400	1.088	5.820	136.6	3.92	98.1	0.699
800	0.325	1.114	6.680	189.0	4.45	137	0.713
1000	0.268	1.140	7.610	250.0	4.95	185	0.719
1200	0.238	1.164	8.450	314.0	5.39	232	0.724
1400	0.204	1.190	9.300	384.0	5.78	282	0.736
1600	0.182	1.220	10.110	459.0	6.16	338	0.740
1800	0.165	1.224	10.820	534.0	6.66	397	0.740

Appendix—Table A.6—Resistivity, Specific Heat and Thermal Conductivity as Function of Temperature [1]

Stainless steel *(19.11 % chromium; 8.14 % nickel; 0.60 % tungsten)*

Temperature (°C)	ρ (Ωm) $\times 10^{-6}$	$c\gamma$ (Ws/m^3 K) $\times 10^6$	λ (W/mK)
20	0.695	4.04	15.9
100	0.776	4.15	16.3
200	0.850	4.24	17.2
300	0.915	4.36	18.0
400	0.976	4.51	19.7
500	1.030	4.83	21.4
600	1.072	4.84	23.0
700	1.111	4.87	24.3
800	1.141	4.90	26.0
900	1.171	4.86	27.0
1000	1.196	4.87	28.0
1100	1.220	4.93	29.0
1200	1.241		29.8
1300	1.257		30.5

Mild steel *(0.23 % carbon)*

Temperature (°C)	ρ (Ωm) $\times 10^{-6}$	$c\gamma$ (Ws/m^3 K) $\times 10^6$	λ (W/mK)
20	0.160	3.65	52.0
100	0.220	3.85	51.0
200	0.290	4.10	49.0
300	0.380	4.40	46.0
400	0.483	4.77	43.0
500	0.610	5.19	39.3
600	0.755	5.66	35.5
700	0.922	6.66	31.5
800	1.095	6.73	26.0
900	1.135	5.94	26.5
1000	1.168	5.09	27.3
1100	1.195	5.10	28.5
1200	1.22		29.7
1300	1.24		31.2
1400	1.26	5.08	32.8

Copper

Temperature (°C)	ρ (Ωm) $\times 10^{-6}$	$c\gamma$ (Ws/m^3 K) $\times 10^6$	λ (W/mK)
20	0.017	3.39	395
100	0.022	3.48	387
200	0.033	3.57	380
300	0.037	3.65	373
400	0.044	3.72	366
500	0.052	3.77	360
600	0.060	3.82	353
700	0.068	3.87	347
800	0.077	3.92	341
900	0.086	3.96	335
1000	0.097	3.99	330
1100	0.215	3.66	324
1200	0.222		319
1300	0.228		313

Aluminium

Temperature (°C)	ρ (Ωm) $\times 10^{-6}$	$c\gamma$ (Ws/m^3 K) $\times 10^6$	λ (W/mK)
20	0.0270	2.52	211
100	0.0364	2.59	219
200	0.0478	2.65	224
300	0.0599	2.71	223
400	0.073	2.78	216
500	0.087	2.84	209
600	0.104	2.89	200
700	0.210	2.51	92
800	0.225	2.52	88
900	0.235	2.54	

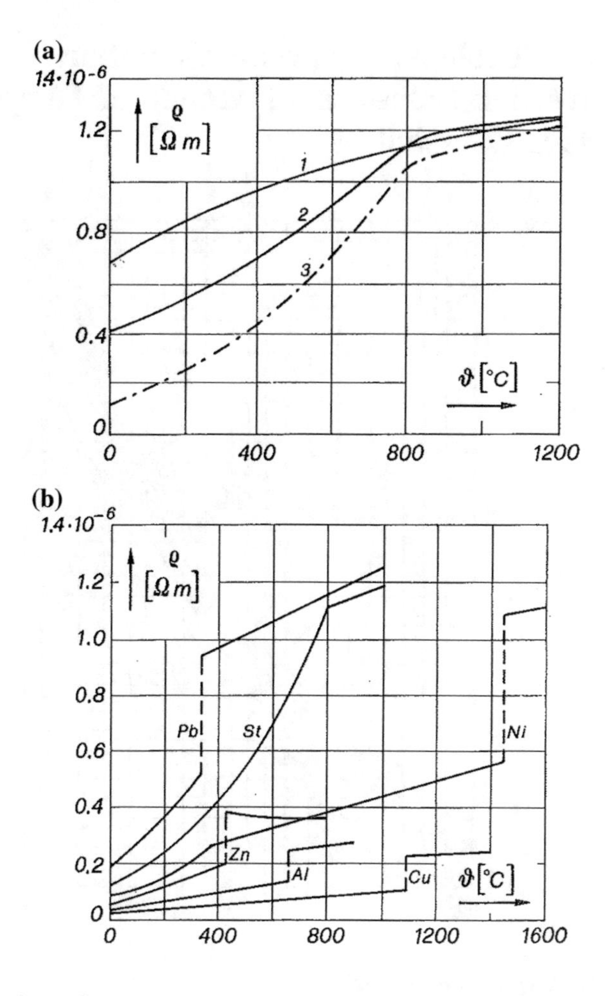

Resistivity of metals as a function of temperature ***a*** *1-Inox 18Cr Ni8; 2-Rapid steel; 3-steel 0.06 % C;* ***b*** *non-magnetic metals*

Appendix—Table A.7—Specific Radiation and Convection Losses as a Function of Temperature ϑ and Surface Emissivity ε

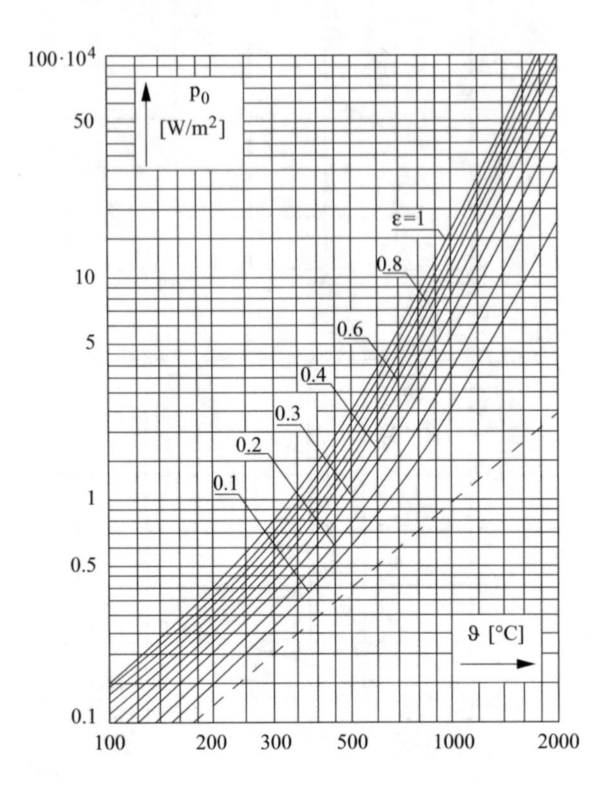

(dashed line: convection losses)

Appendix—Table A.8—Heat Content of Metals

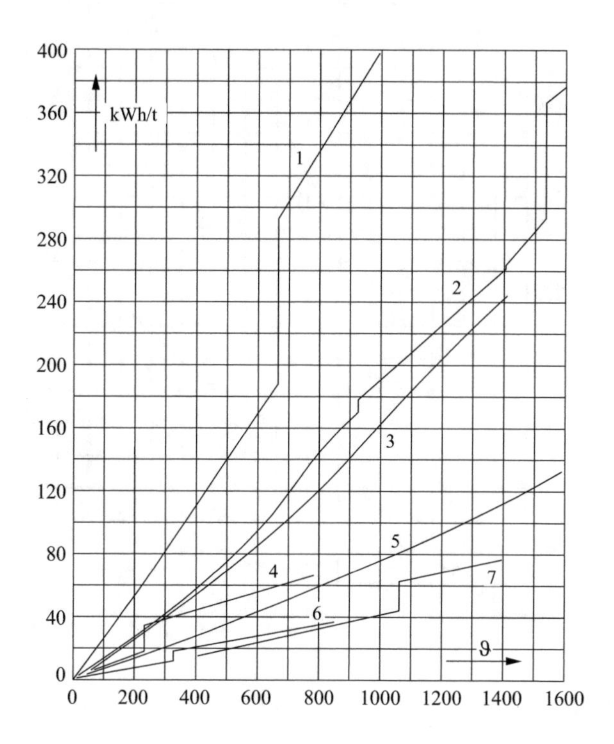

8a 1-Aluminium; 2-pure Iron; 3-Cobalt; 4-Tin 99.92; 5-Molybdenum; 6-Lead 99.75; 7-Gold

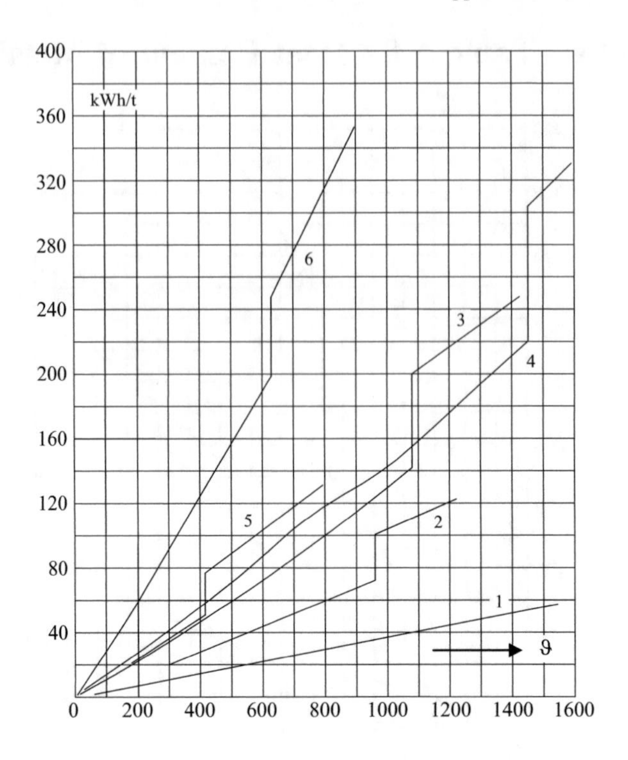

8b 1-Platinum; 2-Silver; 3-Copper; 4-Nickel; 5-Zinc; 6-Magnesium

Reference

1. Davies, E.J.: Conduction and induction heating. Peter Peregrinus Ltd., London (UK), 1990, 385 p., ISBN 0-86341-174-6